THIN FILM TECHNOLOGY HANDBOOK

Electronic Packaging and Interconnection Series

Charles M. Harper, Series Advisor

Alvino • PLASTICS FOR ELECTRONICS

Classon • SURFACE MOUNT TECHNOLOGY FOR CONCURRENT ENGINEERING AND MANUFACTURING

Ginsberg, Schnoor • MULTICHIP MODULE AND RELATED TECHNOLOGIES

Harper • ELECTRONIC PACKAGING AND INTERCONNECTION HANDBOOK

Harper, Miller • ELECTRONIC PACKAGING, MICROELECTRONICS, AND INTERCONNECTION DICTIONARY

Harper, Sampson • ELECTRONIC MATERIALS AND PROCESSES HANDBOOK, 2/E

Lau • BALL GRID ARRAY TECHNOLOGY

Lau • FLIP CHIP TECHNOLOGIES

Lau, Pao • SOLDER JOINT RELIABILITY OF BGA, CSP, FLIP CHIP, AND FINE PITCH SMT ASSEMBLIES

Licari • MULTICHIP MODULE DESIGN, FABRICATION, AND TESTING

Hwang • MODERN SOLDER TECHNOLOGY FOR COMPETITIVE ELECTRONICS MANUFACTURING

Related Books of Interest

Boswell • SUBCONTRACTING ELECTRONICS

Boswell, Wickam • SURFACE MOUNT GUIDELINES FOR PROCESS CONTROL, QUALITY, AND RELIABILITY

Byers • PRINTED CIRCUIT BOARD DESIGN WITH MICROCOMPUTERS

Capillo • SURFACE MOUNT TECHNOLOGY

Chen • COMPUTER ENGINEERING HANDBOOK

Coombs • ELECTRONIC INSTRUMENT HANDBOOK, 2/E

Coombs • PRINTED CIRCUITS HANDBOOK, 4/E

Di Giacomo • DIGITAL BUS HANDBOOK

Di Giacomo • VLSI HANDBOOK

Fink, Christiansen • ELECTRONICS ENGINEERS' HANDBOOK, 3/E

Ginsberg • PRINTED CIRCUITS DESIGN

Juran, Gryna • JURAN'S QUALITY CONTROL HANDBOOK

Jurgen • AUTOMOTIVE ELECTRONICS HANDBOOK

Manko • SOLDERS AND SOLDERING, 3/E

Rao • MULTILEVEL INTERCONNECT TECHNOLOGY

Sze • VLSI TECHNOLOGY

van Zant • MICROCHIP FABRICATION

THIN FILM TECHNOLOGY HANDBOOK

Aicha A. R. Elshabini-Riad

Fred D. Barlow III

McGraw-Hill

New York San Francisco Washington, D.C. Auckland Bogotá
Caracas Lisbon London Madrid Mexico City Milan
Montreal New Delhi San Juan Singapore
Sydney Tokyo Toronto

Library of Congress Cataloging-in-Publication Data

Elshabini, Aicha.
Thin film technology handbook / Aicha Elshabini, Fred D. Barlow.
p. cm.—(Electronic packaging and interconnect series)
Includes index.
ISBN 0-07-019025-9 (hardcover)
1. Thin film devices—Design and construction. 2. Thin films.
3. Electronic packaging. I. Barlow, Fred D. II. Title.
III. Series.
TK7872.T55E47 1997
621.3815′2—dc21 96-36938
CIP

McGraw-Hill

A Division of The ***McGraw·Hill*** *Companies*

1 2 3 4 5 6 7 8 9 0 DOC/DOC 9 0 2 1 0 9 8 7 0

ISBN 0-07-019025-9

The sponsoring editor for this book was Steve Chapman, the editing supervisor was Bernard Onken, and the production supervisor was Clare Stanley. It was set in Times Roman by Graphic World, Inc.

Printed and bound by R. R. Donnelley & Sons Company.

This book is printed on recycled, acid-free paper containing a minimum of 50% recycled de-inked fiber.

To our families
Aicha A. R. Elshabini
Fred D. Barlow III

CONTENTS

CONTRIBUTORS

Fuad Abulfotuh *National Renewable Energy Laboratory, Golden, CO* (CHAPS. 4, 6)

Fred D. Barlow III *The Bradley Department of Electrical and Computer Engineering, Virginia Polytechnic Institute and State University, Blacksburg, VA* (CHAPS. 1, 3, 5)

Richard Brown *Richard Brown Associates, Inc., Huntington, CT* (CHAPS. 1, 10)

Aicha A. R. Elshabini *The Bradley Department of Electrical and Computer Engineering, Virginia Polytechnic Institute and State University, Blacksburg, VA* (CHAPS. 1, 3, 5)

Bradley A. Fox *Kobe Steel USA Inc., Research Triangle Park, NC* (CHAP. 7)

Lynn F. Fuller *Rochester Institute of Technology, Rochester, NY* (CHAP. 2)

Philip Garrou *Dow Chemical, Research Triangle Park, NC* (CHAP. 9)

Lawrence L. Kazmerski *National Renewable Energy Laboratory, Golden, CO* (CHAPS. 4, 6)

David C. Keezer *Georgia Institute of Technology, Atlanta, GA* (CHAP. 11)

Angus Macleod *Thin Film Center, Inc., Tuscon, AZ* (CHAP. 8)

PREFACE

The *Handbook of Thin Film Technology* addresses the major advances in the thin film technology, as related to device fabrication, electronic packaging of microcircuits, and optical applications. The topics discussed include the various important classes of thin film deposition techniques with an emphasis on the principles and practices of film deposition, thin film material synthesis and characterization, design guidelines, and applications.

This handbook has been prepared by experts in the field who bring an in-depth knowledge of thin film processes and applications. Each chapter has been written as an independent source of information on a particular topic of thin film technology or a particular application.

The *Handbook of Thin Film Technology* consists of 11 chapters. The first three chapters provide the core technologies and the foundation upon which many of the later chapters build. Chapter 1 is an introduction to thin film deposition techniques used in various forms to create thin films. Chapter 2 provides an in-depth study of photolithographic techniques used to fabricate thin films with intricate patterns. Chapter 3 presents an overview of thin film materials and their various properties and attributes. Chapter 4 analyzes the nature of semiconductor thin films that form the backbone of modern microelectronics. Chapter 5 outlines design considerations and guidelines for the use of thin film materials, particularly for use in microelectronic assemblies. Chapter 6 is a compendium of characterization techniques used to evaluate and measure the properties of thin films, devices, and structures. The emphasis of this chapter is on semiconductor films; however, these techniques are generally applicable to any thin film material. Chapter 7 provides an in-depth study of diamond thin films, examining the myriad of applications and great potential of this unique material. Chapter 8 focuses on the optical properties of thin films, as well as their applications for anti-reflection coatings, filters, and other electro-optical applications. The rapidly growing arena of electronic packaging is the special focus of both Chapters 9 and 10. Chapter 9 provides a detailed look at thin film multichip modules, while Chapter 10 discusses RF and microwave microcircuits. The final chapter (Chapter 11) is a detailed look at testing and reliability of electronic assemblies, in particular thin film modules, which play an increasingly important role in microelectronics.

Aicha A. R. Elshabini
Fred D. Barlow III

CHAPTER 1
FILM DEPOSITION TECHNIQUES AND PROCESSES

Fred Barlow
Aicha Elshabini-Riad
The Bradley Department of Electrical and Computer Engineering
Virginia Polytechnic Institute and State University

Richard Brown
Richard Brown Associates

1.1 INTRODUCTION

Modern thin film technology has evolved into a sophisticated set of techniques used to fabricate many products. Applications include very large scale integrated (VLSI) circuits; electronic packaging, sensors, and devices; optical films and devices; as well as protective and decorative coatings. The subject of thin film deposition cannot be contained in a single chapter; indeed, whole texts have been dedicated to the myriad of deposition techniques and practices in use today. What this chapter seeks to provide is a basic applied foundation for thin film deposition techniques, with the emphasis on the deposition techniques themselves. The applications will be highlighted in subsequent chapters.

There are three categories of thin film processes: physical vapor deposition (PVD), chemical vapor deposition (CVD), and chemical methods. Figure 1-1 illustrates the classification of the different deposition processes. This chapter focuses on the most common and important techniques in use today. Topics discussed include vacuum systems, evaporation, sputtering, plating, molecular beam epitaxy (MBE), CVD, laser ablation and sol gels. The reader is strongly encouraged to review the comprehensive references for a deeper understanding of these and other deposition techniques. There are a number of complete texts available which focus on thin film deposition in general or on a particular thin film technique [1–6].

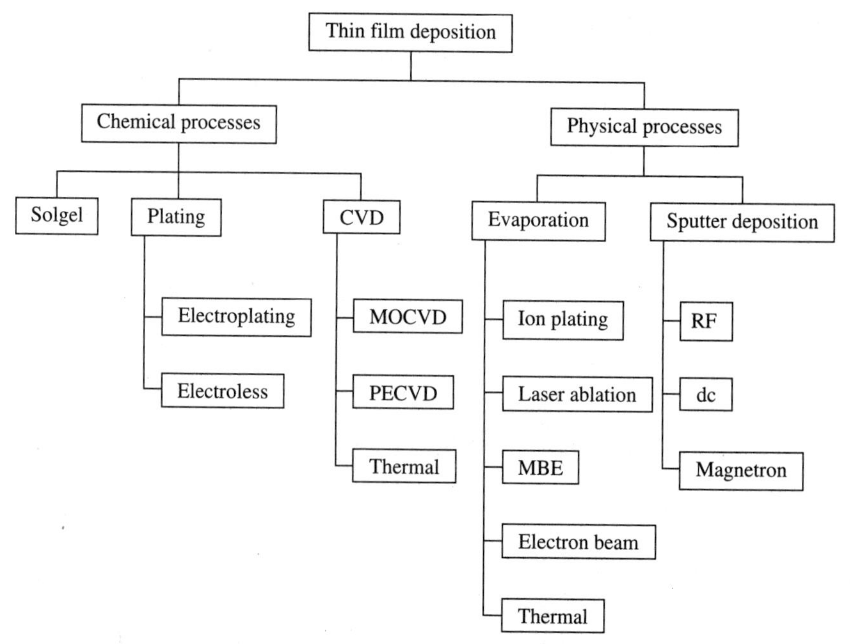

FIGURE 1-1 Classification of the most common deposition processes.

1.2 VACUUM SYSTEMS

1.2.1 Basic Vacuum Concepts

Many prevalent deposition techniques used in modern thin film technology require the creation of a vacuum for the deposition to occur. As a result, an entire industry and a new branch of science have developed to meet this requirement. For many of these techniques, the quality and consistency of the final product depends on the vacuum conditions during deposition. As a result, a basic understanding of vacuum technology is essential to the practice of these techniques.

A number of reasons exist for the deposition of materials in a vacuum environment. Vacuum conditions increase the mean free path for atoms, eliminate the presence of gases that could react with the deposited material, reduce the vapor pressure (thus lowering the evaporation temperature of materials), and provide the ultimate clean environment. Several systems of units have been developed to measure and quantify the quality of a vacuum, including the torr, also referred to as mm of mercury (mm Hg), Pascal (Pa), inches and cm of water, inch of mercury (in Hg), millibars (mbars), pounds per square inch (psi), as well as dynes/cm^2. Unfortunately, this wide range of units can make comparison of vacuum conditions rather confusing. Table 1-1 is included to provide a convenient conversion of units of vacuum measure. The standard international (SI) unit of pressure, Pa, will be the primary unit of measure in this text. However, a conversion to torr (mm Hg) will usually be provided because it is a very common unit of measure.

A vacuum is the absence of material, including the gases, moisture, and particles which normally fill our environment. In a deposition system, a container is provided and the majority of these elements are removed to create a vacuum. The complete and total removal

TABLE 1-1 Conversion factors for units of vacuum measure

		Convert to			
	Pascal (N/m^2)	Torr (mm Hg)	in Hg	mbars	atm
Convert from: Pascal (N/m^2)	—	7.5×10^{-3}	2.95×10^{-4}	10^{-2}	9.87×10^{-6}
Torr (mm Hg) (0°C)	133.332	—	3.94×10^{-2}	1.33	1.32×10^{-3}
in Hg (0°C)	3386.389	25.4	—	33.85000	3.342×10^{-2}
mbars	100	0.75	2.954×10^{-2}	—	9.87×10^{-4}
atm	101,325	760	29.9	1,013	—

of gases from any vessel is impossible. As a result, the degree or quality of vacuum required depends on the application. Normally, vacuum conditions are classified as low vacuum for pressures in the range of 10^5 to 10^3 Pa (760 to 25 torr), medium vacuum for 10^3 to 10^{-1} Pa (25 torr to 0.75 milli-torr), high vacuum for 10^{-1} to 10^{-4} Pa ($\approx 10^{-3}$ to 10^{-7} torr), and very high vacuum for 10^{-4} to 10^{-7} Pa (10^{-7} to 10^{-12} torr). Most practical deposition processes occur in the medium-to-high vacuum ranges.

Gases in a vacuum system result from leaks in the containment vessel, evaporation of moisture and other materials from the walls, and outgassing from the materials that make up the vessel. Leaks occur from any seam or connection between two materials or parts of a vacuum system. Connections have been developed for a variety of applications, all of which vary in the amount of leakage, cost of implementation, and convenience. Normally, the quality of vacuum required dictates the connections used. Evaporation from the chamber walls results from a coating of water vapor, particulates, and gases created by exposure of the components that make up the vacuum system to the environment during manufacture and every time the chamber is opened during maintenance or use. These contaminants then evaporate from the chamber walls, resulting in contamination and degradation of the vacuum quality. Outgassing from the materials that make up the vacuum vessel is the result of moisture and gases, trapped inside the materials, which are absorbed during manufacture or subsequent exposure to the environment and are released into the vacuum. The rate of outgassing from a surface can be expressed as

$$O_r = qA \tag{1}$$

where O_r = the rate of outgassing
q = the specific outgassing rate for the material
A = the surface area

The specific outgassing rate is unique for a given material, as can be seen from Table 1-2. The types of materials used in a vacuum system, the quality of its seals, and the general cleanliness condition can all limit the maximum obtainable vacuum.

1.2.2 Vacuum Chambers

To maintain a vacuum, a chamber is needed to keep out normal atmospheric pressure. Normally, stainless steel or glass chambers are used because they do not corrode, are easily cleaned, are nonmagnetic, and have good outgassing characteristics. Stainless steel is preferred for large systems because of its relatively high strength, ease of welding and machinability. Figure 1-2 illustrates a basic vacuum system used for thin film deposition.

TABLE 1-2 Specific outgassing rates for common materials

Material	q (one hour)	q (10 hours)
Pyrex (new)	7.35×10^{-9}	5.5×10^{-10}
Pyrex (aged 1 month in air)	1.16×10^{-9}	1.6×10^{-10}
Butyl Rubber	1.5×10^{-8}	4.0×10^{-9}
Viton	1.14×10^{-8}	—
Teflon	6.5×10^{-8}	2.5×10^{-8}
Steel	5.4×10^{-7}	5.0×10^{-8}
Stainless Steel	9.0×10^{-8}	2.0×10^{-8}
Stainless Steel Sanded	8.28×10^{-9}	1.04×10^{-9}
Copper	4.0×10^{-8}	4.15×10^{-9}
Aluminum	6.3×10^{-9}	6.0×10^{-10}

This system consists of a vacuum chamber, pumping system, deposition sources, and monitoring equipment. As illustrated, the vacuum chamber provides the basic support for all of the other components, and typically has several ports to connect deposition sources, pumps, and measurement sensors. Two basic configurations exist. The first is a glass or stainless steel enclosure (or belljar) sealed to a metal base or door by a gasket. The second is a stainless steel chamber with several ports.

The first (and oldest) of the two configurations is the bell jar, which has been in use since the beginning of vacuum technology. Typically, glass or stainless steel bell jars are sealed to stainless steel base plates with rubber gaskets, as depicted in Fig. 1-3. The base plate normally provides a large port for a pumping system and an array of smaller ports,

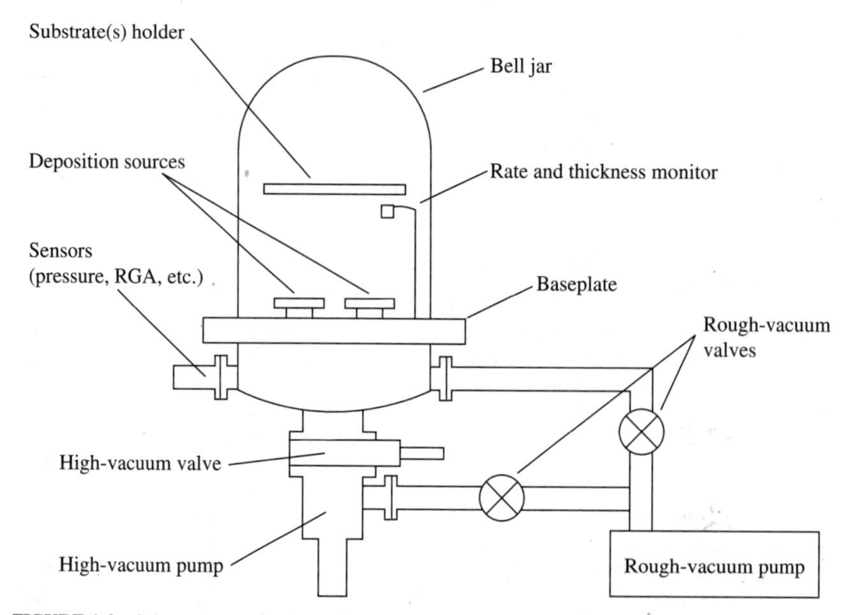

FIGURE 1-2 Schematic representation of a typical vacuum system used for thin film deposition.

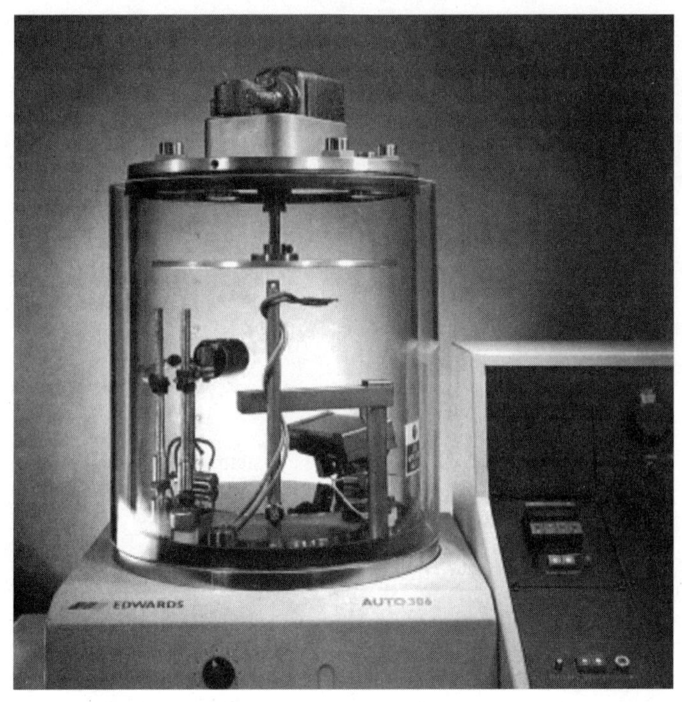

FIGURE 1-3 A deposition system based around a belljar sealed to a stainless steel base. *(Courtesy of Edwards High Vacuum International.)*

or feedthroughs, for deposition sources and vacuum components. A hoist is usually provided to raise and lower the bell over the base plate. This approach has the advantage of allowing easy access to the interior of the vacuum chamber. The disadvantage is that the gaskets used to seal the bell and the base plate together create a large area with a relatively high leakage rate. As a result, bell jar systems normally cannot reach the low vacuum pressures that other solid stainless steel vessels can obtain. On the other hand, the bell jar vacuum system is less expensive to construct and maintain. Many deposition techniques do not normally require pressures lower than 10^{-7} torr. Thus, this approach is favorable for many such techniques, particularly PVD methods.

A popular variation on this theme is the box coating system. Box coaters consist of a stainless steel box with a gasket-sealed door. The interior of the box supports the deposition sources, substrate holders, and monitoring equipment, and has ports for pumping system attachments, as illustrated in Fig. 1-4.

The second type of vacuum chamber has stainless steel components, which are attached together to form a closed vacuum system. Normally, a large chamber is used or a set of large chambers are bolted together. These main chamber sections may include any number of ports with a wide variety of configurations. This type of vacuum system has the advantage of flexibility and can produce very high-quality vacuum conditions (as low as 10^{-11} torr). This increased ability is largely because the system can be configured with higher quality seals. In addition, the whole system can be heated periodically to a few hundred degrees (C); this is usually referred to as baking the system or bakeout. This process evaporates and removes any impurities and moisture remaining inside the system. This

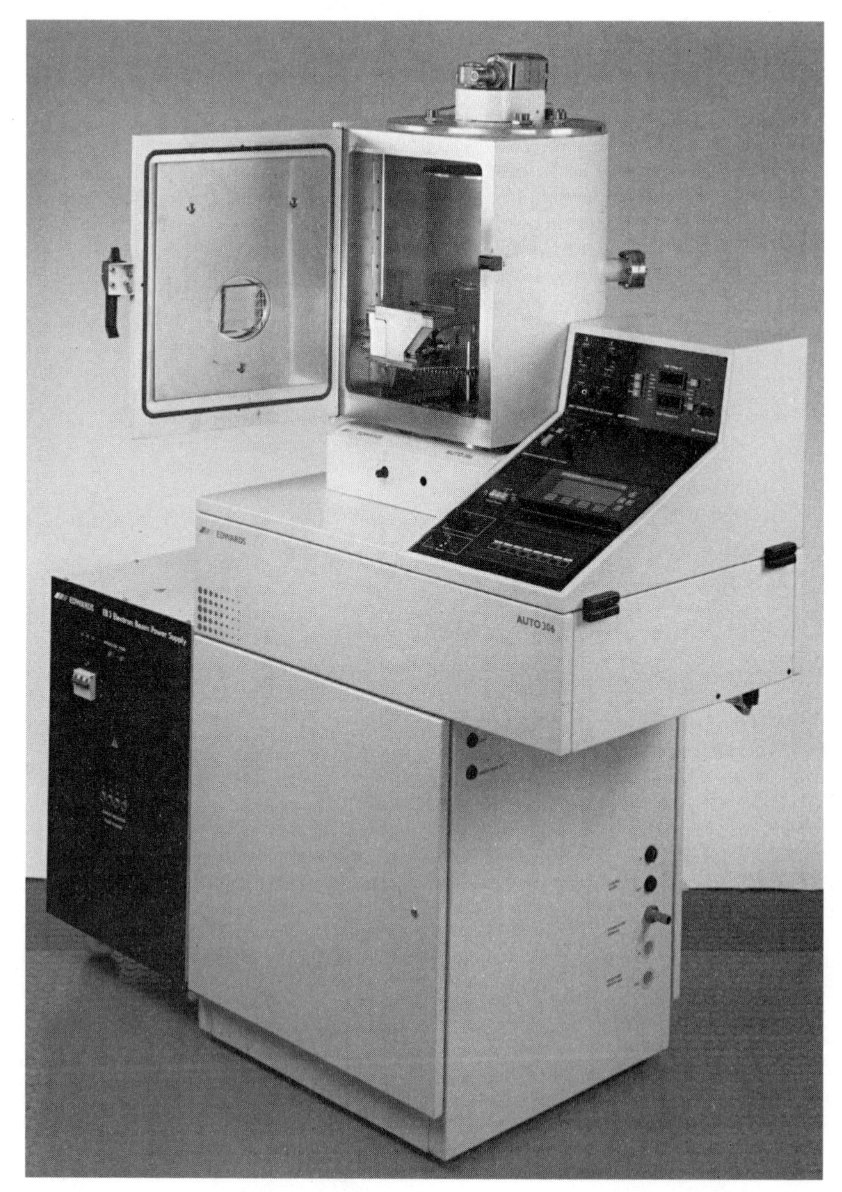

FIGURE 1-4 An electron beam evaporation system. This unit uses a box coating configuration and has a stainless steel box and a gasket-sealed door. *(Courtesy of Edwards High Vacuum International.)*

type of vacuum system is usually more expensive to construct and maintain because all of the components must withstand high-temperature bakeout.

A key issue with any vacuum system is the quality of the seals between various components. Because any interface between two components of a vacuum system has the potential to leak, great care must be taken to ensure the seal is properly made. Several standard vacuum fittings are widely used in the industry today. These fall into two primary classes: seals which use rubber o-rings or gaskets, and seals which use metal gaskets.

Vacuum seals that rely on rubber o-rings normally cannot withstand vacuum pressures as low as seals with metal gaskets, and cannot endure as high a temperature bakeout. However, this type of seal is adequate for most applications. It has the advantages of low cost and ease of assembly, because the o-ring or gasket can be sealed and unsealed hundreds of times. Figure 1-5 illustrates several common o-ring and gasket vacuum seals. Figure 1-5*a* shows a cross-section of a quick flange (QF Flange), which uses a metal clamp to hold an o-ring between two flanges. The o-ring is seated on a metal centering ring to ensure proper alignment. Figure 1-5*b* depicts an international standards organization (ISO) flange that uses bolted clamps and an o-ring. The ISO o-rings also use centering rings; Figure 1-5*c* illustrates a gasket seal between a bell jar and a base plate. This type of seal relies only on the weight of the bell and the internal vacuum to hold the two pieces together. A key point to understand with all vacuum seals and vacuum systems in general is that cleanliness is the key to maintaining seals with a minimum amount of leakage.

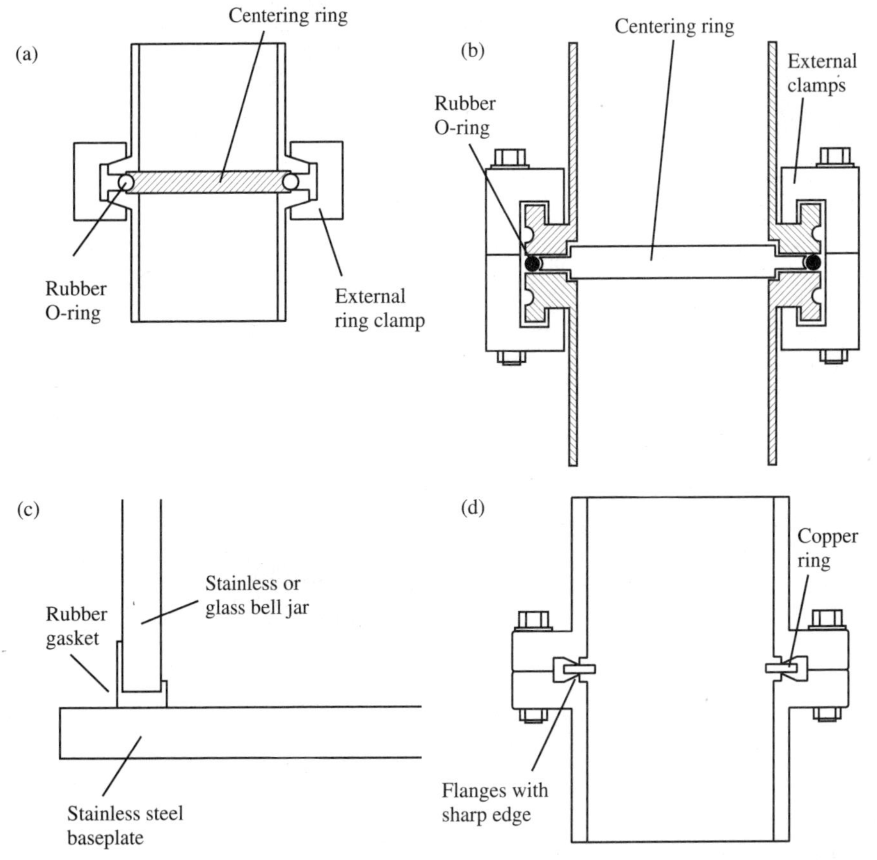

FIGURE 1-5 Cross section of several different vacuum seals: *(a)* quick flange; *(b)* ISO flange; *(c)* bell jar gasket seal; *(d)* knife edge.

Seals that use metal gaskets are more reliable and have reduced leakage when compared to o-ring seals. The two most common of these seals are the wire seal and the knife-edge seal (Figure 1-5*d*). In both cases a metal ring or wire is compressed between two stainless steel mating surfaces and cold flows to provide the seal. A new metal ring or wire is needed each time the seal is removed because compression of a fresh seal is the key to successful use of this type of flange. Knife-edge flanges are the most popular; however, wire seals are more appropriate for large flanges because of their reduced size and mass.

1.2.3 Pumping Systems

Two key characteristics of vacuum pumps are the displacement, or pumping rate, and the ultimate pressure that can be achieved. These characteristics are normally expressed in a performance curve, as illustrated in Fig. 1-6, for a rotary vane vacuum pump. Pressure pumping rates, like vacuum pumping rates, are expressed using a wide range of units. Conversion factors are provided in Table 1-3; note that this text uses the litre·s^{-1} (L·s^{-1}), but provides a conversion to cubic feet/minute (cfm). Different pumping systems have pumping rates that depend on the inlet pressure. In this particular example, this rotary vane pump is only effective down to a few Pa. In addition, other factors, such as the pump's ability to handle gas loads and its potential for oil contamination, must be considered.

Rough pumps and high-vacuum pumps form the two basic classes of vacuum pumps. Rough pumps, as the name implies, are used to achieve a rough-vacuum, or low-vacuum, condition. They are capable of very high pumping rates. However, these low-vacuum pumps have a limited maximum obtainable vacuum. As a result, a second kind of pumps, high-vacuum pumps, is used to create a high vacuum. High-vacuum pumps usually have lower pumping rates than rough pumps, but they can reach very high-vacuum conditions. High-vacuum pumps are normally used in combination with a rough pump. The high-vacuum pump draws on the chamber, while the rough pump draws gases from the high-vacuum pump. The most common rough pumps include the oil-sealed rotary vane, rotary piston, the roots blower, cryogenic absorption pump, and, more recently, the dry mechanical pumps. Oil diffusion, turbomolecular, titanium sublimation, and ion pumps are examples of high-vacuum systems.

The most common rough pump is the rotary vane mechanical pump. This pump consists of a set of vanes embedded in a shaft. As the shaft is rotated by an external motor,

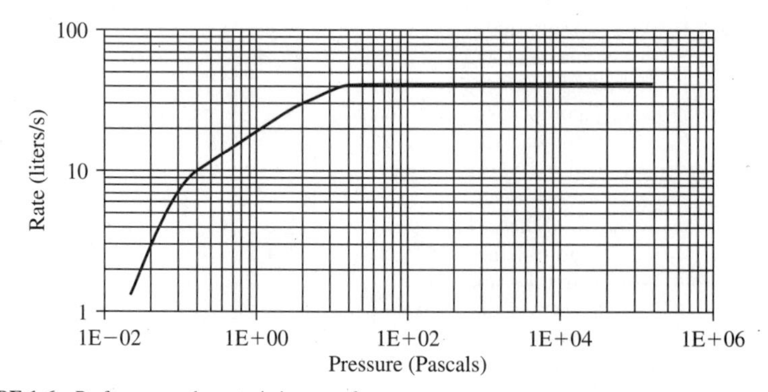

FIGURE 1-6 Performance characteristic curve for a rotary vane vacuum pump.

TABLE 1-3 Conversion factors for units used to measure pumping rates

Convert from: Unit	Convert to: liters/s	liters/min	ft^3/min	m^3/hour	cm^3/s
liters/s	—	60.0	2.12	3.6	1000
liters/min	0.016	—	0.0353	0.06	16.67
ft^3/min	0.47	28.32	—	1.699	471.95
m^3/hour	0.278	16.67	0.589	—	277.8
cm^3/s	0.001	0.06	0.0021	0.0036	—

any gas in the large cavity of the pump is compressed and expelled through the exhaust. This compression generates a vacuum in the cavity, which is then presented to the inlet of the pump as the shaft continues to rotate. Gases from the chamber rush in through the inlet to fill the vacuum in the pump and are subsequently compressed and expelled through the exhaust as the shaft rotation continues. In this way, gases in a vacuum chamber are compressed and expelled out of the system. The cavity of this type of mechanical pump is filled with oil; this provides lubrication, cooling, and a positive seal between the spring-loaded vanes and the walls of the pump. This simple mechanical process can be implemented on a large scale to provide high pumping rates in efficient, compact pumps. The limitations of this configuration are largely the result of the vapor pressure of the oil, which limits the maximum obtainable vacuum. In addition, backstreaming of oil vapors into the vacuum system can be a key concern if filters are not installed.

Rotary piston pumps operate similarly to rotary vane pumps; however, the maximum achievable vacuum (1 millitorr) is an order of magnitude lower than that of a rotary vane pump (1×10^{-4} torr). This pump consists of a piston connected to an eccentric on a rotating shaft. Gases from the inlet of the pump flow through the piston into the chamber created by the stator walls. As the shaft rotates, the piston makes contact with the walls of the pump chamber at one location. This action compresses the gases which are then expelled out the pump's exhaust. The cycle is then repeated. Rotary piston pumps have high pumping rates ($>$150 L/s or 318 cfm) compared to rotary vane pumps (maximum rate $\approx$90 L/s or 190 cfm).

In many cases, the pumping speeds and vacuum generated by rotary vane or piston pumps are not adequate. In such cases, a rotary lobe or roots blower pump can be used in addition to increase the vacuum and boost the pumping rate. This apparatus mounts directly to a mechanical pump and consists of a set of lobes rotating on two independent shafts. One lobe of the blower rotates clockwise while the other rotates counterclockwise. The resulting action compresses the gases entering the pump. These are then expelled out to a rotary mechanical pump. Using this technique, significantly higher rates can be obtained while achieving higher vacuums. For example, a commercially available mechanical pump capable of 42 L/s (90 cfm) can be boosted to a total rate of 240 L/s (510 cfm) with the addition of a roots blower.

One concern about all of these pumping systems is that they all, in varying degrees, use oil. Oil can be a serious problem if allowed to contaminate the deposition chamber, and can be devastating to film adhesion. As a result, oil must be trapped by installing a filter between the pump and vacuum chamber to prevent contamination. An alternative to oil-based pumps is the dry pump, which uses no oil. These types of pumps are gaining popularity for ultraclean film deposition systems, and are readily available from commercial sources.

High-vacuum pumping traditionally has been accomplished through the use of oil diffusion. As illustrated in Fig. 1-7, a heating element boils a quantity of oil, which then expands and is directed downward through a set of orifices at high velocity. This action effectively traps gas molecules in the downward oil stream, which can then be drawn out by a companion mechanical pump. This type of mechanism is very effective and can be used to provide pumping speeds greater than 1200 L/s (2544 cfm), with maximum obtainable vacuums approaching 10^{-6} Pa ($\approx 10^{-8}$ torr). Again, as with any oil-based pump, oil backstreaming into the deposition chamber is a concern. Normally, baffles and liquid nitrogen traps are used to prevent oil contamination. Liquid nitrogen traps can be used to assist in the pumping action, by condensing water vapor. As a result, the maximum vacuum of a diffusion pump can often be boosted to 10^{-8} Pa ($\approx 10^{-10}$ torr) with the addition of a liquid nitrogen trap.

Another very popular type of high-vacuum pump is the turbomolecular pump. This design uses a series of rotating blades (rotors) and stationary blades (stators), alternating in pairs for approximately ten sets of blades, which behave in a manner similar to a jet engine turbine. The rotors are accelerated to a very high speed (typically 27,000 rpm) by a motor built into the pump assembly. These blades impart momentum to gas molecules in a vacuum chamber. The resulting molecular flow of gas through the pump causes a high level of compression, which is a very effective pumping mechanism. Turbomolecular pumps are small and require a mechanical or dry pump to provide backing vacuum at the pump exhaust. In combination with a pressure controlled throttle valve, these

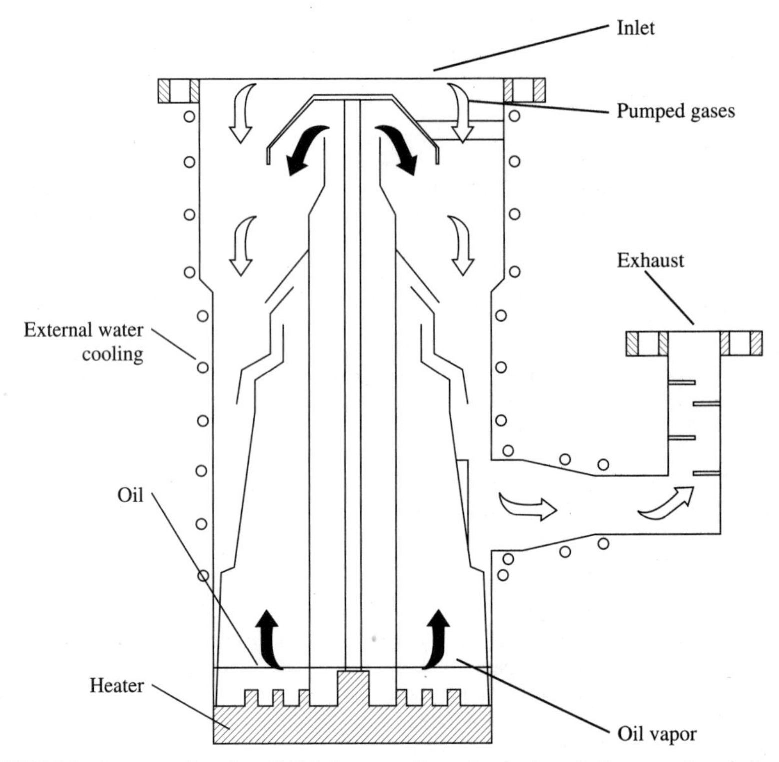

FIGURE 1-7 A cross section of an oil diffusion pump illustrating the flow of oil vapor and pumped gases.

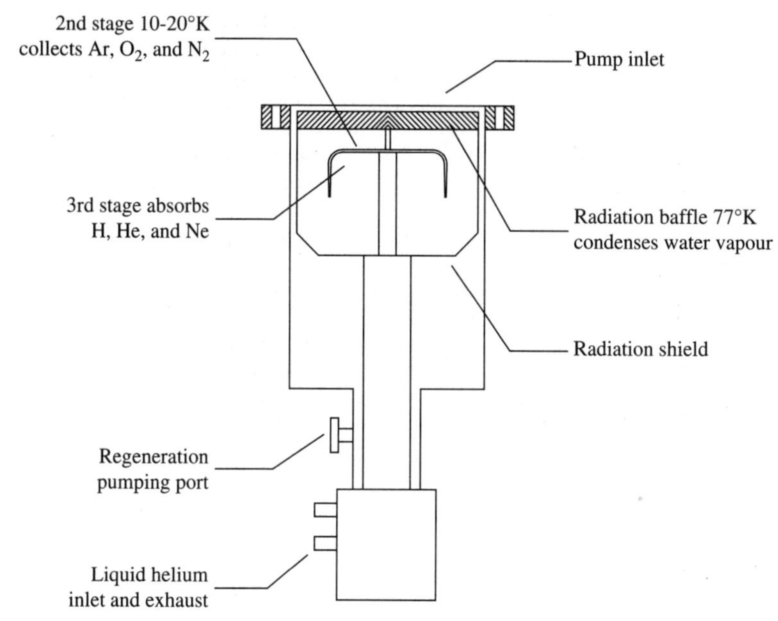

FIGURE 1-8 Cross section of a basic cryopump illustrating three stages and the gases collected by each stage.

pumps are ideal for sputtering systems where high gas loads can damage diffusion pumps. Turbomolecular pumps are also commonly used to evacuate load locks because small-sized pumps can be produced. The range of operation of a turbomolecular pump is almost identical to commercial available diffusion pumps with capacities from 50 L/s to 2000 L/s.

Dry pumping systems that do not use oil are becoming more popular because of the increasing demand for improved film quality and purity. The primary pumping system used for this type of application is the cryogenic absorption pump. These devices are extremely simple to operate. Cold surfaces are provided to condense gases and water vapor. Unlike most pumping systems, the pumped gases are not expelled from the pump, but are trapped inside on the cooled surfaces. Figure 1-8 illustrates a typical cryogenic pump with an external compressor until that delivers liquid helium (He) in a closed loop to surfaces inside the pump. The liquid He has a temperature of 4.2°K (−268.8°C), and is sometimes used in combination with liquid nitrogen (N) at 77°K to cool the internal surfaces of the pump. This three-stage process provides pumping action for three different sets of gases. The first area is maintained at a temperature of approximately 77°K to collect water vapor and any other gases that condense at this temperature. The second area is maintained at a temperature of 10°K to 20°K (−263 to −253°C), and condenses argon (Ar), oxygen (O), N and similar gases. The final phase of the process uses a porous material (charcoal is often used) maintained at the same temperature as the second phase 10 to 20°K (−263°C to −253°C). Because Hydrogen (H), He, and Neon (Ne) will not condense even at very low temperatures, cryogenic absorption in the cooled porous surface is used to effectively eliminate these gases from the vacuum chamber. The advantage of this system is the lack of internal moving parts that need lubrication, eliminating the risk of contamination that is associated with it. It is important to note that these units store all of the pumped gases, and as a result, a periodic purging is required to remove the condensed gases. This purging is referred to as regeneration. It is accomplished by introducing an inert gas, such as nitrogen while drawing on the cryogenic pump with a rough mechanical pump. In this way, all of the condensed vapor and gases are flushed out of the cryogenic pump.

1.2.4 In Situ Analysis and Monitoring Equipment

Several systems often are included in a thin film deposition system to allow an operator to monitor the deposition rate or the properties of the deposited material. A common device is a quartz crystal monitor, which allows the operator to determine the deposition rate and film thickness. These devices have a resonant quartz crystal with one surface exposed to the deposition source. As material is deposited onto the surface of this device, the resonant frequency changes because of the added mass of the deposited film. These monitors then use the density and acoustic impedance of the source material to determine the deposition thickness and rate. Careful calibration is normally required because these two quantities are unknown and because the monitor is normally in a slightly different physical position than the substrate.

In addition to systems which are designed to monitor deposition conditions, several devices and systems can be used to manipulate and control the temperature of the substrates. The goal of these systems is to improve uniformity of film deposition and to promote adhesion of the resulting films. Manipulation of substrates or wafers is often accomplished by rotation or orbital motion, so that any nonuniformity in the deposition rates resulting from position will not be transferred to the deposited films. Substrate heaters are often used to increase the surface mobility of deposited films and thus improve film adhesion, control grain structure, and minimize surface roughness.

Advanced deposition systems often include capabilities to monitor the film growth as well as residual gases in the deposition environment. Analysis techniques used include electron spectroscopy for chemical analysis (ESCA), auger electron spectroscopy (AES),[7] ellipsometry,[8] and reflection high energy electron diffraction (RHEED).[9] These techniques provide information on the chemical composition and crystal structure of deposited films and the level of impurities and film uniformity; these are discussed in detail in Chapter 6. The advantage of in situ analysis is the ability to use real-time information as feedback to process control. This approach can help prevent the loss of yield caused by incorrect deposition conditions.

1.3 EVAPORATION

1.3.1 Introduction

Perhaps the simplest film deposition technology, vacuum evaporation is a versatile technique that can be used to deposit most materials. This is a PVD process and has the advantage of high deposition rates, simplicity, and relative ease of use. Vacuum evaporation is used for deposition of conductor materials in electronic circuits and devices, for the application of dielectric and optical coatings, and for developing technologies, such as high-temperature superconductors.

The first attempts at vacuum evaporation were done by Faraday using metal wires, which he exploded in a vacuum. These attempts led to the discovery of thermal evaporation in the late 1880s. Although these techniques were used to study the properties and effects of thin films for many years, most of the applications of this technique did not arise until the 1940s and 1950s. Several excellent reviews of the history of this technique are available [1,2].

The next section discusses the basic concepts, mechanisms, and equipment associated with vacuum evaporation, and explains how vacuum evaporation is used in modern thin film applications.

1.3.2 Deposition Mechanism

Vacuum evaporation is a deposition technique used to deposit a variety of materials by heating a source material under vacuum until it evaporates or sublimes. This evaporant is then deposited or condensed onto a substrate surface to form a film. The source material melts into a liquid and subsequently evaporates into a gaseous vapor or sublimes directly into a gaseous state. Evaporation takes place in a vacuum where the mean free path of atoms in the evaporant material is much longer than the distance from the source to the substrate. The rate of deposition can be expressed by the Hertz-Knudsen equation:

$$\frac{\partial N}{\partial t}\frac{1}{A} = \frac{\alpha(p'' - p)}{\sqrt{2\pi mkT}} \tag{2}$$

$\frac{\partial N}{\partial t}$ = rate of deposition from a source with surface area A (with respect to time)
α = evaporation coefficient
m = molecular weight of the evaporant
k = Boltzmann's constant
T = temperature
p'' = vapor pressure at the evaporant surface
p = hydrostatic pressure acting on the source's surface.

It is interesting to note that the evaporation coefficient is larger for clean surfaces and lower for contaminated surfaces.

Vacuum evaporation is a low-energy process, because the deposited material condenses onto the substrate with very little kinetic energy. Material incident from an evaporation source to the intended substrate typically possesses 0.5 eV of kinetic energy compared to 10 to 100 eV of kinetic energy for sputter deposition. Also, the film deposition is almost strictly line of sight since the vapor condenses onto the exposed surfaces adjacent to it and does not coat edges perpendicular to the source. This is often referred to as a lack of step coverage. Film deposition thickness can be controlled by the quantity or rate of generated vapor material as well as the distance from the source to the substrate. Rates depend greatly upon the substrate-to-source geometry, and the deposition rate can vary across large substrates because of this strong function of distance. The variation in deposition rate resulting from position relative to the source is given by Knudsen's cosine law, $\cos\theta/r^2$, where r is the radial distance from the source and θ is the angle between that radial vector and the normal to the receiving surface. This can be used to express the thickness variation on a coated surface centered a distance h below or above a source as

$$\frac{t_x}{t_o} = \left[\left(1 + \frac{x}{h}\right)^2\right]^{-3/2} \tag{3}$$

where t_o = thickness at the center of the coated surface directly above or below the source
t_x = thickness of deposition at some distance x from the center of the coated surface

When sources are very small in comparison to the source to substrate distance (h), the source can be treated as a point source and the expression reduces to:

$$\frac{t_x}{t_o} = \left[\left(1 + \frac{x}{h}\right)^2\right]^{-2} \tag{4}$$

since the variation as a function of position becomes $\cos^2 \theta/r^2$.

1.3.3 Evaporation Sources

There are two primary types of evaporation sources in widespread use today: thermal evaporation and electron beam evaporation. Both of these techniques have advantages which make them desirable for different applications.

One of the most common evaporation sources uses direct thermal heating to evaporate the source material. A crucible, boat, or wire coil (see Fig. 1-9) is used to hold the source material, and heat is produced by passing an electrical current through the wire, boat, or an external wire coil. The temperature of the evaporant can be directly controlled by manipulating the current through the heating element. As a result, the evaporation rate can be carefully adjusted. Thus direct thermal heating has the advantage of simplicity as well as the low cost of implementation. A disadvantage of this technique is that the entire source material, as well as its containment structure, such as a boat or crucible, must be uniformly heated to the temperature required to evaporate the source material. As a result, the maximum temperature which can be obtained is limited, and the rate of deposition cannot be rapidly changed. In addition, the material used to con-

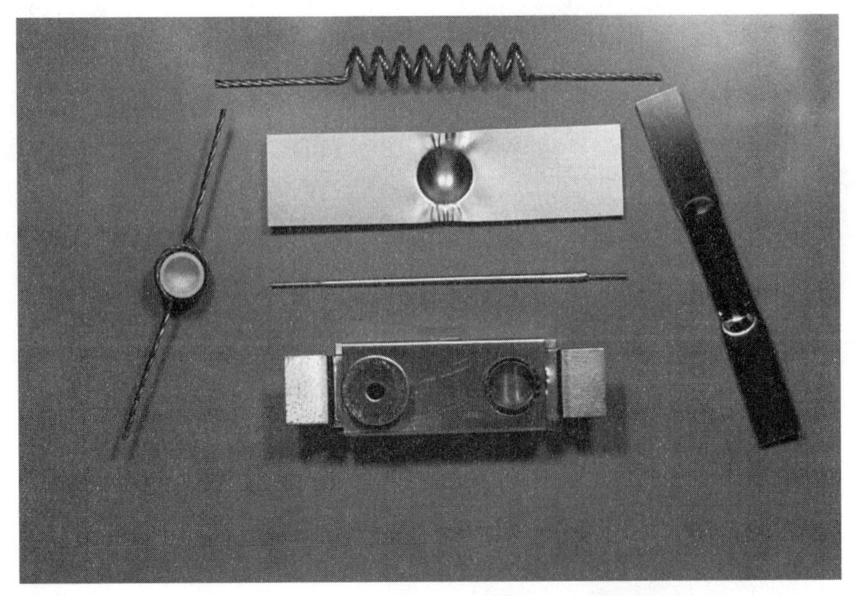

FIGURE 1-9 A variety of thermal evaporation sources. *Far right:* Crucible source with external tungsten coil heater. *Middle from the top:* Coil heater used for evaporation of metal wires, a dimple tungsten boat used for shot or powders, a chromium rod, and a specialized source for evaporation of SiO_2. *Far left:* A tungsten boat used for metal shot or powders.

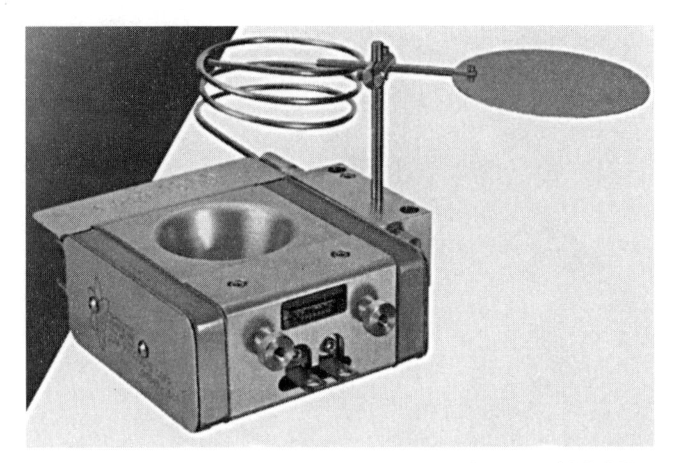

FIGURE 1-10 A typical small-volume electron beam source. *(Courtesy of MDC Vacuum Products Corp.)*

struct this containment structure must be carefully selected to avoid reactions or alloying with the molten source material. The cleanliness and purity of the source material and boat are critical for the production of high purity films because impurities may also evaporate and contaminate the final film structure. Crucible sources are commonly used for high-volume applications where a large quantity of source material is needed, whereas wire coils or metal boats are used for low-volume applications. A set of thermal evaporation sources is often combined into one film deposition system. This approach allows a series of film depositions to occur without the need to break the vacuum. A switching mechanism is used to allow a single power supply to consecutively operate several sources.

Electron beam sources are a vast improvement over conventional thermal sources for many applications. This type of source uses a high-energy beam of electrons, which heats the evaporate charge only in a small localized area. As a result, alloying with container materials is avoided, as is potential contamination. In addition, because the high temperature is generated only in a small area, a much larger range of materials can be deposited. Figure 1-10 illustrates a typical electron beam source consisting of an electron source, a water-cooled cavity or set of cavities (often called hearths), and an electromagnet for directing the electron beam. The highly focused beam of electrons heats the source material in a small localized position. This approach allows for fast modifications of the rate of deposition as well as the production of high-purity films. Many of these systems are also capable of steering or oscillating the beam to consume the source material more uniformly. Multiple hearth sources, as shown in Fig. 1-11, allow the deposition of several layers of material without breaking vacuum because the hearths can be rotated to deposit as many as six materials in succession.

1.3.4 Process Implementation

Although there are several methods to evaporate films, all of these techniques share similar support systems that provide the necessary conditions for the sublimation or evaporation to occur, and may also provide information on the evaporation rate and resulting film

FIGURE 1-11 Multiple electron beam hearths used to deposit successive layers. Each hearth can be rotated into position and used as a source. Multiple layers can be deposited in this manner without breaking vacuum. *(Courtesy of MDC Vacuum Products Corp.)*

structure. In particular, providing a high vacuum environment is critical because heated material undergoes a transition from solid to gaseous state at a much lower temperature if the atmospheric pressure is reduced. As a result, evaporation near or even below the source material's melting point is normally performed, rather than at the boiling point, which is much higher. In addition, high-vacuum conditions prevent the reaction of volatile materials with air, which could transform the deposited film into an undesirable material, such as an oxide caused by the presence of oxygen.

High-vacuum systems for evaporation are normally used initially to evacuate the film deposition chamber to 10^{-6} to 10^{-8} torr. The deposition process itself increases the pressure because of the addition of the vapor from the source material, and in the case of thermal sources, heating of the vacuum vessel. Figure 1-12 illustrates a typical evaporation system.

Because most evaporation processes are used to produce a high deposition rate, oil diffusion pumps, in conjunction with mechanical backing pumps, are often used because they provide high pumping rates. Turbomolecular pumps backed by mechanical pumps as well as cryogenic pumps have also been used. Dry pumps, which, unlike oil diffusion pumps, do not use oil, are gaining popularity because they usually produce cleaner vacuum systems. The choice of the vacuum system is largely determined by the source or sources used, the volume required, and the nature of the material to be deposited.

1.3.5 Deposition Conditions

The optimum conditions for the deposition of a material depend on the application. For example, aluminum films are used as conductors or contacts as well as reflective coatings. It is necessary to optimize different material properties for both these applications, in one case electrical conductivity and, in the other, optical reflection, even though both are aluminum films. In addition, the selection of the substrate material may also affect the prop-

FIGURE 1-12 A typical low-volume evaporation apparatus. *(Courtesy of Kurt J. Lesker Company.)*

erties of the deposited films, such as variations in surface roughness, material interactions, and adhesion. The optimum conditions may also depend on the apparatus used, because the source geometry, substrate geometry, and vacuum conditions vary. As a result, it is impossible to determine the ideal deposition conditions for a given material. A set of controlled experiments is normally used to determine the optimum desired property or properties for a given deposition system and material.

However, it is useful to have some guidelines. A table is included that lists the vapor pressure, melting points, and suggested source materials and configurations for most materials encountered in microelectronics and optical applications (see Appendix A).

1.3.6 Applications

Evaporation has been used for a very broad range of applications, including: metallization deposition for electronic assemblies, devices, and sensors; magnetic coatings [10]; optical coatings [11]; and optoelectronic devices [12,13]. This technique is one of the most common methods currently used to deposit thin film metals and dielectrics for electronic applications, such as VLSI circuits [14] and multichip modules. In recent years, sputtering has replaced evaporation in a number of applications, largely because of the increased consistency and adhesion of sputtered films. However, several applications, such as lift-off patterning and electronic device fabrication, still primarily use electron beam evaporation.

1.4 MOLECULAR BEAM EPITAXY

1.4.1 Process Overview

Epitaxy is a special type of film deposition characterized by a continuation of crystal structure from the substrate to the film. This type of film is essentially single crystalline, and it can be grown one monolayer (a single atomic layer) at a time. Highly order crystalline films can be prepared in this way and used for various applications, such as electronic devices. In contrast, non–epitaxial-deposited films are amorphous or polycrystalline, and the substrate and film lattice are mismatched. As a result, this type of film often is not well-suited to the manufacture of devices, particularly very small submicron devices and integrated circuits that are often smaller than the grain size of polycrystalline films.

Molecular beam epitaxy (MBE) is a highly developed technique used to perform a wide variety of epitaxial depositions, from III-V semiconductors to high-temperature superconductors. The basic film growth concept is essentially a refinement of vacuum evaporation, in which thermal or electron beam sources are used to generate and deposit a beam or flux of atomic or molecular species. There are several sources with varying flux rates that can be used to grow complex compounds with precise elemental compositions. MBE sources are focused on a substrate, which is heated to provide a high degree of surface mobility for incident species. The resulting condensation of molecular or atomic flux results in the formation of a crystalline film, one monolayer at a time. Unlike the majority of thin film deposition techniques, MBE is performed in an ultra-high vacuum environment that is often greater than 6×10^{-9} Pa (5×10^{-11} torr). This allows the deposition of contaminant-free films.

1.4.2 Deposition Systems

The requirements of epitaxial layer film deposition are far more stringent than those of other types of deposition. It is difficult to create and reliably maintain ultra-high vacuum, and strict cleanliness standards are required for monolayer deposition. As a result, modern MBE system designs have become modular, with separate chambers to perform individual functions, such as film analysis, substrate preparation, and the actual growth or deposition process. These separate modules are made from 304 or 316 stainless steel alloys, which exhibit low outgassing characteristics and have been prepared with highly polished internal surfaces to minimize absorption of gases and moisture by the chamber walls. The individual modules are then connected so that wafers can be transported from module to module while under constant high-vacuum conditions. Once a wafer is introduced into an MBE system, the system is normally kept at a minimum vacuum of 6×10^{-8} Pa (5×10^{-10} torr). Figure 1-13 illustrates a typical configuration for a modern MBE system, consisting of an entry module, an MBE growth module, and a wafer preparation chamber. These systems can also include additional chambers for metallization that use conventional sputtering or evaporation techniques.

A typically process flow for MBE deposition consists of the following:

- The wafer(s) are prepared. This includes a rigorous cleaning followed by the growth of a passivation layer
- The wafer or wafers are introduced into the deposition system and the entry module is subsequently evacuated. Wafers are often loaded into cassettes that can be heated during the initial evacuation process to expel any moisture or absorbed gases.
- Desorption of the passivation layer is performed by heating the wafers in high vacuum conditions. The resulting surface is very clean and pure, and it is ideal for epitaxial deposition.
- Film is grown under very-high-vacuum conditions using several beam sources at rates up to one μm per hour. The flux from each source is controlled using fast shutters that allow for monolayer-level control of the deposition. In situ analysis using optical pyrometers and RHEED is often used to control process parameters and the rate of deposition.

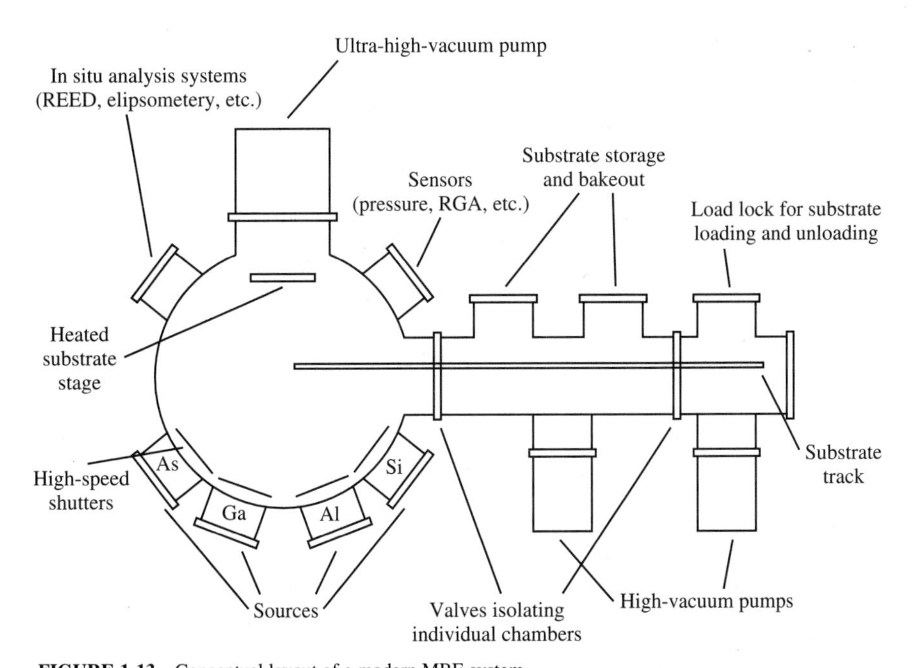

FIGURE 1-13 Conceptual layout of a modern MBE system.

- MBE systems often have separate analysis modules that allow film to be analyzed before exposure to the atmosphere. This method allows quality control analysis, process verification, and feedback without contamination of the films caused by environmental exposure.

The substrates normally used for this technique are single crystalline wafers because one of the key advantages of MBE is its ability to deposit highly crystalline films. Before deposition, the wafers are degreased and chemically etched to remove any surface contamination. A subsequent oxide passivation layer is applied to protect the surface of the wafer and prevent reaction between atmospheric gases and the wafer material [15].

Wafers are bonded by an indium or gallium metallurgic bond or are clipped to wafer holders that provide support for the wafer and attachments for manipulation systems within the MBE chamber. Individual wafers are then loaded into cassettes, which are subsequently introduced into the deposition system. Because it normally requires several days to achieve ultra-high-vacuum conditions, these types of deposition systems are kept at a constant vacuum; only a small entry chamber is exposed to the external atmosphere. Once a cassette has been loaded into the entry chamber, the chamber is evacuated to high-vacuum condition while isolated from the rest of the MBE system by a valve. Thus, only a small amount of contaminates are injected into the system, and the constant high-vacuum condition of the system is maintained.

Wafers are then transported along a track into a preparation chamber. Here, the passivation layer is removed to reveal a clean, pure surface layer. Removal of the passivation layer is accomplished by heating the wafer in an ultra-high-vacuum environment to desorb the oxygen in the passivation film. This task is usually performed at 500°C to 550°C for GaA wafers and 700°C to 900°C for silicon wafers. This step also completely desorbs any moisture or gases that have been absorbed during wafer fabrication or subsequent exposure to the outside environment. Sputter or plasma cleaning can also be used to remove the top layer of the substrate and expose a fresh, pristine surface [16].

The growth stage begins with the transportation of the wafer into the growth chamber. There it is attached to the heated substrate stage. Precise thermal control and wafer rotation during growth are critical to the uniformity and the final film composition and structure. As previously discussed, a variety of thermal or electron beam sources can be used to deposit simultaneously a number of elemental or molecular species to form the desired compound. RHEED analysis, described in detail in Chapter 6, optical pyrometery, or elipsometry [17] can be used to determine the rate of deposition and the substrate temperature.

Most MBE systems have the capability to perform chemical, structural, and crystallographic analysis in either the growth chamber or a separate chamber. With separate chambers, the final wafer is transported to the analysis module, and film composition and crystal structure are determined. This information is used for quality control and process feedback to the growth system.

1.4.3 Applications

The primary applications of MBE are based on its ability to deposit films with precise crystal structures, complex compounds, and highly controlled layer structures. For example, multiple thin layers of alternating III-V compounds can be manufactured with monolayer precision. Thus, MBE has proven invaluable for optoelectronic applications such as LEDs, diode lasers, photodetectors, and solar cells [18–22]. These same attributes are also ideal for the manufacture of high-speed devices using GaA and related compounds, which are faster and have a greater bandwidth than conventional silicon devices [23]. Many advances have been made in heterojunction bipolar transition (HBT), high electron mobility transistor (HEMET), and metal semiconductor field effect transistor (MESFET) devices for high-frequency and or high-speed applications, such as microwave low-noise amplifiers [24–28]. Other applications exist for high-temperature superconductors and diamond films [29–31]. Because the electrical properties of these materials are directly related to the film crystal structure, MBE film deposition is an excellent choice.

1.5 SPUTTER DEPOSITION

1.5.1 Introduction

Sputtering is a PVD process involving the removal of material from a solid cathode. This is accomplished by bombarding the cathode with positive ions emitted from a rare gas discharge. When ions with high kinetic energy are incident on the cathode, the subsequent collisions knock loose, or sputter, atoms from the material. The process of transferring momentum from impacting ions to surface atoms forms the basis of sputter coating. Sputtering was originally developed to deposit refractory metals, which could not be deposited using the thermal evaporation techniques of that time. Today, sputtering has developed into a versatile deposition technique that is able to deposit most materials.

A typical sputter deposition system consists of a vacuum chamber, a sputter source, vacuum sensors, a substrate holder, and a pumping system (see Fig. 1-14). Unlike most other vacuum deposition techniques, sputtering occurs in a so-called "windy" vacuum, thus named because of the constant flow of rare gas into the chamber. Deposition pressure is controlled by the rate of gas passing into the chamber and by a throttle valve placed between the vacuum pump and deposition chamber. By adjusting the gas flow into the system and the throttle valve opening that controls the pumping speed, a constant

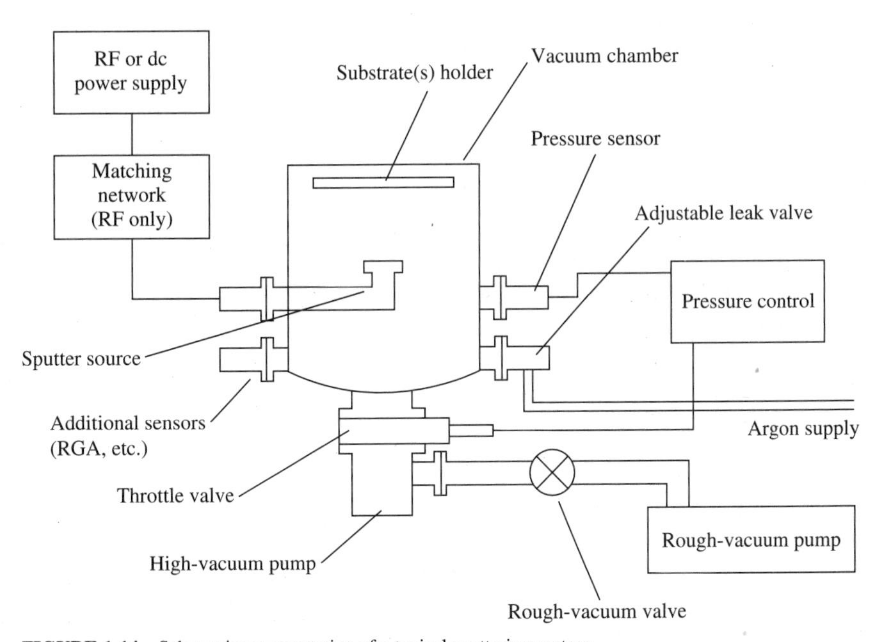

FIGURE 1-14 Schematic representation of a typical sputtering system.

pressure of approximately 13.3 mPa to 13.3 Pa (0.1 to 100 millitorr) is maintained. This supply of rare gas is then ionized using large potentials at the source, resulting in the generation of a plasma and sputtering from a target material onto the substrate and chamber walls.

1.5.2 Sputter Sources

The two most common types of sputter sources are diodes and magnetrons. Both of these configurations can be operated with direct current (dc) or radio frequency (RF) potentials to generate a plasma through the ionization of a rare gas.

A basic diode sputter source consists of a target in the shape of a flat disk bonded to a water-cooled plate. An external potential is applied from an outside power source, charging the target to a high negative voltage (3 to 5 kV). A rare gas, usually Argon, is introduced into the vacuum chamber between the target and the grounded substrate and chamber walls. The large difference in potential forms a plasma, caused by ionization of the Ar atoms in the intense electric field. This ionization results in a negatively charged electron and positively charged ion pair, whereas the plasma itself retains a net neutral charge. Positively charged ions are attracted to the negatively charged target and accelerated by the electric field, resulting in a collision with the target material. Bombardment of the target with these high energy ions leads to sputtering of the target atoms, forming a coating on the substrate and chamber walls. In this type of configuration, the target, substrate holder, and grounded vacuum vessel can be considered a parallel plate capacitor. This creates a number of advantages:

- The plasma or ionized gas is created uniformly between the cathode and anode, causing a uniform consumption or sputtering of the target.
- Diode sources are the cheapest and least-complex sputter source to use.
- Targets for this type of source consist of a flat disk. They can be easily fabricated without the need for complex shapes often required for magnetrons. As a result, the targets are lower in cost, which can be a critical issue for the deposition of precious metals.
- Diode sources can be fabricated in large sizes for production-scale deposition equipment.

Based on these advantages, diode sources are widely used in manufacturing for the metals and metal alloy of films. However, there are several limitations:

- Diode sources are inefficient and do not make use of the generated electrons to maintain the plasma.
- Electrons generated in the plasma irradiate the substrate, which can result in damage to the substrate surface, or lattice, for highly crystalline substrates.
- Deposition rates from diode sources are much lower than those from magnetron sources.

Note that the major limitations of the diode source stem from the lack of control and use of electrons generated in the plasma.

Magnetron sources overcome the limitations of diode sources by using a magnetic field to control the motion of electrons. This approach centers around a magnetic field's ability to exert a force on a charged particle in motion. Lorentz's law states that the force F on a particle with a charge of q and a velocity v from an incident magnetic field B is given by:

$$F = q\vec{v} \times \vec{B} \tag{5}$$

As a result of this basic physical law, a magnetic field applied to a plasma causes the charged particles to move in a helical path with a radius of

$$r = \frac{mv_\perp}{qB} \tag{6}$$

where r = radius of the helix
m = mass of the charged particle
q = charge of the particle
$v_\perp$ = component of the particle's velocity normal to the applied magnetic field B.

Because electrons and ions in a plasma have identical charges but radically different masses, the radius of their helical paths is quite different. Low-mass electrons are highly affected by the magnetic field and move with a radius that is usually much less than the dimensions of the plasma and deposition system. By comparison, the much-heavier ions travel with a radius much greater than the dimensions of the plasma and deposition system; as a result, the magnetic effect on the ions is not significant. Magnetron sources use the properties defined in this physical law to control the electrons generated in the plasma and apply them toward plasma regeneration.

A typical planar magnetron source is depicted in Fig. 1-15. The key difference between a planar magnetron source and the diode source is that a permanent magnet is placed behind the target with its north pole at the center of the target and its south pole in a band at the edge of the target. The resulting magnetic field confines the electrons to a circular path on the surface of the target disk. These energized electrons further ionize the gas molecules through collision, resulting in a large increase in plasma density at the target surface and an increase in sputtering rate from the target, with reduced irradiation of the substrate and chamber walls. Because the sputtered target atoms are relatively massive and neutrally charged, they are not affected by the magnetic field, and they migrate to coat the substrate and exposed chamber surfaces. The key advantages of this type of source include the following:

- Vast improvement in efficiency over conventional diode sources is exhibited.
- Sputter deposition rates are improved because of the increased plasma density.
- A reduction in substrate bombardment by energetic electrons occurs.

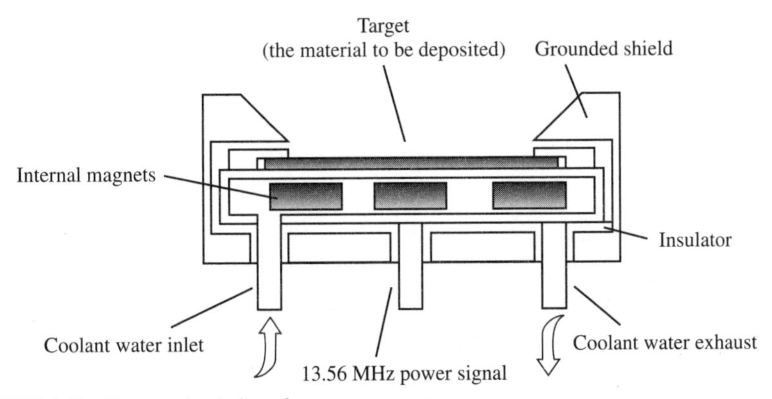

FIGURE 1-15 Cross-sectional view of a magnetron sputter source.

The primary problems of magnetron sources are

- These sources are more complex and therefore more expensive.
- Uneven consumption of the target material results from the confinement of the electrons around a ring at the center of the target. In particular, this is expensive for precious metal deposition.

As mentioned previously, both types of sources can be operated with dc and RF potentials. The first technique used was dc operation, and it is very effective for the deposition of conductive materials. However, targets composed of insulators will build up a significant positive charge, resulting in collection of positive ions that cannot be drained away. As a result, methods using RF sputtering were developed.

RF sputtering uses a 13.56 megahertz (MHz) sinusoidal voltage to drive the source. The substrate and chamber walls are held at ground potential. Using this approach, the charge that builds up on a dielectric target is dissipated through the second half of the cycle. RF sputtering is perhaps the most versatile technique and can be used to deposit virtually any material. The key difference between an RF system and a dc deposition system is the need for a matching network and an RF source rather than a dc source. The matching network is needed to match the impedance of the source to the chamber in order to maximize the power transfer from source to load. A bias can also be applied to the substrate to increase irradiation, causing resputtering of the deposited films [32]. This process is selective such that material at elevated locations on the substrate is selectively removed. In this way, sputtering can also be used for epitaxial liftoff since the normally excellent step coverage is reduced. An added benefit of irradiation, can also be increased film adhesion due to the additional kinetic energy imparted by the irradiating particles. A typical RF sputter deposition system is illustrated in Fig. 1-16.

1.5.3 Applications

Because the deposition mechanism of sputtering is mechanical in nature, refractory materials such as tungsten, tantalum, and molybdenum can be easily deposited at temperatures well below their melting points. As a result, sputtering is the method of choice for a wide range of refractory materials used for metallization of circuits and electronic packages, diffusion barriers, and optical coatings [33,34]. In addition, it is very useful for alloys or compound materials (such as high temperature superconductors) because the resulting film structure can often be tailored to closely match the composition of the starting target material [35,36]. For additional information on sputter deposition the reader is referred to a detailed discussion [37].

1.6 CHEMICAL VAPOR DEPOSITION (CVD)

So far, this chapter has discussed deposition techniques that use some physical means of delivering atoms or molecules of a material to a substrate surface to create a thin film. In contrast, CVD relies on reactive carrier gases to transport precursors of the desired material to the substrate surface. Here they react with other gases or decompose to produce stable reaction products, which are deposited on the substrate. CVD methods are among the most versatile deposition techniques because a wide range of chemical reactants and reactions can be used to deposit a large number of different types of films for a wide range of applications.

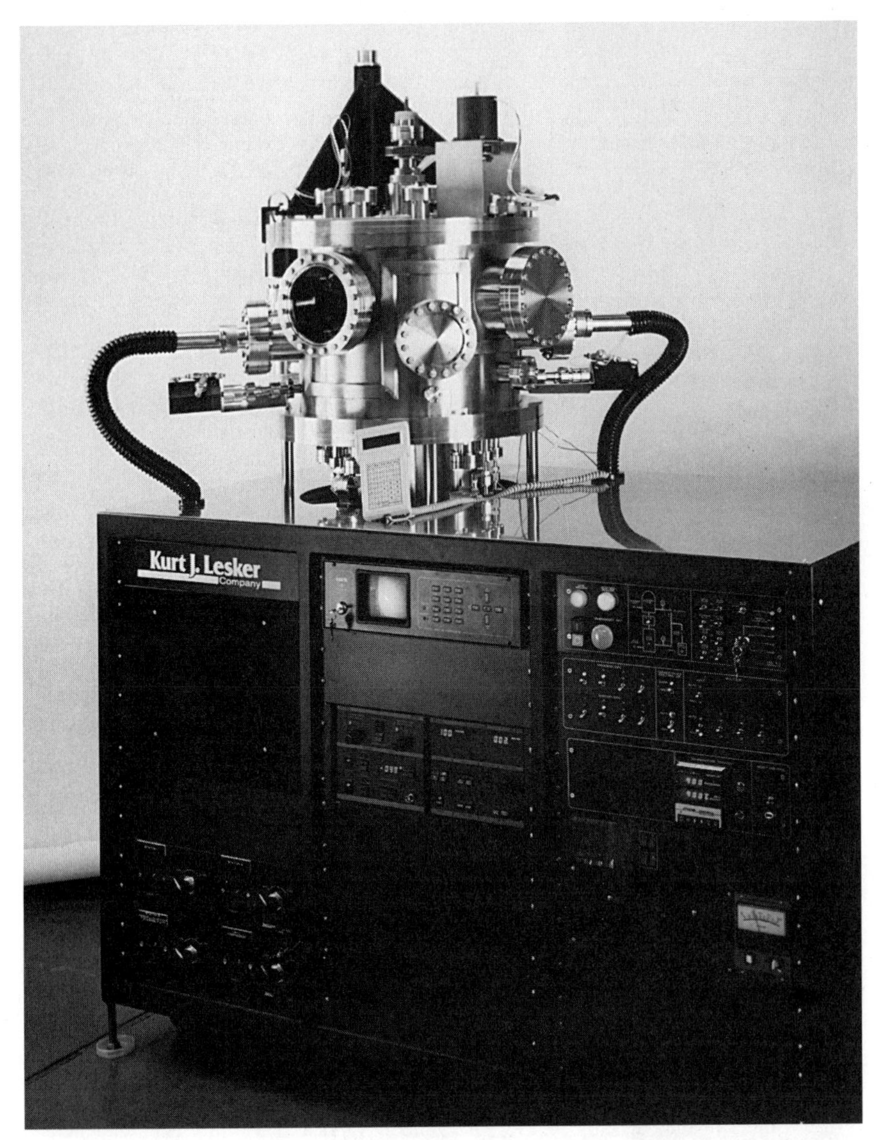

FIGURE 1-16 A typical low-volume sputter deposition system. *(Courtesy of Kurt J. Lesker Company.)*

CVD is normally performed at rough vacuum levels using a reactor, which consists of the following, as depicted in Fig. 1-17: a reaction chamber; thermal heating or plasma energy sources that provide the needed energy for the chemical reactions to occur; gas carriers; and source materials. The carrier gas flows through the source materials and picks up some quantity of the material. These materials are then mixed in a mixing manifold and injected into the reaction chamber. In the reaction chamber, the reactive agents in the carrier gas flow across the substrate(s) and a chemical reaction or set of reactions takes place

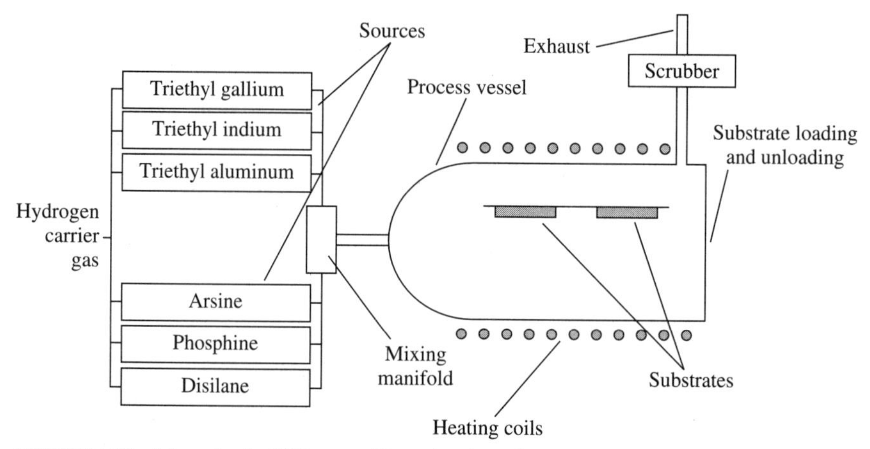

FIGURE 1-17 Schematic of a CVD system illustrating thermal reaction chamber, gas carrier, and an array of common source materials.

as a result of energy provided by some external source, such as thermal heaters or a plasma discharge. Because the exhaust products from this process are often toxic or corrosive, special procedures are often required to ensure that these materials are removed before venting the exhaust to the atmosphere. Scrubbers are normally employed, and they can be tailored to the particular reaction products present in the exhaust. In principle, large substrates can be used because the gas flow can be achieved over a very large area. However, the ability to provide a uniform flow over the surface, which is mandatory for uniform film deposition, is a practical limitation.

A number of variations on and subcategories of CVD techniques have been developed. Some of the more common techniques include:

- Metalorganic CVD (MOCVD): This technique uses a thermally heated reaction chamber and sources that are organometallic in nature. Some examples of this technique are the deposition of III-V and II-VI semiconductors for device applications.
- Plasma-enhanced CVD (PECVD): This technique uses a plasma discharge to provide the excitation necessary for chemical reaction to occur. PECVD is often used for diamond deposition, which is discussed in more detail in Chapter 7. Other applications of this technique include the deposition of silicides, silicon, and refractory metals.
- Atmospheric pressure CVD (APCVD): This technique does not require a vacuum, and it can be performed in a continuous process with a belt transporting a continuous flow of substrates in and out of the deposition chamber. Applications of this technique include oxides and silicate glasses.

Many variations on these basic processes have been developed for and adapted to the deposition of a wide variety of materials.

A typical process flow for a CVD consists of the following:

1. The reactants are transported to the deposition reactor in a carrier gas flow.
2. Gas phase reactions are initiated in the reaction chamber by an energy source, such as a plasma discharge or thermal activation. The products of these reactions are the film precursors.

3. The film precursors are transported to the heated substrate surface in the carrier gas flow, and they are condensed or absorbed by the substrate surface.
4. Deposited species diffuse across the substrate surface to growth sites on the surface.
5. The deposited film grows and reaction by-products are desorbed and transported away from the deposition region by the carrier gas.

For more in-depth information on CVD and the wide range of variations on this basic deposition process, the reader is referred to the literature [38,39].

1.7 LASER ABLATION

Laser ablation uses a high-power laser pulse to vaporize a small area of target material. The cloud of the target material generated is then deposited onto a nearby substrate. This type of process is usually performed in a high-vacuum chamber, which provides a window for the laser pulse from an external laser source. Ultraviolet excimer lasers delivering 0.1 to 1 Joule (J) pulses of 15 to 45 nanoseconds (ns) duration are commonly used with 1 to 100 hertz (Hz) repetition rates to produce deposition rates of 0.1 to 100 nanometers per second (nm/s) [40]. By scanning the laser beam across the target and rotating the substrate, the target is consumed in a uniform manner and a uniform film can be produced.

A typical example of a laser ablation process is depicted in Fig. 1-18*b*. In this case, an external laser is positioned to provide high energy laser pulses directed toward a ceramic target, resulting in a plume of ceramic liberated from the target that consists of a variety of molecular species of the target materials. This plume of material condenses on a nearby substrate, which is heated to promote surface diffusion and film adhesion.

The primary advantage of this process is its ability to reproduce the composition of the target in the vapor status and, subsequently, in the resulting films. This property makes this type of process ideal for the deposition of complex, multicomponent materials. Another advantage of this technique is its capacity to deposit virtually any material because very high temperatures are generated at the point of impact between the laser beam and the target. Unfortunately, the requirements of a UHV system and the high cost of excimer lasers currently limit the widespread use of this deposition technique. However, this process has been used to deposit a number of materials, including high-temperature superconductors [41–43], graphite [44] and electro-optical materials [45].

1.8 PLATING

1.8.1 Introduction

Plating is used for a variety of applications in microelectronics, optics, and related fields. The basic process involves the deposition of a metal film using a chemical reaction. There are two types of plating: electroplating and electroless plating. The difference between these techniques, as the names suggest, is that electroplating requires an applied electrical voltage and electroless plating does not. Electroplating can only be performed on conductive substrate or films, but can be used to produce films with thicknesses up to tens of micrometers (μm). In contrast, electroless plating can be performed on dielectrics and conductors; however, it is usually used for thin protective coatings. Both of these techniques are described in detail in the following sections.

(a)

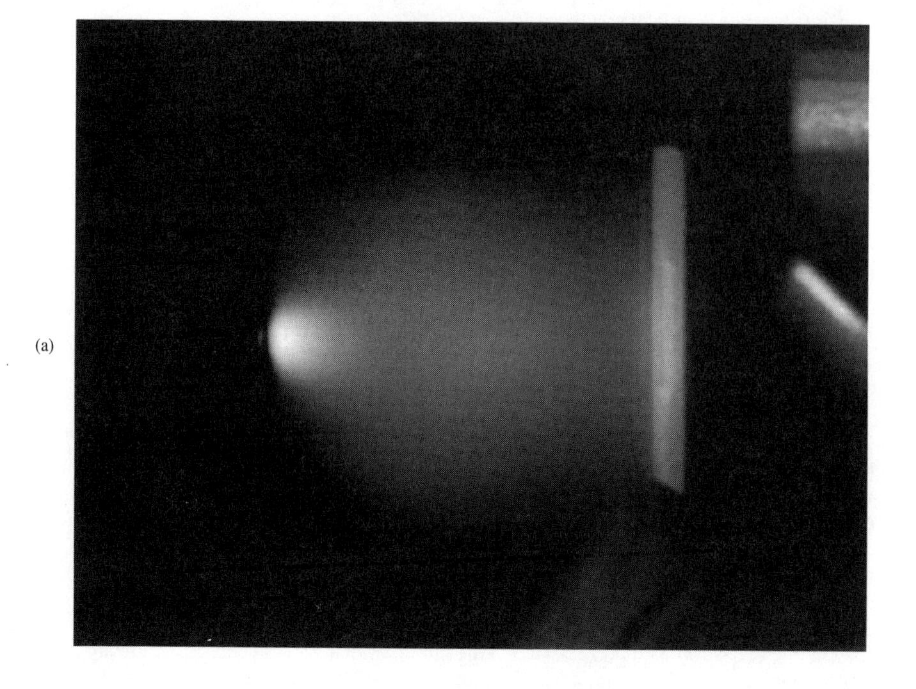

(b)

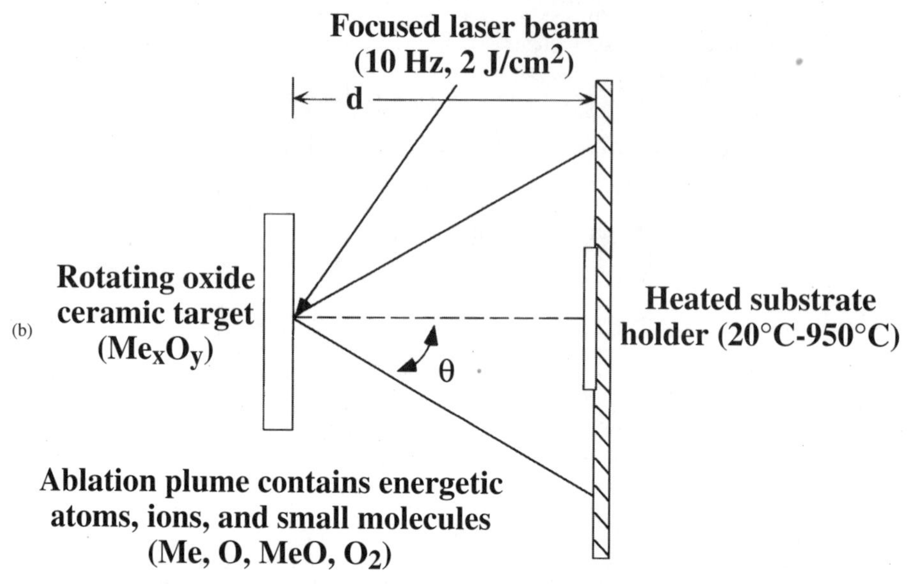

FIGURE 1-18 *(a)* A laser ablation system in action, illustrating a laser beam striking a target and a plume of material being deposited on an adjacent substrate. *(b)* Schematic layout of a laser ablation system used to deposit metal oxides. *(Courtesy of Dr. D. Chrisey, Naval Research Laboratory.)*

1.8.2 Electroplating

Certain spontaneous chemical reactions produce electrical energy under proper conditions. An example of this is the storage battery. Other reactions require the addition of electric energy to proceed. When electric current is used to effect chemical change, the process is called electrolysis. An example is the separation of water into its component parts with the application of an electric current.

With thin film deposition, a more practical application of electrolysis is electroplating. Under proper conditions, electroplating can provide high aspect ratios, dense packing and accurate pattern reproduction. The requisites for successful electroplating include:

- A conducting surface, often supplied by a vacuum-deposited seed layer.
- Good resist adhesion is required, because the plating presses against the resist when selective area plating is used.
- The importance of clean substrates, solutions, and containment vessels cannot be understated. Even low impurity levels result in poor-quality films.
- The plating solution must have easy access to recesses. This is often accomplished with agitation.
- The plating solution must be compatible with resist and substrate material. Many plating solutions are caustic and or reactive.
- Land and void areas should be uniform to maximize uniform current distribution.

The basic apparatus is relatively simple. It consists of a power supply, anode, cathode, and electrolyte (see Fig. 1-19). The electrode connected to the positive terminal is called the anode. It serves as the electron source. Electrons flow from the anode metal to the power supply. The positively charged metal cations which remain are dissolved into

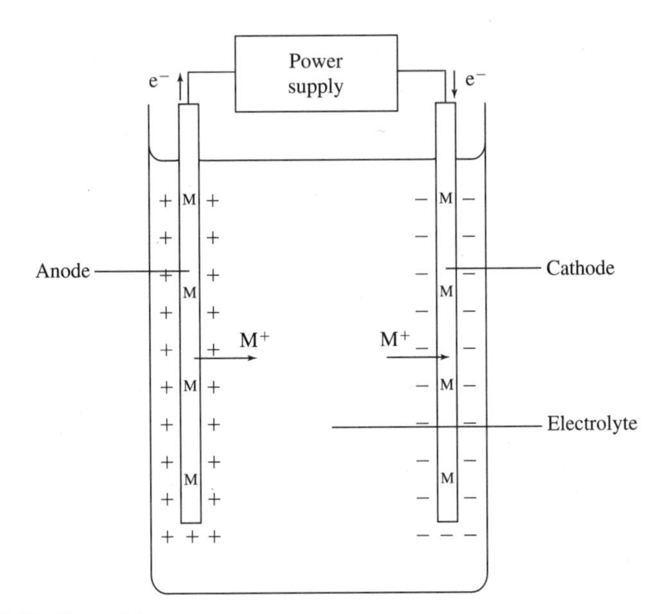

FIGURE 1-19 Electroplating apparatus.

the plating bath. The removal of electrons at the anode creates a potential between anode and cathode. The charged cations diffuse to the negatively charged cathode (substrate), where the solvated metal ions are reduced to the metallic state by accepting one or more electrons. They attach themselves to the cathode, and they are removed from the solution. Removal of metal cations from the plating bath creates a concentration gradient adjacent to the cathode, causing neighboring ions to move toward and diffuse into the cathode. For each electron accepted by the positive ion, one is donated at the anode, creating a flowing current to complete the electric circuit. The net result is that a metal is deposited (electroplated) onto the cathode (substrate) using electrical energy to produce chemical change.

The morphology and composition of electrodeposits are generally more varied than those obtained by PVD, and may depend on the following:

- Current density
- The presence of impurities
- The nature of the anions or cations in solution
- Power supply current waveform
- Physical and chemical nature of the substrate surface
- Bath temperature
- Solution concentration.

Clearly, in the general case, the deposition will be influenced by several of the aforementioned factors. Current density and solution composition are particularly important to minimize ohmic losses in the plated films.

The relationship between the weight of the deposited material and the various plating parameters can be stated as follows:

1. The weight of the deposit is proportional to the magnitude of current.
2. The weight of material deposited by the same quantity of electricity is proportional to the electrochemical equivalent, E. For example, gold (Au) is 2.043 milligrams per coulomb (mg/C), copper (Cu) 0.329 mg/C, and nickel (Ni) 0.304 mg/C.

This relationship can be expressed as

$$\frac{T}{\tau} = \frac{JE\alpha}{\rho} \tag{7}$$

where T = thickness of the deposit
τ = time of deposition
J = current density
E = electrochemical equivalent
α = current efficiency ratio of actual/theoretical weight deposited
ρ = density of deposit

If the plating parameters remain unchanged, the deposition rate will increase proportionally with current density. For this to occur, no secondary reactions can take place; all too often, though, this does occur. As a consequence, competing reactions caused by the presence of addition agents, impurities, or hydrogen evolution reduce the efficiency of the plating process.

The growth of the deposit can be influenced by several of the previously discussed factors, although any one factor may be sufficiently influential to control the growth form. The current density and solution purity seem to be particularly important. When deposition is carried out with low concentrations of simple ions in solutions, surface irregularities may increase [46]. Many baths have organic additives that aid in leveling. In this case, surface irregularities are ironed out during the deposition. Inclusion of these organics in the electroplated films in excessive amounts increases ohmic losses and reduces adhesion in subsequent bonding operations. Thus, adequate metal ions must be available for high-purity deposits with high conductivity and smooth surfaces. For plating baths with expendable anodes (that is, anodes which slowly dissolve, saturating the bath with metal ions), this poses little problem. Although the bath should be monitored for purity, the metal ion concentration remains stable. With Au plating baths, in which the anode is insoluble, the Au ion concentration is depleted with use. Precise ion concentration is very important in this system.

It is important to recognize that the conductivity of electroplated films depends on the bath purity. Small amounts of impurities can drastically increase the resistivity of the deposit. Weisberg tabulated the effect of impurities of only 1% on the resistivity of Au. His results show that 1% Ni or Zn increase the resistivity by 132%, while 1% Cu results in a 63% increase [47]. Thus, very small amounts of unwanted material greatly increase the resistivity. For example, parts preplated with Cu and Ni must be carefully rinsed before Au plating to reduce contamination.

Deposit purity and current density are directly related. Current densities above or below the optimum value can result in an increase of the bath's cation constituents, as illustrated in Fig. 1-20*a*. This cyanide Au bath uses thallium (Tl) as a grain refiner. Plots such as this emphasize the importance of using accurate current densities and maintaining good bath control. Excessive plating rates, usually used in the interest of saving time, can deplete the grain refiner and lead to rough, large grain, porous deposits. They also can lead to co-deposition of unwanted impurities in the film. A plot for Cu, Fig. 1-20*b*, shows that reducing the current density in this simple acid-Cu bath below the recommended value of 200 milliamperes per square inch (mA/in^2) results in an increase in magnitude of the impurity level almost one order. However, the impurity level is still almost 2 orders of magnitude lower than that of the Au deposit because of the very high cathode efficiency of Cu baths as compared to that of gold (the high-90s for Cu as opposed to the mid-50s for Au).

In general, conventional dc is the most widespread plating technique. The time–current curve is shown, along with other rectification schemes, in Fig. 1-21. The uppermost method depicted in Fig. 1-21 is the most often used, in which a dc rectifier supplies a constant current. Here, the rate of arrival of metal ions depends on their diffusion coefficient. Electrode-to-part spacing and electrolyte agitation are important factors. Temperature has some effect, with lower temperatures resulting in larger grains. Areas with high fields will preferentially plate because the supply of ions is limited. Other rectification techniques widen the latitude of plating conditions to minimize porosity, contamination, and surface roughness. They interrupt the plating cycle and change the growth pattern of the grains, enhancing leveling and improving density.

Asymmetric dc is a variation of pulsed dc. The polarity is reversed for a fraction of the cycle, intentionally removing some material, exposing a fresh surface, and, to some extent, leveling the surface. This occurs because this part of the cycle is essentially dc, and high fields generated at surface high points selectively deplete material in those areas. [48].

Pulse plating is a special type of periodic reverse plating. The cathodic current pulses are interrupted by zero current pulses instead of the anodic pulses. Additionally, the current is allowed to drop to low values, allowing ions to diffuse to the cathode. A number of

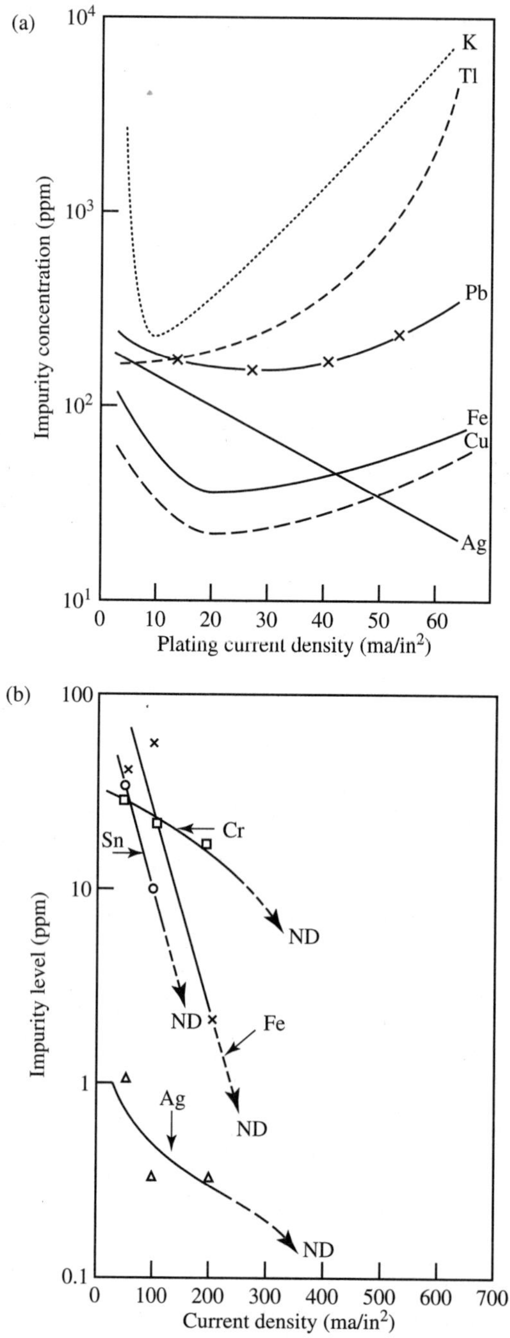

FIGURE 1-20 Effect of current density on impurity concentration in gold (*a*) and copper (*b*) electrodeposits.

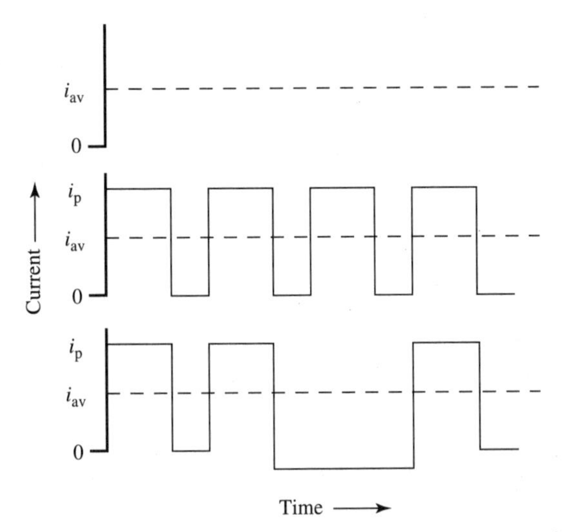

FIGURE 1-21 Current waveforms used for pulse electroplating. *Top:* Direct Current (dc) plating. *Middle:* dc pulse plating. *Bottom:* Asymmetric dc pulse plating.

studies have compared the advantages of pulse plating versus those of conventional dc plating [49–51]. They found the following benefits of pulse plating:

- Decreased porosity and increased density caused by reduction in grain size
- Improved adhesion, because the peak voltage can be several times greater than dc voltage
- Improved deposit distribution, because localized high current-density "burning" associated with dc is significantly reduced
- Reduced stress in the electrodeposit, because the lattice imperfections, impurities, voids, and nodules usually found with dc plating are generally reduced with pulse plating.

Morrissey compared the efficiency versus current density obtained with direct and pulsed-plated current from the same Au bath. He found pulse-plating efficiency was lower than dc efficiency. The difference was greater at low current densities and longer *off* times. It appears that increasing off time provides an effect similar to increasing agitation. This reduced current efficiency may also contribute to a leveling effect by inhibiting the formation of high current-density areas around surface asperities. Bahr and Lamb used pulse plating of Au to manufacture thin film chokes and antenna circuitry [52]. They were able to fabricate conductor lines 18 μm thick and 12 μm wide. They also reported that the amplitude and duty cycle of the pulse did not affect the thickness of the Au pulse over the range tested. Chang fabricated fine lines with improved conductivity for stripline applications [53].

Electrodeposition of metal onto a patterned surface is widely used to manufacture conductors with a high degree of shape fidelity and replication with high aspect ratios. Uniformity of deposit thickness is a prime concern. Ideally, the current distribution should be uniform over the entire surface. Realistically, thickness distribution depends not only on anode-cathode configuration, bath properties, and deposition conditions, but also on the resist pattern used, in the case of lift-off deposition. It is not unusual for circuitry to be dense in some areas and sparse in others, resulting in variation in resist density over the substrate surface. Krongelb and Mehdizadeh modeled this phenomena defining two current densities [54,55]. They assigned a superficial current density, i, which applies Ohm's law and the field equations everywhere in the system with the exception of the close vicinity of the patterned surface. Here, the lines of current squeeze into those areas not covered by resist. It is this higher surface current density, i_{surf}, which enters the plating kinetics. The ratio of i_{surf}, to i' depends entirely on the resist density, θ. As a result, plating in narrow, isolated resist openings tends to be thicker. This problem may be somewhat ameliorated by using lower cathode efficiency processes, such as pulse plating or lower anion concentration.

Another problem that can be attributed to the photoresist coating during electrodeposition is plating blockout. Reagan and Qutub define this phenomena as "the formation of a blocking layer that prevents electroplating deposits from forming on the target material" [56]. They ascribe the source of this layer to solid impurities such as dirt particles, gas bubbles, or residual contamination, such as finger oils or unremoved resist. Entrapped gas usually results from the decomposition of water, which forms H_2 at the cathode due to low bath conductivity or excessive current density. Gas sparging used for agitation may also contribute to gas entrapment. The gas tends to collect in narrow photoresist openings, and it effectively isolates that area of the substrate from the electroplating solution.

1.8.3 Electroless Plating

In the electronics industry, electroless plating has historically been relegated to printed circuit (PC) boards and soft substrates, particularly for plated through holes (PTH). In part, this is because the favorable surface structure of these materials allows good mechanical bonding. In electroless plating (autocatalytic processing), electrons are supplied by a reducing agent, such as alkali borohydrides, alkali diboranes, hypophophorous acid, or formaldehyde. Electroless technology has a number of advantages over electrodeposition:

- No electrical contact is needed (as is necessary with electroplating).
- It is possible on insulating surfaces as well as conductive surfaces, provided the surfaces are first sensitized.
- It is readily adaptable for three-dimensional coverage.
- No field lines are present; this enhances deposit uniformity.

Recently, electroless plating has complimented conventional thin film electrodeposition technologies. Although electroless films have been substituted for vapor-deposited films as seed layers for subsequent additive plating, the surfaces of ceramic substrates are characteristically ill-suited for electroless plating because of their dense, tightly-knit grain structure. To use electroless plating on ceramic surfaces, two options are available: roughen the surface or chemically modify it [59,60].

Table 1-4 lists the properties of the commonly used electroless baths. For most electronic applications, Cu and Au use is more prevalent. Nickel containing 10% to 12% phosphorous finds applications in power modules, in which aluminum wire bonds make Au finishes undesirable as a result of Au-Al intermetalics.

For copper a colloidal conductive layer, usually palladium (Pd) is catalytically deposited onto a nonmetallic surface. Copper salts are then reduced on the Pd seed layer to form a continuous Cu film. Originally, two separate steps were used: first the sensitizer, then a catalyzer. Most systems now use a combined palladium chloride-stannous chloride type catalyst/accelerator. The reaction proceeds by first hydrolyzing the surface absorbed species, so that

$$Pd[SnCl^{3-}]_n \xrightarrow{\text{water}} Pd + Sn(OH)_2 + Sn(OH)_4 \qquad (8)$$

TABLE 1-4 Comparison of electroless processes

Material	Copper	Nickel (P)	Nickel (B)	Gold
Reducing agent	Formaldehyde	Sodium hypophosphite	(1) Alkali borohydride (2) DMAB*	(1) Alkali borohydride (2) DMAB*
Bath pH	13	4.5	(1) 14 (2) 6	(1) 7-10 (2) 8-14
Bath temperature (°C)	60	90	65	70-90
Deposition rate (μm/hr)	3	10-15	(1) 8 (2) 6	1.5-2.0
$\rho/\rho o$	1.6	11-20	2-4	~1
Hardness†	Soft	Slightly hard	Hard	Soft

*DMAB = dimethylamineborane; †depending on heat treatment.

The Sn is converted into water-soluble compounds and removed by reaction between the hydrated Sn and selective acids and bases. When the Sn has been mostly removed, electroless deposition of Cu can take place. This reaction proceeds by reducing the Cu salt to Cu metal using formaldehyde. In an alkaline environment (pH 11 to 13), formaldehyde is oxidized, viz

$$2\,CH_2O + 4\,OH^- \rightarrow 2\,HCO_2^- + H_2 + 2e^- \quad (9)$$

Copper (II) is reduced to Cu metal by the following reaction:

$$Cu_2^+ + 2e^- \rightarrow Cu\ (metal) \quad (10)$$

The formaldehyde only functions at the catalyzed surface and does not spontaneously reduce all the cupric ions in solution providing the bath is properly maintained. The high pH of the electroless bath ($\approx$11) precludes the use of resists for additive plating because they are readily attacked at high pH levels.

Nickel may also be autocatalytically deposited from either hypophosphite- or boride-based baths onto a seed layer that is first deposited by hydrolyzing the surface as with Cu. Nickel deposition by hypophosphite is usually represented by the following equations [61,62].

$$Ni^{++} + H_2PO_2^- + H_2O \rightarrow Ni\ (metal) + H_2PO_3^- + 2\,H^+ \quad (11)$$

and

$$H + H_2PO_2^- \rightarrow P + H_2O \quad (12)$$

The physical and electrical properties of electroless Ni depend to a great extent on the P concentration, which may range from 2% to 12% in the deposited nickel and increase with decreasing pH. The pH of the bath decreases as the orthophosphite concentration increases unless buffers and stabilizers are used.

Nickel deposition may also be accomplished using either the borohydride ion or an amine borane as the reductant [63,64]. In acid or neutral solutions, the borohydride ions rapidly hydrolyze and form nickel boride (Ni_2B) in the presence of Ni ions [63,64]. As such, the reaction occurs between a pH of 12 to 14. Below pH of 12, spontaneous solution decomposition occurs. Above pH of 12, Ni_2B formation is suppressed and the Ni element is the primary product. Electroless nickel deposition may be represented by two equations:

$$BH_4^- + 4\,Ni^{++} + 8\,OH^- \rightarrow 4\,Ni\ (metal) + BO_2^- + 6\,H_2O \quad (13)$$

and

$$2\,BH_4^- + 4\,Ni^{++} + 6\,OH^- \rightarrow 2\,Ni_2B + 6\,H_2O + H_2 \quad (14)$$

Amine boranes are generally limited in electroless Ni applications to two compounds: N-dimethylamine borane (DMAB) and N-diethylamine borane (DEAB). N-dimethylamine borane is generally preferred because it is readily soluble in aqueous solution. On the other hand, DMEB requires short-chain aliphatic alcohols to enhance its solubility. The overall reduction reaction may be summarized by the following two reactions:

$$DMAB + 3\,Ni^{++} + 3\,H_2O \rightarrow 3\,Ni\ (metal) + (CH_3)_2H_2N^+ + H_3BO_3 + 5\,H^+ \quad (15)$$

and

$$2\ DMAB + 4\ Ni^{++} + 3\ H_2O \rightarrow Ni_2B + 2\ Ni\ (metal) + 2\ [(CH_3)_2H_2N]^+ + H_3BO_3 + \tfrac{1}{2}\ H^+ + 6\ H^+ \quad (16)$$

As opposed to the hypophosphite baths, the amine boranes are effective over a wider pH range. The (B) content, like the P content, increases with decreasing bath pH. The boron content from DMAB-reduced baths can vary from 0.2% to 4.0% B.

Alkali borohydrides are also used as reducing agents for the autocatalytic deposition of Au after first depositing a seed layer similar to Cu and Ni. Okinaka showed that excess free cyanide (CN^-) is necessary in the bath as a stabilizing agent. As in pure water, alkali borohydride readily hydrolyzes at room temperature with the evolution of H_2 [65]. He showed that the reaction of stabilized borohydride is a two step process, and it is the BH_2OH^- species that reduces $Au(CN)_2^-$, as follows:

$$BH_2OH^- + 3\ OH^- \rightarrow BO_3^- + \tfrac{3}{2}\ H_2 + 2\ H_2O + 3\ e^- \quad (17)$$

The above reaction coupled with

$$Au(CN)_2^- + e^- \rightarrow Au\ (metal) + 2\ CN^- \quad (18)$$

yields

$$BH_2OH^- + 3\ Au(CN)_2^- + 3\ OH^- \rightarrow BO_3 + \tfrac{3}{2}\ H_2 + 2\ H_2O + 3\ Au\ (metal) + 6\ (CN)^- \quad (19)$$

When Cu and Ni are used, Okinaka observed, a small amount of these metals dissolved in the solution. When using these solutions, very thin layers of Cu and Ni should be avoided.

1.8.4 Applications

Plating is used for a wide range of applications including anticorrosion coatings, interconnect metallization formation, via formation in electronic circuits and devices, and Au and Ni coatings for wire bonding [66,67]. In addition, the current trend in the microelectronics industry toward the use of flip chip and tab attachment has caused growth in the use of plating techniques to form gold bumps for device interconnections.

1.9 SOL-GEL COATINGS

Sol-gel coatings result from spin coating, dipping, or spraying a chemical solution onto a substrate. This solution is a stable mixture of suspended precursor particles known as a sol gel. Once applied to a substrate, a transition or destabilization of the sol occurs. This is marked by a significant increase in the viscosity of the coating, resulting in the formation of a tacky gel. The gel is hardened to form a film by drying, typically in air, at 100°C to 125°C. Although the films are deposited at room temperature, a high temperature heat treatment is usually required to create a dense adherent film. Sintering temperatures are typically in excess of 300°C. Final film thickness is determined by the application method, but films are typically 50 nm to 1 mm thick.

The chemical composition of the sol is determined by the material to be deposited, because this process relies on a chemical reaction. One of the more common types of sols is based on a metal alkoxide $M(OR)_X$, with the R representing an alkoxy group. The metal alkoxide is dissolved in a solvent with water and is destabilized by hydrolysis, resulting in the formation of a metal oxide film. Various catalysts may be added to modify the rate of the reaction and, hence, the properties of the resulting film.

The advantages of this process include high purity and a very-high degree of homogeneity over large areas. The primary disadvantages are the high cost of the materials and the narrow range of materials that can be deposited with this technique. A number of applications have been demonstrated for this process. These include sensors, optical coatings, and transparent conductors [68–71]. Several complete reviews of this technique have been published that have more extensive discussions of this topic [72–76].

REFERENCES

1. L. I. Maissel and R. Glang, *Handbook of Thin Film Technology,* McGraw-Hill, New York, 1970.
2. Kiyotaka Wasa and Shigeru Hayakawa, *Handbook of Sputter Deposition Technology,* Noyes Publications, Park Ridge, NJ, 1992.
3. J. L. Vossen and W. Kern, *Thin Film Processes,* Academic Press, New York, 1978.
4. J. L. Vossen and W. Kern, *Thin Film Processes II,* Academic Press, New York, 1991.
5. K. K. Schuegraf (ed.), *Handbook of Thin Film Deposition Processes and Techniques,* Noyes Publication, Park Ridge, NJ, 1988.
6. D. L. Smith, *Thin-Film Deposition: Principles and Practice,* McGraw Hill, New York, 1995.
7. R. Roth, J. Schubert, M. Martin, and E. Fromm, "Effect of process parameter changes on the composition of magnetron sputtered and evaporated TiN and AlN films measured by UHV in-situ techniques," *Thin Solid Films,* vol. 270, no. 1–2, Dec. 1, 1995, pp. 320–4.
8. John A. Woollam and Scott Heckens "In-Situ Ellipsometry on Sputtered Dielectric and Magneto-Optic Thin Films," *Thin Solid Films,* vol. 270, no. 1–2, Dec. 1, 1995, pp. 65–8.
9. S. R. Das, L. LeBrun, P. B. Sewell, and T. Tyrie, "UHV RHEED System for In Situ Studies of Sputtered Films," *Thin Solid Films,* vol. 270, no. 1-2, Dec. 1, 1995, pp. 314–9.
10. Tao Pan and Geoffrey W. D. Spratt, "Reactively Evaporated Co Thin Film for Tape Media," *IEEE Transactions on Magnetics,* vol. 31, no. 6 (pt. 1), Nov., 1995, pp. 2851–3.
11. Y. Sripathi, L. K. Malhotra, and G. B. Reddy, "GaGeTe Films as Phase-Change Optical Recording Media," *Thin Solid Films,* vol. 270, no. 1–2, Dec. 1, 1995, pp. 60–4.
12. R. Islam, H. D. Banerjee, and D. R. Rao, "Structural and Optical Properties of $CdSe_xTe_{x-1}$ Thin Films Grown by Electron Beam Evaporation," *Thin Solid Films,* vol. 266, no. 2, Oct. 1, 1995, pp. 215–8.
13. T. Sakaguchi, A. Katsube, T. Honda, F. Koyama, and K. Iga, "MgO/SiO_2 Dielectric Multilayer Reflectors for GaN-based Ultaviolet Surface Emitting Lasers," *Conference Proceedings—Lasers and Electro-Optics Society Annual Meeting,* vol. 2, 1995, IEEE, Piscataway, NJ, no. 95CH35739, pp. 102–3.
14. Kazuyoshi Torii, Hirosi Kawakami, and Keiko Kushida, Fumiko Yano, Yuzuru Ohji, "Ultra-thin Fatigue-free Lead Zirconate Titanate Thin Films for Gigabit DRAMs," *Digest of Technical Papers—Symposium on VLSI Technology 1995,* IEEE, Piscataway, NJ, no. 95CB35781, pp. 125–6.
15. M. Kayambaki, R. Callec, G. Constantinidis, Ch. Papavassiliou, E. Loechtermann, H. Krasny, N. Papadakis, P. Panayotatos, and A. Georgakilas, "Investigation of Si-Substrate Preparation for GaAs-on-Si MBE Growth," *Journal of Crystal Growth,* vol. 157, no. 1–4, Dec. 2, 1995, pp. 300–3.
16. W. Hansch, I. Eisele, H. Kibbel, U. Koenig, and J. Ramm, "Device quality of In Situ Plasma Cleaning for Silicon Molecular Beam Epitaxy," *Journal of Crystal Growth,* vol. 157, no. 1–4, Dec. 2, 1995, pp. 100–4.

17. V. S. Varavin, S. A. Dvoretsky, V. I. Liberman, N. N. Mikhailov, and Yu. G. Sidorov, "Controlled Growth of High-Quality Mercury Cadmium Telluride," *Thin Solid Films,* vol. 267, no. 1–2, Oct. 15, 1995, pp. 121–5.

18. J. P. Praseuth, "Molecular Beam Epitaxy of AlGaInAs on Patterned InP Substrates for Optoelectronic Applications," *Microelectronics Journal,* vol. 26, no. 8, Dec. 1995, pp. 841–52.

19. B. J. Wu, L. H. Kuo, J. M. DePuydt, G. M. Haugen, M. A. Haase, and L. Salamanca-Riba, "Growth Characterization of II-VI Blue Light-Emitting Diodes Short-Period Superlattices," *Applied Physics Letters,* vol. 68, no. 3, Jan. 15, 1996, pp. 379–81.

20. V. P. Evtikhiev, I. V. Kudryashov, V. E. Tokranov, A. F. Ioffe, J. S. Yu, S. K. Yang, G. Pak, and T. I. Kim, "Performance of 980 nm Pump Laser Diodes with GaAs/AlAs Graded Short-Period Superlattice Waveguides," *Conference Proceedings—Lasers and Electro-Optics Society Annual Meeting,* vol. 1, 1995, IEEE, Piscataway, NJ, no. 95CH35739, pp. 255–6.

21. A. L. Gutierrez-Aitken, K. Yang, X. Zhang, G. I. Haddad, P. Bhattacharya, and L. M. Lunardi, "Wide Bandwidth InAlAs/InGaAs Monolithic PIN-HBT Photoreceiver," *LEOS Summer Topical Meeting 1995,* IEEE, Piscataway, NJ, 2 pp.

22. C. Hardingham, T. A. Cross, J. Burrage, and C. Goodbody, "P-n and n-p GaInAsP Solar Cells: Technology and Material Analysis," *1994 IEEE First World Conference on Photovoltiac Energy Conversion—Conference Record of the IEEE Photovoltaic Specialists Conference,* vol. 2, 1994, IEEE, Piscataway, NJ, no. 94CH3365-4, pp. 2181–4.

23. Achim Strass, "Nano-MOSFETs for Future ULSI Applications," *Solid State Technology,* vol. 39, no. 1, Jan., 1996, pp. 65–74.

24. F. M. Yamada, A. K. Oki, D. C. Streit, D. K. Umemoto, L. T. Tran, K. W. Kobayashi, P. C. Grossman, T. R. Block, M. D. Lammert, S. R. Olson, J. C. Cowles, M. M. Hoppe, S. B. Bui, D. M. Smith, K. Najita, et. al., "Reliable ICs Fabricated Using a Production GaAs HBT Process for Military and Commercial Applications," *Proceedings of the 1995 Military Communications Conference (MILCOM), San Diego, CA,* vol. 2, 1995, IEEE, Piscataway, NJ, no. 95CB35750, pp. 760–4.

25. A. Schueppen, A. Gruhle, H. Kibbel, and U. Koenig, "Mesa and Planar SiGe-HBTs on MBE-Wafers," *Journal of Materials Science: Materials in Electronics,* vol. 6, no. 5, Oct., 1995, pp. 298–305.

26. K. W. Kobayashi, D. K. Umemoto, T. R. Block, A. K. Oki, and D. C. Streit, "Wideband HEMT Cascode Low-Noise Amplifier with HBT Bias Regulation," *IEEE Microwave and Guided Wave Letters,* vol. 5, no. 12, Dec., 1995, pp. 457–9.

27. Kevin W. Kobayashi, Donald K. Umemoto, Tom R. Block, Aaron K. Oki, and Dwight C. Streit, "Novel Monolithic LNA Integrating a Common-Source HEMT with an HBT Darlington Amplifier," *IEEE Microwave and Guided Wave Letters,* vol. 5, no. 12, Dec., 1995, pp. 442–4.

28. M. Lagadas, Z. Hatzopoulos, F. Karadima, G. Konstantinidis, N. Kornilios, C. Papavasiliou, A. Christou, "Improved Performance of MESFET Devices with L.T.GaAs Buffer Layers," *Defect and Impurity Engineered Semiconductors and Devices Materials Research Society Symposium Proceedings Materials Research Society,* vol. 378, 1995, Pittsburgh, PA, pp. 789–94.

29. A. Brazdeikis, A. Vailionis, A. S. Flodstrom, C. Traeholt, "Molecular Beam Epitaxy Growth and Microstructure of Thin Superconducting $Bi_2Sr_2CaCu_2O_x$ Films," *Physica C: Superconductivity,* vol. 253, no. 3–4, Nov. 1, 1995, pp. 383–390.

30. Hideomi Koinuma, "Crystal Engineering of High-T_c and Related Oxide Films for Future Electronics," *Bulletin of Materials Science,* vol. 18, no. 4, Aug., 1995, pp. 435–445.

31. T. Nishimori, H. Sakamoto, Y. Takakuwa, and Sh. Kono, "Diamond Epitaxial Growth by Gas-Source Molecular Beam Epitaxy with Pure Methane," *Japanese Journal of Applied Physics; Part 2: Letters,* vol. 34, no. 10A, Oct. 1, 1995, pp. L1297–L1300.

32. A. A. Adjaottor, E. I. Meletis, S. Logothetidis, I. Alexandrou, and S. Kokkow, "Effect of Substrate Bias on Sputter-Deposited TiC_x, TiN_y, and TiC_xN_y Thin Films," *Surface & Coatings Technology,* vol. 76–77, no. 1–3, pt. 1, Nov., 1995, pp. 142–8.

33. G. Sade and J. Pelleg, "Co-Sputtered TiB_2 as a Diffusion Barrier for Advanced Microelectronics with Cu Metallization," *Applied Surface Science,* vol. 91, no. 1–4, Oct., 2, 1995, pp. 263–8.

34. T. Minami, S. Takata, T. Kakumu, and H. Sonohara, "New Transparent Conducting $MgIn_2O_4$-$Zn_2In_2O_5$ Thin Films Prepared by Magnetron Sputtering," *Thin Solid Films,* vol. 270, no. 1–2, Dec. 1, 1995, pp. 22–26.

35. Y. Matsunaga, J. G. Wen, and Y. Enomoto, "Fabrication and Characterization of A-Axis/C-Axis-Oriented $YBa_2Cu_3O_y$ Bilayer Thin Film," *Physica C: Superconductivity,* vol. 256, no. 1–2, Jan. 1, 1996, pp. 81–9.

36. H. E. Horng, J. C. Jao, H. C. Chen, H. C. Yang, H. H. Sung, and F. C. Chen, "Critical Current in Polycrystalline Bi-Ca-Sr-Cu-O Films," *Physical Review B,* vol. 39, no. 13, May, 1989, pp. 9628–30.

37. R. V. Stuart, *Vacuum Technology, Thin Films, and Sputtering,* Academic Press, New York, 1983.

38. A. Sherman, *Chemical Vapour Deposition for Microelectronics,* Noyes Publication, Park Ridge, NJ, 1987.

39. S. Wolf and R. N. Tauber, "Silicon Processing for the VLSI Era," *Process Technology,* vol. 1, Lattice Press, Sunset Beach, CA, 1987.

40. J. Dieleman, E. Van de Riet, and J. C. S. Kools, *Japanese Journal of Applied Physics,* vol. 31, no. 6B, pt 1, 1992, pp. 1964–71.

41. A. J. Basovich et al., *Thin Solid Films,* vol. 228, 1993, pp. 193–5.

42. F. Sanchez et al., *Applied Surface Science,* vol. 69, 1993, pp. 221–4.

43. A. Giardini Guidoni and I. Pettiti, *Applied Surface Science,* vol. 69, 1993, pp. 365–9.

44. C. Germain, C. Girault, J. Aubreton, and A. Catherinot, *Applied Surface Science,* vol. 69, 1993, pp. 359–64.

45. G. A. Petersen and J. R. McNeil, *Thin Solid Films,* vol. 220, 1992, pp. 87–91.

46. A. R. Despic et al., *J Electrochem Soc,* vol. 115, 1958, p. 507.

47. A. M. Weisberg, in F. H. Reid and W. Goldie (eds.), *Gold Plating Technology,* Electrochemical Publications Ltd, Ayr, Scotland, 1974, p. 14.

48. A. J. Avila and M. J. Brown, *Plating,* vol. 57, 1970, p. 1105.

49. C. L. Faust et al., *Plating,* vol. 48, 1961, p. 605.

50. N. Ibl, *Surface Technology,* vol. 10, 1980, p. 81.

51. C. J. Raub and A. Knodler, *Gold Bulletin,* vol. 10, April, 1977, p. 2.

52. T. K. Bahr and G. M. Lamb, *J Electrochem Soc,* vol. 126, no. 9, Sept., 1979, p. 1514.

53. W. H. Chang, *Proc AES 2nd Intern Symp on Pulse Plating, Rosemount, IL,* Oct. 6–7, 1981.

54. S. Krongelb et al., *SPIE Intern Conf on Advances in Intercon and Pkg,* vol. 1389, 1990, p. 249.

55. S. Mehdizadeh et al., *J Electrochem Soc,* vol. 139, no. 1, Jan., 1992, pp. 78–91.

56. J. Regan and O. Qutub, *Plating and Surface Finishing,* vol. 78, no. 1, 1991, p. 29.

57. H. Honma and Y. Kouchi, *Plating and Surface Finishing,* vol. 77, no. 6, 1990, p. 54.

58. W. Kinzy-Jones, *Proc 1988 Intern Symp Microelectron, Seattle, WA,* 1988, p. 164.

59. K. Nishiwaki et al., *Proc 1989 Intern Symp Microelectron, Baltimore, MD,* 1989, p. 305.

60. J. Fudala and M. Beke, *1990 Proc 40th Electron Comp Techn Conf, Las Vegas, NV,* May 20–23, 1990.

61. A. Brenner and G. Riddell, *NBS J Res,* vol. 37, 1946.

62. G. Gutzeit, *Plating and Surface Finishing,* vol. 47, 1960, p. 63.

63. R. W. Hoke, U.S. Patent 2,990,296, 1961.

64. T. Berzins, U.S. Patent 3,045,334, 1962.

65. Y. Okinaka, *Plating,* vol. 60, Sept., 1974.

66. S. Bhansali, D. K. Sood, "Novel Technique for Fabrication of Metallic Structures on Polyimide by Selective Electroless Copper Plating Using Ion Implantation," *Thin Solid Films,* vol. 270, no. 1–2, Dec. 1, 1995, pp. 489–92.

67. Hajime Yoshiki, Kazuhito Hashimoto, and Akira Fujishima, "Area-Selective Electroless Copper Deposition Using Zinc Oxide Thin Films Patterned on a Glass Substrate," *Metal Finishing,* vol. 94, no. 1, Jan., 1996, pp. 28–9.

68. B. D. MacCraith, C. M. McDonagh, G. O'Keeffe, A. K. McEvoy, T. Butler, and F. R. Sheridan, "Sol-Gel Coatings for Optical Chemical Sensors and Biosensors," *Sensors and Actuators, B: Chemical,* vol. B29, no. 1–3, Oct., 1995, pp. 51–7.

69. Yoshiro Moriya, Hidenori Shimoda, and Ryoji Hino, "Control of Pore Diameter of Porous Silica by Sol-Gel Coating," *Journal of the Ceramic Society of Japan* vol. 103, no. 1203, Nov., 1995, pp. 1201–4.

70. N. J. Arfsten, *Journal of Non-Crystalline Solids,* vol. 63, 1984, pp. 243.

71. H. Dislich, *Journal of Non-Crystalline Solids,* vol. 57, 1983, pp. 371.

72. C. J. Brinker and G. W. Scherer, *Sol-Gel Science,* Academic Press, Orlando, FL, 1990.

73. D. Segal, *Chemical Synthesis of Advanced Ceramic Materials,* Cambridge University Press, Cambridge, 1989.

74. L. C. Klein, (ed.), *Sol-Gel Technology for Thin Films, Fibers, Preforms, Electronics and Specialty Shapes,* Noyes Publications, Park Ridge, NJ, 1988.

75. L. C. Klein, *Ceram Eng Sci Proc,* vol. 5, 1984, pp. 379.

76. H. Dislich, *Journal of Non-Crystalline Solids,* vol. 80, 1986, pp. 115.

APPENDIX A

Material	MP (°C)	g/cm³	Temperature (°C) for given vapor pressure (Torr)			Evaporation Techniques					Sputter	Comments
			10^{-8}	10^{-6}	10^{-4}	E-beam	Crucible	Coil	Boat	Basket		
Aluminum	660	2.70	677	812	1010	Ex	TiB_2-BN, ZrB_2, BN	W	TiB_2, W	W	RF, dc	Alloys and wets; use stranded tungsten
Aluminum nitride	>2200	3.26	—	—	~1750	F	—	—	—	—	RF, RF-R	Decomposes; reactive evaporation in 10^{-3} Torr nitrogen with glow discharge; **S**
Aluminum oxide	2072	3.97	—	—	1550	Ex	—	—	W	W	RF-R	Sapphire excellent in E-beam; n = 1.66
Barium titrate	—	6.02	—	—	—	—	—	—	—	—	RF	Co-evaporate with two sources or sputter; n = 2.40; **D**
Beryllium oxide	2530	3.01	—	—	1900	G	—	—	—	W	RF, RF-R	Toxic; no decomposition with E-beam; n = 1.72
Boron nitride	~3000	2.25	—	—	~1600	P	—	—	—	—	RF, RF-R	Decomposes with sputtering; use reactive sputtering; **S**
Carbon	~3652	1.8–2.1	1657	1867	2137	Ex	—	—	—	—	RF	E-beam preferred; **S**
Chromium	1857	7.20	837	977	1157	G	VC	W	‡	W	RF, dc	Films very adherent; **S**
Chromium-silicon monoxide	—	†	†	†	†	G	—	—	W	W	dc, RF	Flash evaporation; **S**
Copper	1083	8.92	727	857	1017	Ex	Al_2O_3, Mo, Ta	W	Mo	W	dc, RF	Poor adhesion; use adhesion layer of Cr
Gold	1064	19.32	807	947	1132	Ex	Al_2O_3, BN, VC	W	W§ Mo§	W	dc, RF	Poor adhesion
Inconel	1425	8.5	—	—	—	G	—	W	W	W	dc, RF	Use fine wire wrapped on tungsten; low rate required for smooth films
Indium	157	7.30	487	597	742	Ex	Gr, Al_2O_3	—	W, Mo	W	dc, RF	Wets tungsten and copper; use Mo or Mo liner

Indium (I) oxide	~600	6.99	—	—	650	—	—	—	—	—	RF	Decomposes under sputtering; **S**
Indium (III) oxide	850	7.18	—	—	~1200	G	Al_2O_3	—	W, Pt	—	—	—
Indium tin oxide	1800	—	—	—	—	—	—	—	—	—	—	**S**
Lead	328	11.34	342	427	497	Ex	Al_2O_3, Q	W	W, Mo	W, Ta	dc, RF	Toxic
Magnesium oxide	2852	3.58	—	—	1300	G	C, Al_2O_3	—	—	—	RF, RF-R	Evaporates in 10^{-3} Torr oxygen for stoichiometry; tungsten gives volatile oxides; n = 1.7
Molybdenum	2610	10.2	1592	1822	2117	Ex	—	—	—	—	dc, RF	Careful degas required
Nichrome IV	1395	8.5	847	987	1217	Ex	Al_2O_3, VC, BeO	W	§	W, Ta	dc, RF	Alloys with refractory metals
Nickel	1455	8.90	927	1072	1262	Ex	Al_2O_3, BeO, VC	W	W	W	dc, RF	Alloys with refractory metals
Palladium	1554	12.02	842	992	1192	Ex	Al_2O_3, BeO	W	W	W	dc, RF	Alloys with refractory metals; evaporates rapidly; **S**
Platinum	1772	21.45	1292	1492	1474	Ex	C, ThO_2	W	W	W	dc, RF	Alloys with metals; poor adhesion
Ruthenium	2310	12.3	1780	1990	2260	P	—	—	W	—	dc, RF	—
Silicon	1410	2.32	992	1147	1337	F	BeO, Ta, VC	—	W, Ta	—	dc, RF	Alloys with tungsten; oxides produced above 4×10^{-6} Torr; E-beam best
Silicon nitride	1900	3.44	—	—	~800	—	—	—	—	—	RF, RF-R	—
Silicon (II) oxide	>1702	2.13	—	—	850	F	Ta	W	Ta	W	RF, RF-R	Use baffled box and low rate; n = 1.6; **S**
Silicon (IV) oxide	1610	~2.65	†	†	1025*	Ex	Al_2O_3	—	—	—	RF	Quartz excellent in E-beam; n = 1.544, 1.553

*__S,__ sublimes; **D,** decomposes; †, function of composition; ‡, Cr-plated rod or strip; §, alumina-coated; C, carbon; Gr, graphite; Q, quartz; Incl, Inconel; VC, vitreous carbon; Ex, excellent; G, good; F, fair; P, poor; RF, RF sputtering; RF-R, reactive RF sputtering; dc, dc sputtering; dc-R, reactive dc sputtering.

APPENDIX A—cont'd

			Temperature (°C) for given vapor pressure (Torr)			Evaporation Techniques						
Material	MP (°C)	g/cm³	10^{-8}	10^{-6}	10^{-4}	E-beam	Crucible	Coil	Boat	Basket	Sputter	Comments
Silver	962	10.5	847	958	1105	Ex	Al_2O_3	W	Ta, Mo	W	dc, RF	
Spinel	—	8.0	—	—	—	G	—	—	—	—	RF	n = 1.72
Tin	232	7.28	682	807	997	Ex	Al_2O_3	W	Mo	W	dc, RF	Wets Mo; use tantalum liner for E-beam
Tin oxide	1630	6.95	—	—	~1000	Ex	Q, Al_2O_3	W	W	W	RF, RF-R	Films from tungsten are oxygen deficient, post-oxidize in air; n = 2.0; **S**
Titanium	1660	4.5	1067	1235	1453	Ex	TiC	—	W	—	dc, RF	Alloys with refractory metals; degas required
Tungsten	3410	19.35	2117	2407	2757	G	—	—	—	—	RF, dc	Forms volatile oxides

*__S,__ sublimes; **D,** decomposes; †, function of composition; ‡, Cr-plated rod or strip; §, alumina-coated; C, carbon; Gr, graphite; Q, quartz; Incl, Inconel; VC, vitreous carbon; Ex, excellent; G, good; F, fair; P, poor; RF, RF sputtering; RF-R, reactive RF sputtering; dc, dc sputtering; dc-R, reactive dc sputtering.

CHAPTER 2
PATTERN GENERATION TECHNIQUES

Dr. Lynn F. Fuller
Motorola Professor of Microelectronic Engineering
Rochester Institute of Technology

2.1 INTRODUCTION

Fabrication of thin film integrated circuits and semiconductor devices requires pattern definition through microlithographic processes, which play a critical role in the overall success of integrated circuit manufacturing. Lithographic processes define substrate regions for subsequent etching removal or materials addition. The size and precision of these regions depends on the capabilities of the particular process. Lithographic processes essentially drive integrated circuit design and fabrication and are responsible for advances in integrated circuit feature size, speed, and packaging density. The trend for integration is to continue decreasing feature size and increasing the number of devices. This trend is responsible for the dramatic decrease in cost per function, although the cost per integrated circuit essentially remains the same.

2.2 PATTERN GENERATION

2.2.1 CAD

Computer aided design (CAD) is used extensively to prepare the layout for integrated devices and circuits. Computer aided design refers to the software and hardware tools used in the design process. For electronic device and circuit design, this includes software for schematic capture, analog and digital circuit simulation, process simulation, layout editors, design rule checkers, and pattern generation data preparation.

2.2.2 Layout Editor

The layout editor is the software system that allows the design of the layout of the various layers in the circuit or device. Each layer of a given process is often represented by a different color or pattern in the layout editor. The layers are shown superimposed on each other to illustrate the correct interlayer alignment. Different amounts of automatic layout are possible depending on the software. Automatic placement of standard cells, automatic routing of cell interconnects, and other time-saving software tools are available. The layout for a thin film resistor and capacitor network created using a process with four lithographic levels, two conductive layers, a resistive layer, and several insulating layers is illustrated in Fig. 2-1.

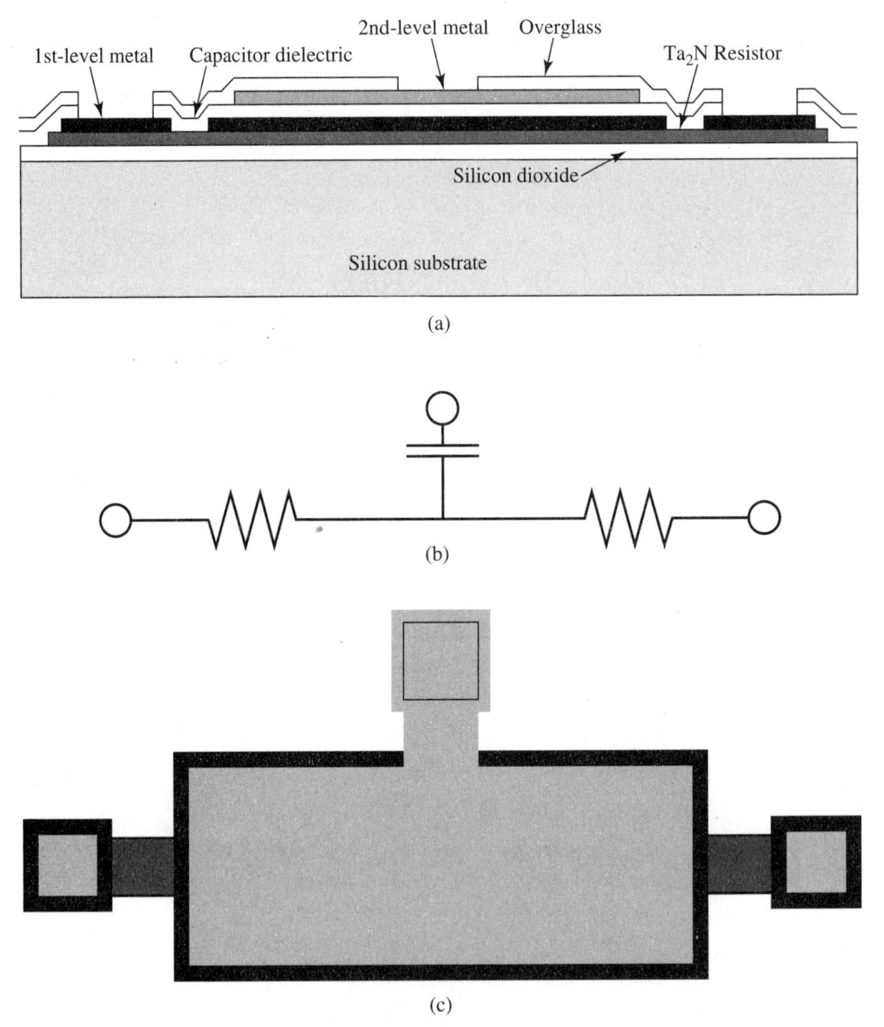

FIGURE 2-1 (*a*) Crossection, (*b*) electrical schematic, (*c*) layout for a thin film resistor and capacitor network.

2.2.3 Data Preparation

Once the layout is completed, the data need to be converted to the correct format for mask preparation. One can visualize the maskmaking machine as a printer that prints on glass instead of paper. The layout editor format may require that objects be referenced from their center or from a corner, as well as the width and height of the objects. Width and height information is all that is required since only box or square objects are used. Data are also sized by 1×, 5×, or some other magnification as required by the photoresist exposure tool that will be used. Boxes can be printed in dark field or light field. Letters, numbers, and other patterns may need to be reflected or rotated to achieve a final pattern on the substrate that reads correctly. Patterns are often biased or bloated to compensate for the expected reduction in pattern size caused by isotropic etching. Patterns are also formed from boolean functions of various layers. For example, in a complementary metal oxide semiconductor (CMOS) process, the N^+ drain and source implant pattern is made from the inverse of the P^+ drain and source pattern. Another example from a complementary metal oxide semiconductor process is the p-well field adjust implant mask which is made from the inverse of the well pattern ANDed with the active pattern. The data are prepared and delivered to the maskmaking tool, they are used to create a clear and opaque pattern.

2.2.4 Maskmaking

Maskmaking is the process of creating a clear and opaque pattern (often chrome on quartz or glass) from the layout data, including the data preparation. The clear and opaque pattern is called the *reticle,* and it includes serial numbers, layer names, bar codes, test structures, fiducial marks (marks to align the reticle to the exposure tool), pellicles, and other features. The mask pattern may be actual size (1×) or magnified (4×, 5×, 10×, etc). The mask may contain only one copy of the pattern or an array of patterns. The pattern may be clear field or dark field. A sketch of a mask is shown in Fig. 2-2 with a 5× pattern, fiducial marks, serial number, layer name, and lettering. Note that the lettering must be read

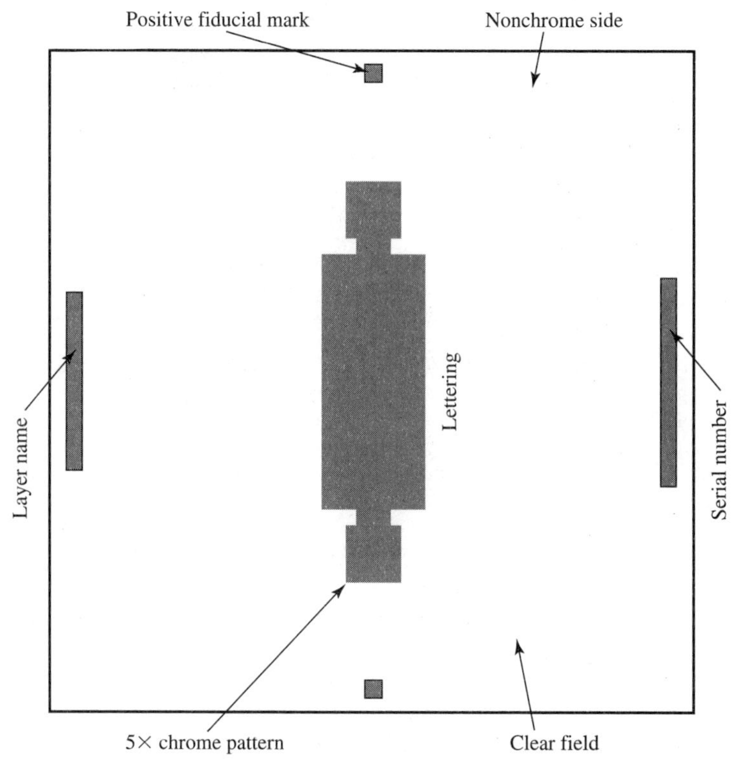

FIGURE 2-2 Sketch of a photomask showing 5× pattern, fiducial marks, serial number, layer name, and lettering.

correctly from the nonchrome side to create correct lettering on the substrate; this is the result of a mirror operation in the data preparation.

The masks are typically made using an electron beam exposure tool. In the semiconductor industry, the mask procurement engineer obtains the mask from the mask vendor. These tools cost several million dollars each and are capable of writing geometries down to 0.1 μm. The mask vendor quotes prices based on time spent on data preparation, writing, inspection, and other variables. The details of a basic chromium photomask process are described in the following list.

Basic chrome photomask process:

1. Deposit chromium (by evaporation, sputtering) onto glass or quartz substrate (1000 Å). Chromium oxide may be added as an antireflective layer (100 Å).
2. Spin coat with electron beam resist (5000 Å). Resist may be either positive or negative acting. Bake in an oven or on a hot plate.
3. Expose to a scanning electron beam with required beam size, address size, and beam current. The exposure dose is measured in coulombs per square centimeter (C/cm^2) and is related to the beam current, scan frequency, beam size, and number of times a pixel site is scanned.
4. Develop in an appropriate developer.
5. Rinse in an appropriate solvent.
6. Etch chromium in chrome etch (ceric ammonium nitrate); rinse in water.
7. Strip resist in solvent or acid (such as sulfuric acid and hydrogen peroxide); rinse in water and dry.
8. Inspect for defects. Measure geometry feature sizes and placement.

Other techniques for maskmaking include laser writing and optical pattern generation.

Laser writing is limited to pattern sizes of 1 μm and larger. Laser writing is similar to electron beam writing, except a laser is used and the chrome mask blank is coated with an optical resist suitable for the laser used. After the pattern is written, the resist is developed and the chrome is etched.

Optical pattern generation is limited to pattern sizes of approximately 2 μm and larger. Optical pattern generation (or photoplotter) uses a tool that places boxes (or other shapes) of various sizes and orientations on either photographic film or chrome mask blanks. If photographic film is used, the light source and optical system will typically be designed for a wavelength in the green region of the visible spectrum. The film can either be an emulsion coating on glass or quartz, or an emulsion coating on acetate film. The final result will be a pattern of black and clear. If the output is a chrome mask on glass or quartz, the exposure source is typically the g-line of a mercury vapor lamp. The chrome mask blank is coated with photoresist and, after development, the chrome is etched. Other techniques use emulsion pattern generation followed by contact print to chrome mask blanks. Older techniques include generating artwork, hand-cut rubylith patterns, and photoreduction on a large format camera. All of these techniques are still generally available in the printing industry.

2.3 MICROLITHOGRAPHY

Photolithography is the production of a three-dimensional relief image based on the exposure to light and subsequent development of light-sensitive photoresist [1]. Microlithography is used to print ultrasmall patterns, primarily in the semiconductor industry. The types of radiation, materials, and tools are important characteristics of a process, but the basic steps are generally the same as those of conventional optical lithography. Radiation-sensitive photoresist is applied to a substrate as a thin film. Image exposure is then transmitted to the photoresist, usually through a mask of clear and opaque areas on a glass substrate. Clear areas within the mask allow exposure of the photoresist material, which photochemically alters the material. Depending on the type of photoresist used, exposure will either increase or decrease the solubility of the exposed areas in a suitable solvent called developer: positive photoresist material will become more soluble in exposed regions, while negative photoresist will become less soluble in exposed regions. The solubility differential of exposed to unexposed areas is responsible for the reproduction of the mask image into the photoresist.

After development, regions of the substrate are no longer covered by the photoresist film. Further subtractive or additive processing can now be performed, either by etching unprotected areas or by depositing layers over exposed layers of the substrate (plating or lift-off). The photoresist, therefore, must be capable of reproducing desired pattern images and providing protection, or resistance, for the substrate in subsequent processes. A schematic of positive and negative photolithographic processes is shown in Fig. 2-3.

2.3.1 Photoresists

Photoresists are generally organic polymer materials with properties tailored to meet specific performance criteria. As discussed, resists may be classified either as positive or

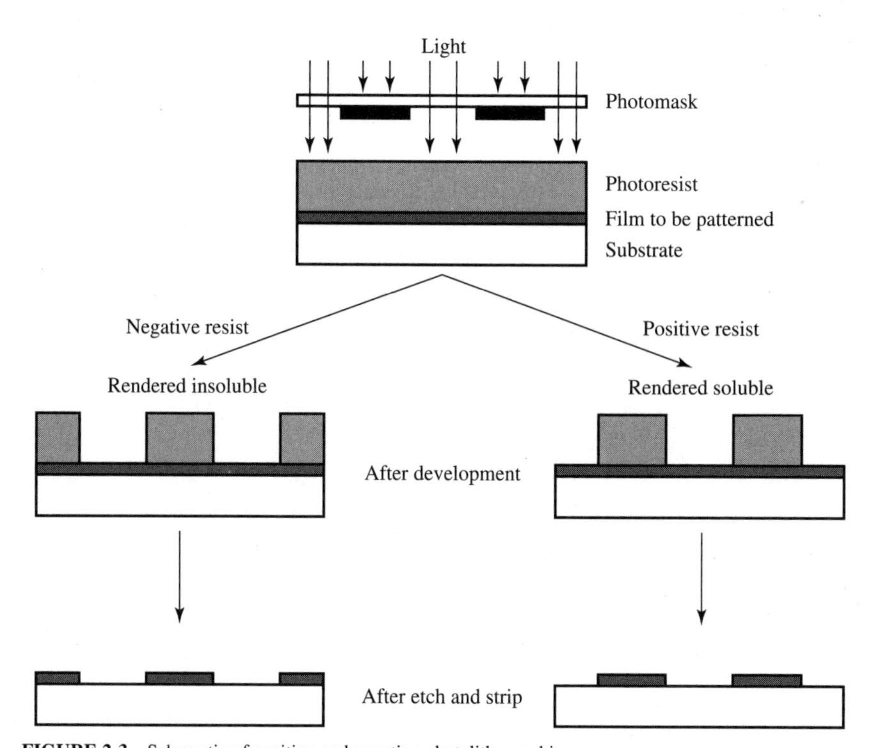

FIGURE 2-3 Schematic of positive and negative photolithographic processes.

negative depending on their response to exposure. In the 1970s, negative-tone resists based on two-component formulations were standard for the microelectronics industry. Limited resolution, resulting from resist swelling, caused these materials to be replaced by two-component positive-tone resists, which offer higher resolution using aqueous processing. In two-component systems, the resist is formulated from a base matrix resin, which serves as a binder for the material, and a sensitizer, which provides appropriate exposure sensitivity. In addition to these components is a casting solvent that keeps the resist in a liquid state until application. One-component and two-component resist hierarchy is illustrated in Fig. 2-4. Recently, new chemically amplified negative resists and image reversal materials (both negative or positive depending on chemical mechanisms) have been introduced into the microelectronics industry.

2.3.2 Positive Photoresists

Positive photoresists are formulated from a base matrix resin such as novolac (about 30%); combined with a sensitizer or photoactive compound (PAC) (about 30%); a casting solvent (about 30%), such as ethylene glycol monoethyl ether; and other additives (about 10%), such as adhesion promoters, surfactants, dyes, antioxidants, and polymerization inhibitors.

Positive photoresists accomplish an image solubility differential upon exposure using changes in the sensitizer component. This PAC undergoes photodecomposition, changing it from a dissolution inhibitor to a dissolution enhancer. Before exposure, the PAC, in combination with a novolac matrix resin, is *less* soluble in an alkali developer solution than the novolac resin alone. Upon exposure, the PAC goes through photochemical rearrangement, and the novolac with exposed PAC becomes *more* soluble in a basic developer than the novolac resin alone. This synergistic effect creates a high-solubility differential, giving the two-component positive resist high contrast and resolution capabilities. Typical resists containing unexposed PAC might dissolve at a rate of 2 nm/s whereas resists with exposed PAC might dissolve at a rate of 200 nm/s. Thus, with a 1.0 μm thick coating, it would take

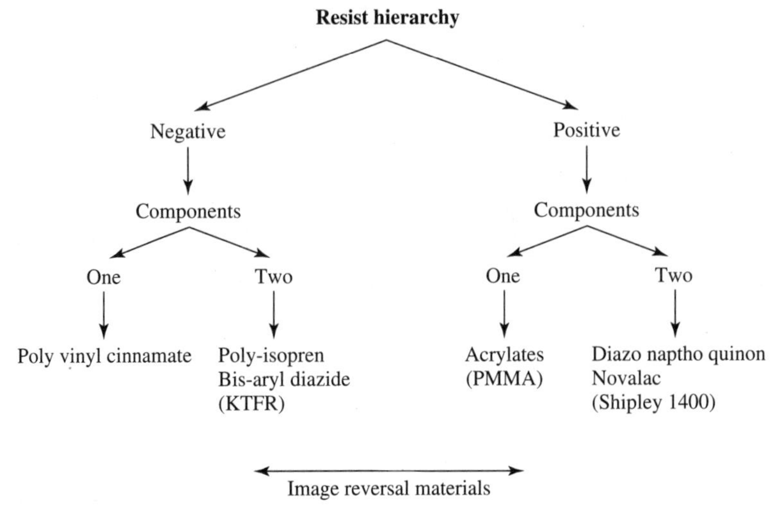

FIGURE 2-4 One-component and two-component resist hierarchy.

500 seconds to develop unexposed regions of resist and 50 seconds to develop partially exposed regions.

The sensitivity of the PAC is caused by the instability of the diazide (N_2) group bonded to an organic molecule—a hydrocarbon ring, adjacent to a ketone (double-bonded oxygen), as shown in Fig. 2-5. This results in a compound known as a benzoquinone diazide, or orthoquinone diazide.

Napthalene forms are alternatives, consisting of a ketone and diazide in a napthalene formation, as shown in Fig. 2-6. This is a *base-insoluble sensitizer* known as a napthoquinone diazide. Exposure to UV light leads to dissociation of the diazide group of the orthoquinone or napthoquinone diazide, rearranging to form a keto-carbene and nitrogen (N_2) illustrated in Fig. 2-7. The keto-carbene is short-lived, and it rearranges to form a more-stable ketene structure, shown in Fig. 2-8.

The ketene reacts with H_2O from the air to form a carboxylic acid, the desired end product, as illustrated in Fig. 2-9. At humidities exceeding 40%, the exposed sensitizer is completely converted to acid[3]. Without water available in the air, a less-soluble ester is formed. At 20% humidity, the dissolution rate is decreased by as much as 50% and the dose to clear is increased by 10% to 15%. The carboxylic acid is a base-soluble photoproduct and will react with an alkaline solution. A dilute alkali (base) solution is capable of developing exposed PAC, while unexposed, insoluble PAC is not developed.

The developer is a dilute alkali solution, such as sodium hydroxide (NaOH) or potassium hydroxide (KOH), and can contain buffers, surfactants, antioxidants, and others. The

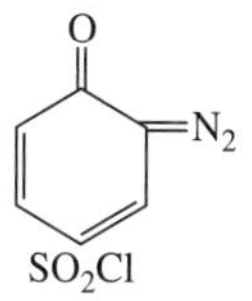

FIGURE 2-5 The organic molecule benzoquinone diazide, or orthoquinone diazide, a PAC.

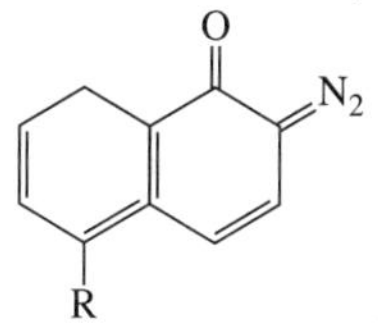

FIGURE 2-6 Napthoquinone diazide, a PAC.

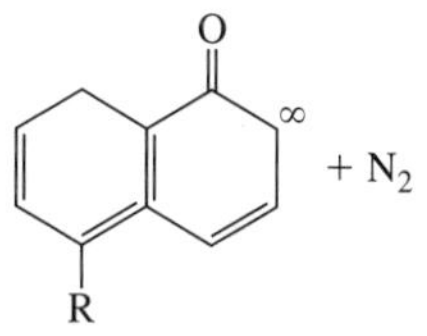

FIGURE 2-7 Keto-carbene and nitrogen (N_2) formed by the dissociation of the diazide group of the orthoquinone or napthoquinone diazide.

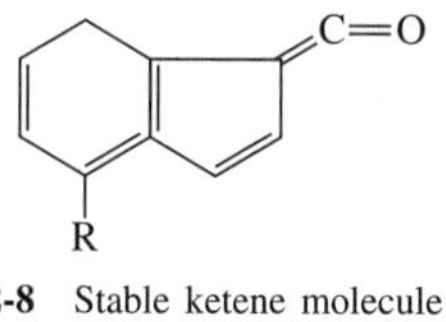

FIGURE 2-8 Stable ketene molecule formed by the rearrangement of the keto-carbene molecule.

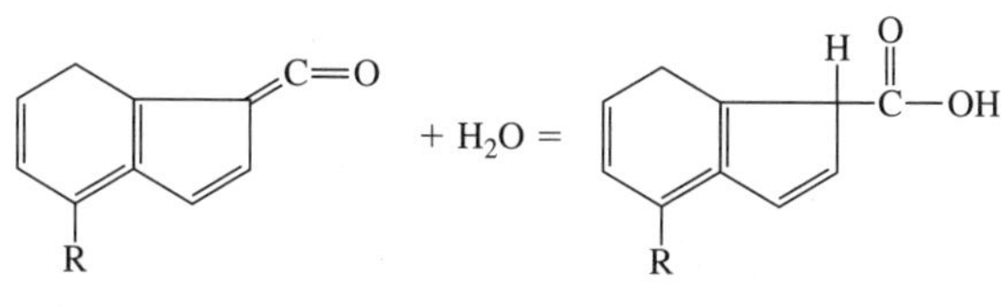

FIGURE 2-9 Base-soluble carboxylic acid formed by the reaction of H_2O from the air and the ketene molecule.

NaOH and KOH are examples of metal ion-containing developers. The metal ion developers give high contrast at the lowest cost. However, they cause concern for manufacturers of metal oxide semiconductor (MOS) devices, because small quantities of metal ions, such as Na^+ and K^+, can cause device parameter shifts. The metal-free developers, such as $(CH_3)_4 N^+OH^-$, $(CH_3)_3N^+CH_2CH_2OH^-$, or $CH_3N^+(CH_2CH_2OH)_3OH^-$, avoid this problem. During development, the developer can combine with the sensitizer in partially exposed resist to create a magenta dye. The dye is used as an indicator to stop development, and any magenta dye is rinsed away with deionized water.

While the resin in a two-component positive resist system is not photosensitive, it has certain other properties: chemical resistance, viscosity, and stability. Solubility is also a factor. The resin must allow exposed areas to dissolve and, thus, must be soluble in alkaline solutions. However, if the resin is too soluble in a base, unexposed areas could be developed, destroying image-forming characteristics. As mentioned previously the binder resin often used in positive photoresist is novolac, illustrated in Fig. 2-10, formed by the condensation polymerization of phenol and formalde-hyde. Novolac resins are soluble in organic solvents, exhibit good film-forming characteristics, and are capable of combining with orthoquinone diazide sensitizers.

OH OH
CH2 CH2 n
CH3 CH3

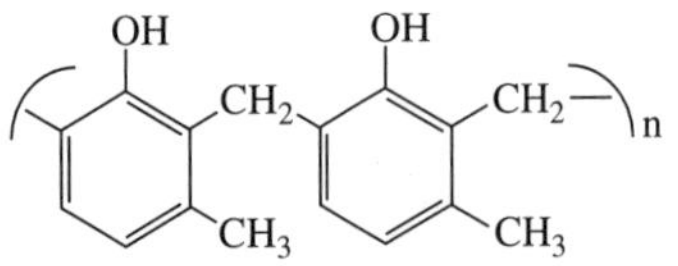

FIGURE 2-10 Novolac resins are soluble in organic solvents, exhibit good film forming characteristics, and are capable of combining with orthoquinone diazide sensitizers.

2.3.3 Chemically Amplified Negative and Positive Photoresists

Chemically amplified negative and positive resists have recently become available. These resists offer the advantage that they do not swell in developer solution like older negative resists. The new resists are capable of resolution below 0.1 μm, can be used at shorter wavelengths, and offer improved sensitivity. Table 2-1 lists the properties of several commercially available resists.

2.3.4 Photosensitive Polyimides

Polyimides are used for interlayer dielectrics in multilayer metal schemes and for multilayer thin film high-performance packages [4,5]. They provide thermal-mechanical buffer and α-particle protection coating for integrated circuit (IC) packages. Polyimides are also used in multilayer lithographic techniques where a planarizing layer is needed or harsh processing, such as ion implantation or plating, is used. The main advantage of polyimide materials is their stability at temperatures from 300°C to 400°C and ability to be processed at low-temperature. Polyimides also have low (3.2 to 3.4) dielectric constant. They can be coated in pinhole-free layers, offer radiation resistance, as well as thermal, mechanical, and chemical stability.

The photosensitive polyimides can be treated like positive photoresists except that the coatings will be thicker and thus longer exposures and developing times are needed. Many companies offer a wide range of polyimide products [6]. A typical process for photosensitive polyimide follows.

Photosensitive polyimide process:

1. Apply adhesion promoter by puddle and spin coating at 5000 rpm for 30 seconds; the adhesion promoter is a proprietary solution provided by the manufacturer of the polyimide.

Text continued on p. 2-14.

TABLE 2-1 Photoresists for microlithography

Company name/resist designation	Resist basic type N (negative) P (positive)	Broadband	G-line	H-line	I-line	Deep UV	e-Beam	x-ray	Sensitivity, 1 mm thick coating	Resolution	Typical contrast	Typical 4000 rpm thickness
Genesis, Santa Clara, CA												
GS 7010	P	X	X	X	X				110	0.25	6	1.2-1.9
Hoechst Celanese Corporation												
AZ 1500	P	X	X	X					100	0.7	2.5	1.0-2.0
AZ 6100	P	X							60	0.5	2.5	1.0-2.0
AZ 6200B	P		X						220	0.45	3	1.0-2.0
AZ 7500	P				X				165	0.35	3	0.9-2.0
AZ 7500T	P			X	X				150[1]	0.75	3	2.0-4.5
AZ DX	P					X			50	0.23	9	.75-1.0
AZ P4000	P	X	X						100	0.8	2.3	1.0-8.0
AZ PF500	P						X	X	35	0.2	>10	1.4
IBM Corp., Technical Products Division, Hopewell Junction, NY												
APEX-E	P					X			15	0.23	8	0.8-1.0
Morton Electronic Materials, Tustin, CA												
EL 2015	P	X	X	X	X				100	0.5	4.2	1.1-2.1
EL 2025												
EL 2026	P	X	X	X	X				100	0.5	4.2	.0-40.0
EL 2026TE												
EPA914EZ	P	X	X	X					60	0.6	2.8	.68-4.4
NOVA 2050	P	X	X	X					80	0.6	2.8	1.2-2.1
NOVA 2070	P	X	X	X	X				100	0.35	3.9	.09-5.0
NOVA 2071	P	X	X	X	X				100	0.35	3.9	.09-5.0
OFPR-800	P	X	X	X					125	1	1.4	0.6-3.0
OFPR-800PG	P	X	X	X					125	1	1.5	1.2-2.4

(Continued)

TABLE 2-1 Photoresists for microlithography (Continued)

Company name/resist designation	Resist basic type N (negative) P (positive)	Broadband	G-line	H-line	I-line	Deep UV	e-Beam	x-ray	Sensitivity, 1 mm thick coating	Resolution	Typical contrast	Typical 4000 rpm thickness
Morton Electronic Materials, Tustin, CA												
OMR-83	N	X			X				30	2.5		
PR 1024 MP	P				X	X		X	90	0.35	5	0.3-0.7
OCG Microelectronic Materials, Inc., West Paterson, NJ												
CAMP-6	P					X			65	0.25	10	0.5-1.0
HIPR 6500	P	X	X	X	X				200	0.4	3	0.5-2.0
HPR 500	P	X	X						100	0.8	2	0.7-3.0
OCG 825	P	X	X	X					160	0.6	3	0.8-3.0
OCG 895	P				X				130	0.4	4	0.5-5.0
OCG 897	P				X				150	0.35		1-2.1
OIR 32	P				X				150-200	0.32	3.5	0.8-1.5
Shipley Company, Inc., Marlboro, MA												
S 1400	P		X						50	0.6	1.8	0.1-2.8
S 1800	P		X						50	0.6	1.8	0.5-2.2
S 3800	P		X	X	X				90	0.45	[7]	1.0-1.8
SAL 601	N						X		[8]	0.1	>3.3	0.2-1.5
SPR 2	P		X						90	0.4	3.3	1.1-1.8
SPR500-A	P				X				80	0.35	>3	0.5-2.6
SPRT500-A	P				X				80	0.35	>3	0.5-2.6
XP-89131	N					X		15-80	0.23	>4	0.5-1.0	
SAL 603	N						X		[9]	0.1	>3	.45-1.8
SAL 605	N						X		[9]	0.1	< 3	.45-1.8
XP-90236	N					X		10	50	0.25	>4	0.5-1.0

Company name/resist designation	Recommended softbake temperature	Recommended hardbake	Softening point (C)	Index of refraction	Index of refraction	Max boron and iron (ppb)	Max total trace metal (ppb)	Solvent type	Filtration	Dyed resist	Shelf life (months)	Purpose of dye	Custom resist available
Toray Industries, Toray Marketing & Sales (America) Inc., San Mateo, CA													
Photomeece[	N	X	X						50-100	2		NA	3.5-3.6
EBR-9	P							[11]	0.2	2		20-0.22	
UCB-JSR Electronics, Inc., Sunnyvale, CA													
PFR GX210	P		X						100	0.58			2
PFR GX250E	P		X						200	0.42			1.2
PFR IX405E	P				X				90	0.4			1.2-1.8
PFR IX500EL	P				X				180	0.34			1.1-1.5
PFR IX560FD	P				X				160	0.36			1.2
PFR IX700	P				X				200	0.32			1.1
PFR IX710D	P				X				200	0.34			1.1
PFR IX715D	P				X				175	0.36			1.1-1.5
System 8	P		X						50	0.5		3	0.5-2.7
Genesis, Santa Clara, CA													
GS 7010	100	160		1.64				EL	0.2	X	6	[14]	X
Hoechst Celanese Corporation													
AZ 1500	100	120	120	1.64	200	200		PGMEA	0.1	X	12	[2]	X
AZ 6100	100	135	130	1.64	200	200		PGMEA	0.1	X	12	[2]	X
AZ 6200B	90	125	120	1.64	200	200		PGMEA	0.1	X	12	[2]	X
AZ 7500	110	110	120	1.64	50	50		PGMEA	0.1	X	8	[2]	X
AZ 7500T	115	115	120	1.64	50	50		PGMEA	0.1	X	8	[3]	X
AZ DX	120	120	130	1.75	50	50		PGMEA	0	X	8	[3]	
AZ P4000	100	120	120	1.64	750	750		PGMEA	0.2		12		X
AZ PF500	110	100	120	1.64	500	500		PGMEA	0.1		12		

(Continued)

TABLE 2-1 Photoresists for microlithography (Continued)

Company name/resist designation	Recommended soft-bake temperature	Recommended hard-	Softening point (C)	Index of refraction	Index of refraction	Max boron and iron (ppb)	Max total trace metal (ppb)	Solvent type	Filtration	Dyed resist	Shelf life (months)	Purpose of dye	Custom resist available
IBM Corp., Technical Products Division, Hopewell Junction, NY													
APEX-E	90	90	130	1.57	50	50		PGMEA	0.1		12		X
Morton Electronic Materials, Tustin, CA													
EL 2015	100	115		1.64	500	500		EL	0.1	X	9	[5]	X
EL 2025													
EL 2026	90	115		1.64	3500	4000		EL	1		6		X
EL 2026TE													
EPA914EZ	110	115		1.64	500	500		PMA	0.1	X	12	[6]	X
NOVA 2050	110	115		1.64	500	500		PMA	0.1	X	12	[5]	X
NOVA 2070	80	130		1.62	100	500		EL	0.1	X	12	[5]	X
NOVA 2071	80	130		1.62	100	500		EL	0.1	X	12	[5]	X
OFPR-800	115	150		1.64	500	500		EGMEA	0.1	X	12	[5]	X
OFPR-800PG	115	150		1.64	500	500		PMA	0.1	X	12	[5]	X
OMR-83	105	140		1.55	[4]	[4]	<500	Xylene	0.2	X	12	[5]	X
PR 1024 MP	100	115		1.62	100	500		EGMEA	0.1		12		X
OCG Microelectronic Materials, Inc., West Paterson, NJ													
CAMP-6	105	115	140	1.61	20	20	20	EEP	0.1		6		X
HIPR 6500	90	130	135	1.68		0[28]	50[28]	EL/EEP	0.1	X	12	[16]	X
HPR 500	115	125	130	1.64	200	300	300	EL	0.2	X	12	[16]	X
OCG 825	100	130	140	1.64	500	500	500	EEP	0.1	X	12	[16]	X
OCG 895	90	120	130	1.62	200	200	200	PGMEA	0.1	X	12	[16]	X
OCG 897	90	120	130	1.7	50	70	50	PGMEA	0.1	X	12	[16]	X
OIR 32	90	120	130	1.7	50	50	30	MMP	0.1	X	12	[17]	X
S 1400	90	115	120	1.64	>200	>200	1000	EGMEA	0.2	X	12		X
S 1800	90	115	120	1.64	500	500	1000	PMA	0.2	X	12		X
S 3800	105	115	120	1.64	500	500	1000	EL	0.1	X	12		X

Shipley Company, Inc., Marlboro, MA												
SAL 601	85	115	>	1.61	>100	>100	1000	EEA	0		6		X
SPR 2	100	120	125	1.64	100	100	1000	EL	0.1	X	12		X
SPR500-A	95	110	115	1.64	15	50	10	EL	0.2	X	9	[5]	X
SPRT500-A	95	110	125	1.64	15	50	10	EL	0.2	X	9	[5]	X
XP-89131	115	125		1.78	50	50	10	PGMEA	0		>6		X
SAL 603	105	110		1.61	>100	>100		PMA	0	X	6	[10]	X
SAL 605	105	110		1.78	>100	>100	>100	PMA	0	X	6	[10]	X
XP-90236	115	125		1.72	50	50	10	PGMEA	0		6		X
Toray Industries, Toray Marketing & Sales (America) Inc., San Mateo, CA													
Photomeece[	80	400	285	1.41	100	100	100	NMP	1		3		X
EBR-9	200	120	133	1.64	100	100	100	MEA	0.1		6		X
UCB-JSR Electronics, Inc., Sunnyvale, CA													
PFR GX210	100	125		1.64	>50	>50	>50	EL		X	6	[18]	X
PFR GX250E	100	125		1.64	>50	>50	>50	EL	0.1		6		X
PFR IX405E	95	125		1.64	>50	>50	>50	MMP/EL	0.1		6		X
PFR IX500EL	100	125		1.64	>50	>50	>50	EL	0.1		6		X
PFR IX560FD	95	125		1.64	>50	>50	>50	EL/MMP	0.1	X	6	[19]	X
PFR IX700	95	125		1.64	>50	>50	>50	MMP	0.1		6		X
PFR IX710D	95	125		1.64	>50	>50	>50	MMP	0.1	X	6	[19]	X
PFR IX715D	95	125		1.64	>50	>50	>50	MMP	0.1	X	6	[20]	X
System 8	105	115	120	1.64	500	500	1000	EL	0.1	X	12		X

2. Apply photosensitive polyimide by puddle and spin coating at 5000 rpm for 30 seconds; coating thickness of 2.5 μm can be achieved.
3. Partially cure the polyimide film in a convection oven at 125°C for 60 minutes.
4. Expose at 200 mJ/cm^2.
5. Develop in a positive photoresist developer containing NaOH or KOH; rinse in water.
6. Oxygen plasma descum may be needed; set power at 100 W for 10 minutes.
7. Final cure at 180°C for 2 hours, 300°C for 60 minutes, or 400°C for 15 minutes.

Nonphotosensitive polyimides can be patterned using photoresist processing similar to the trilayer process (described in section 2.6.1) or in a bilayer process, where photoresist is applied directly to the polyimide without the barrier layer used in the trilayer process. The photosensitive polyimide process is simpler than processing nonphotosensitive polyimides.

Bilayer polyimide process

1. Apply adhesion promoter by puddle and spin coating at 5000 rpm for 30 seconds; the adhesion promoter is a proprietary solution provided by the manufacturer of the polyimide.
2. Apply polyimide by puddle and spin coating at 5000 rpm for 30 seconds; coating thickness of 2.5 μm can be achieved.
3. Partially cure the polyimide film in a convection over at 125°C for 60 minutes.
4. Coat with positive photoresist.
5. Expose with a dose compatible with the type of photoresist used.
6. Develop in a positive photoresist developer containing NaOH or KOH; developer will simultaneously remove polyimide in exposed areas; rinse in water.
7. Oxygen plasma descum may be needed; set power to 100 W for 10 minutes.
8. Strip resist in acetone.
9. Final cure at 180°C for 2 hours, 300°C for 60 minutes, or 400°C for 15 minutes.

2.3.5 Resist Resolution and Contrast

As stated earlier, the resolution and contrast of a resist material are important. The term resolution is used to describe the ability to print minimum-size images consistently under reasonably-varying manufacturing conditions. Contrast is expressed in terms of gamma (γ) and is related to changes in solubility in a resist material. Contrast of a resist directly influences resolution, resist profiles, and line width control. Resists with higher contrast result in better resolution than those with low contrast. If a resist had infinite contrast, vertical resist profiles would result independent of image contrast. For a positive resist, increases in exposure cause decreases in film thickness until the resist is completely removed. The corresponding exposure dose, or dose to clear, is called *Dc*. The minimum exposure dose for the onset of film removal is *D(0)c*. To experimentally calculate contrast, photoresist films are given known amounts of exposure and the thickness remaining after development is measured. The film thickness is normalized and plotted as in Fig. 2-11. Contrast (γ) is determined from the extrapolated slope of the linear portion of the response curve.

Image exposure is delivered to a photoresist in a diffuse (rather than abrupt) manner as a result of diffraction. Within the image, there are areas of varying exposure

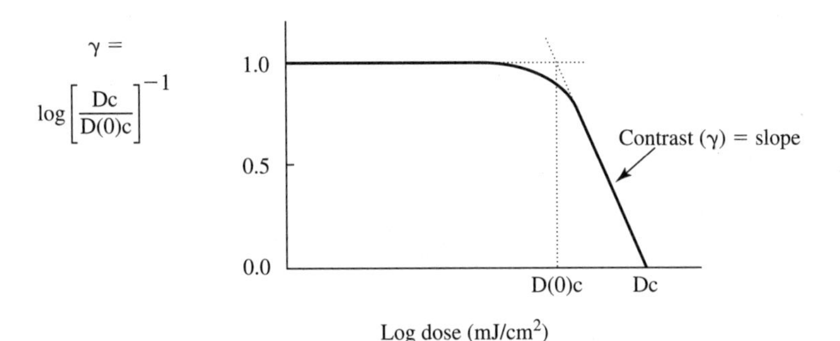

FIGURE 2-11 Plot of normalized film thickness after development showing contrast (γ = slope) of characteristic curve and sensitivity, or dose to clear (*Dc*).

levels (or doses). In areas with exposure greater than Dc, complete removal of the resist will occur. In areas of exposure less than D(0)c, little resist removal will occur. Areas with exposure between these doses will be partially removed. The smaller the range between D(0)c and Dc (that is, the higher the contrast of the resist), the better the resist will be able to resolve the image. These concepts are illustrated in Fig. 2-12.

The aerial image is the local irradiance levels at the surface of the photoresist that result from the imaging of a mask pattern by an optical system. For a mask with clear and opaque areas, the ideal aerial image is zero irradiance in regions corresponding to opaque mask areas, and the maximum irradiance in areas corresponding to clear mask areas. In a real optical system, the aerial image will exhibit maximum irradiance (I_{max}) less than the ideal maximum value and minimum irradiance (I_{min}) values greater than zero. As the spatial frequency increases (the distance imaged between lines and spaces decreases), the available difference between I_{max} and I_{min} decreases. The modulation transfer function (MTF) for the optical system under consideration gives these details. In Fig. 2-13, modulation transfer functions are shown for three different systems. A system illuminated by a point source will exhibit an MTF similar to that of the coherent example. Systems with large illumination sources will have an MTF similar to that of the incoherent example. Most optical systems for microlithography are partially coherent and will exhibit an MTF similar to that of the $\sigma = 0.7$ example depicted in Fig. 2-13.

The ability of a resist to form an image, which results in clear areas and areas with resist, depends on the contrast of the resist (γ). The minimum acceptable image modulation for a resist of contrast γ is defined as critical modulation transfer function $(CMTF)_{resist}$ given by

$$CMTF_{resist} = \frac{10^{1/\gamma - 1}}{10^{1/\gamma - 1}} \qquad (1)$$

and is plotted in Fig. 2-13. If a resist exhibits a contrast (γ) of 3, the required CMTF delivered by the optical system is 0.40, as illustrated in Fig. 2-13. A CMTF of 0.4 corresponds to a spatial frequency of 500 1/mm for the partially coherent system in Fig. 2-13. This means that the combination of optical system and resist can be expected to resolve 1 μm lines and spaces. Increased resist contrast might be achieved

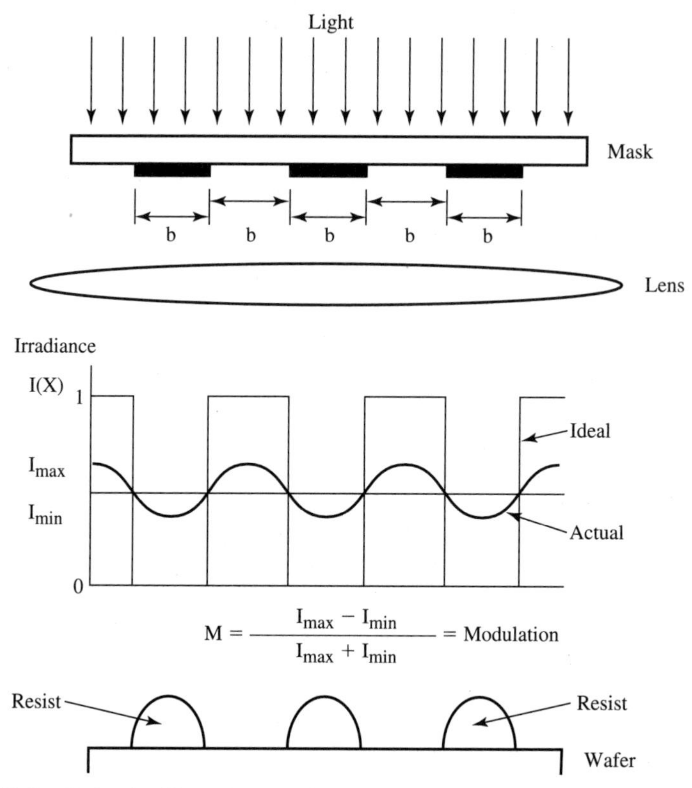

FIGURE 2-12 Ideal and real image formation in an optical system and definition of image modulation.

through process changes, resulting in possible improvements in resolution. Optical system modifications or use of a system with a different CMTF could also improve resolution.

2.3.6 Sensitivity

To provide adequate throughput, a resist must possess relatively high sensitivity to radiation. Sensitivity of a resist is determined by the photoefficiency of the sensitizer, or the PAC for positive resists. Photoefficiency (ϕ) is defined as:

$$\phi = \frac{\text{Number of photo-induced events}}{\text{number of photons absorbed}} \tag{2}$$

Sensitivity is conventionally defined as the exposure dose in milliJoules per unit area in square centimeters (mJ/cm^2) required to achieve a desired photochemical response. For a positive photoresist, sensitivity is determined by the minimum dose required to obtain complete solubility, or Dc, as shown in Fig. 2-11. Resist sensitivity considerations also include matching resist performance with the energy source of the

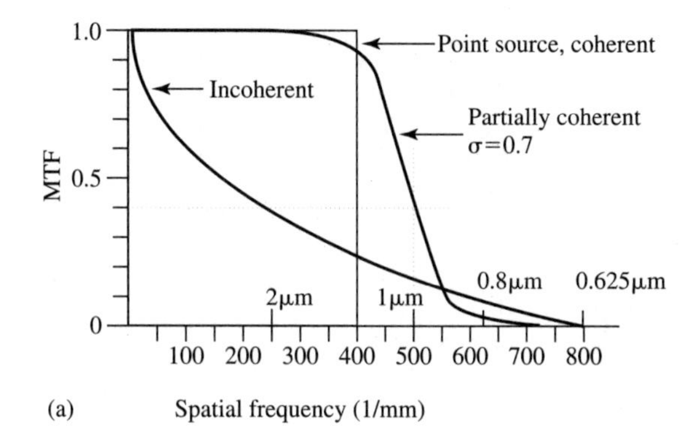

FIGURE 2-13 (*a*) Exposure system MTF, (*b*) CMTF, as a function of a resist contrast (γ).

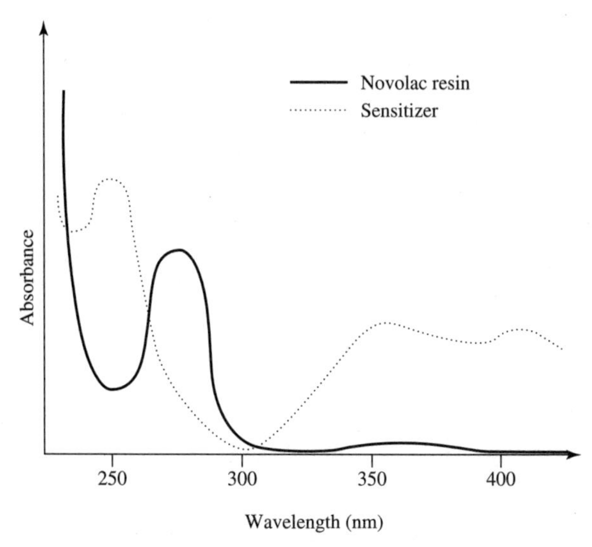

FIGURE 2-14 Absorbance spectrum of a typical diazoquinone sensitizer and a novolac resin.

exposure tool. The effectiveness of a resist/source combination is determined by the resist's spectral sensitivity. A PAC in a positive photoresist exhibits an absorbance spectrum related to its photoefficiency at various wavelengths. The absorbance spectrum of a typical diazoquinone PAC and a novolac resin is shown in Fig. 2-14. It is desirable to have a resin with low absorption and a PAC with high absorption at the desired exposure wavelength. Different PACs have different absorption spectra and are selected to optimize a resist for a particular exposure wavelength. For example, I-line resists may use a different PAC than G-line resists. Two slightly different PAC absorption spectra are shown in Fig. 2-15. Actinic absorbance is the difference between the absorbance of exposed resist and unexposed resist (Fig. 2-16). Exposure of a PAC leads to its destruction and loss of absorption contribution to the resist. This resist property is known as bleaching, and allows incident light to pass effectively through exposed upper regions of the resist film. Resists should have high actinic absorbance at the desired exposure wavelength.

Mercury and mercury-rare gas sources (arc lamps) are generally used for exposure of resist materials in microlithography. Emission lines corresponding to 365 nm (I-line) and 436 nm (G-line) are used for resist exposure, as shown in Fig. 2-17.

It is interesting to note that the glass used for the photomask and optical components in the exposure tool absorbs some of the ultraviolet light needed for exposure. Transmission properties of several optical glasses as a function of wavelength is shown in Fig. 2-18. Note that at 248 nm (KrF), only quartz shows high transmission.

2.3.7 Photoresist Processing

The basic steps involved in positive photoresist processing for microlithography are discussed in the following sections. Each of these steps impacts every other step in some way. This, along with the numerous variables involved with each step, makes careful engineering a necessity for optimizing a process.

Substrate Cleaning. Many problems encountered in microlithographic processing are a result of contamination in coating. Contaminants can lead to adhesion problems. Substrate cleaning often involves high-pressure water scrubbing, followed by a dehydration step. Generally, lithographic processing follows thermal oxidation or deposition steps. It is immediately after these steps that a surface is at its cleanest and contamination can be minimized. Humidity problems also occur, and a dehydrating baking step may be required to drive off any moisture before coating. Coating should be performed immediately after this bake step.

Priming. Adhesion problems often exist with positive photoresists because of the small molecular weight of the polymer molecules. If the substrate is not carefully treated, attachment of the individual molecules may be difficult. An adhesion promoter can be applied before resist application. Adhesion promoters such as hexamethyldisilozane (HMDS), trichloropropyltrimethoxy-silane (TCMS), or vinyltriethoxysilane (VTS) are commonly applied as a liquid or vapor coating. HMDS reacts with the remaining OH molecules, releasing ammonia gas, which creates an organic coating on the silicon surface. The organic nature of the surface improves adhesion of the organic photoresist, as shown in Fig. 2-19.

Spin Coating. After cleaning and priming the wafer with an adhesion promoter, the wafer substrates can be coated with photoresist. This is generally achieved by spin coating resist over the wafer. The spin coating procedure has three stages: (1) dispensing the resist onto the wafer surface; (2) accelerating the wafer to the final rotational speed, and

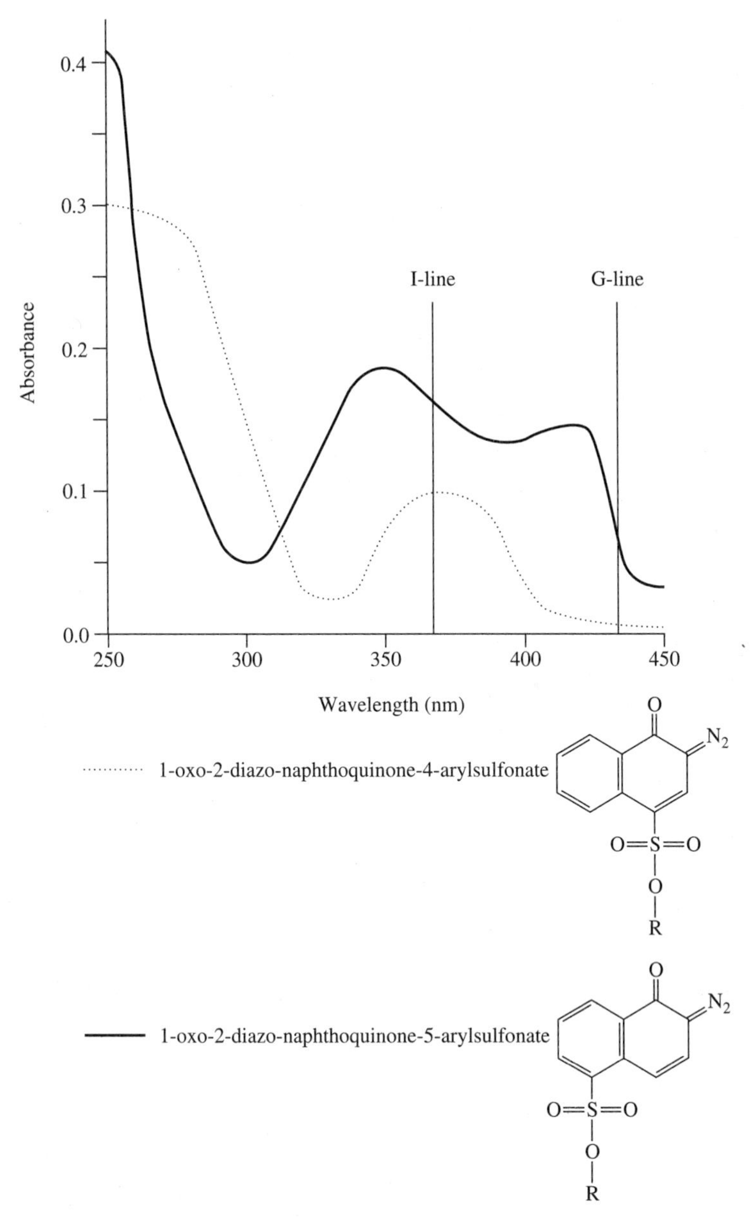

FIGURE 2-15 Absorbance spectrum of a 1-oxo-2-diazo-naphthoquinone-4-arylsulfonate and 1-oxo-2-diazo-naphthoquinone-5-arylsulfonate.

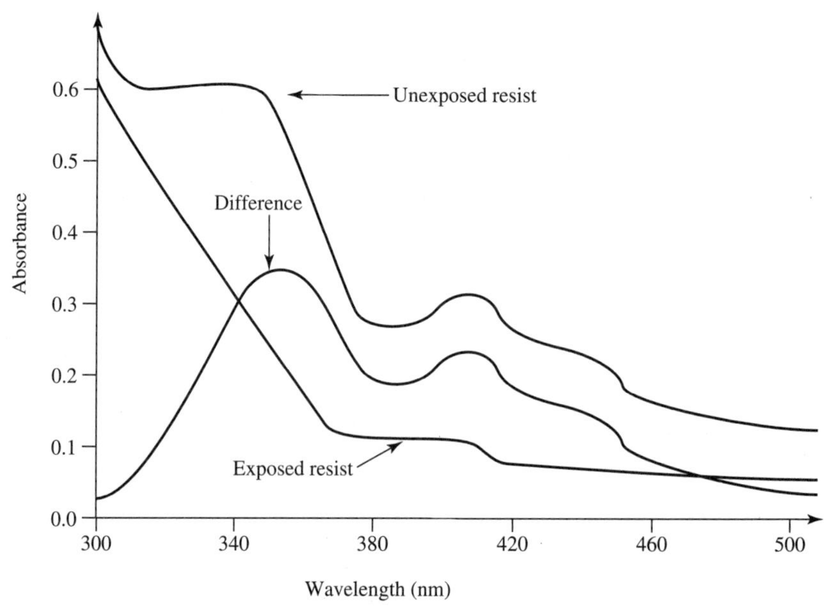

FIGURE 2-16 Actinic absorbance versus wavelength.

(3) spinning the wafer at a constant speed to spread and dry the resist film. After the spinning stage, the coating acquires a relatively uniform, symmetrical flow profile, illustrated in Fig. 2-20. Most spin coating is performed at speeds from 3000 to 7000 rpm for 20 to 60 seconds, producing coatings with uniformities to ± 100 Å over a wafer with a 150 mm diameter. Typical spin speed versus coating thickness is shown in Fig. 2-21. Spin coating can be performed in single-component spinner tools, but automated coating can be achieved with in-line automated tools.

Soft-bake. After the wafers are spin coated and partially dried, some of the resist casting solvent remains in the film. The prebake, or soft-bake steps that follow remove solvents from the resist, reducing them from as high as 30% down to 4% to 7%. Bake temperatures need to be high enough to reach the boiling point of the solvent and the glass transition temperature of the binder, but low enough to avoid destruction of the photoactive compound. In addition to solvent removal, resist adhesion is increased and resist/substrate stresses are reduced. Bake steps can be performed in convection ovens, hot plates, or infrared bake tools, as shown in Fig. 2-22.

Baking in a convection oven takes about 20 minutes because the top layer of the resist is dried first and remaining solvents have to exit the resist through the hardened outer layers. Hot plate baking takes about one minute because the wafers are heated from the back side driving solvents out of the front. Infrared baking will heat the resist throughout the thickness and results in intermediate baking times. Infrared baking is sensitive to the substrate and any layers on the substrate. Convection oven baking is usually used for batch processing of 25 to 50 wafers. Hot plate baking is usually used for single-wafer processing.

Exposure. Exposure of a photoresist is a critical step in that line width size and control is accomplished at this stage. The degree of exposure is controlled by adjusting the energy

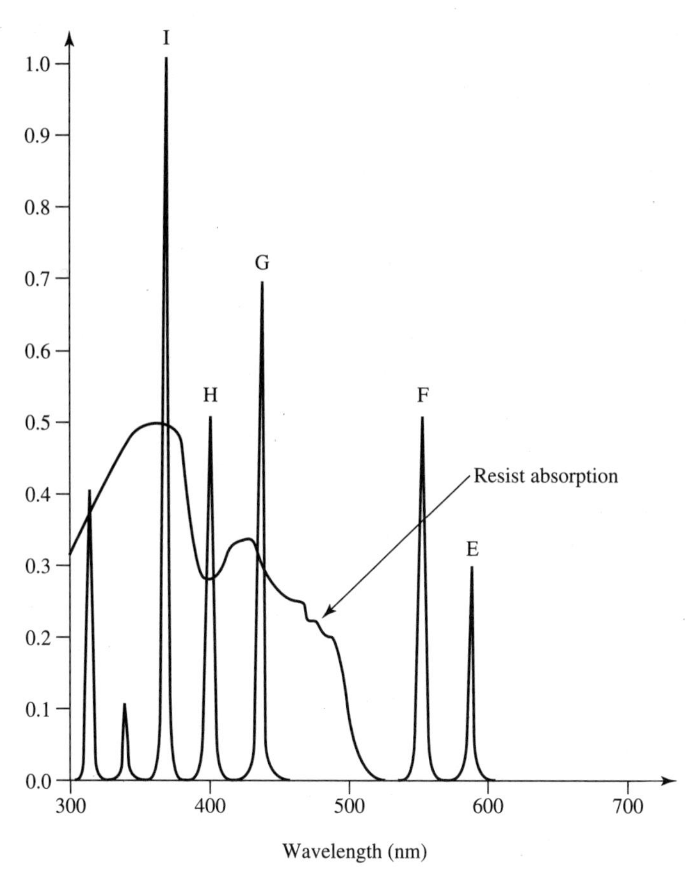

FIGURE 2-17 Positive photoresist response superimposed on spectral emission of high-pressure mercury arc lamp.

dose impinging on a resist. The exposure dose (E, in mJ/cm^2) is equal to the irradiance (I, in mw/cm^2) times the exposure time (t, in seconds)

$$E = I \times t \text{ (mJ/cm}^2\text{)} \tag{3}$$

At humidities exceeding 40%, the exposed sensitizer is completely converted to acid [3]. Without water available in the air, a less soluble ester is formed. At 20% humidity, the dissolution rate is decreased by as much as 50% and the dose to clear is increased by 10% to 15%.

Postexposure Bake. The postexposure bake is used to reduce standing wave effects in resist images and to increase the speed of the resist.

Development. The goals of an optimized development process include (1) minimal loss of resist thickness in unexposed areas of photoresist; (2) relatively short development times (under one minute); (3) minimum pattern distortion caused by absorption of developer solution and swelling; and (4) the specified pattern dimensions should be precisely produced.

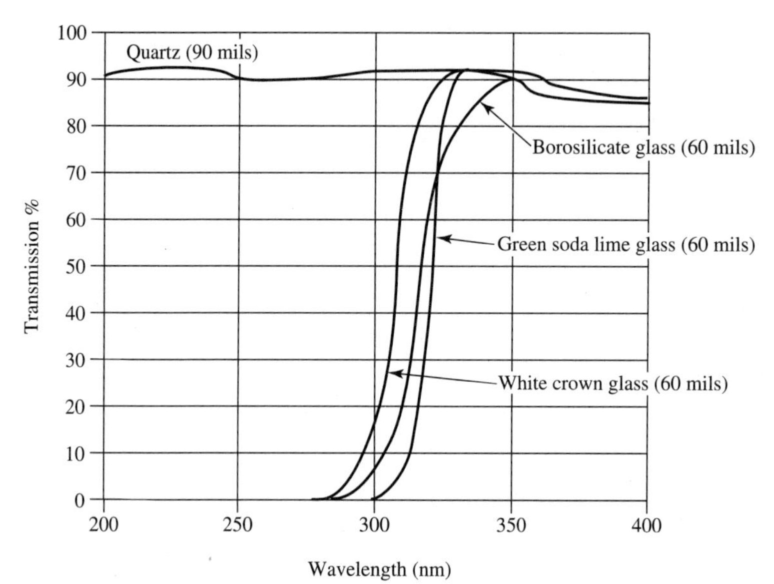

FIGURE 2-18 Transmission properties of several optical glasses as a function of wavelength.

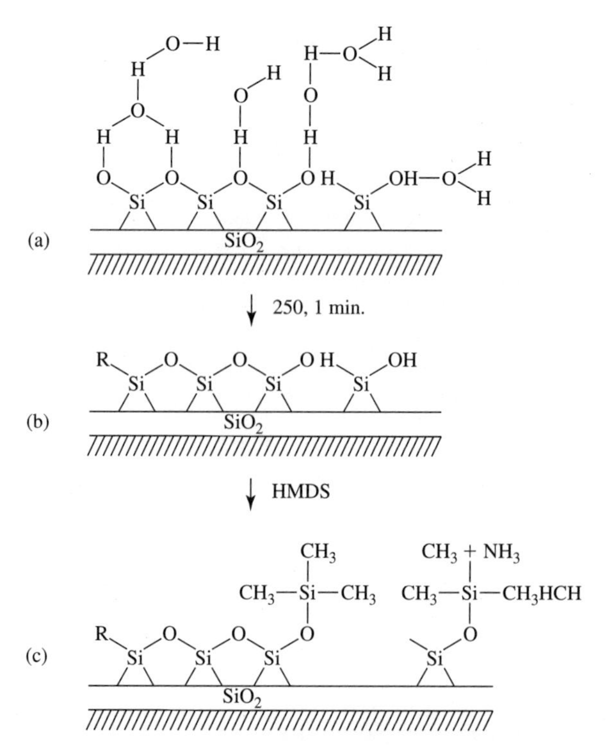

FIGURE 2-19 Wafer surface after (*a*) DI water scrub, (*b*) 250°C, 1 minute dehydration bake, (*c*) HMDS primer spin-coating.

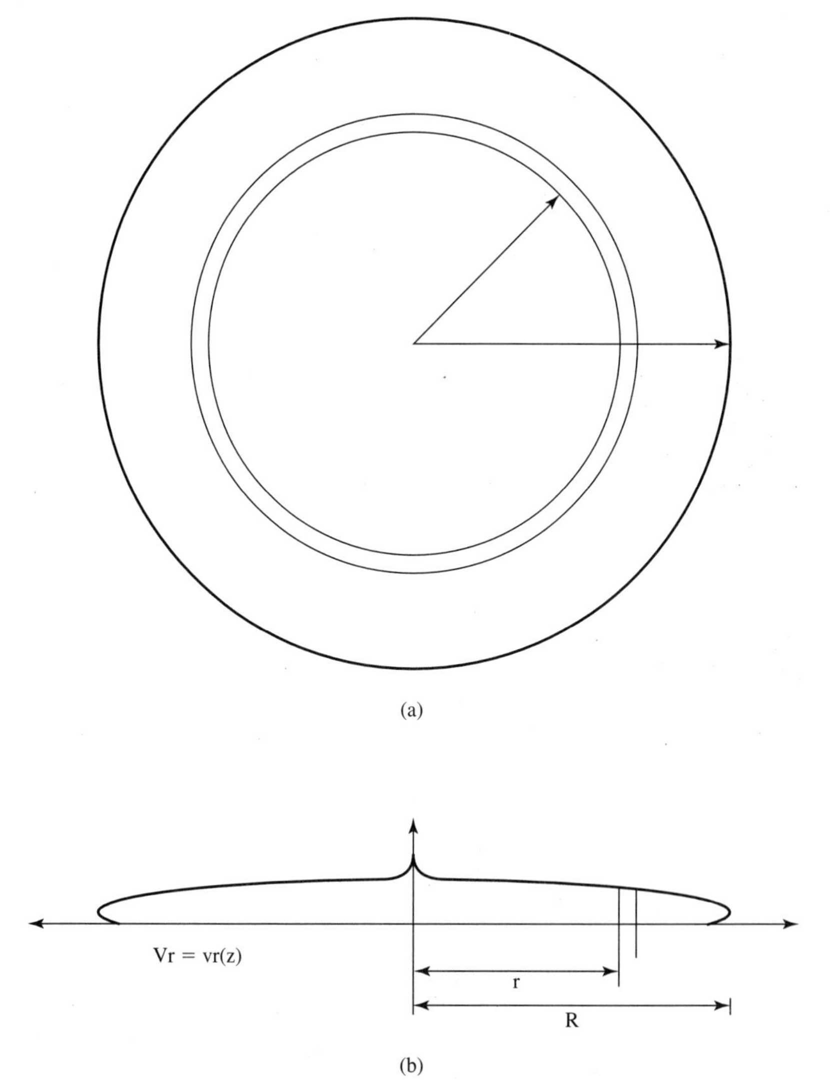

FIGURE 2-20 Photoresist application by spin coating: (*a*) radial symmetry, (*b*) photoresist cross-section.

There are three methods to achieve development: immersion, spray, and puddle development. In each method, development time and temperature must be carefully controlled. Developers for positive photoresist that contain metal ions are a possible source of contamination. For this purpose, there are metal ion-free developers with only small amounts of metal ion content. However, these developers may attack unexposed resist at a higher rate.

Rinse. To stop the development, the developer must be rinsed from the substrate surface with a solvent that is compatible with the developer chemistry. Positive photoresist devel-

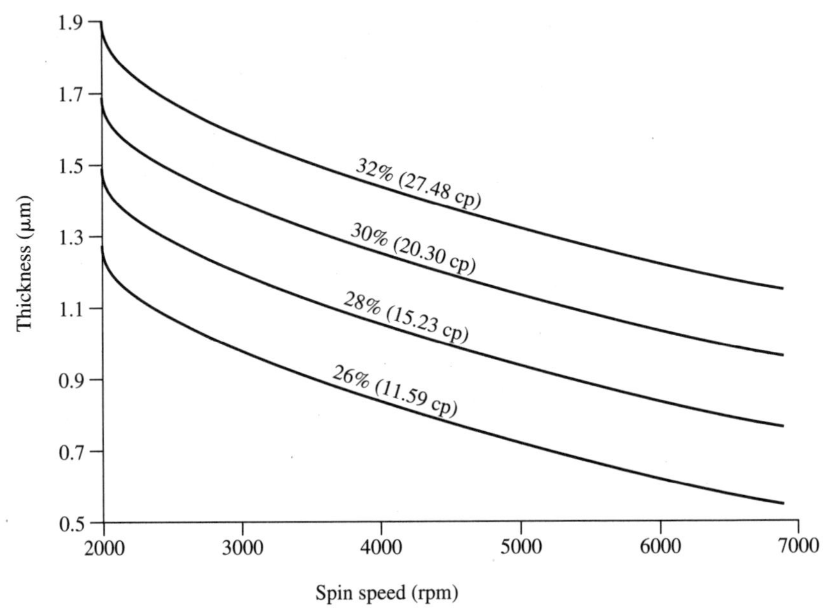

FIGURE 2-21 Photoresist thickness versus spin speed.

opers are alkaline solutions diluted with water, which must be rinsed with water. Solvent-based developers used with poly-methyl-methacrylate (PMMA) or negative polyisoprene resists are rinsed with other solvents that have a higher affinity for the developer.

Hard-bake. A post-development baking step is performed to remove residual solvents, improve adhesion, and increase the etch resistance of the resist. Additionally, this baking step causes some resist flow, which can fill pinholes in the film and reduce stress between film layers. Photoactive compounds are also destroyed, which makes the resist insensitive to radiation. Temperatures must reach the glass transition temperature of the resist, usually around 150°C.

Etching. The etching process selected will result in the removal of material in the regions not protected by photoresist. These etching processes can be done with suitable chemical solutions wet etching in the liquid state or plasma etching in the ionized gas state. Etching techniques are further discussed in the following sections.

Striping. After etching, most of the resist is removed from the substrate surface using chemicals or various plasma ashing techniques. Hot sulfuric acid and hydrogen peroxide can be used if the underlying layers are not affected by these chemicals. Organic solvents, such as acetone, can be used, but these represent a significant fire hazard. Oxygen plasma ashing is often used unless the substrate is sensitive to plasma damage. Downstream oxygen plasma ashers are used when plasma ashing is the preferred technique and plasma damage must be minimized.

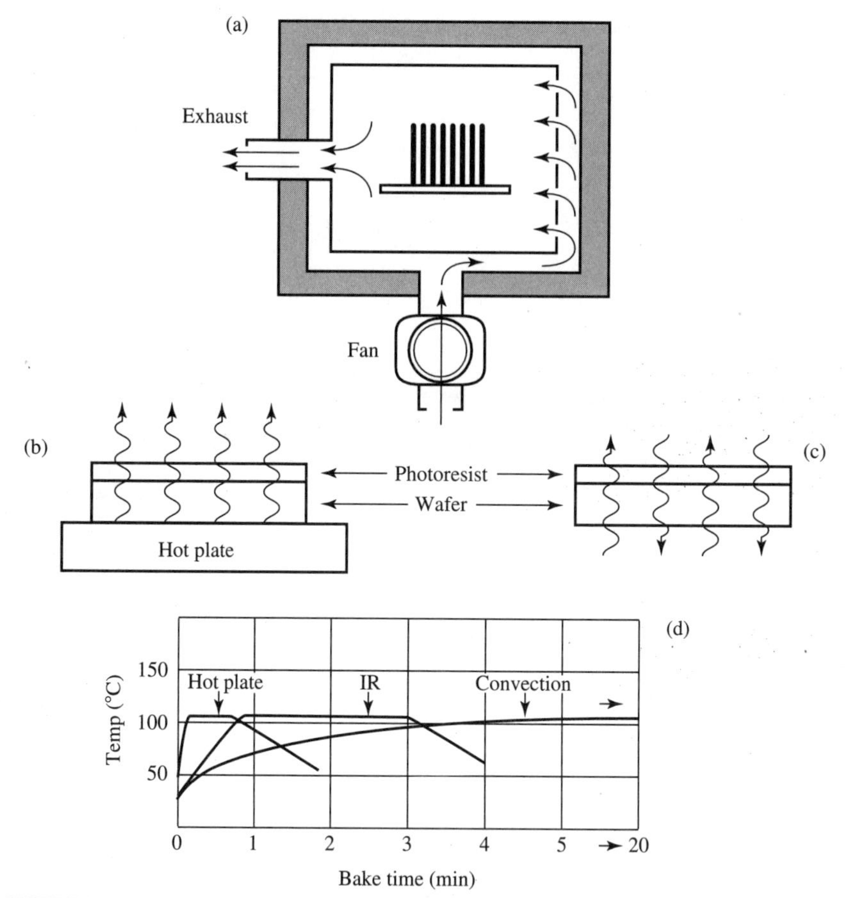

FIGURE 2-22 (*a*) Forced air oven, (*b*) hot plate, (*c*) infrared oven, (*d*) wafer temperature versus time for three bake techniques.

2.4 OPTICAL TOOLS FOR MICROLITHOGRAPHY

The three methods of optically transferring an image onto a resist film are contact, proximity, and projection printing as shown in Fig. 2-23. Each technique uses an illumination system, an optical system, a mask, and a resist-coated wafer. A brief history of optical lithography tools is given in Fig. 2-24.

2.4.1 Contact

Contact printing is the technique with the highest resolution. The major problem with contact printing is that the wafer and mask touch, resulting in increased defects caused by photoresist sticking to the mask. Contact printing and proximity printing are very similar because the mask is separated from the substrate by the photoresist thickness.

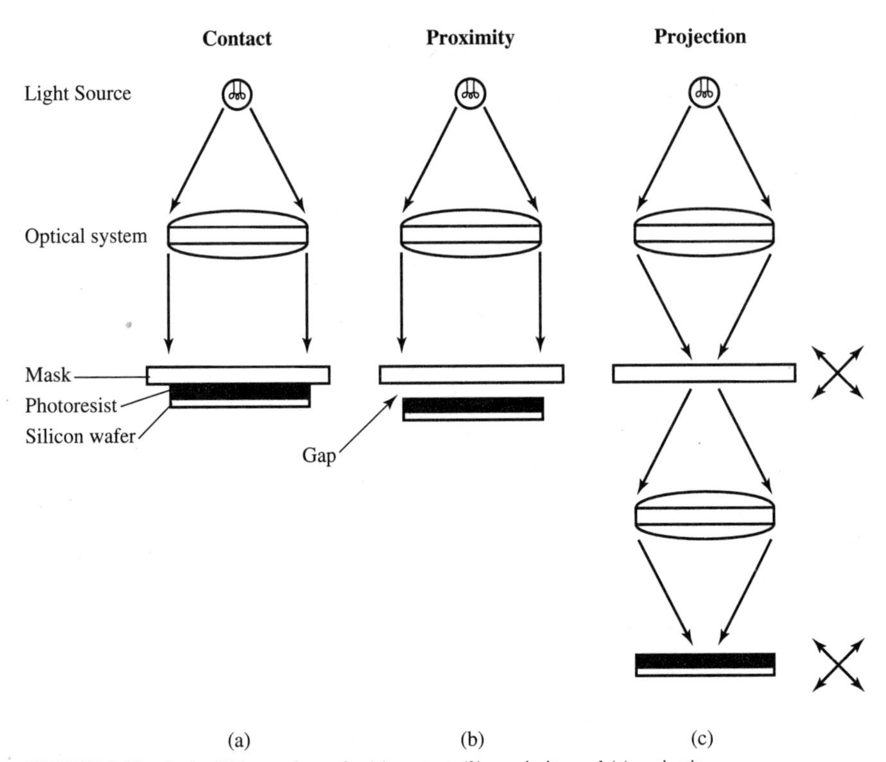

FIGURE 2-23 Optical lithography tools: (*a*) contact, (*b*) proximity, and (*c*) projection.

Contact and proximity printing
- From the start in 1960's

Perkin-Elmer scanner
- Developed by Rod Scott and Abe Offner
- First units in the field in 1973
- More than 3,000 units in use

Step-and-repeat
- Developed from photorepeaters used for mask making
- First introduced in 1978 by GCA with a Zeiss 10-77-82 lens

Ultratech 1:1 stepper
- Developed by Ron Hershel and Martin Lee
- First units delivered in 1979

Deep UV steppers (<248 nm)
- Advanced Wafer Imaging System (AWIS) developed by GCA for VHSIC contract, 1988
- ALS laser step delivered in 1989
- XLS stepper 1992

FIGURE 2-24 History of optical lithography tools.

$$\text{Resolution } (L_{min}) = 1.5\ [(S + d)\ \lambda]^{0.5}$$

where S = spacing between the mask and the resist
d = thickness of the resist
λ = wavelength of light

Example: $S = 0$, $d = 1\ \mu m$, $\lambda = 436$ nm, then $L_{min} = 1\ \mu m$.

2.4.2 Proximity

Proximity printing reduces defects because the mask and wafer never touch. However, the resolution is lower than other techniques.

$$\text{Resolution } L_{min} = 1.5\ [(S + d)\ \lambda]^{0.5}$$

where S = spacing between the mask and the resist
d = thickness of the resist
λ = wavelength of light

Example: $S = 10\ \mu m$, $d = 1\ \mu m$, $\lambda = 436$ nm, then $L_{min} = 3\ \mu m$.

2.4.3 Projection

Projection lithography uses either refractive lenses or reflective mirrors. These optical systems are described by the numerical aperature of the imaging system, NA_O, numerical aperature of the illumination optical system, NA_C, and the wavelength, λ, for which the system was designed. Other factors, such as magnification, field size, coherency, distortion, and depth of focus, are also important optical performance parameters. Examples of lens specifications are shown in Table 2-2.

The numerical aperature of the imaging lens gives an estimate of the resolution capability of the system. The numerical aperature is equal to the index of refraction of the surrounding medium times the sine of the acceptance lens angle. Systems with larger diameter lenses have larger values for NA. Lord Raleigh proposed a criteria for identifying just when two point sources (stars in his case) are resolved. The diffracted image of the point source is called an airy disk; when the peak of one airy disk falls on the first minimum of

TABLE 2-2 Examples of lens specifications

Lens manufacturer	Zeiss	Tropel	Zeiss	Nikon	Cannon
Part number	10-78-56	T22351	10-78-46		UL 1011
Reduction	5×	5×	5×	5×	5×
Wavelength (nm)	365	365	436	436	436
Numerical aperature	0.4	0.35	0.38	0.42	0.43
Field diameter (mm)	20	22	20	21.2	21.2
Square image format (mm)	14.1	15.6	14.1	15	15
Res. limit $L = 0.5\lambda/NA$	0.46	0.52	0.57	0.52	0.51
Depth of focus, $DOF = 0.5\lambda/(NA)^2$	1.14	1.49	1.51	1.24	1.18
Distortion ($\pm\ \mu m$)	0.01	0.1	0.15	0.15	0.15

the other airy disk, then the two point sources can just be resolved. This occurs when the images of two point sources are separated by a distance of

$$L_{min} = 0.61\ \lambda/NA$$

where λ is the wavelength of the light source. This equation gives us an estimate of system resolution and the correct relationship between resolution, wavelength, and numerical aperature. Raleigh also proposed that depth of focus

$$DOF = \lambda/[2(NA)]^2$$

One should be cautioned that Raleigh's criteria was not intended for predicting the performance of projection lithographic systems. However, this criteria allows quick estimates, as shown in Table 2-3.

Refractive. Projection steppers use refractive optics (lenses) to project an image on a wafer. These tools cannot expose the entire wafer in one exposure because the optical field size is smaller than the wafer. The wafer is exposed in sections and the wafer is moved (stepped) in between exposures until the entire wafer has been exposed. The key to this type of system is the precision stage on which the wafer rests and is moved; the overlay accuracy of the system depends on the stage precision. Stage position is measured with a laser interferometer positioning system. The laser interferometer is accurate to $\lambda/4$, $\lambda/8$, or $\lambda/16$, depending on the configuration. For the helium neon laser, $\lambda/8$ equals 0.08 μm. A schematic of a Zeeman interferometer is shown in Fig. 2-25. The laser is helium neon and has two special features. A magnet surrounding the laser causes energy levels associated with electron spin, resulting in two wavelengths of laser light being emitted. The two-frequency Zeeman laser also has a piezoelectric motor to create a small amount of resonant cavity tuning, which makes the two frequencies of laser energy equal in strength. The two circularly polarized frequency components exist in the same beam. One component is left-hand circularly polarized and is approximately 1 MHz from the center operating frequency, f_o of the laser cavity. The other beam frequency component is right-hand circularly polarized and is approximately 1 MHz from f_o in the other direction. The laser beam containing the two circularly polarized frequency components (f_1 and f_2) passes through a quarter-wave plate, $\lambda/4$. This causes the f_1 and f_2 components to become linearly polarized and mutually perpendicular (orthogonal). The two-frequency Zeeman laser beam is split with a beam splitter, forming the reference signal and the measuring beam, with one part containing only the f_1 component and the other containing only the f_2 component. The f_2 component reflects off a beam bender, hits a stationary retroreflector, and returns to the beam splitter, where it is combined with the $f_1 \pm 2 \times \Delta f_1$ beam. A $\lambda/4$ plate is inserted in the f_1 path just after the polarizing beam splitter and before the plane mirror reflector on the moving stage. The $\lambda/4$ plate causes the polarization of the f_1 beam to be rotated through

TABLE 2-3 Examples of estimates of lithographic system performance calculated from the Raleigh criterion

	G-line	I-line	Excimer KrF	Excimer ArF	Excimer ArF
Lambda	436 nm	365 nm	248 nm	193 nm	193 nm
NA	0.28	0.28	0.28	0.28	0.35
L_{min}	0.95 μm	0.800 μm	0.540 μm	0.420 μm	0.340 μm
DOF	2.78 μm	2.33 μm	1.58 μm	1.23 μm	0.788 μm

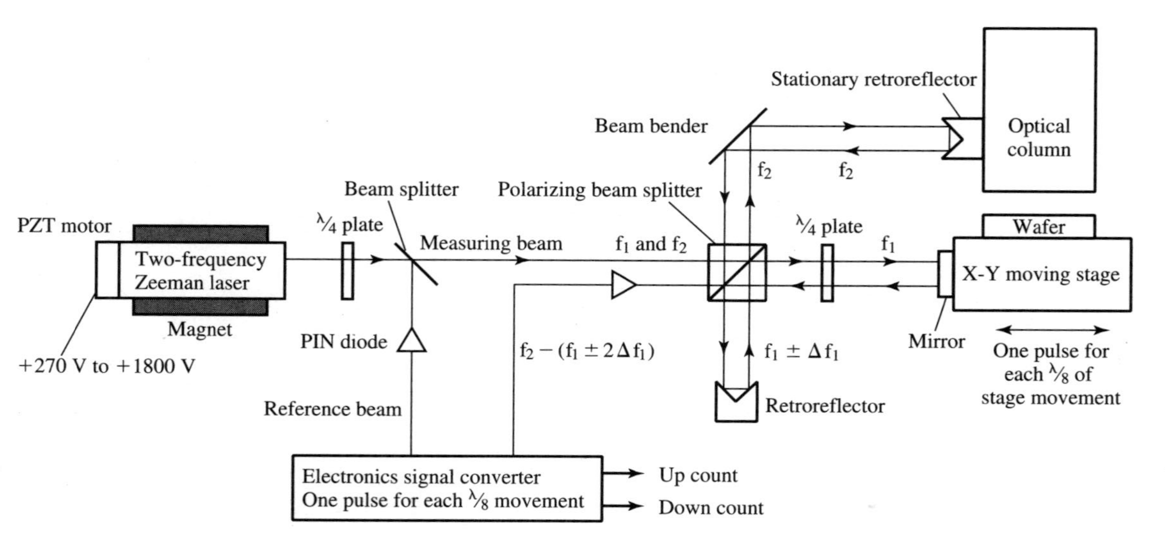

FIGURE 2-25 Schematic of Zeeman two-frequency laser interferometer used for measuring stage position with accuracy of $\lambda/8$.

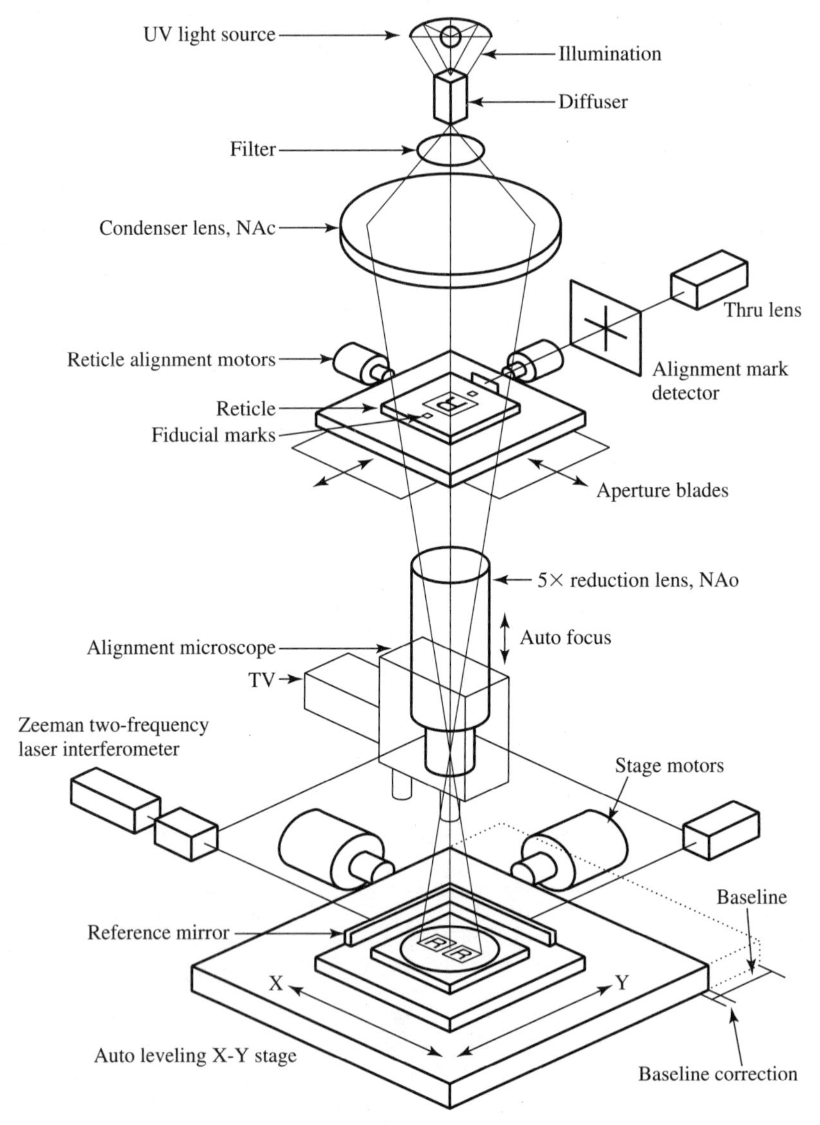

FIGURE 2-26 Schematic of a reduction stepper.

90 degrees, so that $f_1 \pm \Delta f_1$ is reflected out of the polarizing beam splitter, off the bottom retroreflector, and to the plane mirror for a second time, where it is Doppler shifted again. The polarization of the $f_1 \pm 2 \times \Delta f_1$ is rotated through 90 degrees so that it now is transmitted back to the receiver. Resolution doubling is inherent because of the double Doppler shift. The $2 \times \Delta f_1$ part is created if the stage is moving because of a Doppler shift phenomenon. If there is no motion the Δ is zero. The reference signal is converted by PIN photodetectors into electric current representing $(f_2 - f_1)$ and the measuring signal $(f_2 - f_1 \pm 2 \times \Delta f_1)$, which is converted to an electrical pulse for each $\lambda/8$ of stage

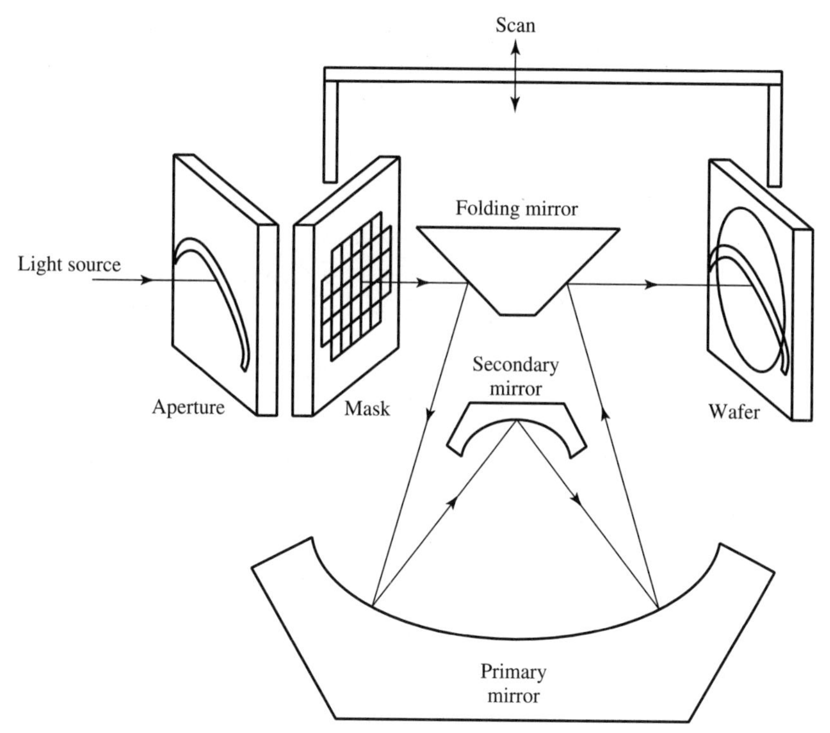

FIGURE 2-27 Schematic of a Perkin-Elmer Microalign 1:1 reflective wafer scanning system.

movement. There is also a signal for up or down count, depending on the direction of stage travel.

A generalized step and repeat system is shown in Fig. 2-26. The system consists of an illuminator and related optics, reduction lens, and stage. In addition, systems for aligning the wafer, aligning the reticle to the optical column, loading wafers and reticles, adjusting focus and leveling the wafer, and measuring stage position are all part of the tool.

Reflective. Projection scanners use noncontact optical imaging, thus reducing defects. The optics in these systems are reflective mirrors and have the advantage of being less sensitive to wavelength. These systems also can expose very large areas. A schematic of the Perkin Elmer Microalign 1:1 reflective scanning system is shown in Fig. 2-27. The system exposes the wafer by moving the mask and wafer simultaneously (scanning) while the optical system remains stationary. Because of the way the mirrors are made and the symmetry of the optical system, only an arc-shaped image can be created at optimum focus, which results in the requirement for scanning. In addition to the wafer-scan system, other variations are possible, including a reduction-step and scan system, (as shown in Fig. 2-28), 1:1 stripe scan, and 1:1 raster scan. The resolution of these systems can be estimated by

$$L_{min} = \frac{K\lambda}{NA}.$$

A value of 0.15 at 436 nm for NA gives resolution of about $L_{min} = 2\ \mu m$; for NA of 0.35 at 248 nm, $L_{min} = 0.50\ \mu m$.

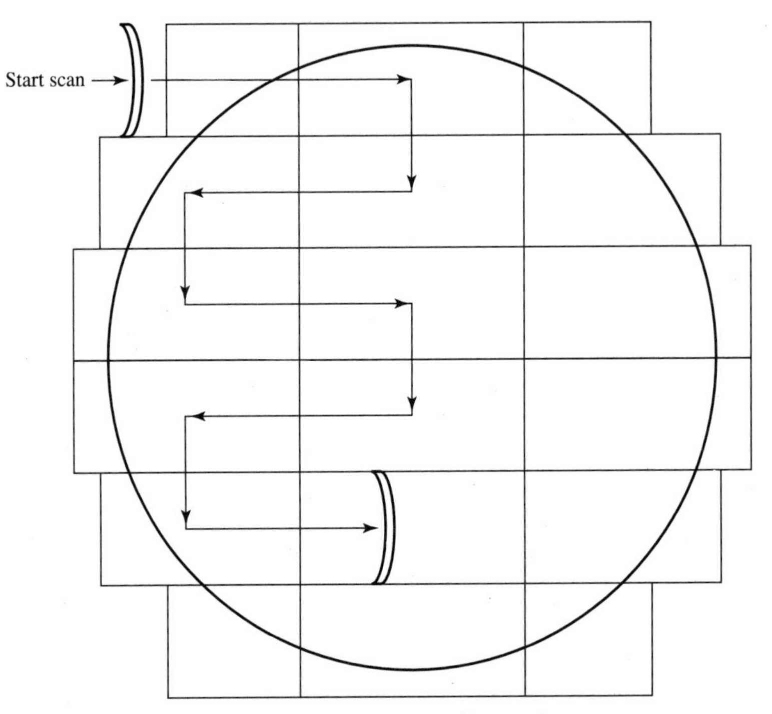

FIGURE 2-28 Step and scan system for reflective exposure of large wafers.

2.5 ETCHING

2.5.1 Wet Etching

In wet etching, the substrate is immersed in a liquid etch bath, usually a strong acid or base solution. The advantage of wet etching is that it is more selective than other techniques. Selectivity refers to the ratio of the etch rate of the film to be etched to the etch rate of the underlying layer. Selectivity to the photoresist is the ratio of the etch rate of the film to be etched to the etch rate of the photoresist. In a good etch process, both selectivities should be high. Positive photoresists are attacked in alkali solutions; as a result, most wet etch solutions used with positive photoresists are based on acidic formulations. The major disadvantage of wet etching is that, for certain films, the selectivity to the photoresist is poor. Films such as silicon nitride and polysilicon are difficult to etch using wet-etch techniques, but they are relatively easy to etch using dry-etch techniques. Another disadvantage of wet etching is that, for most substrates, wet etching is isotropic, meaning that the etch rate is the same in all directions. This results in the lateral etch and vertical etch being almost equal. For small geometries on the order of the film thickness, an anisotropic etch which has a vertical etch rate much higher than the lateral etch rate, is required. Under certain conditions, dry etching can give anisotropic etch profiles.

There are many references containing collections of etch recipes for various situations [8]. The following examples will illustrate several wet-etch processes and the considerations used in developing of these processes. Table 2-4 also illustrates a number of common solutions used for etching.

TABLE 2-4 Wet etching solutions for common electronic materials

Compound to Etch	Etching Solution	Rate	Comments
Aluminum	1 $HClO_4$, 1 $(CH_3CO)_2O$	3 μm/min	
	1 HCl, 4 H_2O		80°C, fine line etching etching
	10 HCl, 1 HNO_3, 9 H_2O	25-50 μm/min	49°C
Copper	$FeCl_3$	50 μm/min	49°C
Gold	4 g KI, 1 g I_2, 40 ml H_2O	0.5-1 μm/min	
	3 HCl, 1 HNO_3	25-50 μm/min	
Nickel	5 HNO_3, 5 CH_3COOH, 2 H_2SO_4	400-500 Å/min	Dilute with H_2O for reduced rate
	$FeCl_3$	12-25 μm/min	
Palladium	1 HCl, 10 HNO_3, 10 CH_3COOH	1000 Å/min	
Platinum	8 H_2O, 7 HCl, 1 HNO_3	400-500 Å/min	85°C
Silver	5 HNO_3, 1 H_2O	12-25 μm/min	49°C
Titanium	9 H_2O, 1 HF	12 μm/min	32°C
Silicon Dioxide	BOE (HF solution)	10-10 nm/min	Buffered oxide etch 25°C
Gallium Arsenide	1 H_2SO_4, 8 H_2O_2, 1 H_2O	5-10 μm/min	20°C etch rate is a function of crystal orientation.

Etching of Aluminum (Al). Pure aluminum can be etched in many acids and bases. An aluminum etch solution is available that contains phosphoric acid, acetic acid, and nitric acid as the major components. This solution does not attack positive photoresist or underlying layers of silicon dioxide, silicon nitride, or silicon (Si). It etches at about 5000 Å/min at 50°C and at lower rates at room temperature. When used below 40°C with Al/1% Si, this etchant will leave behind a residue of silicon. The residue can be removed by etching at 50°C or higher. This aluminum etchant is not recommended for etching Al off of gallium arsenide because it etches gallium arsenide at a high rate. A similar formulation with higher selectivity is available for this application. This illustrates the details that need to be addressed in developing wet-etch processes.

Wet etching of silicon nitride can be useful for unmasked etches (for example, after a LOCOS isolation process). The wet etch will etch the silicon nitride and not the underlying silicon dioxide, unlike dry etching. The process involves a hydrofluoric acid dip to remove the oxy-nitride layer, a rinse in water, then a 165°C phosphoric acid etch, and, finally, a water rinse. The phosphoric acid etch rate is about 55 Å/min. This process could not be used with regular positive photoresist because of the 165°C temperature, but some negative resists or polyimide as a resist could work.

In multilevel metal processes, it is often necessary to etch vias through an insulating interlevel dielectric. When the underlying layer is Al and the insulating layer is glass, the etchant must etch glass but not Al. Straight buffered hydrofluoric acid (BOE) will etch Al. A mixture of 5 parts BOE and 3 parts glycerin will etch glass but not Al [9]. The etch rate is unaffected by the glycerin.

The etching of platinum (Pl) is very difficult. Most wet etchants for Pl destroy the photoresist. In this case, etching can be accomplished using tougher negative photoresists or polyimide as the etch resist. Dry etching is also feasible.

2.5.2 Dry Etching

Dry etching is a process in which the substrate is exposed to gases in an excited state (plasma). Processing is usually done in a vacuum system below atmospheric pressure. Excitation is realized by various means, including high voltages at radio frequency (RF). As a result, the substrate is immersed in ions in the gaseous state; these ions can etch the exposed substrate surface. The main advantage of dry etching is that the substrate can be etched anisotropically (with vertical etch walls). Plasma etching, reactive ion etching, and sputter etching are all similar processes; in each the wafers are exposed to a plasma. The differences are the operating conditions of the plasma and the type of gas in the plasma. The operating pressure is highest for plasma etching and decreases for reactive ion etching, further decreasing for sputter etching. The molecular fragments in plasma etching are chemically active elements such as fluorine (F). In sputter etching, the ions can be inert elements such as argon. The ion energy is lowest in plasma etching and highest in sputter etching. Plasma etching can be selective; sputter etching is the least selective. All three techniques can be anisotropic. Plasma etching can be either isotropic or anisotropic, while both reactive ion etching and sputter etching tend to be more anisotropic.

Plasma Etching. The key to plasma etching is that the chemical formed by the reacting molecular fragments in the plasma is volatile at the temperature of operation [10,11]. For example sulfur hexafluoride (SF_6) plasma contains F fragments. When etching polysilicon, the F reacts with Si to form silicone tetrafluoride (SiF_4), which is volatile at room temperature. The SiF_4 can be removed from the substrate surface, exposing a new Si surface to be etched. Silicone hexafluoride (SF_6) will also etch SiO_2 and Si_3N_4, releasing SiF_4 and O_2 or SiF_4 and N_2, respectively. The ratio of the etch rate of the top film to the etch rate of the underlying film is called the selectivity. For example, the selectivity in SF_6 plasma for polysilicon on SiO_2 is approximately 8, making it difficult to etch polysilicon on top of thin SiO_2 layers without also etching through them.

Plasma etch conditions are difficult to transfer from one type of machine to another. The main controllable variables are chamber pressure during etch, etch time, gas flow (gas flow and pressure interact unless the tool has separate control of pressure with a throttle valve), forward and reverse RF power, gas types, and mixture. The design of the plasma chamber also has a great effect on the etch. Barrel etchers, parallel plate etchers, plate separation, gas inlet position, and exhaust porting make every system different. Etching of polysilicon at pressure of 100 mm Hg, 200 W, 30 sccm of SF_6 and 3 sccm of O_2 may give the desired results in one tool, but might give completely unacceptable results in another. Conditions for particular etch processes can be obtained from vendor application engineers.

Reactive Ion Etching. Reactive ion etching is similar to plasma etching except that, for reactive ion etching, slightly lower pressures and higher ion energies are incorporated. The reactor chamber is a parallel plate design in which the substrate to be etched is in contact with one electrode in the system. This electrode is kept at a negative bias with respect to the other parts of the chamber. Positive ions are attracted toward the substrate and can bombard the surface. If the electrode configuration is parallel and the pressure is low, the bombarding ions are mostly perpendicular to the wafer surface. The etching can be anisotropic when the sidewalls of an etch are protected by a polymer film and the bottom of the etch is free of polymer because of the effect of the bombarding ions.

Sputter Etching. Sputter etching is similar to sputtering except that the sputtering target (the cathode) is replaced in sputter etching by the substrate. The gas used is usually an inert gas such as Ar, and pressure ranges are low (1 to 10 mm Hg). A plasma is created with either a dc or RF power supply. The substrate is the cathode and Ar ions bombard the substrate, physically knocking off atoms. This technique is not selective, meaning that it will remove the masking material (photoresist), the film to be etched, and the substrate at nearly the same rate.

2.6 ADVANCED PROCESSES

There are situations where single-layer photoresist processes do not provide the desired performance in terms of photoresist sidewall profile, resist thickness over topology, or other process parameters. Advanced photoresist processes can sometimes provide a solution. These include bilayer processes such as antireflection coating under positive photoresist or contrast enhancing (slow bleaching) coating on top of positive photoresist. Trilayer, lift-off, and image reversal processes are discussed in detail in the following text.

2.6.1 Trilayer

Trilayer resist processes are used when the substrate has severe topology or whenever a thick photoresist layer is needed. The first layer applied to the substrate is a thick photoresist or polyimide layer (2 to 100 μm). This layer should be thick enough to cover any topology and provide a smooth surface for the next two layers. The second layer is a thin barrier layer (about 0.3 μm) of either spin-on glass or evaporated aluminum. The barrier layer prevents the top layer of photoresist and the botom layer of thick photoresist from mixing; it also will not be etched by oxygen-reactive ion etch. Finally, a top coating of thin photoresist (about 1 μm) is applied and soft-baked. The top layer is imaged and the barrier layer is etched on exposure, exposing the first layer. The substrate is placed in an oxygen-reactive ion-etch tool and the thick polymer layer is etched with vertical walls, resulting in the resist pattern shown in Fig. 2-29.

2.6.2 Lift-Off

Lift-off is a technique in which the thin film being applied to the substrate is added only where the material is desired, as opposed to being removed from regions where the thin film is not desired. The advantage of this method is that etching of the thin film is not necessary; thus, selectivity issues between etching of the thin film and etching of the substrate are avoided. The key to this process is creating a reverse slope photoresist sidewall profile and deposition of the thin film using a technique which does not provide step coverage. A lack of step coverage is defined as line of sight deposition which does not coat vertical walls or slopes, and can be accomplished with evaporation or special sputter deposition techniques, described in Chapter 1. The substrate is coated with photoresist first and imaged so that there is no photoresist where the thin film is desired. Then the thin film is deposited on the substrate. Finally, the photoresist is stripped off in a solvent, lifting off the undesired thin film on the photoresist and leaving the thin film material on the substrate only where desired. A lift-off process is illustrated in Fig. 2-30.

It is important to get a reverse slope photoresist sidewall profile. One technique is to use a trilayer-like process, but to etch the first polymer layer with an isotropic plasma rather than a anisotropic reactive ion etch. Another approach is to use an image reversal process optimized for the desired sidewall slope. This will result in a reverse-slope photoresist sidewall profile.

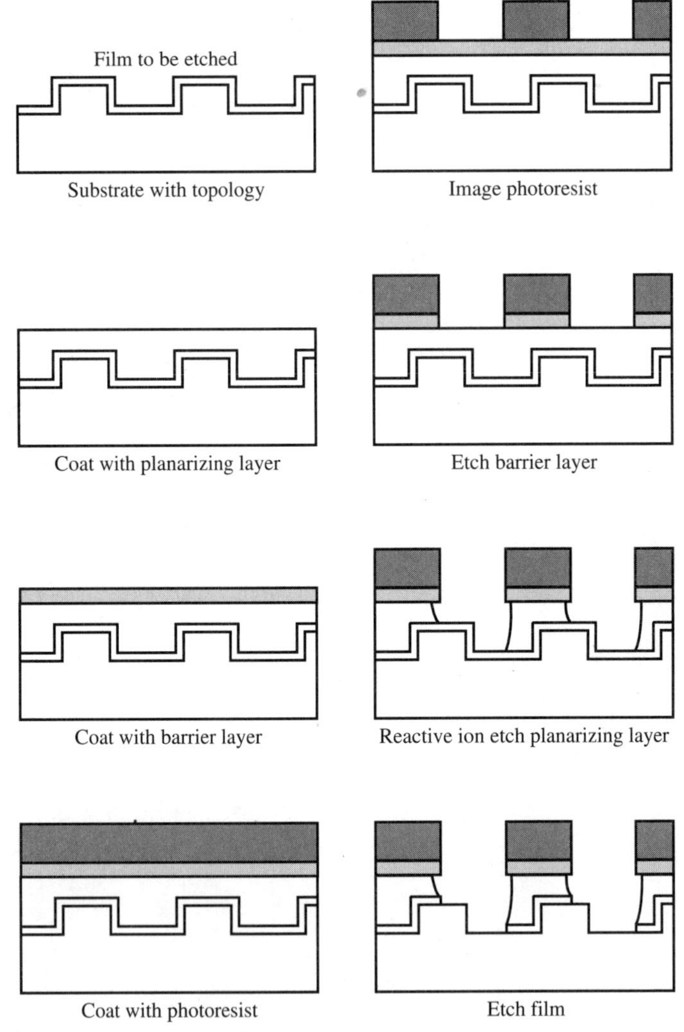

FIGURE 2-29 Trilayer resist process.

2.6.3 Image Reversal

Image reversal processes involve special photoresists that can function not only as a normal positive photoresist but also as a negative photoresist if an additional postexposure bake and flood exposure is added to the photoresist process. A reversal process is useful for providing both negative and positive images from the same photomask. Image reversal photoresists also allow for steeper-slope or even reverse-slope photoresist sidewall profiles. A reverse-slope photoresist sidewall profile can be made by exposing the resist on a high-aperature exposure tool (low depth of focus) with the focus set for the bottom of the photoresist. The exposure at the top of the photoresist will be slightly out of focus and thus

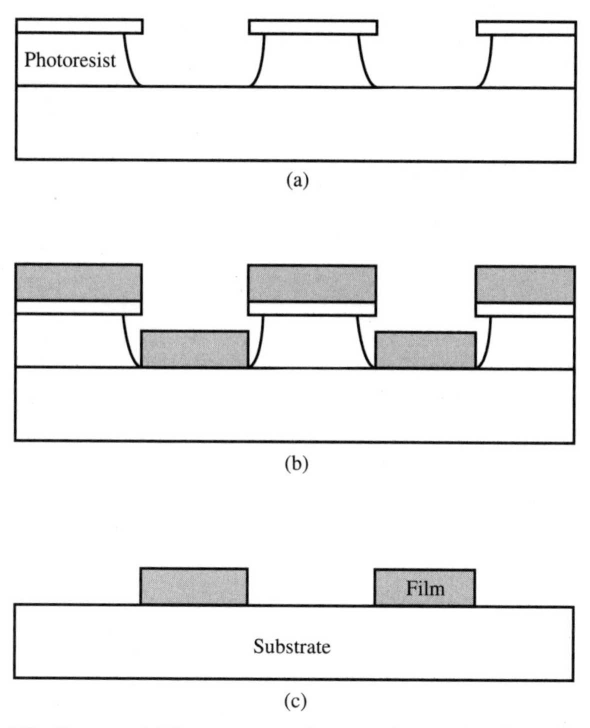

FIGURE 2-30 Lift-off process. (*a*) Create a reverse slope or undercut resist edge profile, (*b*) deposit film by evaporation, (*c*) chemically strip photoresist and lift off film, leaving film in desired pattern.

slightly flared. After image reversal, the flare at the top will create a reverse slope photoresist sidewall profile, as illustrated in Fig. 2-31. The details of an image reversal process are given in the example below.

Image reversal process steps

1. Prepare and clean substrate
2. Dehydration bake at 200°C for 3 minutes
3. Spin coat AZ-5214-E photoresist at 4000 rpm/45 s
4. Soft bake on hotplate at 100°C/45 s
5. Expose 80 mJ/cm^2 (about 10 seconds on Kasper contact aligner)
6. Postexposure bake, image reversal (PEB) on hotplate at 115°C/90 s
7. Flood expose on Kasper aligner for 30 s, 120 mJ/cm^2
8. Develop in Shipley 351 + H_2O (1:4 ratio) in petri dish
9. Mix 100 ml developer with 400 ml H_2O and rinse immediately after developing
10. Blow dry
11. Vapor-deposit metal thin film
12. Lift-off photoresist and metal thin film in ultrasonic bath of acetone.

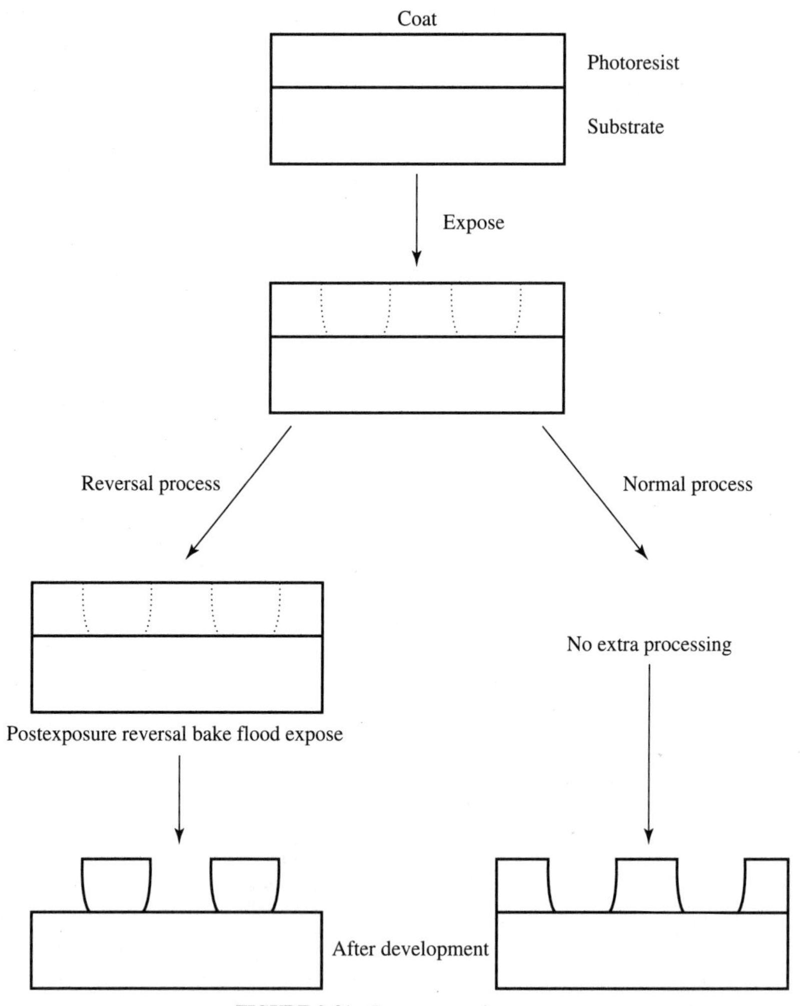

FIGURE 2-31 Image reversal process.

2.6.4 Deep Ultraviolet (DUV)

Microelectronics technology has pushed for devices with dimensions well below 1.0 μm. To obtain these geometries, exposure using shorter wavelength light is necessary. The wavelength of the mercury vapor lamp G-line is 436 nm (0.436 μm). It would be very difficult to make devices at 0.25 μm using exposure systems optimized for G-line or even I-line sources.

Excimer Laser. The excimer laser has emerged as the light source for deep ultraviolet (DUV) lithography. The krypton fluoride (KrF) laser has a wavelength of 0.248 μm and argon fluoride (ArF) has a wavelength of 0.193 μm. It is possible to build optical systems optimized for these wavelengths and thus obtain photoresist images below 0.2 μm.

Top Surface Imaging. One problem with DUV lithography is that many materials that are transparent at visible wavelengths are absorbant in the DUV region, this includes photoresist. (Some of the acylate resists are transparent in the DUV region as well.) The result is that images can be made near the top surface of the resist. One technique that can be used to create a pattern is to expose photoresist at DUV, causing polymerization in the exposed regions. After exposure, the photoresist is exposed to hexamethyldisilozane (HMDS) vapor, which is absorbed in the nonpolymerized regions of the resist. Subsequent oxygen-reactive ion-etch processing can remove the resist in the polymerized areas. When used on silicon substrates, the silicon in the silated areas forms SiO_2, which protects those regions from the oxygen-reactive ion-etch processing.

REFERENCES

1. D.J. Elliott, *Microlithography,* Process Technology for IC Fabrication, McGraw-Hill, New York, 1986.
2. L.F. Thompson, C.G. Wilson, and M.J. Bowden, *Introduction to Microlithography,* 2nd Ed., American Chemical Society, Washington, DC, 1994.
3. J.A. Bruce, S.R. Dupuis, and H. Linde, "Effect of Humidity on Photoresist Performance," OCG Microlithography Seminar, Essex Junction, VT October 29–31, 1995, *IBM Microelectronics.*
4. "DuPont Photosensitive Polyimides For Passivation or as a High Temperature Photoresist," DuPont Publication, 1985.
5. "Photosensitive Polyimide E-38675-50 Preliminary Information Bulletin," DuPont Publication, 1985.
6. P. Burggraaf, "Polyimides in Microelectronics," *Semiconductor International,* Cahners Publishing Company, Des Plaines, IL, 1988.
7. S. Wolf and R.N. Tauber, *Silicon Processing for the VLSI Era,* vol. 1, Lattice Press, Sunset Beach, CA, 1986.
8. M.J. Collie, *Etching Compositions and Processes,* Noyes Data Corporation, Park Ridge, NJ, 1982.
9. J.J. Gajda, IBM System Products Division, East Fishkill Facility, Hopewell Junction, NY.
10. R.A. Morgan, *Plasma Etching in Semiconductor Fabrication,* Elsevier Science Publishers, New York, 1990.
11. B. Chapman, *Glow Discharge Processes,* John Wiley and Sons, New York, 1980.

CHAPTER 3
PROPERTIES OF THIN FILM MATERIALS

Aicha Elshabini
Fred Barlow
The Bradley Department of Electrical Engineering
Virginia Polytechnic Institute and State University

3.1 INTRODUCTION

This chapter focuses on the materials used to fabricate thin film circuitry and multilayer modules: substrates, metallizations, resistor materials, dielectric materials, protective coatings, and advanced thin film materials. There are many materials used in the fabrication of resistor and capacitor elements. Thin film technology and passive components integration is inherent to the technology. Materials selection is closely related to the fabrication process or deposition technique being used [1,4–7,11–19]. In addition, resistors and capacitors with low-voltage ratings can be fabricated in a reliable and consistent way using thin film technology. The reader is referred to Chapter 6 for thin film materials characterization and references [27–37].

3.2 SUBSTRATES FOR THIN FILMS APPLICATIONS

3.2.1 Introduction

The function of the substrate is to provide the base onto which thin film circuits are fabricated and various thin film multilayers are deposited. In addition, the substrate provides the necessary mechanical support and rigidity needed for a reliable circuit, and it has adequate thermal management ability to ensure proper temperature operation and proper electrical insulation to withstand high voltages without breakdown. Dielectric strength, dielectric constant value, dissipation factor, electrical conductivity, thermal conductivity, and flexural or mechanical strength are important substrate properties that are affected by their microstructure, composition, and processing.

3.2.2 Substrate Materials

Commonly used substrate materials for thin film circuits include alumina, glass, beryllia or beryllium oxide-based ceramic, aluminum nitride, silicon (Si), and metals. When using Si, the substrate surface is covered with a silicon dioxide dielectric layer. The use of Si as the substrate material for thin film interconnections offers numerous advantages, including high thermal conductivity, matched thermal expansion coefficient for Si ICs, excellent surface finish, and good dimensional stability. Silicon wafers are necessary for the fabrication of integral SiO_2/Si_3N_4 decoupling capacitors. On the other hand, limited mechanical rigidity dictates the necessity of mounting Si substrates onto a base to provide mechanical support. In addition, Si substrates have a relatively high dielectric constant value that affects their high-frequency operation. Also, warping may result from the deposition of thick (greater than a few microns) polymer dielectrics on Si.

Ceramic materials are often used for thin film applications because ceramics have high thermal conductivity, good chemical stability, and are also resistant to thermal and mechanical shocks. Glazed substrate surfaces provide good surface finish, low porosity, have a low dielectric constant value, and are effective for most thin film deposition techniques with relatively low deposition temperatures. Glazing the substrate surface typically yields a surface roughness of <1 μm/in (about 250 Å). The common choice for substrate material for thin film deposition is 99.6% alumina. Substrate dimensions vary from 1″ × 1″ to 3″ × 3″; thicknesses range from 0.010″ to 0.025″ in increments of 15 mils. The surface finish is 4 microinches (μin) maximum on the front surface and 5 μin maximum on the back surface. The flatness is measured in terms of camber and it is 0.002″/in for 0.010″ substrate thickness, 0.003″/in for 0.025″ substrate thickness, and 0.004″/in for 0.040″ substrate thickness, respectively. The coefficient of thermal expansion (CTE) is 6.3 ppm/°C at room temperature. The density of the ceramic is 3.9 g/cm^3, the dielectric constant is about 9.9 (measured at 1 MHz), and the loss tangent is 0.0001 (measured at the same frequency).

For 99.5% beryllium (Be), the surface finish is 5 μin maximum on the front surface and 10 μin maximum on the back surface, and the flatness is 0.003″/in maximum. For aluminum borosilicate glass, the surface finish is 0.25 μin maximum on both the front surface and the back surface, and the flatness is 0.004″/in maximum.

The CTE mismatch between alumina (6.3 ppm/°C) and GaAs (5.7 ppm/°C) and Si (2.6 ppm/°C) can result in serious stresses. Thus, the use of aluminum nitride (3.5 ppm/°C), thick diamond films (0.9 ppm/°C), or a metal or ceramic matrix composites (2.6 to 5 ppm/°C) are more suitable thin film substrates, especially in that they are more capable of dissipating large amounts of heat.

3.2.3 Requirements of Substrates

The properties of the substrates used to grow defect-free films for thin film circuits include:

- Good surface smoothness (surface finish can vary from about 50 μm/in for 96% aluminum oxide content to 20 μm/in for 99.5% aluminum oxide content)
- Coefficient of thermal expansion (CTE) matched
- Good mechanical strength (>350 MPa)

- High thermal conductivity
- Inertness, or chemical stability
- Porosity
- Low cost
- High electrical resistance ($>10^{14}$ Ω·cm)
- Good uniformity

The CTE is represented in terms of the change in the length of a material per °C to the original length at 0°C. The CTE of most substrate materials is linear over the temperature range of interest. The differential magnitude of the CTE between one material and another (for instance the substrate and the film) is significant, because tensile stresses can result from the mismatch, causing delamination or cracking of the film and/or bowing of the substrate. In addition, the CTE of the substrate should match the CTE of Si (2.6 ppm/°C) to avoid excessive stresses in the die-attach material (especially for large-size die). The CTE of 99.6% alumina is 6.3 ppm/°C, the thermal expansion of 99.5% BeO is 6.9 ppm/°C, and the thermal expansion of aluminum borosilicate glass is 0.46 ppm/°C.

Substrate flatness directly influences the minimum achievable line width and spacing. In general, films are tolerant of surface roughness less than the order of magnitude of the film thickness. To maintain uniform characteristics of the thin film elements, it is important to minimize surface roughness. The flatness, or camber of 99.6% alumina substrate is 0.002″/in for 0.010″ substrate thickness; 0.003″/in for 0.025″ substrate thickness; 0.004″/in for 0.025″ substrate thickness. The camber of 99.5% BeO is 0.003″/in, and the camber of aluminum borosilicate glass is 0.004″/in. The CTE of the substrate should be similar to the deposited film to minimize mechanical and residual stresses in the film. High mechanical strength and thermal shock resistance are required to enable the substrate to withstand the rigors of processing and normal use. Ceramic materials are ideal for this function. High thermal conductivity is required to remove heat, enabling the realization of circuits with high-component densities.

Inertness to chemicals used in circuit processing is a necessary requirement. Ceramic materials possess better chemical stability than glasses, especially at higher temperatures. These materials are not attacked by the etchants used in processing thin film circuits. Low porosity is required to prevent the entrapment of gases, which causes film contamination. These substrate materials must be good insulators at room temperature. Finally, because uniformity of substrate properties must be maintained, control of electrical properties is very important. Similarly, control of substrate purity, density, and surface properties is also necessary.

Substrates for thin films have high resistivity, high dielectric strength, and are chemically resistant or inert. For example, tantalum-film processing requires hydrofluoric acid as the etchant. Therefore, glass or glazed substrates must be covered with a thin layer of tantalum oxide to resist the acid. The volume resistivities of alumina, BeO and aluminum nitride are 10^{15} (or higher), 10^{14}, and 10^{13} Ω·cm, respectively. The dielectric constants (measured at 1 MHz) for alumina, BeO and aluminum nitride are 9.9, 6.8, and 8.9, respectively. The loss tangents are 0.0001, 0.0002, and 0.0001, for alumina, BeO, and aluminum nitride (AlN), respectively. High thermal conductivity is needed for heat dissipation from the active and resistive components deposited on the surface. The thermal conductivity of ceramics is superior to that of glass (about 30 to 33 W/m·°K). A glazed ceramic will have slightly lower thermal conductivity than raw ceramics because of the glass-like glaze layer on its surface. The thermal conductivity of aluminum nitride is 130 to 170 W/m·°K, and that of BeO is 180 to 260 W/m·°K.

TABLE 3-1 Properties of substrates for thin film multilayer interconnections

Material	Density (g/cm^3)	CTE (ppm/°K)	Young's modulus (GPa)	Thermal conductivity (W/m·°K)
Polyimide	1.4	40	2.5	0.15
Si	2.3	2.6	113	148
Al_2O_3 (99.6%)	3.9	6.3-6.7	360	20-35
BeO (99.5%)	2.9-3	6.9	350	251
AlN	3.3	4	340	160-190
SiC	3.1	3.7	400	270
Mo	10.2	4.9	324	138
Cu	8.9	16.8	110	398
Al	2.7	2.5	62	237
Au	18.9	14.3	—	318
Steel (AISI 1010)	7.9	11.3	192	64
Kovar (Fe/Ni/Co)	8.4	6.1	138	16
Cu/Invar/Cu (20/60/20)	8.4	6.4	134	170
Cu/Mo/Cu (20/60/20)	9.7	7	248	208
Cu/Mo/Cu (13/74/13)	9.9	5.7	269	242
CuW (20/80)	17	8	283	186
Natural diamond	3.5	1.1	—	2,000
CVD diamond	3.5	1.5-2.0	890-970	400-1600

The fluctuations in the substrate surface can be described in terms of roughness and waviness. Roughness is a short-range fluctuation (the average deviation of the surface from some arbitrary mean value). The surface roughness can result in an uneven film thickness. Resistive films on rough substrates exhibit a high sheet resistance from an apparent increase in the length between points, and also because of thin spots in the film. These films are also less stable during thermal aging. Additionally, the performance of capacitors is affected by surface roughness, because a premature breakdown of the dielectric film may occur. Raw ceramic materials are not smooth enough for very-fine line widths or for applications requiring very high stability. For this reason, capacitors, high-density interconnects, and high-stability resistors are deposited on polished or glazed ceramic substrates. Waviness is a periodic variation characterized by a peak-to-peak amplitude and a repetition interval. Flatness, or the lack of waviness, is an important requirement if thin ($<50\ \mu m$) and well-defined line widths are to be achieved. The line width of a resistor is widened by the waviness of the substrate surface (even when controlled by the surface mask).

Metal substrates such as aluminum (Al) and copper (Cu) have high thermal conductivity, reasonable cost, ease of machining, proper surface flatness upon lapping and polishing, and may be used as substrates for thin film interconnections. An insulator layer may be deposited first on the surface of the substrate to provide the proper electrical insulation needed to construct the thin film circuitry. The substrates are usually reactive to metal etchants used in thin film processing and may require a protective coating. These metal substrates also possess a high CTE, which makes them poorly matched to Si devices. Aluminum and Cu substrates possess excellent thermal conductivity, they are inexpensive, and can be easily machined; however, these metals have a high CTE that is poorly matched to Si chips. On the other hand, molybdenum (Mo) and tungsten (W) are attactive metal substrates with a low CTE and high elasticity. But these metals are relatively heavy and are difficult to machine.

Alloying metals can provide metal substrates with desirable characteristics. Copper-clad and nickel-clad Mo substrates produced by hot roll pressing, and Cu-W alloys (about 20% Cu) produced by powder metallurgy have a low CTE, high elasticity, high thermal conductivity, and good solderability as potential substrate materials.

Table 3-1 illustrates some important properties of substrates for thin film multilayer interconnections. In addition, the CTE of several materials as well as the thermal conductivity are plotted in Figs. 3-1 and 3-2, as a function of temperature.

3.2.4 Substrate Fabrication

The raw materials for ceramic substrates usually are available as purified oxide powders. The oxides are mixed, reground, and then mixed with organic compounds functioning as plasticizers, binders, or lubricants. A flow chart indicating a number of established methods to form ceramic substrates is shown in Fig. 3-3. These techniques, their advantages and limitations, are briefly reviewed in the following paragraphs.

Powder Pressing. In powder pressing, dry or slightly dampened powder is packed into an abrasion resistant die under sufficiently high pressure (8,000 to 20,000 psi), to form a dense body. This process allows rapid or automatic production of parts with reasonably controlled tolerances because the shrinkage during the sintering process is lower than other ceramic substrate-forming techniques. Substrate features such as holes may be simultaneously formed. However, various limitations exist. Pressure variations from uneven filling of long or complex die configurations to inhomogeneities in bulk properties. As a consequence, these materials are limited to those applications in which lack of uniform properties does not adversely affect circuit performance. Holes cannot be located too close to an outside edge, and the process limits the substrate size. Generally, these substrates are less dense and more porous, with higher surface roughness and lower mechanical strength than other ceramic forming methods. When used for thin films, such substrates must be mechanically polished, and in doing so, they lose their cost advantage over other forming techniques while still suffering from inhomogeneity problems. Powder pressed parts, however, remain the major source of beryllium oxide (BeO) substrate material, particularly for stock thicker than 0.40″. Problems arise in pressing thin substrates of this material because the larger grain size of BeO (as compared to other substrate powders) significantly reduces the amount of powder, or filling fraction, in the die, resulting in increased porosity. Ferrite substrates and plugs are also powder-pressed.

Isostatic Pressing. In contrast to dry pressing, this method applies uniform pressure to the powders. Dry powders are enclosed in an elastic container which is inserted into a cavity, which is first evacuated and then filled with a liquid that surrounds the elastic container. The pressure applied to the cavity is uniformly distributed over the surface of the container, yielding a relatively uniform compact piece. An important advantage of isostatic pressing is that it permits fabrication of pieces with relatively large length-to-width ratios. For example, cylinders one foot in diameter and two feet in length can be made using this method. These cylinders are then sliced into wafers, which are shaped by machinery, ground, and polished to attain desired surface quality for thin film applications. Thus, this finishing method is analogous to that used to make semiconductor single crystal and fused silica substrates. Although capable of making large-sized substrates (such as 4″ × 4″), the

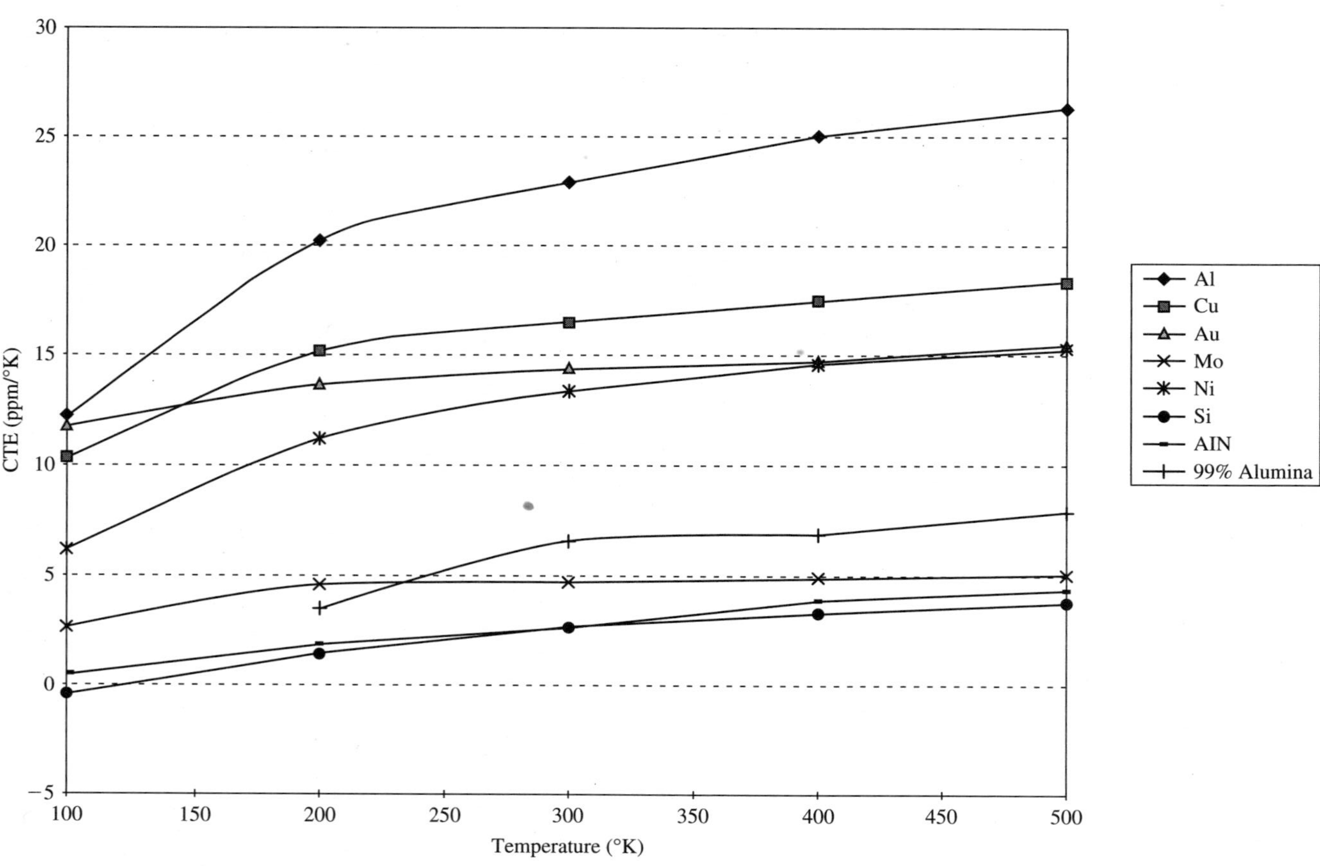

FIGURE 3-1 CTE versus temperature for a variety of substrate and metallization materials.

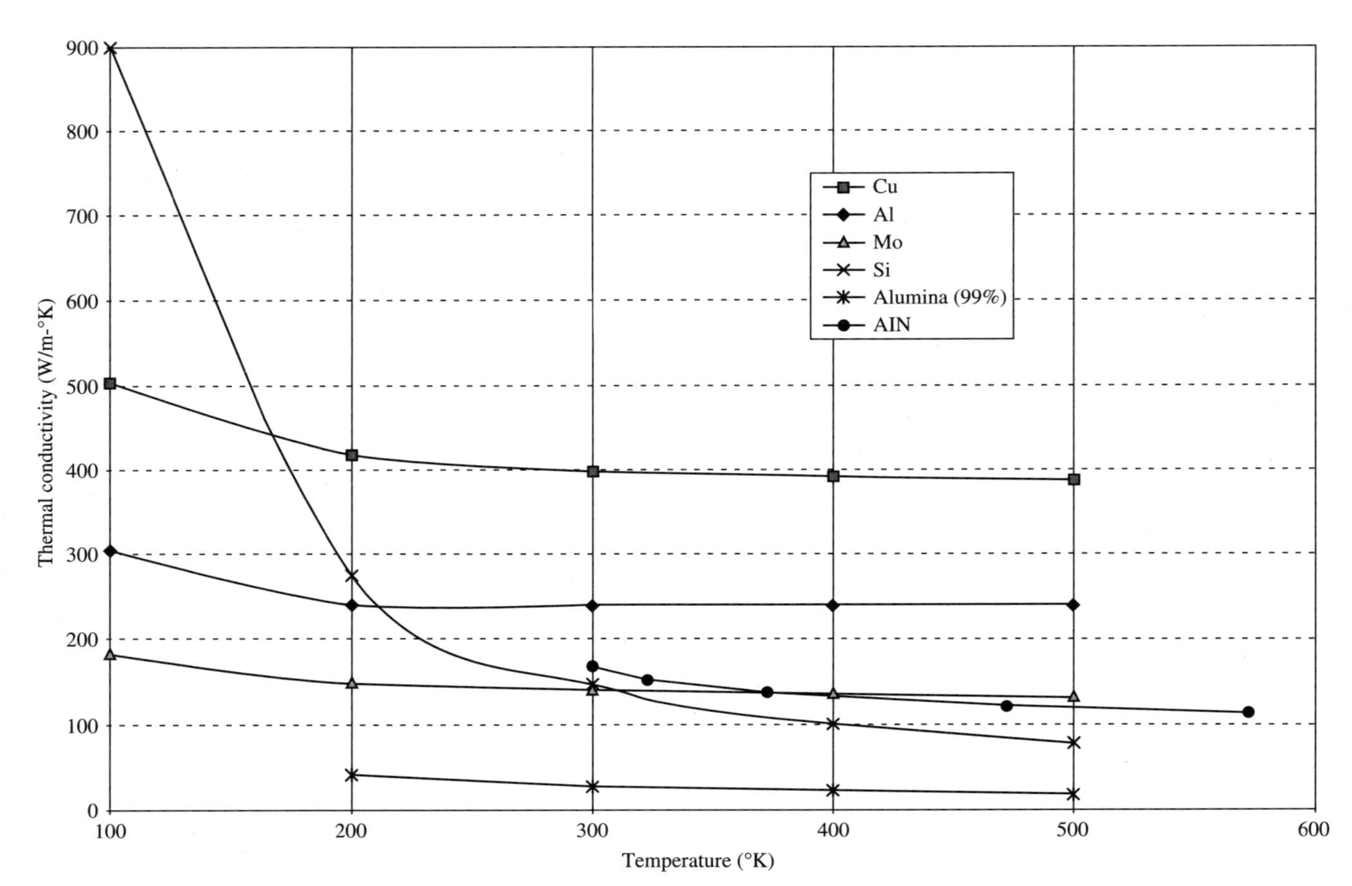

FIGURE 3-2 Thermal conductivity as a function of temperature for a variety of substrate materials.

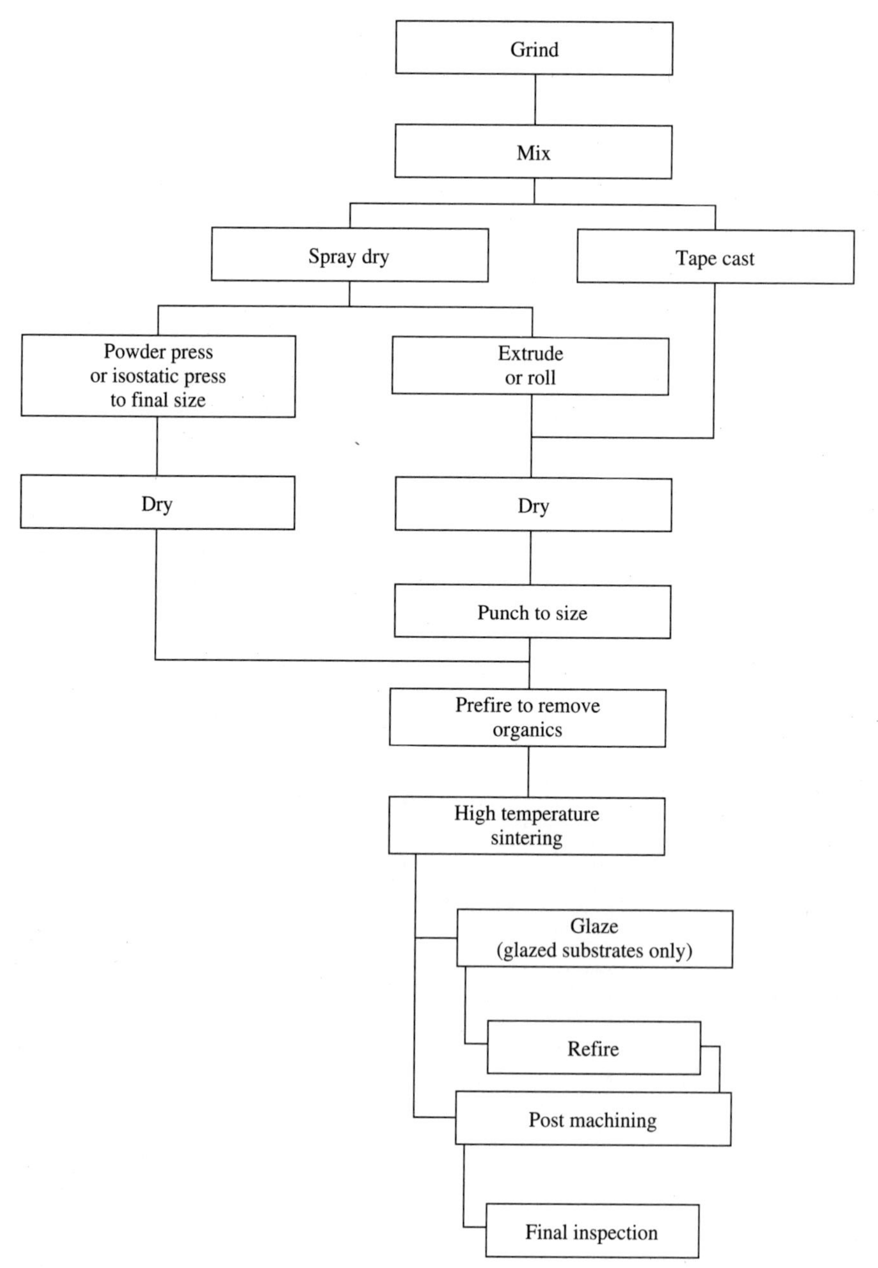

FIGURE 3-3 Process flow of three methods used to produce thin film substrates. *(Courtesy of Richard Brown Associates.)*

amount of material wasted in the cutting operation frequently offsets any advantages of this method. As a consequence, few microwave substrates are made from isostatically pressed stock.

Tape Casting. The most widely used method for producing substrates for thin film hybrid applications, particularly alumina, is tape casting. The manufacturing process consists of the following steps:

Casting. At present, the doctor blade method of sheet or tape formulation offers the best potential for a smooth surface substrate that is relatively free of surface defects. The slurry is spread onto a suitable carrier film (polyester or cellulose acetate), which moves at constant speed under metal knife blades (doctor blades). The blades are positioned a short distance above the film. As the film and slurry move under the blades, a thin sheet of wet ceramic forms. The thickness of this sheet is controlled by adjusting the height of the blade over the carrier film, the speed of the carrier film, the composition of the slurry, and the slurry head in the reservoir.

The resultant ceramic sheet, usually referred to as "green tape" or "in the green state," is air dried to remove the solvents. Typically, a 30% to 50% reduction in green tape thickness occurs during solvent evaporation. In tape cast ceramics, the substrate surface on one side replicates the surface characteristics of the carrier, and the other surface is determined by the doctor blade, the air temperature, and the amount of relative humidity in the air flow of the tape casting machine. By choosing an ultra-smooth carrier, a fine surface finish is facilitated, and by careful attention to process, a similar, but not identical, finish is produced on the air side. Controlled raw materials, slurry properties, and properly designed drying or curing conditions help produce substrates with minimum warp and waviness. This process has been successfully used in the manufacture of Al_2O_3 substrates for several decades. More recently, production runs of BeO and AlN have appeared. While the casting technology using these two materials is not as developed as that of Al_2O_3, the substrates they form show improved surface, mechanical, and electrical characteristics when compared with dry pressed parts. Tape cast BeO substrates are available up to 4.50″ × 4.25″ and 0.35″ thick. Aluminum nitride—a newer material—is available in substrate sizes up to 3.75″ × 4.50″ and 0.40″ thick.

Glazing. During the lapping and polishing operations, which are used to obtain substrate parallelism and flatness, surface grains from polycrystalline ceramics are physically pulled out. The size and frequency of these defects, which interrupt an otherwise flat, smooth surface, depend upon substrate grain size distribution, secondary phase techniques, and machining techniques.

For thin film capacitors, such surface disruptions can significantly reduce component yield. An appropriately smooth surface can be restored by glazing the surface. In addition to providing a smooth surface, the glaze must also be chemical resistant, have good electrical properties, and closely match the expansion of the parent ceramic.

Glazes are formed from finely milled glass in an organic binder (frit). After application by screening or spinning onto the ceramic substrate, the frit is fired in air or nitrogen at temperatures typically between 750°C and 1000°C. The organic binders burn out at temperatures between 500°C and 600°C, and above the flow point of the glass, the particles of glass fuse together forming a continuous glassy layer with a smooth (albeit slightly wavy) surface. The cross-section of a glazed, polycrystalline ceramic shown in Fig. 3-4 clearly demonstrates the glass-like smoothness attained using this method, with smooth, tapered edges allowing uninterrupted metallization over the glaze–ceramic interface.

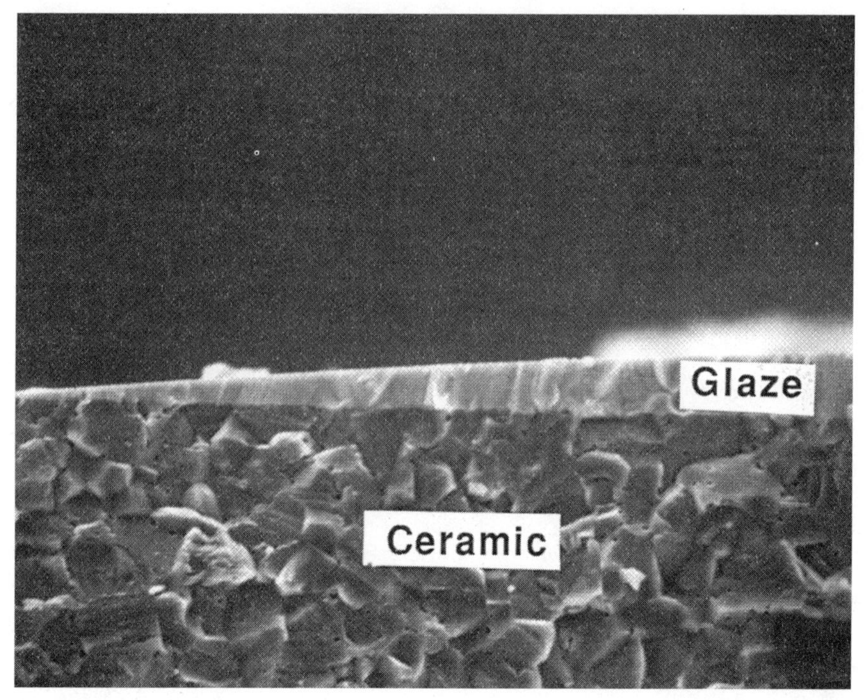

FIGURE 3-4 Cross-section of a glazed substrate. *(Courtesy of Richard Brown Associates.)*

3.3 THIN FILM CONDUCTOR MATERIALS [20, 22, 23]

3.3.1 Introduction and Definition of Various Quantities

Considerations in selecting the proper metallization include its main function, current-carrying capability, interface and adhesion issues, stability, ability to be easily processed, and compatibility with the assembly processes. In addition to these main considerations, there are specific properties to be examined to evaluate the materials under consideration, including the thermal conductivity, CTE, electrical resistivity, and the temperature coefficient of resistance (TCR).

The thermal conductivity of a material presents the time rate of heat conduction per unit area per unit temperature gradient, expressed in cal/cm·s·°K, (1 cal/cm·s·°K = 4.184×10^2 W/m·°C = 0.0001927 Btu·in/ft^2·s·°F).

The CTE is a measure of the amount a material expands as the temperature changes, in 1/°K, (1/°K = 5/9.1/°F).

The electrical resistivity is the inverse of the electrical conductivity, in Ω·cm. The resistivity significantly changes with a temperature change. The resistivity has a high correlation to the purity of the material and the perfection of the crystal lattice, especially for metals. Thus the resistivity can vary greatly with the presence of even small amounts of impurity in the material. Also, because thin films have more imperfections and lattice strain than bulk materials, these films possess generally higher resistivity values than the corresponding bulk values. Annealing at higher temperatures can reduce the number of lattice imperfections, thereby decreasing the resistivity of these films.

The TCR is expressed in ppm/°C or ppm/°K. In general, the TCR of thin film materials is lower than the TCR for the corresponding bulk materials. Thicker film deposition and heat treatment of the film can result in a film with TCR values approaching the corresponding TCR values of bulk materials.

3.3.2 Buffer and Adhesion Layers [45]

Conductor films are typically formed from several metals to achieve the desired properties of a thin film metallization. These properties include

- High conductivity
- Good adherence to both the substrate and the other deposited films
- Resistance to etchants
- Capacity to be etched to line widths of a few microns using conventional photolithographic technology
- Ease of bonding
- Compatibility with other materials and processes

Metals with good conductivity usually have poor adherence to substrate material. The adherence property can be enhanced by providing intermediate layers of titanium (Ti), chromium (Cr), or nichrome. These interface metals are used to achieve good adhesion between the main metallization (such as Cu) and polymer dielectrics, or to prevent chemical attack of the metal. Most of these interface metals are deposited by sputtering to achieve a thickness of 100 to 300 nm.

3.3.3 Copper Metallization

Copper is a good electrical conductor (0.596×10^6 S/cm), resistivity equal to 1.7×10^{-6} Ω·cm, and also has high thermal conductivity. The thermal conductivity of Cu is 398 W/m·°K (0.9508 cal/cm·s·°K). It is often evaporated and can be deposited by sputtering or electroplating processes to form the required metallization. Pure Cu has a sheet resistance (R_s) of 0.2 Ω/square as deposited. Cu and its alloys are nonmagnetic, have excellent solderability, and are readily wetted by lead and tin solders.

Passivation of the Cu surface is required because Cu oxidizes in air at low temperatures. In addition, Cu exhibits poor adhesion to most dielectrics. Thus inert coatings such as SiC can usually protect the Cu from degradation resulting from high temperatures and corrosive environments. In addition, Cu interaction with silicon dioxide-based dielectrics, which form the active portion of the insulating structure between interconnect lines, is also an important issue. During the process of chemical vapor deposition, nucleating Cu onto the various metal oxides and silicon dioxide is difficult because of the lack of chemical bonding between the species, thus lowering the Cu deposition rates on these surfaces [45]. To enhance the adhesion of Cu, researchers have used Cr, Ti, niobium (Nb), Al, or manganese (Mg) as interface layers to promote the adhesion of Cu to these oxide substrate surfaces and to achieve the desired mechanical bonding. Another technique for improving Cu adhesion to oxide layers is to add small amounts of various alloying elements to Cu. Both Ti and Cr have been alloyed with Cu to improve adhesion to dielectrics, especially after heat treatment or annealing.

3.3.4 Gold Metallization

Pure gold (Au) electrodeposits are classified in terms of MIL-G-45520B specifications. Various types of pure Au deposits are obtained from a hot cyanide bath solution containing Au in the form of potassium gold cyanide, with potassium cyanide, carbonate, and phosphate in varying amounts. Higher contents for Au are usually used for thicker Au electrodeposits.

Gold can also be either evaporated or sputtered, and it is used as a top layer to protect other metal layers. It can also be used to achieve the required metallization or to form the capacitor electrodes, possessing basically a high electrical conductivity (0.452×10^6 S/cm). The R_s of Au is 2 to 20 Ω/square as deposited, and 0.45 to 4.95 Ω/square when annealed. Gold has excellent corrosion resistance and it is compatible with wire-bonding attachment techniques. An adhesive underlayer of Ti or Cr is often used because Au does not adhere well to most substrate materials. Gold is more difficult to solder to because it usually alloys with soft solders, resulting in mechanically poor connections. The thermal conductivity of Au is 318 W/m·°K (0.753 cal/cm·s·°K).

3.3.5 Aluminum Metallization [48]

Primary Al is produced by direct current electrolysis of aluminum oxide (alumina) dissolved in a molten sodium fluoride and aluminum fluoride bath at temperatures between 940°C and 980°C. Aluminum is refined to remove impurities that degrade electrical conductivity, corrosion resistance, and electrochemical characteristics. Physical properties consisting of density, CTE, electrical conductivity, and thermal conductivity can be changed by addition of one or more alloying elements. Heat treatment can also substantially alter some of these properties, such as density and conductivity.

Pure unalloyed Al is a single-phase material, and its optical microstructure is composed of only grains and grain boundaries. The tensile yield strength of high-purity Al in its annealed state is 10 MPa (1.5 ksi); on the other hand, some heat-treated, high-strength Al alloys exceed 550 MPa (80 ksi).

Aluminum is a soft, machineable metal that can be evaporated and is commonly used for conductors and capacitor electrodes. In addition, large optical telescope mirrors are often coated using evaporated Al. Aluminum has an acceptable degree of adhesion to various substrate materials and good adhesion to polymer dielectrics. One major drawback is its tendency to react with many types of crucibles during the evaporation process. Also, Al has low electrical conductivity (0.377×10^6 S/cm), and its thermal conductivity is 236 W/m·°C (0.566 cal/cm·s·K). Aluminum has an R_s of 0.41 Ω/square as deposited, and 0.36 Ω/square when annealed.

Traditionally, Al has been regarded as a difficult element to sputter deposit. This can be attributed to its low atomic mass and high reactivity, resulting in the collision of atoms with the argon gas, which leads to atoms being scattered widely. This can cause the Al atoms to return to the target, or to become oxidized. An ultra-high-voltage magnetron sputter system designed and constructed by Wang et al, with a very low sputtering pressure, a small distance between the target and the substrate (32 mm), and the highest purity target has been shown to improve the purity and quality of sputtered Al film deposition on glass, silicon, and sapphire substrates [48].

Both Al and its alloys are used as interconnecting materials. Aluminum alloys are often used because of a large difference in CTE between Al and Si substrate that causes thermal stresses. Copper can be added as an alloying element in Al interconnect lines in concentrations varying between 0.5 to 4 weight % in order to reduce the electromigration failure rate. In addition, Si can be added to Al as an alloying element in concentrations up to 1 weight % to minimize the nonuniform interdiffusion of Si into Al in the contact regions.

3.3.6 Palladium Metallization

Palladium (Pd) and its alloys can also be evaporated, and it is used as a protective layer to prevent surface oxidization of other metals. An adhesive underlayer of Ti or Cr is usually used to improve the adherence of Pd to glass and ceramic substrates. Titanium can be evaporated and is mainly used as an adhesive underlayer material. Palladium has a resistivity equal to 11×10^{-6} Ω·cm, with an R_s of 1.3 Ω/square as deposited.

Palladium is electrodeposited from baths based on diamino-dinitrito-palladium, typically buffered with ammonium nitrite and sodium nitride. Electrodeposits of Pd have hardness and wear resistance comparable to Au; because of this, they often replace Au in many electronics applications. The Pd solution does not react with printed circuit boards, and thicker layers of Pd are possible and less expensive.

3.3.7 Nickel and Chromium Metallization

Nickel (Ni) is a hard, tough metal with good resistance to corrosion. Nickel possesses a thermal conductivity of 90 W/m·°K (0.216 cal/cm·s·°K). The R_s of Ni is 11 to 28 Ω/square as deposited, and 1.8 to 10 Ω/square when annealed. Nickel is often alloyed with Cr to form Nichrome (NiCr). Nichrome is used as an interface material (to promote the adhesion of some metals) or as a thin film resistor material. Nichrome possesses high resistivity of 100 $\mu\Omega$·cm or higher, has poor solderability, and can be easily oxidized. Chromium is used as an adhesive because its adhesion to the substrate and other films is very high. Chromium possesses a resistivity of 15×10^{-6} to 80×10^{-6} Ω·cm, and its R_s is 19 to 170 Ω/square as deposited, and 10 to 60 Ω/square when annealed.

3.3.8 Silver Metallization

In general, electrodeposition of silver (Ag) is performed mainly for decorative purposes, such as the making of jewelry, ornaments, and gifts. The Ag deposit is applied with a thickness of 5 to 50 μm for these uses. Silver electrodeposition with thickness varying between 0.1 to 1.5 mm is used for making contacts and switches and for coating bearing surfaces.

Silver electrodeposition can be performed with baths based on potassium silver cyanide. High-speed bath solutions with higher Ag contents, higher pH level, and elevated temperature are often used. Silver can also be evaporated, and it is often used to presensitize dielectric material surfaces. The thermal conductivity of Ag is 427 W/m·°K (1.02 cal/cm·s·°K.)

3.3.9 Conductor Properties

Table 3-2 provides a list of several metals with their most important properties at room temperature. In addition, the CTE of several conductors is plotted in Fig. 3-1, and the conductors' resistivities are shown as a function of temperature in Fig. 3-5.

It is important to realize that these values, in addition to mechanical properties (yield stress, tensile strength, hardness), may depend on the impurity content (that is, they are sensitive to impurities), and the presence of lattice defects.

Metals, especially precious metals, are used for their good oxidation and corrosion resistance. Copper has good electrical conductivity, good thermal conductivity, high strength, and high resistance to fatigue. Copper can be attacked by common reagents and environments. Although pure Cu resists attack quite well under most corrosive conditions, some Cu alloys have limited usefulness in certain environments because of stress corrosion cracking or hydrogen embrittlement. Copper (and most Cu alloys) containing little or no zinc are not susceptible to stress corrosion cracking. The annealing process prevents the stress corrosion cracking by relieving residual stresses. Hydrogen embrittlement occurs when tough-pitch Cu (Cu alloy containing cuprous oxide) is exposed to a reducing atmosphere. In this case, Cu alloys are deoxidized and are not subject to hydrogen embrittlement.

Aluminum and most of its alloys have good corrosion resistance in natural atmosphere and in many chemicals and solutions. This resistance to corrosion is the result of the presence of a very thin (about 50 Å) aluminum oxide adherent film on the metal surface. This film is formed in air under normal conditions. This oxide film is soluble in alkaline solutions and in strong acids.

Gold does not oxidize at any temperature. Gold is also resistant to hot HNO_3 and H_2SO_4. It is also resistant to HCl dry gas, especially at high temperatures, although it offers poor resistance to Cl at 80°C and higher temperature.

Palladium forms an oxide film in the range 400°C to 800°C. Palladium offers poor resistance to highly oxidizing environments.

Nickel is highly resistant to a variety of corrosive media because of its intrinsic oxide coating. Nickel can be alloyed with elements such as Cu, Cr, and iron to produce alloys that are resistant to corrosion in various environments.

Silver results in a stable oxide near room temperature and in an unstable oxide at elevated temperatures. Silver is generally resistant to hot concentrated solutions of many organic acids.

3.4 RESISTORS FOR THIN FILMS APPLICATIONS

3.4.1 Background and Resistor Materials Properties

The selection of thin film resistor materials depends on basic considerations such as TCR, voltage coefficient of resistance (VCR), electrical noise figure, resistance stability,

TABLE 3-2 Properties of metal films for thin film applications (300°K)

	Melting point in K and (°C)	Resistivity (Ω·cm)	TCR bulk (ppm/°K)	TCR film (ppm/°K)	Density (g/cm^3)	Specific heat (cal/g·°K)	Thermal conductivity (W/m·°K)	Thermal exp. coeff (CTE) (1/°K)
Copper (Cu)	1356.6	1.73×10^{-6}	3,930		8.94	0.09118	398	16.12×10^{-6}
Gold (Au)	1337.58 (1064)	2.249×10^{-6}	3,670	1600-2800	19.3	0.03088	3.18	14.2×10^{-6}
Aluminum (Al)	933.52 (660)	2.73×10^{-6}	4,190	2,800	2.71	0.211	237	23.2×10^{-6}
Palladium (Pd)	(1552)	11×10^{-6}			12.02	245 J/kg·k	70	11.8 μm/m·k
Nickel (Ni)	1726 (1453)	7.16×10^{-6}	5,550	3000-5000	8.9	0.107	90	13.5×10^{-6}
Silver (Ag)	1235.08 (962)	1.628×10^{-6}	3,730		10.5	0.0560	427	19.0×10^{-6}
Tantalum (Ta)	3269 (2996)	13.48×10^{-6}	3,540	100	16.6	0.03306	57	6.3×10^{-6}
Chromium (Cr)	2130 (1875)	13.5×10^{-6}	2,300	100-600	7.16	0.1077	90	5.1×10^{-6}
Nicrome (NiCr)	1670	100×10^{-6}		127	8.4	0.11	13	12×10^{-6}

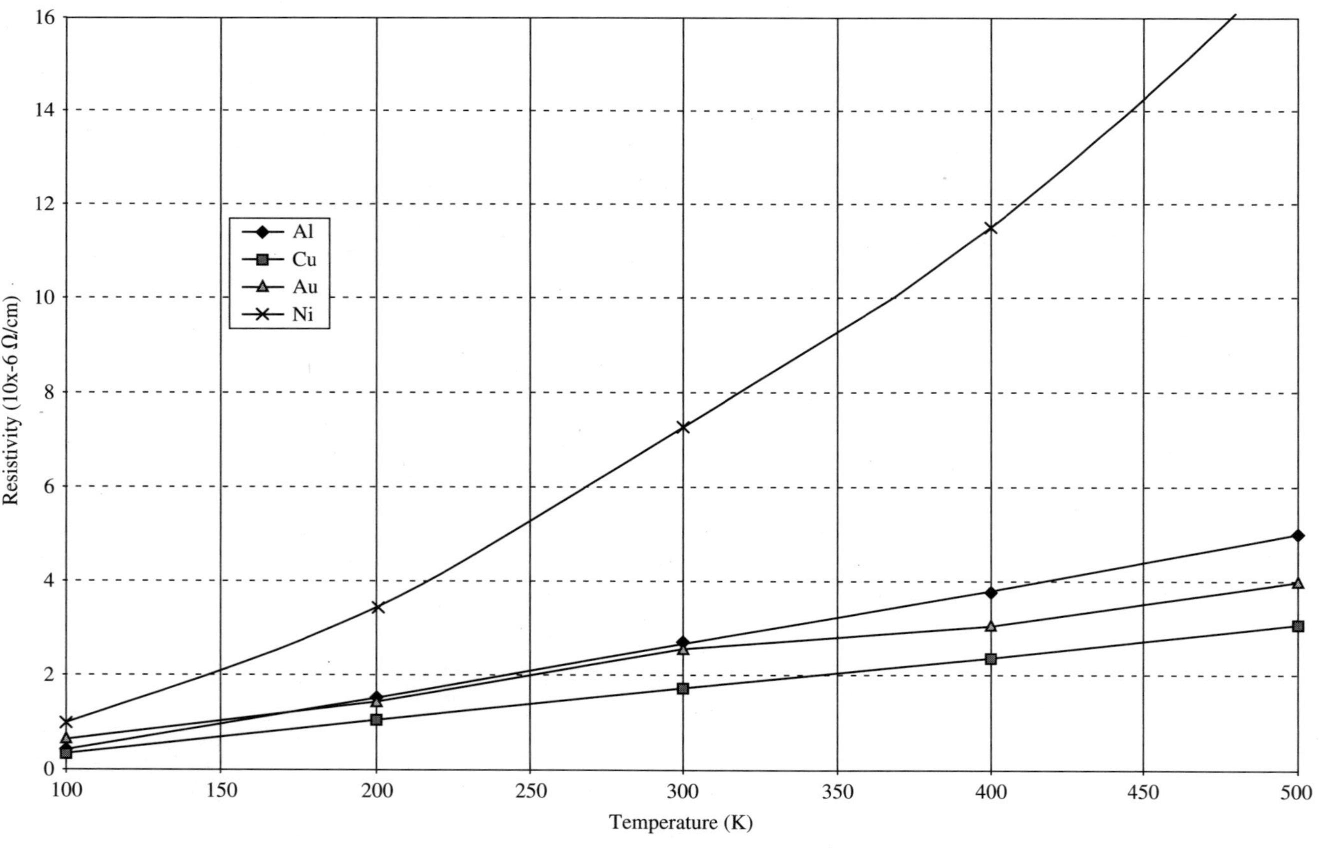

FIGURE 3-5 Resistivity of metallization materials as a function of temperature.

achievable R_s, allowable power density, compatibility with other materials, and film deposition method used.

The R_s of a thin film resistor is a basic material property. Sheet resistance (in Ω/square) refers to the volume or bulk resistivity normalized with respect to the thickness of the film. The 4-point probe measurement technique is used to accurately measure the R_s of a thin film without any etching process. Typical thin film R_s are in the range of 10 to 2000 Ω/square with an initial tolerance of ±10%. Trimming, as discussed in Chapter 5, brings these resistance values to tighter tolerances (0.01%) if desired. The TCR provides information useful in characterizing the properties of the deposited thin film. Temperature coefficient of resistance (in ppm/°C) is defined as $(1/R)(\Delta R/\Delta T) \times 10^6$, where ΔR and ΔT refer to a change in the resistance value R caused by a change in the ambient temperature, T (room temperature represents the reference point in this case). This parameter is especially important in some applications that require circuit operation in unstable ambient temperature that affects the circuit performance. In general, thin films have very low TCR values as compared to bulk materials. Very low values of TCR, approaching zero, can also be achieved in some cases. Typical TCR values are well within the range of ±100 ppm/°C. Resistor stability is closely related to substrate inertness, substrate surface smoothness, and substrate thermal conductivity. Therefore, thin film resistor design and performance are greatly affected by the choice of substrate material. Thin film resistors exhibit good stability over time under different environmental conditions. The VCR describes the dependence of the resistance value on the voltage applied across the thin film resistor. Small VCR values are normally observed in thin films (10 ppm/volt or less). The improved performance of thin film resistors as compared to wire-wound resistors have made these elements very reliable.

A typical thin film resistor structure is discussed in Chapter 5. A thin resistive film of a few hundred Å to a few microns in thickness is deposited on an insulating substrate. Metallic-end contacts are then deposited. Thin film resistor materials can be metals, metal alloys, and metal compounds. Metals in bulk form possess low resistivity values that range between 0.1 and 200 μΩ·cm. When deposited in film form, these metals achieve significantly higher resistivities, depending on the thickness achieved.

3.4.2 Tantalum Films

Tantalum (Ta) films (tantalum nitride and tantalum oxynitride resistors), Ni-Cr alloy, Cr film, chromium silicon-monoxide-film, and chromium-cobalt film will be covered in this section. Properties of interest to the design engineer include R_s, TCR, stability, and stability after trimming.

Tantalum Films. Tantalum is known as a "valve" metal, a name given to metals forming tough self-protective oxides through anodization (anodic oxidation) or heat treatment in the presence of oxygen. Tantalum is commonly used for thin film resistors because of its high stability with respect to aging and temperature. Bulk Ta has resistivity of 15 μΩ·cm, while Ta thin films possess a higher resistivity at the same temperature.

Thin films of Ta are usually deposited by sputtering because of its refractory nature. Special precautions are taken during deposition because the sputtered films tend to become contaminated. Some doping impurities, such as nitrogen (N) or oxygen (O), are allowed in the Ta films to create specific properties. These sputtered films normally possess a high resistivity as deposited. Anodizing these films also provides a means of trimming these resistors to tight tolerance.

3.4.3 Tantalum Nitride Films

Nitrogen-doped Ta films have resistivities of ~400 $\mu\Omega\cdot$cm and TCR of -100 to -200 ppm/°C. The doping is accomplished by introducing a controlled amount of N into the sputtering chamber. These films also have higher stability under various environmental conditions than Ta films. Resistors composed of Ta_2N are more stable than those formed with TaN. A stabilizing treatment of film anodization, followed by heat treatment, and then another anodization at higher voltage promotes the stability of these elements. Tantalum-nitride films with thicknesses of 200 to 1000 Å produce R_s values of 300 Ω/square to 50 Ω/square. Also, these nitrogen-doped films can be anodized like pure Ta. The sputtering voltage, bias voltage, N_2 partial pressure, and temperature are critical parameters that affect the resistivity of Ta films.

Tantalum-Oxynitride Films (TaO_2N). Tantalum-oxynitride films can be made by adding O and N in the reactive sputtering process of Ta. The additional N and O are used to provide higher resistivity films. Temperature coefficients of resistance can be greatly modified by varying the pressure of both the O and the N in the sputtering chamber during the sputtering process.

3.4.4 Nickel Chromium Films (Ni-Cr)

A common resistive film composed of 80% Ni and 20% Cr alloy provides the desired properties of TCR and R_s for some applications. In general, the ratio of Ni:Cr can vary from 40:60 to 80:20. Bulk resistivity of 110 $\mu\Omega\cdot$cm and TCR of 100 ppm/°C have been cited. The TCR depends on the Ni:Cr ratio as well as specific deposition conditions. The TCR is more negative for films with high Cr content and high R_s. The 80:20 composition has an increased TCR and a higher R_s. Nickel-chromium films are commonly obtained by sputtering, but they can also be deposited by vacuum evaporation. Sputtering provides improved composition control of NiCr films over a wide range of R_s values. Special care should be taken in the evaporation process to provide a film with uniform composition. The difficulty arises from the fact that Cr and Ni possess different vapor pressures. (Chromium has a higher vapor pressure than Ni.) The difference in vapor pressure decreases as the temperature increases. Some of the techniques to obtain uniform film composition include using a different source composition from the desired film deposition, using a large source, or using two separate sources. If evaporation is used as the technique for NiCr film deposition, then several requirements need to be satisfied to produce successful results. These requirements include relief of internal stresses in substrates by keeping them within a certain temperature range during deposition, adjustment of residual gas pressure and rate of deposition to control oxidation, and post-annealing in air. Partial oxidation of Cr constituent during deposition is a serious problem. Nickel-chromium resistors possess R_s values in the range of 1 to 500 Ω/square. The incorporation of additives such as Cu and Al modifies the characteristics of these films. In addition, heat treatment affects the R_s. Maissel and Glang [20] illustrated the percent resistance change from heat treatment as a function of R_s for both vacuum-evaporated and sputtered NiCr. The sputtered films show a larger resistance increase than evaporated films. Nickel-chromium films are also known to be unstable under high humidity; therefore, a coating layer, such as silicon monoxide, is provided to increase the stability of these films. This protective technique improves the resistor stability in a wide variety of atmospheric conditions. Polyimide coatings are also used to improve the stability and passivate resistors.

3.4.5 Chromium Films

Chromium is a simple single-component system. It is not a particularly refractory metal. Therefore, temperature limitations for continuous operation are not relevant. Heat treatment of Cr films in vacuum may cause a decrease in resistance values. Pure Cr films have higher resistivities than NiCr films with achievable R_s values of 200 Ω/square within a tight tolerance without trimming. Much higher R_s values can be achieved with thinner layers but an increased tolerance. Chromium films are used for thin film resistors as well as underlay for thin film conductor metals. Chromium films are deposited by either vacuum evaporation or sputtering. The film properties are extremely sensitive to deposition conditions. The TCR of Cr films are not as good as NiCr films; however, the adhesive strength of Cr to substrates is superior, making it an attractive metal for resistive films.

3.4.6 Rhenium Films

High resistivity can be achieved with rhenium (Re) films. The films are deposited by evaporation onto ceramic, glass, or silica substrates. Sheet resistance values of 250 to 300 Ω/square with TCR close to zero may be achieved. Higher R_s values of 10,000 Ω/square with TCR around −400 ppm/°C are also obtained. A silicon monoxide (SiO) passivation layer assures the stability of this type of resistor.

3.4.7 Cermets

Cermets constitute an important class of materials for resistor fabrication (metal-dielectric systems such as the chromium-silicon monoxide system, Cr-SiO). Several techniques or processes are used to deposit films of these materials, such as evaporation and R_f sputtering. The cermet composition greatly affects the film resistivity. Cermets of 20% SiO and 80% Cr have resistivity of 550 $\mu\Omega\cdot$cm (TCR of −180 ppm/°C) after deposition (200°C), and 400 $\mu\Omega\cdot$cm (TCR of 0 ppm/°C) after annealing at 400°C. High resistivities of 10^3 to 10^5 $\mu\Omega\cdot$cm and good stability can be obtained with reasonable TCR values. Varying the percentage of SiO and Cr greatly affects both the resistivity and the TCR values.

Table 3-3 summarizes the various important materials used for thin film resistors.

TABLE 3-3 Characteristics of thin film resistor materials

Film material	Deposition technique	Sheet resistance R_s (Ω/square)	TCR (ppm/°C)
Cr-Ni (20:80)	Evaporation	10-400	100-200
Cr-Si (24:76)	Evaporation	100-4000	±200
Cr-Ti (35:65)	Evaporation	250-600	±150
Cr-SiO (70:30)	Evaporation	Up to 600	−15 to −200
SnO_2	Spray pyrolysis/CVD	Up to 10^4	−50 to −200
W, Mo, Re	Sputtering	10-500	−20 to −100
Ta	Sputtering	Up to 100	±100
Ta_2N, TaN	Reactive sputtering	10-100	−85

3.5 THIN FILMS DIELECTRICS

Silicon monoxide, tantalum oxide, manganese oxide-tantalum oxide, silicon monoxide-tantalum oxide, and barium titanate are used as common dielectric materials for thin film capacitors fabrication. In adding silicon dioxide and polymer dielectrics are common materials used to produce deposited multichip modules and thin film interconnection structures.

3.5.1 Background and Capacitor Materials Properties [47]

A thin film capacitor has three layers: a bottom electrode, a dielectric material, and a top electrode. The insulating material may contain two different types of dielectrics. Characteristics of capacitor elements include capacitance per unit area, temperature coefficient of capacitance (TCC), dissipation factor, dielectric constant, and breakdown voltage. Other properties of interest are leakage current and maximum working stress. High-quality thin film capacitors have low pinhole density, little leakage of current, low dissipation factor, and low TCC in addition to reasonably high capacitance per unit area. Because of low manufacturing yields, few companies use parallel-plate thin film capacitors. The low yields are caused by the inherent difficulties in avoiding pinhole formation in dielectric films. Thin film capacitors fabricated using interdigitated conductor geometries are often used in the fabrication of high-frequency circuits.

Common dielectric materials used for thin film capacitor formation include SiO, tantalum oxide (pentoxide Ta_2O_5), and organic polymers. Capacitor electrodes are usually made from Al, Au, Cu, or Ta, depending on the dielectric material used.

3.5.2 Silicon Monoxide Capacitors

Deposition of amorphous SiO films by vacuum evaporation results in capacitance values between 10 and 1.8 nanoFarads (nF)/cm^2 for dielectric film thicknesses between 0.5 and 3 μm. The tolerance of the capacitance value can be controlled to within $\pm 5\%$. The properties of the deposited dielectric film depend on the degree of oxidation during deposition and annealing (around 400°C). To ensure the proper amount of oxygen to be added, specially designed evaporation sources are used in the process as well as an oxygen partial pressure in the evaporation system. The vapor pressure of SiO dielectric increases considerably as the temperature increases. This property enables the evaporation of the material at much lower temperatures than either Si or silicon dioxide.

The SiO dielectric material has a dielectric constant of about 6, a leakage current <1 nÅ, TCC between +100 and +300 ppm/°C, a dissipation factor of approximately 0.01 to 0.1 at 100 kHz, and a maximum allowable stress of 2×10^6 V/cm. These properties depend on the deposition conditions and the nature of electrode material.

Aluminum is commonly used for both the base electrode and the counter electrode for SiO capacitors, because it forms a smooth layer and can vaporize around pinholes by discharge currents. Copper metallization is also used for the base electrode and either Cu, Al, or Au are used and for the counterelectrode.

3.5.3 Tantalum Oxide Capacitors

Tantalum pentoxide (Ta_2O_5) dielectric is formed by anodizing Ta metallization [46]. The resulting dielectric material is amorphous and chemically resistant. The Ta metallization for the base electrode is achieved through sputtering (depositing a layer of Ta or tantalum nitride of thickness of about 0.5 μm). After anodization, a counterelectrode of Al or Au is deposited by evporation or sputtering. The dielectric constant of the resulting oxide is 25, the capacitance per unit area is about 0.3 $\mu F/cm^2$ (depending on thickness of anodization), the maximum working stress is 1.5×10^6 V/cm, and the dissipation factor is 0.002 at frequencies in the kilohertz range. The TCC falls in the range +160 to +250 ppm/°C. To eliminate pinhole formation in the oxide, constant-current anodization is used, followed by constant-voltage anodization, in addition to the use of an anodic etch.

Oehrhein conducted a comprehensive study of the high-frequency capacitance voltage properties of Al/Ta_2O_5/p-Si storage capacitors used in dynamic random access memories [46]. After cleaning boron-doped Si wafers, Ta films can be deposited by electron beam evaporation from a high-purity Ta source. The Ta layers deposited are polycrystalline. The Si substrates are heated to 150°C during Ta evaporation by heating the wafer holder and establishing a temperature equilibrium between the Si substrate and the wafer support. The Ta can be oxidized in a dry oxygen atmosphere for an hour at a temperature of 430°C to 675°C. The dielectric constant is in the range of 22 to 33, depending on the oxidation temperature.

Sasaki et al. reported the use of anodized film of Ta_2N formed at 80 V as a dielectric material for anodized capacitors with a dielectric constant of about 12 [47]. This resulted in improved heat resistance in terms of the loss tangent, improved TCC, and lower leakage current upon thinner dielectric formation, as compared to thin film capacitor with Ta-anodized film as the dielectric material. The anodization of the Ta_2N film was achieved by a constant current–constant voltage technique in a boric ammonia solution. No thermal degradation of loss tangent or TCC were observed even during a 400°C heat treatment.

3.5.4 Manganese Oxide Tantalum Oxide Capacitors

A high-conductivity semiconductor manganese oxide (MnO_2) film is deposited by reactive sputtering or evaporation on top of an anodized Ta film. In this capacitor structure, Ta is used as the base electrode, and Au or Al is used as the counter, or top, electrode. The composite dielectric structure enhances the dielectric properties of the capacitor element, eliminating weak spots and thus enabling the use of thinner tantalum oxide layers. The resulting dielectric constant is 22, maximum working stress is 2×10^6 V/cm, and capacitance per unit area is 1 $\mu F/cm^2$.

3.5.5 Silicon Monoxide Tantalum Oxide Duplex Capacitors

The combination of these two dielectric materials enables one to use thinner dielectric thicknesses with a low probability of pinhole formation occurring. An SiO film of 2000 Å thickness and Ta_2O_5 film of 500 Å thickness may be used. Tantalum or tantalum nitride is used as the base electrode and Al is used as counterelectrode for this capacitor structure. A dielectric constant of 6, a maximum allowable working stress of 2.5×10^6 V/cm, and a capacitance per unit area of 0.02 $\mu F/cm^2$ typically result from this type of structure.

3.5.6 Barium Titanate Capacitors and Other Thin Film Capacitor Materials

Other dielectric materials for thin film capacitor formation include Al_2O_3, Si_3N_4, MgF_2, and $BaTiO_3$. Table 13-4 summarizes the dielectric properties of thin insulating films of various materials for capacitor applications [43]. Aluminum (or Au) films are generally used as electrodes for these measurements.

TABLE 3-4 Properties of thin film dielectric materials

Film material	Dielectric constant	Dissipation factor	Frequency (kHz)	Breakdown V (V/cm)	TCC (ppm/°C)	Thickness (μm)	Deposition technique
Al_2O_3	9	0.01	1		300	1.5	Ev
$BaSrTiO_3$	3-5	0.004-0.02	0.1-1000	3×10^5	300	2	Ev
SiO	6	0.015-0.02	1-1000			1	Ev
SiO_2	3-4	0		$\approx 10^6$		1	Rs
Si_3N_4	5.5			107		0.03-0.3	CVD
Ta_2O_5	25	<0.01	0.1-50	6×10^6	250	1	An
TiO_2	50	0.01	1		300	1	An
W_2O_5	40	0.6				1	An
CaF_2	≈ 3	0.05	0.1			1	Ev
LiF	≈ 5	0.03	0.1			1	Ev
Nb_2O_5	39	0.07	1			1	R_s
Polystyrene	2.5	0.001-0.002	0.1-100			1	GDP
Polybutadiene	2.5	0.002-0.01	0.1-100			1	UVP
$BaTiO_3$	200	0.05	1			1	Ev

Note: Ev, evaporation; R_s, reactive sputtering; CVD, chemical vapor deposition; An, anodization; GDP, glow-discharge polymerization; and UVP, ultraviolet polymerization.

3.5.7 Silicon Dioxide Dielectric

The SiO_2 dielectric layers with relative permittivity (ϵ_r) equal to 4.0 for thin film interconnections are deposited by a plasma chemical vapor deposition (CVD) process to produce dielectric films 10 to 20 μm thick. Silicon dioxide is relatively simple to process compared to polymer dielectrics. It also has a thermal conductivity higher than polymers (by about a factor of 10). In addition, this dielectric material does not absorb moisture and has high reliability with thermal changes. On the other hand, SiO_2 lacks planarization and is susceptible to pinhole formation.

3.6 THIN FILMS MAGNETICS

Use of thin film inductors is limited to high-frequency circuits because of the low permeability value of the substrate [49–53]. Practical physical dimensions of the spiral limit the inductance value to 10 microHenry (μH) with a quality factor of 50 to 100 at about 15 MHz. Higher inductance values can be achieved by depositing the inductor on a ferrite substrate or by using multilayer structures. The use of ferrites is also common in tapes for magnetic recorders and electromagnetic wave absorbers.

Nickel-zinc ferrite ($Ni\text{-}Zn\text{-}Fe_2O_4$) and manganese-zinc ferrite ($Mn\text{-}Zn\text{-}Fe_2O_4$) constitute two ferrimagnetic compositions with improved permeability and quality [49,50]. Ni-Zn ferrites are used in high frequencies because of their high resistivity, mechanical hardness, low porosity, and chemical stability. Improvement in the permeability is achieved by substituting a small amount of cobalt (Co) for Ni-Zn and Mn-Zn to produce Ni-Zn-Co and Mn-Zn-Co ferrites, respectively, at the expense of reducing the frequency bandwidth property [51–53]. For frequencies below 1 MHz, Mn-Zn ferrites are preferred. On the other hand, Ni-Zn ferrites provide better performance above 1 MHz. In addition, constituent ferrite powders should have low porosity and fine grain size (1 μm or less), and have nearly zero anisotropy.

These ferrites are prepared in polycrystalline form by high temperature solid state reaction method. The NiO, ZnO, and Fe_2O_3 high purity oxides are heated for a long period with grinding in between to achieve homogeneous powder at 1000°C and then heated (sintered) for a few hours at 1300°C for complete chemical reaction. The heat treatment of the samples in alumina are carried out in air and are followed by a slow cooling.

The effect of Fe ions replacement by rare earth ions (R) in a small amount of 0.5 mol% on the properties of $Ni_{0.7}\,Zn_{0.3}\,Fe_2\,O_4$ ferrites has resulted in enhancement of the intrinsic properties comprising of the magnetic and electrical properties. This replacement enables these materials to be used at higher frequencies of operation. In this case, R is Yb, Er, Tb, Gd, Dy, Sm, and Ce. The polycrystalline samples are prepared by mixing after sieving powder of the starting materials consisting of pure nickel oxide (NiO), Zinc oxide (ZnO), iron oxide (Fe_2O_3), and rare earth oxides (R_2O_3). The powders are mixed in a ball mill and then pressed into toroidal forms at a pressure of 5×10^7 N/m^2. The mixture is sintered at 1300°C in air for a few hours, followed by a slow cooling in the furnace [54].

3.7 POLYMER DIELECTRICS

The dielectric material for thin film interconnection must have good thermal and chemical stability during processing of materials to produce the electrical module, as well as during use. The CTE of the dielectric should be closely matched to the parent substrate to minimize stresses in the deposited films and substrate warping. Low CTE (5 to 15 ppm/°C) polyimides possess anisotropic dielectric constants [3.9(x, y), 2.9(z)]. It is important to deposit the dielectric in thicknesses reaching 25 to 30 μm without pinhole formation or stresses. A low dielectric constant with a low dissipation factor is also a prime factor in achieving low resistance.

Most polyimides possess CTE in the range 30 to 65 ppm/°C. During the polyimide cure, the loss of solvent and the release of water take place. Improper curing can result in a pronounced effect on the electrical and mechanical properties of the film.

In addition, moisture uptake severely affects the electrical characteristics of polymers, namely the dielectric constant and the loss tangent. Uncontrolled outgassing of absorbed moisture from underlying layers of dielectric during subsequent high temperature processing can result in blistering and delamination of the thin film structures being fabricated. Solutions to remedy the problem may include bake out cycles, use of meshed ground planes, and use of gridded signal traces. Reduction of dielectric constant and water absorption can be also achieved by incorporation of fluorine containing groups into the polyimide backbone (fluorinated polyimides). Low dielectric constant value is desired in order to allow faster propagation velocity, lower line capacitance per unit length, and higher characteristic impedance. Therefore, wider lines can be designed and smaller dielectric thickness can be used, lowering the line resistance and the electrical crosstalk.

Polymer dielectrics, particularly polyimides, are the common dielectric materials used for thin film multilayer interconnections. These polymers have high thermal stability, low dielectric constants of 2 to 3.5, and relatively low dissipation for minimum conversion of electrical energy to heat and small overall power loss. They are inert to process solvents and chemicals, and can be easily deposited from a solution using either spin coating or spray coating, making them excellent candidates for low-impedance interconnections. These polyimides are the most common polymer dielectrics, characterized by aromatic groups and an amide ring in the repeat structure. Curing the solution at temperatures up to 450°C removes the solvents and converts the precursor to a stable, insoluble polyimide. When cured properly, polyimides become stable with temperature. They are inert to chemicals and solvents, and they are mechanically tough when deposited with enough thickness.

Before polymer deposition, an organosilane-based adhesion promoter is applied by spin coating to improve the adhesion of the polymer coat to the substrate surface. Polymer films can be deposited from a solution using either spin coating, spraying, or dipping. The polymer solution flows freely and spreads on the surface because of the low viscosity, thus achieving a good planarization. The thickness, uniformity, and planarization of the polymer film depend on the viscosity and solids content of the solution. Multiple coatings may be necessary to achieve film thicknesses with the specified impedance interconnections. An improvement in the planarization is achieved with every deposited polymer layer. Polymer-to-polymer adhesion depends on the degree of cure (usually a lower cure temperature is recommended) and the glass transition temperature of the underlying film. Surface tension developed in the solution causes a polymer build-up near the edge, inhibiting patterning near these edges.

Polymer films are heat-cured in an oven or in a furnace after deposition to evaporate the solvents contained in the solution and to convert the polyamic acid reaction into a stable, insoluble polyimide. The final curing temperature (varying from 250°C to 450°C), curing rate, and curing atmosphere (usually nonoxidizing) result in a precise viscosity change. This results in the film flowing on the surface in a certain way, conforming to the underlying topography. During the polyimide cure, the loss of solvent and the release of water take place. Improper curing can result in a pronounced effect on the electrical and mechanical properties of the film. Relatively high moisture absorption and a large CTE on the order 30 to 65 ppm/°C can be disadvantages with some polymers, resulting in delamination of films, warpage of substrate, and stresses induced in the films. In addition, moisture uptake severely affects the electrical characteristics of polymers, namely the dielectric constant and the losts tangent. Table 3-5 illustrates properties of MCM-D dielectrics (or polymers) as compared to ceramics.

Polymers exhibit good thermal and electrical properties as well as high resistance to chemical attack from acids and solvents. In addition, polyimides can withstand higher processing temperatures than epoxies without degradation of their electrical and physical properties. Polymers also display a good thermal stability up to a temperature of 400°C, low dielectric constant (~3.5), low dissipation factor (a key parameter in high speed digital and microwave electronics, $\tan\delta$ is on the order of 0.002 for polyimides and 0.0008 for benzocyclobutenes), ability to planarize interconnect structures and irregular substrate features, and mechanical toughness. One attractive feature of polymers is that the addition of reactive groups to the polymer backbone can tailor the material properties to meet stringent electrical and mechanical requirements. For instance, this feature can be realized in polyimides by the selection of the diamine and dianhyride components used as the starting materials. In addition, polyimide precursor solutions can be etched using chemicals common to the fabrication of printed circuits for microelectronics.

3.7.1 Photosensitive Polyimides

Photosensitive polyimides contain photoreactive species or groups to result in photocrosslinking with adjacent polymer chains when exposed to ultraviolet light. This addition leads to a differential solubility with respect to the exposed and unexposed portions of the film. Covalent bonding or ionic bonding represent the two basic chemical methods of photosensitivity presence in a polyimide structure. Photosensitivity is found to be higher for covalent material, resulting in better resolution, especially for layers that are thicker than 10 μm. On the other hand, the ionic material demonstrates better mechanical properties once fully cured (about 20 hours), tensile strength, and elongation are a factor of 10 better, with superior peel strength. Thus, the patterning process becomes quite simple. Photosensitive polyimides may experience some problems: poor shelf life, cracking during development (especially related to film thickness deposited), and limited resolution caused by swelling. Because newly developed photosensitive polyimides overcome most of these problems, the polymer is widely used by various industries (such as NTT, NEC, Toshiba, and Mitsubishi).

TABLE 3-5 Properties of MCM-D dielectrics as compared to ceramics

Properties	PIQs	Ceramics	PI	Fluoropolymers	BCB	PPQ
ϵ_r	3.2-3.4	5-9	2.5-2.8	1.9-2.6	2.6-2.8	2.8-3
Tg (°C)	300-400	N/A	300-400	160-320	310-350	361
CTE (ppm/°C)	3-58	3-7	2-60	90-300	65	40
Percent water uptake at 95% RH	80-100	~0	25-40	1-10	25	10
Percent planarization (ha-hb)/hb	20-70	N/A	5-50	≈0	up to 95	40-50

Note: ha, average height of features before planarization; hb, average height of features after planarization; PI, polyimides; BCB, benzocyclobutenes; PPQ, polyphenylquinoxaines; PIQ, polyimide iso-indoloquinazolinediones.

3.8 ADVANCED THIN FILM MATERIALS

Advanced thin film materials include superconductor materials, diamond thin films, and optical thin film materials. Superconductor thin film materials will be discussed in this chapter. Diamond thin films are discussed in Chapter 7 of this handbook [24,25,26]. Optical thin film materials are discussed in Chapter 8 of this handbook.

3.8.1 Superconductor Thin Film Materials [2, 3, 8–10]

The pioneering discovery by Bednorz and Muller of superconductivity above 30°K in the La-Ba-Cu-O system has generated an unprecedented amount of research on superconducting oxides [21]. This includes the discovery of a superconducting transition above 77°K in the Y-Ba-Cu-O, Bi-Sr-Ca-Cu-O, and Tl-Ba-Ca-Cu-O systems. Although other materials exist, these three systems are the most important for many applications, because they have the highest critical temperatures. These material systems are very complex and have many phases. For example $Y_1Ba_2Cu_3O_7$ is a high-temperature superconductor (HTS) with a critical temperature of 91°K, although $Y_2Ba_1Cu_1O_X$ is an insulator. The principle phases of these systems and their associated transition temperatures are listed in Table 3-6.

The most common methods used to make superconducting films include sputtering, evaporation, and laser ablation. The most popular method used to produce high-temperature superconductive films is RF-sputtering. The deposited films have little structure and are insulating. Postdeposition annealing treatments in an oxygen atmosphere at temperatures above 800°C are normally necessary for the films to become superconducting at high temperatures. Inadequate heat treatments, such as annealing at temperatures below 750°C, usually make films semiconductive before the superconducting transition. Annealing treatment is an important process that determines the final oxygen content and the resulting properties of the film. The critical processing variables in this technique are annealing temperature, oxygen partial pressure, and quenching rate.

Substrate reactions of the superconducting films are a major factor in degradation. Diffusion of substrate material into an HTS results in the creation of insulating phases. These insulating phases reduce the current density of the overall film, and may result in an insulating film if the reaction is severe. Certain substrate materials, such as yttrium-stabilized zirconia (YSZ), do not react with most of the superconductors, although other materials require a buffer layer to prevent substrate reactions. These buffer layers prevent physical contact between the superconductive material and the substrate.

Most of the applications for these materials are based on their low loss characteristics. These materials have significantly lower loss than conventional metals particularly for high-frequency applications. Additionally, these materials have lower propagation delays than conventional metals, and they are prime candidates for use as high-speed circuit interconnections.

TABLE 3-6 High-temperature superconductors and transition temperatures

Material	Critical temperature (°K)
$Y_1Ba_2Cu_3O_7$	91
$Bi_2Sr_2Ca_2Cu_3O_X$	110
$Tl_2Ba_2Ca_2Cu_3O_{10}$	125

3.9 PROTECTIVE COATINGS [39]

Thin films can be thermally stabilized and protected against environmental effects by depositing the proper protective layers to cover the elements.

3.9.1 Resistor Passivation [38,44]

Thin film resistor materials include Ta films, NiCr films, Cr films, Re films, and cermets. Tantalum nitride films possess higher stability under various environmental conditions. A stabilizing treatment with film anodization, followed by heat treatment and another anodization at higher voltage, further promotes stability of these elements. Although NiCr films provide the desired properties of TCR, R_s, and adhesion, they are known to be unstable under humid conditions. Therefore, a coating layer (SiO) is provided to maintain acceptable stability.

A technique for passivating thin film circuits using NiCr resistors includes the use of polyimides. The method consists of spinning the polyimide on the substrate, and then baking it at a low temperature to form the tentative film. The low-temperature cure is necessary to remove the solvents before the high-temperature cure. Patterning this passivation material is accomplished through the application of positive photoresist, exposure, and development to etch the resist as well as the polyimide. The film is then baked at moderate temperature (250°C to 300°C) to form the polyimide film. The polyimide protects thin film resistors under different operating and environmental conditions (under load, high humidity, and high temperature). The polyimide material results in no effect on the microwave performance of these microcircuits. For Re films, an SiO protective overlayer helps assure good resistor stability.

3.9.2 Surface Preparation [40–42]

Surface preparation is necessary to produce a suitable top layer interconnection that can resist corrosion and be compatible with the dielectric, in addition to bonding pads and solderable areas on the substrate. A barrier metal such as Ni is deposited first on the surface, usually by electroplating, to prevent diffusion of the Cu conductor. Also, Ni withstands pressure applied during the wire-bonding process. Chromium or TiW can also be used as barrier metals, and they are deposited by a sputtering process. A thin film sputter deposition of an Au layer of 1 to 3 μm thickness usually follows the Ni deposition. Soft Au with few impurities is needed to achieve successful bonds and proper die attachment sites.

REFERENCES

1. J. L. Vossen and W. Kern, (eds.), "Thin Film Processes II," Academic Press, San Diego, CA, 1991, pp. 501–522.
2. C. Stolzel, M. Huth, and H. Adrian, "C-Axis oriented thin $Bi_2Sr_2CaCu_2C_{8+}$ films prepared by flash evaporation," *Physica C,* vol. 204, 1992, pp. 15–20.
3. T. Hato, Y. Takai, and H. Hayakawa, "High Tc Superconducting Thin Films Prepared by Flash Evaporation," *IEEE Transact Magnet,* vol. 25, no. 2, March 1989, pp. 2466–2469.
4. C. De Las Heras and C. Sanchez, "Characterization of Iron Pyrite Thin Films Obtained by Flash Evaporation," *Thin Solid Films,* vol 199, 1991, pp. 259–267.

5. H. E. Horng et al., "Critical Current in Polycrystalline Bi-Ca-Sr-Cu-O Films," *Phys Rev B,* vol. 39, no. 13, May 1989, pp. 9628–9630.
6. P. Wagner, H. Adrian, and C. Tome-Rosa, *Physica C,* 198, 1992, p. 258.
7. J. Dieleman, E. Van de Riet, and J. C. S. Kools, *Jpn J Appl Phys,* vol. 31, no. 6B, pt. 1, 1992, pp. 1964–71.
8. A. J. Basovich et al., *Thin Solid Films,* vol. 228, 1993, pp. 193–95.
9. F. Sanchez et al., *Appl Surface Sci,* vol. 69, 1993, pp. 221–224.
10. A. Giardini Guidoni, I. Pettiti, *Appl Surface Sci,* vol. 69, 1993, pp. 365–369.
11. C. Germain et al., *Appl Surface Sci,* vol. 69, 1993, pp. 359–364.
12. G. A. Petersen and J. R. McNeil, *Thin Solid Films,* vol. 220, 1992, pp. 87–91.
13. N. J. Arfsten, *J of Non-Crystalline Solids,* vol. 63, 1984, pp. 243.
14. H. Dislich, *J of Non-Crystalline Solids,* vol. 57, 1983, pp. 371.
15. C. J. Brinker and G. W. Scherer, "Sol-Gel Science," Academic Press, Orlando, FL, 1990.
16. D. Segal, "Chemical Synthesis of Advanced Ceramic Materials," Cambridge University Press, Cambridge, United Kingdom, 1989.
17. L. C. Klein, (ed.), "Sol-Gel Technology for Thin Films, Fibers, Preforms, Electronics and Specialty Shapes," Noyes Publications, Park Ridge, NJ, 1988.
18. L. C. Klein, *Ceram Eng Sci Proc,* vol. 5, 1984, pp. 379.
19. H. Dislich, *J of Non-Crystalline Solids,* vol. 80, 1986, pp. 115.
20. Maissel and Glang, (eds.), "Handbook of Thin Film Technology," McGraw-Hill, New York, 1970.
21. J. G. Bednorz and K. A. Muller, *Z Phys B,* vol. 64, 1986, p. 189.
22. L. Kempfer, *Materials Engineering,* vol. 107, no. 5, May 1990, pp. 26–29.
23. R. Messier et al., *J of Metals,* September 1987, pp. 8–11.
24. W. Zhu et al., *Proc IEEE,* vol. 79, no. 5, May 1991.
25. Y. Hirose and M. Mitsuizumi, *New Diamond,* vol. 4, 1988, pp. 34–35.
26. A. H. Deutchman and R. J. Partyka, *Adv Materials & Processes,* ASM Int., June 1989, pp. 29–33.
27. I. Bahl and P. Bhartla, "Microwave Solid State Circuit Design," John Wiley and Sons, New York, 1988.
28. P. A. Stenmann and H. E. Hintermann, *J Vac Sci Tech,* A7, 1989, p. 2267.
29. J. Valli, *J Vac Sci Tech,* A4, 1986, p. 3007.
30. V. E. Cosslett, "Fifty Years of Instrumental Development of the Electron Microscope," in *Advances in Optical and Electron Microscopy,* R. Barer and V. E. Cosslett, (eds.), vol. 10, Academic Press, San Diego, CA, 1988, pp. 215–267.
31. J. I. Goldstein et al., *Scanning Electron Microscopy and X-Ray Microanalysis,* Plenum Press, New York, 1981.
32. G. Thomas and M. J. Goringe, *Transmission Electron Microscopy of Materials,* John Wiley and Sons, New York, 1979.
33. M. Ohring, "The Materials Science of Thin Films," Academic Press, Orlando, FL, 1992, p. 276.
34. L. E. Davis et al., *Handbook of Auger Electron Spectroscopy,* Physical Electronics Industries Inc., Eden Prairie, MN, 1976.
35. G. E. McGuire, *Auger Electron Spectroscopy Reference Manual,* Plenum Press, New York, 1979.
36. M. T. Bernius and G. H. Morrison, "Mass Analyzed Secondary Ion Microscopy," *Rev Sci Instrum,* 58, October 1987, pp. 1789–1804.
37. J. R. Bird and J. S. Williams, (eds.), *Ion Beams for Materials Analysis,* Academic Press, Sydney, 1989.
38. S.C. Miller, "A Technique for Passivating Thin Film Hybrids," *Proc of the 1976 International Symposium of Hybrid Microelectronics (ISHM),* 1976, p. 39.

39. G. Lu, "CVD Diamond Films Protect Against Wear and Heat," *Adv Mat Proc ASM Int.,* vol. 144, no. 6, December 1993, pp. 42–43.
40. B. Banks, "Modify Surfaces with Ions and Arcs," *Adv Mat Proc ASM Int.,* vol. 144, no. 6, December 1993, pp. 22–25.
41. B. Holtkamp, "Ion Implantation Makes Metals Last Longer," *Adv Mat Proc ASM Int.,* vol. 144, no. 6, December 1993, pp. 45–47.
42. H. A. Naseem et al., "Metallization of Diamond Substrates for Multichip Module Applications," *Proc 1993 Intern Conf Exh on Multichip Modules,* April 14–16, 1993, pp. 62–67.
43. C. D. Iacovangelo and E. C. Jerabek, "Metallizing CVD Diamond for Electronic Applications," *Proc of the Intern Symp of Hybrid Microelectronics (ISHM),* 1993, Dallas, pp. 132–138.
44. M. Okuyama, S. Kambe, and D. L. Piron, "A Proposed Mechanism on Ni^{2+} Dissolution During Growth of Passive Films on Nickel in a Neutral Solution," *Mat Sci Forum,* vols. 192–194, 1995, pp. 489–496.
45. S. W. Russell et al., "Enhanced Adhesion of Copper to Dielectrics Via Titanium and Chromium Additions and Sacrificial Reactions," *Thin Solid Films,* vol. 262, 1995, pp. 154–167.
46. G. S. Oehrhein, "Capacitance Voltage Properties of Thin Ta_2O_5 Films on Silicon," *Thin Solid Films,* Electronics and Optic, vol. 156, 1988, pp. 207–229.
47. K. Sasaki et al., "Heat-Proof Properties of Ta_2N Anodized Thin Film Capacitors Prepared at Low Anodization Voltage," *Electronics and Communications in Japan,* vol. 78, part 2, no. 9, 1995, pp. 97–103.
48. C. H. Wang et al., "Correlation of Roughness, Impurity, Infrared Emissivity and Sputter Conditions for Aluminum Films," *Mat Res Soc Symp Proc,* vol. 354, 1995, pp. 523–528.
49. J. Smit, "Magnetic Properties of Materials," McGraw-Hill, New York, 1971, Ch. 3, pp. 76–78.
50. E. C. Snelling and A. D. Giles, "Ferrites for Inductors and Transformers," John Wiley & Sons, United Kingdom, 1983, Ch 3.
51. J. G. M. De Lau, "Influence of Chemical Composition and Microstructure on High Frequency Properties of Ni-Zn-Co Ferrites," Philips Research Reports Supplements, The Netherlands, no. 6, 1975, Ch 4, pp. 42–87.
52. E. E. Riches, "Ferrites," Mills & Boon Limited, United Kingdom, 1972, Ch 3, pp. 17–33.
53. E. C. Snelling, "Soft Ferrites," CRC Press, Cleveland, OH, 1969, Ch 3.
54. N. Rezlescu and E. Rezlescu, "The Influence of Fe Substitutions by R Ions in a Ni-Zn Ferrite," *Solid State Communications,* vol. 88, no. 2, 1993, pp. 139–141.

CHAPTER 4
PRINCIPLES AND PROPERTIES OF SEMICONDUCTOR THIN FILMS

Lawrence L. Kazmerski
Fuad Abulfotuh
National Renewable Energy Laboratory

4.1 INTRODUCTION

Understanding and control of the electro-optical properties of thin films have gained significance with the evolution of thin film device technologies. The literature is abundant with books, references, and reviews dealing with the physical, structural, electrical, optical, and thermal characteristics of metal, semiconductor, and insulator thin films [1–10]. This chapter focuses on semiconductor thin films; delineates fundamental mechanisms; integrates and compares recent developments with earlier work; and provides a general basis for understanding thin semiconductor film properties, parameters, and phenomena. The chapter provides an introduction and overview of the basic properties encountered in thin film research and development. The authors present a logical progression from simple concepts to more complex concepts, dealing with the electrical transport and optical characteristics of single-crystalline, polycrystalline, and amorphous thin film semiconductors. For further information, the reader's attention is directed to the references at the end of this chapter.

4.2 TRANSPORT MECHANISMS IN POLYCRYSTALLINE SEMICONDUCTORS

The electrical properties of thin films depend on the physical nature of the films, including their structural characteristics, crystallinity, impurity levels (or doping), defect characteristics, uniformity, and stoichiometry. The effects of each characteristic on the general flow of positive and negative carriers explain the associated transport mechanisms. This section develops and presents various conductivity models that describe transport in thin film materials by using the evolution of the films over various growth and processing stages as the basis for the electro-optical considerations.

4.2.1 Discontinuous Thin Films

Growth of thin films have regions with distinct physical characteristics and electrical properties [11–12]. During the initial deposition process, nucleation usually occurs at many sites on the substrate surface [13]. Small clusters of material develop and increase in area until a continuous film is formed. The discontinuous film-growth areas, or islands, provide a useful starting point for understanding the transport properties of thin films.

Physical and Electrical Observations. The interdependence of the effective island size (r) and the interisland separation (d) on the average film thickness is illustrated in the electron micrographs of Fig. 4-1, and the representative data in Fig. 4-2. Here, the average film thickness is defined as the effective thickness for the deposited mass if the film was continuous. Note that the island size increases with thickness, whereas the interisland distance decreases—except for a region over which d can remain approximately constant. (The island size distributions are approximately gaussian, and the mean size is designated

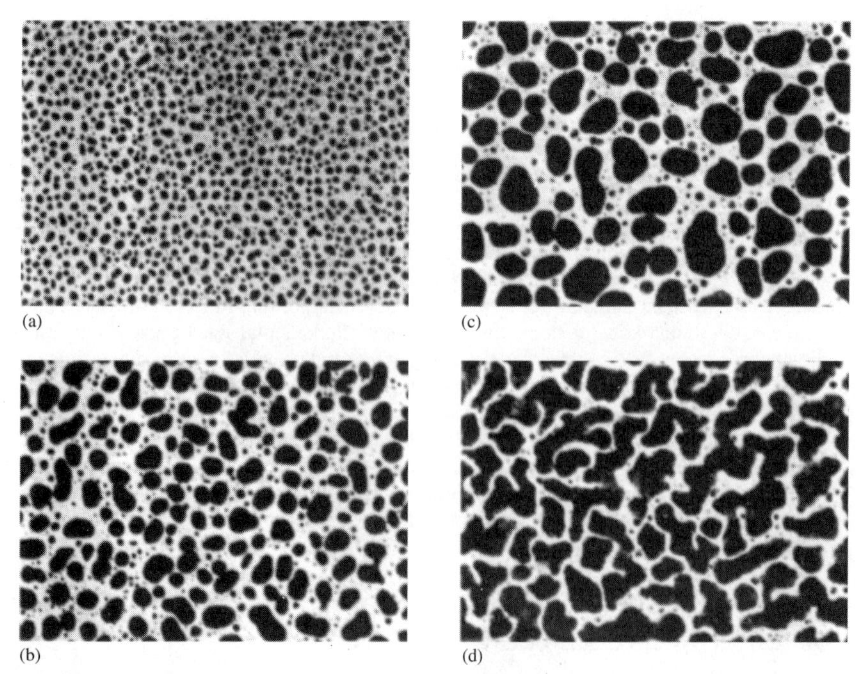

FIGURE 4-1 Transmission electron micrographs of ultra-thin Au films. Average thicknesses are (a) 10 Å, (b) 40 Å, (c) 60 Å, and (d) 150 Å. Films are deposited under identical conditions, and show progression from discontinuous to network stages.

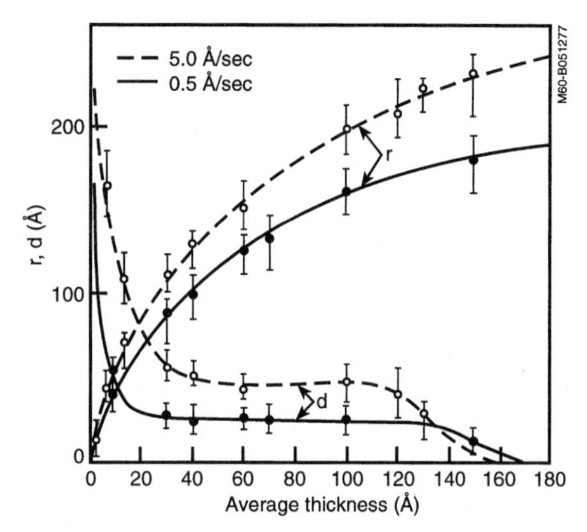

FIGURE 4-2 Dependence of island cross-sectional dimension r and inter-island separation d on average film thickness for Au films deposited at rates of 0.5 Å/s and 5.0 Å/s. *(From L. L. Kazmerski and D. M. Racine,* J Appl Phys, *vol. 46, 1975, p. 791.)*

in Fig. 4-2.) These physical data are typical for metallic and most semiconductor thin films grown on foreign substrates. The films typically become continuous at average thicknesses that depend on the material, the substrate, and the growth conditions, as well as any in situ or external processing.

Films grown at higher deposition rates typically have larger island sizes (Fig. 4-1), and the effective island size increases with substrate temperature (Fig. 4-3). These characteristics are associated with the enhanced surface mobility of adatoms and clusters with increased surface energy. The electrical properties of the films are usually examined first for the effect of temperature on conductivity. These data are represented in Fig. 4-4 for discontinuous gold (Au) films having different thicknesses and, therefore, different r and d values. Proximal probe microscopy has allowed the observation of similarly grown processes for semiconductors on single-crystal substrates [14–16]. The conductance generally exhibits linear dependence on inverse temperature for various temperature ranges. This observed experimental electrical characteristic forms the basis for deriving the models for the electrical transport of these films.

Electrical Transport. Several models have been introduced to explain the electrical behavior of these films in their initial stages of growth. The earliest models considered thermionic emission between the material islands (particles), with the conductivity depending on the separation between the particles and the temperature [17,18]. Charge transfer by quantum-mechanical tunneling was also considered, but predicted the conductivity to be nearly temperature independent [19–21].

The work of Neugebauer and Webb provides a simple model based on activated tunneling of the film conductivity and includes not only the observed temperature dependence but also the relationship of the electrical properties to the physical parameters of the films [22]. The model includes thermal activation, which involves the transfer of charge from one initially neutral island to another island that is separated from it. The activation energy is simply $q^2/\epsilon r$ (in which q includes the magnitude and polarity of the charge and

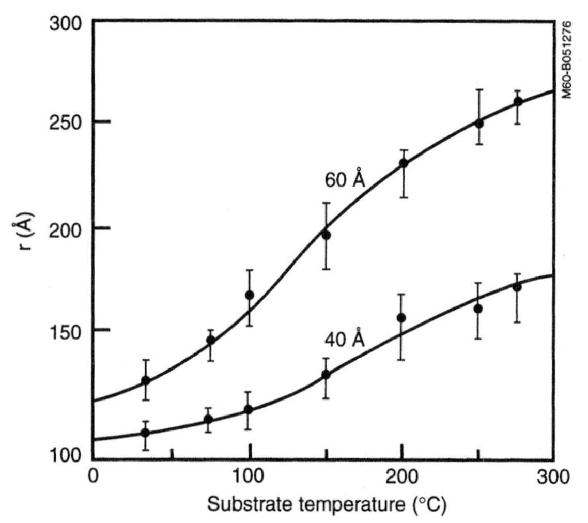

FIGURE 4-3 Cross-sectional dimension r as a function of substrate temperature for Au films 40-Å and 60-Å thick. Mean value of dimension is indicated within range bars. *(From L. L. Kazmerski and D. M. Racine,* J Appl Phys, *vol. 46, 1975, p. 791.)*

ϵ is the permitivity), and only electrons or holes excited to states of at least this energy from the Fermi level can tunnel from the one neutral island to another. Neugebauer and Webb proposed an activated transfer mechanism in which the charge carrier (created by thermal activation over the electrostatic potential barrier) is transported by tunneling from one neutral island to another.

This model considers the net probability of a transition between islands in the applied field direction and calculates this probability in terms of the transmission coefficient. This coefficient is a function of the potential barrier ($q\varphi$) between the islands and the distance between them. The resulting conductivity σ is given by

$$\sigma = C(d/r)\exp(A/\epsilon rkT - Bd) \quad (1)$$

where C and $B \propto q\varphi^{1/2}$

A = function of q^2

ϵ = an effective dielectric constant of the media between islands.

This model predicts the observed temperature dependence. In addition, the relative dependence of the conductivity on the island size and separation (Fig. 4-4) is also included in this formulation.

In the Neugebauer-Webb model, the activation energy E is expressed as

$$E = (q^2/\epsilon)(d/r[r + d]) \quad (2)$$

This dependence of E on the physical size of the islands and island separation is confirmed by the data of Fig. 4-5 showing the linear dependence on the $d/r(r + d)$ parameter for two different deposition cases. The model, though simple, does a good job of predicting the relationships among the conductivity and the physical nature of the film and the film's temperature.

In general, models are used as indicators of behavior, and counter examples of observed characteristics introduce limitations that require closer examination and alteration of the analytical treatments. For example, deviations from the linear conductivity-

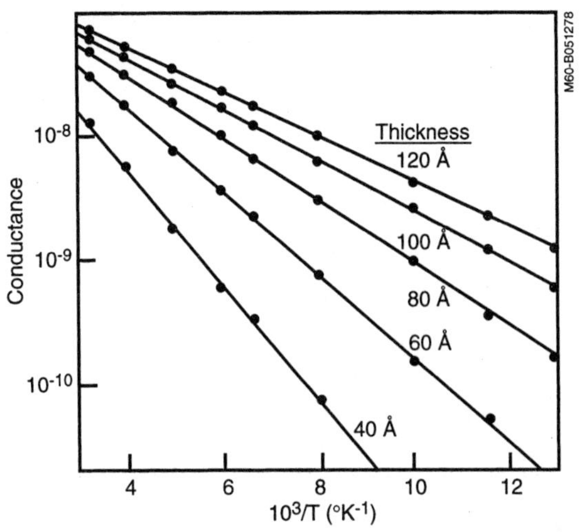

FIGURE 4-4 Conductance as a function of inverse temperature for various ultrathin Au film thicknesses.

temperature dependence are observed for the discontinuous Au films (Fig. 4-6), which is behavior not explicitly covered by the Neugebauer-Webb approach. These data show a deviation from the linear dependence predicted by Eq. 1 at higher temperatures for films grown under poorer vacuum conditions. In the development of these models, it is sometimes useful to examine the simple model more closely and modify it to include possible parameter variations, which is the case in this chapter. The enhancement in the conductivity can be explained in terms of a quantum-mechanical consideration for tunneling via substrate and traps [21].

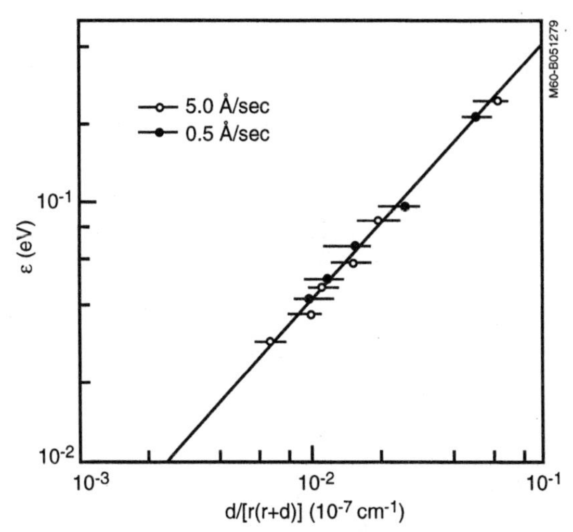

FIGURE 4-5 Activation energy as a function of the dimensional parameter $d/[r(r+d)]$, showing linear dependence for ultrathin Au films. *(From L. L. Kazmerski and D. M. Racine,* J Appl Phys, *vol. 46, 1975, p. 791.)*

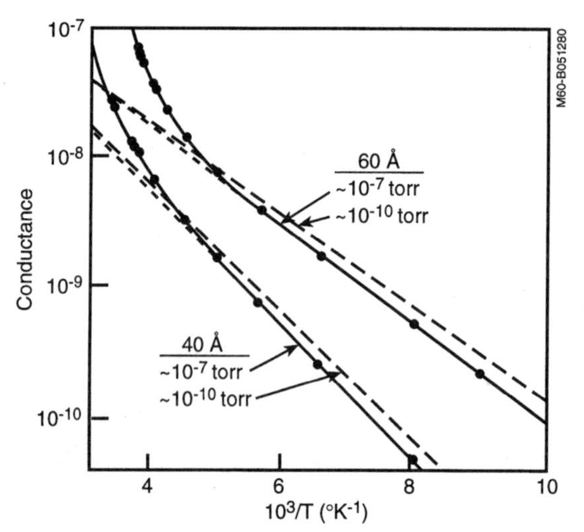

FIGURE 4-6 Conductance as a function of inverse temperature for Au films, indicating anomalous conductivity effects for films deposited at higher base pressures. This is ascribed to quantum-mechanical tunneling between islands at higher temperatures. *(From L. L. Kazmerski and D. M. Racine,* J Appl Phys, *vol. 46, 1975, p. 791.)*

Rewriting the basic Neugebauer-Webb formulation of Eq. 1,

$$\sigma \approx (d^2/r)D \exp(-q^2/\epsilon rkT) \tag{3}$$

where D is the transmission coefficient. Using the Wentzel-Kramers-Brillouin method [22] to calculate D for two separated metallic particles, the expression for the conductivity becomes

$$\sigma = C(d/r)\,[\sin(GkT) - 1]\exp(A/\epsilon rkT - Bd) \tag{4}$$

where G is a constant that depends on the thin film material. The quantum-mechanical tunneling term quantitatively accounts for the increased conductivity at higher temperature observed in the experimental data of Fig. 4-6. In this case, the difference is caused by contamination of the discontinuous films and the substrate because of the initial adsorption of oxygen at the film's surface. Equation 4 predicts that the tunneling process is enhanced when the effective distance between the islands is small, or when $d_{\text{eff}} < d$. This effective tunneling distance is caused by the absorption of the contaminant. Such anomalous effects have been noted for other materials, including semiconductors [24,25]. The point is that the simple model explaining general conductivity usually does not lead to a general or universal prediction for electrical transport. Finally, the expressions for conductivity (Eqs. 3 and 4) provide some insight into the transport mechanisms for continuous polycrystalline films in which defects (such as grain boundaries) might dominate. That is, the conductivity is controlled by the activation of the carriers over the potential barrier (perhaps derived from carrier depletion) associated with the defects.

4.2.2 Crystalline Films

The next stage in the evolution of thin-film electrical transport involves a continuous, perfect, single-crystal film in which the electrical transport properties are essentially the bulk properties, but are controlled or altered only by the surfaces. In this stage, the conductivity is influenced by the surfaces (perfect and real) on the carrier transport. Both rigorous and simplified analyses are used to consider the effects of band bending, and experimental results (primarily Hall-effect parameters) are compared to the developed analytical predictions.

The surface of a thin film affects the transport properties of a semiconductor by limiting the transversal of the charge carriers and their mean-free paths. Because of the large surface-to-volume ratios in thinner layers, these surface effects are more pronounced in thin films than in thick or bulk layers. When the thickness of the film becomes comparable to the mean-free path of the carriers, the scattering of the electrons and holes from the film surfaces have measurable effects on the carrier transport properties and can dominate the electrical characteristics. The extent of these surface events depends on the nature of the scattering mechanisms involved. The two limiting cases are:

- *Specular reflection.* During the scattering process, the carriers have only their velocity component perpendicular to the surface reversed; their energy remains constant. Because no losses occur, there is no effect on the conductivity. The surface represents a perfect reflector; the scattering is *elastic,* which is expected from an ideal surface.
- *Diffuse reflection.* After scattering, the carriers emerge from the surface with velocities independent of their incident ones. This process is indicative of *inelastic* or random scattering. The change in momentum leads to a related change in conductivity. Real surfaces exhibit disorder, resulting in some degree of diffuse scattering.

The extent of diffuse scattering is determined by the type, density, and cross sections of the surface defects. Major mechanisms, primarily surface charge impurities and electron phonon interactions, that determine the extent of diffuse scattering have been discussed in detail [26–31]. Quantitatively, total diffuse scattering can lower effective film conductivities and mobilities more than an order of magnitude below their single-crystal bulk values. For a real surface, both partial specular and partial diffuse scattering mechanisms exist, and the resultant electrical properties usually lie between those predicted by either scattering mode.

For the perfect semiconductor thin film, the effective transport properties are the bulk properties, altered by the surface conditions. The initial, most simple model involves energy bands constant to the surface—the *flat-band* condition shown in Fig. 4-7*a.* This situation is somewhat artificial for real surfaces, but provides a basis for understanding the various scattering phenomena and their effects. It can be used to predict the general transport properties and electrical behavior of films. The more complicated model includes band bending at the surface. The cases for surface depletion, inversion, and accumulation are shown in Fig. 4-7*b-d.* For the purposes of discussion, the z direction is defined in this chapter as perpendicular to the film surface, and the total film thickness is d.

Transport Mechanisms. The standard of comparison is the bulk crystal [32]. If the bulk crystal was perfect, the electrons and holes could flow unimpeded in the perfect periodic potential. In the absence of external fields, each carrier maintains its velocity and wave vector indefinitely. In the real bulk-crystal lattice, vibrations, impurities, and defects cause deviations from the ideal behavior. The carriers experience a nonzero scattering probability which provides for random movement and continual velocity change for the carriers. A drift current results when an electric field is imposed.

Each of the various scattering processes can be characterized by a fundamental relaxation time, τ, defined as the average time required for a disturbance in the electron distribution to fade by the random action of the scattering. Two scattering processes are especially important for bulk-crystal behavior: 1.) scattering by lattice vibrations,

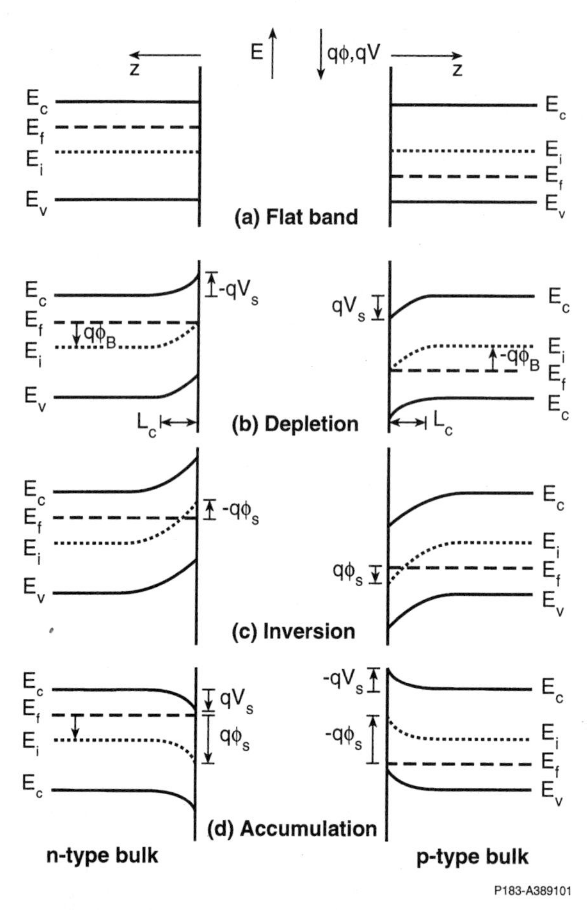

FIGURE 4-7 Representative and structures near the semiconductor surface for (*a*) flatband, (*b*) depletion, (*c*) inversion, and (*d*) accumulation conditions for *n*-type and *p*-type semiconductors.

dominant in chemically pure crystals at normal temperatures; 2.) scattering from impurity centers. For the first case, the relaxation time τ_L is expressed as [33]

$$\tau_L \propto E^{-\frac{1}{2}} T^{-1} \tag{5}$$

where T = temperature
E = carrier energy.

In the second case, scattering occurs from impurity centers (such as ionized impurities) [33]. In this case

$$\tau_i \propto E^{\frac{3}{2}} / N_i \tag{6}$$

where N_i is the density of ionized impurities. All scattering mechanisms act essentially simultaneously in real crystals. The total bulk relaxation time can then be expressed as

$$1/\tau_b = 1/\tau_L + 1/\tau_i + \ldots \tag{7}$$

The transport equations are derived by incorporating the relaxation times into the Boltzmann relation [34]. From this, the current-field relationship is generated, and the relevant transport parameters (conductivity, mobility, and carrier concentration) are derived. The temperature dependencies of the lattice and the ionized impurity-scattering carrier mobilities (μ_L and μ_i) are expressed as [33]

$$\mu_L = CT^{-3/2} \tag{8}$$

and

$$\mu_i = C'T^{3/2} \tag{9}$$

where C and C' are material-related constants. If the scattering types are present and noninteracting, the mobilities add according to the relaxation times (Eq. 7), with the effective bulk crystalline mobility μ_b expressed by [35]

$$1/\mu_b = 1/\mu_L + 1/\mu_i + \ldots \tag{10}$$

By examining Eqs. 8 and 9, it is apparent that lattice scattering dominates at high temperatures, and impurity scattering at lower temperatures. The transition from domination by one type to domination by the other depends on the material, the carrier type, and the impurity concentration. It has been experimentally observed that mobility variations are usually confined to the dependence range from $T^{-3/2}$ to $T^{3/2}$ for most semiconductors [36,37].

Surface Effects. In general, the bulk-transport case from the previous section must be modified further for a thin film when the number of carriers scattering from the surface exceeds the number of those not scattered. Starting with the case of flat bands and noninteracting scattering mechanisms, the total relation time τ' for a nondegenerate semiconductor can be intuitively derived using Matteissen's rule [35]:

$$1/\tau' = 1/\tau_b + 1/\tau_s + \ldots \tag{11}$$

where τ_s represents the average time a carrier requires to collide with the surface toward which it is moving, and τ_b is expressed by Eq. 7. The mean-free distance of a carrier from a surface is approximately the film's half thickness ($d/2$). The average surface scattering time relates to this mean-free distance through the unilateral mean velocity, v_z, the average over negative or positive z-direction velocity component of all carriers [38], or

$$\tau_s = (d/2)/v_z \tag{12}$$

But the unilateral mean-free path, λ, is defined as [38]

$$\lambda = \tau_b\, v_z \tag{13}$$

from which

$$\tau_s = (d/2\lambda)\tau_b \tag{14}$$

Relating this to the mobility through Eq. 10, μ' is defined as

$$\mu' = \mu_b/(1 + 2\lambda/d) \tag{15}$$

This simple derivation assumes that the surface scattering is specular. If only some fraction, p, of these carriers undergo specular reflection, then $1 - p$ are scattered diffusely. The expression for τ_s is adjusted by this parameter; that is, τ_s becomes $(1 - p)v_s$, and the mobility can be written as [38]

$$\mu' = \mu_b/[1 + (2\lambda/d)(1 - p)] \tag{16}$$

A rigorous approach was used, leading to essentially the same result. Starting with the Boltzmann equation, and incorporating the distribution for the carriers undergoing surface scattering, the derived mobility is of the form [39]

$$\mu'/\mu_b = 1 - (1 - p)(2\lambda/d) + (1 - 2p)(2\lambda/d)\Gamma_{2\lambda/d} + (2p\lambda/d)\Gamma_{\lambda/d} \tag{17}$$

where

$$\Gamma_{n\lambda/d} = \int_0^\infty \exp[-E - (d/n\lambda)(\pi E)^{-1/2}]dE \quad \text{and} \quad n = 1 \text{ or } 2. \tag{18}$$

Figure 4-8 shows the functional dependencies of this rigorous solution for both degenerate and nondegenerate semiconductors. The excellent agreement between this more complex treatment and the simplified approach leading to Eq. 17 is also shown, with the dots representing the prediction for the simple case.

If the scattering processes from the upper and lower surfaces of the film are different, the contributions from each surface can be averaged to a first-order approximation. Therefore, the parameter p can be replaced by $(p + q)/2$, where p and q are the specular scattering coefficients from the upper and lower film surfaces, respectively. Such differences might be expected because of the different environments in which each film surface is exposed, the relative roughness of the surfaces, and the differences in defect or impurity levels. Therefore, Eq. 16 becomes

$$\mu' = \mu_b/[1 + (2\lambda/d)(1 - (p + q)/2] \tag{19}$$

This somewhat artificial formulation is used to accentuate two film features: (1) the surface influences the transport properties for thinner layers, and (2) the contributions to the scattering from either surface might be considerably different because of physical and chemical influences.

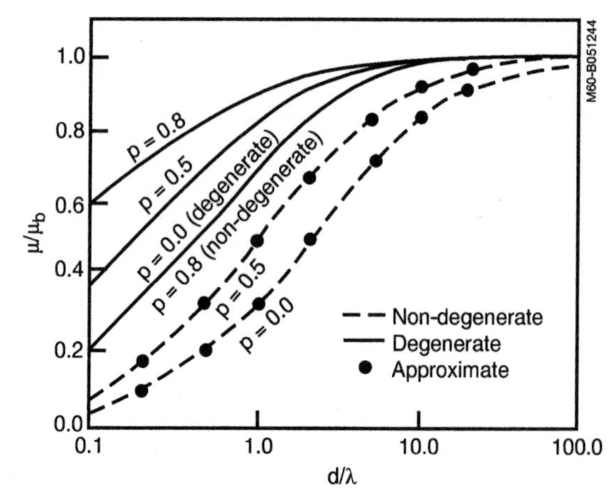

FIGURE 4-8 Ratio of film mobility μ to bulk mobility μ_b as a function of the ratio of the film thickness d to the mean scattering length λ, using both rigorous Fermi solutions (*solid lines*) for degenerate semiconductor and Boltzmann solutions (*broken lines*) for nondegenerate case from Eqs. 17. The effects of specular and diffuse scattering are included in the specularity factor (p), and the dots represent the approximate solution given by Eq. 16. *(From J. C. Anderson,* Advan Phys, *vol. 19, 1970, p. 311.)*

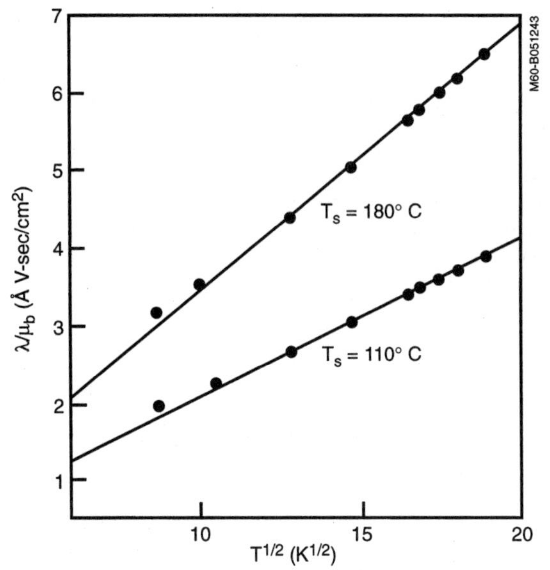

FIGURE 4-9 The ratio of the mean surface scattering length to the bulk mobility as a function of $T^{1/2}$ for substrate temperatures T_s of 383 K and 453 K, confirming relationship of Eq. 22 for these CdS thin films. *(From L. L. Kazmerski,* Thin Solid Films, *vol. 21, 1974, p. 273.)*

Surface Scattering. The parameter λ is a measure of the influence of the surface on the carrier transport. The surface-scattering length is defined in terms of the mean velocity of the carriers, or by [38–40]

$$\lambda = v_z \tau_b \tag{20}$$

The velocity component is measured perpendicular to the film surface over which it is averaged. This mean velocity is expressed as

$$v_z = (kT/2m^*)^{1/2} \tag{21}$$

Combining Eq. 20 with Eq. 21, and noting that $\mu_b = q\tau_b/m^*$, the mean-surface-scattering length is [40]

$$\lambda = (m^*k/2\pi q^2)^{1/2}\, \mu_b T^{1/2} \tag{22}$$

Figure 4-9 shows the dependence of λ/μ_b on temperature, indicating the predicted dependence of Eq. 18 for CdS thin films as an example.

The magnitude of λ is material dependent. For carrier concentrations comparable to the CdS ($\approx 10^{16}/\text{cm}^3$), $\lambda = 1.95\ \mu\text{m}$, $0.85\ \mu\text{m}$, and $0.12\ \mu\text{m}$ for Si, GaAs, and Ge, respectively. At a given film thickness, surface scattering affects the carrier transport in Si far more than GaAs or Ge [41]. For a 1 μm-thick film, with no other considerations, the mobility of Ge is expected to be only 5% less than its bulk value (although that value for Si is about 55% less).

Band Bending: Real Surfaces. Some degree of band bending takes place at real surfaces, and therefore the flat-band assumption made in the previous sections must be modified. The deviations from the constant, bulk-energy-band condition are illustrated in Fig. 4-7. In depletion, the majority carriers are trapped in acceptor-type surface states in an n-type semiconductor (donor-like states in a p-type) at energy levels below (or above) the Fermi level. For an n-type semiconductor, electrons are repelled from the surface region to make it less n-type. The band edges bend up at the surface away from the Fermi level. If the band bending is sufficient, the surface can become p-type, which is the condition of inversion. Analogous conditions exist for a p-type semiconductor. The band bending is equivalent to applying a positive bias to the surface of the n-type semiconductor, or a negative bias to the p-type material. The situation of accumulation is converse to depletion. Donor states at the surface of an n-type semiconductor (acceptor states for a p-type) contribute additional majority carriers to the conductor (or valence bands). The band edges will bend toward the Fermi level. The analogy in this respect is the application of a negative bias to an n-type (or negative to a p-type) semiconductor. Generally, the free surface of an impurity semiconductor is in depletion in the absence of an applied external field and is the condition considered in this chapter.

Surface Space Charge Region. The width of the surface-depletion region, L_c, depends upon the condition of the surface (surface charge, surface potential) and the condition of the bulk semiconductor (impurity doping, Fermi-level position, intrinsic concentration) [41–43]. Several analytical and experimental research efforts have demonstrated the significance of L_c, including those of Waxman et al. in the case of CdS [44]. In this calculation, the change in L_c is predicted to be more than an order of magnitude for a corresponding 0.2 eV change in qV_s. This dependence is further illustrated using a solution of the one-dimensional Poisson's equation [38]:

$$d^2v/dx^2 = -(\rho/\epsilon kT) \tag{23}$$

$$= -(q^2/\epsilon kT)[n_b - p_b + p_b\exp(-qV/kT) - n_b\exp(qV/kT)]$$

where x = directional parameter
n_b and p_b = bulk carrier concentrations
V = the potential barrier (defined as the potential at any point in the space-charge region with respect to the value in the bulk, $V = \varphi - \varphi_b$).

For small perturbations ($|qV/kT| < 0.5$), with the boundary condition that $[d(qV/kT)/dz]z = 0$,

$$dv/dz = \pm F(qV/kT, u_b)/L \tag{24}$$

where L = effective Debye length
F = electric field
$u_b = q\varphi_b/kT$

The negative sign signifies $qV/kT > 0$, and the positive sign, $qV/kT < 0$, and $F(qV/kT, u_b)$ is expressed as

$$F(qV/kT, u_b) = 2^{1/2}\,[\cos h(u_b + qV/kT)/[\cos h(u_b) - v \cdot \tan h(u_b) - 1]^{1/2} \tag{25}$$

The integration of this equation yields the potential profile and is shown as a function of z/L in Fig. 4-10. The flatband condition exists when $qV/kT\,(0) = 0$. Band bending continues until $qV/kT\,(0) = -2u_b$, in which condition the semiconductor is quasi-intrinsic because the minority-carrier density equals the majority-carrier density. As the band bending continues ($|qV/kT\,(0)| > 2u_b$), the total inversion condition is reached (that is the surface undergoes a change in majority-carrier type).

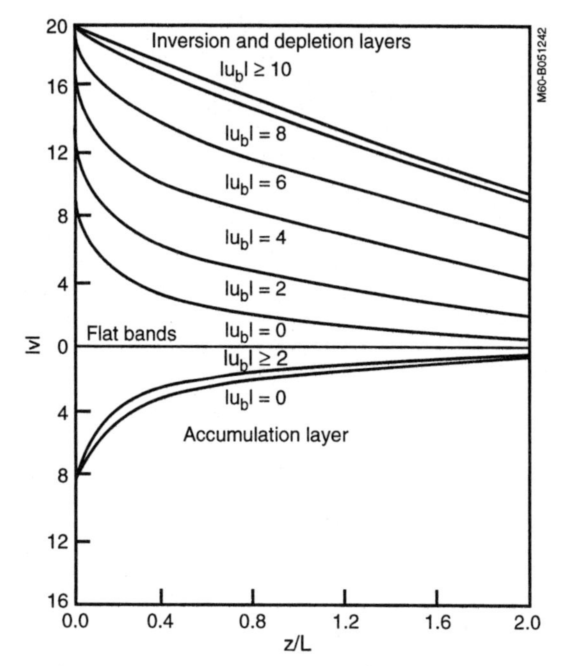

FIGURE 4-10 Near-surface barrier potential shape as a function of the Debye-length normalized distance from the surface for values of the bulk potential. *(From A. Many, Y. Goldstein, and N. B. Grover,* Semiconductor Surfaces, *North-Holland, Amsterdam, 1965, Ch. 2.)*

Surface Transport. Surface-band bending generates an excess or deficiency of mobile carriers within the surface region. These carriers are defined as [38]

$$\Delta N = \int_0^\infty (n - n_b)\, dz \tag{26a}$$

and

$$\Delta P = \int_0^\infty (p - p_b)\, dz \tag{26b}$$

where $p = p_b \exp(-v)$
$n = n_b \exp(v)$
ΔN and ΔP = surface parameters with per-unit-area units.

Eqs. 26*a* and *b* have been solved numerically for accumulation, depletion, and inversion conditions. In turn, these parameters can be related to the carrier transport because ΔN and ΔP relate to a change in the surface conductivity through the relation

$$\Delta\sigma = q(\mu_{ns}\Delta N + \mu_{ps}\Delta P) \tag{27}$$

where ΔN and ΔP = surface area concentrations
μ_{ns} = electron hole surface mobility
μ_{ps} = hole surface mobility.

The surface conductivity change can be measured, although it depends greatly on the nature and magnitude of the surface potential. Although it is straightforward to measure the surface conductance, it is impossible to separate the product $\mu_{ns}\Delta N$ (or $\mu_{ps}\Delta P$) without some further considerations.

The surface mobilities can be calculated and their relationships to the surface predicted. The general case of nonparabolic bands for a nondegenerate and degenerate semiconductor has been solved for either the depletion or accumulation cases [45]:

$$\mu_{ns}/\mu_b = 1 - \lambda n_b/\Delta N\, H_n(v) \text{ (depletion)} \tag{28}$$

and

$$\mu_{ns}/\mu_b = 1 - \lambda p_b/\Delta p\, H_p(v) \text{ (accumulation)} \tag{29}$$

where $H_n(b)$ and $H_p(v)$ are functions that reduce to the Γ functions (Eq. 18), and v is defined as the reduced potential, qV/kT. Figure 4-11 shows the dependence of μ_{ns}/μ_p on surface potential for both degenerate and nondegenerate cases. For the nondegenerate semiconductor, μ_{ns}/μ_p decreases upon an increase in v_s as expected. However, a maximum of μ_{ns}/μ_b is not predicted for the $v_s = 0$ (no band bending) situation. In this limit, μ_{ns} corresponds to the surface mobility with normal diffuse scattering. For the degenerate film, the mobility cusp is predicted by other analytical approaches [27,46–48].

Anderson presented one approach to understand the behavior with the model shown in Fig. 4-12 [49]. The surface region (width L_c) is approximated by two independent mean-scanning times. The first is associated with bound carriers; these carriers are constrained to move in the surface-potential well and to scatter diffusely at the surface ($z = 0$), but specularly at the boundary ($z = L_c$). The second mean-scattering time is associated with unbound electrons, having energies above the well and scattered at the surface under flat-band conditions modified by the surface potential. The mean surface-scattering time is

$$1/\tau_s = 1/\tau_s\text{(bound)} + 1/\tau_s \text{ (unbound)} \tag{30}$$

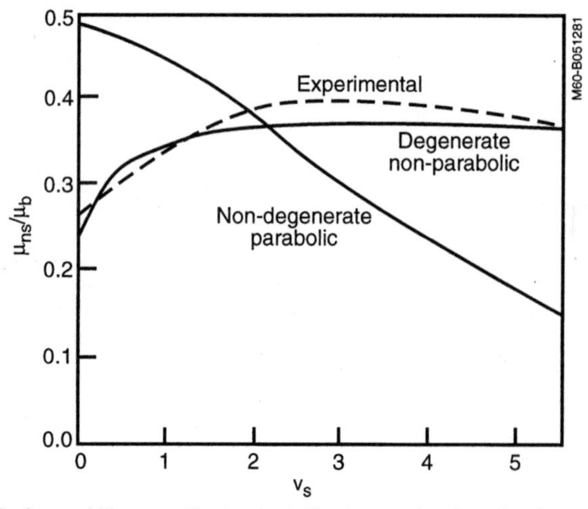

FIGURE 4-11 Surface mobility normalized to the bulk value as a function of surface potential for degenerate and nondegenerate semiconductors. Data for degenerate InSb thin film are shown by the dashed line. *(From C. Juhasz and J. C. Anderson,* Radio Electron. Eng. *vol. 33, 1967, p. 223.)*

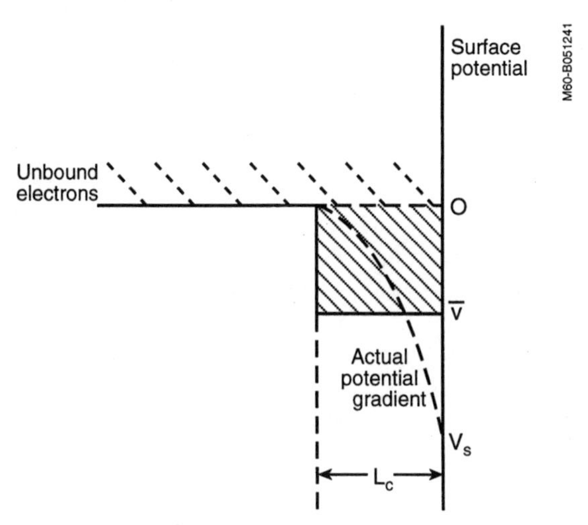

FIGURE 4-12 Anderson potential well model for evaluating surface scattering and surface transport. *(From J. C. Anderson,* Advan Phys, *vol. 19, 1970, p. 311.)*

which can be written

$$\frac{1}{\tau_s} = \frac{(1 - p)\lambda(1 + v_s)^{1/2}}{L_c\tau_b} + \frac{(1 - p)\lambda(1 + v_s)^{1/2}}{d\tau_b} \tag{31}$$

With this modified surface-scattering time, the effective mobility (Eqs. 28 and 29) becomes

$$\frac{\mu'}{\mu_b} = \{1 + (1 - p)\lambda[(1/L) + (2/d)]\,[1 + qV_s/kT]^{1/2}\}^{-1} \text{ (accumulation)} \tag{32}$$

and

$$\frac{\mu'}{\mu_b} = \frac{1 + (1 - p)(2\lambda/d)[1 - \exp(qV_s/kT]}{1 + (1 - p)(2\lambda/d)} \text{ (depletion)} \tag{33}$$

These equations are consistent with the flat-band situation. Because as $V_s \Rightarrow 0$, and $L_c \Rightarrow \infty$, these relations become identical to Eq. 16. Figure 4-13 shows the effects of the surface potential on the mobility.

Galvanomagnetic Effects. The common experimental approaches to determine transport parameters use magnetic fields (Hall effect and Van der Pauw). The measured carrier concentrations are surface scattering and thickness dependent when the number of carriers being scattered from the surface is much greater than the number not being scattered. Amith used the relationship of this scattering mechanism and the magnitude of the Hall constant to solve the Boltzmann equation for an extrinsic semiconductor with an additional drift field resulting from the applied magnetic field [50]. The result indicates that the effective Hall constant (R'_H) is related to the crystalline value (R_{Hb}) through the expression

$$R'_H = \eta(\lambda/d)R_{\mathrm{Hb}} \tag{34}$$

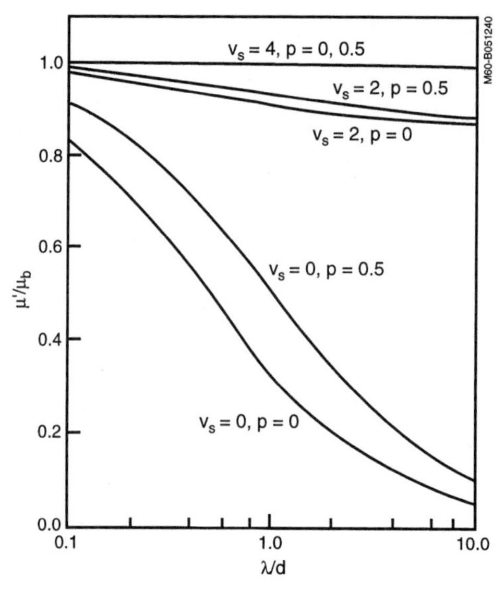

FIGURE 4-13 Ratio of film mobility (μ') to bulk mobility (μ) dependence on film thickness normalized surface scattering length λ for various surface potentials ($v_s = qV_s/kT$), using the nondegenerate model of Eq. 33. Specular and diffuse scattering effects are represented.

where the solution of the Boltzmann equation yields

$$\eta(\lambda/d) = \frac{1 - 4\lambda/d + 4(\lambda/d)\Gamma_1(\lambda/d) + \Gamma_3(\lambda/d)}{\{1 - 2\lambda/d + (2\lambda/d)\Gamma_1(\lambda/d)\}^2} \tag{35}$$

where $\Gamma_1(\lambda/d)$ is defined in Eq. 35, and $\Gamma_3(\lambda/d) = \int_0^\infty (E\pi)^{-1/2}\,\Gamma_1(\lambda/d)dE$. Values of Γ_1 and Γ_3 are evaluated by numerical integration, and the general dependence of $\eta(\lambda/d)$ on the mean-scattering length and thickness is shown in Fig. 4-14.

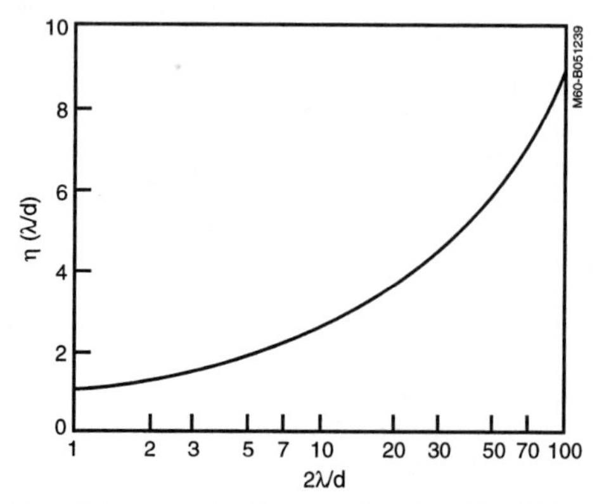

FIGURE 4-14 The Hall coefficient correction factor as a function of the thickness-normalized mean surface scattering lengths. *(From A. Smith,* J Phys Chem Solids, *vol. 14, 1960, p. 271.)*

Zemel et al. evaluated the relative mobilities for a semiconductor film with a surface space-charge region with and without an applied magnetic field [51–54]. With no magnetic field

$$\mu'/\mu_b = 1 - \exp(\alpha^2)\text{erfc}(\alpha) \tag{36}$$

where $\alpha = \lambda_c(2mkT)^{1/2}/q\tau V_s$. The application of a magnetic field perturbs the surface potential, and

$$\mu'/\mu_b|_{RH} = [1 - 2\alpha\pi^{-1/2} - (1 - 2\alpha^2)\exp(\alpha^2)\text{erfc}(\alpha)]^{1/2} \tag{37}$$

Figure 4-15 compares Eqs. 36 and 37, and the difference in mobilities is very small. For minimal band bending, V_s approaches zero and α becomes large. As expected, $\mu' = \mu_b = \mu'|_{RH}$ for this flat-band condition. As the surface potential well gets larger (that is V_s becomes large and surface band bending exists), α approaches zero. In this situation, $(\mu'/\mu_b)|_{a \Rightarrow 0} = 2\alpha p^{-1/2}$, and $(\mu'/\mu_b)R_H|_{\alpha c \Rightarrow \infty} = \alpha$. The mobilities differ by only about 12% under this extreme condition.

Measurement of resistive and Hall coefficients is necessary to determine both the mobility and carrier concentrations in thin films. However, errors in measurements can arise from specimen contours (electrode size, geometry, and position), electron symmetry, and spatial and thickness variations. These potential sources of error for galvanomagnetic measurement have been reviewed in the literature [55–58].

4.2.3. Polycrystalline Films

Most films grown by physical vapor deposition, sputtering, or evaporation do not have the perfect (defect-free) layers described in the previous sections. Even epitaxial films have defect concentrations (point defects and dislocations) that can affect carrier transport. Polycrystalline thin films are common in many technologies, from integrated circuits to solar cells. In these examples, the conduction mechanism is dominated by the inherent intercrystalline rather than the intracrystalline characteristics. Building on the results of the previous section, we consider defects in the films in terms of their effect on the carrier

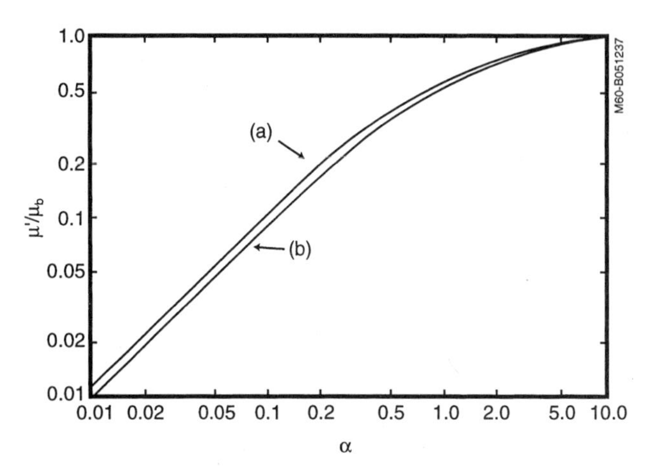

FIGURE 4-15 The ratio of the film mobility μ' to the bulk mobility μ_b as a function of the parameter α (equal to $\lambda_c(2mkT)^{1/2}q\tau V_s$) for (*a*) the effective conductivity mobility from Eq. 36, and (*b*) the effective Hall mobility from Eq. 37. *(From C. Anderson,* Advan Phys, *vol. 19, 1970, p. 311.)*

transport. This section focuses on the relationship between the defects found in polycrystalline semiconductor films and the resulting electrical characteristics. The intercrystalline boundaries (grain boundaries) are emphasized because they are the dominant and most interesting complexities of these films. To provide insight, modeling methods that can predict behavior accurately only under a specific set of circumstances are described. The evolution of various general and pioneering models (Volger, Petritz, and Berger) are detailed and compared. Compound semiconductor films are different from elemental ones and are considered separately. The combined effects of surface scattering with the defect-dominated phenomena are presented.

Boundary Scattering: Initial Concepts. The most common representation of a polycrystalline thin film has nearly regular, columnar, and cylindrical grain structure. This geometry has been observed in many thin film systems. For several films, deviations are encountered. For example, grains are many times smaller near their initial growth base than at the top surface. The columnar growth pattern can be interrupted, and grain boundaries can exist along the z-direction of the film. One rule of thumb is that for most physical deposition processes, the grain cross section can be about the same as the film thickness for as-deposited layers. Modeling generalizes most of these observed physical properties of real polycrystalline films to predict electro-optical behavior.

The simplest approach considers a film's grain boundaries with a columnar structure to have the major, controlling effect on the carrier transport from grain to grain [59]. Therefore, the carriers collide at and with the boundary interruptions, and have an effective mean-free path λ_g and a mean relaxation time τ_g in steady state. The mobility associated with this mechanism can be expressed as

$$\mu_g = q\tau_g/m^* \qquad (38)$$
$$= (q\lambda_g)/m^*v$$

where v is the mean thermal carrier velocity. For a nondegenerate semiconductor

$$\mu_g = q\lambda_g[(9\pi/8)m_i^*kT]^{-1/2} \qquad (39)$$

and for the degenerate semiconductor case

$$\mu_g = (q\lambda_g/h)(3\pi^2 n)^{-1/3}(m_d^*/m_i^*)^{1/2} \qquad (40)$$

where n = free-carrier density
m_d^* = the density of state's effective mass
m_i^* = inertial or conductivity mass of the carrier.

The temperature sensitivities in Eqs. 39 and 40 depend on the semiconductor material. If the bandgap is small, the temperature sensitivities of m_i^* and m_d^* are enhanced, and the mobility of a nondegenerate semiconductor is more likely to deviate from the $T^{-1/2}$ characteristic for a low-bandgap. On the other hand, the temperature sensitivity of the degenerate semiconductor mobility depends on the ratio of the effective masses. Because the temperature dependence of each of these is approximately the same, only small variations in μ_g are expected with T. For nondegenerate and degenerate semiconductors having larger bandgaps, the masses are relatively temperature independent. Thus, μ_g follows the $T^{-1/2}$ dependence for nondegenerate semiconductors and is independent of T for degenerate semiconductors. These generalizations have been supported for various semiconductors, including III-V, binary, and ternary semiconductors, and have been reported in the literature [60–83].

Grain Boundary Potential Barrier Models. The simplest modeling of the polycrystalline semiconductor correlates the transport properties with the effect of the space-charge regions at the intragrain (or grain boundary) interfaces. At these regions, band bending results in potential barriers to the charge transport, as shown in Fig. 4-16 [84–88]. In general, these barriers reduce the conductivity of the semiconductors as compared to those of single-crystal analogues. This section considers the evolution of models to account for the carrier transport mechanisms.

Transport Models: Evolution and Mechanisms. Among the earliest analytical approaches evaluating the conductivity of polycrystalline semiconductor films is the Volger model [89]. This approach considers an inhomogeneous conductor consisting of series-connected, separately homogeneous domains of high conductivity and low conductivity in which no space-charge regions exist. The width of the low-conductivity (grain boundary) regions is negligible with respect to the high-conductivity (grain) domains. The polycrystalline semiconductor is simulated by ohmic transport of the carriers. With I_1 as the grain diameter and I_2, the grain boundary width, Volger derived the Hall coefficient based primarily on geometric factors.

$$R_H = R_{H1} + c(I_2/I_1)^2 R_{H2} \tag{41}$$

where c is an unspecified constant, and R_{H1} and R_{H2} are the Hall coefficients in the grain and boundary regions, respectively. This is related to the mobility

$$\mu_g = \mu_1\{[1 + (I_2/I_1)\exp(q\varphi_b/kT)]^{-1} + (I_2/I_1)\} \tag{42}$$

where μ_1 is the grain mobility, and φ_b is the barrier potential (a function of the grain carrier concentrations).

One of the pioneering and more frequently cited analyses of transport mechanisms in polycrystalline thin films was presented by Petritz [90]. The emphasis of these modeling efforts was to develop a theory of photoconductivity in compound, polycrystalline semiconductors, but the straightforward modeling of the transport applies to a more general framework. The Petritz model differs from the earlier Volger approach in that it is based on the thermionic emission of carriers from grain to grain. A major postulation of the Petritz approach is the averaging of parameters over many grains using an analysis based on a single grain and single-grain boundary or barrier region. The total resistivity for these two regions is written as

$$\rho_g = \rho_1 + \rho_2 \tag{43}$$

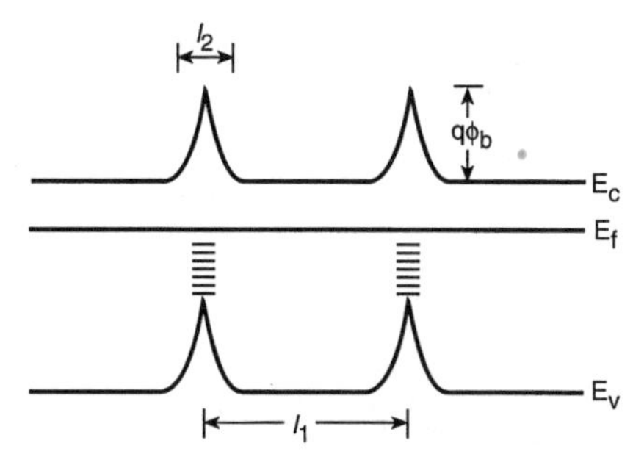

FIGURE 4-16 Energy band representation of an *n*-type polycrystalline semiconductor thin film with grain size I_1, grain boundary width I_2, and grain boundary barrier potential $q\phi_b$.

where the subscripts 1 and 2 refer to the grain (sometimes termed crystallite) and boundary, respectively. If it is assumed that $\rho_2 \gg \rho_1$, then the current density-voltage relationship can be written analogous to simple diode theory as

$$j = Mn_1 \exp(-q\varphi_b/kT)[\exp(-qV_b/kT) - 1] \tag{44}$$

where n_1 = mean majority carrier density in the grains
φ_b = potential height of the barriers
V_b = voltage drop across the barrier
M = a factor that is barrier dependent, but independent of φ_b.

For a film having many such barriers, the voltage drop across any one barrier is small compared to kT/q, and Eq. 44 can be rewritten as

$$j = Mn \exp(-q\varphi_b/kT)\ [qV_b/kT] \tag{45}$$

If V is the total voltage drop across the film, and there are n_c grains per unit length along the film (having a total length L), then

$$j = [qm_o n \exp(-q\varphi_b/kT)] \tag{46}$$

where $\mu_o = M/n_c kT$, E is the electric field, and the quantity in the brackets is the conductivity. The exponential term in Eq. 46 provides the essential characterization of the barrier. Petritz interpreted the case of this barrier. With $\rho_1 \ll \rho_2$, the carrier concentration is not reduced by the exponential factor; instead, all carriers take part in the conduction process, but with reduced mobility. That is, with $\mu_o = \mu_b(T)$

$$\mu_g = \mu_b \exp(-q\ \varphi_b/kT) \tag{47}$$

where μ_b = $C\mu_{cryst}$
C = a constant
μ_{cryst} = the perfect crystallite value of the mobility [39].

Therefore, μ_b is the bulk representation of the grain or crystallite mobility, and it includes contributions from inherent defects or impurities. The expression in Eq. 47 encompasses significant scattering events within the grain. The general temperature dependence of the mobility predicted by the Petritz model is shown in Fig. 4-17 for p- and n-type $CuInSe_2$ films having various grain boundary activation energies.

This model was extended to other transport parameters, and the Berger model [91,92] demonstrates that the Hall coefficient and carrier concentrations can also exhibit the exponential dependencies. That is

$$R_H = R_b \exp(-E_n/kT) \tag{48}$$

where the magnitude of the activation energy E_n depends on the relative concentrations in the grain and the boundary region. Another step was provided by Mankarious and others [93–97], who observed that the conductivity term can be written in a similar and more general fashion as

$$\sigma_g \propto \exp(-E_\sigma/kT) \tag{49}$$

where E_σ is the conductivity activation energy. From the relationships between the conductivity and the carrier concentration, the n and p relationships can be derived (consistent with the Berger model) according to the relations

$$n \propto \exp(-E_n/kT) \tag{50}$$

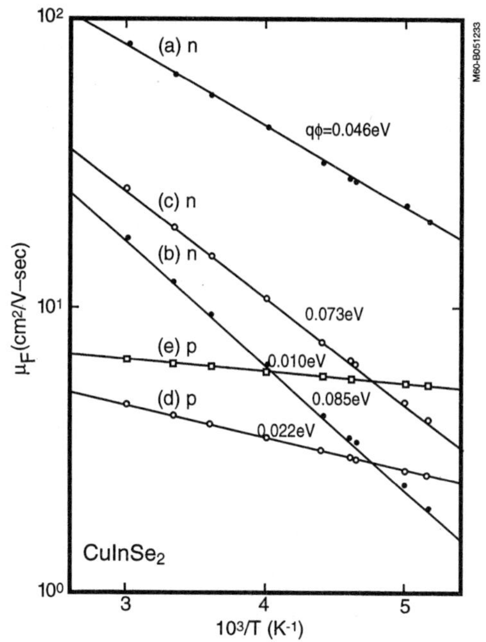

FIGURE 4-17 Hall mobility dependence on inverse temperature for $CuInSe_2$ films deposited at various substrate temperatures: (*a*) 500°C, annealed 15 min in Ar; (*b*) as-deposited film, process as in (*a*); (*c*) 520°C, as-deposited; (*d*) 520°C, annealed 15 min in Ar/H_2Se; (e) 520°C annealed 80 min in Ar/H_2Se. Annealing temperature is 427°C. Slopes are used to determine the barrier heights from the Petriz model. *(From L. L. Kazmerski, M. S. Ayyagari, and G. A. Sanborn,* J Appl Phys, *vol. 46, 1975, p. 4685.)*

and

$$p \propto \exp(-E_p/kT) \tag{51}$$

where E_n and E_p are the carrier activation energies for n-type and p-type films, respectively. Since $\sigma = n$ (or $p)q\mu_{n(\text{or } p)}$, the model predicts

$$E_\sigma = E_{n(\text{or } p)} + q\varphi_b \tag{52}$$

That is, for the Petritz model, with $\rho_1 \ll \rho_2$, the mobility activation energy is identical to the conductivity activation energy, and R_H is constant. Figures 4-18*a* and *b* present one verification of this activation-energy interdependence for CdS and $CuInTe_2$ thin films, respectively.

The Berger model expands Petritz's further, using a more general and precise representation of the polycrystalline film. The grain size and resistivity are represented by l_1 and ρ_1, respectively, and the boundary parameters by l_2 and ρ_2. Similar to the Volger approach, the resistivity can be written as

$$\rho_g = \rho_1 + (l_2/l_1)\rho_2 \tag{53}$$

and the effective mobility can be similarly derived:

$$\mu_g = \mu_1/[1 + c\,\exp(-\Delta E/kT)] \tag{54}$$

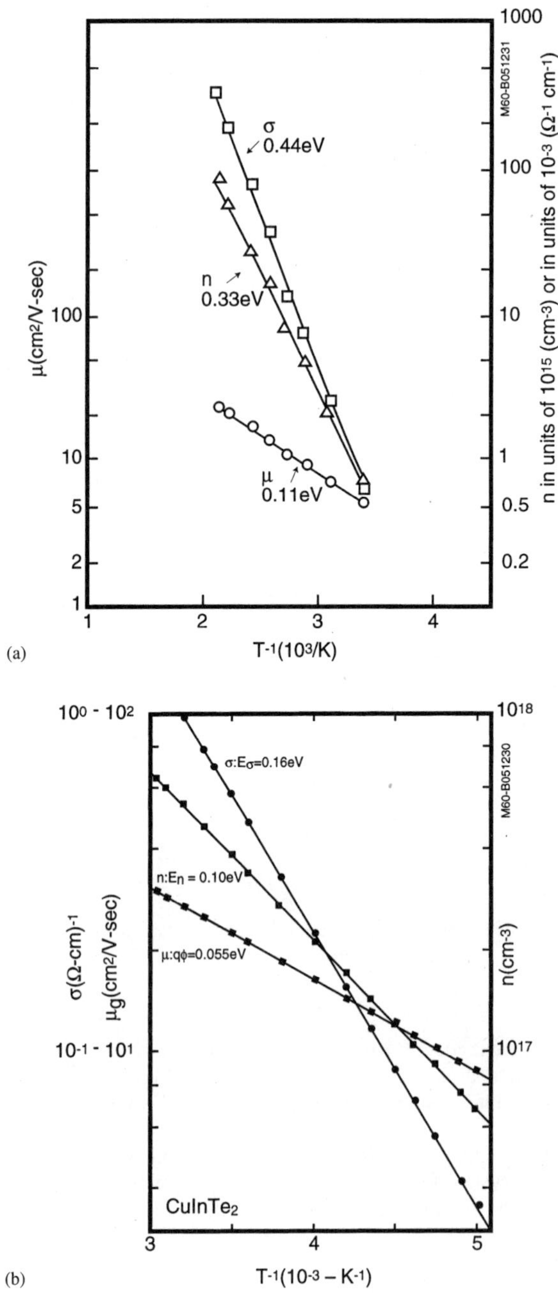

FIGURE 4-18 Verification of the interdependence of the grain boundary, conductivity, and carrier concentration activation energies for (*a*) CdS thin film *(from R. G. Mankarious,* Solid State Electron, *vol. 7, 1964, p. 702)* and (b) $CuInTe_2$ thin film predicted by various thin film models *(from L. L. Kazmerski et al.,* J Vac Sci Technol, *vol. 13, 1976, p. 769).*

where $c = (I_2/I_1)(n_{o1}/n_{o2})$
n_{o1} and n_{o2} = the characteristic carrier concentrations at infinite temperature in the grain and boundary regions
μ_1 = grain region mobility.

Because the magnitude of $[c \exp(-\Delta E/kT)]$ is usually $\gg 1$, the mobility can be written in a format similar to that derived by Petritz:

$$\mu_g = \mu_1[(I_2/I_l)(n_{o2}/n_{o1}) \exp(-\Delta E/kT)] \tag{55a}$$

Comparison of Eqs. 47 and 55*a* leads to two interpretations of the activation energies and the values of μ_o and μ_1. The limiting cases for these two models is

$$\mu_o = \mu_1(I_2/I_1)(n_{o2}/n_{o1}), \tag{55b}$$

and μ_1 is the mobility of the bulk, single-crystal semiconductor. If the grain diameter is large enough (that is greater than the mean-free path of the carriers), scattering within the grain becomes important. In this case, Eq. 55*b* reduces to $\mu_o = \mu_1 = \mu_b$, and Eq. 55 is satisfied. It should be emphasized that all of these models are based on assumptions that must be interpreted and applied to the particular case and that several demonstrations have been made [98]. These demonstrations are not universal and general models, and therefore, the boundary conditions and postulations must be recognized.

Potential Barriers, Barrier Heights, and Activation Energies. The activation energies used in the various transport models are comparable. For the Volger and Petritz models, the barrier heights φ_b [89,90] are of the form

$$\varphi_b = kT\ln(n_1/n_2) \tag{56}$$

where n_1 and n_2 are the carrier concentrations in the grain and grain boundary regions, respectively. By comparison, Berger represented the same barrier as

$$\Delta E = E_1 - E_2. \tag{57}$$

This difference is determined by

$$\Delta E = \ln[(N_A/N_D)|_{\text{grain boundary}} - (N_A/N_D)|_{\text{grain}}] \tag{58}$$

where N_A and N_D are the acceptor and donor densities, respectively.

Noting that $n_1 = n_{o1}\exp(-E_1/kT)$ and $n_2 = n_{o2}\exp(-E_2/kT)$ for Eq. 56 are more general, this equation becomes

$$q\varphi = kT\ln(n_{o1}/n_{o2}) + (E_2 - E_1) \tag{59}$$

Finally, substituting this result into Eq. 47 leads to

$$\mu_g \propto (n_{o1}/n_{o2}) \exp[-(E_1 - E_2)/kT] \tag{60}$$

which is consistent with the Berger representation.

4.2.4 Elemental Semiconductors

The inherent characteristics of elemental polycrystalline semiconductor films are developed through the incorporation of impurity atoms rather than through compositional or stoichiometric effects [99–127]. The modeling and interpretation of transport mechanisms are made through early models (segregation and grain boundary trapping) and through the Seto model [133,134]. Assumptions are identified for each of these models to ensure that proper application of the analytical result is clearly understood. In this section, the extent and limitations of theses models are discussed, especially in terms of a wealth of experimental results on Si-polycrystalline thin films. The *segregation model* is based on the concept of segregation of impurities, which become electrically inactive at the boundary region [128,129]. This model fails to account for the temperature dependence of the resistivity and predicts a negative temperature coefficient of resistance—both of which are not observed experimentally. The *grain boundary trapping model,* on the other hand, is based on active trapping sites at the grain boundary that capture free carriers [130–133]. These charge states become potential barriers (analogous to the Berger and Petritz models) which limit the transport of carriers among grains. The mechanism is dominated by thermionic emission. It is this model that is most applicable to the case of polycrystalline elemental semiconductors.

The evolution of the potential at the grain boundary is, in itself, an interesting phenomenon. The grain boundary barriers are formed when the boundary region has a lower electrochemical potential for minority carriers than the grains, which provides for the influx of electrons or holes. A space-charge region is thereby formed that inhibits further redistribution of the carriers. The Fermi level is located near the center of the bandgap if the barriers are formed in both n- and p-types of a semiconductor. This is the case for Si [130–132]. In general, the Fermi level is not in the midgap region, and potential barriers can be created in only one majority carrier type of the semiconductor. This is the case for Ge, in which grain boundary potential barriers form only in n-type material [134–137].

Grain-Boundary Trapping. Seto provided the most comprehensive analysis of grain boundary trapping based on the physical, charge, and energy-band structures [133–134]. This model is based on the following assumptions:

- Monovalent trapping is the dominant mechanism (that is only one type of impurity atom is present).
- The impurity is distributed uniformly with a concentration of N/cm^3.
- The grains are identical, with a cross section σ.
- The grain boundary thickness is negligible and has Q_t/cm^2 traps located at energy E_t with respect to the intrinsic level, E_i.
- The traps are initially neutral and become charged by trapping a free carrier. Therefore, all mobile carriers which are in a region $l/2 - d$ (Fig. 4-19) from the grain boundary are trapped, forming a space charge region. Mobile carriers are neglected in this region.

The solution of the one-dimensional Poisson equation for a p-type semiconductor yields

$$V(x) = (qN/2\epsilon)(x - \delta)^2 + V_{vo} \qquad \delta < |x| < l/2 \tag{61}$$

where V_{vo} is the potential at the valence band edge. Two possible conditions exist for the trap densities and these conditions depend on the doping or impurity concentrations. First, for $Q_t > lN$, the grain is completely depleted of carriers and the traps are partially filled. The potential is obtained for $\delta = 0$ from Eq. 61. Because $p(x) = N_v \exp[-qV(x) - E_f/kT]$, the average concentration is obtained by integration over the region $-l/2 < x < l/2$, or

$$p_{\text{av}} = (N_v/ql)(2\pi\epsilon kT/N)^{1/2} \exp[(E_b + E_f)/kT] \operatorname{erf}[ql(N/8\epsilon kT)^{1/2}] \tag{62}$$

where n_i is the intrinsic concentration of the single crystal (grain) and $E_b = q\varphi_b = q(ql^2N/8\epsilon)$.

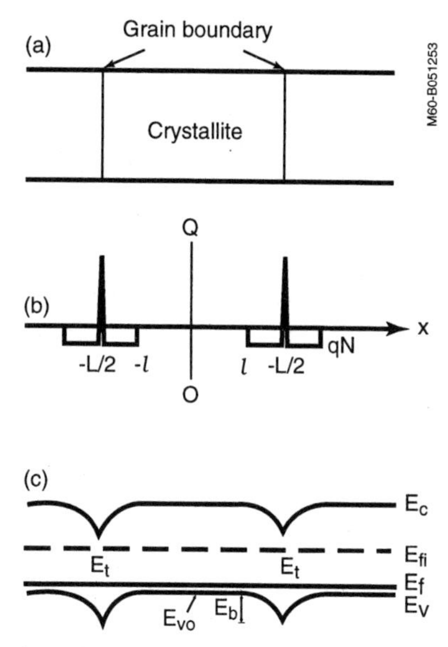

FIGURE 4-19 Model for electronic structure of a polycrystalline film region in the vicinity of a single grain: (*a*) generalized geometry, (*b*) charge distribution, and (*c*) energy band representation.

Second, $Q_t < lN$, only a portion of the grain is depleted of carriers and $\delta > 0$. In this case,

$$p_{av} = p_b[(1 - Q_t/lN) + (1/q)(2\pi\epsilon kT/N)^{1/2}]\ \mathrm{erf}[(qQ_t/2)(2\epsilon kTN)^{-1/2}] \qquad (63)$$

where $p_b = N_v \exp[(E_{vo} - E_f/kT]$ (analogous to doped single-crystal Si).

The variation of the doping concentration and trap density adjusts or changes the barrier height, $q\varphi_b$. This interrelationship is shown in Fig. 4-20. The variation results from a dipole layer that is created by impurities and by trap filling. As the impurity concentration increases, so does the strength of the dipole region. After filling all the traps, both the width of the dipole layer and the magnitude of the barrier height decrease, while the total charge

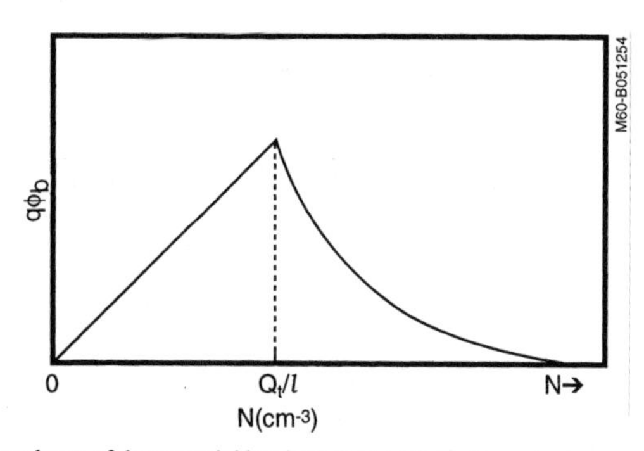

FIGURE 4-20 Dependence of the potential barrier height on impurity or doping concentration for the Seto model. *(From J. Y. W. Seto,* J Appl Phys, *vol. 46, 1975, p. 5247.)*

in the region remains constant. By using Petritz's thermionic emission current, the conductivity can be determined by

$$j = qp_{av}(kT/2\pi m^*)^{1/2} \exp(-q\ \varphi_b/kT)[\exp(qV_B/kT) - 1] \quad (64)$$

where V_B is the voltage applied across the grain boundary. For small $V_B(\ll kT/q)$, the conductivity (from Eq. 64) is

$$\sigma_g = q^2/p_{av}(2\pi m^* kT)^{-1/2} \exp(-q\varphi_b/kT) \quad (65)$$

Two solutions are possible, corresponding to the different doping-density regimes. Combining Eqs. 62 and 63 with Eq. 65, the conductivity becomes

$$\sigma_g \sim \exp[-(E_g/2 - E_f/kT)^{-1/2} \qquad \text{for } Q > lN \quad (66)$$

and

$$\sigma_g \sim T^{-1/2} \exp(-q\varphi_b/kT) \qquad \text{for } Q < lN \quad (67)$$

In either case, the effective mobility (from Eq. 65) is

$$\mu_g = q/(2\pi m^* kT)^{-1/2} \exp(-q\varphi_b/kT) \quad (68)$$

Seto provided the experimental validation of these conductivity relationships for polycrystalline Si as shown in Fig. 4-21 [133]. The straight line dependences, as well as the magnitudes of the slopes, are verified. Figure 4-22 presents the mobility dependence on inverse temperature. As predicted, a straight-line dependence is observed, with a slope $(-q\varphi_b)$ and a deviation when $q\varphi_b < kT$. The Seto model accounts for the Hall mobility, resistivity, and carrier concentration dependence on doping concentration for polycrystalline Si films, as shown in Figs. 4-23, 4-24, and 4-25. In each case, the model is represented by a solid line, demonstrating the excellent correspondence.

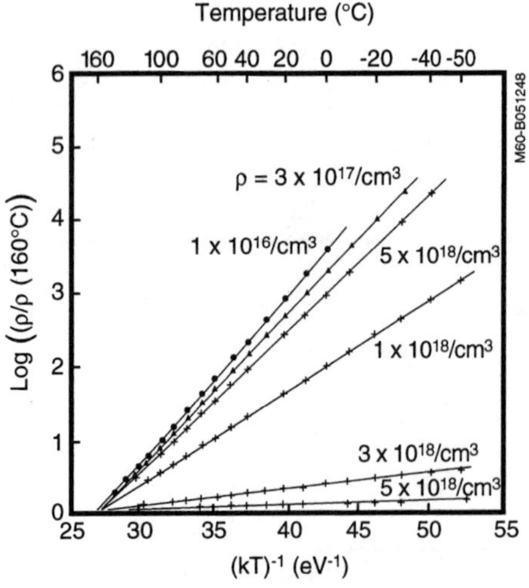

FIGURE 4-21 Dependence of the temperature (160°C) normalized resisitivity on inverse temperature for Si films having different doping concentrations. *(From J. Y. W. Seto,* J Appl Phys, *vol. 46, 1975, p. 5247.)*

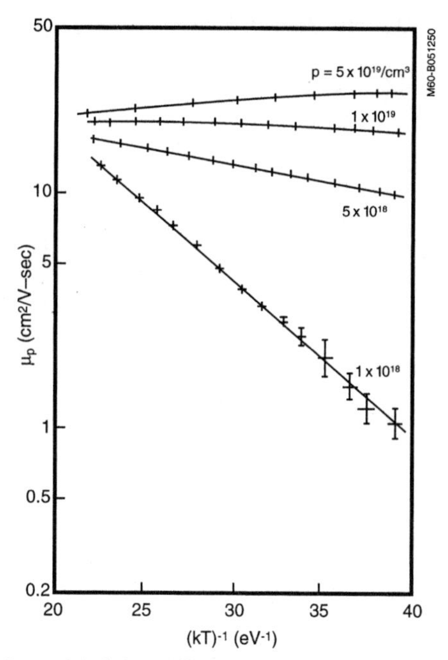

FIGURE 4-22 Dependence of the hole mobility μ_p on inverse temperature for various doping concentrations in polycrystalline Si. *(From J. Y. W. Seto,* J Appl Phys, *vol. 46, 1975, p. 5247.)*

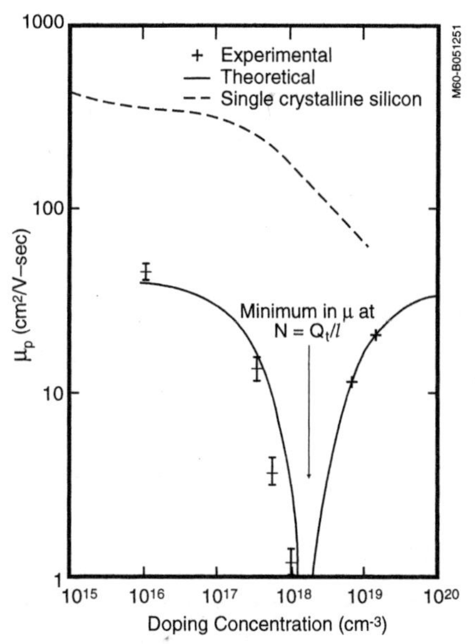

FIGURE 4-23 Room-temperature Hall mobility as a function of doping concentration for *p*-type polycrystalline Si, comparing experimental results with the Seto model. The broken line is for single-crystal Si. *(From J. Y. W. Seto,* J Appl Phys, *vol. 46, 1975, p. 5247.)*

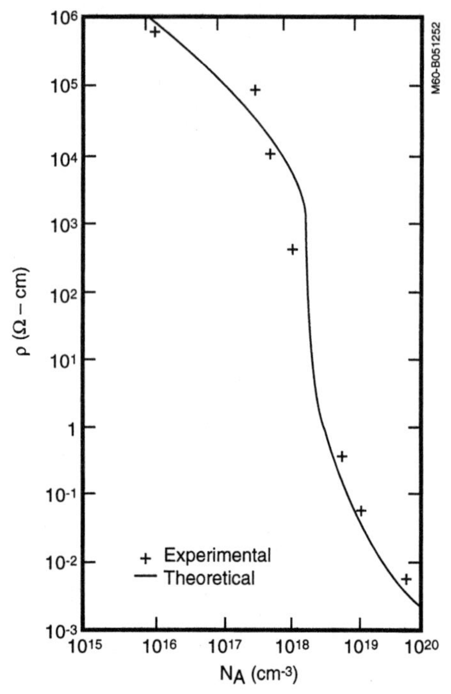

FIGURE 4-24 Room-temperature resistivity as a function of doping concentration for polycrystalline Si, comparing experimental results with the Seto model. *(From J. Y. W. Seto,* J Appl Phys, *vol. 46, 1975, p. 5247.)*

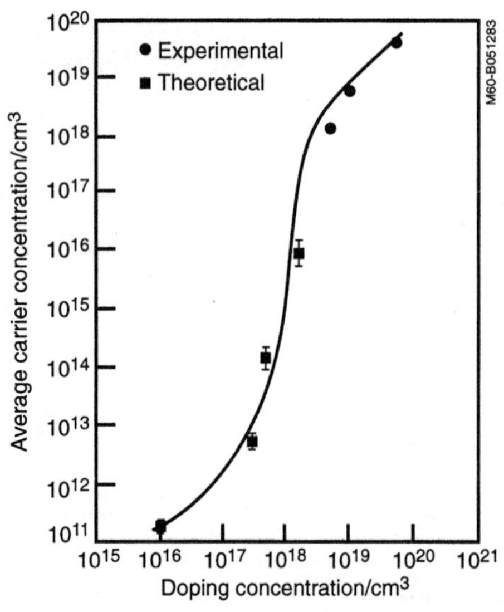

FIGURE 4-25 Room-temperature carrier concentration dependence on doping concentration for polycrystalline Si, comparing experimental results with the Seto model. *(From J. Y. W. Seto,* J Appl Phys, *vol. 46, 1975, p. 5247.)*

Model Limitations and Refinements. The Seto grain boundary-trapping model has several significant limitations [132,133,138–140]. These limitations have led to some alterations in the model to address specific cases. The constraints include:

- *Grain resistivity.* In the Seto model, the resistivity of the grain is assumed to be insignificant compared to the resistivity of the boundary region itself. However, for large grain sizes and high doping levels, the grain resistivity should be considered. Kamins, for example, treated this case for polycrystalline Si doped in excess of $7 \times 10^{18}/cm^3$ [130,141–145].
- *Discrete versus distributed energy states.* It is possible to have trapping states with energies distributed over some energy range as opposed to the fixed value assumed in the Seto model. Such distributions have been reported for the Si surface [146], as well as at interfaces (such as, SiO_2/Si) [147]. This influences the mobility, the carrier concentration, and the activation energy, especially if $N < Q_{tl}$ (Fig. 4-20).
- *Free-carrier density in depletion region.* In selected cases, such as large-grain polycrystalline Si), the carrier concentration in the space charge layer can be appreciable, leading to inaccurate values of the barrier heights calculated through the Seto approach [148]. Significant variance can occur because μ_g depends exponentially on the barrier height. On the other hand, the carrier concentration is affected less by changes in the magnitude of the barrier height and more by the shape of the barrier potential.
- *Energy level of interface states.* The midgap position of the trapping state is excluded in the Seto considerations. The trapping states are located in either the upper or lower portion of the bandgap [140,142].
- *Population of traps.* The major assumption of the Seto model is that the grain boundary traps in the energy gap are always filled. This is not always the case, and examples have been reported in the literature.

Modifications to the Seto model have been reported. The work of Baccarani et al. clarifies the model for the intermediate range of impurity concentrations [140]. Two cases are considered in this alteration: monovalent trapping states and continuous distribution of trapping states within the energy gap. These complex modifications to the Seto model provide further insight into the relationship between the transport in these polycrystalline films and the fundamental film parameters.

Monovalent Trapping States at the Grain Boundary. Baccarani et al. assume the existence of N_t acceptor states in an n-type semiconductor, each state at energy E_t with respect to E_i at the interface [140]. For a given set of $\mathcal{G}$, N_t, and E_t, there is an impurity concentration N_D' such that if $N_D < N_D'$, the grains are completely depleted. Two possible impurity conditions exist.

First, for *complete depletion* of the grains (that is $N_D < N_D'$), the energy barrier is of the form

$$E_b = q^2\mathcal{G}^2N_D/8\epsilon \tag{69}$$

and the conductivity becomes

$$\sigma_g = \frac{q^2\mathcal{G}^2N_cN_Dv}{2kT(N_t - \mathcal{G}N_D)} \exp(-E_a/kT) \tag{70}$$

where $v = (kT/2\pi m^*)^{1/2}$ and the activation energy is $E_a = (E_g/2) - E_t$.

Second, for *incomplete depletion* of the grains (that is $N_D > N_D'$), the energy barrier is of the form

$$E_b = E_f - E_i + kT \ln[(qN_t/2\epsilon N_DE_b)^{1/2} - 1)]. \tag{71}$$

The solutions to Eq. 71 must be obtained by iterative techniques. However, the conductivity can be evaluated according to two specific energy regions:

$$\sigma_g = (q^2 \vartheta N_c^2 v n_o / kT) \exp(-E_a/kT) \qquad \text{for } E_f - E_t - E_b \gg kT \text{ with } E_a = E_b \qquad (72a)$$

and

$$\sigma_g = qN_c^2 v(2\epsilon N_D^{-1} E_b)^{1/2}(kTN_t)^{-1} \exp(-E_a/kT) \qquad \text{for } E_t + E_b - E_f \gg kT \text{ with } E_a = E_g/2 - E_t \qquad (72b)$$

where n_o is the concentration in the neutral region, neglected in the Seto model. The energy region of Eq. 72 is also not specified in the Seto treatment.

The calculated activation energy as a function of the donor-doping level is shown in Fig. 4-26 for selected trap densities (for an assumed grain size of 10 μm). The *broken line* indicates the boundary between the two energy regions and their respective solutions (Eqs. 71 and 72*a,b*). For lower values of N_t, an abrupt transition in E_a occurs between $E_g/2$ and E_b at the onset of complete depletion. However, for larger N_t, E_a becomes more or less continuous. For either incomplete or complete depletion, E_a approaches $E_g/2 - E_t$ for the conditions leading to Eq. 71.

Continuous Trapping State Distribution. In the comparative situation of a continuous energy distribution of interface states, with N_i expressed in units of cm^{-2}/eV^{-1}, donors are assumed to be uniformly distributed in the lower half of the energy bandgap and the acceptors in the upper half. Baccarani et al. solved the charge-neutrality equation for this case to determine the effective depletion region width and the position of the Fermi level in terms of the fundamental parameters.

For complete depletion of the grains ($N_D < N_D'$), the conductivity is

$$\sigma_g = (q^2 \vartheta N_c v/kT) \exp(-E_a/kT) \qquad \text{with } E_a = E_g/2 - \vartheta N_D/N_i \qquad (73)$$

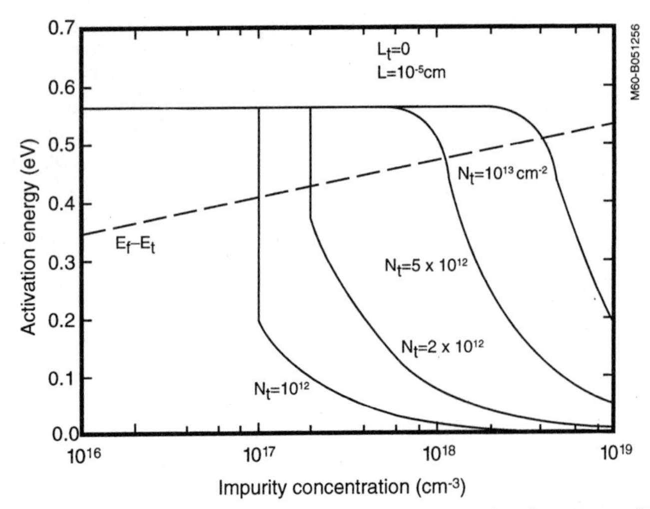

FIGURE 4-26 Activation energy as a function of impurity concentration for polycrystalline Si (grain size = 10 μm). Reference is made to the monovalent trapping states at midgap. *(From G. Baccarani, B. Ricco, and G. Spandini,* J Appl Phys, *vol. 49, 1978, p. 5565.)*

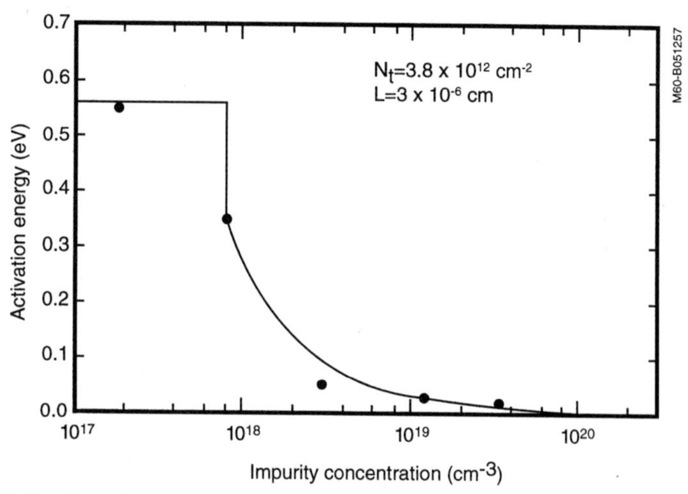

FIGURE 4-27 Comparison of theoretical and experimental values of activation energy as a function of impurity concentration in polycrystalline Si. *(From G. Baccarani, B. Ricco, and G. Spandini,* J Appl Phys, *vol. 49, 1978, p. 5565.)*

For *incomplete depletion* of the grains ($N_D > N_D'$) with a depletion width $< \mathcal{G}/2$, the corresponding conductivity solution yields

$$\sigma_g = (q^2 \mathcal{G} N_D v/kT)(N_c/N_D)^{-\xi} \exp(-E_a/kT) \quad \text{with } E_a = E_g/2 - [(4\epsilon N_D)/(q^2 N_i)^2]\,(\xi - 1) \quad (74)$$

where $\xi = (1 + q^2 N_i^2 E_g / 4\epsilon N_D)^{-1/2}$.

The calculated activation energy as a function for the donor-doping level for the distributed interface case (with the same fundamental parameters as in Fig. 4-26) is shown in Fig. 4-27. For lower impurity densities, E_a approaches the midgap energy, and for higher concentrations, it is proportional to N_D^{-1}. These two limits correspond to monovalent trapping states. However, the abrupt transition between complete and partial depletion always occurs, even at high interface state densities.

Problems: Barrier Height and Concentrations. Three problems limit the general application of the previous transport models to semiconductor thin films. First, the actual doping of the film is uncertain. This is true for physical and chemical vapor deposition in which uniform control is very difficult. Even for the ion-implanted films, the postprocessing, high-temperature heat treatments required to minimize the high-energy particle damage can lead to impurity diffusion, especially at the grain boundaries [140]. Second, the uncertainty in the impurity concentration spatial distribution is high because of variations with film thickness [130,149]. And third, contamination of the grain boundary by gases inherent to the film-growth procedure is possible. Although such contamination can be minimized by careful process control, the large grain boundary, surface-to-volume ratios in most polycrystalline thin films enhance the problem.

Seager and Castner analyzed and characterized the electrical transport properties of neutron-transmutation-doped polycrystalline Si, and sought to avoid these impurity distribution problems [96]. The samples in their studies had larger grain sizes (minimum diameter of 25 μm). The electrical measurement techniques included four-probe and traveling-potential probe analyses performed as close to zero-bias as possible. The resistivity was linearly dependent on inverse temperature (as previous studies reported) below a doping level of about $2 \times 10^{13}/cm^3$ (phosphorous) for uniform samples [130–132]. The activation energy was nearly the midgap value (0.55 eV). In contrast, for doping levels above $2 \times 10^{13}/cm^3$, deviations from the linear dependence were observed (see Figs. 4-28 and 4-29). Potential probe measurements indicated that a large range of grain boundary impedances exist (that is, a variety of grain boundary barrier heights are present).

Seager and Castner [139] were able to analytically represent the grain boundary potential under three separate conditions for the interface defect state density N_i:

1. *Energy-independent* N_i:

$$q\varphi_b = \Delta E_f [1 + \alpha/2\Delta E^f - (\alpha/\Delta E_f + \alpha^2/\Delta E_f^2)^{1/2}] \tag{75}$$

   where $\Delta E_f = E_{fg} - E_{fb}$, is the separation between the Fermi levels in the grain and boundary regions, respectively, (measured from the valence band edge), and $\alpha = 8\epsilon N_D/q^2 N_t$ (with N_t = the two-dimensional density of defect states and N_D = the donor (doping) density).

2. *Monoenergetic* N_i: In this case, a closed-form solution is not possible, but

$$[8eN_D q\varphi_b/(q^2 N_t^2)]^{1/2} = (\alpha q\varphi_b)^{1/2} \tag{76}$$

   which can be solved using iterative techniques.

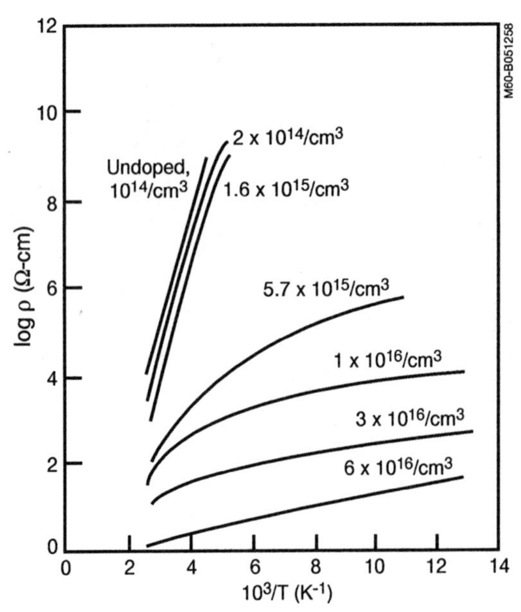

FIGURE 4-28 Dependence of the resisitivity on inverse temperature for neutron transmutation-doped polycrystalline Si. *(From C. H. Seager and T. G. Castner,* J Appl Phys, *vol. 49, 1978, p. 3879.)*

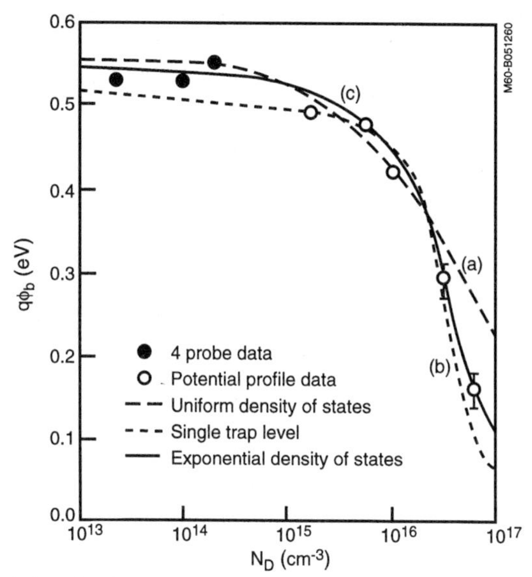

FIGURE 4-29 Grain boundary barrier height as a function of doping density for neutron transmutation-doped polycrystalline Si. Open circles are data from small-area potential profile experiments, and solid circles are four-probe resistivity data. *(From C. H. Seager and T. G. Castner,* J Appl Phys, *vol. 49, 1978, p. 3879.)*

3. *Exponentially dependent* N_i: A closed form solution is again not possible, and

$$(8\epsilon N_D/q^2)^{1/2}\,(q\varphi_b)^{1/2} = E_o[N_{To}\exp(-E_{fb}/E_o)]\{1 - \exp[q\varphi_b - \Delta E_f)/E_o]\} \qquad (77)$$

where $N_t = N_{to}\exp(-E_{fb}/E_o)$, and N_o and E_o are adjustable parameters.

These three cases are summarized in the band diagram of Fig. 4-30, with the parameters from Eqs. 75 to 77 used to fit the specified zero-bias data. The exponential density-of-states model fits the data somewhat better than the single-trap or the energy-independent N_i models. This comparison indicates that the largest grain boundary-state densities correspond to about $6 \times 10^{11}/cm^2$, located within 0.2 eV of the midbandgap position. This is lower than the corresponding value used by either Seto or Baccarani ($3 \times 10^{12}/cm^2$) and results from the extrinsic quality (likely contamination) of the grain boundary states in those cases.

Seager and Castner correlated their results with the effects of doping concentrations on the electrical transport in polycrystalline Si[149]. The conclusions are defined in three regions:

1. $N_D < 10^{14}/cm^3$: The barriers primarily exhibit $q\varphi_b = 0.55$ eV (the midgap value), and the resistivity is dominated by this activation energy.
2. $10^{14}/cm^3 < N_D < 2 \times 10^{15}/cm^3$: A substantial population of the barriers is < 0.55 eV, but the largest $q\varphi_b$ still dominate the resistivity-temperature characteristics [150].
3. $N_D > 2 \times 10^{15}/cm^3$: A range of barrier heights exists, and the magnitude of the resistivity activation energy depends on the physical features of the sample and analysis technique. A major problem of the shape of the grain boundary–density-of-states function limits the interpretation. The evaluation requires an extensive, difficult determination of the current-voltage characteristics of a large number of grain boundaries.

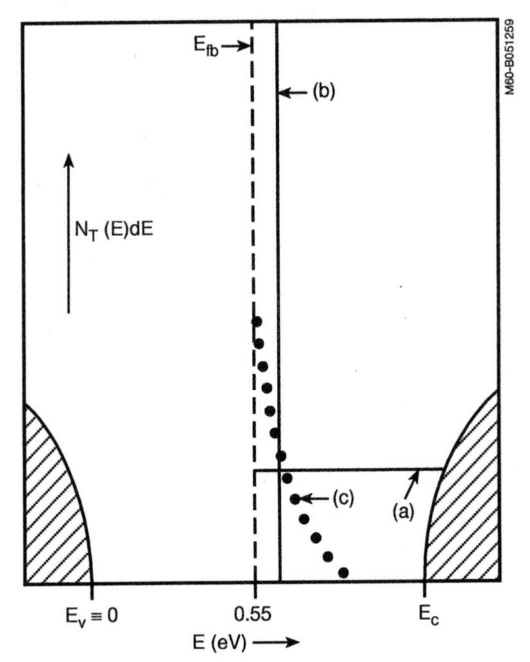

FIGURE 4-30 The densities of states and Fermi level in barrier region for polycrystalline Si for three cases: (*a*) $N_t = 3.3 \times 10^{12}$/cm²·eV; (*b*) $N_t = 5.4 \times 10^{11}$/cm²·eV at −0.626 eV above E_v and zero elsewhere, and (*c*) $N_t = 1.89 \times 10^{15}$/cm²·eV exp(−E/0.099). *(From C. H. Seager and T. G. Castner,* J Appl Phys, *vol. 49, 1978, p. 3879.)*

4.2.5 Grain Boundary Model

A microscopic examination of the electronic structure of the grain boundary provides information for macroscopic measurements, as well as for understanding the electro-optical behavior and modeling of thin film, polycrystalline semiconductors [151–161]. A typical microscopic band diagram of a charged grain boundary in a *p*-type semiconductor is represented in Fig. 4-31. The holes are captured in interface states above the Fermi level. This positive charge is compensated for by the donors in the space charge region. The thermionic current flows from left to right in this diagram, and the j_{ss} is the current resulting from the dynamic equilibrium between the capture and emission of holes.

Thermionic Considerations. This is the basic model used for polycrystalline semiconductor modeling [80,82,94,138,139]. For the charged-grain boundary in Fig. 4-31, the net current j_{th} of the majority holes is given by

$$j_{th} = A^*T^2 \exp[-(\zeta + \varphi)/kT]\,[1 - \exp(-V/kT)] \tag{78}$$

where A^* is the Richardson constant, and V is the applied voltage. The forward biased (left) barrier is $q\varphi$, and $q\zeta$ is the Fermi level in the crystalline grains. This latter term is a function of the doping or concentration level.

Displacement Currents and Frequency Effects. In addition to the thermionic current, j_{th}, there is a second current caused by the interface states permanently capturing holes from the valence band and reemitting holes to the valence band [162]. This current is the

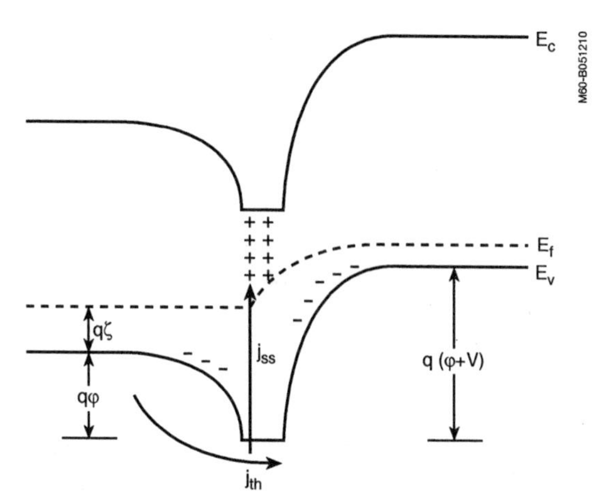

FIGURE 4-31 Energy band diagram of a charged grain boundary in a *p*-type semiconductor. Holes are captured in interface states above the Fermi level. This positive charge is compensated for by the negative acceptors in the space charge region. The thermionic current flows from left to right.

difference between the capture and the emission rates of holes and can be described by the Shockley-Read-Hall statistics. This displacement current is labeled j_{ss} in Fig. 4-31.

The displacement currents are frequency dependent. Under conditions near thermodynamic equilibrium (that is, $qV \ll kT$ and zero bias), j_{ss} is 0. With the application of a small AC voltage, $\delta V \ll kT/q$, with the dc voltage ($V_{dc} \geq kT/q$), the time-periodic change of the barrier is

$$j = j_{dc} - \delta\varphi \tag{79}$$

where $\delta\varphi \leq kT/q$. In addition to the thermionic current in Eq. 79, two other displacement currents exist.

1. *Space-charge-layer displacement current.* This results from the high-frequency capacitance of the depletion layer, and is expressed in complex form as

$$j_{sc} - i(2\pi\nu)C_{HF}\,\delta V \tag{80}$$

where ν is the frequency and C_{HF} is the depletion layer capacitance at high frequency.

2. *Capture/emission displacement current.* This is a result of the periodic barrier change, the hole that leads to the hole concentration at the interface also varying periodically in time. This change drives a periodic capture and emission of holes by the interface states.

The difference between the capture and emission rate is the current j_{ss}, represented in Fig. 4-31. It is phase-shifted with respect to $\delta\varphi$ because the capture and emission processes at the interface are not infinitely fast. This displacement current is written as

$$j_{ss} = Y_{ss}\,\delta\varphi \tag{81}$$

where Y_{ss} is the characteristic admittance of the traps. This parameter depends on the trap distribution in energy and space and on the capture cross section. The definition of this admittance is

$$Y_{ss} = G_{ss} + i(2\pi\nu)C_{ss} \tag{82}$$

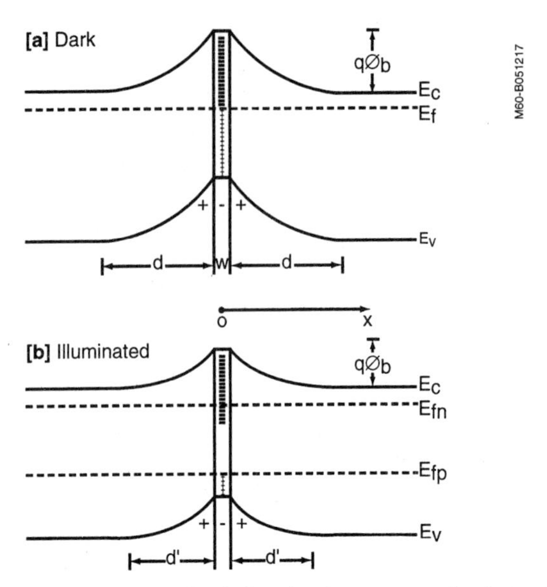

FIGURE 4-32 Energy band diagram of grain boundary in a *p*-type semiconductor: (*a*) dark case, and (*b*) illuminated case.

where G_{ss} and C_{ss} are the conductance and capacitance, and the admittance is analogous to the physics of the metal-oxide-semiconductor interface.

Minority-Carrier Processes. Many semiconductor devices are controlled by the minority-carrier transport. As a result, the problems and mechanisms of recombination at grain boundaries in polycrystalline semiconductor films have been the focus of interest and investigation [151–155,163–169]. Such an understanding is important in improving thin film device performance.

Card and Yang developed one of the first comprehensive treatments of minority-carrier properties, with a focus on polycrystalline Si [152]. They considered an ensemble of parameters, including doping concentrations, grain size, and interface-state densities that were important in determining the minority-carrier lifetime. This work has led to numerous other reports, in which models have been refined and extended to other semiconductors [153–169].

The approach considers the band diagram for the grain boundary both in the dark and under illumination (Fig. 4-32). Considering the width of the grain boundary to be much less than the width of the depletion region, the charge balance equations give $Q_i = Q_d$ where the first term is the net charge contained in the interface states, and the second term is the charge in the depletion region. From this, the diffusion potentials are calculated and are shown in Fig. 4-33 for the case of *p*-type $CuInSe_2$ as a function of interface-state density and for various concentrations.

More important, the model leads to expressions for the recombination velocity, v_r, which is based on a careful consideration of the recombination currents J_r [152,153], where $J_r = qv_rp$, and

$$v_r = \tfrac{1}{4}\,\sigma v(E_{fn} - E_{fp})N_i \exp(q\varphi_b/kT) \tag{83}$$

illustrating that the grain boundary diffusion potential enhances the recombination at those defects. This is further illustrated in Fig. 4-34, showing the dependence of the

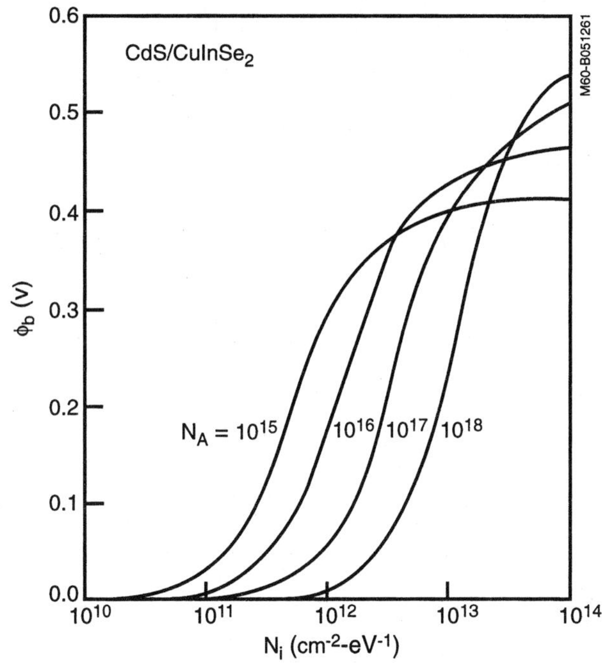

FIGURE 4-33 Dark diffusion potential dependence on interface state density for various carrier concentrations in p-type $CuInSe_2$. *(From A. Rockett and R. W. Birkmire,* J Appl Phys, *vol. 70, 1992, p. R81.)*

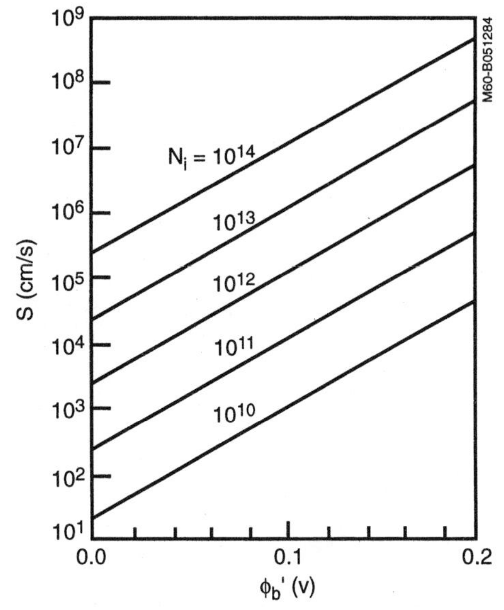

FIGURE 4-34 Dependence of the recombination velocity on the illuminated diffusion potential and interface state density in $CuInSe_2$. *(From A. Rockett and R. W. Birkmire,* J Appl Phys, *vol. 70, 1992, p. R81.)*

recombination velocity on the illuminated diffusion potential and interface-state density for $CuInSe_2$.

From this relationship, the minority-carrier lifetime can be estimated. Card and Yang considered films with columnar grain geometries. For a *p*-type semiconductor, the resulting lifetime is

$$\tau_n = 1/\sigma v N_r \qquad (84)$$
$$= I \exp(q\varphi_b/kT)/\sigma v N_i(E_{fn} - E_{fp}),$$

where I is the grain diameter, and E_{fn} and E_{fp} are the positions of the quasi-Fermi levels. This simple expression has been used to model the properties of several polycrystalline semiconductors. Such data are shown in Fig. 4-35 for τ as a function of grain diameter for $CuInSe_2$. This result is consistent with the physical representation of the grain boundary in terms of interface states (that is, high-angle grain boundaries with $N_i > 10^{13}/\text{cm}^2\cdot\text{eV}$, medium-angle grain boundaries with $10^{11} < N_i < 10^{13}/\text{cm}^2\cdot\text{eV}$, and low-angle grain boundaries with $N_i < 10^{11}/\text{cm}^2\cdot\text{eV}$) [170].

Yee has used this analysis to further show the behavior of minority carriers, specifically addressing the minority-carrier mobility [171,172]. In this approach, the depletion region is divided into two distinct parts, which lead to expressions for the carrier concentrations and the diffusion currents. The electron and hole mobilities are then derived, with

$$\mu_n = [qI/2\pi m^* kT)^{1/2}] \exp(-q\varphi_b/kT) \qquad (85)$$
$$= \mu[I/(I + L^*)]$$

where $L^* = (2L_p)/[\alpha \tanh(I/2L_p)]$
$L_p = [(kT/q)\mu\tau]^{1/2}$
τ = lifetime associated with μ.

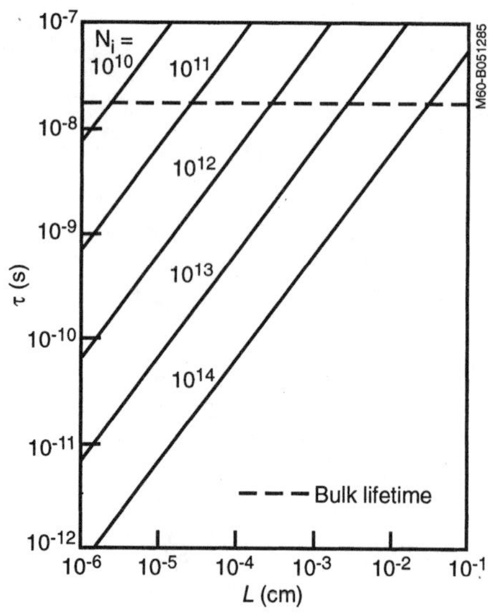

FIGURE 4-35 Dependence of the minority carrier lifetime on the grain diameter for various interface state densities in $CuInSe_2$. The dashed line indicates a single-crystal value.

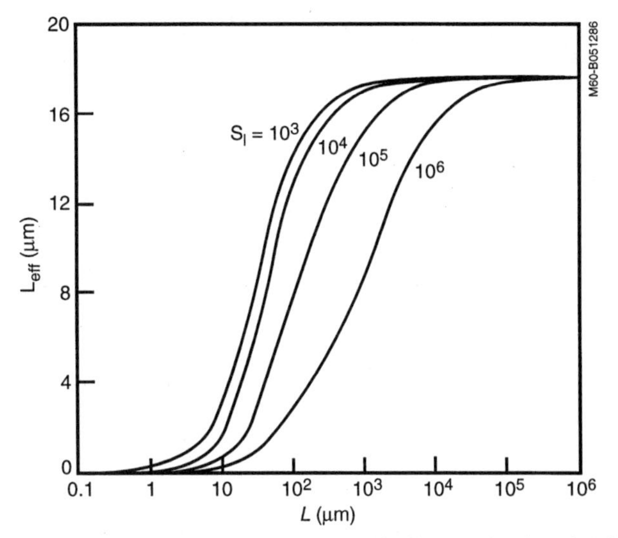

FIGURE 4-36 Diffusion length of polycrystalline CdTe thin film as a function of grain diameter for several grain boundary recombination velocities. *(From J. H. Yee,* The Systematic Computation of the Performance of Photovoltaic Cells Based on First Principles, *DOE Report W-9405-ENG-48, 1979; and J. Y. Leong and J. H. Yee,* Appl Phys Lett, *1979.)*

Yee's treatment is used to approximate the effects of grain size and interface recombination velocity on the minority-carrier diffusion length and the minority-carrier lifetimes. Such an analysis is shown in Fig. 4-36 for polycrystalline CdTe [171,172].

General models for electronic transport in semiconductor thin films have been published [173–175]. These include three-dimensional considerations of the grain structures, as well as in-depth considerations of the minority-carrier properties. In most cases, these modeling efforts are complex, but are of significant use in specific applications.

4.3 PROPERTIES OF AMORPHOUS THIN FILMS

Amorphous thin films have significantly different characteristics than crystalline thin films [176–178]. Therefore, the physics governing the transport mechanisms must be treated differently. This section introduces some of the concepts that distinguish the conductivity and current behavior of amorphous semiconductors. Amorphous Si is used as the major example because it has received the most technical attention and currently has the most significant set of device applications, ranging from sensors and transistors to displays and solar cells [178]. The reader is directed to a number of excellent references for more discussion of transport mechanisms, and for more specific treatments of the physics of this special semiconductor [176–181].

4.3.1 Comparisons of Crystalline and Amorphous Semiconductors

The differences between the crystalline and amorphous semiconductors permeate their structural, electronic, optical properties, and device attributes and behavior. The emphasis

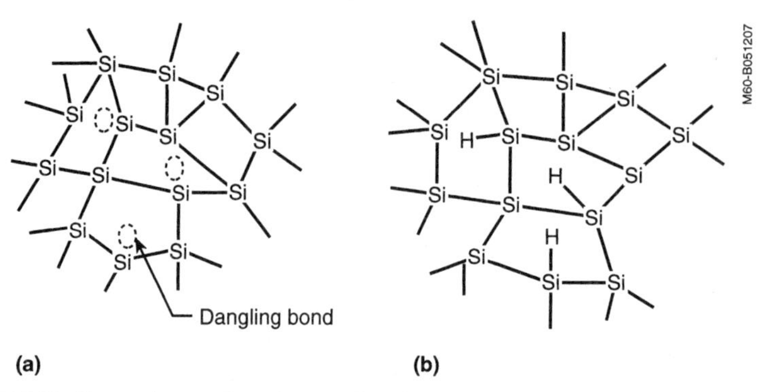

FIGURE 4-37 Structural comparisons for amorphous Si, showing (*a*) amorphous Si containing no H, and (*b*) amorphous Si containing H.

in this case lies in the fact that the physics governing the amorphous semiconductors are much different from that developed for the "classical" crystalline counterpart. In these comparisons, Si is used as an example.

Physical Structure. *Crystalline* Si has a diamond-type atomic structure, with its four atoms positioned at each apex of regular tetrahedra, covalently bonded with an Si atom positioned at the center of the tetrahedra. The crystal structure is regular, with a lattice constant of 5.421 Å. *Amorphous* Si (a-Si) does not have longer range periodicity or symmetry in its structure. Although the atoms are not positioned regularly, an individual atom must be covalently bonded with other atoms around it. The number of bonds (coordination number), bond angles, and bond lengths of each atom is similar to single crystals, but over several atoms, the periodicity is not preserved [178]. Because of randomness, some of the atomic bonds are unsatisfied, or "dangling" [178,179]. This comparison is illustrated in Fig. 4-37.

Energy Band Structure. Crystalline Si has an indirect energy gap of 1.1 eV. Amorphous Si has fluctuations in the Si–Si bond, leading to a difference in gap, especially in the tailing of the edges. Therefore, the exact magnitude or extent of bandgap tailing is difficult to determine. The levels present in the forbidden gap of a-Si are called localized states. Comparisons between the crystalline Si and a-Si are shown in Figs. 4-38, 4-39, and 4-40.

Conduction between these localized states is accomplished by hopping, which results in a suppressed or decreased mobility of carriers. If the mobility is expressed as a function of energy, a region in which the mobility shows a sharp decrease can be observed. This region is called the mobility gap [181], but it is not measurable. The determination of the absorption coefficient is used to determine the bandgap of a-Si in most cases.

Amorphous Si, used in devices, is actually a hydrogenated amorphous semiconductor (that is a-Si:H), with some 5% to 15% atomic-percent hydrogen. The bonding of H to Si ties up the dangling bonds, which causes the defect levels to move into the conduction and valence bands. The density of states in the bandgap is reduced. The addition of H also relaxes local strain in the amorphous materials. The overall effect of adding H to a-Si is to increase the bandgap of the material. The bandgap of a-Si (with about 10% H) is about 1.7 eV, and the absorption characteristics show that the transitions are direct [182].

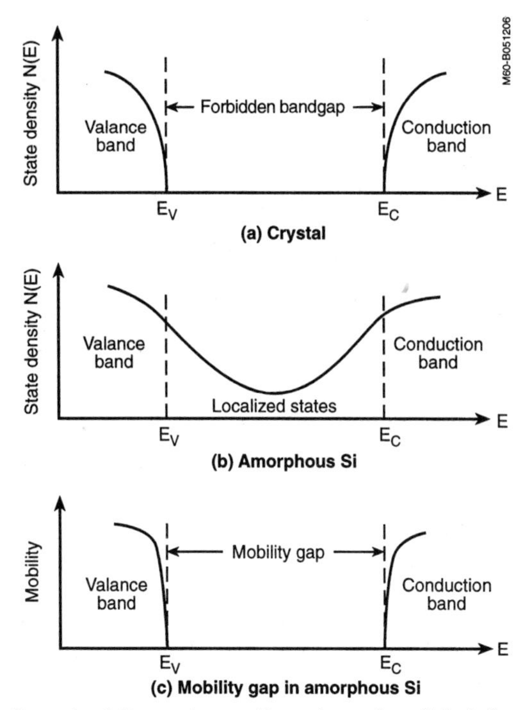

FIGURE 4-38 Energy band diagrams in crystalline and amorphous Si, including representation of the mobility gap.

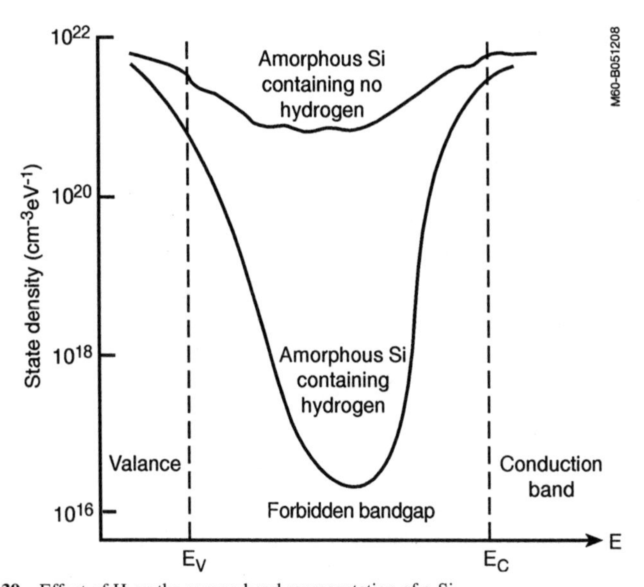

FIGURE 4-39 Effect of H on the energy band representation of a-Si.

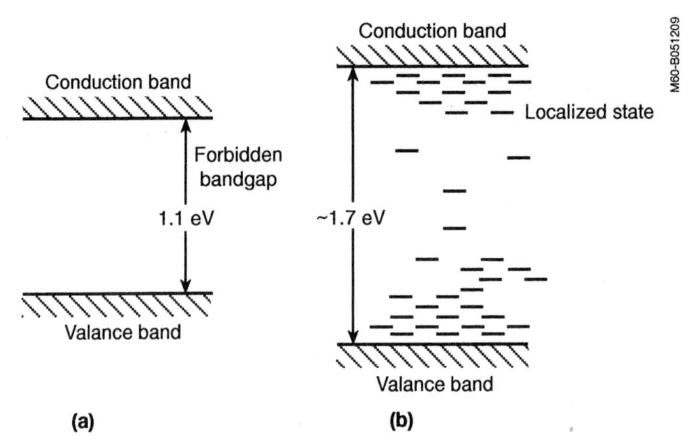

FIGURE 4-40 Energy band structure of (*a*) crystalline Si and (*b*) a-Si:H.

4.3.2 Conductivity and Band Gap

An examination of the recent literature indicated that the conductivity and carrier transport in amorphous semiconductor films (a-Si, Ge, GaAs, GaS_2, CdTe) remains an area of intensive investigation [183–200]. The conductivity has been attributed to several major mechanisms, as summarized in this section.

General Observations. The conductivity of amorphous films is critically dependent on the deposition technique and conditions of the deposition. This is especially true of a-Si:H films in which the substrate temperature (during deposition) variation can show a dark resistivity variation from about 10^{11} Ω·cm at T_{sub} = 100°C to 10^5 Ω·cm at T_{sub} = 550°C in films produced in glow discharge. With increased substrate temperature, undoped films become more *n*-type, indicating that the density of states in the gap is sensitive to the deposition conditions [201].

Another very interesting observation seen in the conductivity in a-Si:H films is the vast difference in the magnitudes of the dark conductivity (σ_d) and the photoconductivity (σ_{ph}). It has been experimentally demonstrated that the thickness of the a-Si:H layer is critical [202]. If the thickness is less than about 0.3 μm, the conductivity will be controlled only by the surface (recall the previous discussion on surface scattering and surface electronic structure), making the measured value lower than anticipated. If the layer is very thick, the photons for the photoconductivity will not sufficiently penetrate into the a-Si:H to be effective because of the relatively high absorption coefficient. Some general rules for device-quality material used by device engineers are as follows [202–205]:

- If $\sigma_{ph} > 10^{-5}$ Ω/cm, there is a high percentage of Si–H_2 bonds (dihydride bonding), and the material is of poor quality for device applications.
- For $5 \times 10^{-5} < \sigma_{ph} < 5 \times 10^{-4}$ Ω/cm and $1 \times 10^{-10} < \sigma_d < 1 \times 10^{-8}$ Ω/cm, the bonding is primarily Si–H (monohydride), and the material is more suitable for device applications.

Photoconductivity. For comparison, since the surface reflection and the film thickness vary from sample to sample, the most consistent method by which to model the photoconductivity is to calculate the photon density (N_{ph}), which is actually injected and absorbed in the a-Si:H, using light of long wavelength and assuming it is uniformly absorbed by the film. The $\eta\mu\tau$ product can then be calculated from consideration of the light-generated current as [206]:

$$j_{ph} \sim qN_{ph}\eta\mu\tau V/L \tag{86}$$

where $\nu \sim 1$
V = applied voltage
L = sample length.

Dark Conductivity. This conductivity is descriptive of the semiconductor case in which the mean-free path is of the order of the lattice spacing of the material. The dark conductivities are very small in undoped amorphous semiconductors. A number of theoretical and modeling efforts have provided background toward understanding the conduction and transport mechanism in these amorphous semiconductors. Several of the major ones are highlighted here, but the reader is referred to a number of excellent books and reviews dealing with this subject area [1,171–183].

Kubo-Greenwood Formulation. This is the simplest approach to depicting the conductivity. It assumes that, because of the short mean-free path I_{mfp} of the carriers, the single scattering mechanism on which the Boltzmann formula is based breaks down. This formulation calculates the conductivity as a function of frequency, then takes the limit of that conductivity as the frequency goes to zero [207], as in

$$\sigma = C_o q^2 k^2 I_{mfp}/h \tag{87}$$

This simple representation has relevance only to cases in which the carrier mean-free paths are extremely small. The difficulty in establishing the validity of this criterion makes the application of this model difficult. However, the equation does predict the very small conductivity encountered in amorphous films.

Anderson Localization and the Mobility Edge. This approach considers a crystalline array of potential wells. The tight-binding approximation is used, with the overlap between wave functions considered negligible except for wells that are nearest other wells. Starting with the Schrödinger equation, the solution is of the form

$$\Psi = \Sigma \exp(ik \cdot a_n)\, \Psi(|a_r - a_n|) \tag{88}$$

where a is the spatial vector.

Anderson introduced a random potential at each well, with depths in the range V_o. If $V_o < B$, the effect is to introduce a mean-free path, I_{mfp}. The resulting conductivity, termed Ioffe-Regel (I-R) conductivity [208], is expressed as

$$\sigma_{I\text{-}R} = Cq^2/3ha \tag{89}$$

This simple result has to be corrected for semiconductors. Because I_{mfp} is small, multiple scattering is not negligible. The results in the derivation by Kaveh and Mott [209] give

$$\sigma = \sigma_B[1 - (C/k_F I_{mfp})^2(1 - I_{mfp}/L] \tag{90}$$

where L is the length of the sample and σ_B is the bulk material conductivity. Furthermore, the case for the $V_o > B$ reduces the density of states. If V_o/B is not large enough to give localization thoughout the band, the states in the band tails would be localized and the energies of localized and extended (nonlocalized) states would be separated by a sharp energy E_c (known as the *mobility edge*). The existence of a mobility edge in a semiconductor means that the lowest states have become traps, and conduction by carriers will be excited normally to the mobility edge. Thus the conductivity behaves [210,211] as

$$\sigma = \sigma_o \exp[-(E_c - E_f)/kT] \tag{91}$$

Hopping Conduction. If the Fermi level is below the mobility edge, the conduction will be of two types:

1. *By excitation to the mobility edge,* where

$$\sigma = Cq_2/\eta L_i \tag{92}$$

   given that L_i is the inelastic diffusion length, a result of collisions with phonons and Auger processes [212]. The carriers (electrons) lose energy to other carriers that have energies below E_f.

2. *By thermally activated hopping,* which occurs when the number of states near the Fermi level is finite. This is a process in which an electron in an occupied state with energy below E_f receives energy from a phonon, enabling it to move to a nearby state above E_f. This is illustrated in Fig. 4-41 [213].

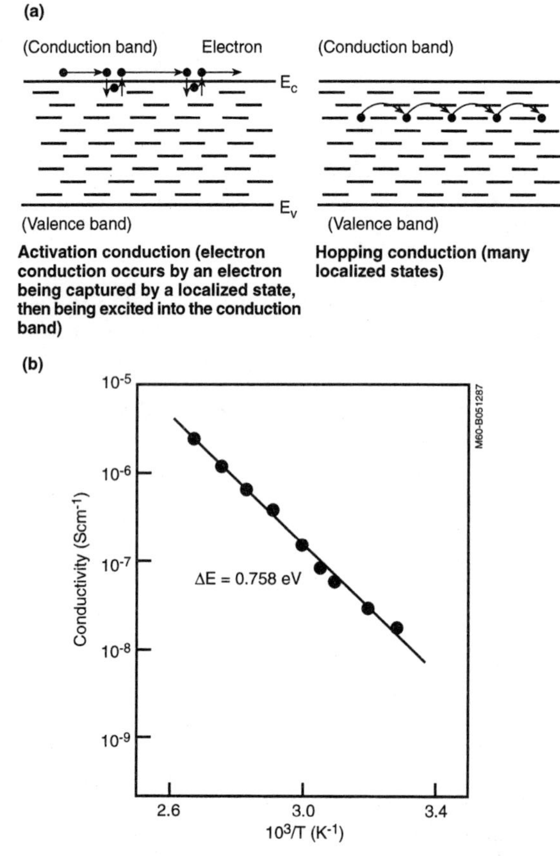

FIGURE 4-41 Amorphous hydrogenated Si: (*a*) models of conduction and localized state density, (*b*) typical dependence of σ on inverse temperature for supporting hopping mechanisms.

This treatment was further developed and refined by Mott and Davis to show that the conductivity is of the form [214]

$$\sigma = A \exp(-E/kT) \tag{93}$$

This is termed *variable-range hopping,* and applies to both doped crystalline and amorphous semiconductors. The temperature dependence of the dark conductivity of undoped films is expressed as

$$\sigma = \sigma_o \exp(-E_a/kT) \tag{94}$$

where $0.2 < E_a < 0.8$ eV, typically. Many research groups report single activation energy over the temperature range $180°K < T < 220°K$. Also, a transition at approximately 250°K suggests hopping through localized states in the gap to be the transport mechanism below this temperature [215,216].

Meyer-Neldel Rule. Both doped and undoped a-Si:H thin films follow the Meyer Neldel rule, which modifies the hopping model by extending the expression for the pre-exponential conductivity term [217]. That is,

$$\sigma = \sigma_o \exp(-E_a/kT), \tag{95}$$

where $\sigma_o = \sigma_{oo}\exp(E/kT_o)$, with σ_{oo} and T_o representing constants which are a function of the material properties, and E is the applied electric field.

This model specifically addresses high-field regimes in which a-Si:H films exhibit non-ohmic behavior. This mechanism has been identified as Poole-Frankel conduction [218].

Mobility, Lifetime, and Density of States. Typical drift mobilities (majority carrier) in n-type and p-type material (over the range $200°K < T < 400°K$) are $\mu_{dn} = 2$ to 5×10^{-2} cm^2/V·s and $\mu_{dp} = 5$ to 6×10^{-4} cm^2/V·s. Variations of the Hall mobility with electron lifetimes are deduced from photoconductivity measurements and range from milliseconds to hundreds of nanoseconds in undoped films. Hole lifetimes are as high as 20 ns. These low values are indicative of the short-range structural order in this material. The lifetimes are illumination-level dependent.

The density of states in device quality a-Si:H is on the order of 10^{17}/eV·cm^3. Figure 4-42 shows the density of states for a range of samples [219,220]. Both undoped and lightly doped samples are represented. The exponential edge is shown, and E_c,E_v are the two mobility edges. E_x and E_y are the positions of the two dangling bond states. The density of states in glow-discharge a-Si:H has shown some consistency in regard to the method used to make the measurements. Figure 4-43 presents such data from field effect, deep-level transient spectroscopy (DLTS) and space-charge limited current methods [221].

4.3.3 Light-Induced Instabilities

When a-Si:H is illuminated by a strong light with an intensity in the range of 80 to 100 mW/cm^2, the photoconductivity (σ_{ph}) is lowered and the dark-state conductivity decreases. This effect is partially reversible with heat treatment, reverse bias annealing, and storage of films in the dark for prolonged periods of time. This phenomena was first observed in the early 1980s as an unwanted property affecting photovoltaic devices, and it was named the Staebler-Wronski effect for its discoverers [222–225]. An example of the Staebler-Wronski effect is shown in Fig. 4-44, demonstrating the change in the conductivity (σ_d and σ_{ph}) as a function of time with illumination. The $\sigma - (1/T)$ characteristics for

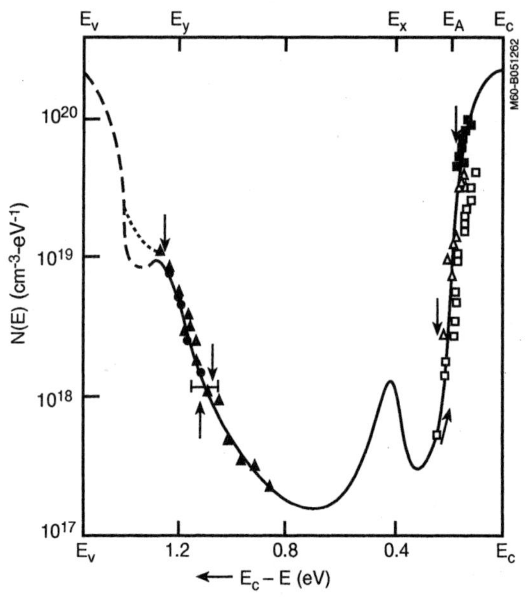

FIGURE 4-42 Density of states of a-Si:H deposited at high temperature. The points represent lightly doped samples. The exponential edge is shown, and E_c and E_v are the two mobility edges. E_x and E_y are the positions of the two dangling bond states *(From W. E. Spear and P. G. LeComber, in* Hydrogenated Amorphous Silicon, *vol. 1, J. D. Joannopoulous and G. Lucovsky [eds.], Springer-Verlag, New York, 1984,pp. 63–118; also, K. D. MacKenzie, P .G. LeComber, and W. E. Spear,* Phil Mag, *vol. B46, 1982, p. 377.)*

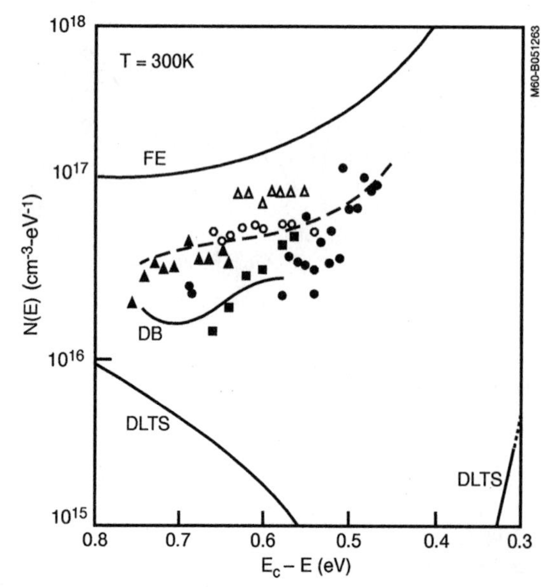

FIGURE 4-43 Density of states in a-Si:H (glow-discharge) measured from field-effect (FE), space-charge limited current (DB), and deep-level transient spectroscopy (DLTS).

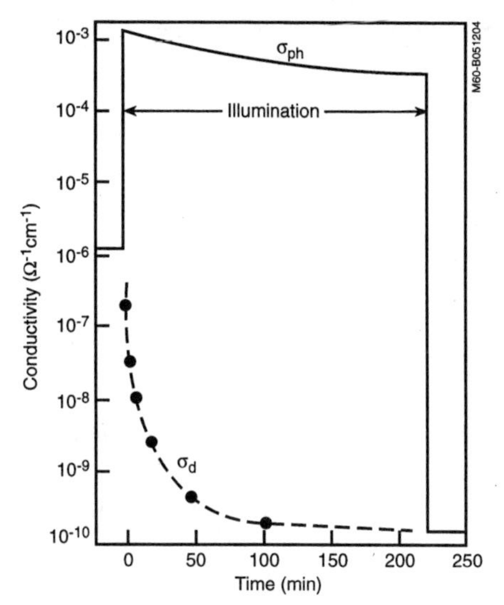

FIGURE 4-44 Staebler-Wronski effect in undoped a-Si : H under illumination. Effect on σ_d and σ_{ph} is shown for illumination of 200 mW/cm^2, for film grown at substrate temperature of 320°C and 0.7 μm thickness.

initial and degraded films are shown in Fig. 4-45, indicating that the activation energy is also affected.

Several major observations about the Staebler-Wronski effect have been reported.

- The instability is significant in undoped films, but does not occur in doped films with high concentrations of dopants (P, B, or As) [226].

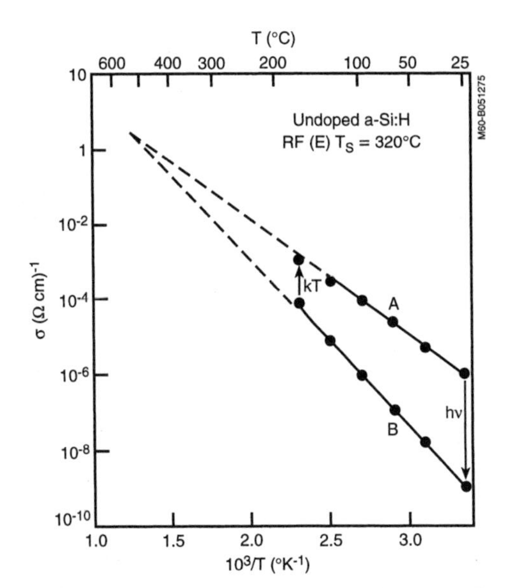

FIGURE 4-45 Conductivity-inverse temperature characteristics for a-Si : H film **A** in dark, and **B** illuminated, showing Staebler-Wronski effect.

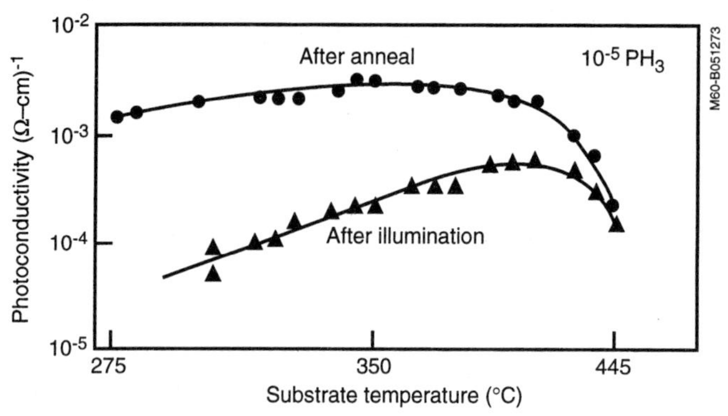

FIGURE 4-46 Recovery of Staebler-Wronski light-generated degradation by heating in dark at 150°C.

- The initial value is almost recovered upon heat treatments, typically for about 1 hour at 150°C (shown in Fig. 4-46) or when left in the dark for 4 to 8 hours [227].
- The extent of the alteration of σ_{ph} depends on the decrease in the $\eta\mu\tau$ product of the thin film [228–229].
- The optical properties of the films are unaffected [230].
- The change in σ_d signifies a shift of the Fermi level toward the center of the bandgap [231].
- The dependence of σ_{ph} on intensity (F): $\sigma_{ph} = F_\gamma$, where the γ parameter is initially $0.5 < \gamma < 0.9$. After illumination, $\gamma = 0.9$ [232].
- The Staebler-Wronski effect is small in fluorinated amorphous Si of low H content [233–234].
- The thinner the film, the less significant the effect [235,236].

Several theories of the Staebler-Wronski effect have been reported [235–240]. The effect has been ascribed to both bulk (defects and impurities) and surface effects. One representation of the possible effect of oxygen involvement is illustrated in Fig. 4-47 [241]. Before illumination, the oxygen bridges the bonding between two Si atoms. After illumination, the O–Si bond is broken, leaving a hydroxide and several dangling bonds and increasing the density of states in the gap. This problem is a limitation of the class of amorphous Si devices intended to operate under some

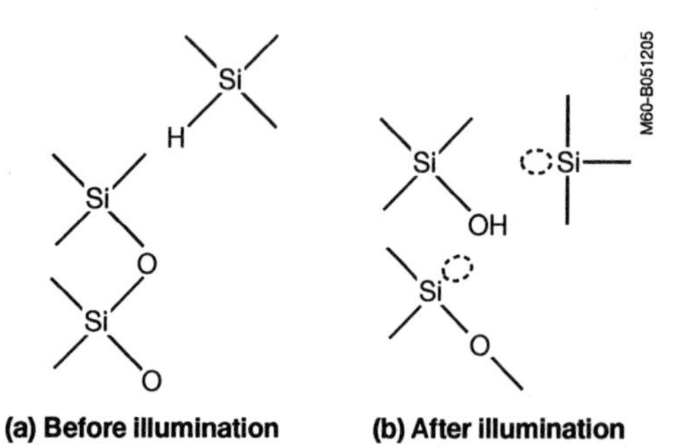

FIGURE 4-47 Possible model accounting for oxygen involvement in Staebler-Wronski effect.

type of radiation (such as solar cell) [242–244]. Such changes in device performance are indicated in Fig. 4-48 for an a-Si:H cell. However, recent advances have limited the light-generated degradation by manipulating the device structure and materials (with improvements shown in Fig. 4-48); solar cells and modules with stabilized efficiencies above 10% have been reported [245,246].

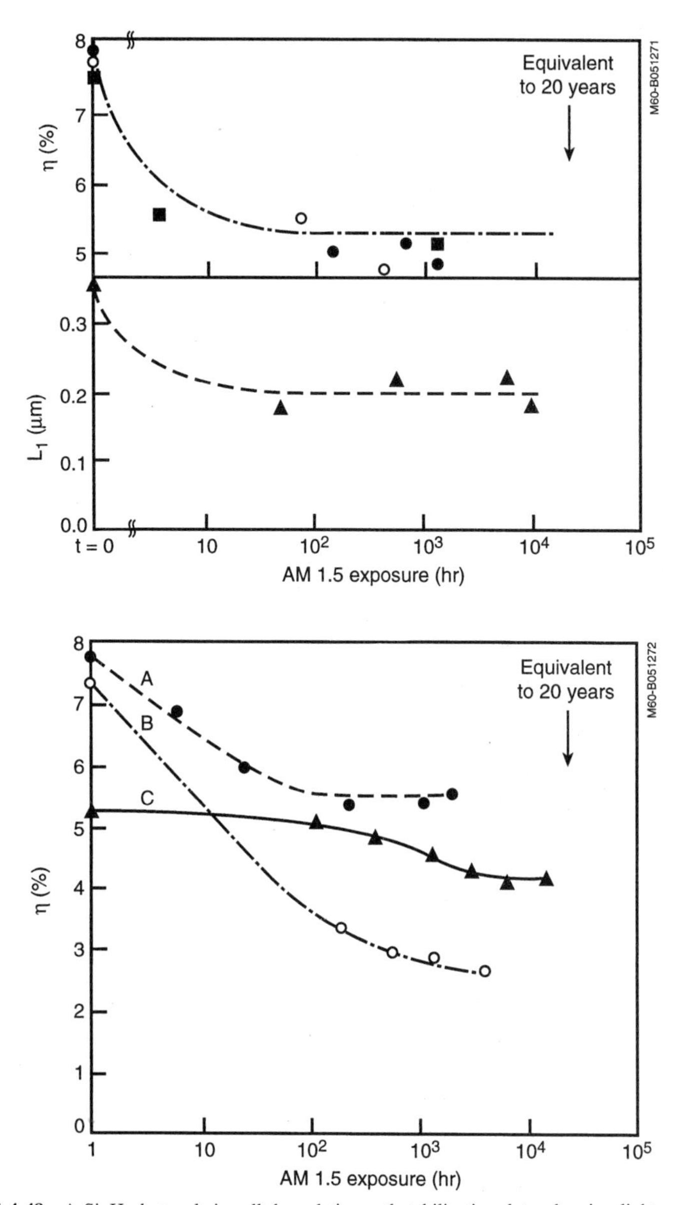

FIGURE 4-48 A-Si:H photovoltaic cell degradation and stabilization data, showing light-generated instability (initial loss in performance) and region of stabilization.

4.4 OPTICAL PROPERTIES OF SEMICONDUCTOR FILMS

Semiconductor films are important because of their absorbing and reflecting properties and are used for their photoconductivity, electro-optical, or magneto-optic effects. The range of optical phenomena in semiconductor films is considerable, and a comprehensive treatment would include infrared properties determined by lattice vibrations, inelastic scattering (that is, Raman and Brillouin scattering), photoconductivity, and various forms of luminescence (such as stimulated emissions), in addition to treatment of the optical constants. The reader is referred to several other sources, books, and monographs which address these phenomena in more detail [247–249].

4.4.1 Definitions of Optical Parameters

This section discusses the optical constants of semiconductor films, developing the definitions and relations between the various optical properties. It also discusses some measurement techniques (optical properties in polycrystalline and amorphous films, specifically). Finally, the relationships between the most common semiconductor thin film optical constants are presented. The complex index of refraction, n_c, is defined as

$$n_c = n - ik \tag{96}$$

and is related to the velocity (v) of the propagation of light in the material by

$$v = c/n_c \tag{97}$$

where c = velocity of light in vacuum
n = index of refraction
k = extinction coefficient.

The intensity I of a light wave in an absorbing medium decreases as

$$\mathrm{I} = \mathrm{I}_o \exp(-\alpha x) \tag{98}$$

where x is the distance into the sample from the surface under illumination. The quantity α is termed the absorption coefficient and is related to k through the expression

$$\alpha = 4\pi k/\lambda \tag{99}$$

where λ is the wavelength of the light in vacuum. The dielectric constant can be introduced in two ways. One can define the dielectric constant ϵ to be real and describe any losses by a conductivity (σ), which is frequency-dependent and is usually not identical to the dc conductivity. Then,

$$n^2 - k^2 = \epsilon \tag{100a}$$

and

$$nk = \sigma/\upsilon \tag{100b}$$

where υ is the frequency.

Alternatively, ϵ can be defined in complex form as

$$\epsilon \equiv \epsilon_1 - i\epsilon_2 \tag{101}$$

Then, combining these expressions,

$$n^2 - k^2 = \epsilon_1 \tag{102a}$$

and

$$2nk = \epsilon_2 \tag{102b}$$

Clearly, a knowledge of n and k determines ϵ_1 and ϵ_2, and vice versa. The quantity ϵ_2 is usually easy to calculate, though experimental results often appear in terms of α. The α-photon energy characteristic has the same general shape as the ϵ_2 versus photon energy characteristic because n usually does not vary much with energy.

In addition, for a linear medium, n and k (as well as ϵ_1 and ϵ_2) are related in that if the complete frequency dependence of n is known, one can calculate k, and vice versa. This dependence is known as the Kramers-Kronig relation [250], and they are a consequence of causality.

4.4.2 Experimental Routes for Determining Optical Constants

The experimental techniques used to determine the optical constants provide a very good basis for understanding optical properties. The choice of a particular technique depends on the accuracy or precision required, the wavelength range of interest, the type of semiconductor film, and the physical qualities of the film.

Transmission and Reflection: Normal Incidence. The simplest case is a film surface and substrate that are perfectly flat, are of uniform film thickness, and possess homogeneous optical constants. The incident monochromatic radiation is adjusted to impinge normally on the absorbing film that is grown on a transparent substrate. The transmission (T) and reflection (R) of the system depend on a number of critical parameters, including:

- The optical constants n_2 and k_2
- The film thickness
- The wavelength of the light
- The indices of refraction (n) of the substrate and the medium above the film

Two distinct wavelength regions generally occur for semiconductors. At long wavelengths (photon energies less than the bandgap), T and R exhibit oscillations from interference effects in the transparent film. At short wavelengths (photon energies greater than the bandgap), T rapidly decreases to zero, while R approaches a constant magnitude. Such a typical transmission-wavelength characteristic is presented in Fig. 4-49.

The actual expressions for R and T are not derived here, but have had numerous treatments [251–253]. The most useful format for these relationships are

- *At long wavelengths* (with $n_2 \gg k_2$, $n_2 > n_3, n_1$)

$$T^{\max}_{\min} = \frac{C(1 - R_o)(1 - R_1)\exp(-\alpha d)}{[1 \pm (R_o R_1)^{1/2} \exp(-\alpha d)]^2} \tag{103}$$

where
$$R_o^{1/2} = r_{21}$$
$$R_1^{1/2} = r_{23}$$
$$R_1'^{1/2} = r_{32}$$
$$R_3^{1/2} = r_{13}$$

$$r_{ij} = n_i - \frac{n_j}{n_i + n_j}$$

$$\alpha = 4\pi k_2/\lambda$$

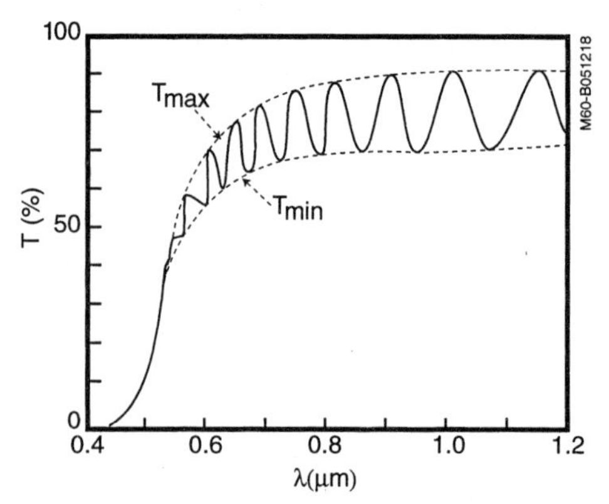

FIGURE 4-49 Transmission spectrum at 300 K for SnO_2 thin film, with thickness 2.03 μm (±0.08). *(From J. C. Manifacier et al.,* Thin Solid Films, *vol. 41, 1977, p. 127.)*

The notation $T_{\min}^{\max}$ refers to the value of the maximum or minimum in T (Fig. 4-49). The value of n_2 can be determined in the long wavelength region, because the maxima satisfy the Bragg's relationship, expressed as

$$2n_2(\lambda)d = m\lambda \tag{104}$$

where $m = 0, 1, 2, \ldots$, and the minimum satisfy the same equality with m = ½, ³⁄₂, . . .

At shorter wavelengths ($\alpha d \gg 1$)

$$T = \frac{C\,|\tilde{t}_{12}|^2\,|\tilde{t}_{23}|^2}{n_3 \exp(-\alpha d)} \tag{105a}$$

and

$$R = |\tilde{r}_{12}|^2 \tag{105b}$$

where the complex quantities $\tilde{r}_{ij}$ and $\tilde{t}_{ij}$ are defined by

$$\tilde{r}_{ij} = (\tilde{n}_i - \tilde{n}_j)/(\tilde{n}_i + \tilde{n}_j) \tag{106a}$$

$$\tilde{t}_{ij} = 1 - \tilde{r}_{ij} \tag{106b}$$

There are a number of approaches for using these relations to obtain n and k for the semiconductor film. If the thickness is determined independently (using an interferometer or a topographic profiling technique), n is usually determined in the low-absorption region using Eq. 104. The value of n in the short wavelength region may be obtained from T and R data (Eq. 105) or by extrapolating n from the long wavelength region. This may be done graphically or by fitting n to some reasonable function [254]. It is then possible to obtain k or α from Eqs. 103 and 105.

It should be emphasized that there is usually a significant difference in accuracy between measurements of T and R. Measurements of T are limited only by the linearity and sensitivity of the detection system. An accuracy of better than ~0.1% can be obtained with reasonable precautions. The determination of R, however, is more difficult to achieve with high accuracy. An accuracy of around 1% is realistic, unless exceptional methodologies are followed. Thus although self-consistent treatments of R and T values over the entire spectral range may, in principle, be used to determine n and k, it is far more common to adopt a procedure that weights the T data more heavily.

Manifacier et al. have developed a consistent, straightforward procedure for determining n and d from T, under the condition $k^2 \ll n^2$ [255,256]. The envelope functions T_{max} and T_{min} are constructed (Fig. 4-49), and these in turn are treated as continuous functions of λ. Then

$$n = n_2 = [N + (N^2 - n_3^2)^{1/2}]^{1/2} \tag{107}$$

where

$$N = (1 + n_3^2)/2 + 2n_3(T_{max} - T_{min})/T_{max}T_{min} \tag{108}$$

The thickness is calculated from two maxima (or two minima) according to

$$d = M(\lambda_1\lambda_2/2)[n(\lambda_1)\lambda_2 - n(\lambda_2)\lambda_1] \tag{109}$$

where M is the number of oscillations between the two extrema occurring at λ_1 and λ_2. With n and d determined, n can be extrapolated into the high-absorption region, from which k can be determined as before. This method is attractive because d is determined precisely in the region of the film in which the transmission is measured. This method is best applied to sufficiently thick (1-2 μm) films in the visible and near infrared spectral regions, so that the interference peaks are closed spaced, thus defining T_{max} and T_{min} with best precision.

With careful R and T measurements, n can be determined within a few percent. The absorption coefficient (α) for a film thickness of 1 μm may be determined in the range $10^2/cm^{-1}$ to $10^6/cm^{-1}$. One must generally include the effects of surface roughness to obtain the two absorption coefficients in the small α range (discussed later in this section). Figures 4-50 and 4-51 present n and α, determined from R and T measurements, for an a-Si film [251].

Ellipsometric Considerations. When the angle of incidence of the incoming light is not zero, the reflected intensity is polarization-dependent. The reflectance of light with the electric field parallel to the plane of incidence (R_p) is different from the reflectance with the electric field perpendicular to the plane of incidence (R_s). In addition, there is usually a phase shift (Δ) between the two components. Figure 4-52 presents the geometry under consideration.

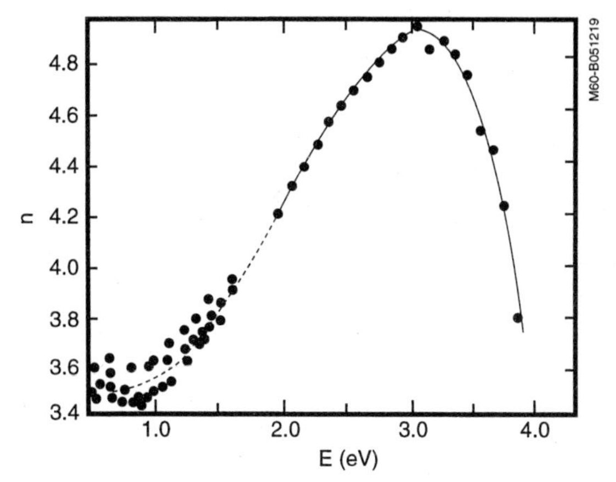

FIGURE 4-50 Index of refraction as a function of photon energy for a-Si:H. *(From G. D. Cody et al.,* Solar Cells, *vol. 2, 1980, p. 227.)*

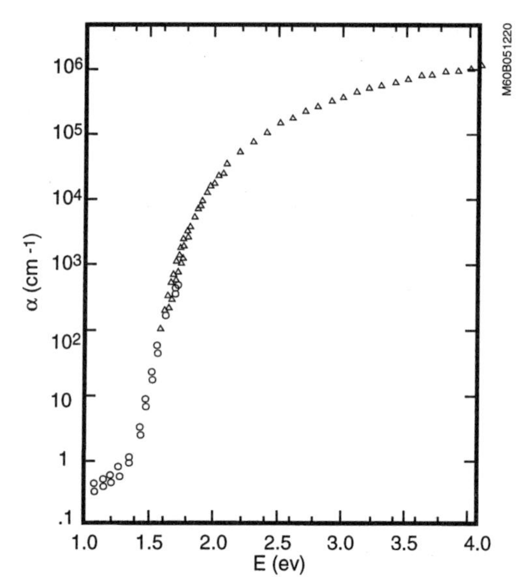

FIGURE 4-51 Absorption coefficient for a-Si: H, derived from optical absorption and photoconductivity. *(From G. D. Cody et al.,* Solar Cells, *vol. 2, 1980, p. 227.)*

In principle, R_p, R_s, or R_p/R_s may be measured as a function of the angle of incidence, and the optical constants of the thin film can be extracted. Bennett and Bennett discuss the various configurations used in this experiment [257]. The precision obtained in a specific configuration depends on the nature of the thin film system being investigated.

Ellipsometry refers to techniques for measuring reflectance amplitudes rather than intensities. Therefore, both phase and magnitude are determined. The complex reflectance ratio is defined as

$$\tilde{\rho} \equiv \tilde{r}_p \tilde{r}_s^{-1} \equiv (\tan\Psi)\exp(i\Delta) \tag{110}$$

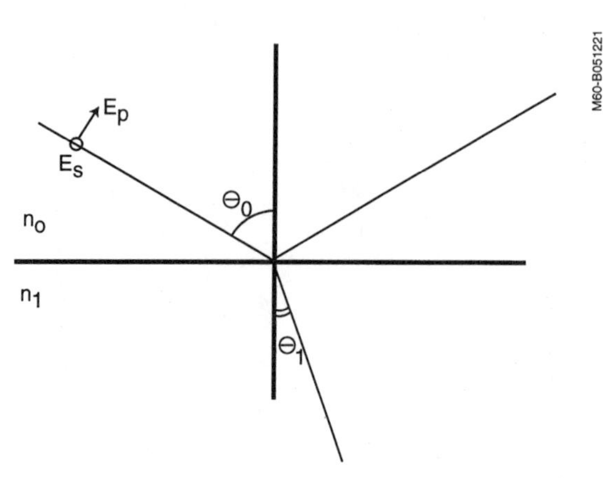

FIGURE 4-52 Definitions corresponding to reflectance at interface. *(From J. B. Theeten and D. E. Aspnes,* Am Rev Mater Sci, *vol. 11, 1981, p. 97.)*

Therefore, the angles Ψ and Δ correspond to the relative amplitude and phase of the two reflections, respectively.

Ellipsometry has rapidly developed, particularly in the areas of spectroscopic ellipsometry (in contrast to the fixed-wavelength technique) [258,259]. In Fig. 4-52, the relevant results for this two-phase model are

$$\tilde{n}_o \sin \Theta_o = \tilde{n}_1 \sin \Theta_1 \tag{111}$$

$$\tilde{r}_p = (\tilde{n}_1 \cos \Theta_0 - \tilde{n}_o \cos \Theta_o)(\tilde{n}_1 \cos \Theta_0 + \tilde{n}_o \cos \Theta_o)^{-1} \tag{112}$$

and

$$\tilde{r}_s = (\tilde{n}_o \cos \Theta_0 - \tilde{n}_1 \cos \Theta_1)(\tilde{n}_o \cos \Theta_0 + \tilde{n}_1 \cos \Theta_1)^{-1} \tag{113}$$

Therefore, ellipsometry can be used to unambiguously determine n and k of a material if $\tilde{n}_o$ is known. Most semiconductors, however, have a near-surface region that is different from that of the bulk.

Figure 4-53, for example, shows a three-phase model approximating bulk Si and its native oxide. If the optical constants of the substrate are known, only three unknowns remain—n_1, k_1, and k_2, with two ellipsometric quantities, Ψ and Δ. As the number of layers increases, one clearly cannot determine all the parameters directly from the ellipsometric measurement. However, if the measurements are made as a function of wavelength and angle of incidence, the results are modeled, and the confidence limits are established rigorously, the parameters can be accurately determined [259].

Ellipsometry instrumentation is described in detail in other readings [258]. Null ellipsometers use a compensator to produce a phase shift exactly opposite of that produced by reflection from the sample. Such systems yield high accuracy, but are a bit more cumbersome than conventional ellipsometric systems. By rotating the polarizer or analyzer (there is no compensator), an ac signal may be generated that contains Y and D in the Fourier components. Alternatively, an elector-optical or piezo-optical compensator plate can be used in a null ellipsometer configuration. State-of-the-art ellipsometer systems can follow changes in Ψ and Δ on a scale smaller than milliseconds.

Ellipsometry is particularly useful to in situ studies of semiconductor thin film growth as discussed in Chapter 1. Except in high vacuum or ultrahigh vacuum environments, ion and electron probes are not suitable for in situ work. Various optical techniques can be applied, such as Raman scattering and photoluminescence. Ellipsometry is one of the more powerful in situ approaches.

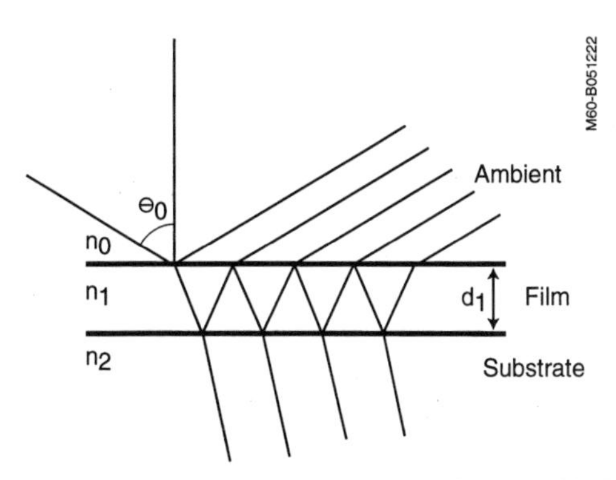

FIGURE 4-53 Definitions for the 3-phase model. *(From J. B. Theeten and D. E. Aspnes,* Am Rev Mater Sci, *vol. 11, 1981, p. 97.)*

An example of an in situ study is presented in Figs. 4-54 and 4-55. The complex refractive indices of GaAs and AlGaAs layers grown by chemical vapor deposition are determined using a He-Ne laser source. In Fig. 4-54, the results of growing $Al_{0.25}Ga_{0.75}As$ on GaAs are shown. Point 1 in Fig. 4-54*a* corresponds to the GaAs substrate; point 2 corresponds to the thick AlGaAs layer. The results show that the transition to AlGaAs is not abrupt, but rather occurs over several hundred Å. On the other hand, growing GaAs on AlGaAs under these same conditions produces a more abrupt junction, as illustrated in Fig. 4-55. This could be interpreted in terms of a compositional profile in a well-characterized system. Thus in situ ellipsometry is capable of compositional profiling on a scale comparable to Auger electron spectroscopy, without the artifacts induced by sputter etching. Furthermore, with spectroscopic (wavelength-adjustable) ellipsometry, one can tune the profile to examine particular chemical or structural features.

Surface Roughness Effects. The preceding discussions have assumed a perfectly smooth film on a perfectly flat substrate. This assumption is implicit in the majority of literature relating to optical properties of thin films. The effects of surface roughness on

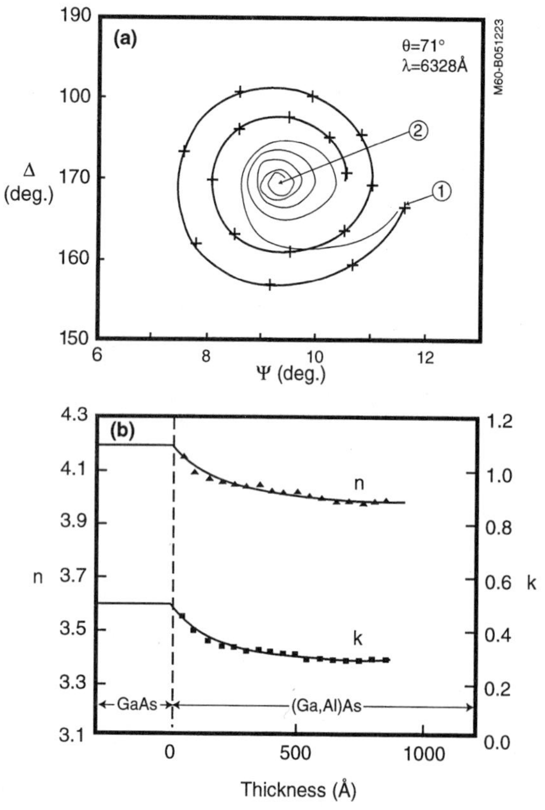

FIGURE 4-54 GaAs to GaAlAs transition. (*a*) Experimental trajectory in the (Ψ,Δ) plane. The calculated trajectory (assuming a 3-phase model) is shown (*solid line*), with crosses corresponding to 100 Å increments in thickness. (*b*) Refractive index profile calculated from data. *(From J. B. Theeten and D. E. Aspnes,* Am Rev Mater Sci, *vol. 11, 1981, p. 97.)*

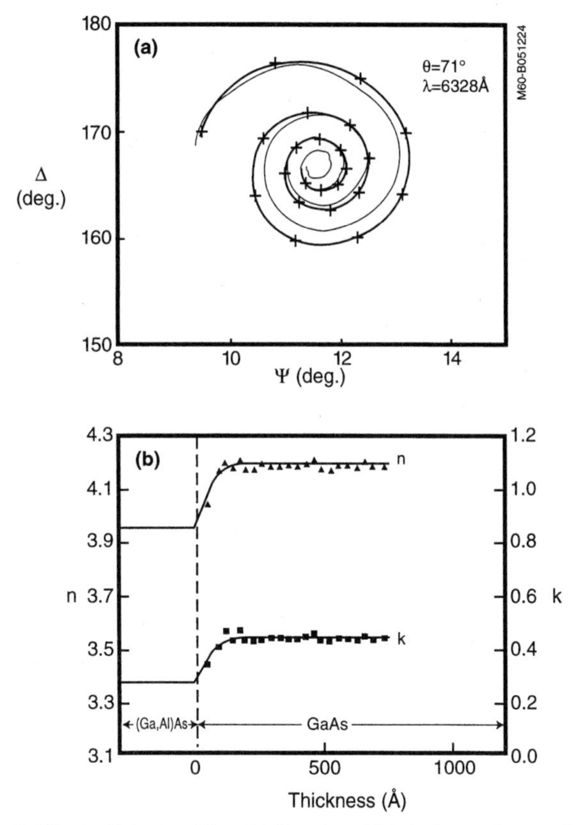

FIGURE 4-55 GaAlAs to GaAs transition. (*a*) Experimental trajectory and comparison with 3-phase model (*solid line*) with crosses corresponding to 100 Å increments in thickness. (*b*) Refractive index profile calculated from the data. *(From J. B. Theeten and D. E. Aspnes,* Am Rev Mater Sci, *vol. 11, 1981, p. 97.)*

the reflectance and transmission (coherent and incoherent) of bulk surfaces and thin films have been documented in the literature [253,257,260–262]. For a normal incidence, the ratio of reflectance from a rough surface relative to a smooth surface of the same material is given by

$$R/R_o = \exp[-(4\pi\sigma/\lambda)^2] + \{1 - \exp[-(4\pi\sigma/\lambda)^2]\}\,\{1 - \exp[-2(\pi\sigma\alpha/m\,\lambda)^2]\} \tag{114}$$

where σ = rms height of the surface irregularities
λ = wavelength
m = rms slope of the irregularities
α = half acceptance angle of the instrument.

This expression is valid for $\sigma/\lambda \ll 1$. The first term is a result of coherently reflected light, and the second term results from that part of the incoherently reflected light recorded by the instrument. Bennett and Bennett have discussed the effect of surface roughness on the accuracy of reflectance measurements [257]. They show that to obtain 0.1% accuracy in a reflectance measurement, σ must be ~10 Å for visible light wavelengths. In general, special polishing and surface preparation techniques are required to achieve such optical flatness. Alternately, one can attempt to determine the roughness and correct for that factor when determining the reflectance.

Szczyrbowski and Czapla have studied the effects of surface roughness on R and T for thin films on a smooth transparent substrate [253]. Their analysis includes the effects of multiple reflections. They then apply these results to a determination of n and k for polycrystalline InAs film. As an example, Fig. 4-56 shows the absorption coefficient for a 3.5 μm-thick layer, first assuming $\sigma = 0$, then using a best-fit value, $\sigma = 0.10$ μm. The effect of surface roughness is significant for low absorption coefficients. It must be considered to understand small effects, such as those resulting from grain boundaries in polycrystalline films and disorder-induced states in amorphous films.

A direct method of accounting for the effect of surface roughness on absorption is to measure the total sample reflectance (specular plus diffuse) with an integrating sphere. Coupled with a transmission measurement, absorptance of the sample can be determined. For measuring very low absorption coefficients ($\alpha < 10^2$/cm) in thin films, different techniques are required.

Low Absorption Coefficients. The determination of the absorption coefficient in semiconductor films of approximately 1 μm thickness is limited to about 10^2/cm using conventional techniques. The problem is twofold. First, the measurement involves the difference between two nearly equal quantities, so that instrument limitations have a great effect. Second, light scattered by the roughness of the substrate surface produces an effective absorption that can be significant if not properly considered. A better approach in the low-absorption region is to use a technique in which the response is more directly related to the absorption.

Two techniques are highlighted here. The first is photoacoustic spectroscopy (PAS) [262,265]. In PAS, the absorbed light generates acoustic energy that is detected with a microphone or piezoelectric transducer. The detected signal is related to α; for small α and thin samples, it is nearly linear in α. Figure 4-57 shows a block diagram of a typical PAS measurement system. Figure 4-58 presents the absorption coefficient of undoped

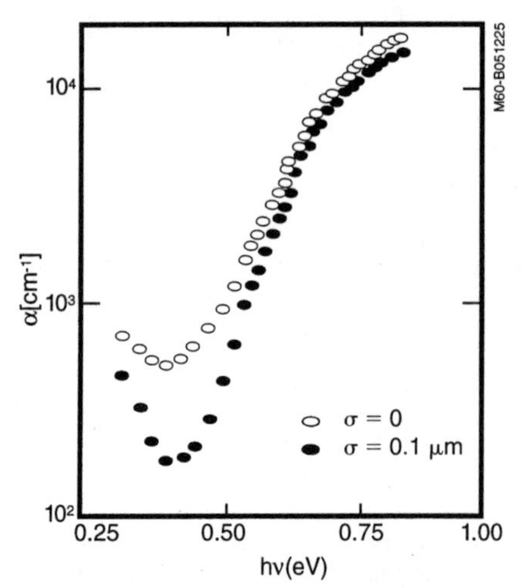

FIGURE 4-56 Absorption coefficient for a 3.51 μm sputtered InAs film, showing the effects of neglecting surface roughness. *(From J. Szczyrbowski and A. Czapla,* Thin Solid Films, *vol. 46, 1977, p. 127.)*

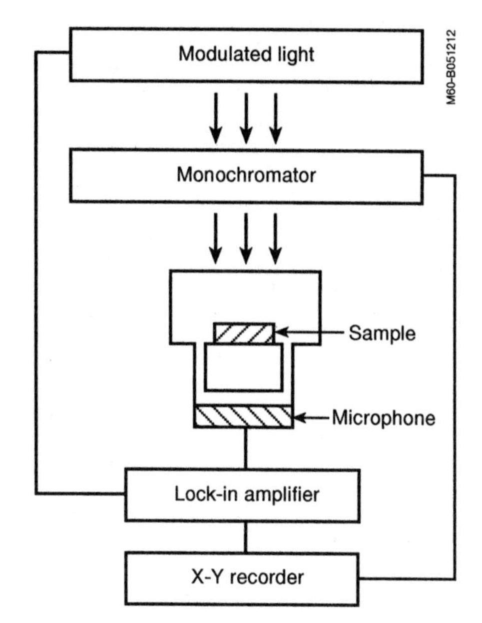

FIGURE 4-57 Block diagram of a PAS measurement system.

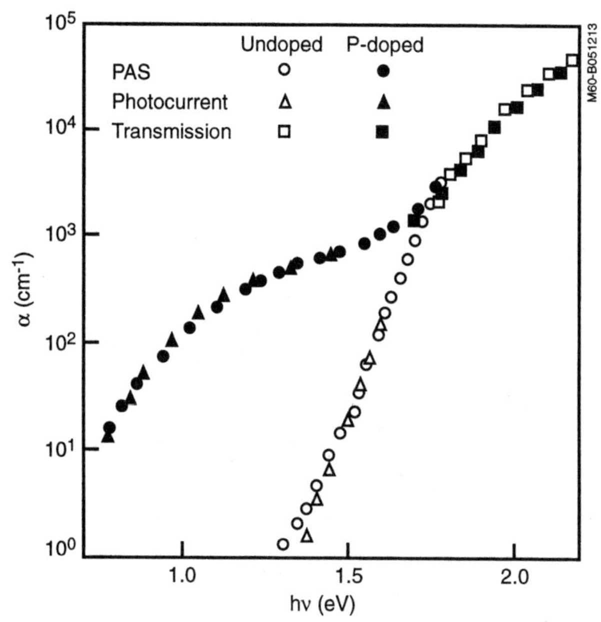

FIGURE 4-58 Absorption coefficient for undoped and P-doped a-Si:H films derived from PAS, secondary photocurrent, and transmission measurements. *(From S. Yamaski et al., in* Tetrahedrally Bonded Amorphous Semiconductors, *R. A. Street, D. K. Biegelsen, J. C. Knights [eds.], AIP Conference Series, no. 73, American Institute of Physics, New York, 1981, p. 258.)*

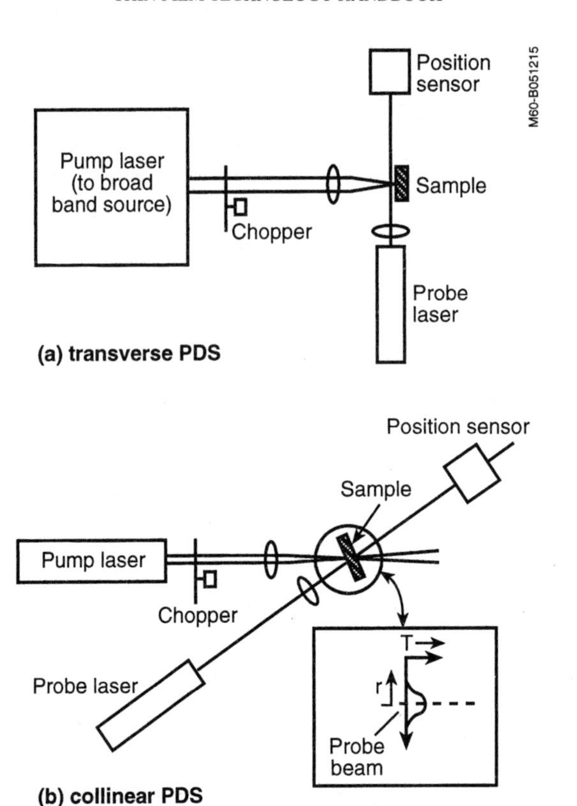

FIGURE 4-59 Experimental system for: (*a*) transverse PDS; (*b*) collinear PDS. *(From W. B. Jackson et al.,* Appl Optics, *vol. 20, 1981, p. 1333.)*

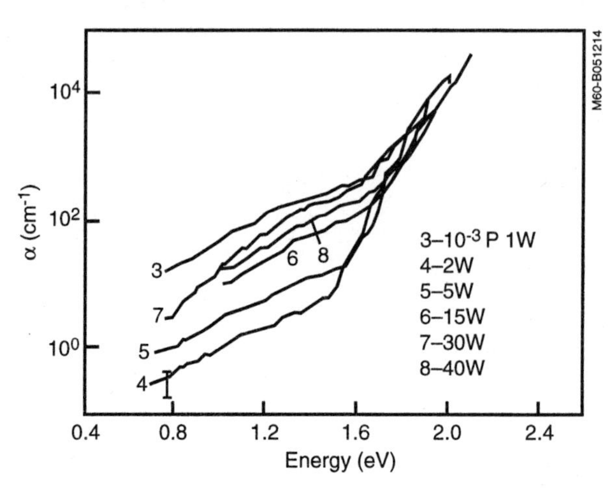

FIGURE 4-60 Absorption coefficient of a-Si:H as a function of RF power, determined by PDS. *(From W. B. Jackson et al.,* Appl Optics, *vol. 20, 1981, p. 1333.)*

and doped a-Si:H films. The second technique is photothermal deflection spectroscopy (PDS), which provides somewhat better results, in most cases, than a PAS system. In this approach, the light absorbed in the film causes a temperature gradient in the adjacent medium (typically an organic solvent). This gradient is detected by monitoring the deflection of a second beam with a position-sensitive detector. Figure 4-59 shows two experimental configurations, and Fig. 4-60 presents absorption coefficient determinations for a-Si:H films.

4.4.3 Models of the Optical Properties

This section addresses specifically the optical properties of semiconductor films, including single-crystal, amorphous, and polycrystalline structures. Some focus is directed on the spectral region near the absorption edge. In these discussions, films are assumed to be homogenous.

Absorption in Single-Crystal Semiconductors. The dominant feature of the energy dependence of the absorption coefficient in semiconductors, $\alpha(h\nu)$, is the onset of absorption near the region of interband transitions from valence to conduction bands. Only the fundamental principles are provided here; further information has been documented thoroughly in the literature [247,267,268].

The energy dependencies of a near-the-band edge for band-to-band and exciton transitions are shown in the following:

- *Allowed direct transitions* (neglecting exciton effects)

$$\alpha(h\nu) \sim (h\nu - E_g)^{1/2}/h\nu \tag{115}$$

  where E_g is the energy gap at $k = 0$

- *Forbidden direct transitions* (neglecting exciton effects)

$$\alpha(h\nu) \sim (h\nu - E_g)^{3/2}/h\nu \tag{116}$$

- *Indirect transitions* (neglecting exciton effects)

$$\alpha(h\nu) = \alpha_e(h\nu) + \alpha_a(h\nu) \tag{117}$$

  where

$$\alpha_e(h\nu) = A(h\nu - E_g - E_p)^2/[1 - \exp(-E_p/kT)] \tag{118}$$

  For $h\nu > E_g - E_p$, α_e corresponds to the emission of a phonon of energy E_p to conserve momentum, and

$$\alpha_a(h\nu) = A(h\nu - E_g + E_p)^2/[\exp(E_p/kT) - 1] \tag{119}$$

- *Exciton effects:* In sufficiently high quality, pure semiconductors, the intrinsic absorption is dominated by exciton formation, giving rise to sharp line structure in direct-gap semiconductors at low temperatures. These in turn lead to a more complex expression for $\alpha(h\nu)$ at all temperatures. The explicit dependence of α on $h\nu$ is not treated here, but the reader is referred to several literature sources [247,249,252]. Excitons in disordered semiconductors are more important, although not well understood.

In addition to band-to-band absorption, a single crystal may exhibit impurity effects in its absorption spectrum. These effects include acceptor to conduction band, valence band to donor level, and possibly, acceptor level to donor transitions, all on the low-energy side of the absorption edge. For low to moderate carrier concentrations ($<10^{18}/cm^3$), these effects will be negligible compared to band-to-band absorption. For heavily doped semiconductors, complications arise because of the onset of degeneracy, impurity bending, and potential fluctuations that generally affect the shape of the main absorption edge [249]. In amorphous and polycrystalline semiconductors, the disorder-induced state plays a role analogous to that of impurities.

At energies above the absorption edge, there is generally structure in the optical constants resulting from transitions involving the critical point in the band structure. A point in the Brillouin zone where

$$\nabla_k E(k) = 0 \tag{120}$$

corresponds to a singularity in the density of states and produces a large contribution to the optical properties (that is, reflection or absorption). The determination of such singularity points is important in understanding the band structure; several modulation spectroscopies have evolved for making such determinations [268]. Such techniques are not as easy to apply to disordered semiconductors in which k is not a good quantum number.

Absorption in Amorphous Semiconductor Thin Films. The absorption characteristics in most amorphous semiconductors are described by three regions, as depicted schematically in Fig. 4-61. Region A corresponds to transitions between extended states. Region B likely relates to the disorder-induced tail states. Region C corresponds to transitions to or from states strongly localized deep in the gap.

The interband absorption region (A) is treated using a parabolic density of states, constant matrix elements, and relaxation of k-conservation in the transition [269,270]. It is this

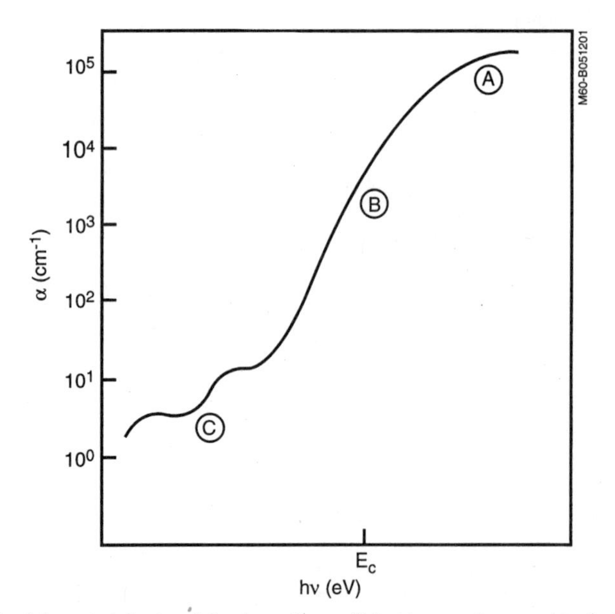

FIGURE 4-61 Schematic behavior of the absorption coefficient in amorphous semiconductors.

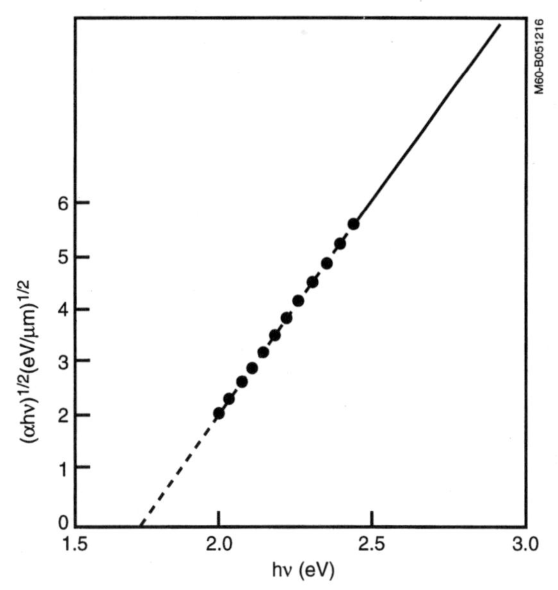

FIGURE 4-62 Absorption characteristic for a-Si:H.

last assumption that is unique to disordered systems and that results in a different energy dependence than the one derived by Eq. 119. The absorption coefficient is written

$$\alpha(h\nu) \sim (1/n)\int N_v(E)N_c(E + h\nu)/h\nu \, dE \qquad (121)$$

where the valence band edge is zero energy.

For a parabolic density of states, that is, $N_c \sim (E - E_c)^{1/2}$, $N_v \sim (-E)^{1/2}$, neglecting the small energy dependence of *n,* the absorption coefficient is

$$\alpha(h\nu) \sim G\,(h\nu - E_c)^2/h\nu \qquad (122)$$

where G is a materials-related constant.

Thus the amorphous semiconductors absorb more strongly in this energy region than do their crystalline counterparts. In crystalline semiconductors, only slices of reciprocal space corresponding to the appropriate *k*-conservation condition can contribute to the absorption. In amorphous semiconductors, any state can contribute to absorption if energy is conserved in the transition. Amorphous Si solar cells, for example, are more efficient absorbers than crystalline Si cells, even though crystalline cells have superior performance characteristics.

Figure 4-62 presents the dependence of the absorption coefficient as a function of $h\nu$ for an a-Si:H film. The intercept defines an optical gap for the amorphous semiconductor, although this should be treated more as an operational definition than as the actual bandgap. Another operational definition is the energy at which the magnitude of α is 10^4/cm. The constant in Eq. 122 can also be modeled to extract the minimum metallic conductivity [271].

The parabolic density of states above the optical gap should not be assumed to have fundamental significance. First, the graph of most $(\alpha h\nu)^{1/2}$ versus $h\nu$ characteristics are not straight lines. Second, the parabolic density of states for crystalline semiconductors comes from a boundary condition on *k,* which can be discounted as a quantized variable. Third, even if the density of states were parabolic near the band edge, it would not be expected to remain parabolic for more than a few tenths of electron volts. This behavior, including the alternative interpretations of the absorption characteristics, has been discussed in detail by Mott and Davis [270].

Region B in Fig. 4-61 is depicted by an exponential dependence of α upon $h\nu$. This behavior is observed in a number of semiconductors, both crystalline and amorphous, and is generally referred to as an Urbach edge [270]. The general form of the dependence is

$$\alpha = \alpha_o \exp\{-[\gamma'(E_o - h\nu)]/kT\} \tag{123}$$

where γ' is a material's related coefficient. (However, the temperature dependence of Eq. 123 is not always observed.)

The origin of the Urbach edge is a matter of discussion. One possible source in amorphous semiconductors is the disorder-induced tail states [270,272]. Other sources cited include exciton broadening and electric-field broadening. (These models are considered again in discussions of polycrystalline semiconductors in the next section.) A relationship in a-Si:H has been reported between the characteristic width of the exponential edge and the optical gap, with lower gaps also giving wider edges. This suggests that H affects the gap only indirectly, through its effect on disorder. It is not yet clear, however, whether this behavior is true of a-Si:H produced by all experimental methods.

Region C (Fig. 4-61) has been explored with PAS and PDS, which allow the probing of the mid-gap region (Fig. 4-60). Theoretical analysis of these optical transitions is more difficult, because the localized nature of the states complicates the treatment of matrix elements for the transition. Optical spectroscopy in this region is an important tool for determining defect levels that significantly affect device performance.

Absorption in Thin Film Polycrystalline Semiconductors. The optical properties of polycrystalline semiconductors are the subject of numerous publications in the literature. However, only a small portion of these actually address the effects of grain boundaries. The implicit assumption is that such effects on the optical properties are small.

It is hard to isolate grain boundary optical effects because it is usually difficult to produce polycrystalline films that differ from a single-crystal analogue by only the grain boundaries. Grain boundaries tend to be a source of impurities. For example, gases such as O_2 may be incorporated (preferentially) at the defects during growth. Additionally, compound semiconductors may have regions of large stoichiometry deviation or even mixed phases. Two models, relating to optical absorption, that consider grain boundary contributions are discussed below.

Mixed Amorphous-Crystalline Semiconductors. Szczyrbowski and Czapla have attempted to describe the optical absorption of sputtered InAs films using a model in which the film is composed of micrograins imbedded in an amorphous matrix. This model also has application to a-Si:H [253,273]. The assumption that the grain boundary region is amorphous is a tenuous one; what is required for optical absorption in this model is that the transition be nondirect (that is, k conservation is not important). The grain boundary could be sufficiently disordered such that the wave vector is not an accurate quantum number, even though the boundary might not be strictly in an amorphous state. Impurities incorporated into the boundary region would modify the effective energy gap.

In this model, the absorption data is fitted to the expression

$$\alpha = \beta\alpha_{\text{crys}} + (1 - \beta)\,\alpha_{\text{amor}} \tag{124}$$

where β is the volume fraction of the crystalline phase. This expression follows from the effective medium approximation if $n_1 \sim n_2 \gg k_1, k_2$. In Eq. 124, α_{cryst} is adapted from Eq. 117, modified to account for degeneracy and nonparabolicity. Equation 123 gives α_{amor}. Figure 4-63 shows a best fit for one of the InAs films, with $\beta = 0.95$, $E_{g(cryst)} = 0.35$ eV, and $E_{g(amor)} = 0.33$ eV. The analysis also yields an electron concentration of $7.5 \times 10^{18}/\text{cm}^3$, which is in reasonable agreement with that obtained from the plasma-resonance measurement ($6.0 \times 10^{18}/\text{cm}^3$).

Bagley et al. have also applied the amorphous-crystalline mixture model to the optical properties of low-pressure chemical vapor deposition (LPCVD) Si [274]. They used the effective medium approximation to account for ϵ_1 and ϵ_2 at energies well above the fundamental absorption edge. In contrast to the InAs studies, this system is known from structural investigations to contain crystalline and amorphous regions. The authors found that the best fit for undoped material was for a mixture of 64% amorphous a-Si and 20% crystalline a-Si with 16% voids.

The Dow-Redfield Model. In a series of papers, Dow and Redfield have shown that one important contribution to the Urbach tail is caused by a Franz-Keldysh effect arising from electric fields present in the semiconductor [275–278]. The electric fields arise from a variety of sources, such as charged impurities, external surfaces, and grain boundaries. Typically these fields are approximately 10^5 V/cm. This effect may be the origin of the Urbach tail in amorphous semiconductors, because electric fields resulting from density and compositional fluctuations may be present.

The Franz-Keldysh effect is basically photon-assisted tunneling. In the presence of an electric field, a photon with energy less than E_g can cause a transition to the conduction band. The absorption coefficient dependence as a function of the field E is

$$\alpha = \mathrm{I}(E)\, E^{1/3} \tag{125}$$

where

$$\mathrm{I} = (h\nu)^{-1} \int A_i^2(z)\, dz \tag{126}$$

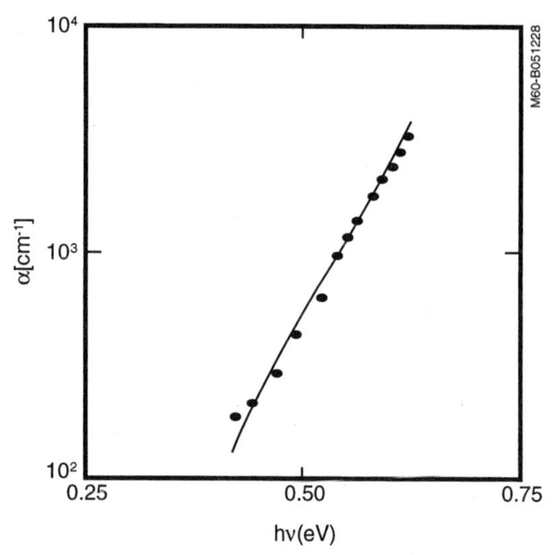

FIGURE 4-63 Absorption coefficient for the sample used in Fig. 4-56, showing correspondence to crystalline-amorphous model. *(From G. D. Cody et al.,* Solar Cells, *vol. 2, 1980, p. 227.)*

and

$$y = \gamma^{1/2} E^{-2/3} (E_g - h\nu) \tag{127}$$

A_i is the Airy function, whereas γ and K are constants depending on properties of the material.

These relations do not account for exciton effects. The Coulomb attraction between the electron and hole causes a significant reduction in the tunneling barrier [279]. Numerical solutions show that the subband gap absorption is enhanced by several orders of magnitude by the inclusion of exciton effects. Despite this, the shape of the absorption characteristic is approximately the same for $E_g - h\nu$ greater than about 1 to 2 Rydbergs (0.035 eV in CdS and 0.055 eV in ZnS). Because the models are normalized to the absorption at the energy gap, neglecting the exciton effects is not critical.

Although the electric fields responsible for this enhanced absorption are generally nonuniform and random, the absorption coefficient for a given energy must be averaged over the field. In the effective medium approximation, the averaging is done over the field distribution:

$$\alpha(h\nu) = \int_o^\infty P(E)\alpha(h\nu,E)dE \tag{128}$$

where P(E) is the probability that there is a field E in the material. Alternatively, the integration can be performed over the volume (V) as

$$\alpha(h\nu = (1/V)\int_V \alpha(\tilde{r},E)dV \tag{129}$$

For polycrystalline CdS films, Gujatti and Marcelja have provided an analysis of the Dow-Redfield formulation [280]. Because their evaporated films were exposed to air during deposition, the presence of oxygen increased the charge on the grain boundaries. The samples were divided into two groups: those composed of grains 0.1 to 0.5 μm in size and those with smaller grains. Potential barriers of approximately 0.3 eV were measured in the films of the first group.

In approximating the effect of the electric field, Bujatti and Marcelja assumed spherical grains with some distribution P(r) of the radius of the spheres. Equation 129 is then written [280]

$$\alpha = (4\pi N/V)\int_o^\alpha P(R)dR\int_{r_m}^R \alpha(r,F)r^2dr, \tag{130}$$

where N is the number of grains in the volume (V) and r_m is the minimum radius, measured from the center of the grain, at which absorption can occur for a given field and photon energy. Two limiting cases are then considered:

1. The number of acceptor-like surface states is so large (and the grain diameter so small) that the donors in the grain are fully ionized, giving a constant charge density. This results in a radially increasing field within the grain.
2. The grains are sufficiently large that only donors near the surface are ionized. In this case, the field decays exponentially from the surface. The field-dependent absorption coefficient (Eq. 125) is then calculated and suitably averaged.

A comparison of the model with the experiment is shown in Fig. 4-64. Group 1 is composed of films with an unusually high population of large grains that almost cover the entire surface. Group 2 is composed of films with only a few such large grains. Group 3 is composed of films with only very small grains (<0.05 μm).

The films in Group 2 can be analyzed. Using the exponential field approximation and Debye length in the range of 10^{-5} to 10^{-6} cm, a surface charge density of $10^{12}/cm^2$ to

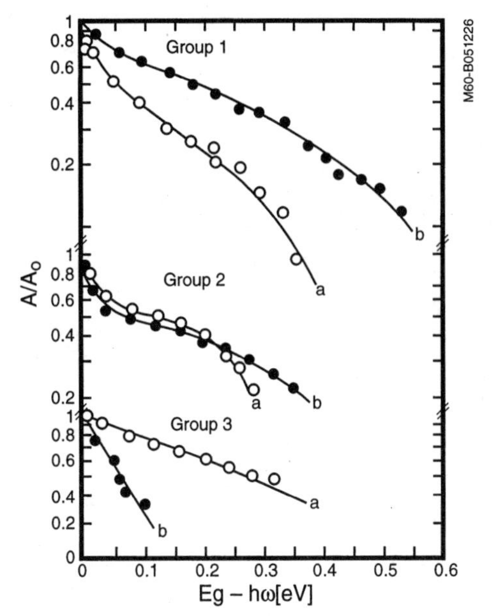

FIGURE 4-64 Absorption coefficient for various CdS polycrystalline films. The absorption coefficient is normalized to unity at the energy gap. The solid lines represent the Dow-Redfield model. *(From J. Callaway,* Phys Rev, *vol. 134, 1964, p. A998.)*

$10^{13}/cm^2$ and potential barriers of 0.3 to 0.4 eV are determined. The films in Groups 2 and 3 can be fit using a sum of exponential and linear field terms to account for the smaller grains. The parameters extracted for these analyses do not yield quantities that are particularly meaningful.

A similar analysis by Bugnet was reported on polycrystalline ZnS thin films [281]. These results are shown in Fig. 4-65. The excess absorption at the bandgap (3.7 eV) could be attributed to the Franz-Keldysh effect with a surface potential of 1.36 eV at the grain

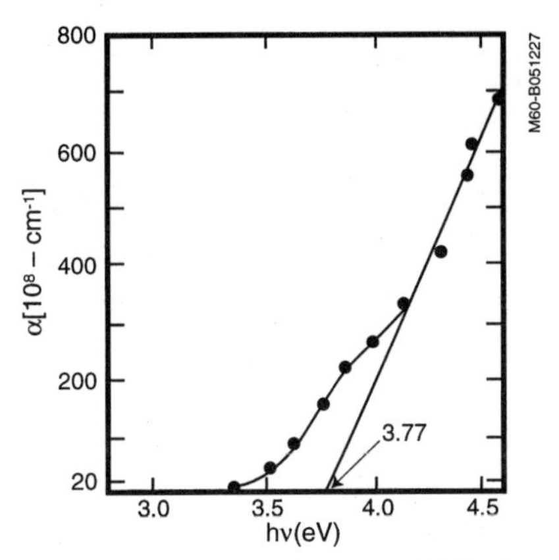

FIGURE 4-65 Absorption coefficient characteristic for sputtered ZnS polycrystalline thin films. *(From M. Bujatti and F. Marcelja,* Thin Solid Films, *vol. 11, 1972, p. 249.)*

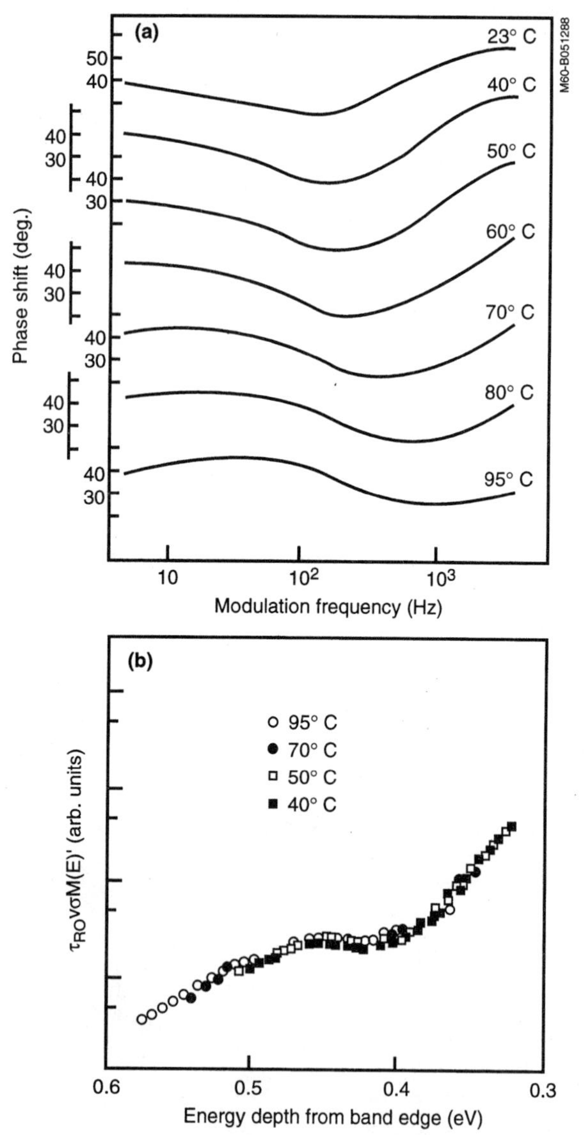

FIGURE 4-66 CdS film photoconductivity data: (*a*) modulation frequency dependence of the phase shift; and (*b*) the distribution of the localized density of states derived from data in (*a*). *(From D. L. Greenway and G. Harbeke,* Optical Properties and Band Structure of Semiconductors, *Pergamon, Oxford, 1968.)*

boundary. This potential seems high, but it is not inconsistent with the value of 1.2 eV reported by Swank from contact potential measurements [282]. At longer wavelengths, the excess absorption cannot be explained by this effect. Furthermore, the *e*-wavelength absorption is very sensitive to preparation conditions when compared to absorption near the band edge. Bugnet suggested that this additional absorption is caused by other defects and impurities [281].

4.4.4 Photoconductivity

Photoconductivity effects in semiconductors have been valuable for both understanding materials properties and for the development of device technologies [283]. For example, photoconductivity methods are important for the investigation of the density of states. This includes the evaluation of the photoconducivity-time decay characteristic and monitoring of the position of the quasi-Fermi level at which the density of states is determined [284–286]. Another approach is to examine the phase shift of the photocurrent with respect to the input excitation source [287–288]. The rate equations are used to determine both the generation and recombination of the photoelectrons and the energy-gap density of states. Such is the case in Fig. 4-66*a,* which shows the phase shift at various modulating frequencies for a CdS thin film. The energy distribution of the energy state density (Fig. 4-66*b*) is derived from the data in figure 4-66*a* [288].

The optical properties of both amorphous and polycrystalline semiconductor films are clearly far from being completely understood. The models presented here should be considered with some caution, because they represent reasonable first attempts to understand the optical properties of rather complex materials systems.

4.5 SUMMARY

With the evolution of modern device technologies, thin films (especially semiconductor thin films) are becoming increasingly important. The applications of thin films range from the nanoscale (quantum dots and computer memory) through the macroscale (displays and solar cells). Because of the intimate link between the understanding of these thin layer films and the operation, performance, and reliability of the devices they are applied to, research on thin films will continue to lead to the production of better components and new technologies.

REFERENCES

1. See references accompanying this chapter.
2. L. L. Kazmerski, ed., *Properties of Polycrystalline and Amorphous Thin Films and Devices,* Academic Press, New York, 1980, Ch 3–6.
3. T. J. Coutts, in *Active and Passive Thin Film Devices,* T. J. Coutts, (ed.), Academic Press, New York, 1978, Ch. 3.
4. L. I. Maissel, *Handbook of Thin Film Technology,* L. I. Maissel, (ed.), McGraw-Hill, New York, 1970, Ch. 13. Also, J. G. Simmons, Ch. 14.
5. C. Anderson, *Advan Phys,* vol. 19, 1970, p. 311.
6. R. H. Bube, *Annual Rev of Mater Sci,* vol. 5, 1975, p. 201.
7. J. L. Vossen and W. Kern, "Introduction" and "Sputter Deposition Processes," in *Thin Film Processes II,* Academic Press, New York, 1991, Ch. 1 and 5.
8. R. V. Stuart, *Vacuum Technology,* "Thin Films, and Sputtering" Academic Press, New York, 1991.
9. S. K. Ghandhi, in S. K. Ghandhi, (ed.), *VLSI Fabrication Principles,* John Wiley and Sons, New York, 1983, Ch. 4.
10. W. Kern, *J Electrochem Soc,* vol. 137, 1990, p. 1982.
11. J. E. Morris and T. J. Coutts, *Thin Solid Films,* vol. 47, 1977, p. 3.

12. L. L. Kazmerski and D. M. Racine, *J Appl Phys,* vol. 46, 1975, p. 791.
13. V. N. E. Robinson and J. L. Robbins, *Thin Solid Films,* vol. 20, 1974, p. 155.
14. G. Binnig et al., *Phys Rev Lett,* vol. 50, 1983, p. 120.
15. L. L. Kazmerski, *J Vac Sci Technol,* vol. B 9, 1991, p. 1549.
16. L. He, E. I, Z. Q. Shi, *J Vac Sci Technol,* 1996, in press.
17. W. B. Phillips, E. A. Desloge, and J. G. Skofronick, *J Appl Phys,* vol. 39, 1968, p. 3210.
18. J. G. Simmons, *J Appl Phys,* vol. 35, p. 2472, 1964.
19. H. Mayer, R. Nossek, and H. Thomas, *J Phys Radium,* vol. 17, 1956, p. 204.
20. T. E. Hartman, *J Appl Phys,* vol. 34, 1963, p. 943.
21. A. A. Milgram and C. Lu, *J Appl Phys,* vol. 37, 1966, p. 4773.
22. C. A. Neugebauer and M. B. Webb, *J Appl Phys,* vol. 33, 1962, p. 74.
23. R. M. Hill, *Proc R Soc Lond,* vol. A 309, 1969, p. 377.
24. S. L. Wu, C. L. Lee, and T. F. Lei, *Appl Phys Lett,* vol. 62, 1993, p. 3491; also, M. L. Knotek and T. M. Donovan, *Phys Rev Lett,* vol. 30, 1973, p. 652.
25. L. L. Kazmerski, D. M. Racine, and M. S. Ayyagari, *J Appl Phys,* vol. 46, 1975, p. 2658.
26. R. F. Greene, *J Phys Chem Solids,* vol. 14, 1960, p. 291.
27. R. F. Greene, *Surf Sci,* vol. 2, 1964, p. 101.
28. R. F. Greene, *Phys Rev,* vol. 141, 1966, p. 687.
29. B. Tavger and W. Kogan, *Phys Rev Lett,* vol. 19, 1965, p. 353.
30. B. Tavger and E. Erukhimov, *Zh Eksper Teor Fiz,* vol. 51, 1966, p. 528.
31. B. Tavger, *Phys Stat Solidi,* vol. 22, 1967, p. 31.
32. See, for example, K. L. Chopra, *Thin Film Phenomena,* McGraw-Hill, New York, 1975, pp. 55–108.
33. A. Many, Y. Goldstein, and N. B. Grover, *Semiconductor Surfaces,* North-Holland, Amsterdam, 1965, Ch. 2.
34. J. L. Moll, *Physics of Semiconductors,* McGraw-Hill, New York, 1964, Ch. 2–4.
35. A. Matthiessen, *Rep Brit Ass,* vol. 32, 1862, p. 144.
36. H. F. Wolf, *Semiconductors,* Wiley-Interscience, New York, 1971, pp. 281–283.
37. K. L. Chopra, *Thin Film Phenomena,* McGraw-Hill, New York, 1975, Ch. 5.
38. A. Many, Y. Goldstein, and N.B. Grover, *Semiconductor Surfaces,* North-Holland, Amsterdam, 1965, Chs. 2 and 4.
39. L. L. Kazmerski, W. B. Berry, and C. W. Allen, *J Appl Phys,* vol. 43, 1972, p. 3515.
40. L. L. Kazmerski, *Thin Solid Films,* vol. 21, 1974, p. 273.
41. E. E. Namimanov, *Fiz I Tek Pol,* vol. 29, 1995, p. 235.
42. Y. Alpern and J. Shappir, *J Appl Phys,* vol. 64, 1988, p. 4987.
43. N. H. Nickel, N. M. Johnson, and C. G. Van de Walle, *Phys Rev Lett,* vol. 72, 1994, p. 3393.
44. C. C. Chen, A. H. Clark, and L. L. Kazmerski, *Thin Solid Films,* vol. 32, 1976, p. L5.
45. C. Juhasz and J.C. Anderson, *Radio Electron Eng,* vol. 33, 1967, p. 223.
46. D. R. Frankl, *Electrical Properties of Semiconductor Surfaces,* Pergamon Press, New York, 1967, pp. 42–45.
47. A. Y. Cho, *J Vac Sci Technol,* vol. 13, 1979, p. 139.
48. R. Hezel and V. Siemens-Forsch, *Entwickl,* vol. 3, 1974, p. 160.
49. J. C. Anderson, *Advan Phys,* vol. 19, 1970, p. 311.
50. A. Amith, *J Phys Chem Solids,* vol. 14, 1960, p. 271.
51. J. N. Zemel, *Phys Rev,* vol. 112, 1958, p. 762.
52. J. N. Zemel and J. O. Varela, *J Phys Chem Solids,* vol. 14, 1960, p. 142.

53. J. N. Zemel, J. D. Jensen, and R. B. Schoolar, *Phys Rev,* vol. 140, 1965, p. A 330.
54. J. N. Zemel and R. L. Petritz, *J Phys Chem Solids,* vol. 8, 1959, p. 102.
55. H. H. Weider, *Thin Solid Films,* vol. 31, 1976, p. 123.
56. V. Snejdar and J. Jerhot, *Thin Solid Films,* vol. 37, 1976, p. 303.
57. L. J. Van der Pauw, *Philips Res Rept,* vol. 13, 1958, p. 1.
58. L. J. Van der Pauw, *Philips Tech Rev,* vol. 20, 1958–59, p. 220.
59. J. D. S. Vlachavas, R. C. Pond, *Inst Phys Conf Ser,* vol. 60, 1989, p. 159.
60. R. F. Egerton, Ph.D. Thesis, University of London, 1969.
61. A. Bourret, *J de Phys,* vol. C4, 1983, p. 44.
62. J. Werner and H. Strunk, *J de Phys,* vol. C1, 1982, p. 99.
63. G. E. Pike, in *Grain Boundaries in Semiconductors,* H. J. Leamy, (ed.), North-Holland, Amsterdam, pp. 2–14 1982; also, *Phys Rev B,* vol. 30, 1984, p. 795.
64. D. Bhattacharyya, S. Chaudhuri, and A. Pal, *Mater Chem and Phys,* vol. 40, 1995, p. 44.
65. B. B. Ismail and R. D. Gould, *Internl J Electron,* vol. 78, 1995, p. 261.
66. T. Hayashi et al., *Japn J Appl Phys,* vol. 30, 1991, p. 2449.
67. B. B. Ismail, Ph.D., Thesis, Univ. Keele, United Kingdom, 1990.
68. J. C. Bernede et al., *J de Phys III,* vol. 4, 1994, p. 677.
69. A. J. Simons and C. B. Thomas, *Phl Mag,* vol. B 68, 1993, p. 465.
70. A. Jankowska-Frydel and M. Lepek, *Cryst Res and Technol,* vol. 24, 1989, p. 567.
71. D. Bhattacharyya, S. Chaudhuri, and A. Pal, *Mater Chem and Phys,* vol. 40, 1995, p. 44.
72. B. B. Ismail and R. D. Gould, *Internl J Electron,* vol. 78, 1995, p. 261.
73. T. Hayashi et al., *Japn J Appl Phys,* vol. 30, 1991, p. 2449.
74. B. B. Ismail, Ph.D., Thesis, Univ. Keele, United Kingdom, 1990.
75. N. V. Gorev, E. F. Prokhorov, and A. T. Ukolov, *Russian Microelectronics,* vol. 23, 1994, p. 23.
76. D. P. Saha, P. S. Basak, and R. P. Singh, *Indian J Pure and Appl Phys,* vol. 32, 1994, p. 171.
77. A. K. Collins, M. A. Pickering, and R. L. Taylor, *J Appl Phys,* vol. 68, 1990, p. 6510.
78. S. Martinuzzi et al., *Rev Phys Appl,* vol. 22, 1987, p. 645.
79. J. Palm, D. Steinbach, and H. Alexander, in *Polycrystalline Semiconductors III,* H. P. Strunk et al., (eds.), Scitech Publ, Switzerland, 1994, pp. 183–188.
80. M. Acciarri et al., *Phys Stat Sol,* vol. A 138, 1993, p. 451; also, M. Acciarri et al., in *Polycrystalline Semiconductors III,* H. P. Strunk et al., (eds.), Scitech Publ, Switzerland, 1994, pp. 219–224.
81. S. Martinuzzi et al., *Proc 13th European Photovoltaic Solar Energy Conf, Nice,* Kluwer Scientific, The Netherlands, 1995; also, S. Martinuzzi, I Perichaud, and M. Stemmer, in *Polycrystalline Semiconductors III,* H. P. Strunk et al., (eds.), Scitech Publ., Switzerland, 1994, pp. 361–366.
82. J. W. Orton et al., *J Appl Phys,* vol. 53, 1982, p. 1602.
83. J. Palm, D. Steingach, and H. Alexander, in *Polycrystalline Semiconductors III,* H. P. Strunk et al., (eds.) Scitech Publ, Switzerland, 1994, pp. 183–188.
84. O. Ka, in *Polycrystalline Semiconductors III,* H. P. Strunk et al., (eds.) Scitec Publ, Switzerland, 1994, pp. 201–212.
85. L. M. Fraas, *J Appl Phys,* vol. 49, 1978, p. 871.
86. C. Donolato, in *Polycrystalline Semiconductors,* Springer-Verlag, 1984, pp. 138–154.
87. C. Donolato, *J Appl Phys,* vol. 54, 1983, p. 1314.
88. R. W. Glew, *Thin Solid Films,* 1977, pp. 52, 59; also, J. Mannhart et al., *Phys Rev Lett,* vol. 61, 1988, p. 2476.
89. J. Volger, *Phys Rev,* vol. 9, 1950, p. 1023.

90. R. L. Petritz, *Phys Rev,* vol. 104, 1956, p. 1508.
91. H. Berger, W. Kahle, and G. Janiche, *Phys Stat Solidi,* vol. 28, 1968, p. K97.
92. H. Berger, *Phys Status Solidi,* vol. 1, 1961, p. 739.
93. R. G. Mankarious, *Solid-State Electron,* vol. 7, 1964, p. 702.
94. L. L. Kazmerski, M. S. Ayyagari, and G. A. Sanborn, *J Appl Phys,* vol. 46, 1975, p. 4685.
95. L. L. Kazmerski et al., *J Vac Sci Technol,* vol. 13, 1976, p. 139.
96. L. L. Kazmerski and Y.J. Juang, *J Vac Sci Technol,* vol. 14, 1977, p. 769.
97. L. L. Kazmerski et al., *J Vac Sci Technol,* vol. 14, 1977, p. 65.
98. R. Kassing and W. Bax, *Japn J Appl Phys,* suppl. 2, 1974, p. 801.
99. V. Damodara Das and C. Bahulayan, *Solid-State Comm,* vol. 93, 1995, p. 949.
100. B. A. Mansour and M. A. El-Hagary, *Phys Stat Sol,* vol. A 146, 1994, p. 669.
101. Y. Tang, R. Braunstein, and B. von Roedern, *Appl Phys Lett,* vol. 63, 1993, p. 2393.
102. B. R. Sethi et al., *Phys Status Sol,* vol. A 134, 1992, p. 151.
103. S. K. Viswanathan et al., *Proc Conf Phys and Tech of Semicond Dev and Integ Circuits,* SPIE, Washington, 1992, pp. 453–455.
104. M. Y. Satyanarayana, N. B. Srinvasulu, and R. P. Jayarama, *Phys Stat Sol,* vol. A 115, 1989, p. 175.
105. A. K. Collins, M. A. Pickering, and R. L. Taylor, *J Appl Phys,* vol. 68, 1990, p. 6510.
106. F. Mansour et al., *Thin Solid Films,* vol. 261, 1995, p. 12.
107. W. K. Chai, *Phys Stat Sol,* vol. A 147, 1995, p. K17.
108. N. Yamauchi, N. Kakuda, and T. Hisaki, *IEEE Trans Electron Dev,* vol. 41, 1994, p. 1882; also, N. Lifshitz and S. Luryi, *IEEE Electron Dev Lett,* vol. 15, 1994, p. 274.
109. C. A. Dimitriadis et al., *Acta Phys Hung,* vol. 74, 1994, p. 155.
110. T. Serikawa et al., *Japn J Appl Phys (Lett),* vol. 33, 1994, p. L409.
111. E. A. Schiff et al., *Amorphous Silicon Technology Mater Res Soc,* Pittsburgh, 1993, pp. 545–550.
112. H. Sehil et al., *Solid-State Electron,* vol. 37, 1994, p. 159.
113. S. Marco et al., *Sensors and Actuators A,* vol. 68 1993, pp. A37–A38.
114. C. A. Dimitiadis, J. Stoemenos, P. A. Coxon, S. Friligkos, J. Antonopoulos, and N. A. Economou, *J Appl Phys,* vol. 73, 1993, p. 8402.
115. H.-J. Gossmann and F. C. Unterwald, *Phys Rev B,* vol. 47, 1993, p. 12618.
116. P. S. Basak and D. P. Singh, *Indian J Pure and Appl Phys,* 31, 1993, 271.
117. N. Sakakibara, S. Fujino, H. Muramoto, T. Hattori, and K. Goto, *J Appl Phys,* vol. 73, 1993, p. 2590.
118. S. Lombardo, S. U. Campisano, and F. Baroeeo, *Microelectron Eng,* vol. 19, 1992, p. 601.
119. M. Yazaki, S. Takenaka, and H. Ohshima, *Japn J Appl Phys,* vol. 31, 1992, p. 206.
120. K. Shimizu et al., *Japn J Appl Phys,* vol. 30, 1991, p. 3704.
121. V. G. Kobka et al., *Neo Mat,* vol. 26, 1990, p. 1362.
122. G. Queirolo et al., *J Electrochem Soc,* vol. 137, 1990, p. 967.
123. A. Anagnostopoulos, G. L. Bleris, and I. Y. Yanchev, *Phys Stat Sol,* vol. B 158, 1990, p. K43.
124. M. R. I. Ramadan and M. M. Elsherbiny, *Solar and Wind Technol,* vol. 7, 1990, p. 107.
125. S. Verghese et al., *IEEE Trans Electron Dev,* vol. 36, 1989, p. 1311.
126. R. D. Black, *J Appl Phys,* vol. 63, 1988, p. 2458.
127. M. Ada-Hanifi et al., *J Appl Phys,* vol. 63, 1988, p. 2311.
128. M. E. Crowder and T.O. Sedgewick, *J Electrochem Soc,* vol. 119, 1972, p. 1565.
129. A. L. Fripp, *J Appl Phys* vol. 46, 1975, p. 1240; also, A. L. Fripp and L. R. Slack, *J Electrochem Soc,* vol. 120, 1973, p. 145.

130. T. I. Kamins, *J Appl Phys,* vol. 42, 1971, p. 4357.
131. K. Sagara and E. Murakami, *Appl Phys Lett,* vol. 54, 1989, p. 2003.
132. P. Rai-Choudhury and P. L. Hower, *J Electrochem Soc,* vol. 120, 1973, p. 1761.
133. J. Y. W. Seto, *J Electrochem Soc,* vol. 122, 1975, p. 701.
134. J. Y. W. Seto, *J Appl Phys,* vol. 46, 1975, p. 5247.
135. W. E. Taylor, N. H. Odell, and H. V. Fan, *Phys Rev,* vol. 88, 1952, p. 867.
136. R. K. Mueller, *J Appl Phys,* vol. 32, 1961, p. 635.
137. R. K. Mueller, *J Appl Phys,* vol. 32, 1961, p. 640.
138. G. Baccarani, B. Ricco, and G. Spandini, *J Appl Phys,* vol. 49, 1978, p. 5565.
139. C. H. Seager and T. G. Castner, *J Appl Phys,* vol. 49, 1978, p. 3879.
140. G. Baccarani et al., *Proc 3rd Internl Symp Silicon Mater Sci and Technol,* Philadelphia, 1977.
141. T. I. Kamins, *IEEE Trans Parts Hyb Packag,* VHP-10, 1974, p. 221.
142. J. Manoliu and T. I. Kamins, *Solid-State Electron,* vol. 15, 1972, p. 1103.
143. B. Raicu, T. Kamins, and C. V. Thomson, in *Polysilicon Thin Films and Interfaces, Mater Res Soc,* vol. 59, Pittsburgh, 1986.
144. T. I. Kamins, *J Electrochem Soc,* vol. 127, 1980, p. 686; also, vol. 126, 1979, p. 838.
145. M. Mandurah et al., *J Appl Phys,* vol. 51, 1980, p. 5755.
146. A. Many, V. Goldstein, N. B. Grover, *Semiconductor Surfaces,* North-Holland, Amsterdam, 1965, Ch. 5 and 9.
147. S. T. Pantelides, (ed.), *The Physics of SiO_2 and Its Interfaces,* North-Holland, Amsterdam, 1965.
148. N. C. C. Lu et al., *IEEE Trans Electron Dev* ED-30, 1983, p. 137; also, C. Y. Wu and N. D. Kern, *Solid-State Electron,* vol. 26, 1983, p. 675.
149. C. H. Seager and T. G. Castner, *Bull Am Phys Soc,* vol. 22, 1977, p. 434.
150. J. W. Cleland et al., *Proc Natl Workshop on Low-Cost Polycrystalline Silicon Solar Cells,* Dallas, 1977, pp. 113–117.
151. M. Kohyama and R. Yamamoto, in *Polycrystalline Semiconductors III,* H. P. Strunk, et al., (eds.), Scitec Publ., Switzerland, 1994, pp. 55–66; also, A. Broniatowski, ref 162, pp. 95–117.
152. H. Hwang, H. C. Card, and E. S. Yang, *Appl Phys Lett,* vol. 36, 1980, p. 3155; also, H. C. Card and E. S. Yang, *IEEE Trans Electron Dev,* vol. ED-24, 1977, p. 397.
153. L. L. Kazmerski, *Solid-State Electron,* vol. 21, 1978, p. 1545.
154. A. Rockett and R. W. Birkmire, *J Appl Phys,* vol. 70, 1992, p. R81.
155. P. T. Landsberg and C. M. Kimpke, *Proc 13th IEEE Photovoltaics Spec Conf,* New York, 1978, pp. 665–666.
156. L. M. Fraas, *J Appl Phys,* vol. 49, 1978, p. 871.
157. C. H. Seager, *J Appl Phys,* vol. 52, 1981, p. 3960; also, C. H. Seager and G. E. Pike, *Appl Phys Lett,* vol. 37, 1980, p. 747.
158. C. H. Seager et al., *J Vac Sci Technol,* vol. 20, 1982, p. 430.
159. C. H. Seager, in *Grain Boundaries in Semiconductors,* North Holland, New York, 1982, pp. 85–98.
160. L. L. Kazmerski, *Surf Sci Reports,* vol. 19, 1993, p. 1969.
161. H. J. Möller, *Solar Cells,* vol. 31, 1991, p. 77.
162. J. Werner, in G. Harbeke, (ed.), *Polycrystalline Semiconductors,* Springer-Verlag, New York, 1985, pp. 76–94; also, J. Werner and H. Strunk, *J de Phys,* vol. C1, 1982, p. 99.
163. C. Feldman et al., *J Electron Mat,* vol. 7, 1978, p. 309.
164. R. Singh et al., *J Vac Sci Technol,* vol. 13, 1979.
165. D. J. Thompson and H. C. Card, *J Appl Phys,* vol. 54, 1983, p. 1976.

166. C. H. Seager et al., *J Phys,* vol. 43, 1982, pp C1–103.
167. C. Habler et al., *Phys Stat Sol,* vol. A 173, 1993, p. 463; also, A. Barhdadi et al., in *Polycrystalline Semiconductors III,* H. P. Strunk et al., (eds.), Scitech Publ, Switzerland, 1994, pp. 189–194.
168. U. Creutzburg et al., *Proc Ninth EC Photovoltaic Solar Energy Conf,* Kluwer Publ, The Netherlands, 1988, pp. 9–12.
169. A. A. Zahab, M. Abd-Lefdil, and M. Cadene, *Phys Stat Sol,* vol. A 117, 1990, p. K103.
170. A. Bourret and J. Baccman, *Rev Phys Appl,* vol. 22, 1987, p. 53; also, *Polycrystalline Semiconductors,* Springer-Verlag, New York, 1984, pp. 2–26.
171. J. H. Yee, *The Systematic Computation of the Performance of Photovoltaic Cells Based on First Principles,* DOE Report W-7405-ENG-48, 1979.
172. J. Y. Leong and J. H. Yee, *Appl Phys Lett,* 1979.
173. K. M. Doshchanov, *Fiz I Tek Pol,* vol. 28, 1994, p. 1645.
174. J. H. Werner, *Diffusion and Defect Data, Part B,* pp. 37–38, vol. 213, 1994.
175. V. Snejdar and J. Jerhot, *Thin Solid Films,* vol. 37, 1976, p. 202.
176. See the series of bibliographies of amorphous silicon material, published in *Solar Cells* (Elsevier Scientific, Switzerland): vol. 4, 1981, pp. 205–448; vol. 7, 1983, pp. 347–425; vol. 10, 1984, pp. 315–376; vol. 13, 1985, pp. 319–422; vol. 17, 1986, pp. 397–504.
177. R. A. Street and N. F. Mott, *Phys Rev Lett,* vol. 35, 1975, p. 1293; also, R. A. Street, in *Semiconductors and Semimetals,* J. I. Pankove, (ed.), Academic Press, New York, 1984, pp. 191–199.
178. N. F. Mott, *Adv Phys,* vol. 16, 1976, p. 49.
179. S. T. Pantelides, *Phys Rev Lett,* vol. 57, 1986, p. 2979.
180. N. M. Johnson et al., *Appl Phys Lett,* vol. 53, 1988, p. 1626.
181. S. G. Bishop, U. Strom, and P. C. Taylor, *Phys Rev,* vol. B 15, 1977, p. 2278.
182. A. Madan, in *Silicon Processing for Photovoltaics I,* C. P. Khattak and K. V. Ravi, (eds.), North-Holland, Amsterdam, 1985, pp. 331–376.
183. D. G. Cahill, M. Katiyar, and J. R. Abelson, *Phys Rev,* vol. B 50, 1994, p. 6077.
184. A. Nagy et al., *J Non-Cryst Sol,* vol. 529, 1993, pp. 164–166.
185. A. I. Yakimov, N. P. Stepina, and A. V. Dvurechenskii, *Zhur Eksp I Teor Fiz,* vol. 102, 1992, p. 1882.
186. M. Y. Jung, S. S. Yoo, and J. Jang, *J Kor Instit Telematics and Electron,* vol. 29A, 1992, p. 463.
187. H. Uchida et al., *Japn J Appl Phys,* vol. 30, 1991, p. 3691.
188. G. N. Parsons, *IEEE Trans Electron Dev Lett,* vol. 13, 1992, p. 80.
189. M. S. Aida et al., *Thin Solid Films,* vol. 1, 1992, p. 207.
190. H. Miki et al., *Japn J Appl Phys,* vol. 30, 1991, p. 2740.
191. J.-L. Lin, W.-J. Sah, and S.-C. Lee, *IEEE Electron Dev Lett,* vol. 12, 1991, p. 120.
192. J. Kanicki et al., *Proc Amorph Si Technol,* Mater Res Soc, Pittsburgh, 1989, pp. 239–46.
193. S. S. Hegedus and J. M. Cebulka, *J Appl Phys,* vol. 67, 1990, p. 3885.
194. K. R. Kumar and K. Sathianandan, *Indian J of Phys,* vol. A 63A, 1988, p. 501.
195. N. Lustig and W. E. Howard, *Solid-State Comm,* vol. 72, 1989, p. 56.
196. M. N. Makadsi and M. F. A. Alias, *Phys Rev,* vol. B 38, 1988, p. 6143.
197. G. Micocci, R. Rella, and A. Tepore, *Thin Solid Films,* vol. 172, 1989, p. 179.
198. M. Kumru, *Thin Solid Films,* vol. 198, 1991, p. 75.
199. R. D. Gould and B. B. Ismail, *Phys Stat Sol,* vol. A 134, 1992, p. K65.
200. N. Lustig and W. E. Howard, *Solid-State Comm,* vol. 72, 1989, p. 56.
201. W. Beyer and H. Overhof, in *Amorphous Semiconductors,* vol. 21 Part C, J. L. Pankove, (ed.), Academic Press, New York, 1984, pp. 258–308.

202. B. Divon and F. S. Barnes, *Solar Cells,* vol. 6, 1982, p. 125.

203. M. Konagai et al., *16th IEEE Photovoltaic Spec Conf,* IEEE, New York, 1982, pp. 1321–1326.

204. Z. H. Jan, R. H. Bube, and J. C. Knights, *J Appl Phys,* vol. 51, 1981, p. 3278.

205. A. Madan et al., *Appl Phys Lett,* vol. 37, 1982, p. 826.

206. P. B. Kirby et al., *Phys Rev B,* vol. 28, 1983, p. 3635.

207. H. Fritzsche, *Solid-State Comm,* vol. 9, 1971, p. 1813; also, R. Kubo, *Phys Rev,* vol. 86, 1952, p. 929.

208. M. Gurvitch, *Phys Rev B,* 24, 1981, 7404; also, A. A. Gogolin, *Zeits fur Phys B,* vol. 52, 1983, p. 19.

209. N. F. Mott and M. Kaveh, *Phil Mag B,* vol. 61, 1990, p. 147.

210. P. W. Anderson, *Phys Rev Lett,* vol. 34, 1975, p. 953.

211. P. W. Anderson, *Rev Mod Phys,* vol. 50, 1978, p. 191.

212. M. H. Cohen, H. Fritzsche, and S. R. Ovshinsky, *Phys Rev Lett,* vol. 22, 1969, p. 1065.

213. M. G. Grunewald and P. Thomas, *Phys Status Sol B,* vol. 94, 1979, p. 125; also, V. Ambegaokar, B. I. Halperin, and J. S. Langer, *J Non-Cryst Solids,* vol. 497, 1972, pp. 8–10.

214. N. F. Mott, *Adv Phys,* vol. 16, 1967, p. 49; *Phil Mag,* vol. 19, 1969, p. 835.

215. P. G. LeComber, D. I. Jones, and W. E. Spear, *Phil Mag,* vol. 35, 1977, p. 1173.

216. N. F. Mott and E. A. Davis, *Electron Processes in Noncrystalline Materials,* Claredon Press, Oxford, 1971.

217. W. Meyer and H. Neldel, *Z Tech Phys,* vol. 18, 1937, p. 588.

218. See, for example, P. Blood and J. W. Orten, *The Electrical Characterization of Semiconductors: Majority Carriers and Electronic States,* Academic Press, New York, 1992 pp. 435–436.

219. W. E. Spear and P. G. LeComber, in *Hydrogenated Amorphous Silicon I,* J. D. Joannopoulos and G. Lucovsky (eds.), Springer Verlag, New York, 1984, pp. 63–118; also, K. D. Mackenzie, P. G. LeComber, and W. E. Spear, *Phil Mag B,* vol. 46, 1982, p. 377.

220. D. V. Lang, J. D. Cohen, and J. P. Harbison, *Phys Rev B,* vol. 25, 1982, p. 3285; also, J. Kocha et al., *Proc European Photovoltaic Solar Energy Conf,* Kluwer Scientific, The Netherlands, 1982, pp. 433–447.

221. W. Luft, B. Stafford, and B. von Roedern, in *Photovoltaic Advanced Research & Development Project,* R. Noufi, (ed.), AIP, New York, 1992, pp. 347–356.

222. D. L. Staebler and C. R. Wronski, *Appl Phys Lett,* vol. 31, 1977, p. 292.

223. D. L. Staebler and C. R. Wronski, *J Appl Phys,* vol. 51, 1980, p. 3262.

224. C. R. Wronski, in P. I. Pankove, ed. *Semiconductors and Semimetals,* vol. 21, pt C, Academic Press, New York, 1984, pp. 317–362.

225. D. L. Staebler, R. S. Crandall, and R. Williams, *Appl Phys Lett,* vol. 28, 1981, p. 105; also, D. Staebler, *J Non-Cryst. Solids,* vols. 35 and 36, 1980, p. 387.

226. R. S. Crandall, *Phys Rev B,* vol. 24, 1981, p. 7457; also, A. E. Delahoy and R. W. Griffith, *Proc 15th IEEE Photovoltaic Spec Conf,* IEEE, New York, 1981, pp. 704–707; also, P. E. Vanier, A. E. Delahoy, and R. W. Griffith, *J Appl Phys,* vol. 52, 1981, p. 5235; also, R. A. Street, D. K. Biegelsen, and J. C. Knights, *Phys Rev B,* vol. 24, 1981, p. 969; also, R. Crandall, *Phys Rev Lett,* vol. 44, 1980, p. 749.

227. S. T. Pantelides, *Phys Rev B,* vol. 36, 1987, p. 3479.

228. R. Arya, A. Catalano, and J. Newton, *Appl Phys Lett,* vol. 49, 1986, p. 1089; also, V. Dalal et al., in R. Noufi, (ed.), *Photovoltaic Advanced Research & Development Project,* Conf. Series 268, Amer. Instit. of Phys., New York, 1992, pp. 388–394.

229. Y. Tang, and R. Braunstein, *Appl Phys Lett,* vol. 66, 1995, p. 721.

230. K. Shimakawa, *J Non-Cryst Solids,* vol. 43, 1981, p. 229.

231. R. A. Street, *Phys Rev Lett,* vol. 49, 1982, p. 1187.

232. I. Hirabayashi, K. Morigaki, and S. Nitta, *Japn J Appl Phys,* vol. 19, 1980, p. L357.

233. A. Madan et al., *J Phys,* vol. 42 C, 1981, pp. 4–463.

234. J. D. Cohen, in *Amorphous Semiconductors,* vol. 21, pt C, J. L. Pankove, (ed.), Academic Press, New York, 1984, pp. 9–98.

235. Y. Uchida et al., *Solar Cells,* vol. 9, 1983, p. 3.

236. H. Fritzsche, J. Kakalios, and D. Bernstein, *Optical Effects in Amorphous Semiconductors,* American Institute of Physics, New York, 1984, pp. 229–233.

237. M. Stutamann, W. B. Jackson, and C. C. Tsai, *Phys Rev B,* vol. 32, 1985, p. 23.

238. M. S. Brandt and M. Stutzmann, *J Non-Cryst Sol,* vol. 211, 1985, pp. 137–138.

239. D. E. Carlson et al., ref. 236, pp. 234–241.

240. R. A. Street, J. Kakalios, and C. C. Tsai, *Phys Rev B,* 35, 1987, p. 1316; also, vol. 37, 1988, p. 4209.

241. D. Adler, *J Solid State Chem,* vol. 45, 1982, p. 40; also, R. Grogorovici, *Solar Energy Mat,* vol. 8, 1982, p. 177; also, S. Keleman, Y. Goldstein, and B. Abeles, *Surf Sci,* vol. 116, 1982, p. 488.

242. L. Mrig, Y. Caiyem, and D. Waddinton, in *Photovoltaic Advanced Research & Development Project,* R. Noufi, (ed.), AIP, New York, 1992, pp. 429–444.

243. T. Glatfelter, *Proc Photovoltaic Perf and Reliability Workshop,* NREL/CP-411–7414, 1994, p. 125.

244. S. Guha, *Final Subcontract Report,* NREL/TP-411-7190, 1994, p. 38.

245. W. A. Nevin, H. Yamagishi, and H. Tawada, *Appl Phys Lett,* vol. 54, 1989, p. 1226.

246. B. von Roerdern and L. Mrig, *Proc 13th European Photovoltaics Solar Energy Conf,* Kluwer Scientific, Nice, France, The Netherlands, 1995.

247. J. Pankove, *Optical Processes in Semiconductors,* Prentice-Hall, New Jersey, 1971.

248. J. S. Blakemore, *Solid State Physics,* Sanders, Philadelphia, 1974.

249. A. Rose, *Concepts in Photoconductivity and Allied Problems,* Krieger, New York, 1978.

250. H. A. Kramers, *Physica,* vol. 1, 1934, p. 182; also, L. Ley, in *The Physics of Hydrogenated Amorphous Silicon,* T. D. Joannopoulos and G. Lucovsky, (eds.) Springer-Verlag, New York, 1983, pp. 62–64.

251. G. D. Cody et al., *Solar Cells,* vol. 2, 1980, p. 227.

252. O. S. Heavens, *Optical Properties of Thin Solid Films,* Butterworths, London, 1955.

253. J. Szczyrbowski and A. Czapla, *Thin Solid Films,* vol. 46, 1977, p. 127.

254. J. C. Manifacier et al., *Thin Solid Films,* vol. 41, 1977, p. 127.

255. T. S. Moss, *Optical Properties of Semiconductors,* Butterworths, London, 1959.

256. J. C. Manifacier, J. Gassiot, and J. P. Fillard, *J Phys,* vol. E 9, 1976, p. 1002.

257. H. E. Bennett and J. M. Bennett, *Physics of Thin Films,* G. Hass and R. E. Thun, (eds.), vol. 4, Academic Press, New York, 1967.

258. R. M. A. Azzam and N. M. Bashara, *Ellipsometry and Polarized Light,* North-Holland, Amsterdam, 1977; also, N. M. Bashara and S. C. Assam, (eds.), *Ellipsometer,* North-Holland, Amsterdam, 1976.

259. J. B. Theeten and D. E. Aspnes, *Am Rev Mater Sci,* vol. 11, 1981, p. 97.

260. P. S. Hauge, *Surf Sci,* vol. 96, 1980, p. 108.

261. I. Filinski, *Phys Status Solidi B,* vol. 49, 1972, p. 577.

262. Y. H. Pao, (ed.), *Optoacoustic Spectroscopy and Detection,* Academic Press, New York, 1977.

263. S. Yamasaki et al., *Tetrahedrally Bonded Amorphous Semiconductors,* R. A. Street, D. K. Biegelsen, J. C. Knights (eds.), AIP Conf. Series, no. 73, American Institute of Physics, New York, 1981, p. 258.

264. W. B. Jackson et al., *Appl Optics,* vol. 20, 1981, p. 1333.

265. J. C. Murphy and L. C. Aamodt, *J Appl Phys,* vol. 51, 1980, p. 4580.

266. A. C. Boccara, D. Fournier, and J. Badoz, *Appl Phys Lett,* vol. 36, 1980, p. 130.
267. D. L. Greenaway and G. Harbeke, *Optical Properties and Band Structure of Semiconductors,* Pergamon, Oxford, 1968.
268. F. Bassani and G. Pastori-Panavicini, *Electronic States and Optical Transitions in Solids,* Pergamon, Oxford, 1975.
269. J. Tauc, *The Optical Properties of Solids,* in F. Abeles, (ed.), North-Holland, Amsterdam, 1970.
270. N. F. Mott and E. A. Davis, *Electronic Processes in Non-Crystalline Materials,* Clarendon, Oxford, 1979.
271. J. Tardy and R. Meajdre, *Solid State Comm,* vol. 39, 1981, p. 1031.
272. G. D. Cody et al., *Phys Rev Lett,* vol. 47, 1981, p. 1480; also, G. D. Cody, in J. Pankove, (ed.), *Semiconductors and Semimetals,* Academic Press, New York, 1984.
273. J. G. Hernandez and R. Tsu, *Bull Am Phys Soc,* vol. 27, 1982, p. 206.
274. B. G. Bagley et al., *Appl Phys Lett,* vol. 38, 1981, p. 56.
275. D. Redfield, *Phys Rev,* vol. 130, 1963, p. 916; also, *Solid-State Comm,* vol. 44, 1982, p. 1347.
276. D. Redfield and M. A. Afromowitz, *Appl Phys Lett,* vol. 11, 1967, p. 138.
277. J. D. Dow and D. Redfield, *Phys Rev B,* vol. 1, 1990, p. 3358.
278. J. D. Dow and D. Redfield, *Phys Rev B,* vol. 5, 1972, p. 594.
279. J. Callaway, *Phys Rev,* vol. 134, 1964, p. A998.
280. M. Bujatti and F. Marcelja, *Thin Solid Films,* vol. 11, 1972, p. 249.
281. P. Bugnet, *Rev Phys Appl Fr,* vol. 9, 1974, p. 447.
282. R. K. Swank, *Phys Rev,* vol. 153, 1967, p. 44.
283. R. H. Bube, *Photoconductivity in Solids,* John Wiley, New York, 1960; also, R. S. Crandall, in J. L. Pankove, (ed.), *Amorphous Semiconductors,* vol. 21, pt A, Academic Press, New York, 1984.
284. H. B. deVore, *RCA Review,* vol. 20, 1959, p. 79.
285. R. H. Bube, *J Appl Phys,* vol. 38, 1967, p. 3515.
286. V. den Boer, *J de Phys,* vol. 42, 1981, p. 451.
287. H. Oheda, *J Appl Phys,* vol. 52, 1981, p. 6693.
288. K. Abe, *Phil Mag B,* vol. 58, 1988, p. 171.

CHAPTER 5
DESIGN GUIDELINES FOR THIN FILM COMPONENTS

Aicha A. R. Elshabini
Fred Barlow
The Bradley Department of Electrical and Computer Engineering
Virginia Polytechnic Institute and State University

5.1 INTRODUCTION

The design of thin film components and circuits uses insulating substrates on which conductor paths for interconnects and passive components are deposited by means of vacuum evaporation, sputtering, or chemical vapor deposition techniques. Attached components and devices may include resistors, capacitors, inductors, and various semiconductor devices.

Thin film hybrid circuits use very little space on the substrate surface because of the nature of fine-line realization. They also perform well compared to the etched, printed circuit boards and the custom monolithic integrated circuits. This is because of the shorter circuit paths, tighter thermal coupling, superior control of parasitics, and tighter component tolerances that can be achieved with PCBs. In addition, thin film technology is highly reliable because of the small number of interconnections, and quick turnaround for prototypes that are readily adaptable to any design modifications. Troubleshooting thin film hybrids is a reasonable task that can be successfully achieved because of the functional blocks. In fact, thin film technology allows for rework and possesses a good ability for repair. Thin film hybrids are very advantageous in high frequency applications (exceeding 500 MHz). Deposited multichip modules (MCM-D) evolve as a natural development of thin film hybrid circuitry.

This chapter presents design guidelines to realize resistor, capacitor, and inductor elements, in addition to transmission lines and resonators. It also includes design guidelines for MCM-D technology. Thermal considerations, component attachment, and trimming techniques are included in the discussion.

The subjects of thin film deposition, thin film hybrids, and MCM-D technology are covered in Chapter 9 and they are well documented in the literature [1–36].

5.2 THIN FILM DESIGN GUIDELINES

5.2.1 Resistor Design Rules

A thin film resistor consists of a resistive path deposited on an insulating substrate. Deposited metal end terminations form the required contacts. The resistance value of a rectangular or bar pattern in ohms is (R) is of the form

$$R = \rho L/tW \tag{1}$$

where ρ = the specific or bulk resistivity of the material in Ω·cm
L = length of the resistor film in cm
W = width of the resistor film in cm
t = thickness of the resistor film in centimeter.

The thickness t that can be achieved varies typically from 50 to 10,000 Å. Uniform thickness can be obtained on smooth substrate surfaces (such as glass) as compared to ceramic surfaces. Equation 1 can be rewritten as

$$R = R_s L/W \tag{2}$$

where R_s refers to the sheet resistivity of the resistor ($R_s = \rho/t$) in Ω/square, and L/W refers to the aspect ratio of the resistor and defines the number of squares needed to realize a certain resistance value (see Fig. 5-1).

For a certain thin film substrate, one sheet resistivity is normally used to achieve all resistance values with different numbers of squares, thus determining different lengths of straight or serpentine resistor lines. Sheet resistivity is a widely used parameter for

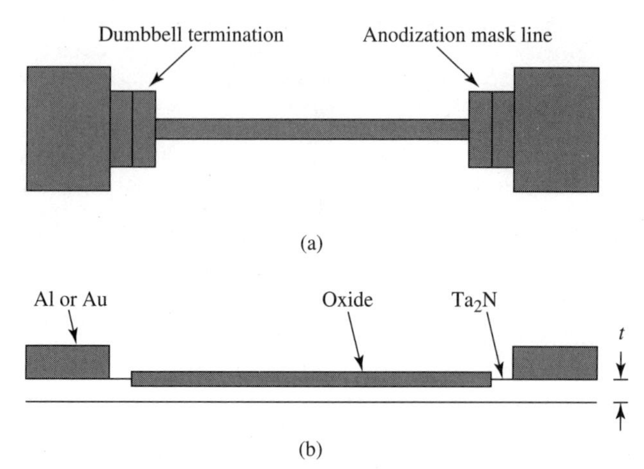

FIGURE 5-1 Typical thin film resistor structure. *(a)* Top view; *(b)* cross section.

comparing films, particularly those films of the same material deposited under similar conditions.

To achieve a high resistance value in a small area, it is desirable to have resistor films with high sheet resistance values. On the other hand, one can observe that the higher the sheet resistivity value of the thin film resistor material, the more difficult it is to control its stabilization.

Resistor patterns can be of rectangular shape or meandering shape or a combination of the two patterns. The resistor design must consider the trimming technique that is subsequently used, otherwise an adverse effect can occur on the resistor stability and also the probe pad locations. Figure 5-2 illustrates both the rectangular (or bar) and the meandering or zig-zag pattern. A meandering pattern is characterized by a high square count. Each corner of the meandering pattern can be considered one-half square because of current crowding, and the Dumbbell terminations (widening of the path closer to the termination) each contribute to a half a square. The transition section between the film resistor and its associated conductor pattern is usually wider than the resistor path. This widened section, referred to as Dumbbell termination, is used to minimize the effect on resistor stability of the narrow area of resistor material at each end of the resistor which must be left unanodized during processing.

With the proper control of deposition processes, thin film resistors can be achieved to a $\pm 10\%$ tolerance without trimming. Control of final dimensions is necessary and critical to achieve a specific resistor value to tight tolerances. Resistor tolerances as tight as $\pm 0.1\%$ can be attained using thermal stabilization and trim-anodization or laser trimming action. Conductor lines possess a minimum line width of 50 μm and a minimum space width of 50 μm. The minimum resistor dimensions are 2 mils by 2 mils. Conductor-resistor overlap is 1 mil minimum and resistor underlap is 2 mil minimum from each side of the conductor. Figure 5-3 and the corresponding

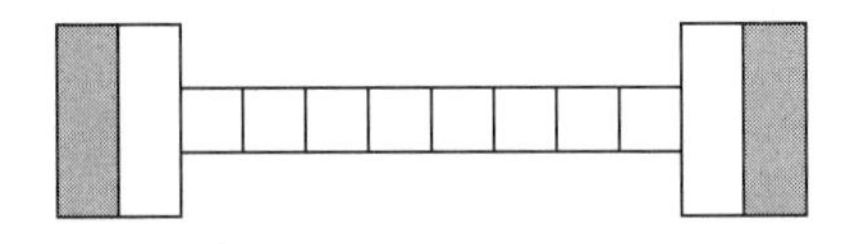

Number of squares = 8 + 2 (0.5)

(a)

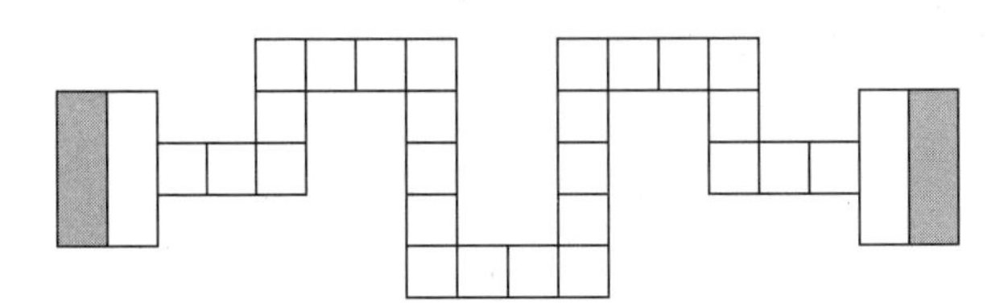

Number of squares = 18 + 10 (0.5)

(b)

FIGURE 5-2 Resistor designs. *(a)* Rectangular or bar pattern; *(b)* meander pattern.

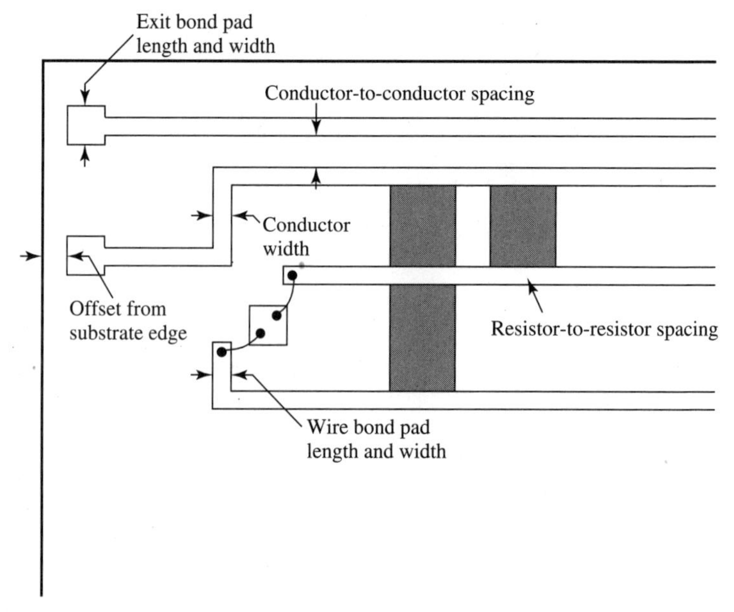

FIGURE 5-3 Thin film design guidelines.

Table 5-1 illustrate various thin film design guidelines for resistor elements and metallization interconnects.

Substrate flatness directly influences the minimum achievable line width and spacing. In general, films are tolerant of surface roughness less than the order of magnitude of the film thickness. When noticeable surface roughness exists, the characteristics of the thin film elements in turn vary. Thin film resistors have an apparent higher sheet resistance value when deposited on rough substrate surfaces. In addition, resistor elements demonstrate more stability on flat surfaces. Thin films can be thermally stabilized and protected against environmental effects by depositing the proper protective layers to cover the elements.

TABLE 5-1 Thin Film Design Guidelines

Factor/quantity	Minimum value in mm (mils)	Typical value in mm (mils)
Conductor width	0.048 (2)	0.24 (10)
Conductor spacing (edge to edge)	0.048 (2)	0.12 (5)
Offset from substrate edge	0.24 (10)	0.30 (12.5)
Conductor to resistor spacing	0.048 (2)	0.12 (5)
Exit bond pad width	0.24 (10)	0.30 (12.5)
Exit bond pad length	0.24 (10)	0.30 (12.5)
Wire bond pad (width and length)	0.12 (5)	0.24 (10)
Resistor to resistor spacing	0.48 (2)	0.12 (5)
Resistor width	0.12 (5)	0.24 (10)
Resistor length	0.12 (5)	May vary

The design of thin film resistors involves the choice of a suitable film, determination of film thickness, choice of the pattern with the required number of squares, and the selection of the line width and spacing that results in an acceptable power density. The choice of film material is based on considerations such as the temperature and voltage coefficients of resistance, the noise coefficient, the stability of resistance with time, the allowable power density, and the method of film deposition.

Resistor Passivation [37]. Thin film resistor materials include tantalum (Ta) films, nickel-chromium (nichrome) films, chromium (Cr) films, rhenium (Re) films, and cermets. Tantalum nitride films possess higher stability under various environmental conditions. A stabilizing treatment consisting of film anodization, followed by heat treatment and another anodization at higher voltage, further promotes the stability of these elements. Although nichrome films provide the desired properties of temperature coefficient of resistance, sheet resistivity value, and good adhesion, they are known to be unstable under high humidity condition. Therefore, a coating layer of silicon monoxide (SiO) is provided to maintain an acceptable stability when using these films.

A technique for passivating thin film hybrids using nichrome resistors uses polyimides. The method consists of spinning the polyimide on the substrate, then backing it at a low temperature to form the film. The low-temperature cure is necessary to remove the solvents before the high-temperature cure. Patterning of this passivation material then takes place through the application of positive photoresist, exposure, and development to etch successfully the resist as well as the polyimide. The film is then baked at a moderate temperature (250°C to 300°C) to form the polyimide film. The polyimide can satisfactorily protect thin film resistors under different operating and environmental conditions (under load, high humidity, and high temperature). The polyimide material has no effect on the microwave performance of these hybrid circuits. For Re films, an SiO protective overlayer helps ensure good resistor stability.

5.2.2 Capacitor Design Rules

Figure 5-4 illustrates a top view and a cross-sectional view of a thin film, parallel-plate capacitor structure. The structure has a dielectric film sandwiched between two conductive layers. The capacitance value (C) is of the form

$$C = \epsilon_o \, \epsilon_r A/t \tag{3}$$

where A = common area of the overlapping electrodes
ϵ_o = permittivity of the free space ($\epsilon_o = 8.85 \times 10^{-14}$ F/cm)
ϵ_r = dielectric constant (relative permittivity)
t = thickness of the dielectric film.

The capacitance density (c) is defined as

$$c = C/A \tag{4}$$

A maximum value of c is desirable. The capacitance density is used to determine the area needed to achieve a certain capacitance value. Fringing effects at the edges of the thin film capacitor are negligible because of the thickness of the dielectric compared

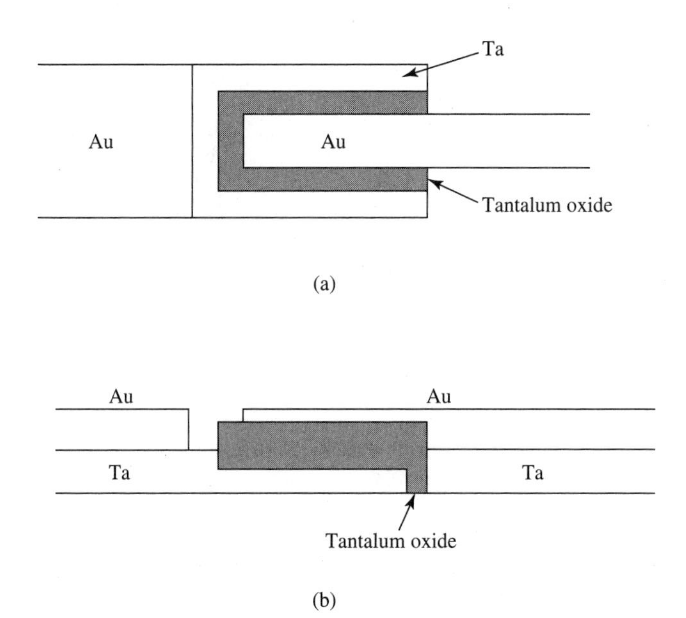

FIGURE 5-4 Typical thin film capacitor structure. (*a*) Top view; (*b*) cross section.

to the lateral dimensions. There exists an upper limit for the capacitance density value achievable for a particular dielectric material. The reason can be attributed to a very thin dielectric film (below 1000 Å) severely affecting the dielectric strength because of pinholes and other defects in the thin film. Thus, the upper limit for capacitance density is 0.01 $\mu F/cm^2$, 0.1 $\mu F/cm^2$, and 0.2 $\mu F/cm^2$ for SiO, tantalum pentoxide (Ta_2O_5), and titanium oxide (TiO_2) dielectrics, respectively.

For most dielectric film materials, the dielectric strength ranges between 1×10^5 and 1×10^7 V/cm. For SiO film, capacitor breakdown has been observed when electric fields ranging between 4×10^5 and 4×10^6 V/cm are applied, even for films of several thousands Å dielectric thickness. The presence of defects in the oxide film results in a certain variability of the breakdown field for capacitors deposited on the same substrate. The breakdown field depends also on the substrate material, and it can be increased significantly by precoating the substrates with a 1 μm layer of SiO. In addition, the metal selected for the capacitor electrodes affects the breakdown characteristics of the capacitor structure.

The dielectric film must be continuous and free of pinholes, thus dictating a lower limit on achievable film thickness. The lower limit is 3000 Å for practical evaporated dielectric film and 100 Å for anodic film. Other characteristics of interest include stored energy, leakage current, dielectric breakdown, maximum working stress, and dielectric loss. Polarity is a factor in thin film capacitor design. It is important in the prevention of dielectric breakdown under large reverse bias. To realize a nonpolar capacitor structure, two capacitors are formed in series; thus, a large area of the substrate is required.

In addition to the parallel plate capacitors, interdigitated capacitors are also commonly used. In this design type, a separation width (W) of 25 μm for finger width and spacing is common. The capacitance expression (C) is of the form

$$C = (2\,N_f - 1)\,K_1\,(H + 2.1\,W) \tag{5}$$

where N_f = number of adjacent finger cells
W = width and spacing of fingers
$K_1 = (\epsilon_r + 1)(0.05)$ pF/cm
ϵ_r = dielectric constant of the substrate material
H = finger cell length

H may be selected in a convenient manner to provide an integer value of N_f. Simple crossovers are achievable by thermocompression wire bonding or ribbon between bonding pads (typically 120 μm × 120 μm).

5.2.3 Inductor Design Rules

Thin film spiral inductors can be either square or circular in shape. These inductors can be realized by depositing metal either by evaporation or sputtering. Thin film deposited inductors are limited to approximately 100 nH. The inductance value (L) of a square spiral is of the form

$$L = 0.216\, S^{1/2}\, N^{5/3} \qquad (\text{in } nH) \tag{6}$$

where S refers to the surface area of the coil in mils^2 and N is the number of turns of the coils. The inductance value of a circular spiral is of the form

$$L = 0.8\, a^2\, N^2/(6a + 10c) \qquad (\text{in } nH) \tag{7}$$

where N = number of turns of the coil
a = average radius ($a = do/4 + di/4$)
c = breadth of the coil ($c = do/2 - di/2$)

do and di refer to the outer and the inner diameters of the coil, respectively. The quality factor Q of a thick film spiral inductor is of the form

$$Q = w\, L_s[1 - (f/f_r)^2]/R_s \tag{8}$$

where f = frequency of interest
L_s = low frequency inductance value
R_s = series resistance of the spiral inductor
C_s = distributed capacitance value
f_r = self-resonant frequency

We can express f_r as

$$f_r = 1/(2\pi\sqrt{L_s C_s}) \tag{9}$$

5.2.4 Transmission Lines

Transmission lines are used to carry information or energy from one location to another in microwave circuits. Passive elements in conventional microwave circuits are commonly distributed and use sections of transmission lines and waveguides. Transmission lines are characterized in terms of the characteristic impedance (Z_o) (the ratio of the voltage to the current at any point along a perfectly terminated line), and the complex propagation constant (γ); comprising a real part representing the attenuation constant (α), and an imaginary part representing the phase constant (β). The strip line, the microstrip line, the slot line, and the coplanar line represent typical planar transmission lines. Expressions for the characteristic impedance and the attenuation constant of these lines are summarized in reference [38] and presented in Chapter 10.

5.2.5 Fabrication Sequence for Thin Film Resistor-Conductor Circuits

Nickel-Chromium (Nichrome) and Tantalum Nitride System. The realization of planar resistors consists of depositing thin films of lossy metal on a dielectric base. Nickel-chromium (nichrome) and tantalum nitride are the most popular and useful film materials in the manufacturing of thin film resistors. Consider a structure consisting of a tantalum nitride-resistive film and a conductor structure of titanium (Ti), palladium (Pd), and gold (Au).

Substrate Preparation. For glass and ceramic-glazed substrates, a tantalum oxide film is placed between the substrate and the resistive film (resistant to hydrofluoric acid as tantalum etchant). This is achieved by sputtering a thin layer of Ta, followed by an oxidation of the Ta material.

Thin Film Integrated Components: Resistors and Capacitors. In this section, the four fundamental thin film processes are used to make thin film resistors and capacitors using Ta technology as an illustrative example.

The resistor material is tantalum nitride. The oxide layer, resulting from anodization of the resistor to obtain the final desired value, grows at the expense of the underlying film. It is the consumption of the Ta material that decreases the resistor cross section, increasing the resistor value to the desired level.

One of the most critical factors in a thin film circuit is the resistor film temperature. The resistor film temperature is affected by the resistor geometry, the total circuit power dissipation, the influence of other dissipative elements, the substrate used in the application, and the type of external heat sink used. If not enough information is provided for power dissipation, then it is reasonably safe to assume a value of 50 W per square inch and a maximum film operating temperature of 150°C when designing these thin film resistors. (Refer to section 5.4 regarding thermal considerations.)

A reasonably high capacitance per unit area can be obtained with Ta material. Anodically grown Ta_2O_5 is used as the capacitor dielectric (dielectric constant of ~25, with a very high dielectric strength). Very thin films of Ta_2O_5 are produced (<1000 Å). The Ta_2O_5 dielectric is sandwiched between two Ta electrodes. The Ta film adheres well to the substrate. But these capacitors cannot be used in precision networks because of their high sensitivity to environmental conditions (such as humidity). The problem is solved by using a counterelectrode consisting of a thin layer of nichrome followed by Au deposition.

Gold System. For Au deposition (as opposed to Ta deposition), the process flow for thin film substrate fabrication consists of vacuum deposition of nichrome followed by etching. Next, Au is deposited. Electroplating the Au may follow (in order to attain a certain thickness), then etching the Au metallization. An annealing process may also be required.

Following the thin film processing for the interconnects and elements to be integrated on the substrate, passive add-on components (including resistors, capacitors, and inductors) are attached through solder or adhesives. This requires a low-temperature cure. Active components are attached to the proper metallization, and wirebonding is performed to connect active windows in the chips to the appropriate interconnects. Eutectic bonding may be used for semiconductor die attachment when lower contact resistance is desirable.

Integrated Termination and Decoupling Components. Typical physical termination resistance is 50 or 75 Ω for some applications, while decoupling capacitance can assume a typical value in the range 200 to 2000 pF. Therefore, termination resistor materials should have resistivity in the range 1.6 to 1.8 Ω/cm^2 to produce the desired resistance termination values. Nichrome is a potential resistor material to achieve this objective. This resistor system can be deposited by either sputtering or evaporation, and the resistivity can be adjusted to a specific value by changing the ratio of Ni to Cr in the film. The nichrome film can be defined by an etching process. In addition, the system has a low-temperature coefficient of expansion (120 ppm/°C) for the traditional ratio of 80% Ni to 20% Cr, making this system suitable for military and some commercial applications that experience a significant change in temperature.

For achieving the proper decoupling capacitance values, a dielectric system with a dielectric constant value of $\epsilon_r = 12$ is usually desired to avoid the use of both excessive film area on the substrate or pinhole formation in the dielectric resulting from very shallow dielectric film deposition. Tantalum pentoxide ($\epsilon_r = 25$) with 0.1 μm thickness of deposition can achieve a 1000 pF capacitance value with about 5×10^5 μm^2 area, while $BaTiO_3$ ($\epsilon_r = 2000$) can achieve this same capacitance value in an area of about 625 μm^2.

5.2.6 Chip Mounting to Thin Film Substrates

Add-on thin film resistor and capacitor chips are available in sizes of 20 $mils^2$ area and larger. Capacitor values accommodated in thin film hybrids are 1000 pF, 0.01 μF, and 0.1 μF with NPO temperature characteristic (±30 ppm/°C maximum). Resistor values accommodated in thin film hybrids extend from a few ohms to a few mega ohms. Chip inductors may have values in the range of tens of nH to one hundred nH. Finally, the ceramic cover or the leaded package for enclosure is fabricated. Table 5-2 provides the guidelines for thin film chip mounting.

TABLE 5-2 Thin Film Design Guidelines for Chip Mounting

Factor/quantity	Minimum value in mm (mils)	Typical value in mm (mils)
Active die mounting pad beyond chip	0.24 (10)	0.30 (12.5)
Die to die spacing	0.24 (10)	0.30 (12.5)
Passive chip to bonding pad beyond chip	0.75 (30)	0.84 (35)
Passive chip mounting pad beyond chip	0.48 (20)	0.60 (25)
Passive chip to conductor overlap	0.24 (10)	0.30 (12.5)
Passive chip to wire bonding pad spacing	1.0 (40)	1.25 (50)
Spacing between passive chips	1.0 (40)	1.25 (50)

5.2.7 Substrates for Thin Film Deposition [39]

The thin film substrate should have a clean, flat, and highly polished surface to accommodate thin film deposition of uniform films. In addition, the substrate should be inert to the various processing chemicals, environmental changes, and temperature variation. The thin film substrate should withstand handling, thermal cycling, and thermal shock. Thermal conductivity is important factor in the substrate performing a successful thermal management. Additionally, the coefficient of thermal expansion is an important criteria for the substrate to meet in order to match the thin film deposited materials onto the surface without developing stresses. Chapter 3 discusses the various attributes of the thin film substrate.

The most common substrate material for thin film deposition is 99.6% alumina. Substrate dimensions normally vary from 1 in × 1 in to 3 in × 3 in with thicknesses ranging from 0.010 to 0.050 in. The surface finish is ~4 μin maximum on the front surface and ~5 μin maximum on the back surface. The flatness is measured in terms of camber and it is ~0.002 in/in for 0.010 in substrate thickness, 0.003 in/in for 0.025 in substrate thickness, and 0.004 in/in for 0.040 in substrate thickness. The thermal coefficient of expansion is 6.3 ppm/°C at room temperature. The density of the ceramic is 3.87 g/cm^3. The dielectric constant is ~9.9 (measured at 1 MHz), and the loss tangent is 0.0001 (measured at the same frequency).

For 99.5% beryllium (Be), the surface finish is ~5 μin maximum on the front surface and ~10 μin maximum on the back surface, and the flatness is 0.003 in/in maximum. For aluminum borosilicate glass, the surface finish is ~0.25 μin maximum on both the front surface and the back surface, and the flatness is 0.004 in/in maximum. Additional thin film design guidelines are available in the literature [40].

The coefficient of thermal expansion (CTE) mismatch between alumina (6.1 ppm/°C) and GaAs (5.7 ppm/°C) and Si (2.6 ppm/°C) can result in serious stresses; thus, the use of aluminum nitride (3.5 ppm/°C) or thick diamond films (0.9 ppm/°C) present more suitable candidates for thin film substrates, especially in view of their superior capacity for dissipating large amounts of heat.

The use of Si as the substrate material for thin film interconnections can indeed offer numerous advantages, among which are high thermal conductivity, matched thermal expansion coefficient for interconnecting chips, good surface finish, and good dimensional stability. Silicon wafers allow the fabrication of integral SiO_2/Si_3N_4 decoupling capacitors as well as active driver circuits. On the other hand, limited mechanical rigidity dictates the necessity of mounting Si substrates onto a base to provide the needed mechanical support. In addition, Si substrates have a relatively high dielectric constant value that affects the high-frequency operation.

Often, silicon dioxide (SiO_2) is used as the dielectric medium on Si substrates. These dielectric layers are deposited by a plasma chemical vapor deposition (CVD) process, producing films up to 20 μm thick.

5.3 MULTICHIP MODULES DEPOSITED (MCM-D) [41–45]

5.3.1 Concept

MCM-Ds are multichip modules constructed using conventional thin film techniques with unreinforced low dielectric constant ($k < 5$) materials that are adjacent to the signal plane, with base substrates (including ceramic, Si, copper (Cu), glass reinforced laminate, or metal) required for dimensional stability. Conductors are fabricated by sputtering or plating high-conductivity metals (such as Al, Cu, Au, or Pd) onto the low-dielectric materials. The metals are subsequently patterned with photolithography techniques. Cost, conductivity, and material compatibility are the prime factors defining a specific metallization. The thin film elements are formed by sequentially evaporating or sputtering resistive and conductive materials from a source or target onto a substrate in a vacuum environment. The deposition process is subtractive, with the resistive material deposited under all conductor patterns and usually exposed by etching the conductor to form the various resistive elements. The most common dielectric materials used in MCM-D applications are polymer materials, in particular polyimides. Polyimides possess thermal, chemical, and mechanical stability, as well as low dielectric constant values. Currently, of all the multichip module technologies, MCM-D technology produces the highest circuit density and smallest size. On the other hand, the processing techniques and materials used to construct these modules make it a costly technology. The following sections describe the existing materials and processes for MCM-D technology, as well as some advances in materials and processes that include the use of photodefineable polymers, integrating components within the base substrate, plating instead of vacuum deposition for conductor fabrication, and spray coating instead of spin coating for the deposition of polymer dielectric materials. Also included is the use of advanced materials such as diamond and high-temperature superconductors.

5.3.2 Process Sequence

Additive approach. In this process, a buffer layer followed by deposition of metal (such as Cu) on the substrate takes place. Conductor interconnects are defined and electroplated to build up thickness in certain areas. Vias are then defined and the vias posts are electroplated for internal connection between the layers. An additional buffer layer may be deposited to prevent the interaction of metallization with the dielectric material. Polyimide is then applied to coat the surface area, and the top of vias is then exposed. To achieve additional layers, the sequence is repeated as needed, but the multilayer structure may be limited to four metal layers for most practical applications.

Subtractive approach. The metallization is evaporated or sputtered onto the entire substrate, and the thickness is added through an electroplating process. The conductor pattern is then defined. The dielectric layer is then deposited and patterned with openings for vias. The vias are etched by a wet process. A second layer of metal is deposited with the metal conforming to the sidewalls of the via hole. Next, layers of metal and vias are then produced and defined. The process is repeated as needed.

5.3.3 Materials Properties

Substrates. Base substrates should have such properties as high thermal conductivity, good electrical insulation, high mechanical strength, high temperature resistance, low weight, and minimal cost. A substrate with thermal expansion coefficient (CTE) close to that of Si is usually desirable for flip-chip bonding or attachment of large dice. The most commonly used substrates tend to be ceramics, metals, Si, Cu, or glass-reinforced laminates.

Ceramic substrates include alumina (Al_2O_3), aluminum nitride (AlN), and beryllium oxide (BeO). Alumina is widely used and can be found in many sizes and thicknesses. Aluminum nitride offers improved heat dissipation over Al_2O_3. Also, it offers CTE-matched materials, it has high bending strength, and high strength. Aluminum nitride is not as readily available as Al_2O_3, and it also costs more. Similar to AlN, BeO also offers an improved heat dissipation over Al_2O_3 for efficient heat removal from the substrate. Beryllium also is harder to obtain than Al_2O_3 and has a significantly higher cost, but the most significant disadvantage is that it is highly toxic and presents a danger if handled incorrectly.

Metal substrates are mostly suitable for harsh environments. They possess high strength, are resistant to shock, and also act as an effective ground plane. Metal substrates also have low cost and a good thermal conductivity. Copper presents a good metal substrate for MCM-D applications. It is usually alloyed with other materials to obtain a good CTE match with Si and it can be coated with an insulating layer to prevent reaction with most metal etchants.

Recent trends indicate that Si is the substrate of choice for applications requiring extremely high circuit density. Higher circuit density is acquired by incorporating components into the substrate itself. Modules have been fabricated with resistors, capacitors, diodes, npn transistors, and pnp transistors. Having the capability to incorporate components into the substrate, especially passive components integration, significantly increases circuit density and frees up vital surface areas for conductor line patterns, active devices, and integrated circuits. Surface area is crucial because of the limited number of conductor layers that can be processed (~5 to 10 layers). In addition, Si has a high thermal conductivity and it is stable, allowing extremely fine lines.

Although Si substrate matches mounted silicon chips perfectly, and it is extremely flat, smooth, and dimensionally stable, it possesses poor flexural strength and a relatively high dielectric constant, in addition to a warpage caused by the polymer dielectric and metallization stresses imposed on the wafer. Some advanced research in substrate materials has been in the area of CVD diamond films. Synthetic diamond substrates can offer improved properties over Al_2O_3, AlN, and BeO substrates, including a lower dielectric constant, higher thermal conductivity, chemical susceptibility, and a good CTE match with Si.

The thermal conductivity, with values in excess of 2000 W/m·°K, is the greatest asset of diamond substrates. Incorporating diamond substrates into MCM-D technology can greatly improve the thermal management performance by enhancing heat removal from chips. Synthetically produced CVD diamond substrates with high thermal conductivity have addressed potential problems in large-size MCM applications, namely thermal management problems spreading and dissipating large transient power pulses as well as removing large steady heat loads. Researchers have described several methods for the provision of reliable metallization processes for CVD of diamond. These methods include transition metal bond coats, diffusion barriers, patterning, and solder die attachment. Diamond substrates can also be doped to have semiconductor properties that provide the capability to integrate components directly into the substrate. Because diamond is expensive, diamond thin films can be used as an alternative at a reduced cost. Diamond thin films can be applied to a highly thermal conductive substrate, such as AlN, to improve heat dissipation with a much smaller cost then that of diamond substrates.

The CTE for diamond is considerably different than gallium arsenide or indium phosphide, and structural integrity problems can occur when a module fabricated with these materials is subjected to temperature extremes. Despite the disadvantage, diamond has characteristics that can greatly enhance the MCM-D technology performance.

Dielectrics. Dielectrics for use in MCM-D applications should have a low dielectric constant, low loss tangent, low moisture absorption, good adhesion to the metallization and base substrate, good planarization, high thermal conductivity, and a CTE comparable to that of the die and base substrate. Polyimides and benzocyclobutenes (BCBs) possess these vital requirements and are often used for MCM-D applications.

Most recent research has involved polyimides: fluorinated polyimides, low stress polyimides with low CTE value in the xy direction, low dielectric constant value, low water absorption, photosensitive polyimides to reduce the number of processing steps, and polyphenylquinoxaline (PPQ). Conventional processing techniques (including dry etching, wet etching, and reactive ion etching) have been simplified by the development of photosensitive polyimides.

This comparison indicates that photosensitive polyimide processing requires fewer steps than other techniques, therefore reducing considerably the processing time and cost. To address current problems (including corrosion caused by direct use of Cu over polyimides when deposited from acid base), five layers of metal deposition are used. An additional problem involves proper adhesion of Cu to polyimides; adhesion promotion is necessary in this case.

Benzocyclobutane have basically the same properties as polyimides. They have low dielectric constant values and low loss factors. They also have a low moisture absorption and adhere well to ceramic, Cu, and Al, and they have superior surface smoothness. High-density multilayer interconnect is usually formed over an Si substrate with BCB as the dielectric material. Metal layers are deposited, followed by photoresist patterning, etching, and then a coating of BCB polymer, and etching of vias in the polymer by reactive ion etch after photolithographic definition of the vias. Conventional processing techniques, however, have limited BCB advancement in MCM-D applications. Limitations include methods with processing vias, which require the use of metal or inorganic masks, and processing methods that use photoresist or wet-etch methods, which are relatively complicated, time-consuming, and costly. Recent improvements with processing techniques have been made, reducing considerably the curing time. The most recent advances have been in the development of photosensitive BCBs. Photosensitive BCBs transcend some of the limitations that exist with previous BCBs by incorporating photolithography and wet-etching techniques. They generally have better properties than polyimides (such as a lower conductor resistance, lower propagation delay, and improved planarization). Table 5-3 illustrates properties of MCM-D dielectrics as compared to ceramics.

Metallization. Three types of metallization are used in conjunction with MCM-D technology: Cu, Au, and Al. Copper is the most widely used because of low cost, high stability, and high conductivity. The disadvantage of Cu is that it is not compatible with many polyimides; therefore, a buffer layer must be applied between the Cu and the polyimide for good adhesion and interface. This additional layer increases the number of processing steps to fabricate the module. Aluminum is also widely used, although not as frequently as

TABLE 5-3 Properties of MCM-D Dielectrics as Compared to Ceramics

Properties	PIQs	Ceramics	PI	Fluoro	BCBs	PPQs
ϵ_r	3.2-3.4	5-9	2.5-2.8	1.9-2.6	2.6-2.8	2.8-3
Tg (°C)	300-400	N/A	300-400	160-320	310-350	361
CTE (ppm/°C)	3-58	3-7	2-60	90-300	65	40
% water uptake at 95% RH	80-100	0	25-40	1-10	25	10
% Planarization $(ha - hb)/hb$	20-70	N/A	5-50	~0	≤95	40-50

Note *ha,* Average height of features before planarization; *hb,* average height of features after planarization; PI, polyimides; BCBs, benzocyclobutenes; PPQs, polyphenylquinoxaines; PIQs, polyimide iso-indoloquinazolinediones.

Cu. It is compatible with most polyimides, but its conductivity is lower than that of Cu. Gold is an alternative to Cu and Al, possessing a good conductivity and good stability, but at a higher cost. Additional metal layers are usually deposited on the main metal. Some of these layers act as barrier layers to promote adhesion and to realize the proper interface layer. Photoresist layers (with thickness ranging from 2 μm to 5 μm) are deposited to define the conductor material. The polyimide dielectric layers (2 to 3 layers) are then spin- or spray-coated and cured. Vias are realized in the polymer layer. The process repeats and usually involves five metal layers.

In summary, the high cost of fabricating MCM-Ds has limited its use to high-performance and high-speed applications, such as computer, workstation, and military applications. Current applications mostly use photodefineable polymers for the dielectric material with Cu metallization, although Al metallization is occasionally used as an alternative. A variety of substrates are being used in current applications that include Al, Al_2O_3, AlN, and Si. Silicon substrates have been the choice for high-speed applications because of capacity to integrate components into the substrate.

5.3.4 Processing Considerations and Design Guidelines

Overall Process Sequence. MCM-D can be fabricated using either the *chips-first* technique or the *chips-last* technique. Chips are mounted in cavities in the base substrate using a chips-first technique, such as the General Electric high-density interconnect overlay approach. Then dielectric and metallization layers are deposited to form interconnects. A Kapton™ (polyimide) film is laminated to the front of the chips, and a laser is used to etch vias for contact to the chip bonding pads. A thin film multilayer interconnect structure is built on the Kapton™ overlay. Copper metallization is sputter-deposited, the thickness is increased through plating, and the metallization is patterned by wet-etching. In this technique, all dice can be tightly positioned, and the thermal management is very effective because these dice are in direct contact with the substrate. Neither thermal vias nor special chip bonding process are required with these techniques. On the other hand, rework is a serious problem. With the chips-last technique, the dielectric layers and metal layers are deposited first on the base substrate, and chips are mounted last to interconnects on the top layer. Figure 5-5 illustrates these two techniques.

Design Guidelines. Minimum line width and spacing is ~0.06 mm (0.0025 in) with a nominal value ~0.25 mm (0.010 in). The metallization is kept at a minimum distance of the ~0.25 mm (0.010 in) from the edge of substrate. The dielectric constant of the substrate material is often high, while the dielectric constant of the separating dielectric layers is normally in the range of 3 to 6. In general, a final thickness of Au metal ranging between 0.0025 to 0.005 mm (0.0001-0.0002 in) is needed to provide conductivity and wirebonding. The dielectric layers are deposited by spray or spin coating. Vias can be produced by wet etching, laser patterning, or through the use of photosensitive polyimides. Vias can be filled with metals using wet chemical plating. Metal layers can be applied by electroplating, sputtering, evaporation, and enhanced ion-plating (EIP), and are patterned using reactive ion etch (RIE), or wet etch (for nonphotosensitive materials). Photosensitive dielectrics can be used to eliminate many processing steps, and involve applying and developing the dielectric material through photolithography and then applying metal.

Interconnection Design and Process Characteristics for MCMs Operating at High-System Clock Rates. A general set of design guidelines for MCM structures fall within five frequency ranges: range 1, up to 100 MHz clock rates (2 to 5 ns rise times); range 2,

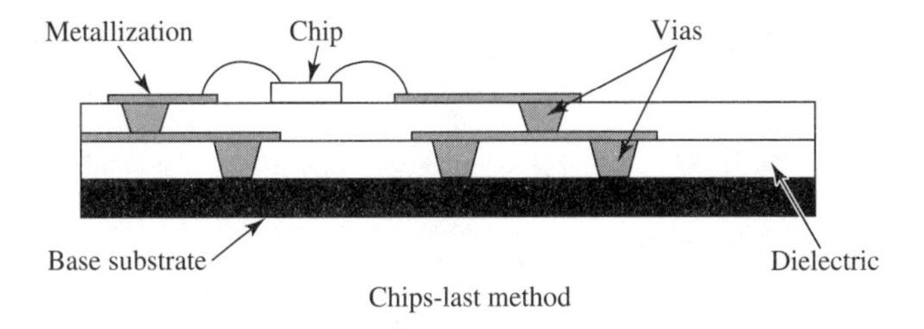

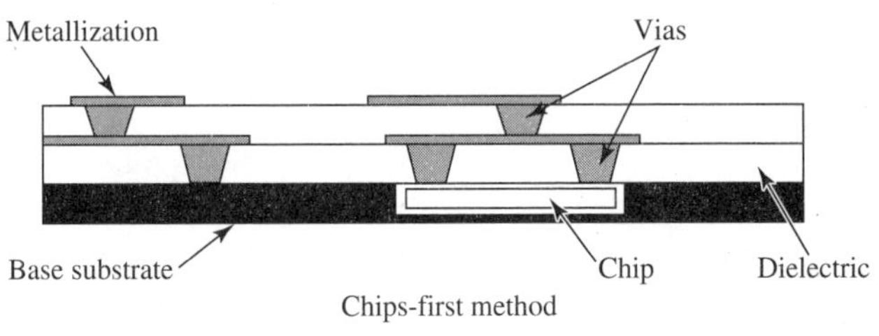

FIGURE 5-5 Chips-first and chips-last methods.

100 to 500 MHz clock rates (0.5 to 2 ns rise times); range 3, 500 MHz to 1.5 GHz (0.2 to 0.5 ns rise times); range 4, 1.5 GHz to 3.0 GHz (0.07 to 0.2 ns rise times); and range 5, >3.0 GHz (rise time <0.07 ns).

5.3.5 New Trends in the Technology

MCM makers (such as IBM) deposit thin film on top of low-cost multichip module laminates. This provides a higher circuit density of interconnects for signal lines on an MCM-D module, in conjunction with power and ground circuitry on the laminates. Also, MCM-D/C, or sometimes called MCM-D in ceramic technology, involves patterning the thin film multilayer interconnections directly on a package base, such as cofired ceramic. The technology eliminates the substrate to package bonds. Power and ground distribution layers can also be incorporated into the substrate. The technology has been implemented in large mainframe computers. The use of photosensitive polyimides replacing conventional polyimides has simplified the pattern definition process in multilayer thin film structure.

5.3.6 Limitations of the Technology

Some curing reactions of organic dielectrics are associated with weight loss due to initial solvent loss. At relatively high temperatures, additional polymer weight loss can occur as a result of the outgassing of residual solvent or degradation reactions, such as cross-linking of the polymer. This reaction can lead to embrittlement, lowered elongation at break, a noticeable change in the glass transition temperature, and adhesion loss.

Polymer dielectrics may suffer high relative permittivity and high loss tangent, especially at high frequencies of operation, as well as under high humidity conditions. Water absorption and film shrinkage are key concerns with polymer films. In addition, at lower temperatures, deterioration occurs in air through radical initiated oxidation, and at high humidity through hydrolysis. Poor planarization in individual signal layers can result from variations in the photoresist thickness. It is also known that the degree of planarization is dependent on the film thickness compared to the feature thickness, the number of coatings, the feature's dimensions and size, the curing speeds and the temperature profiles, the shrinkage of the cured film, and the molecular weight of the materials.

The packing density is usually limited by the component footprints. The routing density determines the minimum package size.

Finally, most polymers suffer from having a relatively high CTE of the order 50 ppm/°C, making them difficult to match semiconductor chips.

5.4 THERMAL CONSIDERATIONS

5.4.1 Thermal Management

Thermal management is a key issue in the design of thin film hybrid circuits and MCM-D units. The proper design ensures that the peak temperatures remain within a specified operating range to produce reliable modules. The main objective of thermal analysis and simulation of circuits and modules is to maintain the semiconductor device junction temperature below the maximum temperature specified by the device manufacturer. Design challenges included in heat removal are higher circuit densities, close proximity of adjacent devices or components, low thermal conductivity substrate, thinner layers of metal forming interconnects, and thermal resistance of the heat sink system.

Generally, two thermal problems are associated with a hybrid module. The first problem concerns the heat transfer by conduction from the heat sources to the package surface. This heat is defined in terms of an internal thermal resistance. Heat transfer by conduction occurs when a hot region in the solid has rapidly vibrating molecules, transfering its energy to a cooler region with fewer vibrating molecules. The heat flow from the junction of a chip to its case or package is a typical example. The total internal resistance from the chip to the case is the sum of the individual resistances (from the junction to the epoxy or solder attachment, and from the epoxy or solder attachment to the packaging case). The second problem concerns the removal of heat from the package surface by convective exchange with an ambient medium. This heat removal may be totally free, or may be achieved through either forced air flow or a liquid coolant. This term is defined in terms of an external resistance. The external resistance in the model represents the bond to the heat sink, the heat sink, and the convection to the cooling medium or air. Heat transfer by convection occurs as a result of a change in the density of the medium, basically causing a medium motion. In the case of forced air or fluid flow, this convection is referred to as forced convection. Typical air-cooled heat sinks can reduce the external resistance to approximately 15 degrees Kelvin per Watt (°K/W) in natural convection, to 5°K/W for moderate forced velocities. Both the internal and external thermal resistances are important in die attachment techniques.

5.4.2 Thermal Resistances

The junction temperature is a function of the sum of the thermal resistances between the junction and the ambient air, the amount of heat being dissipated, and the ambient air temperature. The maximum junction temperature of an operating chip or a discrete device is typically limited to about 150°C or lower to ensure a reliable and functional circuit. All power-consuming elements, devices, and resistors act as individual heat sources. The overall thermal performance of the module can be expressed in terms of the thermal resistance θ_{ja} or R_{jc}. This thermal resistance is equal to the sum of all thermal resistances between the junction and the ambient atmosphere, and it should be minimized. This thermal resistance can be measured in terms of the difference in temperature between the chip junction temperature T_j and the ambient air temperature T_a, per watt of power produced by the chips

$$\theta_{ja} \text{ or } R_{jc} = (T_j - T_a)/W \qquad (\text{in } ^\circ K/W) \tag{10}$$

Thermal resistance (R_{jc}) can vary from ~80°K/W for a plastic package with no heat spreader, to ~20°K/W for a similar package with heat spreader, and to ~5°K/W for a ceramic package.

The thermal environment for a hybrid substrate includes the bond between the substrate and the heat sink (or case); the heat-sink configuration; the natural or forced convection cooling of the heat sink; the cooling medium conditions; and the altitude and atmosphere, or air temperature. The factors determining the thermal resistance include the chip size, the chip attach or die attach, the substrate material, the substrate attach, and the package material.

The module must provide thermal passage to release the heat generated from the various chips and devices integrated within the module. The thermal passage can be from the chip to the substrate and to the ambient; from the chip to the lid and to the ambient; from the chip to the lead frame to the ambient; from the chip to the substrate to the heat sink to the ambient; or a combination of the above passages. The selection of a substrate with enhanced thermal conductivity dictates the realization of thermal vias and the use of metal heat spreaders, usually at the expense of routing density.

5.4.3 Thermal Model and Thermal Analysis

Typically, a nondimensional thermal model is defined, where nondimensional parameters produce results that represent a general case using the minimum number of variables. The thermal model can be solved using a thermal analysis program. The thermal analyzer for multilayer structures (TAMS) thermal analysis program is described by Ellison [47]. The program uses Fourier series expansions to solve the full three-dimensional heat conduction equation in multilayer structures of differing thermal conductivity and thickness. In this program, surface boundary conditions permit simulation of both convection and radiation of heat. The results provide information concerning the peak temperature level for different junctions and simulated heat contour maps to illustrate the temperature gradient. These findings are critical to properly conduct electrical design with optimum thermal performance. Other thermal analysis and simulation programs exist (such as Flowtherm, TEMDIS, and TRATEM).

5.4.4 Thermal Paths

The heat produced within a semiconductor chip must be dissipated efficiently through the metallization or interconnect to the substrate and the package and out to the ambient atmosphere. The technique as well as the materials used to connect the chip to the ceramic substrate for first-level packaging interconnect must provide: the electrical function for signal and power distribution; the structural function through the strong attachment to the substrate; and establish the thermal function or thermal path through the provision of effective heat transfer or heat removal with low thermal resistance.

In chips attached with wire bonding and tape-automated bonding (TAB), the heat generated within these chips flows through an Au eutectic layer or an epoxy layer to the metallization, substrate, package, and heat spreader. When flip-chip designs are used, heat can be removed from the chip with solder bumps or TAB interconnect by pressure contact with a metallization or conduction through epoxy or solder bond.

5.4.5 Temperature Cycling and Power Cycling

Temperature cycling of components and modules (typically in the range of −55°C to +125°C through alternating heating and cooling) provides a common technique for the evaluation of component/module integrity and reliability. This test, simulating fluctuations likely to be encountered during component operation and storage, can induce nearly isothermal changes in the entire electronic packaging assembly.

On the other hand, power cycling tests (chips are periodically activated for a certain duration of time) can provide internal temperature variations expected to result from these fluctuations. A combination of both temperature and power cycling are necessary to truly simulate the thermal structure of the package/module.

5.4.6 Various Effects

Substrate Material. A large reduction in temperature rise can be achieved with an increase in substrate thermal conductivity by substituting AlN or BeO for standard Al_2O_3. The reduction in temperature rise is not as large as the ratio of substrate thermal conductivities when the effect of the thermal environment on heat flow spreading in the substrate is included.

Metallization Layer. Adding a high-conductivity metal layer, such as Cu, to Al_2O_3 results in a reduction in temperature rise. This reduction gradually diminishes as the layer thickness increases. The relative enhancement from the metallization layer is not as great for high-thermal conductivity substrates such as AlN and BeO. The metallization slows the reduction in temperature rise more with large external resistance in the thermal environment. For the case of an isothermal heat sink, adding a metallization layer of any thickness to AlN or BeO can increase the temperature rise. In this case, adding a Cu layer to Al_2O_3 reduces the thermal resistance almost as much as switching to a bare AlN substrate.

Device Size. A large device produces a higher temperature rise than a small device when compared on the basis of equal heat dissipation per unit area. The metallization layer provides a larger reduction in temperature rise for small devices as compared to large devices.

Thermal Environment. The thermal resistance external to the substrate from the heat sink bond, heat sink, and convection to the cooling medium has a direct effect on the heat flow paths and thus spreading resistance with the substrate itself. An estimate of the external thermal resistance is needed to obtain actual device junction temperature rather than just temperature rise above the heat sink temperature, and should be included in the analysis of the substrate. The practice of decoupling the substrate from the thermal environment by assuming an isothermal interface at the heat sink is usually not valid. A design example using TAMS for thermal simulation shows that this method leads to a large underprediction of temperature rise as compared to the coupled analysis. Clearly, the resistance of the thermal environment should be included when evaluating substrate spreading resistance. As the substrate conductivity increases, the constant external resistance represents a larger resistance relative to the substrate. The heat flow tends to spread out more in the substrate, causing the decrease in temperature rise to be less than expected. To greatly improve the thermal performance, the use of heat spreaders, heat sinks, forced air flow, and liquid cooling may apply to various thin film hybrid packages or MCM-D modules.

Device Spacing. Device spacing affects heat dissipation and heat interaction. The spacing between devices needed to eliminate the thermal interaction depends on the substrate material, the heat sinking of the substrate, and the thickness of the metal layer. In general, larger device spacings are required for lower substrate thermal conductivity, higher heat sink thermal resistance, and thinner metal layers. For closely spaced devices, the increase in temperature rise upon a decrease in device spacing is a measure of thermal interaction. Thus, lower substrate conductivity, lower heat sink resistance, and thinner metal layers lead to more thermal interaction. The reader is referred to other sources [46–50] for additional information on the subject.

5.5 COMPONENT ATTACHMENT

This section discusses the first-level connection technology that attaches the die to the substrate or to the multichip module. These connection technologies include wire-bonding, TAB, and flip-chip solder bump.

5.5.1 Wirebonding Technology

Wirebonding Process. The semiconductor die is attached to the substrate with epoxy or solder, and the bond pads of the chip are connected to the corresponding pads on the substrate with a fine wire. Wirebonding necessitates bringing the bonding tool to a first-search level where the tip of the tool and the surface to bond are clearly seen under magnification. The wire bonder brings down the bonding tool, and applied force enables the wire to bond to the bonding surface. The bonding tool retracts, releasing the wire in a manner that permits the wire to flow freely. The wire bonder again moves the bonding tool to a second-search level. Once more, applied force moves the wire to bond to the bonding surface. The wire is then separated and the machine is ready for another bond. In addition to a visual examination of the quality of the bond, a pull test can confirm the bond strength (Mil-Std-883 requires a force gram of 3 g for 1.25 mil or 32 μm, 99% Al to 1% Si wire bond). It is important to have a clean interface at the bond site for the bond or weld to properly occur. A plasma cleaning of the substrate before wirebonding can enhance the reliability of the attachment process.

Thermocompression Wirebonding. The bonding equipment contains a microscope, a heated stage; a heated wedge or capillary to apply pressure to the wire at the interface of the bonding surface; and a wire-feed mechanism. The thermocompression bonding concept relies on heat and pressure to deform a gold wire against the heated bonding surface to form a metallurgical weld with the bonding surface. The high temperatures and the long bonding times can affect the performance of some of the semiconductor chips.

The thermocompression ball-bonding sequence starts with a ball formed at the end of the wire with a hydrogen torch burning the end of the wire. The capillary tip is held at a temperature of 150°C to 200°C, and the substrate is heated at 300°C to 400°C. The bonding tool or capillary brings the ball down to the substrate with a distant clearance from the surface to make a ball bond. This is called the first-search level. The bonder is activated and the bonding tool moves down. In thermocompression bonding, heat and pressure are applied to make the first bond, basically deforming the ball and establishing the desired contact or bond. The capillary or bonding tool is retracted to allow the wire to feed out and to move to the second-bond position and second-search level. Heat and pressure are applied once more to make the second bond. The wire is cut with a hydrogen torch and a new ball is formed. The tail is also removed.

The capillary is made of quartz, tungsten carbide, ceramic, or Ti thick-walled tube. Thermocompression bonding can be either ball bonding or wedge bonding to target small bonding areas. The high temperatures combined with long bonding times can affect semiconductor dice performance and reliability. Some difficulties may be encountered in wire bonding, such as dealing with die that are relatively small with respect to the wire diameter, improper stage or capillary temperatures, contamination of the wire or bonding surfaces, and poor thermal contact between the tool and the substrate.

Ultrasonic Wirebonding. In ultrasonic wirebonding, the ultrasonic energy and the acoustical high-frequency movement of the Al wire against the deposited metal breaks the refracting oxides, displaces contaminants, and removes organics around the wire. This vibration energy can cause a temperature increase at the interface between the wire and the conductor to enable the wire to bond to the surface, establishing the desired contact (temperature rise can reach half the melting degree of the metal). The combination of the bond time and power is very crucial in producing successful bonds.

In comparison with thermocompression wirebonding, ultrasonic wirebonding is considered a relatively cold process. Special care must be provided to optimize the wirebonding parameters to avoid intermetallic formation. Ultrasonic wedge bonding results in producing a fine-pitch bonding at both the first- and second-bond sites (pitch of 0.04 mm). The elongated nature of ultrasonic wirebonding necessitates the alignment of the line of flight of the wire between the chip and the substrate pads so as not to cause a sidewise tearing motion on the already deformed Al wire.

The Al bonding wire is composed of 99% Al to 1% Si alloy, because pure Al is very soft. This feature strengthens the bonding Al wire without affecting its conductivity or boundability. In addition, this feature makes it possible to manufacture small diameter wires (<2 mils in diameter).

The narrow, elongated wedge bond does not completely encapsulate the chip pad metallurgy with noble metal as does the Au ball bond, decreasing the ionic corrosion resistance of the interconnection and the Al wire. On the other hand, rework is facilitated by the narrower wedge bond. In addition, the ultrasonic action tends to break up and displace contaminants, especially brittle oxides, making the Al wedge process more tolerant in manufacturing processes.

Thermosonic Ball-Bonding. In thermosonic ball-bonding, Au wire with diameter less than 0.7 mils is used. An electric flame-off is used to melt a small portion at the end of the wire extending beneath the capillary tip to form the ball. The bonding action is achieved by compressing the wire while the capillary tool is allowed to vibrate at a certain frequency to generate the necessary heat at the bonding site. Some heat may be applied to bring the substrate to a certain temperature to promote the bonding action. Figure 5-6 illustrates the thermosonic ball-bonding process.

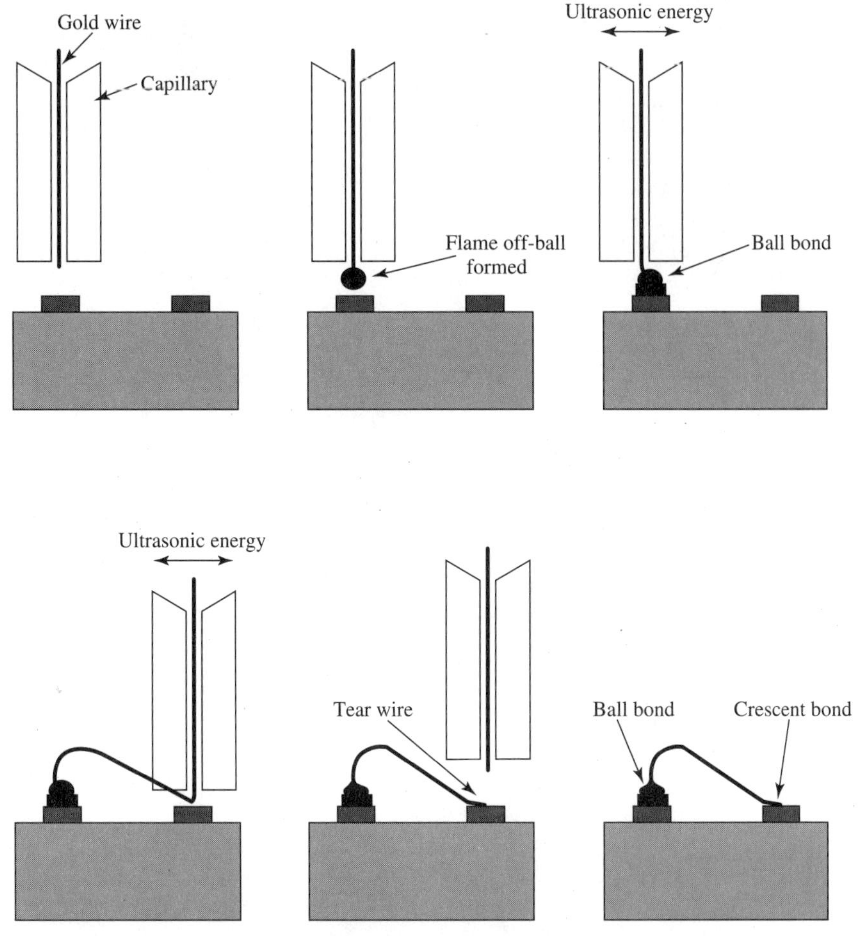

FIGURE 5-6 Thermosonic ball bonding process.

Wire-Bonding Equipment. A constant increase in the bonding rate and a decrease in the bond-pad pitch are the main objectives in producing high-density interconnects. Currently, speeds reaching 4 wires per second for wedge bonding and 10 wires per second for ball bonding are attained. Palomar, Kulicke and Soffa, and Orthodyne are among the major companies in the United States that produce wirebonder equipment for today's market. Figure 5-7 illustrates typical bonding equipment. Figure 5-7*a* is a wedge bonder and Fig. 5-7*b* is a ball bonder.

Reliability and Performance. The wirebonding technique selected is basically a function of the nature of the module designed; the desired wire density; current-carrying capabilities; metallurgical properties; characteristics of the bonding wire; type of application; and environmental conditions. Few necessary precautions and special conditions have to be satisfied to obtain successful bonds. First, the alignment of the line of flight of the wire between the bonding pad within the diffused window of the chip and the metallization pad

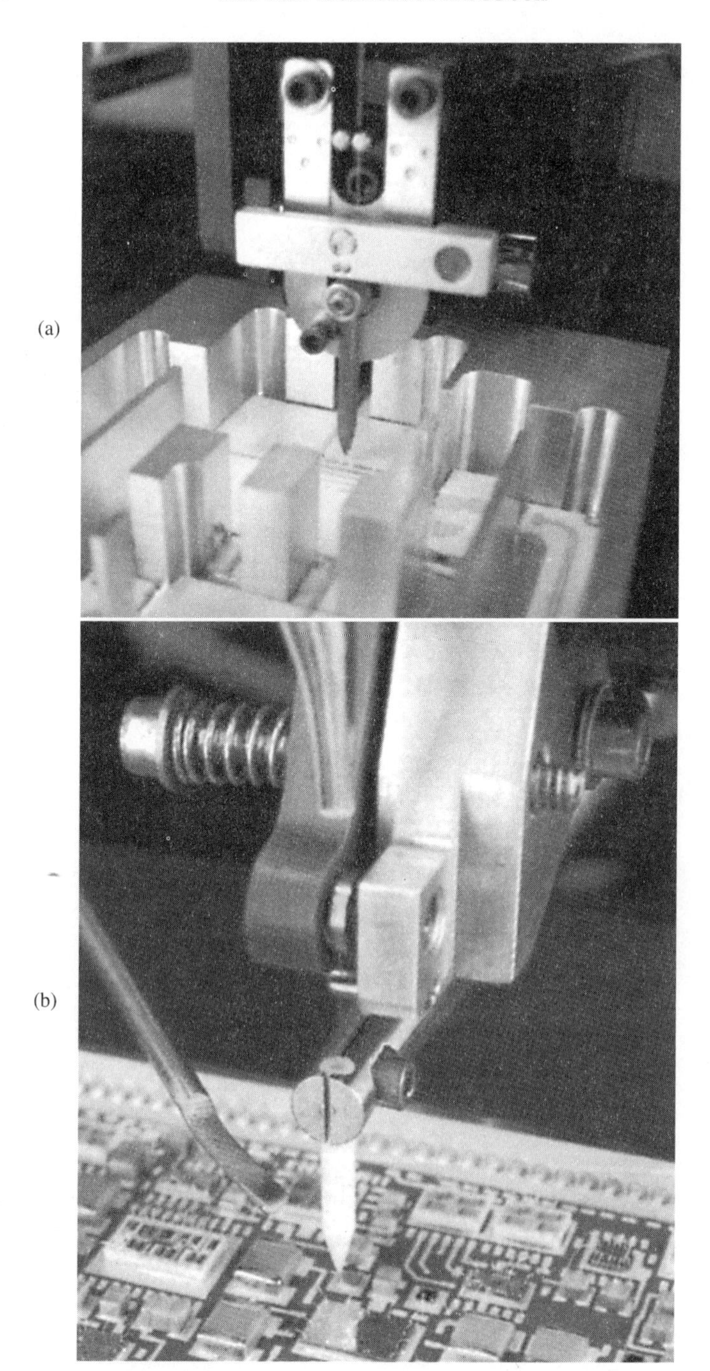

FIGURE 5-7 (*a*) Wedge bonding machine and (*b*) ball bonding machine. *(Courtesy of Palomar Products Inc.)*

on the substrate has to be ensured. In this case, a sidewise tearing motion on the deformed wire is avoided by simple rotation and movement of the workstation or substrate holder. Second, contamination prevention has to be ensured to create successful bonds. Contamination can be oxides formed on the bonding interfaces or chemical contamination (such as finger oils from improper handling). Some measures of cleanliness have to be reinforced and restricted to ensure no contamination. Third, surface irregularities of the metals involved can affect the formation of a good bond (this factor depends on the deposition techniques and the thickness of metal).

Strength and quality of the wirebonds are tested with a pull test or shear test. The most common techniques for testing wirebond integrity and quality are the pull test for wedge bonding, and the shear test for ball bonding. The reader is encouraged to read further about reliability and yield problems of wirebonding [51].

5.5.2 Tape-Automated Bonding (TAB) Technology [52]

Bonding Process: Basic Concept. Tape-automated bonding is an attachment technique that connects a semiconductor die to the substrate by bonding a patterned conductor to the corresponding I/O pads on the die and package. In TAB, the inner lead bonds are connected to the bond pads or bumps on the devices, and the outer lead bonds are attached to the proper metal connections. The TAB concept and TAB features are illustrated in Fig. 5-8. Figure 5-8*a* is a schematic illustration of the TAB concept. Figure 5-8*b* illustrates important features of a TAB tape design. Figure 5-8*c* is an actual TAB-bonded device on an MCM module. Tape-automated bonding is classified into conventual TAB, flip-TAB, and cavity-TAB configurations, Fig. 5-9. Advantages of TAB include an ability for the die to be directly bonded to a printed wiring board or to an MCM substrate; a good electrical performance; and an assembly of high-lead count, especially for fine-pitch interconnects. Thus TAB gives high-circuit density, the ability to test devices before package interconnection, and repairability. On the other hand, disadvantages of TAB may include inadequate infrastructure for advanced equipment and materials needed, with a lack of necessary standards. Various companies are pursuing advances in this basic technology, including Shinko and NEC.

Basic Process Sequence for TAB. The basic sequence for TAB consists of wafer bumping; tape processing; inner-lead bonding; encapsulation; test on tape and burn-in; outer-lead bonding; and final test. Wafer bumping results in the formation of metal standoffs on the I/O pads of the chip or integrated circuit before the wafer sawing by the manufacturer. These metal standoffs, usually gold plated or coated with solder with Ti-W as a diffusion barrier, provide the proper connection for the inner-lead bonds, preventing the TAB lead from shorting to the edge of the die and protecting the metallization underneath it.

TAB Tape. The TAB tape is manufactured in reels or mounted into individual slide carriers. The width of the tape varies typically between 10 to 50 mm and the dielectric thickness is ~0.005 in (0.127 mm). The tape is a patterned conductor matching the I/O pattern on the semiconductor die, laminated to a dielectric layer. Copper is the common choice for the metal material of the tape, and the Cu form is either rolled and annealed or electrodeposited. Although the annealed and rolled Cu has a flat and elongated grain structure with a grain size of about 1 μm, the electrodeposited Cu has a grain size of about 5 μm. Both forms of Cu have typical or similar properties, except for the percentage of elongation (about

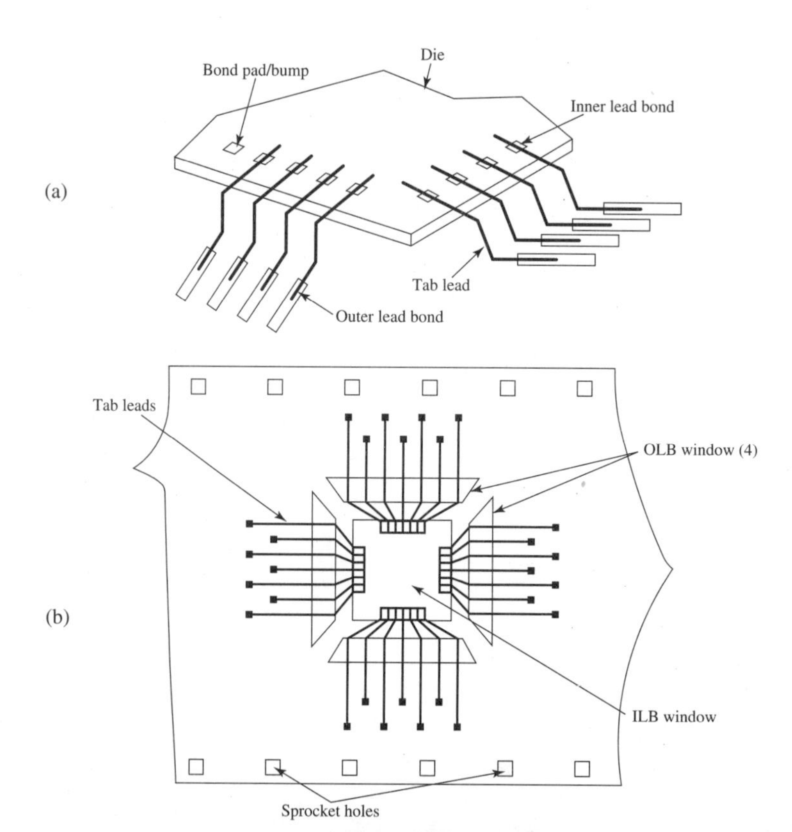

FIGURE 5-8 (*a*) Schematic illustration of the TAB concept; (*b*) important features of a TAB tape design; (*c*) example of TAB attachment in a signal processor MCM. *(Courtesy of IMI, Inc.)*

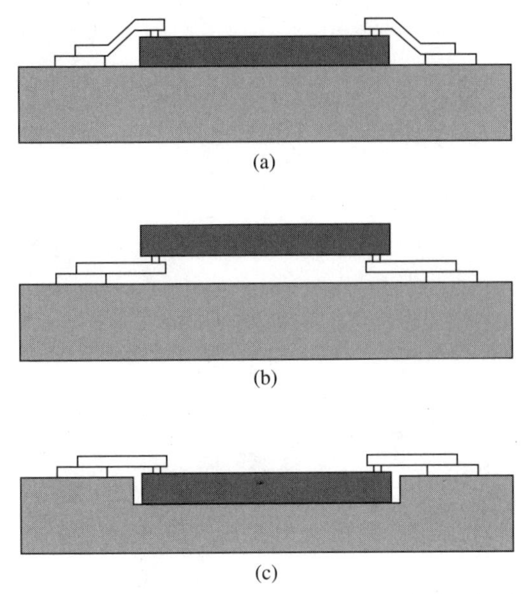

FIGURE 5-9 TAB assembly configurations: *(a)* conventional TAB; *(b)* flip-TAB; *(c)* cavity TAB.

20% for rolled and annealed Cu and about 12% for electrodeposited Cu); and tensile strength (35-40 MPa for rolled and annealed Cu and 45-55 MPa for electrodeposited Cu).

Polyimides have been the common choice for the dielectric material of the tape. These materials are resistant to high temperatures, high moisture, high frequency, and are mechanically stable. Polyimides may widely vary in values for CTE, tensile strength, elastic modulus, elongation, and moisture absorption. Adhesives, based on epoxy materials, are used to adhere the Cu conductor to the dielectric material.

TAB inner-lead spacings range from approximately 0.25 mm to a lower limit of 50 μm, although TAB outer-lead spacings range from approximately 0.50 mm to a lower limit of 100 μm.

Inner-Lead Bonding. Inner-lead bonding (ILB) can be achieved using either single-point or gang-bonding. One bond is heated at a time in single-point bonding with a pulse tip thermocompression for Au tape and bump metals, or thermosonic with Au tape and either Al or Au bumps. Single-point bonding usually requires less force per bond and eliminates the temperature uniformity and planarity problems, in addition of achieving a fine pitch of ~0.004 in. On the other hand, all of the leads are simultaneously bonded to the die in gang-bonding. Thus, the attachment is achieved in a one-step process for a few seconds per die, regardless of number of leads. Planarity of the die, tape and thermode, and temperature uniformity across the thermode (especially for large die size) are issues of concern in gang-bonding. In this type of bonding, the tool is very flat and is evenly heated around the desired bonding area. Thermocompression bonding provides constant heat thermodes. In thermocompression bonding, heat and pressure are applied to form the metallurgical bond between the TAB tape and the bump, causing a softening of the Cu (with copper-coated leads or pads) or Au (with gold-plated leads or pads). Typical bonding temperatures vary in the range 450°C to 550°C with a pressure greatly depending on the equipment used. On the other hand, eutectic bonding and reflow provides hot bar thermodes for pulse-heated hot bars. Selection of tape and bump metallurgies may include gold-plated tape on tin-coated bumps, or tin-plated tape on gold bumps. Both thermocompression bonding and eutectic bonding and reflow are common attachment techniques in gang-bonding.

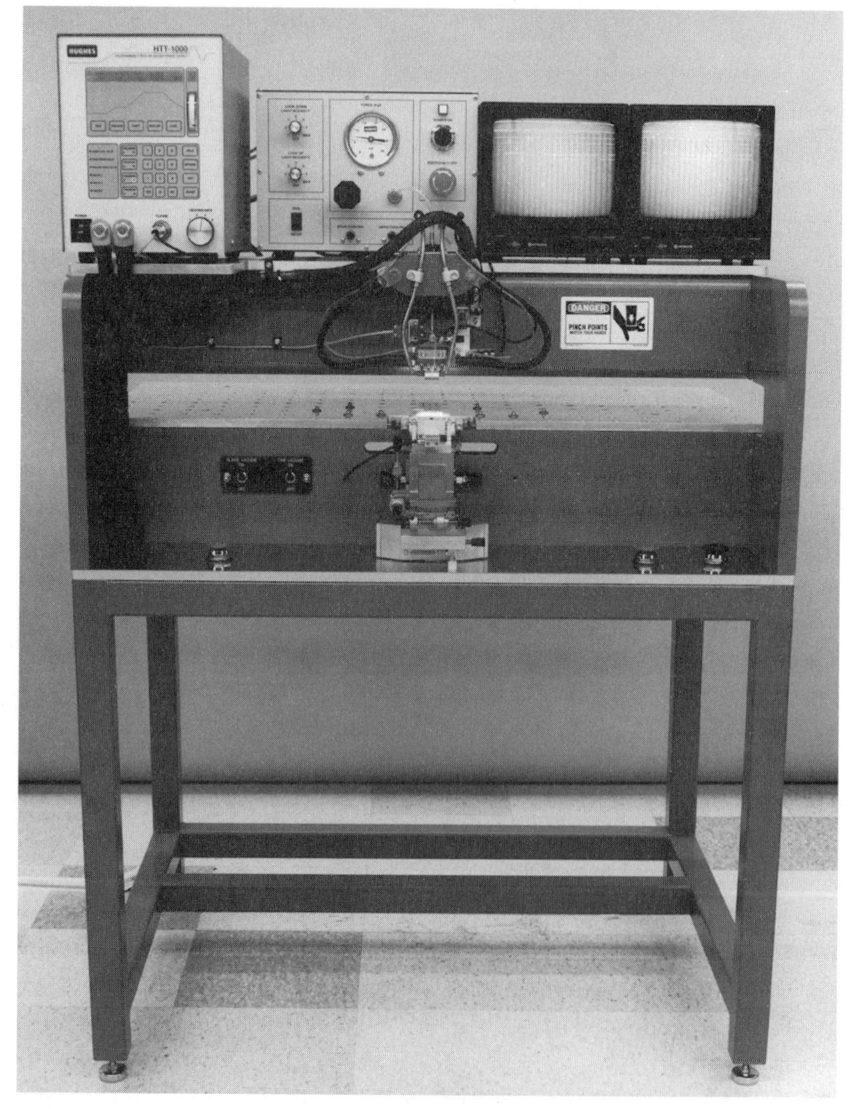

FIGURE 5-10 Typical TAB bonder. *(Courtesy of Palomar Products Inc.)*

Next, the integrated circuit is encapsulated using epoxy or silicon to protect the chip both chemically and mechanically. The encapsultant material must have temperature stability, low moisture absorption, stress resistance at low temperatures, and CTE matching the die.

Test-on-tape and burn-in tests an inner-lead bonded and encapsulated die before attachment to the MCM. Figure 5-10 illustrates a typical TAB bonder.

Outer Lead Bonding. The outer-lead bonding (OLB) process provides a connection of the die to the package. The OLB consists of excise (the ILB chip is removed from the tape before placement); form (lead forming brings the leads from the plane of the inner-lead bonding to the plane of the outer-lead bonding); fluxing and die attach; alignment (to align the die precisely; bonding; and cleaning (cleaning removes the flux residue that was added to enhance solderability). Finally, the packages are inspected and electrically tested.

Thermal Consideration. Copper leads present in TAB technology greatly enhance the thermal management. Conducting thermal analysis and simulation demonstrates that the effective thermal resistance of a typical Cu lead (length, 50 mils; width, 5 mils; and thickness, 1.5 mils) constitutes a factor about 30 times smaller than a Au wire-bond loop of 100 mils span with 1 mil wire diameter.

5.5.3 Flip-Chip Bonding, Attachment Technology

Flip-Chip Bonding. Flip-chip technology allows an assembly approach for active chips integration similar to their original form on the wafer, thus enabling these devices to perform well electrically and thermally while achieving a fine pitch and close-chip proximity.

The process involves flipping the chip and directly attaching its I/Os via connecting bumps, either on the substrate or the chip or on both. Location of bumps depend on several factors, among them is ease of assembly, testing, and repair; the yield percentage; and the temperature of attachment. Figure 5-11 illustrates a schematic of a solder bump before and after reflow. Figure 5-12 illustrates the bump's location to connect the chip to the substrate. A typical bump profile is illustrated in Fig. 5-13. Solder bump interconnects were first introduced by IBM almost three decades ago, and the technology was known then as controlled-collapse chip connections (C4). Solder bump reflowing enables the solder to undergo a round transition between a solid state and a liquid state, allowing the chip connection to the substrate. A wettable area underneath the bump constitutes the bonding interface to the solder. The I/O pad of the chip is located directly below this wettable area. On the other hand, a nonwettable area exists in the vicinity of the bump to confine the solder within the allowed region. The flip-chip solder bumps are made using thermal evaporation or electroplating of lead-tin (Pb-Sn) solder materials, 95 Pb/5 Sn for a reflow temperature of about 330°C. Often a long period of time is needed to deposit a solder thickness of a few microns, either in the form of an alloy or two-elemental charges for lead and tin constituents. The different vapor pressures of the two elements cause one to deposit before the other, depending on the form of the evaporation sources.

Materials Selection. The base metals of the wettable area underneath the bump consist of successive thin film deposition by sputtering or evaporation of Ti or Cr for adhesion to

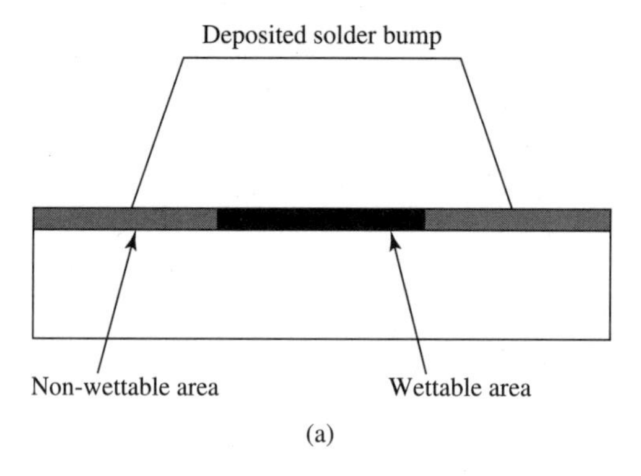

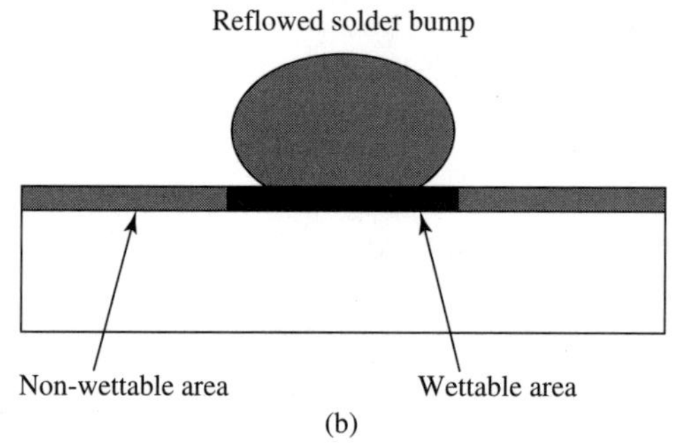

FIGURE 5-11 Schematic of a solder bump (*a*) before and (b) after reflow.

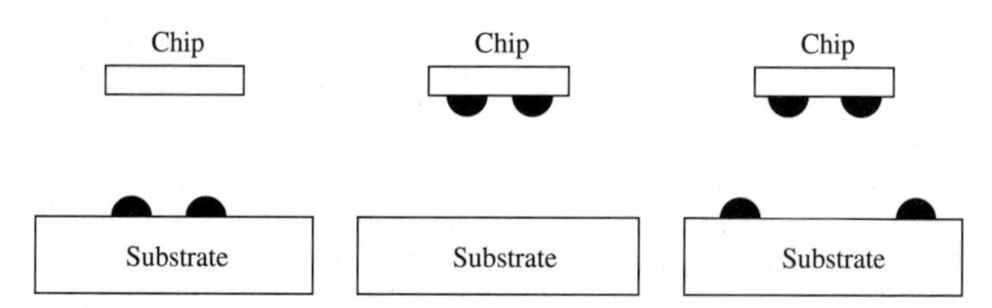

FIGURE 5-12 Several configurations for the locations of bumps connecting chips to substrates.

the underlying surface (about 1000 to 2000 Å); Cu, followed by Ni or Ag for surface preparation; and a barrier metal (about 10,000 Å) by thin film deposition; followed by plating; and finally a top surface layer of Au for good surface properties enabling excellent solderability and nonoxidation (about 1000 to 2000 Å).

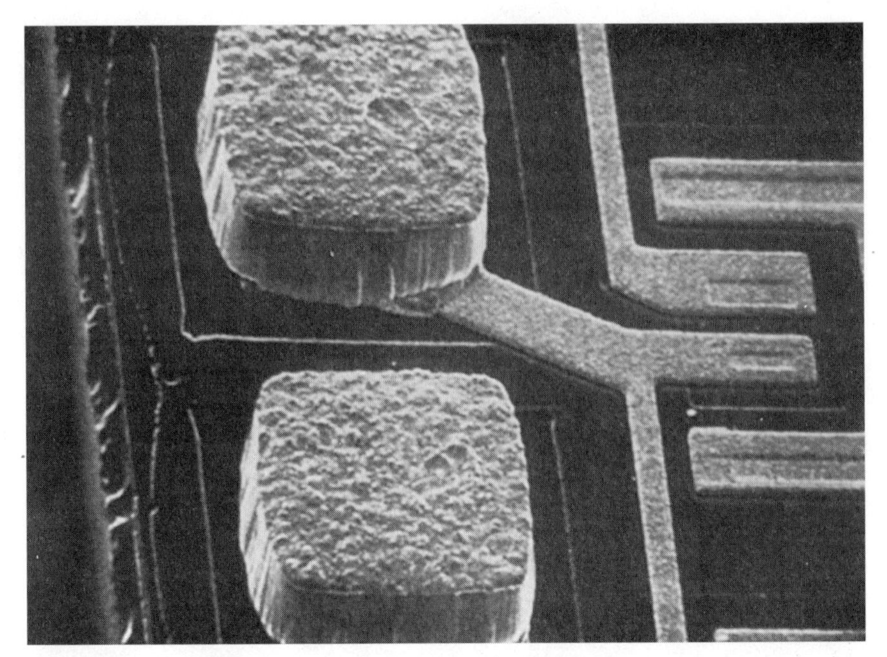

FIGURE 5-13 A typical solder bump profile. *(Courtesy of IMI, Inc.)*

The nonwettable area for the formation of a solder dam is often composed of a dielectric material incapable of interacting with molten solder. Patterning to form the solder bumps following a desired geometry can be either additive (a pattern deposition followed by selective removal) or subtractive (entire deposition followed by patterned etching). Chapter 2 covers in detail the patterning technique. The reader is referred to other sources [52,53] for more information on flip-chip technologies. Flip-chip technology is actively used in electronic products by IBM, Delco, and Hitachi.

5.5.4 Die Attach

Another key area of the assembly process is the attachment of the semiconductor devices to the substrate metallizations. This is only an issue for wirebonded circuits, since flip-chip and TAB include the die attachment. For high reliability and power applications, the nature of the bond between the die and substrate can be critical due to thermal loss and long-term mechanical strength.

Die attach is performed by an automated die bonder or "pick in place" which picks a die from a waffle pack that is used for storage and places the die on the substrate in a precise location. Once all the dice have been placed, the substrate is then heated in order to reflow the solder or set the adhesive. An automatic die bonder is illustrated in Fig. 5-14.

Normally, the attachment media is an adhesive or solder and care must be taken to select the correct material for a given application. Some of the issues involved in the die attach include compatibility of the metallizations with a given solder, void formation, the

FIGURE 5-14 Automatic Eutectic Die Bonder. *(Courtesy of Palomar Products Inc.)*

mechanical strength, electrical resistivity, and thermal conductivity of the attachment media.

Adhesives are available in thermally conductive and electrically conductive varieties and are normally composed of silver or ceramic-filled epoxies. These materials are typically used in applications where the component does not have a high power dissipation because the thermal conductivity of these materials ranges from 0.5 to 3 W/m·°K. They do have the advantage of low temperature cure and ease of assembly, because a measured quantity of the adhesive can be dispensed at the appropriate location on the substrate.

Solder and other eutectic attachments have been used since the inception of microelectronics for the attachment and joining of a wide variety of materials. They offer the advantage of high strength and ease of rework (because the joint can be disassembled by heating the solder to its melting point). Solders with a wide range of properties are in use today, and several of these materials and their properties are summarized in Table 5-4.

Solders are used in paste form or in preforms. Pastes are suspensions of solder and flux and can be applied to the substrate by stencil printing. Preforms are solid ribbons or strips of solder which can be cut or punched out into intricate shapes and are used in conjunction with a liquid flux. Solder attachment can be more complex than adhesives because the metallurgy of the substrate, solder, and die must be compatible. In addition, the application of flux must be carefully monitored with large die sizes because the reactive gases released by the flux can become trapped and form voids in the die attach, as illustrated in Fig. 5-15.

TABLE 5-4 Properties of a few common solder attachment materials.

Solder composition (%)	Liquidus (°C)	Solidus (°C)	Electrical conductivity (% of Copper)	Thermal conductivity (W/m·°C)	Tensile strength (psi)
62 Sn 36 Pb 2 Ag	179	179	12	50	6380
63 Sn 37 Pb	183	183	12	50	7500
97 In 3 Ag	143	143	23	73	800
60 Sn 40 Pb	191	183	12	49	7600
80 Au 20 Sn	280	280		57	40,000
95 Pb 5 Sn	312	308	8	35	4000
90 Pb 10 Sn	302	275	8	36	4400

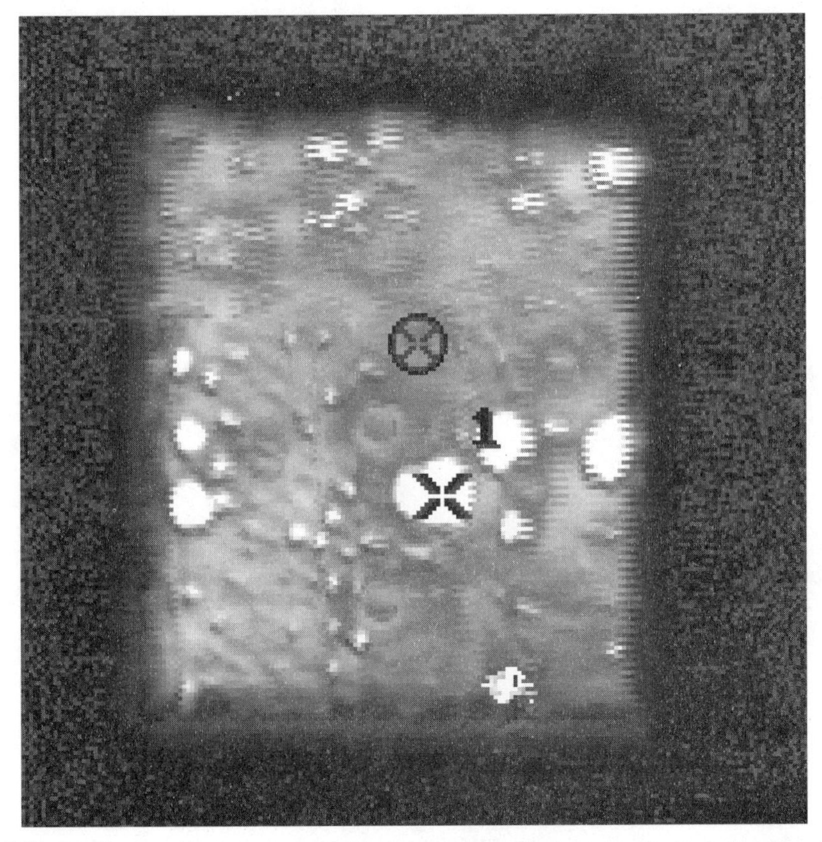

FIGURE 5-15 Acoustic microscope image illustrating void formation in a power die attached to a Au metallization using an 80% Au 20% Sn preform.

5.6 TRIMMING TECHNIQUES

5.6.1 Introduction

The inaccuracies and variations in the thin film component deposition techniques result in resistance values close to ±10% of their designed values. Resistor trimming is an effective tool in producing accurate and reliable electrical elements, with minimum effect on power handling capability, element value variation with time, temperature, and voltage, and electrical noise. Accurate resistor trimming depends on various factors such as available resistor surface area, available resistor length between the two end-metallizations, resistor material, trimming capability, material interaction at the resistor-metallization interface, resistor passivation method, resistance value desired, and the tight tolerance desired. Laser trimming provides an effective tool for adjusting value of thin film resistors to a tight tolerance. Trimming a capacitor structure involves controlling both the capacitance density of the structure and the overlap area of the electrodes. In addition, very tight tolerances can be achieved for these capacitors by adjusting the capacitance value in discrete steps.

5.6.2 Resistor Trimming

Resistor trimming can be achieved using several methods, including anodization, heat trimming, chemical trimming, and laser trimming.

Anodization. In trim-anodizing technique of a thin film resistor, the thickness of the conductive metal is uniformly reduced by the growing layer of insulating tantalum oxide, at the expense of the parent film in a Ta system. Thus, a reduction of the effective thickness of the film occurs. The anodic film is also used as the dielectric film for thin film capacitors. The minimum anodization voltage for Ta resistors is 30 V. Tantalum oxide grows at the rate of 17 Å/V while consuming 6.3 Å of Ta per volt. The anodization is conducted at a constant current until the correct resistance value is measured. The resistor is anodized at one value of current until the resistance is close to the final value. Then the current is reduced to a lower value to reduce the anodization rate. The process is repeated until the desired tolerance is achieved.

Successive anodizing and measurement of the resistor value are achieved. The resistance of the film must alternate with the anodization. Resistive films are made at least 5% lower in value than specified. Low tolerances of 0.01% can be obtained. Anodization is the preferred method for trimming Ta-film resistors because it is the advantage of providing a protective layer on top of the resistor. The interference color of the film changes as the anodization progresses. Thus, the formed oxide layer shows an even color. Large resistance values may cause anodization to form a thicker oxide layer near the termination connected to the power supply. Lower current density and slower rates of anodization usually overcome this problem.

Heat Trimming. Heat trimming involves localized heating to adjust the resistance of the film. The process involves heating in the air at temperatures of several hundred degrees centigrade. This process is a treatment used for stabilization of resistance values as well as trimming. The method is not a very precise one. Heat sources include pulsed lasers, infrared lamps, and carbon arcs.

Electrical Trimming. Electrical trimming involves an appropriate current passing through the resistor film, generating the required heat.

Mechanical trimming. Mechanical trimming involves removal of a portion of the resistor material by sandblasting, vaporization, or cutting without causing any weak spots where local heating may occur. Usually, the mechanical trim involves the reduction of the line width of the resistor uniformly along its length.

Chemical Trimming. Chemical trimming includes the selective application of conductive inks. In this case, the resistance decreases in value by trimming.

Laser Trimming. In this trimming method, the film properties are unchanged, but the geometry varies. The bar, top-hat, ladder, and loop-resistor designs represent different shapes for laser trimming.

5.6.3 Capacitor Trimming

Successful trimming of a thin film capacitor structure involves two factors: controlling the capacitance density of the structure and controlling the overlap area of the electrodes. These two factors determine the achieved capacitance value. As an example, the capacitance density of a tantalum oxide structure can be controlled accurately by controlling the thickness of the anodic oxide through the anodization voltage control. The overlap area of the capacitor electrodes depend on the control of the width of each of the electrodes and their precise alignment.

Tolerances as tight as $\pm 0.1\%$ can be achieved for thin film capacitors by adjusting the capacitance value in discrete steps. The pattern type consists of a main capacitor in parallel with a set of binary weighted trimming capacitors, which can be removed by trimming. Laser trimming of the final value of an interdigitated capacitor may be accomplished by cutting through an exact number of fingers (or discrete steps of capacitances).

5.6.4 Background for Laser Trimming [54–75]

Current trends use laser trim considerations in the design of trimmable thin film hybrid components. Efforts in determining and reducing the effects of laser parameters on the heat-affected zone and resulting resistor stability are underway. In addition to using lasers for trimming resistors and capacitors, laser systems are also used to cut through fired and unfired ceramics, pattern thin films, and, in some cases, repair open circuits in specially deposited thin film circuits.

Methods for resistor trimming are based on abrasive or sandblasting techniques and laser systems; namely, carbon dioxide (CO_2) lasers and acousto-optic Q-switched neodymium:yttrium-aluminum-garnet (Nd:YAG) lasers operating at a wavelength of 1064 nm. Laser-trimmed resistors are less stable and more noisy as compared to abrasive-trimmed resistors and untrimmed resistors. The main reasons are high temperatures and thermal gradients associated with the laser beam, causing various stresses and compositional changes in the resistor material. This causes small cracks and melting at the end-of-Kerf of the trim material. Different laser trimming techniques and properly processed resistor elements can remedy or alleviate some of these effects. In fact, these trimming techniques and the position of the various cuts can be adjusted to yield optimum stability. Sometimes, a second harmonic generator is added to provide green wavelength (532 nm) in the visible spectrum to improve trimming capability by providing better alignment. Laser trimming can also be used while the circuit containing the element is in operation. This technique is referred to as active laser trimming. Active trimming reduces the trimming time by providing in-situ circuit response to the trim.

5.6.5 Laser Trimming Thin Film Resistors

Laser trimming of thick film resistors is an important function in manufacturing hybrid circuits on ceramic substrates. The laser spot is only a few mils in diameter, and the kerf is a few mils. The laser uses a light beam to remove the material deposited on the ceramic substrate in a very short time duration (<1 ms). The laser beam has a high-intensity coherent light pulse that impacts the resistor material. The absorption of the light energy causes the resistor material to heat and vaporize. The process depends on the intensity of the laser pulse (power level), the focus of the laser pulse, and the material composition. As the cut progresses into the main part of the resistor, the resistance value changes continuously. Because laser trimming adjusts resistors to higher values, resistors that require trimming need to be designed for values lower than those specified. One important requirement for precision laser trimming is that the measurement system of the laser must be fast enough to stop the laser beam between pulses once the specified value is reached.

A pulsed CO_2 laser (long pulse width with high energy per pulse) causes vaporization of the deposited film. Damage to the resistive material and the underlying dielectric can occur as a result of CO_2 laser application. On the other hand, the Q-switch provides a Nd:YAG laser with a two-way optical switch, blocking oscillations between laser mirrors (with *on* position), and letting the laser beam pass through (with *off* position). With the acousto-optic Q-switch, the Nd:YAG laser is able to produce short pulses of high peak power for rapid vaporization, with precise control and minimum heat flow into neighboring areas. As a result, well-defined kerfs, free of residues with minimum microcracking in the surrounding material and stable resistors, are achieved with a Nd:YAG laser system.

Excimer lasers are a high-resolution class of lasers (1 mm) that emit high-intensity, short duration pulses of ultraviolet light. These devices can emit up to 1 J of energy in a single pulse of 20 ns duration. Lasers of 100 W are currently available for precision ablation and etching. In addition, excimer lasers are also used for submicron lithography and for producing patterned materials based on various deposition techniques.

5.6.6 Trimming Techniques (Shapes of Cuts)

Various conventional trimming techniques consist of performing cuts into the resistor element; single straight cut, L-cut, and multiple straight cuts.

The single straight cut starts at one end of the resistor element and proceeds as a cut in the element, perpendicular to the current flow, resulting in a rapid increase in the resistance value. An overshoot often occurs with this trimming technique caused by the time constant of the control system. Poor precision, element heating, and width expansion are among the factors that influence this trimming technique and affect the final resistance value.

The L-cut is an effective trimming method, where the resistance value experiences a rapid increase as the trimmer is cutting perpendicular to the current flow until the value is within a specified tolerance. Then the trimmer changes direction and cuts parallel to the current flow, causing a much smaller change in the resistance value, thus achieving a tight tolerance. This action allows the trimming control system enough time to shut the trimmer off while the resistance value is within close tolerance. The control system is adjusted to monitor the change in the resistance value as well as the resistance-value itself. Difficulty in trimming with L-cut technique often occurs with high-aspect ratio resistors (that is, when resistor length is much greater than the resistor width), where cutting perpendicular to the current flow can sometimes cause the cut to exceed half the width of the resistor.

The multiple straight cut method is based on the principle of the single straight cut. It uses multiple cuts to overcome the problem of cutting too deep on any straight cut, assuming that the spacing between the trims is kept to a certain minimum. Double straight cut, double reverse cut, and serpentine cut are standard types of multiple-straight cuts. The double straight cut uses two cuts, with each cut originating on the same side of the resistor. The first cut quickly brings the resistor to within a tolerance closer to the final desired value. The second cut, usually shorter in length than the first cut, causes a slower variation of the resistance value because little current flow occurs in that location. The double reverse cut uses two cuts that begin from opposite sides of the resistor. Finally, the serpentine cut uses a number of cuts that alternate in the origination sides. The serpentine cut uses between four and six cuts to trim resistors of larger areas to a specified resistance value. One fact remains applicable to these trimming techniques; the accuracy of resistor trimming is higher when the resistance value dictates a small amount of trim.

5.6.7 Trimming Tab for Resistors

In addition to the bar design for thin film resistors, top-hat, ladder, and loop-resistor designs represent different shapes for laser trimming. In most cases, a trimming tab is provided to the resistor geometry to enable the laser to trim the element properly. For the top-hat resistor design, the trimming tab may be trimmed from the top or the bottom with a straight cut or L-cut, providing a more significant effect.

5.6.8 Control of Oxide Thickness for Capacitors as Well as Adjustment of Capacitance Value in Discrete Steps

Trimming a thin film capacitor structure involves controlling both the capacitance density of the structure and the overlap area of the electrodes. These two factors determine the capacitance value. As an example, the capacitance density of a tantalum oxide structure can be accurately achieved (to better than 1%) by controlling the thickness of the anodic oxide through the anodization voltage. The overlap area of the capacitor electrodes depends on the control of the width of each of the electrodes and their precise alignment. Tolerances as tight as ±1% can be achieved for thin film capacitors by adjusting the capacitance value in discrete steps. The pattern type consists of a main capacitor in parallel with a set of binary-weighted trimming capacitors that can be removed or separated through the trimming process. Laser trimming of the final value of an interdigitated capacitor may be accomplished by cutting through an exact number of fingers.

5.6.9 Technology Trends

Researchers have evaluated the stability of laser-trimmed thin film resistors on AlN substrates. With suitable laser power (1 W maximum for Nd:YAG laser at the interface between the resistor and AlN), the resistance values can be controlled to within ±1% of their designed values. These laser-trimmed elements have resistance changes <±4% undergoing thermal cycling test (−55°C to 150°C, 1000 cycles) and high-temperature storage test (150°C, 1000 hr). In general, monoxide ceramics have poor chemical stability with laser beam irradiation, manifested in the formation of minute solidified substrates as well as a degradation of electrical insulation. This is associated normally with high-power laser irradiation.

A fast laser trimming speed of 100 mm/s (4 in/s) to produce stable, thick film resistor (using DuPont resistor series QS87) with 1% tolerance has been also applied. An ability to trim thin film precision resistors to tolerances of better than 0.01% using an automated laser system with superior precision measurement and trim software capabilities has been demonstrated. A stability analysis of laser-trimmed thin film resistors has been conducted and the study concluded that the high energy required to vaporize material in the cut also modifies the aging characteristics of the film in the surrounding heat-affected zone (HAZ). For the parameters controlling the post-trim drift, both short-term and long-term drift of laser-trimmed thin and thick film resistors have been investigated and established. Among these parameters are the protective overglaze coating, the resistor dimensions, the sheet resistivity, and the fired resistor thickness. A reduction of post-trim drift of thin film resistors by optimizing the Nd:YAG laser output characteristics, including varying laser pulse width, power, wavelength, and repetition rate has been investigated and realized. Long-term resistance drift, caused by slow annealing in the HAZ, close to the laser kerf, was eliminated by trimming and by link cutting. The trimmed resistance is realized using a number of resistors in series or parallel. The trim is achieved by completely cutting some of these links, as necessary. Clearly, the desired resolution comparable to the continuous trim approach will be on the expense of area for a large number of these links.

5.6.10 Thermal Stabilization

The discontinuous transition between the different materials is unstable. By thermal oxidation (consisting of heating and aging), oxygen atoms diffuse into the Ta at the metal-oxide interface, producing oxygen-doped Ta and Ta-rich tantalum oxide. Thus, a continuous transition results between the different materials. A relatively small increase in the resistance value occurs (the conductivity of Ta at the interface is reduced because of the oxygen diffusion into the Ta). A trim-anodization step is necessary to adjust the value to a tight tolerance.

5.6.11 Protection of Thin Film Resistors

The Ta_2O_5 grown by anodization forms a hard, glassy, uniform film over the resistor to provide protection from damage during handling, and from deterioration of the resistor material by heat or other factors.

5.7 MODIFICATIONS OF SURFACES AND FILMS [76–78]

The objective of surfaces and films modification is to improve the surface and film properties. This objective can be achieved (in terms of films and coatings deposition or modification of existing surfaces) using directed energy sources.

The chemical vapor deposition process permits the deposition of thin films of diamond on a variety of materials used as cutting tools and tribology, thus benefiting from superior diamond resistance to wear and corrosion, as well as its low coefficent of friction. In addition, a deposited diamond film on the back of wafers and substrates can greatly improve the thermal management of hybrid circuitry built on these surfaces, because of the high thermal conductivity of diamond.

The effects of ion-beam techniques on a variety of materials has been demonstrated, as well as a summary of surface texturing technologies or surface modification techniques (such as ion-beam sputter texturing, sputter etching, etching, and simultaneous deposition and etching of surfaces, arc texturing technology, and atomic oxygen texturing) to improve the surfaces' properties of these materials. Ion surface interactions through an ion implantation process can be used to accelerate Ni or Cr ions toward metal surfaces to improve their properties. These charged atoms are embedded beneath the surface at a certain depth on colliding with the surfaces, producing a new alloy in the implanted region with desired properties of enhanced wear resistance and improved corrosion resistance. Steels, stainless steels, carbides, and Al are among the metals used in the ion-implantation process for better performance. Ion implantation has been used successfully with semiconductors to produce a film layer resistant to formation of defects. These implants can be followed by thermal annealing.

5.8 SURFACE PREPARATION

Surface preparation is necessary to produce a suitable top-layer metal interconnection that is resistant to corrosion and compatible to the dielectric, in addition to bonding pads and solderable areas on the substrate. A barrier metal such as Ni is deposited first on the surface, usually by plating, to prevent diffusion of the Cu conductor. Also, Ni acts as a hard surface withstanding pressure applied during the wirebonding process. Chromium or Ti-W can also be used as barrier metals, and they are deposited by a sputtering process. A thin film sputter deposition of a Au layer ~1 to 3 μm thick usually follows the Ni deposition. Soft Au with minimum amount of impurities is usually desirable to realize successful bonds and die attachment sites.

5.9 CONCLUSION

The design of thin film multilayer interconnections is usually determined by a tradeoff between wide and thick metallization interconnects for high speed performance and fine resolution lines equivalent to thin deposited layers for high interconnect density and high operating frequency.

REFERENCES

1. J. L. Vossen and W. Kern, (eds.), *Thin Film Processes II,* Academic Press, San Diego, CA, 1991, pp. 501–522.
2. C. Stolzel, M. Huth, and H. Adrian, "C-Axis oriented thin $Bi_2Sr_2CaCu_2C_8$+ and films prepared by flash evaporation," *Physica C,* vol. 204, 1992, pp. 15–20.
3. T. Hato, Y. Takai, and H. Hayakawa, "High Tc Superconducting Thin Films Prepared by Flash Evaporation," *IEEE Transactions on Magnetics,* vol. 25, no. 2, March 1989, pp. 2466–2469.
4. C. De Las Heras and C. Sanchez, "Characterization of Iron Pyrite Thin Films Obtained by Flash Evaporation," *Thin Solid Films,* vol. 199, 1991, pp. 259–267.
5. H. E. Horng et al., "Critical Current in Polycrystalline Bi-Ca-Sr-Cu-O Films," *Physical Review B,* vol. 39, no. 13, May 1989, pp. 9628–9630.
6. P. Wagner, H. Adrian, and C. Tome-Rosa, *Physica C,* vol. 198, 1992, p. 258.
7. J. Dieleman, E. Van de Riet, and J. C. S. Kools, *Jpn J Appl Phys,* vol. 31, no. 6B, pt. 1, 1992, pp. 1964–1971.
8. A. J. Basovich et al., *Thin Solid Films,* vol. 228, 1993, pp. 193–195.

9. F. Sanchez et al., *Applied Surface Science,* vol. 69, 1993, pp. 221–224.
10. A. G. Guidoni and I. Pettiti, *Applied Surface Science,* vol. 69, 1993, pp. 365–369.
11. C. Germain et al., *Applied Surface Science,* vol. 69, 1993, pp. 359–364.
12. G. A. Petersen and J. R. McNeil, *Thin Solid Films,* vol. 220, 1992, pp. 87–91.
13. N. J. Arfsten, *Journal of Non-Crystalline Solids,* vol. 63, 1984, pp. 243.
14. H. Dislich, *Journal of Non-Crystalline Solids,* vol. 57, 1983, pp. 371.
15. C. J. Brinker and G. W. Scherer, *Sol-Gel Science,* Academic Press, Orlando, FL, 1990.
16. D. Segal, *Chemical Synthesis of Advanced Ceramic Materials,* Cambridge University Press, Cambridge, United Kingdom, 1989.
17. L. C. Klein, (ed.), *Sol-Gel Technology for Thin Films, Fibers, Preforms, Electronics and Specialty Shapes,* Noyes Publications, Park Ridge, NJ, 1988.
18. L. C. Klein, *Ceram Eng Sci Proc,* vol. 5, 1984, pp. 379.
19. H. Dislich, *Journal of Non-Crystalline Solids,* vol. 80, 1986, pp. 115.
20. Maissel and Glang, (eds.), *Handbook of Thin Film Technology,* McGraw-Hill, New York, 1970.
21. J. G. Bednorz and K. A. Muller, *Z Phys B,* vol. 64, 1986, p. 189.
22. L. Kempfer, *Materials Engineering,* vol. 107, no. 5, May 1990, pp. 26–29.
23. R. Messier et al., *Journal of Metals,* September 1987, pp. 8–11.
24. W. Zhu et al., *Proceedings of the IEEE,* vol. 79, no. 5, May 1991.
25. Y. Hirose, M. Mitsuizumi, *New Diamond,* vol. 4, 1988, pp. 34–35.
26. A. H. Deutchman, R. J. Partyka, *Advanced Materials & Processes, ASM Int.,* June 1989, pp. 29–33.
27. P. A. Stenmann and H. E. Hintermann, *J Vac Sci Tech* vol. A7, 1989, p. 2267.
28. J. Valli, *J Vac Sci Tech,* vol. A4, 1986, p. 3007.
29. V. E. Cosslett, "Fifty Years of Instrumental Development of the Electron Microscope," in *Advances in Optical and Electron Microscopy,* R. Barer and V. E. Cosslett, (eds.), vol. 10, Academic Press, San Diego, CA, 1988, pp. 215–267.
30. J. I. Goldstein et al., *Scanning Electron Microscopy and X-Ray Microanalysis,* Plenum, New York, 1981.
31. G. Thomas and M. J. Goringe, *Transmission Electron Microscopy of Materials,* John Wiley & Sons, New York, 1979.
32. Ohring, *The Materials Science of Thin Films,* Academic Press, Inc., Orlando, FL, Table 6-3, p. 276, 1992.
33. L. E. Davis et al., *Handbook of Auger Electron Spectroscopy*, Physical Electronics Industries Inc., Eden Prairie, MN, 1976.
34. G. E. McGuire, *Auger Electron Spectroscopy Reference Manual,* Plenum Press, New York, 1979.
35. M. T. Bernius and G. H. Morrison, "Mass Analyzed Secondary Ion Microscopy," *Rev Sci Instrum,* vol. 58, October 1987, pp. 1789–1804.
36. J. R. Bird and J. S. Williams, (eds.), *Ion Beams for Materials Analysis,* Academic Press, Sydney, 1989.
37. S.C. Miller, "A Technique for Passivating Thin Film Hybrids," *Proceedings of the 1976 International Symposium of Hybrid Microelectronics (ISHM),* 1976, p. 39.
38. I. Bahl and P. Bhartia, *Microwave Solid State Circuit Design,* John Wiley & Sons, New York, 1988.
39. H. Farzanehfard, J. He, and A. Elshabini-Riad, "A Wide-Band Comparison of Aluminum Nitride, Alumina, and Beryllia Microcircuit Substrates," *IEEE Transactions on Instrumentation and Measurement,* vol. 40, no. 2, April 1991, pp. 490–492.
40. *Thin Film Hybrid Design Guidelines,* Tektronix, Hybrid Components Operation, 1985.
41. H.G. Muller and R.F. Miracky, "Laser Writing to Customize Multichip Module Substrates," *Proceedings of the 5th Electronic Materials and Processing Congress, ASM International,* Cambridge, MA, August 24–27, 1992.

42. R. Miracky et al., "Technologies For Rapid Prototyping of Multichip Modules," *Proceedings of The 1991 IEEE International Conference on Computer Design,* Cambridge, MA, Oct. 14–16, 1991, pp. 588–592.

43. I. A. Bhutta et al., "Time Domain Measurement, Modeling and High Frequency Analysis of MIMIC Packages," *ASM International's 5th Electronic Materials and Processing Congress,* Cambridge, MA, August 24–27, 1992.

44. I. A. Bhutta et al., "Electrical Characterization of a Multilayer Low-Temperature Co-Fireable Ceramic Multichip Module," *ASM International's 7th Electronic Materials and Processing Congress,* Cambridge, MA, August 24–27, 1992.

45. P. E. Garrou, I. Turlik, *Multichip Module Technology Handbook,* McGraw-Hill, New York, 1998.

46. G. N. Ellison, "A Thermal Analysis Computer Program Applicable to a Variety of Microelectronic Devices," *Proceedings of 1978 International Symposium on Hybrid Microelectronics,* Minneapolis, MN, 1978, pp. 332–339.

47. G. N. Ellison, *Thermal Computations for Electronic Equipment,* Van Nostrand Reinhold, New York, 1989.

48. Avram Bar-Cohen and Allan D. Krause, (eds.), *Advances in Thermal Modeling of Electronic Components and Systems,* volume 3, ASME Press and IEEE Press, New York, 1993.

49. G. N. Ellison, "TAMS: A Thermal Analyzer for Multilayer Structures," *Electrosoft,* vol. 1, no. 2, Computational Mechanics Publications, 1990, pp. 85–97.

50. M. Hussein, D. Nelson, and A. Elshabini-Riad, "Thermal Interactions of Semiconductor Devices on Copper Clad Ceramic Substrates," *IEEE Transactions on Components, Hybrids, and Manufacturing Technology (CHMT),* vol. 15, no. 5, October 1992, pp. 651–657.

51. G. Harman, *Wire Bonding in Microelectronics Materials, Processes, Reliability, and Yield,* 2d ed, The International Society of Microelectronics (ISHM) and McGraw-Hill, 1997.

52. D. A. Doane and P. Franzon, *Multichip Module Technologies and Alternatives The Basics,* Van Nostrand Reinhold, New York, 1993.

53. B. Romenesko, "Ball Grid Array and Flip Chip Technologies: Their Histories and Prospects," *The International Journal of Microcircuits and Electronic Packaging,* vol. 19, no. 1, First Issue 1996, pp. 64–74.

54. F. Burns, "A Compact Low-Cost Thick and Thin Film Laser Resistor Trimmer," *Proceedings of the 1970 International Symposium Hybrid Microelectronics* (ISHM), vol. 5, no. 3, 1970, pp. 1–9.

55. M. W. Dowley, "Laser Resistor Trimming, Accuracy and Speed," *Proceedings of the 1970 International Symposium of Hybrid Microelectronics* (ISHM), 1990.

56. G. Stone, "Programmable Continuous Trimming, A System Approach to High Speed Laser Trimming of Hybrid Microcircuits," *Proceedings of the 1969 International Symposium of Hybrid Microelectronics (ISHM),* 1969, pp. 177–184.

57. R. Chapman and D. Farbsten, "ND:YAG Laser Fabrication of Fine Geometry Thin Film Substrate Edge Wrap Arounds," *Proceedings of the 1988 International Symposium of Hybrid Microelectronics Symposium (ISHM),* 1988, pp. 567–571.

58. K. Murakami, "Laser Trimming of Polymer Thick Film Resistor," *Proceedings of the 1985 International Symposium of Hybrid Microelectronics Symposium (ISHM),* 1985, pp. 245–252.

59. J. Graves and J. Moorman, "Laser Trimming Techniques for High Accuracy Thermistors," *Proceedings of the 1990 International Symposium of Hybrid Microelectronics Symposium (ISHM),* 1990, pp. 691–698.

60. R. Cote et al., "Factors Affecting Laser Time Stability of Thick Film Resistors, *Proceedings of the 1976 International Symposium of Hybrid Microelectronics Symposium (ISHM),* 1976, pp. 128–137.

61. R. Knapp and S. Wendel, "A Laser System for Active Trimming of Precision Hybrid Devices," *Proceedings of the 1986 International Symposium of Hybrid Microelectronics Symposium (ISHM),* 1986, pp. 83–87.

62. J. Hong and Y. Chiou, "The Surface Leakage Current of Laser Trimmed Thick Film Resistors," *Proceedings of the 1985 International Symposium of Hybrid Microelectronics Symposium (ISHM),* 1985, pp. 369–372.
63. M. Sturrett, "Process Control for Functional Laser Trimming of Thick Film Resistors," *Proceedings of the 1987 International Symposium of Hybrid Microelectronics Proceedings (ISHM),* 1987, pp. 140–146; Also, appeared in *Hybrid Circuit Technology,* 1988, pp. 19–22.
64. J. Brannon, "Excimer Laser Ablation and Etching," *IEEE Circuits and Devices,* vol. 6, no. 5, September 1990, pp. 25–31.
65. Glazer and Subak-Sharpe, *Integrated Circuit Engineering,* Addison-Wesley, Reading, MA, 1979.
66. Y. Kurihara et al., "Laser Trimming of Thick Film Resistors on Aluminum Nitride Substrates," *IEEE Trans on Components, Hybrids, and Manufacturing Technology,* vol. 13, September 1990, pp. 596–602.
67. N. Iwase and K. Iyogi, "A Study for Thick Film Resistors Formed on AlN Substrates," *Proceedings Annual Meeting IEICE,* Japan, 1986, pp. 1–132, (in Japanese).
68. H. Schmidt and B. Couch, "Predicting The Effect of High-Speed Laser Trimming on Resistor Stability," *Proceedings of Electronic Components Conference,* 1987, pp. 585–592.
69. L. Groth, "Laser Trimming Thin Film Precision Resistor Networks with An Automated System," *Electronic Components Conference Proceedings,* 1973, pp. 38–44.
70. A. Bulger, "Stability Analysis of Laser Trimmed Thin Film Resistors," *Electronic Components Conference Proceedings,* 1975, pp. 286–292.
71. J. S. Shah and L. Berrin, "Mechanism and Control of Post-Trim Drift of Laser Trimmed Thick Film Resistors," *Electronic Components Conference Proceedings,* 1977, pp. 252–259.
72. R. Dow et al., "Reducing Post-Trim Drift of Thin Film Resistors By Optimizing YAG Laser Output Characteristics," *Electronic Components Conference Proceedings,* 1978, pp. 87–92.
73. J. Shier, "A Finite Mesh Technique for Laser Trimming of Thin Film Resistors," *IEEE J of Solid State Circuits,* vol. 23, no. 4, 1988, pp. 1005–1009.
74. H. G. Muller et al., "Laser Process for Personalization and Repair of Multichip Modules," *1991 Conference on Lasers in Microelectronic Manufacturing, Proc. SPIE 1598,* 1991, pp. 132–140.
75. Kenneth Lee Venzant, "Analysis of Residual Stresses in Laser Trimmed Alumina Substrates," Masters thesis in Material Science and Engineering, VPI & SU, Blacksburg VA, 1992.
76. G. Lu, "CVD Diamond Films Protect Against Wear and Heat," *Advanced Materials and Processes, ASM Int,* vol. 144, no. 6, December 1993, pp. 45–47.
77. B. Banks, "Modify Surfaces with Ions and Arcs," *Advanced Materials and Processes, ASM Int,* vol. 144, no. 6, December 1993, pp. 22–25.
78. B. Holtkamp, "Ion Implantation Makes Metals Last Longer," *Advanced Materials and Processes, ASM Int* vol. 144, no. 6, December 1993, pp. 45–47.
79. G. Messner et al., *Thin Film Multichip Modules,* The International Society For Hybrid Microelectronics (ISHM), second printing, 1992.

CHAPTER 6

CHARACTERIZATION OF SEMICONDUCTOR THIN FILMS: A COMPENDIUM OF TECHNIQUES

Fuad Abulfotuh
Lawrence L. Kazmerski
National Renewable Energy Laboratory

6.1 INTRODUCTION

The analysis and characterization of thin films complement and guide efforts in film modeling, property engineering, as well as device design and optimization. The precise and accurate determination of important thin film parameters is the key to verifying and realizing thin-film technologies. The ensemble of measurement techniques and tools range from structural and topographic to electro-optical determinations. Many apply directly to the evaluation of bulk properties; others are adaptations to the thickness and size constraints imposed by film technologies; still others are developed especially for thin layer characterization. The evolution of these characterization techniques is driven, directed, and even enhanced by the advancing technologic requirements for higher spatial resolutions, accuracies, and novel component configurations. The constraints placed on thin film measurement scientists challenge their ingenuity as well as the frontiers of measurement science and technology. Thin film characterization is sometimes taken for granted, but, nevertheless, it is a key component of the technology base.

This chapter presents, summarizes, or references several diverse and important techniques used to evaluate thin film materials and devices. The emphasis is on the general representation of the techniques, rather than an exhaustive discussion of the physics, optics, and electronics. The principles and applications of each method are summarized, and their strengths and limitations are stressed. Pertinent examples from the literature relating to elemental and compound semiconductors and devices are provided. The reader is also directed to several excellent sources for a more-detailed analysis and evaluation of various methods discussed generally in this chapter [1–14]. Other chapters in this book also serve as references for further information on thin film characterization approaches. Some optical characterization techniques relating to spectrophotometry and ellipsometry have been covered in Chapter 4. Although this chapter emphasizes the characterization of semiconductor layers, most of the techniques have equally important application to other thin film types.

6.2 ELECTRO-OPTICAL MEASUREMENTS: FILM PROPERTIES

The more fundamental and routine measurements of thin film electrical and optical properties are often taken for granted. The determination of carrier concentrations, mobilities, conductivities, absorption characteristics, and related properties have dominated the literature from the earliest days of film investigations. As the demand for precision and accuracy in reporting has increased, so have the sophistication and complexity of these measurements. Most interpretations depend on a careful analysis of the inherenent characteristics of the particular thin film (such as crystallinity, structure, geometry, carrier type and level, substrate or superstrate, and topography), and they are accompanied by at least some degree of theoretical modeling. This section details a selection of the basic measurements for determining the electrical properties of crystalline and amorphous thin films. Special geometry, carrier concentration, resistivity, contacting, and thickness constraints are summarized, and limitations to appropriate analysis are emphasized.

6.2.1 Hall Effect

- Information: Majority-carrier type, concentration, mobility, and sample resistivity
- Application: Single-crystal, polycrystalline, and amorphous thin films
- Approach: Measurement of transverse voltage across thin film, produced as a result of applied voltage (current) and magnetic field (Lorentz force)

Carrier transport is important in the development of materials and devices. The determination of the conceptually simple parameters, such as resistivity, evolves from basic experimental configurations. One of these is the four-probe method, shown in Fig. 6-1. However, such techniques are limited in the information provided, and possibly in their accuracy and reproducibility. The Hall effect and its derivative measurement schemes provide some of the most important information on the majority-carrier properties of thin films [15–18]. Its elegance is a result of its simplicity, which sometimes leads to misinterpretation because of the sensitivity to contact and sample geometry. However, the physics of this method is presented in shortened detail, and it is important for background and understanding of other techniques.

Analytical Treatment. Consider the experimental configurations for a rectangular-shaped semiconductor shown in Fig. 6-2*a, b*. For a *p*-type semiconductor, the Lorentz force (i.e., $\mathbf{F} = q\mathbf{v} \times \mathbf{B}$, where $\mathbf{v}$ is the velocity of the carriers and $\mathbf{B}$ is the applied magnetic field) causes an accumulation of positive charge on the lower surface of the film [19]. The evolution of the Hall field, E_H, cancels the Lorentz field, E_l, such that no vertical (y direction) current flows. For this simple geometry, the Lorentz force is [19]

$$F_y = qV_xB_z \tag{1}$$

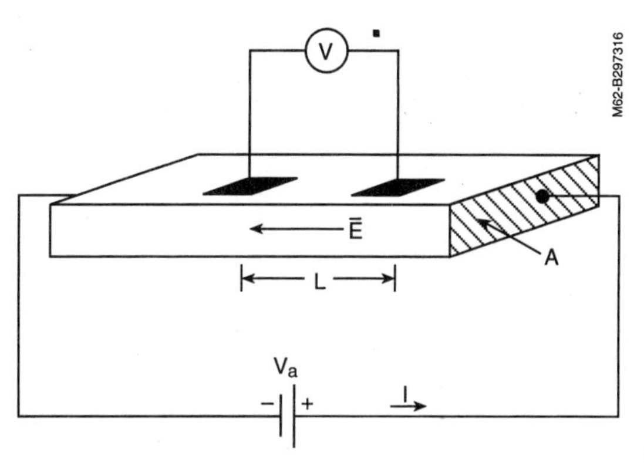

FIGURE 6-1 Configuration for 4-point (contact) method for determination of thin-film resistivity.

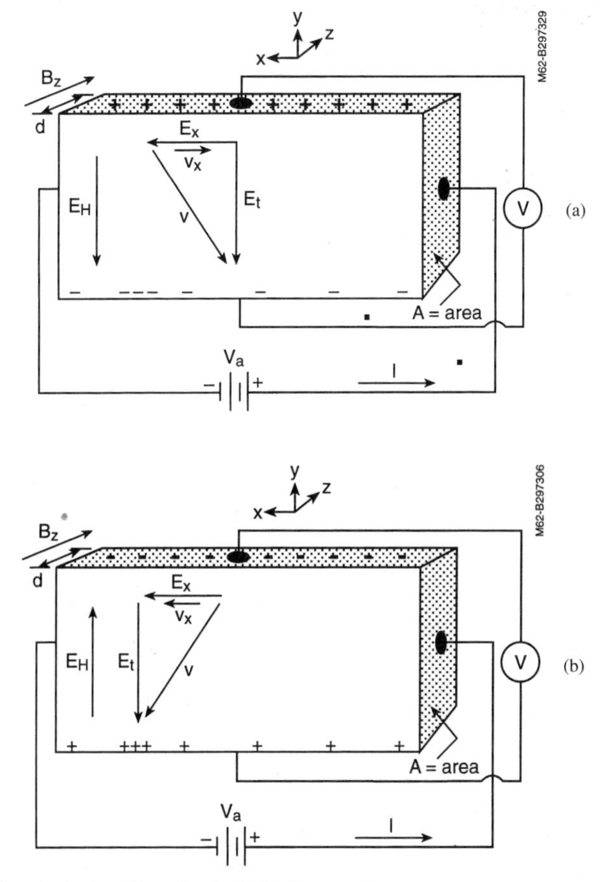

FIGURE 6-2 Experimental configuration for Hall effect measurement on rectangular-shaped sample. *(a)* *n*-type semiconductor; and *(b)* *p*-type semiconductor.

If this positive carrier has a mobility μ_p, then

$$v_x = \mu_p E_x \tag{2}$$

The Lorentz force is then written

$$F_y = q\mu_p E_x B_z \tag{3}$$

and this field drives the positive carrier in the $-$y direction.

At $t = 0$, the current flow is

$$\begin{aligned} J_y &= \sigma E_y \\ &= -\mu_p E_x B_z \end{aligned} \tag{4}$$

The contacts on the z surfaces are blocking contacts because of the high impedance of the measurement device. Therefore, E_H builds and cancels the effect of the Lorentz force. As a result,

$$J_y = \sigma(E_y - E_H) \tag{5}$$

At equilibrium, $J_y = 0$, and

$$|E_y| = |E_H|$$

or

$$E_H = \mu_p E_x B_z \tag{6}$$

Therefore, the mobility can be calculated from the measured Hall field,

$$\mu_p = (E_H/E_x)/B_z \tag{7}$$

where the ratio E_H/E_x $(= \mu_p B_z) \equiv \tan(\Theta_H)$, defining the Hall angle, Θ_H [15–16].

Similar relations can be derived for the analogous n-type semiconductor (with appropriate changes in polarities in the equations).

Measurement Procedure. The method follows these steps:

1. Provide current input along sample's x-direction (J_x).
2. Measure Hall voltage ($V_H = E_H\, d$, where d is the sample thickness).
3. Determine Hall coefficient at equilibrium ($R_H = E_H/J_x B_z$).
4. Determine resistivity ($J_x = \sigma E_x = qp\mu_p E_x = \mu_p E_x/R_H$), Hall mobility ($\mu_p = E_H/E_x B_z$), and carrier concentration ($p = 1/R_H$).

The sign of the Hall angle and Hall voltage reverse, depending on whether the carrier is an electron or hole. In practice, these basic relationships can easily be incorporated into a computer program that controls a set of stable electronics (current and voltage supplies, variable-field magnet, voltage- and current-sensing electronics) to provide automated data acquisition. In fact, commercial Hall effect instrumentation is available that is useful for a large range of semiconductor and other thin film investigations.

Two-Carrier Hall Effect. In a significant number of cases, comparable numbers of electron and holes exist in the semiconductor of interest (e.g., near-intrinsic semiconductors). In the case where $n \sim p$, the simple relationships of the last sections have to be adjusted to accommodate this somewhat more-complex situation. Consider the total current for electrons and holes,

$$J = q(p\mu_p + n\mu_n)E \tag{8}$$

where E is the applied electric field.

The experimental geometry is presented in Fig. 6-3. From this expression for the current, the Hall field can be derived as

$$E_H = E_x B_z \frac{\mu_p \sigma_p - \mu_n \sigma_n}{\sigma_p + \sigma_n} \tag{9}$$

from which

$$E_H = qE_x B_z \frac{p\mu_p^2 - n\mu_n^2}{p\mu_p + n\mu_n} \tag{10}$$

The two-carrier Hall angle can be derived from this equation and is written

$$\tan(\Theta_H) = B_z \frac{p\mu_p^2 - n\mu_n^2}{p\mu_p + n\mu_n} \tag{11}$$

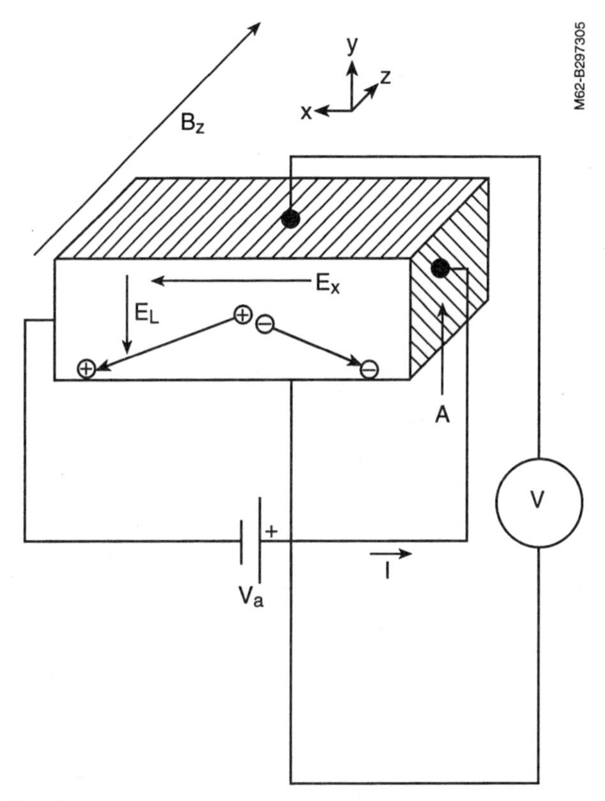

FIGURE 6-3 Experimental geometry for two-carrier Hall effect measurement.

and the two-carrier Hall coefficient becomes

$$R_H = \frac{1}{q}\left[\frac{p - nb^2}{(p + nb)^2}\right] \tag{12}$$

where b is the mobility ratio, μ/μ_p. When $b \gg 1$, electrons dominate the two-carrier Hall effect if the number of electrons is approximately equal to the number of holes. This is usual for photoconductors when the incident light produces electron-hole pairs and the densities are equal. One can manipulate the relative hole and electron densities in Eq. 12 to reduce the expected Hall coefficient to the previously derived single-carrier expressions [16].

6.2.2 Van der Pauw Technique

- Information: Majority-carrier type, mobility, resistivity, and carrier concentration
- Application: Bulk and thin-film samples of arbitrary shape
- Approach: Similar to the Hall effect

In most cases, real samples are not regular, rectangular slabs of semiconductor materials. This complicates the Hall measurement. Irregularly shaped films can be analyzed very easily using the Van der Pauw technique [18], as illustrated in the uniform-thickness representation of Fig. 6-4.

Van der Pauw's Theorem states [18]

$$\exp(-\pi R_{AB,CD} d/\rho) + \exp(-\pi R_{BC,CA} d/\rho) = 1 \tag{13}$$

and this relationship applies to any arbitrary-shaped sample—as long as it is uniform in thickness. The resistance value is measured from

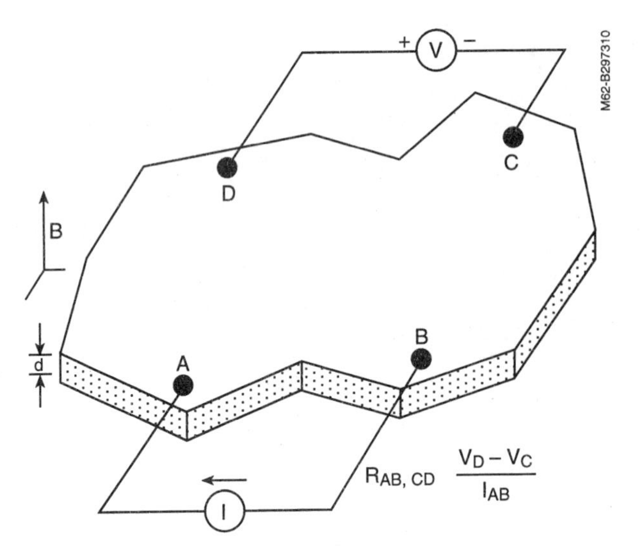

FIGURE 6-4 Van der Pauw measurement for irregularly-shaped, uniform-thickness film, indicating current and voltage contacts.

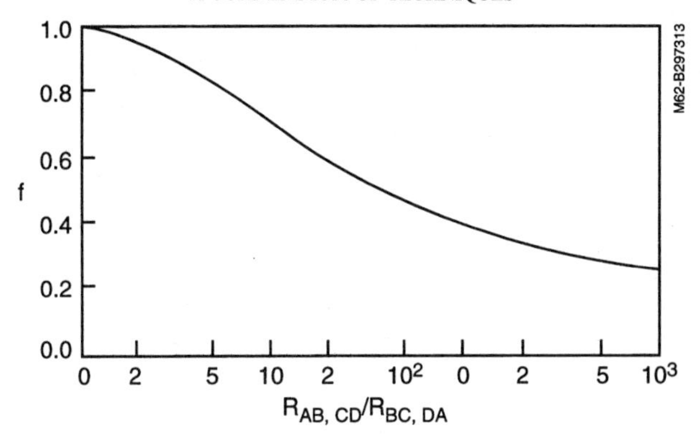

FIGURE 6-5 Van der Pauw coefficient, f, as a function of resistance ratio.

$$R_{AB,CD} = (V_D - V_C)/I_{AB} \tag{14}$$

The resistivity can be defined in terms of two sets of measurements of transverse resistivity. After some algebraic manipulation,

$$\rho = [\pi d/2 \ln(2)](R_{AB,CD} + \pi R_{BC,DA}) f(R_{AB,CD}/R_{BC,DA}) \tag{15}$$

where the function $f(R_{AB,CD}/R_{BC,DA})$ can be determined from a solution of the transcendental equation

$$f = \frac{R_{AB,CD} - R_{BC,DA}}{R_{AB,CD} + R_{BC,DA}} \cdot \frac{2}{\exp[\ln(2/f)]} \tag{16}$$

For symmetrically placed contacts, $f \sim 1$. The resistivity resistance ratio (the first term in Eq. 16) varies between 1.0 and 1.5. In general, the value of f can be found from the solution shown in Fig. 6-5, which presents f as a function of the resistance ratio.

If the resistances $R_{AB,CD}$ and $R_{BC,DA}$ are about equal, the mobility can be calculated from

$$\mu = d\, \Delta R_{BD,AC}/(Bp), \tag{17}$$

where the value $\Delta R_{BC,DA}$ specifies the change in resistance with the magnetic field on and magnetic field off (see Fig. 6-6) [18].

6.2.3 Hall Effect: Polycrystalline Semiconductors

The interpretation of Hall and Van der Pauw results for polycrystalline semiconductor is not as simple as it is for single-crystal materials [20–23]. What is measured by the Hall effect in a polycrystalline semiconductor, in most cases, is not the grain-boundary contribution to the measured Hall coefficient, which is not specifically included and identified. Consider the simple case of a polycrystalline material with random-oriented grains and a single-grain geometry shown in Fig. 6-7. The same current density should flow in both grain and grain boundary because of the random arrangement of the structure. Each current line goes through neutral, depletion, and grain-boundary regions, and the current can be assumed to be nearly uniform.

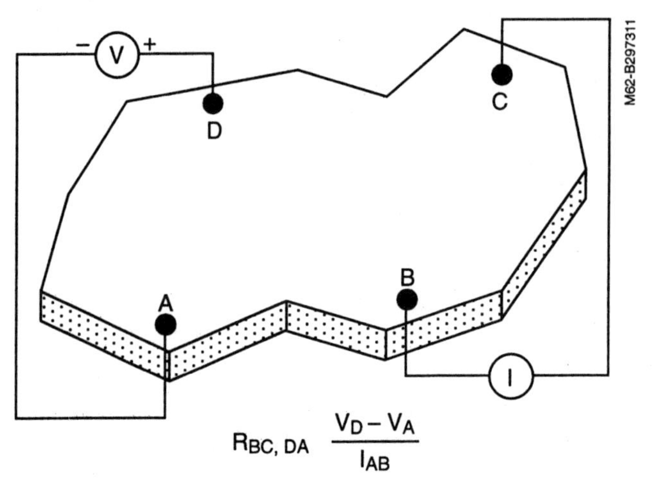

FIGURE 6-6 Experimental geometry showing Van der Pauw configuration for equal resistance measurement.

Consider the general definition of the measured Hall coefficient, R_p, which applies to this case of an inhomogeneous semiconductor, in terms of the conventional Hall coefficient (R) [24]:

$$R_p = <\sigma^2 R>/<\sigma>^2 \tag{18}$$

The brackets indicate spatial averages. To consider specific cases, one must use models for electrical behavior, such as those developed in Chapter 4. In this case, consider the carrier concentration as a function of the coordinates (the Seto model) because of the depletion in the grain, and the mobility has the same value throughout the grain (μ_1) and a constant value (μ_2) in the boundary region. Two cases are possible:

Partial Depletion of the Grain. Using the Seto formulation [25],

$$n = n_o \exp(-q\phi/kT) \tag{19}$$

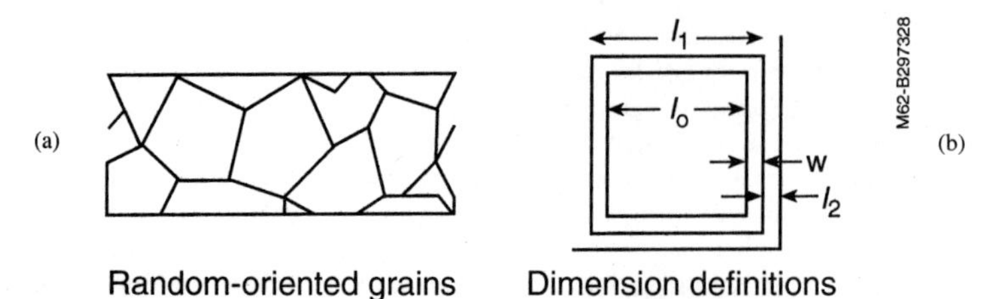

FIGURE 6-7 Polycrystalline thin film, showing *(a)* random grains and *(b)* definitions of dimensions in defined single grain ideal geometry.

where $q\phi = q\phi_b(w - x)w/w^2$. Applying the boundary conditions, the average conductivity is calculated as

$$<\sigma> = \frac{qn_o I_o^2 \mu_1}{(I_1 + I_2)^2} \Big\{ \underbrace{1}_{\substack{\Downarrow \\ \text{Neutral} \\ \text{region}}} + \underbrace{2(w/I_o + w^2/I_o^2)(\pi kT/q\phi_b)^{1/2}\,\mathrm{erf}(q\phi_b/kT)}_{\substack{\Downarrow \\ \text{Depletion} \\ \text{region}}} + \underbrace{[1 + 2\mathrm{w}/I_o + I_2/(2I_o)][2I_2\mu_2/I_o m_1] \cdot \exp(-q\phi_b/kT)}_{\substack{\Downarrow \\ \text{grain-boundary} \\ \text{region}}} \Big\} \tag{20}$$

The $<\sigma^2 R>$, term is obtained from this expression for $<\sigma>$ by substituting μ_1 and μ_2 for μ_1^2 and μ_2^2. With $\mu_1 = \mu_2 = \mu$, $<\sigma> = q<n>\mu$, as expected.

For reasonable values for the parameters (that is, $I_2 = 2$ nm, $I_1 > 10$ nm, $\mu_2/\mu_1 < 1$, and $q\phi_b > 3kT$, the grain-boundary term is not significant. Therefore,

$$<\sigma> = q\mu_1<\mathrm{n}>,$$

and

$$<\sigma^2 R> = qm_1^2<n> \tag{21}$$

From this, it follows that

$$R_p = 1/(q<n>) \tag{22}$$

which is the definition of the Hall coefficient without grain-boundary effects. Thus, the grain-boundary effect is insignificant, and the measured Hall coefficient corresponds to that of the grain.

Complete Depletion of the Grain. In this case, the depletion length is defined $I_D > I_1$. It is assumed that there is little band bending, and the concentration is nearly uniform. The average conductivity is

$$\sigma = \frac{qn\mu_1 I_1^2}{(I_1 + I_2)^2}\,[1 + (\mu_2/\mu_1)(2I_2/I_1 + I_2^2/I_1^2)] \tag{23}$$

and the Hall coefficient is

$$R_p = \frac{1}{qn}\,(1 + I_2/I_1)^2\,\frac{1 + (\mu_2^2/\mu_1^2)(2I_2/I_1)}{[1 + (\mu_2^2/\mu_1^2)(2I_2/I_1 + I_2^2/I_1^2)]^2} \tag{24}$$

Therefore, the measured Hall coefficient (and conductivity) is not just that of the grain, but is modified by the properties and geometry of the grain boundaries. A reduced carrier concentration is measured because of the inhomogeneity in mobility. This corrective factor can be significant, especially for small grain sizes. The interpretation of the Hall results depends significantly on the complexity of the modeling used in the analysis.

6.2.4 Minority-Carrier Techniques

In general, these techniques use electron and photon beams to evaluate the interface recombination velocity, diffusion lengths, and lifetimes of minority carriers at defects, inhomogeneities, and junctions. The focused beam provides a localized source of excess carriers in the semiconductor. If the semiconductor contains an internal field (such as one caused by a *p-n* junction or the space-charge layer of a grain boundary), the injected electrons and holes are driven in opposite directions and produce a current in an external circuit. This signal carries considerable information on the electrical properties of the material and can be used to determine the minority-carrier properties. By scanning over the sample while monitoring the induced current, an image of the sample's minority-carrier characteristics can be obtained. Two similar techniques are summarized in the first part of this section: electron-beam-induced current (EBIC), in which the input is an energetic electron beam, requiring analysis in a vacuum environment; and light-beam induced-current (LBIC), in which the excitation source is a photon beam. LBIC analysis can be accomplished in almost any environment (including air).

Electron Beam-Induced Current (EBIC)

- Information: Minority-carrier diffusion length, diffusion coefficient, carrier loss
- Application: Single-crystal, polycrystalline, and amorphous thin films
- Approach: Measurement of generated current response to spatially resolved electron-beam excitation

The general schematic configuration for EBIC observations is shown in Fig. 6-8*a-d* [26–28]. The electron beam penetrates a depletion region (such as a thin Schottky junction) and generates electron-hole pairs in the semiconductor. These pairs are separated and then collected by the built-in field, and they produce a current that can be used (after amplification and processing) to modulate the intensity of an output device (such as a monitor or printer). The relative current-voltage characteristics are represented in Fig. 6-8*d*. Because the sample and the display are scanned synchronously, the EBIC image gives a map of the current collected at each scanned point. At a defect, the collected current is generally lower, because the locally enhanced carrier recombination is higher. (Note: The related electron beam-induced voltage [EBIV] technique is also depicted in Fig. 6-8*b*. This measures the voltage response to the incident electron beam and provides information about areas of electronic loss or enhancement within electronic materials [29–30].)

The quantitative interpretation of the induced-current images of grain boundaries and other defects requires a model for the contrast mechanism involved. The basic processes are the generation of carriers by the electron beam and the transport and collection of beam-generated carriers, with the recombination of these carriers at defects or inhomogeneities. These two processes are considered separately in the following material.

Carrier Generation. The electron beam is usually on the order of 2 to 40 keV and has a spatial diameter as low as 80 Å. (Recently, nanoscale EBIC has also been demonstrated using 30 eV excitations, with 10-nm-scale "beam" diameters [31].) The electrons penetrate the sample and undergo elastic and inelastic scattering. As a result, they spread laterally, and they simultaneously lose energy by ionizing the sample [32]. This is illustrated in Fig. 6-9*a*. In crystalline semiconductors, this ionization process involves the raising of electrons from the valence to the conduction band, which leaves holes in the valence band. The energy required to form an electron-hole pair (E^*) is approximately 3 times the band-gap energy. For silicon (Si), this is about 3.6 eV. Therefore, a single electron with energy

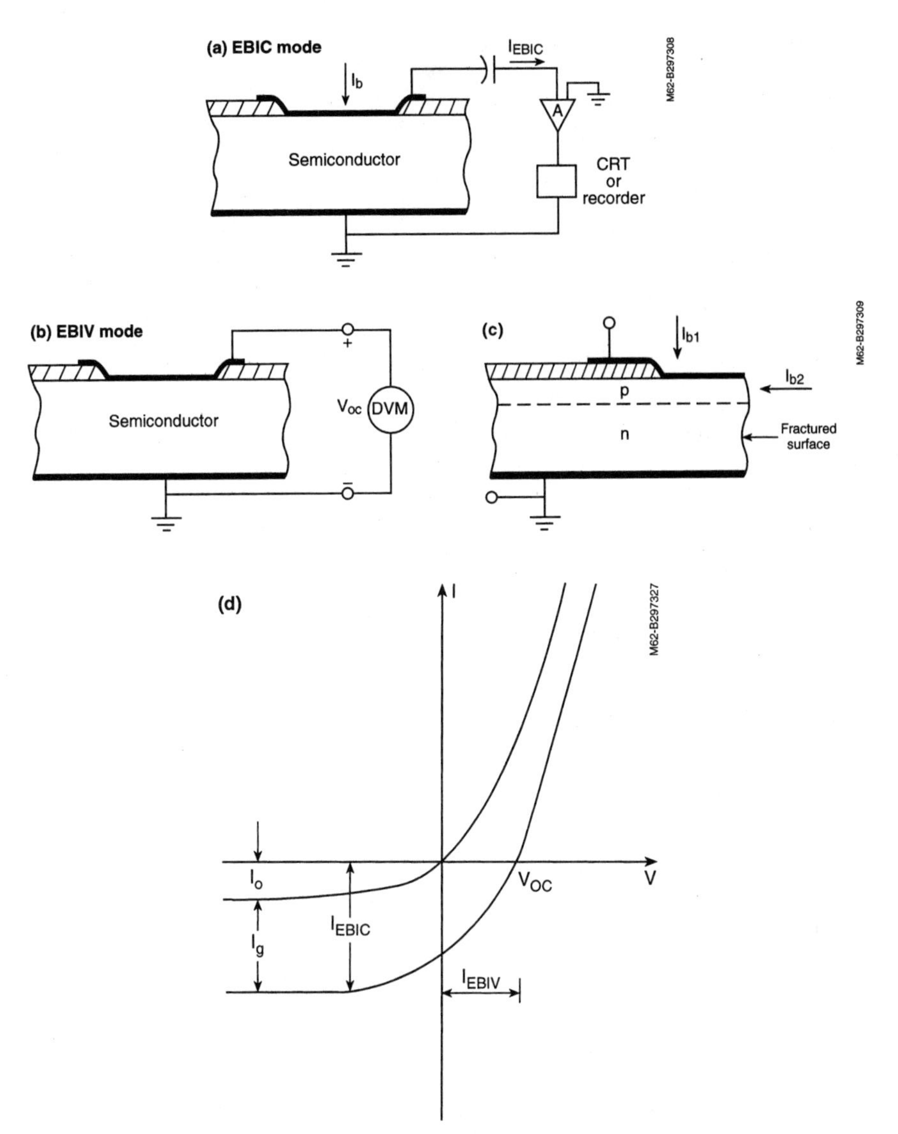

FIGURE 6-8 Schematic representation for *(a)* EBIC (in-plane) measurement; *(b)* EBIV measurements; *(c)* EBIC of fractured cross section; and *(d)* current-voltage characteristics indicating measured EBIC and EBIV signals.

E generates about E/E^* pairs. In Si, the electron range (corresponding to the maximum depth of pair generation) follows the empirical relation [33]

$$R = 0.0171E^{1.75} \tag{25}$$

where R is in units of μm and E is in units of keV. The electron beam generation is not uniform over the generation volume, but is highest closest to the surface and the beam

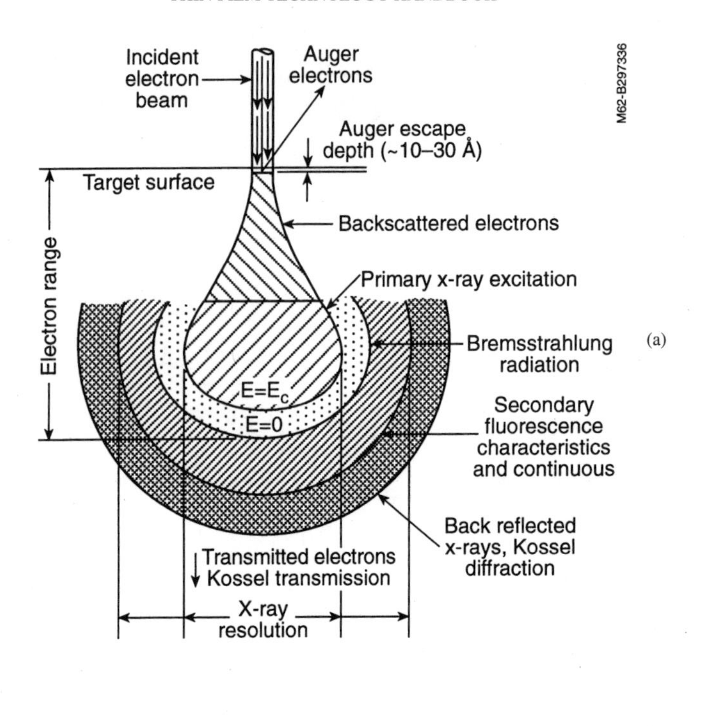

(a)

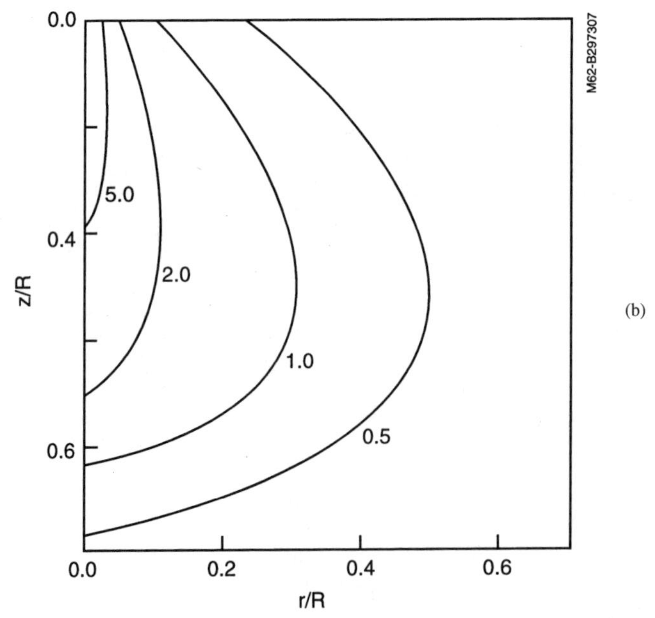

(b)

FIGURE 6-9 Electron beam penetration of solid surface *(a)* representations of various emissions, and *(b)* calculated value of normalized distribution of the generation due to a focused electron beam, with contours of equal ionization scaled to electron range, R.

TABLE 6-1 Electron range (R_e) as a function of incident beam energy (E) for Si and GaAs

E (kV)	1	5	10	15	20	25	30	40	45
R_e-Si (μm)	0.032	0.47	1.49	2.94	4.75	6.89	9.34	15.10	18.38
R_e-GaAs (μm)	0.017	0.25	0.81	1.59	2.57	3.73	5.05	8.17	9.94

axis. Typical values for R as a function of E for Si are presented in Table 6-1 [33], with a graphical representation for z/R as a function of r/R in Fig. 6-9*b*.

By integrating the spatially dependent generation rate over the appropriate volume, the depth-dose function, $g(z)$, is obtained [34]

$$g(z) = g_o/R\ \Lambda(z/R) \tag{26}$$

where g_o is the initial dose and the function $\Lambda(z/R)$ is shown in Fig. 6-10. Knowing the value of $g(z)$ forms the basis of solving the carrier transport equations, which in turn give the fundamental information on the minority-carrier properties [34].

Charge Collection and Minority-Carrier Properties. The generation function for electrons is the basis for finding the EBIC current:

$$\begin{aligned} \mathrm{I} &= g_o \int L(\zeta)\exp(-\zeta R/L)d\zeta \\ &= g_o \eta(R/L) \end{aligned} \tag{27}$$

From this, the EBIC current is given as a function of distance from the depletion region by [35]

$$\mathrm{I}_{\mathrm{EBIC}} = \mathrm{I}_o \exp(-z/L) \tag{28}$$

which gives a first-order, one-dimensional evaluation of the diffusion length of the minority carriers L and assumes that the surface recombination velocity is insignificant. More-exact evaluations require more-complex analytical treatment [35].

Optical Beam-Induced Current (OBIC)

- Information: Minority-carrier diffusion length, diffusion coefficient, recombination velocity, loss sites
- Application: Bulk and thin-film samples of arbitrary shape (large-area capabilities)
- Approach: Evaluation of current-generated response to spatially resolved light input excitation

Optical beam-induced current is termed OBIC; it is also referred to as light beam-induced current (LBIC) [36–37]. Images are formed in a manner similar to the EBIC method, but a focused light beam is used as the source instead of an electron beam. The positive features of this technique are: it is less destructive; it can be used to evaluate relatively large areas; it can provide spectroscopic (wavelength-tunable) data; and it can be operated in almost any environment. However, the spatial resolution is limited by the wavelength of the light, and the interpretations are sometimes more complex. The methods of carrier generation and charge collection are similar to, but distinctly different from, those involved in EBIC.

Carrier Generation. The carrier generation is rather different than that involving an incident electron beam. Visible photon energies are on the order of a few

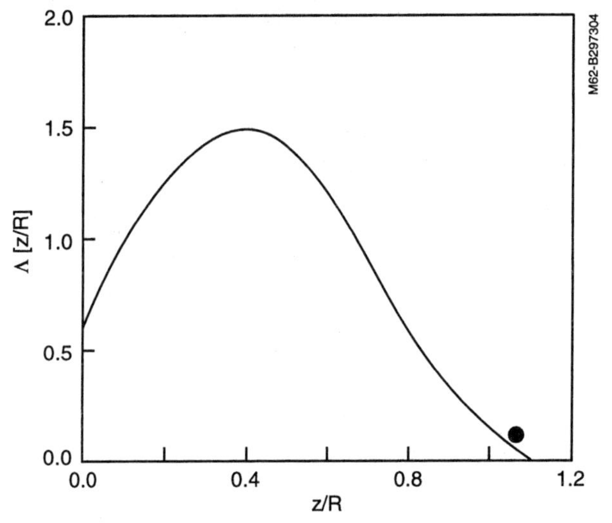

FIGURE 6-10 The universal electron depth-dose, $\Lambda(z/R)$ as a function of normalized depth, z/R.

electron volts (eV), a factor of 10^3 to 10^4 times less than the usual EBIC electron energies. Thus, a photon generates a single electron-hole pair, whereas an electron generates thousands.

It is not possible to define a maximum penetration depth for photons in the semiconductor, because the light absorption follows an exponential law. Accordingly, the generation rate can be written as [38]

$$g(z) = \Phi_o \exp(-\alpha z) \tag{29}$$

where Φ_o is the number of photons incident per $cm^2 \cdot s$ and α is the light absorption coefficient.

Charge Collection and Minority-Carrier Properties. The evaluation of OBIC data is analogous to the EBIC techniques, and the minority-carrier parameters (diffusion length and recombination velocity) can be extracted. However, because of the spatial-resolution limitations, many collection events are integrated into the detected signal. The results can be complex and particularly susceptible to misinterpretations [39].

Minority-Carrier Lifetime Spectroscopy

- Information: Minority-carrier lifetime, diffusion lengths
- Application: Bulk and thin-film samples; special structures sometimes required
- Approach: Excitation with radiative source; measurement of time-dependent response

A number of characterization techniques have been developed to determine the most fundamental parameter used for certifying device-quality material for use in many electronic devices: the lifetime of the minority carriers. The intricacy of these techniques have increased with the complexity of the structures being investigated. In all methods, a very good theoretical model is needed for interpreting the results. Several of these techniques are presented here, and the reader is referred to a number of references that deal with this important, evolving measurement field.

Photoconductive Decay. Initial investigations of techniques used to determine the lifetimes of minority carriers were primarily concerned with photoconductive decay (PCD). For single-crystal wafers and amorphous Si:H films, this method is still a primary characterization tool [40–41].

Figure 6-11 presents a schematic representation of the PCD technique. A constant current is passed through the semiconductor, and the voltage drop is measured with a device such as an oscilloscope. Excess minority carriers are produced from a short photon pulse, with energy or wavelength set to produce adequate penetration of the light. (The current at the surface is negligible if the carriers are produced deep within the semiconductor.) The photon source can be a semiconductor light-emitting diode (LED), a tunable laser, or a broadband source (such as a flash or discharge tube). In the last case, undesired short-wavelength radiation is removed by the use of filters. The transient voltage across the sample is proportional to the density of excess minority carriers. In general, the conductivity resulting from the photon excitation can be expressed (under conditions of low injection) as

$$\begin{aligned}\sigma &= qN_A\mu_p + q\mu_n\Delta n_o \exp(t/\tau)\\ &= \sigma_o + \Delta\sigma(t/\tau)\end{aligned} \tag{30}$$

The voltage measured is

$$\begin{aligned}V(t/\tau) &\sim I/\sigma\\ &I/[\sigma_o + \Delta\sigma(t/\tau)]\end{aligned} \tag{31}$$

From Eq. 31, the magnitude of the voltage can be predicted to increase with the magnitude of the lifetime, assuming that there is no significant surface effect.

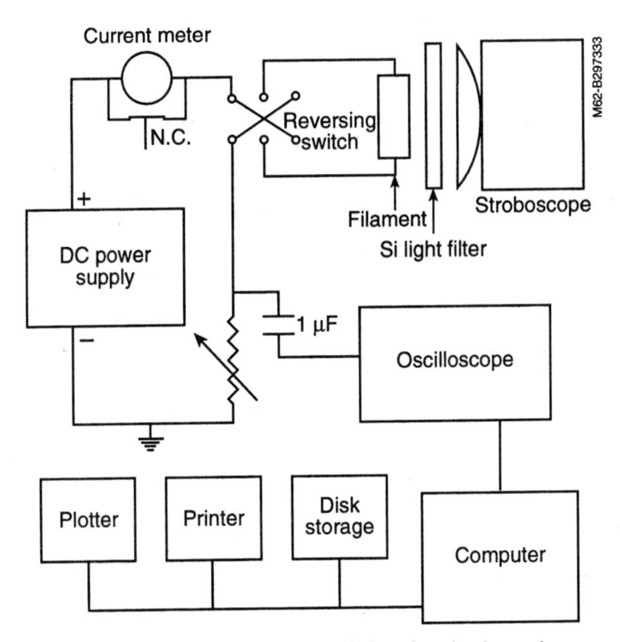

FIGURE 6-11 Schematic representation of the PCD technique for minority-carrier spectroscopy.

Microwave Absorption. This technique is relatively well developed and is available commercially. Excess carriers are generated with a solid-state laser (tuned, for example, to about 880 nm for Si samples), as shown in Fig. 6-12 [42–44]. Simultaneous to this photon excitation, the sample is irradiated with microwaves. As the sample conductivity increases, the microwave reflection increases. The reflected microwave power is detected, and the corresponding lifetime is evaluated. The advantages of this method are that it needs no contacts and can be used in a manufacturing line. It has limitations for thin film samples, in which the surface recombination can be significant because of the large surface-to-volume ratio in the films. Microwave absorption techniques work best with bulk semiconductors rather than with thin films.

Photoluminescence Decay. This technique is becoming routine, with special applications to thin films and compound semiconductors. The methods are based on careful analyses of the recombination mechanisms in semiconductors and thin films [45–48]. A photoluminescence (PL) signal is excited with a pulsed, tunable laser. The pulse durations and separations are short (usually in the picosecond to femtosecond range) and are adjusted for the semiconductor system being investigated. Photoluminescent decay spectroscopy has several advantages over other techniques for determining minority-carrier parameters, especially for semiconductors having large radiative probabilities (such as III-V or II-VI materials). The radiative output is strong in this case. If there is severe trapping in the semiconductor, interpretation using techniques such as PCD is difficult. Photoluminescent decay allows better paths for the separation of the recombination processes. Finally, shorter minority-carrier lifetimes (nanoseconds versus microseconds) require less-expensive instrumentation with narrower bandwidths. The PL-decay electronics and experimental configuration are less complex.

The PL photon generated within the semiconductor is collected with a lens and focused onto the input (slit) of a spectrometer, as shown in Fig. 6-13. The spectrometer is tuned to a wavelength corresponding to the bandgap of the semiconductor and eliminates spurious photons arising from other sources. The detector must be fast and sensitive. A photomultiplier tube (PMT) satisfies most requirements for bandgaps above that for Si. Other III-V and Ge detectors are used for lower-band gap semiconductors. A PL photon produces a

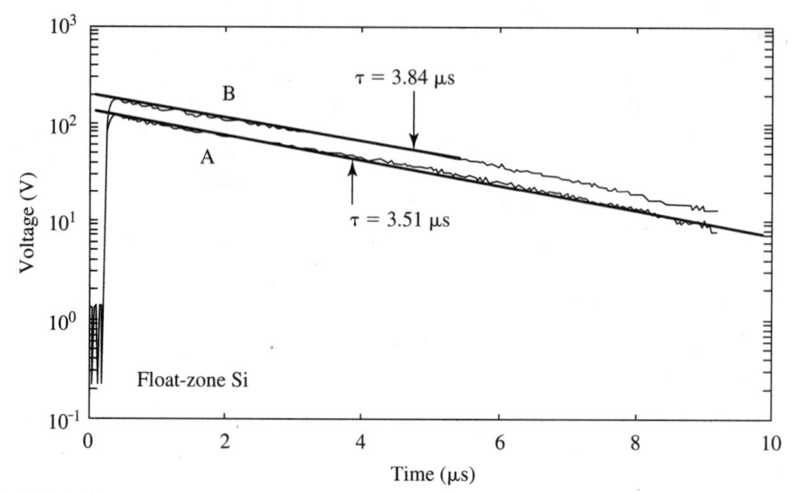

FIGURE 6-12 Excess carriers are generated with a solid state laser (e.g., tuned to about 880 nm for Si samples), as shown in Fig. 6-11.

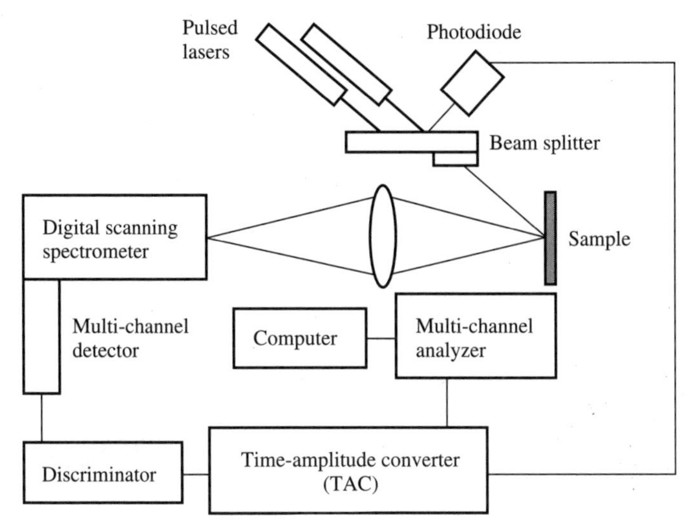

FIGURE 6-13 The PL photon generated within the semiconductor is collected with a lens, and focused onto the input (slit) of a spectrometer.

photoelectron at the cathode of the PMT. Those electrons are multiplied by factors of 10^5 to 10^9 by the PMT dynode chain. To determine the minority-carrier lifetime, the PL photon must be correlated in time relative to the incident laser pulse.

The *time-correlation single-photon counting* is extremely powerful in determining the lifetime [49–50]. Using a beam splitter and photodiode, a portion of the laser pulse is sampled to initiate a start trigger into the delayed coincidence electronics. The start trigger enables a time-to-amplitude converter (TAC). When a photon-initiated signal is passed by the amplitude discriminator, it produces a stop trigger at the TAC. The TAC output is a ramp with amplitude proportional to the time delay between the laser pulse and the arrival of the first photon. The TAC signal is detected by a multichannel pulse-height analyzer, which collects one count per detected photon. This signal is stored in a channel appropriate to the time delay, and thus, one count is recorded for every laser pulse. A profile of the PL decay evolves in terms of counts as a function of time. Some typical characteristics are shown for GaAs in Figs. 6-14 and for CdTe in Fig. 6-15 [51].

Radio-Frequency Photoconductive Decay. This is another contactless method for determining the minority-carrier properties [52–53]. Both radio frequency photoconductive decay (RFPCD) and ultrahigh-frequency photoconductive decay (UHFPCD) can be used for indirect bandgap semiconductors (that is, they do not require photoluminescence), and they are not limited by conductivity and skin depth like microwave techniques.

This technique measures the transient photoconductivity using various pulsed excitation sources to generate excess carriers. The decay of the transient photoconductivity is analyzed in terms of minority-carrier or recombination lifetime. Figure 6-16 shows a schematic of the experimental configuration. The light pulse produces a bridge unbalance because of the photoconductivity of the sample. The RF output signal from the splitter is amplified by about 20 dB and connected to a mixer or demodulator. The second input of the mixer is connected to the same oscillator as the sensing coil. A phase shifter is required to adjust the phase of the input signal and a reference signal coming directly from the oscillator. The doubled-frequency and direct current (dc) components are generated by the mixing circuit when an unbalance signal is present. The dc signal is selected and processed

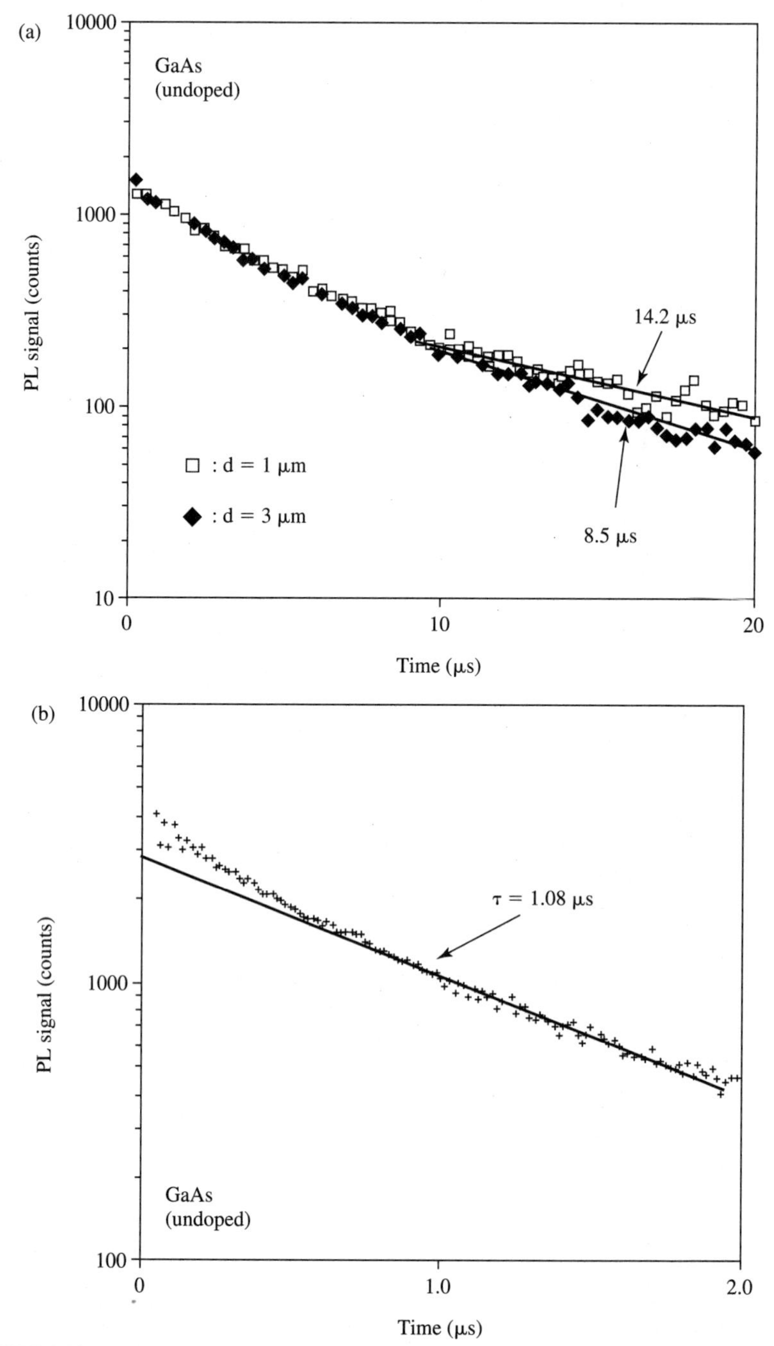

FIGURE 6-14 Typical decay characteristics for epitaxial GaAs thin layer.

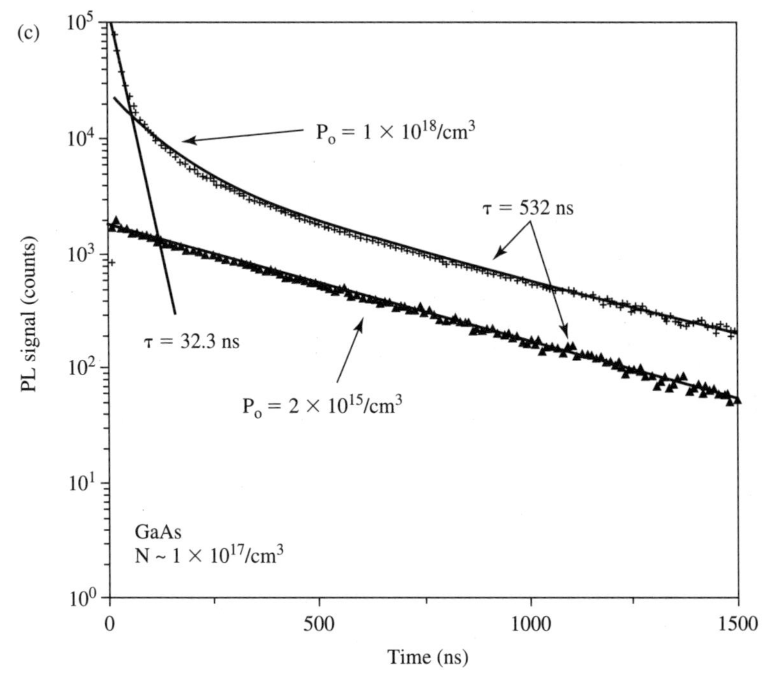

FIGURE 6-14 *(Continued)* Typical decay characteristics for epitaxial GaAs thin layer.

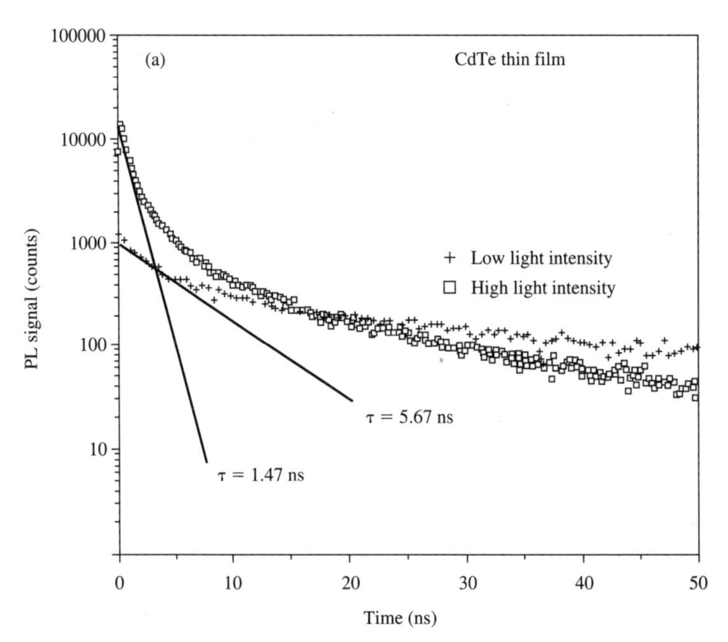

FIGURE 6-15 Typical PL decay characteristics for polycrystalline CdTe thin film.

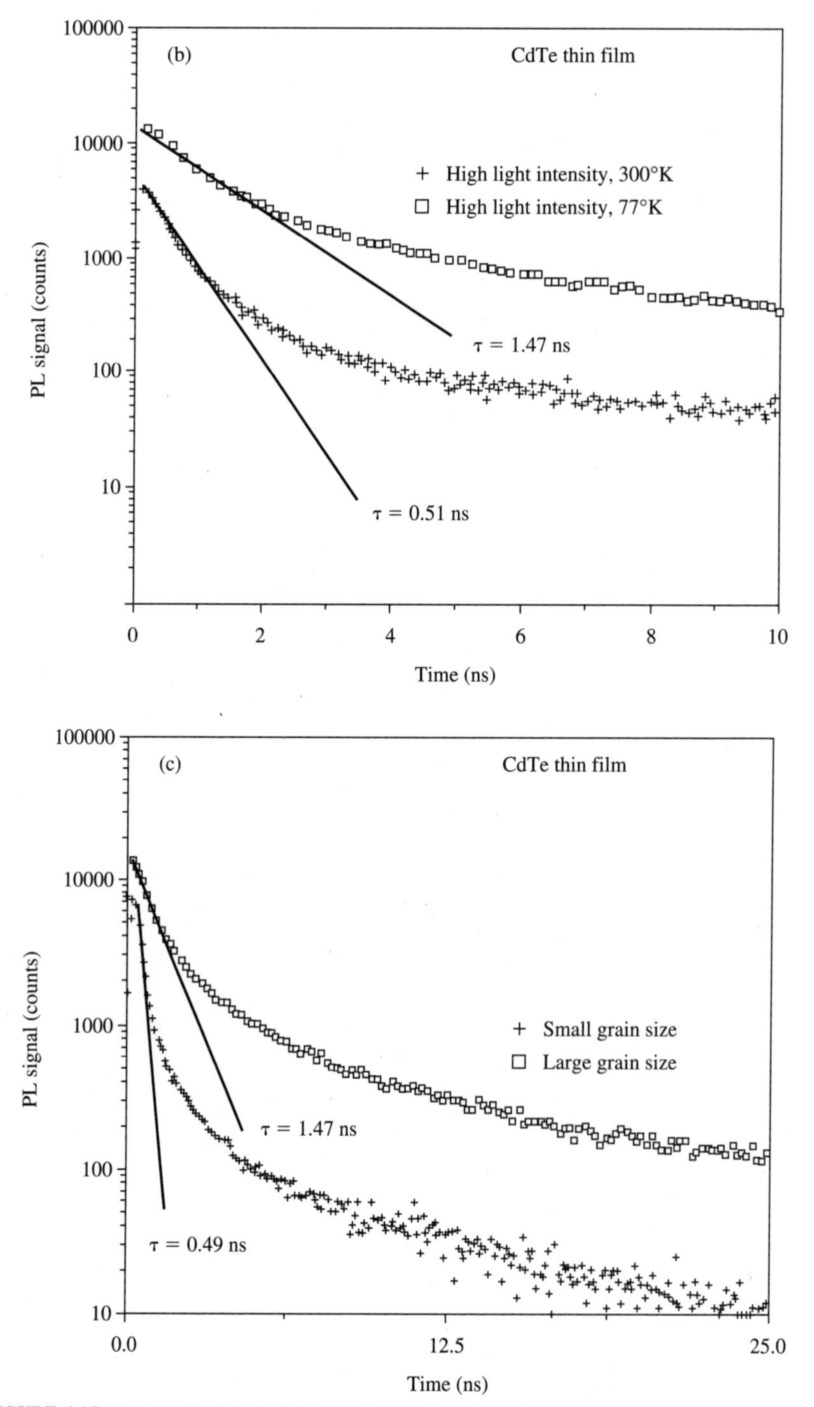

FIGURE 6-15 *(Continued)* Typical PL decay characteristics for polycrystalline CdTe thin film.

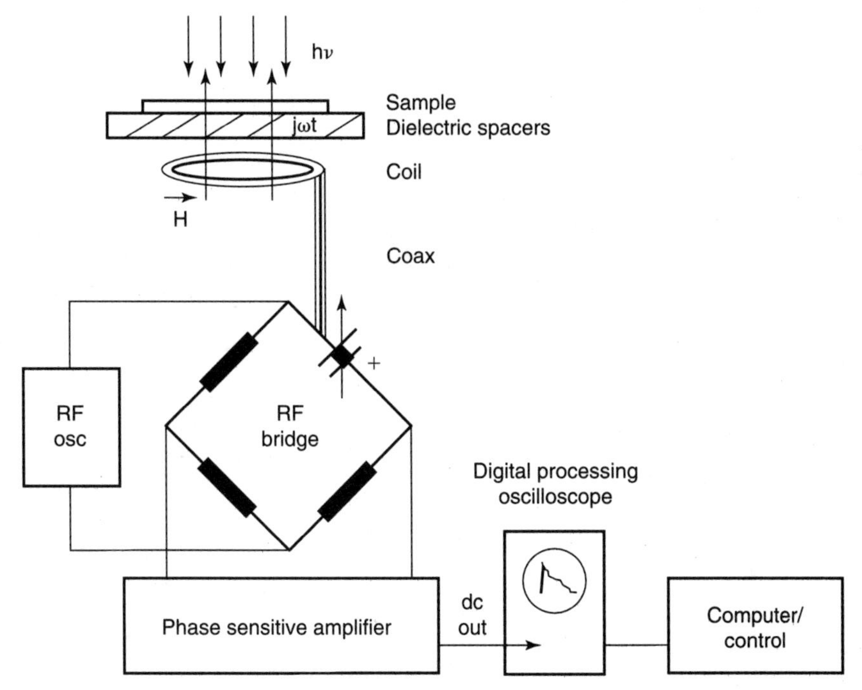

FIGURE 6-16 Experimental configuration for radio-frequency CVD (RFCVD) method for minority-carrier spectroscopy.

accordingly. The mixer output is passed through a low-pass filter (~300 MHz cutoff) and amplified by a 20-dB wideband dc amplifier. The detection circuit is basically a phase-sensitive amplifier. The unbalance or dc signal coming from this system is proportional to the sample photoconductivity. The mixer circuit is linear over a range of offset voltages from zero to about 350 mV. Therefore, the system output signal is linear over a range up to about 3 V. This signal is sent to the digital processing oscilloscope for storage and analysis. This system has been used to determine lifetimes in Si, Ge, GaAs, and InGaAs. Silicon wafers with resistivity ranges from 0.01 to more than 100 Ω·cm have been measured. The system risetime is about 10 ns, and lifetimes have been measured that are close to this time.

In all measurements, the sample is inductively coupled to the ultrahigh-frequency bridge circuit by means of a coil. The high-frequency magnetic field of the coil induces eddy currents in the sample. The bridge is balanced for each sample so that the majority-carrier conductivity is nullified at the bridge output. This procedure maintains operation in the linear range of the mixer. When the excess carriers are generated by the light source, the unbalanced eddy currents induce an additional voltage in the coil that appears across the bridge. Several types of light sources can be used. For Si, a yttrium aluminum garnet (YAG) laser (pulse of 3.0 ns measured at full-width half-maximum [FWHM], 1.064 μm) is used because the absorption is weak at this wavelength. Uniform volume generation of electron-hole pairs occurs in the illuminated volume of the sample. For surface excitation and shorter pulses, a mode-locked, cavity-dumped dye laser (15-ps FWHM) is preferred. For high-resistivity samples, a red light-emitting diode can be used as the excitation source.

One additional advantage of these approaches is that the Si can be immersed in hydrofluoric acid (HF) solutions and inductively coupled to the measurement system [53]. This

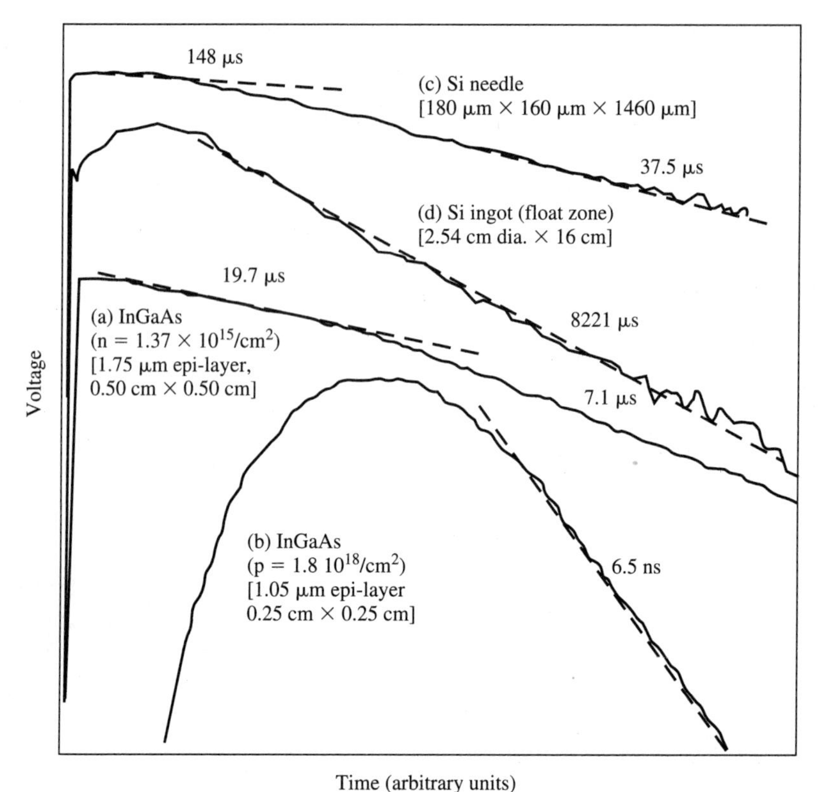

FIGURE 6-17 RFCVD data for thin film polycrystalline Si.

results in preferred surface-state passivation. An example of the application of this technique is presented in Fig. 6-17 for a liquid phase-epitaxy Si thin film, grown on a metallurgical Si substrate. The UHFPCD data are shown at very low and very high injection levels. At lowest injection levels (*A*), the lifetime is about 1.3 μs, but it increases rapidly with laser energy. At highest injection levels (about 1000 mJ/cm^2), the lifetime increases to more than 170 μs. These data clearly illustrate the saturation of defects and may be related to either bulk or surface states.

Transient Spectroscopies

- Information: Minority-carrier and majority-carrier trapping levels; deep electronic levels
- Application: Single-crystal, polycrystalline, and amorphous thin films
- Approach: Measurement of capacitance time-decay as a function of temperature; modeling provides information of electronic level(s)

Deep-level transient spectroscopy (DLTS) uses a transient junction capacitance measurement to detect the concentration and energy levels of deep impurity states. It has had wide application to thin films, but the interpretation can be complicated because of the large concentrations of defects, especially in polycrystalline and compound semicon-

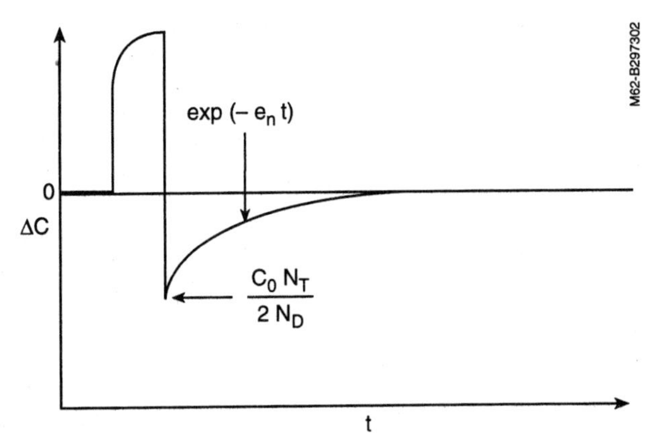

FIGURE 6-18 Transient capacitance data: representative signal.

ductors [54–57]. The experimental technique involves monitoring of the capacitance of a junction device after the space-charge layer has been filled with either majority- or minority-carrier charge. The trapped charge modifies the capacitance of the junction. One approach is to use a fast capacitance bridge to monitor the charge in the junction as the trapped charge is emitted to either band.

The measurement of a capacitance transient is illustrated by the schematic diagram of Fig. 6-18 [54]. The capacitance transient is monitored with a capacitance bridge that is modified for high-speed response. The rise-fall time is better than 50 μs for most applications. The dc component of the capacitance is blocked by a filter with a very low cutoff (typically 0.1 Hz). The output of the bridge is amplified by a wide-bandwidth preamplifier, and the transient is processed by a programmable digitizer. A charge-coupled device (CCD) input stage allows the measurement of the pre-pulse capacitance for a benchmark determination. Typically, the transient is averaged for more than 250 repetitions at a given temperature. The output signal is typically stored for subsequent processing.

The approach is to monitor the transient capacitance data as a function of temperature. This provides a family of curves, represented in Fig. 6-19. At each temperature, the emission rate can be calculated from

$$e_n = V_{\text{th}}\sigma_n \text{N}_{\text{c}} \exp(-\Delta E/kT) \tag{32}$$

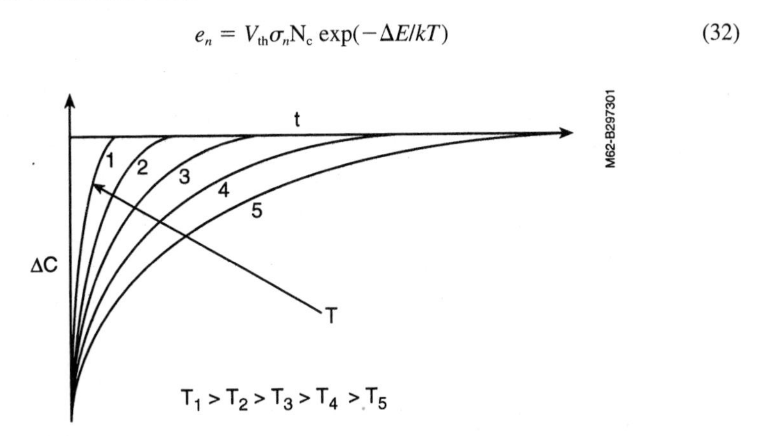

FIGURE 6-19 Family of transient capacitance data characteristics showing changes with temperature.

The Arrhenius dependence of $\ln(e_n)$ has a slope representative of the electronic or trap level (that is, the activation energy ΔE in Eq. 32). Such data are shown in Fig. 6-20*a,b* for *p*-$CuInSe_2$ [54]. The capacitance decay curves are shown for $T =$ 150°K, 174°K, and 210°K. A level at 0.52 eV is calculated from the slope of the characteristic in Fig. 6-20*b*. However, this interpretation is very risky because the film is very defective and the data can also be used to identify many (rather than a single) levels. This calculated level is interpreted as an average of several levels (a continuum rather than a distinct transition). Deep-level transient spectroscopy is best applied to single crystals and low defect-density thin films; otherwise, the interpretation may be misleading. Deep-level transient spectroscopy can be used effectively for monitoring the quality of thin polycrystalline films (qualitative representation) as part of a quality-control effort.

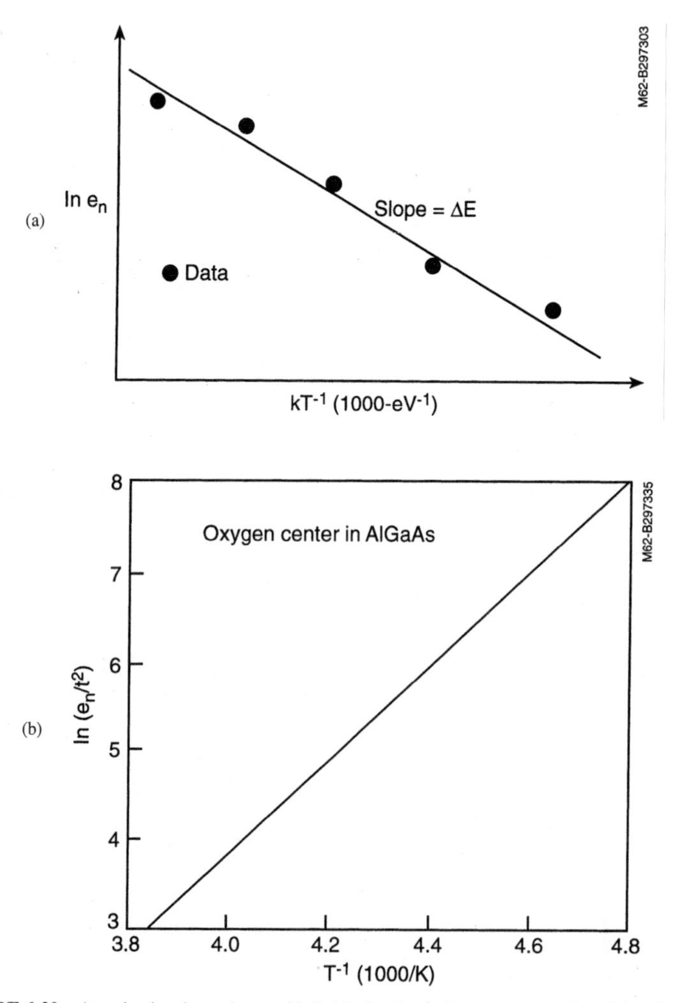

FIGURE 6-20 Arrenhenius dependence of $\ln(e_n)$ indicating a slope representative of the electronic or trap level (i.e., the activation energy ΔE) for *(a)* CdTe and *(b)* AlGaAs thin layers.

6.3 ELECTRO-OPTICAL MEASUREMENTS: GRAIN BOUNDARIES

Determining the majority- and minority-carrier properties involved in the electronic transport across grain boundaries in polycrystalline semiconductors is fundamental to understanding these materials [58–61]. In general, the majority-carrier measurements provide information on the density of states in the grain boundary by evaluating the steady-state (dc) and transient (ac) parameters. The minority-carrier measurement evaluates the interface recombination velocity (carrier lifetime and diffusion length). Some of these methods have been referenced in the previous chapter, and these techniques complement some of the techniques presented there. This section emphasizes the special measurements used to determine the fundamental properties of grain boundaries (ensemble and single-boundary methods). In these discussions, the reader is referred to Chapter 4 for discussions of the models. Specifically, the details of the grain-boundary band-energy diagram have been presented in Fig. 4-31.

6.3.1 Conductivity-Doping Method

- Information: Density of states, barrier heights (activation energies)
- Application: Semiconductors and thin films with large population of grain boundaries; different doping levels
- Approach: Measurement of conductivity versus temperature for a number of different doping levels in samples; may require complementary Hall data

This technique involves a deconvolution method for determining the density of states [62]. That is, the density of states is determined from the measurement and analytical manipulation of a number of parameters. The density of states is expressed as

$$\begin{aligned} N_{ss} &= N_{ss}(E_f) \\ &= \frac{dQ}{dE_f} \\ &= \frac{dQ}{dN} \div \frac{dE_f}{dN} \end{aligned} \tag{33}$$

First consider the change in the Fermi level with doping (the denominator of Eq. 33). As discussed in Chapter 4, the conductivity of polycrystalline films generally can be modeled to follow a thermionic emission mode

$$\sigma = \sigma_o \exp(-E_\sigma/kT) \tag{34}$$

where the activation energy, E_σ, can be extracted from the slope of the Arrhenius characteristics. For a polycrystalline film of length L and average grain size a, the thermionic emission current can be written

$$j_{th} = A^*T^2 \exp[-B(\zeta + \phi)] \cdot [1 - \exp(-\beta Va/L)] \tag{35}$$

where Va/L is the voltage drop per grain boundary. Because $Va/L \ll kT/q$, Eq. 35 can be expanded by a Taylor expansion, and the conductivity is

$$\begin{aligned} \sigma &= j_{th}/(V/L) \\ &= q/kA^*Ta \exp[-\beta(\zeta + \phi)] \end{aligned} \tag{36}$$

From Eqs. 34 and 35, the activation energy is

$$E_\sigma = q[\zeta(\mathrm{T}) + \phi(\mathrm{T})] + kT \ln(qA^*Ta/k\sigma_o) \tag{37}$$

But, $E_f(T) = q[\zeta(T) + \phi(T)]$, and the activation energy can be determined from

$$E_\sigma = E_f(T = 0) \tag{38}$$

That is, the activation energy of the thermally activated conductivity gives a measure for the Fermi level at the grain boundary, extrapolated to zero temperature. The conductivity activation energies, on the other hand, are extremely sensitive to the doping level. The method follows these steps:

1. Measure σ versus $1/T$ for a given doping, N. This gives a straight-line dependence.
2. Determine the slope: determines $E_\sigma = E_f$.
3. Repeat for various other doping levels (other films), and graph the resulting E_f versus N.
4. The slope of this characteristic at each energy is the required denominator of Eq. 33 (dE_f/dN).

Now consider the term in the numerator of Eq. 33, dQ/dN. The boundary charge, Q, can be determined from $q\phi$, because $q\phi = E_\sigma - q\zeta$. The method follows these steps:

1. Use the values for the activation energy (E_σ) for each N (known from the conductivity data).
2. Calculate the values for $q\zeta$ or determine them from Hall measurements.
3. Graph the values of Q for each N; the slope provides the numerator of Eq. 33.

Therefore, the density of states, N_{ss}, can then be calculated for each energy level by using Eq. 33. Data obtained using this method are shown in Fig. 6-21 for polycrystalline Si and GaAs.

This method is very involved and requires a large number of samples that are identical apart from the doping levels. For accuracy, averages of a large number of grain boundaries are taken. Therefore, it is not a method that can easily be used for single grain boundary studies, because it is very difficult to produce a series of identical, isolated grain boundaries that differ only in their doping. Other methods are used for single-grain-boundary determinations.

6.3.2 Current-Voltage Spectroscopy

- Information: Density of states
- Application: Single, isolated grain boundaries and bicrystals
- Approach: Measurement of current-voltage characteristics; measurement of zero-bias conductance or capacitance; deconvolution of data to provide density of states

This technique provides information on single grain boundaries, isolated specifically for this measurement [65]. Ohmic contacts are made on each side of the boundary, usually using some photolithographic process. To increase the precision and reproducibility of the measurement, four contacts are used in a linear configuration. The inner ones are cur-

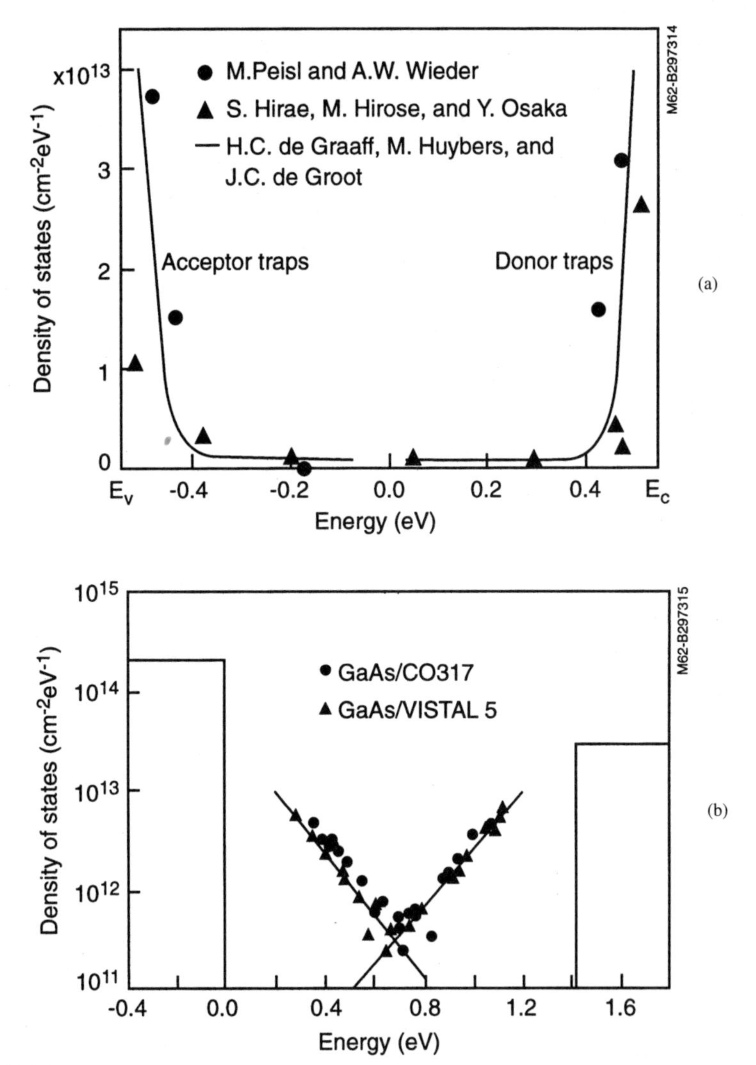

FIGURE 6-21 Density of states determinations. *(a)* Polycrystalline Si thin films; and *(b)* polycrystalline GaAs, showing states within the bandgap.

rent sensors; the outer ones are voltage-input sensors. The current-voltage (*I-V*) characteristic is determined over a large range of voltage (for Si, it is about 0 to 15 V). The current can vary over five orders of magnitude. Figure 6-22 presents a typical characteristic for Si.

The value for the density of states is deduced directly from the *I-V* characteristic. The method follows these steps:

1. Find N_T (or N_{ss}) by equating the total positive charge in the depletion layers (from Poisson's equation) with the total number of excess electrons trapped at the grain boundary (found by integrating over the occupied states).

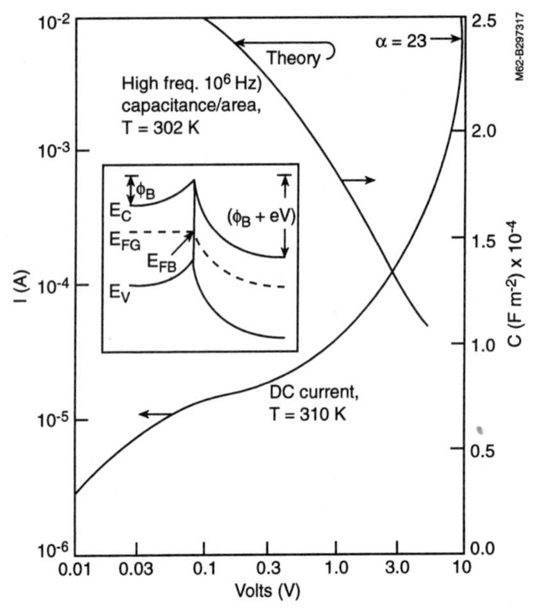

FIGURE 6-22 Illustration of the I-V method for determining the density of states of a Si grain boundary. The insert shows the representation of the grain boundary model. Also presented is the high-frequency capacitance as a function of the voltage, with the theory and experimental data having exact agreement.

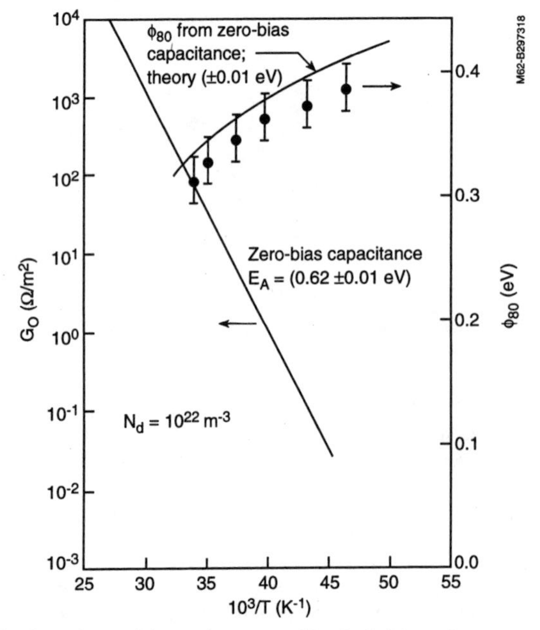

FIGURE 6-23 The dependence of the conductance and barrier height on inverse temperature for the I-V method (Si grain boundary). The zero-bias, dc conductance for this grain boundary is also represented.

2. Differentiate this equation with respect to the applied voltage, giving the rather complicated expression

$$N_T(E) = \frac{1}{kT}\int_0^{E_f} N_T(E)f'(E)dE$$
$$= (\epsilon\epsilon_o N_d/2q^2)^{1/2} \cdot [\phi_B^{-1/2} + (1 + q/\phi'_B)(\phi_B + qV)^{-1/2}]$$

where $\phi'_B = \phi_{Bo}(-kT)\ln(qJ/G_o kT)$ and G_o is the zero-bias conductance. This gives the density of states in terms of the current and the voltage from the *I-V* characteristic. (J is the current density.)

3. Find ϕ_{Bo}. Two methods are: from activation energy measurement from zero-bias conductance; and from capacitance.

4. Solve the equation for $N_T(E)$ with these parameters.

The interpretation of the various regions of the *I-V* characteristic (and the corresponding portions of the density of states versus E) of Fig. 6-22 is provided in Figs. 6-23 and 6-24. For the charged grain-boundary energy-band diagram, the left side of the barrier has a height that decreases under bias, and the right side increases. The current and charge increase, until the Fermi level reaches a decreasing density of states. A rapid breakdown (varistor) of the barrier occurs, and a relatively strong increase of the current results with increasing voltage thereafter.

The determined G and ϕ_{Bo} values are shown in Fig. 6-23, and the determined density of states corresponding to the *I-V* characteristic of Fig. 6-22 is presented in Fig. 6-24.

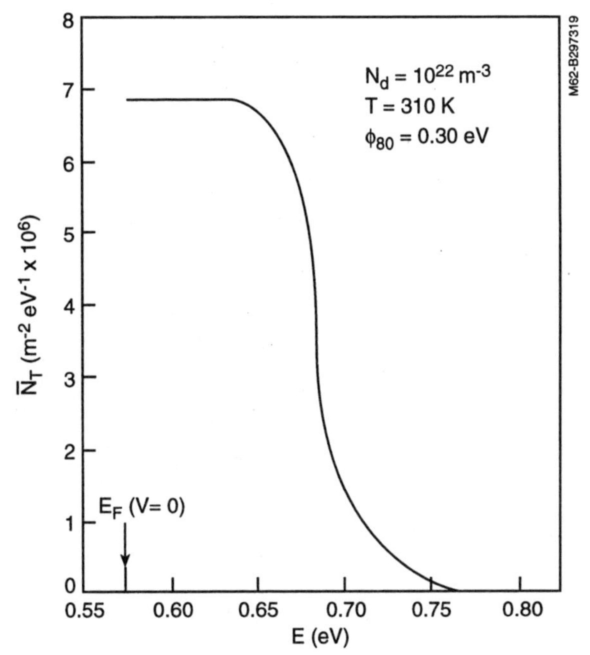

FIGURE 6-24 Density of grain boundary states as a function of energy for a Si grain boundary, determined from the I-V data of Figs. 6-22 and 6-23.

Problems with this technique center around the high fields involved in the measurement, sometimes exceeding 10^5 V/cm, leading to the possibility of impact ionization or heating.

6.3.3 Optical Method

- Information: Density of states
- Application: Single, isolated grain boundaries and bicrystals
- Approach: Measurement of photocapacitance

Capacitive spectroscopy can be used to determine the density of states, including that for single grain boundaries [66]. Again, the grain-boundary defect must be isolated, with ohmic contacts positioned on either side of the defect. The capacitance is measured under zero bias to avoid the application of high electric fields. The sample (specifically, the grain boundary) is simultaneously illuminated. In usual experimental configurations, a tunable light source is used (typically in the infrared region for most semiconductors), and a monochromator is used to select the photon energy.

Figures 6-25 and 6-26 provide representative photocapacitive data for isolated grain boundaries in Si. The onset of the change in capacitance agrees with the energy difference between the dark Fermi level and the edge of the valence band. This indicates a direct op-

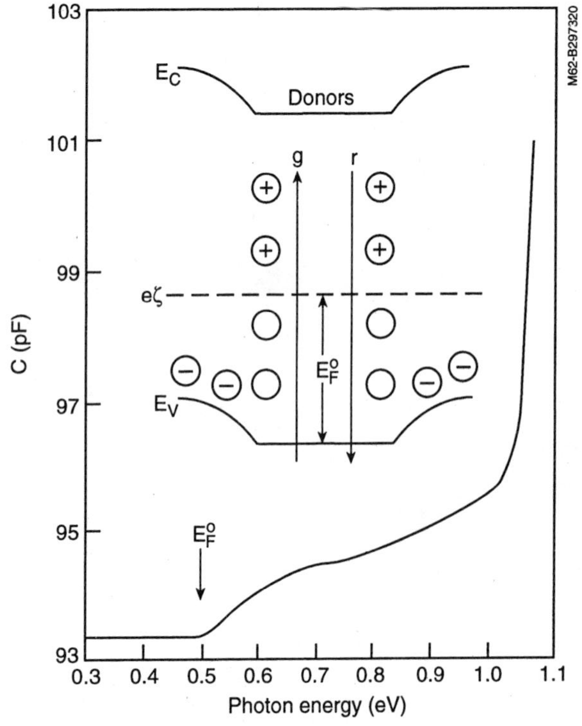

FIGURE 6-25 Illustration of the optical method, showing the capacitance of an isolated grain boundary in *p*-type Si. The photon source provides subbandgap irradiance.

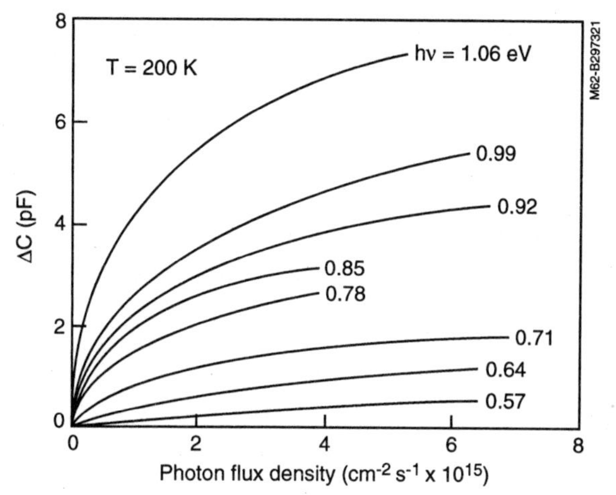

FIGURE 6-26 Photocapacitance as a function of irradiance (intensity) for various photon energies (Si grain boundary).

tical excitation of trapped holes from the forbidden gap into the valence band. When the light energy is held constant (that is, constant wavelength) and the intensity is increased, the capacitance increases to a saturation value, as shown in Fig. 6-26. This saturation value corresponds to a situation in which all traps between the valence band edge and an energy, $h\nu$, are emptied by the optical excitation of trapped holes. The density of states can be obtained by differentiating this remaining charge with respect to the photon energy. This result is shown in Fig. 6-27 for the Si grain boundary.

6.3.4 Thermal Method

- Information: Density of states, grain-boundary barrier height
- Application: Single, isolated grain boundaries and bicrystals

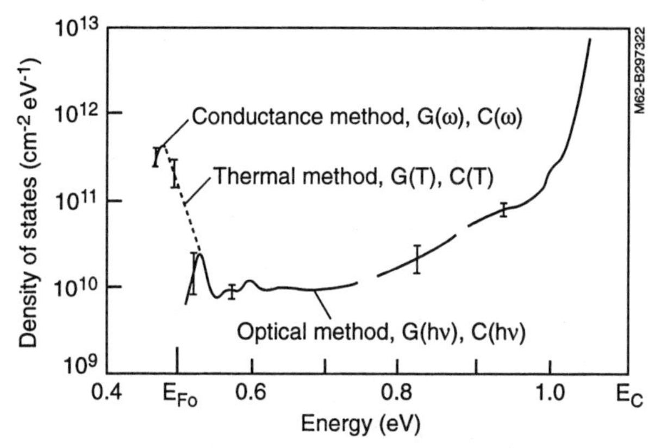

FIGURE 6-27 Determined density of states for a grain boundary in p-type Si.

- Approach: Measurement of the temperature dependence of the conductance or capacitance

The sample is again configured with ohmic contacts on either side of the grain boundary. Two measurements can be made to obtain the identical result [67]:

- Temperature dependence of the zero-bias dc conductance
- Temperature dependence of the high-frequency capacitance

The result from either measurement provides the temperature-dependent, zero-bias barrier, $q\phi_o$. The method follows these steps:

1. Determine $q\phi_o$.
2. Use the Hall effect to determine the Fermi level within the grains ($q\zeta$) as a function of temperature.
3. E_{fo} is the sum of the last two quantities, and yields the temperature-dependent, zero-bias Fermi level at the grain boundary.
4. The temperature-dependence of the trapped grain boundary charge can be calculated from the barrier, $q\phi_o$.
5. The density of states at the grain boundary can be calculated by taking the derivative of the equilibrium boundary charge, Q_o, with respect to the equilibrium Fermi level, E_{fo}:

$$N_{ss} = \frac{dQ_o}{dE_{fo}} = \frac{dQ_o/dT}{dE_{fo}/dT}$$

Typical results are shown in Fig. 6-27, along with those gained from the optical method. The major disadvantage with this technique is its very limited energy resolution.

6.3.5 Admittance Spectroscopy

- Information: Density of states, majority-carrier properties
- Application: Single, isolated grain boundaries (primarily)
- Approach: Measurement of the admittance of the grain boundary region (i.e., the real and imaginary parts of $Y = G + i\omega C$; these terms can be related to the density of states (and other parameters) through various models and deconvolution techniques

This technique is a powerful tool used in grain-boundary experimental analysis [68] that determines three important interface parameters: interface density of states; majority-carrier cross section; and standard deviation of potential fluctuations.

The determination of the admittance involves high-frequency investigation and measurements of the grain boundaries. Additional care is taken to ensure that the contacts are ohmic and that the leads are shielded well. For example, rectifying contacts can be the source of spurious capacitances. It is possible to undertake a study of a grain boundary using these high-frequency methods, only to find that the data gained pertain to the contact to the grains. It is critical, therefore, to ensure that the contacts are proper. To help minimize this contact effect, four-terminal measurements are usually used.

The experimental set-up can involve sophisticated admittance bridges or more simple lock-in approaches. For example, the signal can be delivered to a lock-in amplifier to determine the in-phase and quadrature current components relating to the G and C terms. The measurements are typically taken at 10 kHz or higher. Low-capacitance amplifiers should be used to buffer the signals supplied to the lock-in amplifier.

The measured small signal admittance for a single, p-type grain boundary is

$$Y_{\text{total}} = G + i\omega C \tag{39}$$

and depends on the frequency (as shown in the experimental data of Figs. 6-28 and 6-29). This admittance is given by the derivative of the total current

$$j_{\text{total}} = j_{th} + j_{ss} + j_{sc} \tag{40}$$

Therefore,

$$Y_{\text{total}} = \gamma + \frac{\delta\phi}{\delta V} + Y_{ss}\frac{\delta\phi}{\delta V} + i\omega C_{HF} \tag{41}$$

where $\gamma = q/kA^*T \exp[-\beta(\zeta + j_{DC})] = \beta j_{th}^{\text{dc}}$

The total ac current is

$$J_{\text{total}} = \gamma\delta\phi + Y_{ss}\delta\phi + i\omega C_{HF}\delta V \tag{42}$$

$$= j_{th}^{\text{ac}} + j_{ss} + J_{sc} \tag{43}$$

From Eq. 43, the thermally emitted ac current (j_{th}^{ac}) is determined by the amplitude and phase of $\delta\phi$, which, in turn, depends on j_{ss}.

Equation 43 also provides an electrical engineering equivalent of the grain boundary region. The total current divides into three parts. However, the prevoltage terms are written with respect to two different potentials, $\delta\phi$ and δV. If G_x and C_x are the conductance and capacitance with respect to dV, the equivalent circuit can readily be derived, as shown in Fig. 6-30. The circuit drawn with respect to the potential $\delta\phi$ is shown in Fig. 6-31. These are the equivalent circuits for the total ac current. Contained is the thermally emitted current, j_{th}^{ac}, which is the equivalent to the emitter-collector current of a transistor.

The modeling, therefore, is sometimes referred to as a trap transistor model. The capture-emission current of holes, j_{ss}, which flows between the grain-boundary traps and the valence band, is the base current. The thermionically emitted ac current (j_{th}^{ac}) corresponds to the emitter-collector current. As in the transistor, the transistor action of j_{ss} on j_{th}^{ac} can be expressed in terms of a current amplification factor.

Since

$$j_{th}^{\text{ac}} = \gamma\delta\phi$$

and

$$j_{ss} = Y_{ss}\delta\phi \tag{44}$$

it follows that

$$j_{th}^{\text{ac}} = (\gamma/Y_{ss})j_{ss} \tag{45}$$

This shows that the current j_{th}^{ac} is controlled by the capture-emission current, j_{ss}. The current amplification factor is controlled by the properties of the interface states, which are represented by the admittance, Y_{ss}. This admittance has real and imaginary parts

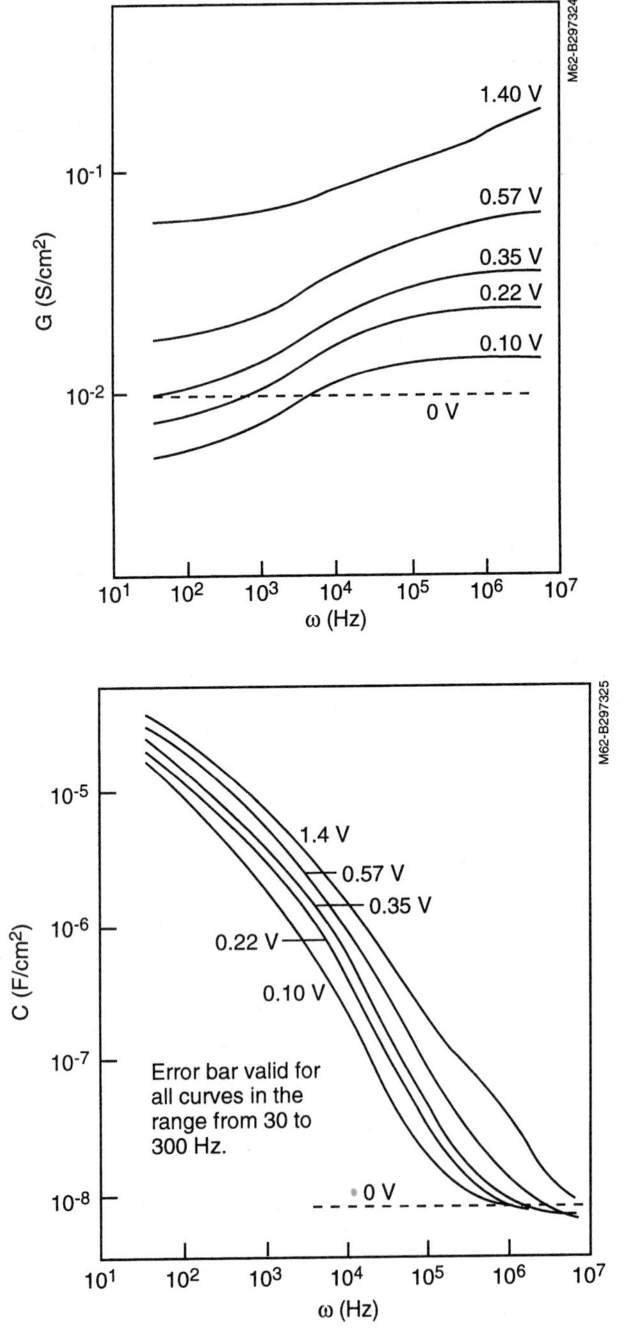

FIGURE 6-28 Dependence of the *(a)* conductance and *(b)* capacitance on frequency for an isolated grain boundary in p-type Si (admittance method).

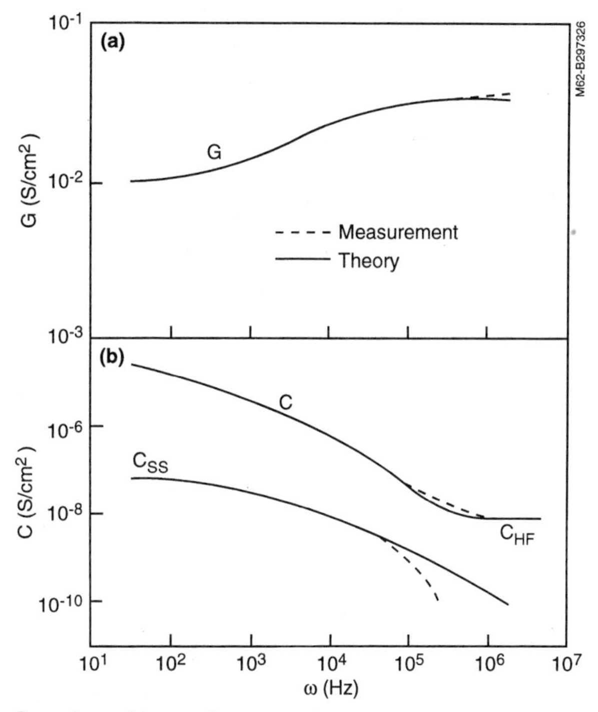

FIGURE 6-29 Dependence of the capacitance *(a)*, and on the conductance *(b)* on frequency for an isolated grain boundary in p-type Si (same as Fig. 6-29 for determination of the equivalent circuit components).

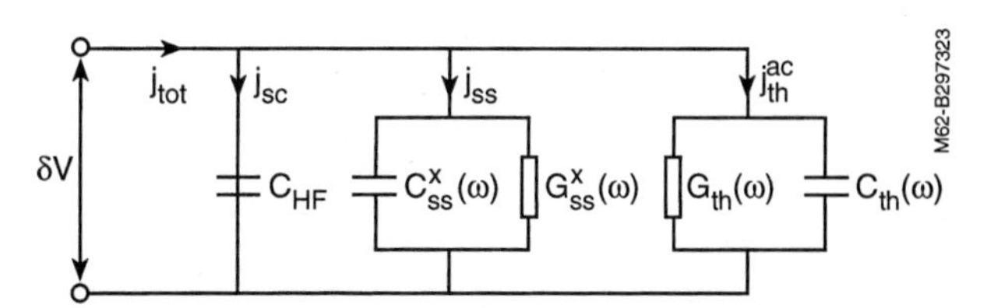

FIGURE 6-30 Equivalent electrical circuit using the conductance and capacitance data with respect to the applied voltage.

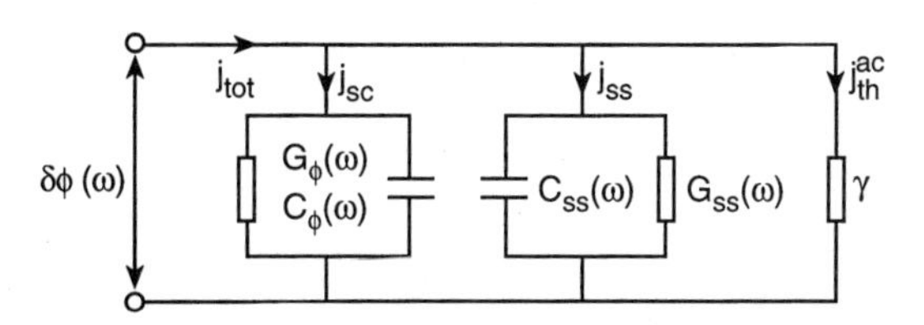

FIGURE 6-31 Equivalent electrical circuit with respect to the grain boundary potential.

(equivalent circuit). Therefore, there is a phase shift between the base current (j_{ss}) and the emitted current, j_{th}^{ac}.

6.4 CHEMICAL, COMPOSITIONAL, AND STRUCTURAL DETERMINATIONS

The range of analytical tools useful to the chemical, compositional, and structural properties of thin films is expansive [69–72]. An exhaustive treatment of all such approaches is not possible in the space available in this chapter. Even the application of a subset of such techniques is formidable. This section provides a brief introduction to some major methods used for such investigations. Analytical approaches include macro- and micro-analysis (methods in which the volume analysis of materials is considered); micro- and nano-analysis (primarily surface and interface techniques); and nano- and atomic-level analysis (proximal probe methods providing the highest spatial resolutions). These techniques encompass microscopies and spectroscopies for ascertaining specific information of film surfaces, subsurface regions, and interfaces.

Table 6-2 provides a tabulation of some common methods for these analyses. Table 6-3 provides more-specific and comprehensive information, including input and output probes, quantity measured, lateral spatial resolution, and probe depth. Spatial resolution is an important parameter for many thin film studies of composition and chemistry.

6.4.1 Micro-Composition Determinations

Transmission Electron Microscopy

- Information: Microscale structural information (defects, atomic arrangements, orientations)
- Application: Single-crystal, polycrystalline, and amorphous thin films
- Approach: Illumination of a sample with a well-focused, high-energy electron beam, and the evaluation of direct and diffracted electron beams transmitted through the sample thickness

Transmission electron microscopy (TEM) is the primary investigative tool for the investigation of the structural properties of crystals, thin films, and interfaces [73]. Combined with ancillary techniques described below, TEM can provide high-spatial-resolution information on chemistry, composition, and electrical properties, as well. Its sister technique—scanning TEM, or STEM—produces images of the internal microstructure of thin samples using a scanned high-energy electron beam. Scanning transmission electron microscopy is used to describe the group of analysis methods collectively assigned as analytical electron microscopies [74–78]. The strength of STEM is its ability to obtain high-resolution imaging combined with microanalysis from almost all solid materials. Transmission electron microscopy is well established and documented, and only a brief introduction will be provided here. The reader is directed to several references that provide extended treatment of the electron optics, electronics, and sample constraints [79–83].

In TEM, a focused high-energy electron beam is incident on a thin sample. The TEM signal is extracted from directed and diffracted electron-beam components that penetrate the sample. Magnetic and electronic lenses focus the emitted beam onto a detector (usually a fluorescent screen, a video camera using a charge-couple device, or a photographic

TABLE 6-2 Compilation of some selected common methods for analyzing specific structural, chemical, and physical properties of surfaces and interfaces of solid materials

Property	Technique
Element identification; composition; purity	Auger electron spectroscopy (AES)
	x-ray photoelectron spectroscopy (XPS); Also termed, electron spectroscopy for chemical analysis (ESCA)
	Secondary ion mass spectrometry (SIMS)
	Time-of-flight SIMS (TOFSIMS or static SIMS)
	Ion-scattering spectroscopy (ISS)
	Scanning-neutral ion mass spectrometry (SNMS)
	Surface analysis by laser ionization (SALI)
	Surface analysis by resonance ionization of sputtered atoms (SARISA)
	Laser ionization mass spectrometry (LIMS)
	Laser microprobe mass analysis (LMMA)
	Rutherford backscattering spectroscopy (RBS)
	Electron energy-loss spectroscopy (EELS)
	Atom probe field ion microscopy (APFIM)
	Nuclear reaction analysis (NRA)
	Particle-induced x-ray emission (PIXE)
	Glow discharge optical spectroscopy (GDOS)
	Glow discharge mass spectrometry (GDMS)
	Surface compositional analysis by neutral and ion impact radiation (SCANIIR)
	Energy-dispersive spectroscopy (EDS)
	Wavelength-dispersive spectroscopy (WDS)
	Electron probe microanalysis (EPMA)
Chemical state characterization	Infrared and Raman spectroscopy
	Fourier-transform infrared spectroscopy (FTIR)
	EELS
	XPS
	SIMS
	Electron-stimulated desorption ion angular distribution (ESDIAD)
	Ellipsometry
	x-ray fluorescence (XRF)
	Laser fluorescence (LF)
	Electroluminescence and cathodoluminescence (EL and CL)
Structural characterization	Low-energy electron diffraction (LEED)
	High-energy electron diffraction (HEED)
	Reflection high-energy diffraction (RHEED)
	Field ion microscopy (FIM)
	Field emission microscopy (FEM)
	Transmission electron diffraction (TED)
	x-ray diffraction (XRD)
	High-voltage electron microscopy (HVEM)
	Analytical electron microscopy (AEM)
	Extended x-ray fine structure spectroscopy (EXAFS)
	ISS
	Ion and electron channeling
	ESDIAD
	Ultraviolet photoelectron spectroscopy (UPS)
	Surface-extended x-ray fine structure (SEXAFS)
	Vibrational EELS

Continued.

TABLE 6-2 Compilation of some selected common methods for analyzing specific structural, chemical, and physical properties of surfaces and interfaces of solid materials *(Continued)*

Property	Technique
Structural characterization *(Continued)*	Raman spectroscopy Neutral scattering spectroscopy (NSS)
Topography	Scanning electron microscopy (SEM) Electron microscopy (EM) Optical microscopy x-ray topography Interferometry (including spectroscopic) Profilometry (mechanical stylus) Scanning tunneling microscopy (STM) Atomic force microscopy (AFM) Near-field scanning optical microscopy (NSOM)
Film growth/material absorbed or deposited	Microgravimetric techniques Ellipsometric techniques Radiotracer techniques Laser/optical techniques Quartz crystal (microbalance) Electron-stimulated desorption (ESD)
Surface area/roughness	Microgravimetric and volumetric absorption EM Optical microscopy Proximal probe microscopies Optical beam-induced current spectroscopy (OBIC)
Inclusions/defects	Scanning auger microscopy (SAM) Secondary ion mass spectroscopy Electron-dispersive spectroscopy with transmission electron microscopy (TEM) Scanning transmission electron microscopy (STEM) OBIC Electron-beam induced-current (EBIC) Electron-beam induced-voltage (EBIV)
Microdefects/features	EM Electron diffraction (ED) Small-angle x-ray spectroscopy (SAXS) STEM AEM Proximal probe microscopies NSOM
Electron density of states/ defect chemistry	UPS Photoluminescence spectroscopy (PL) Deep-level transient spectroscopy (DLTS) Scanning tunneling spectroscopy (STS) Near-field scanning optical spectroscopy (NSOS) EL and CL Ballistic electron-energy microscopy (BEEM)
Atomic-level imaging	FIM STM Scanning force microscopy (SFM) TEM (lattice imaging)

TABLE 6-3 Summary of selected parameters for a variety of structural, chemical, physical, and compositional techniques for the bulk characterization and analysis of materials

	Electron probe microanalysis (EPMA)	Neutron activation analysis (NAA)	Spark source mass spectroscopy (SSMS)	Emission spectrographic analysis	Inductively coupled plasma emission spectroscopy (ICP)	Atomic absorption spectroscopy (AAS)
Sample type	Solid	Solid or liquid	Solid	Inorganic solids (metals/alloys); elements (simultaneously)	Solid (in solution) or liquid	Metallic elements (in solution)
Elemental range	>Li	All materials with reasonable half lives (multielements simultaneously	All elements	50-100 elements simultaneously	Multielements simultaneously or in rapid succession	Metals
Detection limits	0.1 at.-%	10^{-9} g	10^{-3} g	0.002-0.2 μg (typical) 0.001-1.0 at.-%	10-1000 ppb	0.00005-0.1 mg/L
Precision	0.5-10%	1%	300%	—	1000%	—
Analysis volume (minimum or material required)	0.001-1.0 μm^3	mg to μg	mg	1-20 mg	mg	g
Quantitative analysis	Very good (best using standards)	Difficult; requires standards	Range only (semiquantitative)	Semiquantitative	Semiquantitative	Semiquantitative
Spatial resolution	0.1-1.0 μm	Limited (with sample cutting/preparation)	None	None	None	None

film). The magnification can range from ~100× to more than 1,000,000×. The TEM is shown schematically in the diagrams of Fig. 6-32 [82–85]. The spatial resolution is gained by the ability to focus the charged electron beam. For example, 100 keV electrons correspond to wavelengths of 3.7×10^{-3} nm. In general, the higher the operating energy of the TEM, the greater the lateral spatial resolution. Most commercial instruments operate between 300 and 400 keV, with resolutions higher than 0.2 nm.

In general, TEM offers two mechanisms for observing the sample: diffraction and image. In the diffraction mode, the image of the diffracted electrons is obtained from the electron-illuminated sample. The electron diffraction patterns correspond to x-ray diffraction analogues, and they provide information on crystallinity and crystal orientation. Spot patterns correspond to single crystals; ring patterns correspond to polycrystalline samples; and diffuse ring patterns correspond to amorphous structures. Examples of these are shown in Fig. 6-33. The image mode produces a representation of the entire sample depth, and contrast results from several mechanisms, including: diffraction contrast (the scattering of the electrons by structural inhomogeneities); mass contrast

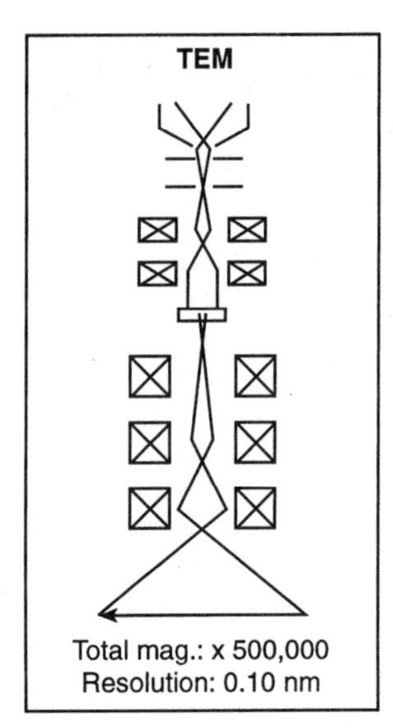

FIGURE 6-32 Schematic representation of transmission electron microscope.

(caused by the spatial separations and orientations of constituent atoms), thickness contrast (caused by nonuniformities in film thickness), and phase contrast (the result of coherent elastic scattering). Each of these modes provides specific information about a selected sample area.

The most important, and perhaps most complex, aspect of TEM involves sample preparation. Imaging may take minutes, but sample preparation can take hours or days. Thinning can be accomplished by ion milling, electropolishing, ball cratering, and polishing. Cross sections (especially important in interface or junction observations) are made from layered samples. An example is shown in Fig. 6-34. These sections are prepared to be viewed along the plane of the layers, and this sample preparation procedure is among the most difficult. The difficult part is to ensure that the defects and other structures observed in the TEM are not imposed by the sample preparation technique itself. The literature contains many descriptions of such procedures used for a variety of thin films and for semiconductor devices [84–86].

Scanning Electron Microscopy

- Information: Microscale topography/topology
- Application: Single-crystal, polycrystalline, and amorphous thin films
- Approach: Spectroscopic evaluation of secondary or backscattered electrons; or x-rays from the sample surface, produced from a highly focused and rastered electron beam source

This instrument is the traditional one for observing topography or topology on the microscale. Scanning electron microscopy (SEM) provides a high-magnification image of the surface of a material and can have resolutions approaching a few nm [87]. The instrument typically operates with electron beams in the range of 20 to 30 keV. Magnification from 10× to more than 400,000× are possible [88–92]. The basic electron optics are shown in Fig. 6-35. Although topographic information is the most common, the SEM can also provide information on the composition of the near-surface region.

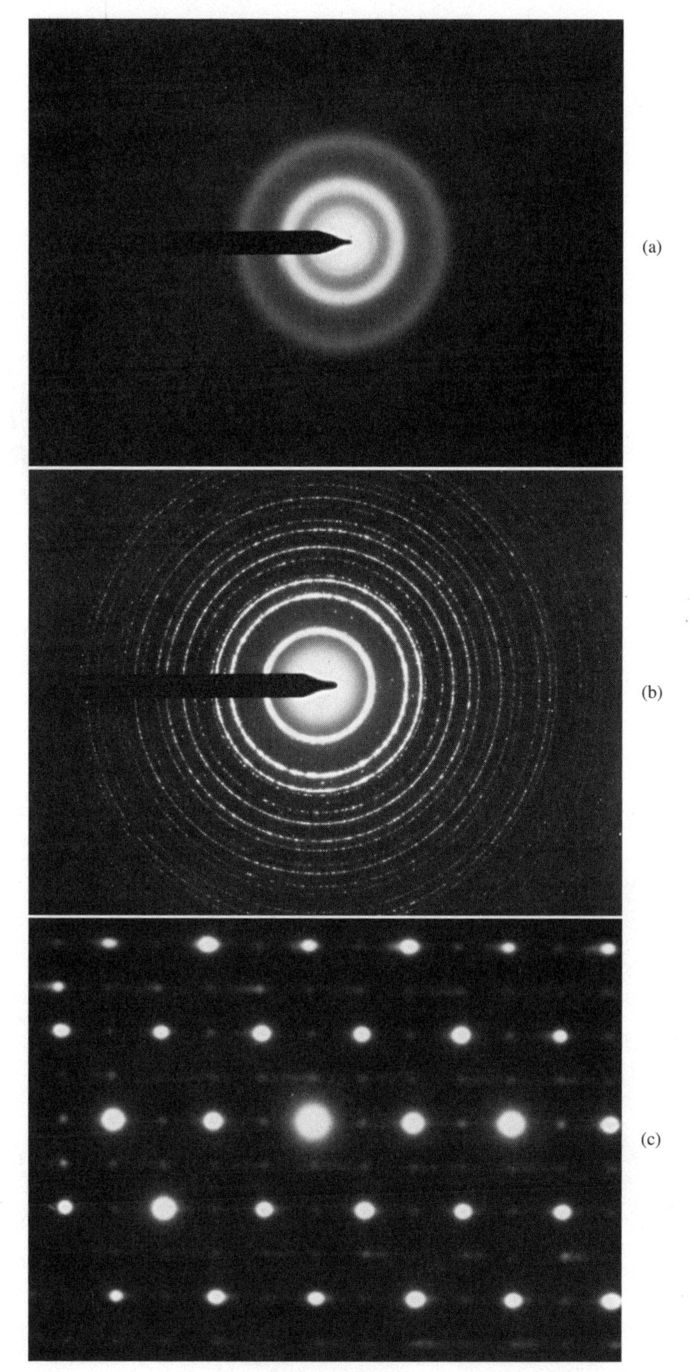

FIGURE 6-33 Electron diffraction patterns for *(a)* amorphous, *(b)* polcrystalline, and (*c*) single crystal thin films.

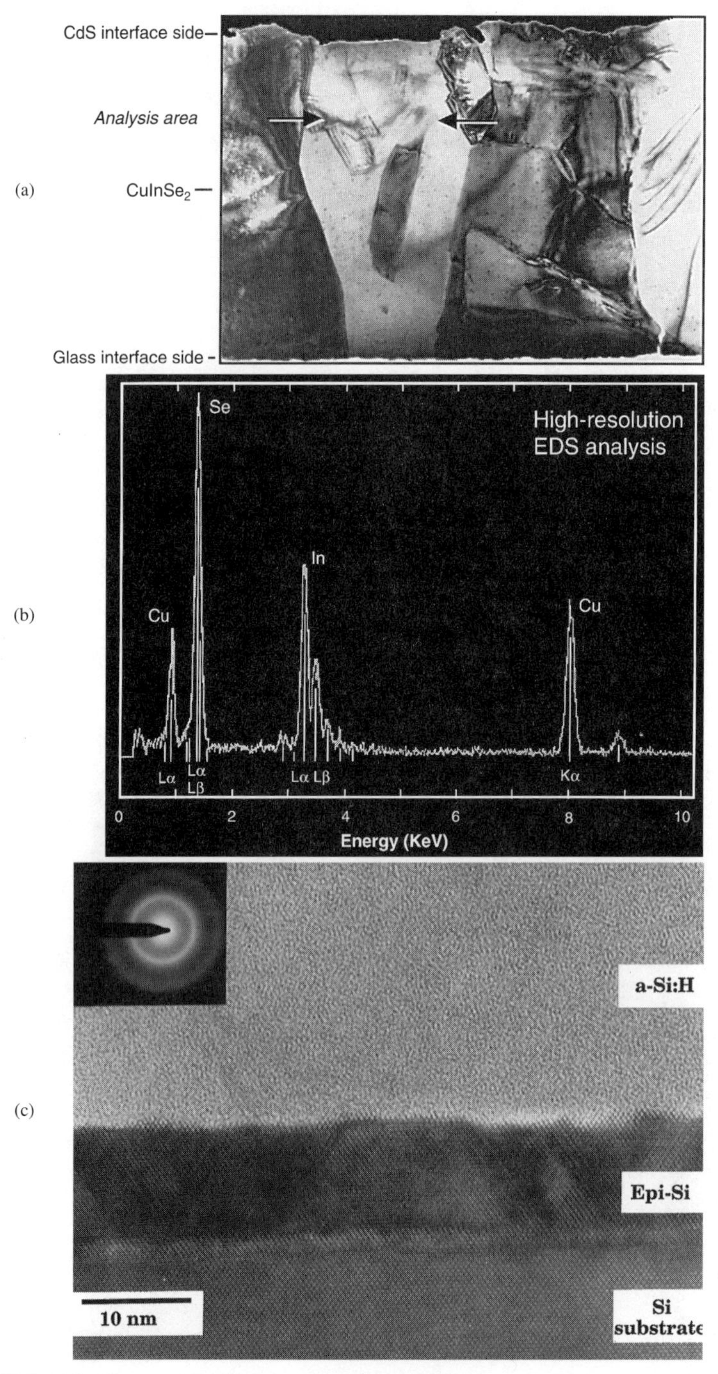

FIGURE 6-34 Cross-sectional TEM image of GaAs junction (*a*), corresponding EDS analysis (*b*), and amorphous silicon/silicon junction (*c*).

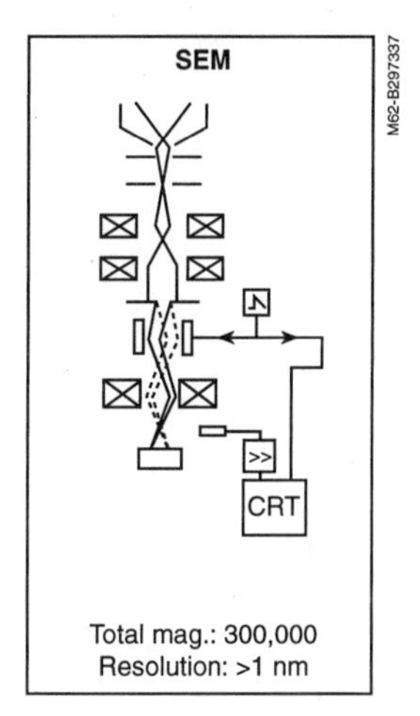

FIGURE 6-35 Schematic representation of scanning electron microscope.

The technique involves focusing a source of electrons on a target in a vacuum chamber. The finely focused electron beam is rastered over the sample surface, and the emitted electrons or photons are used to construct the image of that surface. A large percentage of the electrons emitted are collected by detectors, and the output is used to modulate the brightness of a cathode-ray tube. Three major image types are produced: secondary electron images, backscattered electron images, and elemental x-ray maps. The principles of operation for the instruments have been discussed in a number of sources. However, the SEM is an extremely powerful tool for imaging and forms the basis of discussion for the use of other techniques described in the following sections. Fig. 6-36 presents several SEM images for semiconductor thin films.

Energy Dispersive and Wavelength Dispersive Spectroscopies

- Information: Microscale elemental composition
- Application: Single-crystal, polycrystalline, and amorphous thin films
- Approach: Spectroscopic evaluation of x-rays generated from sample as a result of high-energy electron beam

Microanalytical volume-composition analysis techniques are readily available in the laboratory. The more common are energy dispersive spectroscopy (EDS) and wavelength-dispersive spectroscopy (WDS) [93–96]. Both of these methods involve the generation of x-rays by high-energy electron probes. The analysis is usually performed in a scanning electron microscope or in a more-specialized, dedicated instrument called an electron-probe microanalyzer. In general, these techniques are capable of detecting elemental species to about the 0.1 at.-% (about 1000 ppm) range. Because the input probe is a well-focused electron source, spatially resolved compositional data can be gained. Depth resolution is possible if the energy of the incoming electron beam is varied, which changes the penetration depth of the input probe. Depth information can also be obtained by combining the analysis with ion etching or ball cratering. Finally, because these methods involve a controlled electron beam, useful information, such as topographical features using secondary electron detection or electrical information by EBIC/EBIV detection, can be obtained for a given film analysis.

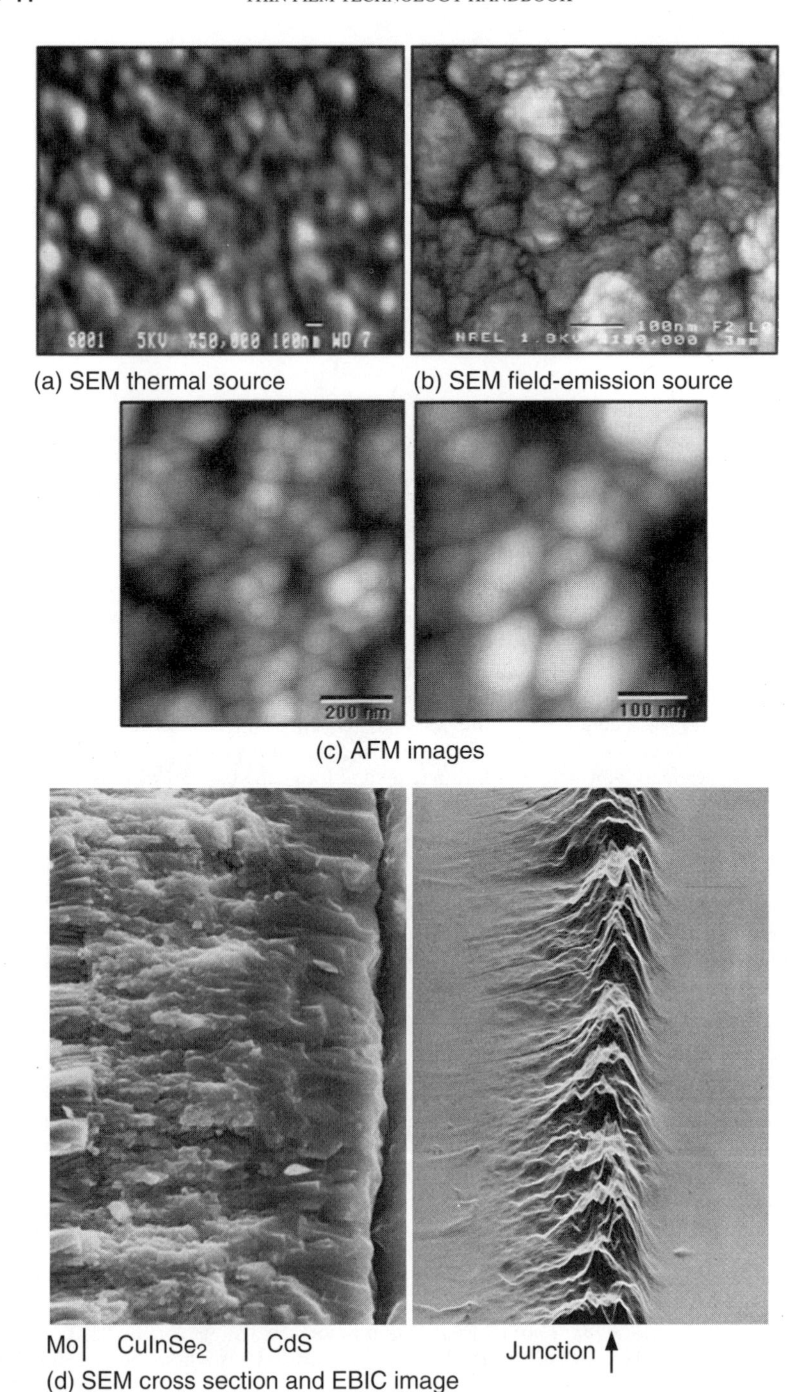

FIGURE 6-36 SEM images of thin films surfaces, illustrating resolution and magnification features.

When a sample is bombarded with an energetic electron beam, both radiative and non-radiative emission processes occur. The non-radiative mechanisms result in the generation of Auger electrons (considered in the surface-analysis discussions later in this chapter). Vacancies are created in the electronic shells of atoms as a result of the input electrons. Electrons from outer shells fill the inner-shell vacancies to keep the atom in a lower energy state. If the energy differences between the transition electron and the vacancy shells are large enough (several thousand electron volts), characteristic radiation in the form of x-rays are emitted. The creation of these x-rays and their detection constitute the basis for both EDS and WDS.

Generation Volume. The generated x-rays originate from subsurface volumes called the generation volumes [97]. These have shape and depth properties that are related to the energy of the incoming electron beam and the atomic number (mass) of the elements being analyzed. The generation volume is one of the more important physical parameters involved in x-ray microanalysis. At low beam energies, or for elements with high atomic numbers, the volume is approximately one hemisphere. As the electron beam energy increases, or the atomic number decreases, the volume becomes characteristically pear-shaped. Figure 6-37 shows the range and origin of a number of the possible emissions from the sample when excitated by an energetic electron beam [97]. Mathematical models have been derived for the physical shape and size of the generation volume [97,98]. Basically, the dimensions of the generation volume are orders of magnitude larger than the lateral dimensions of the incoming electron beam. Therefore, the elemental spatial resolution is more related to the generation volume at lower beam sizes than to the geometrical feature of the electron probe itself. The depth resolution is about one micron for common instrument resolution. For thinner films, the lateral resolution can be enhanced because the beam cannot spread out into the volume of the sample. Therefore, the researcher can use this (or thinning) advantageously to gain lateral resolution for such analyses [99].

Detection. The x-rays measured per-unit-time escaping from the generation volume within the solid angle and at a given take-off angle. Thus, the detection is a function of the geometrical positioning, which is limited primarily by the design of the detector. The methods for detecting the generated x-rays include either evaluating the energy or pulse height (the EDS mode), or determining the wavelength (the WDS mode). In general, WDS detectors have larger areas and are operated further from the x-ray source (the sample), and they inherently have lower collection efficiencies. The EDS system uses a solid-state

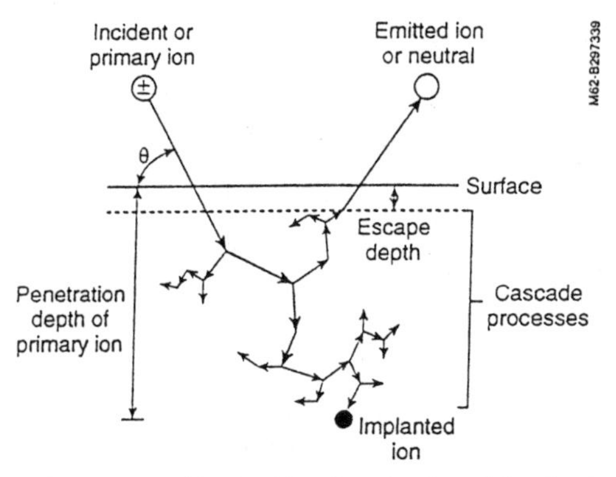

FIGURE 6-37 Electron beam penetration of solid surface showing origin of various emitted species and radiations.

diode and associated electronics to detect, amplify, and sort charge pulses generated by the x-rays. The common detector is *p*-type Si, with Lithium (Li) diffused through the back to form a *p-n* junction (and compensate for other impurities). Because Li can diffuse rapidly at room temperature, these detectors are commonly held at liquid nitrogen temperatures [100].

The WDS system separates the x-rays by wavelength before they reach the detector by diffraction through a crystal. The x-rays of desired wavelength are selected and directed to the detector by the accurate positioning of the crystal and detector with respect to the sample. Crystals with varying lattice spacing are selected to cover the desired x-ray spectrum. Typical EDS and WDS spectra are presented in Fig. 6-38 for CdTe polycrystalline thin films [101]. The superior sensitivity and higher peak resolution capabilities of the WDS method are apparent. Also, the contribution of the background to the WDS spectrum is less detrimental.

Properties and Quantification. The detection limits, energy ranges, and other specifications for these techniques are summarized in Table 6-3. One strength of these tech-

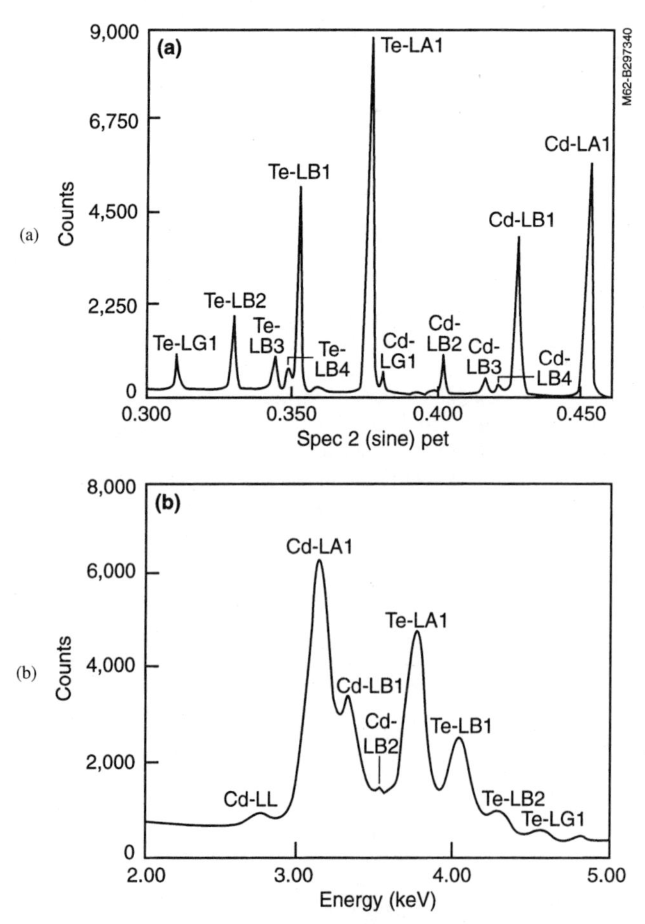

FIGURE 6-38 Compositional data for polycrystalline CdTe thin film, comparing *(a)* WDS, and *(b)* EDS data.

niques is their ability to provide quantitative information, which is especially important for establishing deposition conditions for either optimizing materials or device properties. Several quantitative analysis routines are available, and the more-accurate ones use standards that are near or the same as the films being analyzed. A common quantitative methods is ZAF, which involves calculating the original x-ray intensity by correcting the measured intensity for the atomic number of the x-ray scattering differences between the standard and the unknown atomic number (Z), x-ray absorption (A), and x-ray fluorescence (F) [102]. Thus, the technique compares x-ray intensities between the known composition of the standard with the unknown sample. Details of such quantitative analysis include corrections for background effects and isolation from interference (overlapping) peaks [103].

6.4.2 Surface Spectroscopies and Spectrometries

- Information: Composition and chemistry of near-surface and interface regions
- Application: Single-crystal, polycrystalline, and amorphous thin films
- Approach: Evaluation of emitted electrons, ions, and photons from incident ionizing radiation (electrons, x-rays, ions, photons)

Surface elemental compositions, segregated impurity species at interfaces, doping profiles and gradients, interfacial chemical reaction, and quantitative compositional information are critically important measurements in optimizing thin film parameters and in the design and performance of devices. These properties can be evaluated accurately and with desired sensitivity using one or more of a class of techniques categorized under the generic term "surface analysis." These characterization tools examine the topmost atomic layers using electron-, photon-, and ion-input probes to excite and detect a variety of output or emitted species. These methods, the majority of which are summarized in Table 6-4, provide qualitative and quantitative information on the chemistry and composition of surface and interface regions in electronic materials.

Because of the surface sensitivity of these techniques, they have special and extensive applications to thin-film technologies. Thin films also offer some special challenges in analysis and interpretation because of several factors; the inherent roughness in polycrystalline deposited layers, overlap of elemental and chemical species in the obtained spectra, sensitivity limitations, potential damage or artifact creation from the input probe or associated analysis methods (such as sputtering), and elemental range limitations in detection. The techniques have been extensively refined during the past 20 years, and they offer rapid, accurate, sensitive, and useful analysis of the composition (and possible electronic features) of surface and interface regions—if proper control and understanding of the techniques are used.

These techniques have been treated extensively in the literature, and a minibibliography of such texts, review articles, and special publications is provided at the end of this chapter [104–124]. Therefore, the physics and principles of each of these important analysis methods will not be detailed here. However, some of the key limitations will be discussed regarding the more-common surface analysis tools. This knowledge will help in understanding the application of these methods to thin film analysis.

Sputter-Etching and Ion Bombardment. Common to several of the surface-analysis approaches is the use of an ion beam to obtain elemental or chemical information, either as part of the inherent analysis operation (as in SIMS, INS, ISS) or to provide

TABLE 6-4 Comparison and properties of common surface analysis techniques

	Auger electron spectroscopy (AES)	Electron energy loss spectroscopy (EELS)	x-ray photoelectron spectroscopy (XPS)	Secondary ion mass spectrometry (static SIMS)	Secondary ion mass spectrometry (dynamic SIMS)	Ion-scattering spectroscopy (ISS)	Rutherford backscattering spectroscopy (RBS)	Electron desorption spectroscopy (EDS)
Input probe	Electrons	Electrons	x-rays	Ions	Ions	Ions	Ions	Electrons
Species or range	0.20-20 keV	20-600 eV	1254 eV and 1487 eV	Ar, Xe, He, Ga, (0.05-30 keV)	Ar, Cs, O, O_2, Ga, He, H (1-30 keV)	He, Ne, Ar, Li, Na, K (0.1-5.0 keV)	H^+, He^+	
Detected species	Auger electrons	Electrons	Photoelectrons	Secondary ions	Secondary ions	Scattered ions	Scattered ions	Ions
Range or type	10-2500 eV	0-20 eV	10-2500 eV	+ or −	+ or −			
Elemental range	>Li	>H	>Li	>H	>H	>H	>H	>H
Detection limit	0.1 at.-%	0.1 at.-%	0.1 at.-%	10^{-7}-10^{-3} at.-%	10^{-7}-10^{-3} at.-%	10^{-2}-10^{-1} at.-%	10^{-2}-10^{-1} at.-%	
Spatial resolution (minimum)	~10 nm	~5 nm	~50 μm	0.5 μm	0.5 μm	>5 μm	>1 μm	~100 nm
Depth resolution	~0.3-6.0 nm	~0.3-6.0 nm	~0.3-6.0 nm	0.3 nm	<0.6 nm	<0.1 nm	~15 nm	~0.3 nm
Quantitative analysis capability	Very good	—	Excellent	Good-poor	Good-poor	Poor		Poor
Mapping capability	Excellent	Good	Poor	Very good	Very good	Poor		

depth-compositional profiles simultaneous to the analysis (as in AES, XPS, EDS, UPS). Ion etching is also routinely used to clean the sample surface of oxides or carbon layers that occur during extra-vacuum handling. Any ion bombardment is affected by several important factors. These can be categorized into instrument factors and ion-matrix factors that influence depth resolution and the quality of depth-composition profiles. These processes are represented schematically in Fig. 6-39, and they include the following [125]:

- Crater geometry
- Roughening and non-uniform or preferential sputtering
- Edge/halo effects
- Knock-on effects
- Ion migration (especially for low-conductivity films)
- Secondary ion-extraction optics
- Neutrals and impurities in the ion beam

These effects influence or control the shape of the resulting profile, or contribute to unwanted data detection during the analysis. Figure 6-40 illustrates the influence of several instrument factors (crater-edge and redeposition effects) on the shape of an ideal profile [126]. The knock-on effect can lead to an exponential tail in a secondary ion mass spectrometry (SIMS) profile, as illustrated in Fig. 6-41 [127]. Some uncertainties arise in the interpretation of unexpected shapes of depth-profile curves. Certainly, for compound-semiconductor thin films, the potential for preferential removal of one or more of the inherent species is a problem. This can lead to miscalculation of quantitative information, especially at interfaces located some distance beneath the surface. Such effects have been reported in $CuInSe_2$ and other films [128]. Many times, the preferential sputtering can be minimized by adjusting the experimental conditions, especially the energy of the incoming ion beam, the sputtering gas species (mass), and the angle between the sample surface and the incoming ion flux.

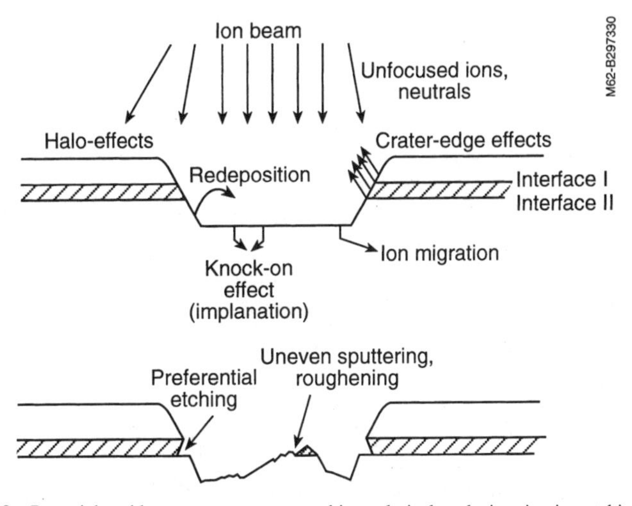

FIGURE 6-39 Potential problem sources encountered in analytical analysis using ion etching.

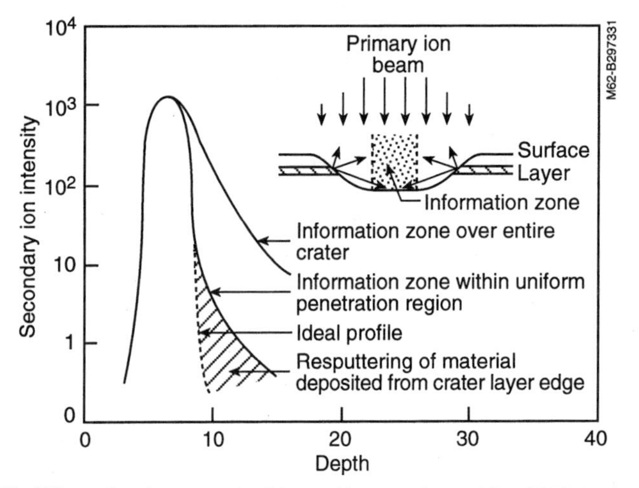

FIGURE 6-40 Effects of various sputter etching problems on the resulting SIMS depth compositional profile.

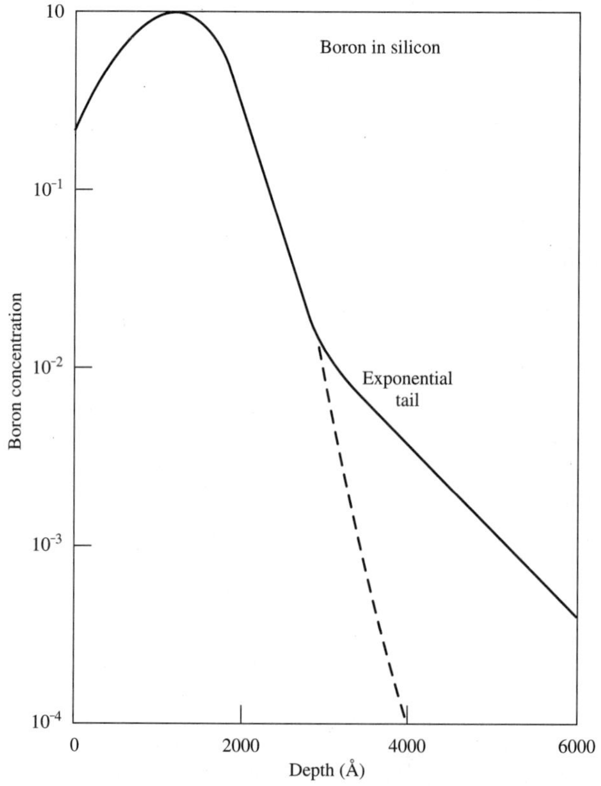

FIGURE 6-41 SIMS depth compositional profile of boron diffused into Si, showing the exponential tail due to knock-on effects.

Any time an energetic ion beam is exposed to a sample surface, damage to that surface and the subsurface region occurs [129–131]. The extent of this damage depends on the physical nature and conditions of the ion beam, and on the properties and chemistry of the sample being analyzed. In any case, the use of ion beams can lead to the generation of artifacts during the analysis. The interpretation of such results depends heavily on the following: the surface analyst's expertise and knowledge of the materials system under investigation; the control of the instrumentation; and a thorough understanding of the techniques. The literature has extensive examples of technique-induced, rather than process-occurring, research results [129–135].

Auger Electron Spectroscopy and Electron-Loss Spectroscopy. These techniques have some common features: the input probe is an energetic electron beam and the detected species is an electron. The differences are in the energy of the input probe (<500 eV for ELS and usually >2000 eV for AES) and in the nature of the species detected (energy loss by electrons for ELS and Auger electrons—produced by radiationless process—for AES) [136–137].

The exposure of a semiconductor surface to an energetic electron beam provides the potential for alteration of the analysis area [138–142]. First, the electrons may introduce damage in the region near the surface. For example, the substantial generation of dislocations and point defects has been reported for Si, for electron energies as low as 100 eV [141,142]. Second, the electron beam can itself cause changes in chemistry at the surface. For example, low-energy electron beams have been shown to oxidize $CuInSe_2$ and Si thin-film surfaces and remove oxide from semiconductors at higher energies [143,144]. The electron beam has also been shown to alter the oxide species on GaAs surfaces. The relationship between the electron beam conditions and the alteration of the $CuInSe_2$ surface is illustrated in Fig. 6-42 [144].

Interference between generated peaks can cause difficulties, although most elemental transitions in normal AES spectra are sufficiently separated [145]. One example is that of indium-tin-oxide (ITO) films [146]. The tin peak is very close to one of the oxygen peaks, and at the low concentrations, it can easily be missed or masked by the oxygen. Quantitative analysis for AES and EDS is relatively straightforward, but it is sensitive to the experimental conditions (beam currents, sputtering, sample orientation, surface roughness).

When using energetic electron beams, it is also difficult to analyze samples having lower conductivities [147]. Charging of the sample can easily occur. In mild cases, the detected transitions can shift in energy position. In the case of insulating films, the charging masks the generation and detection of the AES or EDS signals. There are approaches to overcome this effect in thin films. If the layer is sufficiently thin, the use of higher electron energies will actually punch a higher conductivity region through the film to the conducting substrate, and the charging will be eliminated. Sometimes, very-low-energy beams can be used. It is also possible to provide conducting regions (such as using conducting paints or deposition of metals) near or over the analysis area to lower the path length for conduction. Finally, the use of low-current, pulse-electron beams can also help by minimizing the total amount of charge delivered to the sample. For intrinsic semiconductors, charging remains a problem that needs clever experimental attention.

X-ray and Ultraviolet Photoelectron Spectroscopies. A major application of x-ray photoelectron spectroscopy (XPS) is the identification of the chemical species or chemical state at the surface or interface being investigated. A positive feature of XPS and ultraviolet photoelectron spectroscopy (UPS) is that the techniques produce little sample damage or alteration because of the nature of the input probe [148,149]. This feature, and the fact

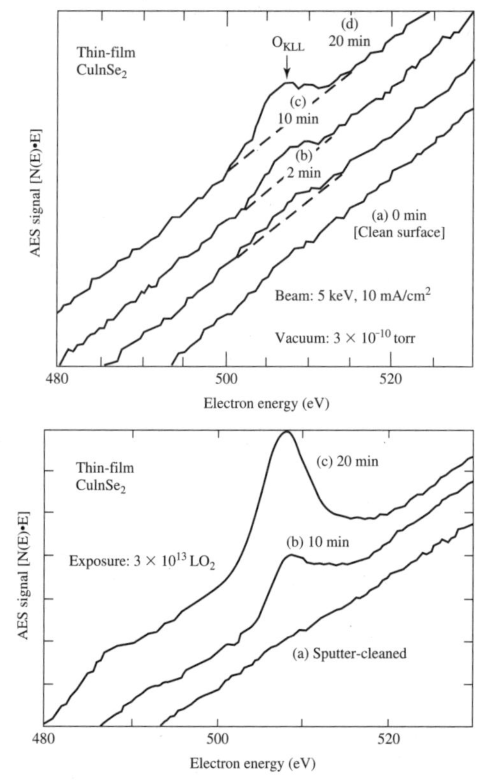

FIGURE 6-42 Effects of electron beam on the alteration of the chemistry of the $CuInSe_2$ surface.

that XPS uses the direct analysis of core electrons, makes XPS the most reliable quantitative surface-analysis method.

Major peaks in the XPS spectra are associated with the binding energies of the electrons in the atoms and the energies of the emitted Auger electrons. These transitions are usually dominant, but the spectra contain other peaks that can confuse the interpretation. The interpretation of many "oxide" or other chemical species is actually attributed to these additional lines. These include the following [150]:

- **Satellite peaks.** The spectrum of the x-ray probe (such as the Mg-Kα) not only contains the characteristic x-ray, but also some minor components at higher photon energies. Thus, each can give rise to a family of peaks in the XPS spectrum at binding energies below the expected transitions. For example, the Mg spectrum contains at least four higher-energy peaks, with about 10 to 60 times less intensity than the major Kα peak, but within 25 eV of the major excitation.
- **Ghost peaks.** Another element, rather than the expected source material, may inadvertently cause unwanted x-rays. This contaminating radiation may come from an impurity in the anode or from the material near the anode (such as the anode base or, even rarely, from within the sample itself). This radiation causes minor lines (called ghost lines) under unusual conditions. Ghost lines are rare, but do exist, and they can lead to misinterpretations.

- **Shake-up lines.** An ion can be left in an excited state, several electron volts above the ground state after the photoelectric process. In this condition, the kinetic energy of the emitted photoelectron is reduced. This results in the formation of a satellite peak at a position a few electron volts higher in binding energy than the major transition peak. These satellite lines can be a real problem for interpretation because their intensities can approach those of the major lines. Positive identification using the Auger line in the XPS spectrum can be useful in identifying the actual chemical state. Of course, if monochromatic x-rays are used (as available with most current instrumentation), the Auger transitions may be suppressed.
- **Valence lines.** Photoelectric emission from the valence bands can produce low-intensity lines in the XPS spectrum in the low binding-energy region. This is usually between the Fermi level and 10 to 15 eV binding energy.
- **Energy-loss lines.** The photoelectron can lose specific amounts of energy because of interactions with other electrons in the surface region of the sample. Energy-loss lines usually occur at binding energies higher than the major transition, and they are located at periodic intervals. The energy between the primary peak and the loss peak is called the plasmon energy. Plasmon or energy-loss lines are usually prominent in very conducting or very perfect samples.

Secondary Ion Mass Spectrometry. The surface sensitivity associated with this technique is gained by removing the top layers of the semiconductor sample and detecting the secondary ions (the positively and negatively charged species) that are ion-sputtered from the sample [151]. It is also the most sensitive of the techniques, and many current instruments can actually detect boron in Si at the level of 10^{12} to 10^{13} per cubic centimeter (ppb range). The problems of the SIMS technique are focused in two areas. The first is the generation of the ionic species involving the sputtering process. As already indicated, these sputtering mechanisms are very complex, destructive, and disruptive on the microscale. They provide a margin of interpretive error for the technique because the results are extremely dependent on the experimental conditions. The recent development of time-of-flight instrumentation has minimized many of these problems by using more-gentle ion sources and more control of the removal process (static SIMS) [152]. Very distinct relationships exist between the ion conditions, the removal rates, and the detection sensitivity, as presented in Fig. 6-43 [153]. These relationships must be kept in mind when requesting ultimate detection with highest spatial resolutions. The second area of concern is that of the detection systems used for analysis. Detection sensitivity to low-level species and mass separation is a function of the instrumentation. High-budget investments are required for highest sensitivities.

Of the surface-analysis techniques, SIMS is perhaps the most operator-dependent. Data acquisition is as important as data interpretation. Table 6-5 provides a summary of some important instrument- and operating-dependent parameters for SIMS.

Rutherford Backscattering Spectrometry (RBS) and Ion-Scattering Spectroscopy (ISS). The RBS technique has increased use for thin film and thin layer analysis. It uses 1 to 3 MeV ions (usually ^4He) to analyze the surface and outer 0.5 to 3.0 μm of semiconductors and other materials [154]. The results produce information on the compositions as well as qualitative and quantitative information about the distribution of atoms within the vicinity of the surface. The RBS technique is limited in its lateral microanalysis and spatial resolution; the analysis of interfaces and layers deeper in the material or device is difficult or impossible. Data interpretation requires extensive experience with the technique and the material system being analyzed.

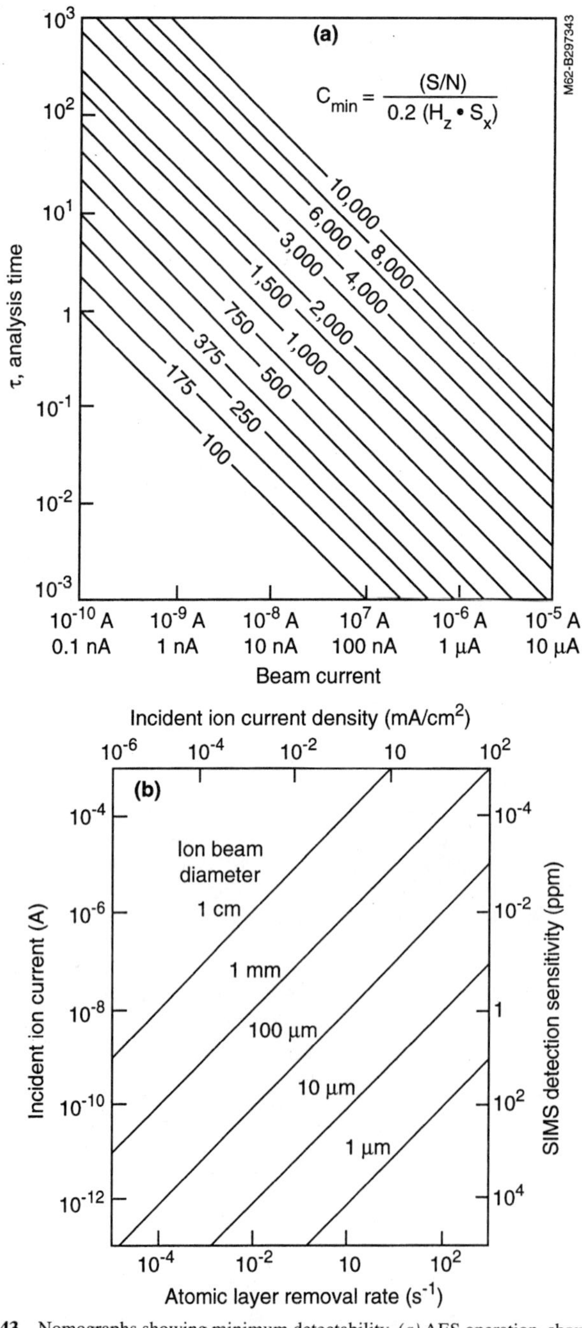

FIGURE 6-43 Nomographs showing minimum detectability. *(a)* AES operation, showing analysis time as a function of beam current for various normalized concentrations; and *(b)* SIMS process. The relationship among the incident ion current density, atomic layer removal rate, diameter of the incident ion beam, and the detection sensitivity is indicated.

TABLE 6-5 Technique, instrument, and operator strengths and concerns for SIMS

Strengths	Limitations
Direct detection of hydrogen	Destructive analysis
Isotope detection	Difficult quantification (especially complex materials systems)
High sensitivity to low level concentrations	Detection sensitivity factors cover wide range (~6 orders of magnitude)
Chemical information from molecular fragments	Mass interferences made identification difficult
Rapid collection of mass data	High mass resolution requires more expensive equipment
Very good depth resolution	Complex experimental apparatus; less user-friendly
Inherent depth profiling	Operator expertise-dependent
Rapid depth compositional data (dynamic)	Multi-parameter control/consideration
Excellent surface sensitivity (static)	
Analysis of fragile materials	
Charging problems minimal during rapid sputtering	
Good spatial resolution (liquid ion sources)	
Integratible with other surface techniques	
Wide range of commercial equipment with wide range of cost	

The ISS technique has similar analysis strengths and problems [155]. Because it is so sensitive to the surface, it is important to minimize contamination or other artifacts in that region. For example, a carbon layer that collects on the surface of a sample may dominate the detected spectrum. On the other hand, ISS can provide fundamental compositional information with extremely high detection sensitivity.

6.4.3 Thickness and Roughness Measurements

- Information: Film thickness and surface topography (roughness)
- Application: Single-crystal, polycrystalline, and amorphous thin films
- Approach: Mechanical and optical characterization of the sample surface

Thin film thickness and thin film roughness have a direct impact on the data acquisition and interpretation of many analyses. Topographical variations can control sputtering and profiling; electron-, ion-, and photon-beam analysis; and charging. The thickness of a film is important for many of the electro-optical measurements and modeling discussed here and in Chapter 4. Related to this is the determination of crater profiles from depth profiling in the surface-analysis techniques to determine not only sputtering rates, but also preferential or non-uniform etching. A wide variety of techniques have been used to evaluate these properties. The determination of surface roughness is provided by a host of mathematical routines that are incorporated into the general computer software of most of these instruments. Scanning electron microscopy is one common approach used to investigate film topography. There are also many mechanical and optical methods that are used for these purposes. The use of these methods depends on the application, including the thickness measurement required and the scale of roughness.

Mechanical Profilometers. These instruments use the movement of a diamond or other hard stylus over a sample surface [156–158]. The force of the stylus on the surface can be adjusted to optimize the load for the surface (material) of interest, with forces in the range of 1 to 50 mg. The signal is digitized and stored. The instrument usually has a microscope or camera for locating the exact region for analysis. The lateral resolution of the profilometer depends on the radius of curvature of the stylus. If the surface curvature exceeds that of the stylus, then the surface cannot be scanned accurately. Typically, the radius of curvature is 1 to 5 μm. By scanning the stylus over the edge of the film, the thickness can be recorded. Likewise, the mechanical profilometer can be used to determine the depth of craters formed by rastered ion beams. The profilometer offers a very rapid technique for evaluating roughness and depth, but with somewhat limited depth resolution. The best profilometers offer minimum step resolutions of about 25 to 40 Å, with an uncertainty of about 5 Å. Maximum steps can exceed 150 μm, although such steps might require special accessories for commercial instruments. Lateral resolutions, important for evaluating roughness and step shape, range from 0.05 to 25 μm, depending mainly on the radius of curvature of the stylus. A wide range of sample thicknesses can be evaluated, although most instruments are adjusted for the semiconductor wafer industry (thickness of 10 to 20 mm). Scanning ranges can be as large as 200 mm. Mechanical profilometers are commercially available and have become common instruments for thin film and surface analysis operations.

Optical Profilers. Interferometry has been a standard technique for determining surface features [159,160]. This method offers some advantages in that no contact is needed. Light is reflected from the surface and interferes with light from an optically flat reference surface. Deviations in the fringe pattern produced by the interferences can be related to the differences in the surface features (such as a step height or film thickness). Two-dimensional and three-dimensional data allow the generation of maps.

Optical interferometers have the same advantages as mechanical profilometers in that no sample preparation is needed for most films, and the analysis is rapid. Disadvantages include problems with surface roughness, which can destroy or interfere with the interference patterns because of excessive scattering. Optical reflectance is important, and sometimes it is necessary to coat the step with a highly reflective thin film—a process that can affect the accuracy of the measurement. Depth resolutions of 1 Å are attainable, with minimum step resolution in the range of 2.5 Å. The largest step or feature that can be resolved is about 15 μm. Lateral resolutions of 0.4 to 10 μm and maximum sample thicknesses of 125 mm are usual with commercial instruments. These instruments are becoming less common than mechanical systems. Software that is now available has lowered interpretive errors (that is, error associated with the operator). However, the optical systems are less convenient for many of the uses that thin film devices and surface-analysis researchers require, especially given that the cost for such systems is typically 10% to 20% higher.

Scanning Force Microscopes. The scanning force microscope (SFM) is a high-resolution analog of the mechanical profiler [161,162]. It is becoming very common in research laboratories for determining the roughness of samples. However, SFM offers depth resolution of 0.1 Å and lateral resolution near 1 Å, and it can provide imaging of wide ranges of samples with high accuracy and spatial resolution. It is very useful for investigating sputter craters: not only can it determine crater depths, but it can also be an excellent tool for investigating the features at the crater bottom [162]. The SFM is discussed in more detail in the next section.

6.5 NANO-SCALE AND ATOMIC-SCALE MEASUREMENTS

- Information: Electronic and spectroscopic imaging; nanoscale topographic and atomic imaging; nanoscale electro-optical characterization; atomic manipulation
- Application: Single-crystal, polycrystalline, and amorphous thin films
- Approach: *Imaging:* application of controlled electronic signals through special tip and measurement of response (e.g., tunneling current for STM, cantilever deflection for AFM); *nanocharacterization:* use of STM or AFM probe to confine signal to nano-area (electrical and optical spectroscopy)

Scanning probe microscopies and spectroscopies are useful analytical techniques that have been introduced during the past 15 years [163]. The scanning tunneling microscope (STM) and the atomic force microscope (AFM), also known as scanning force microscope (SFM) are the most recognizable of a number of sister instruments in this classification [164]. With depth resolutions of 0.01 nm and lateral resolutions of 0.1 nm, the instruments are capable of imaging to atomic dimensions. The probes can also be used for spectroscopic characterization (photoluminescence, EBIC, cathodo- and electro-luminescence, carrier identification, minority-carrier spectroscopy). Finally, these instruments have also been used to manipulate atomic arrangements at surfaces, including the movement of single atoms at semiconductor surfaces. These nanoscale and atomic-scale techniques have excellent representation in a number of recent reviews and books [164–169].

6.5.1 Scanning Tunneling Microscopy (STM)

The scanning tunneling microscope (STM) is a high-spatial-resolution instrument capable of real-space electronic and spectroscopic imaging of surfaces and interfaces on a scale that extends to atomic dimensions [166–170]. The instrument is represented schematically in Fig. 6-44. A specially prepared (sharpened) wire is brought into proximity (that is, 0.3 to 1.0 nm) to a surface using special piezoelectric positioners. These piezoelectric materials can also be used to drive the tip in the x and y lateral directions for mapping the signal over a selected area. A voltage is applied between the tip and the sample, causing a current to tunnel between the tip and the surface. When properly prepared, the electron transfer is from a single atom on the tip to the sample surface, providing excellent spatial resolution. The measured tunneling current provides information about the electronic states in the tip and the surface, and these data can be analyzed to provide images of the topography, as well as the electronic and chemical structure. The STM can be operated in either the constant-current mode or the constant-height mode (for more rapid and higher spatial-resolution operation).

For STM, the sample must be conducting to allow the collection of the tunneling current; insulators and higher-resistivity semiconductor thin films cannot be analyzed using this technique. Surface roughness also provides some limitations, because the instrumentation cannot respond to large changes in the topography. A number of imaging modes can be used to provide various information. Although the STM can be operated in air, highest resolutions are only attained in ultrahigh vacuum (UHV) [172].

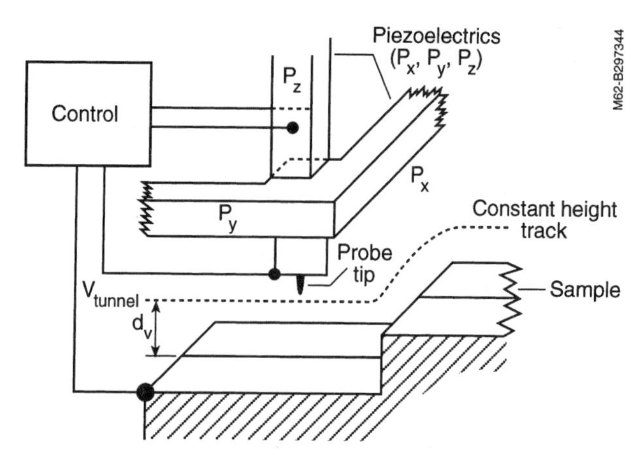

FIGURE 6-44 Schematic representation of STM instrument.

6.5.2 Atomic Force Microscopy (AFM)

The atomic force microscope (AFM) provides nanoscale to atomic-scale microscopy by measuring the tip-surface interactions from forces and translating these to a sensor to provide information on the topographic features of the surface [173,174]. The forces involved can include van der Waals, electrostatic, frictional, and magnetic, and instruments working on each of these force mechanisms are available (such as magnetic force microscopes [MFM] and electrostatic force microscopes [EFM]). Typically, forces measured are in the range 10^{-13} to 10^{-6} N. The AFM probe is typically a tip attached to a cantilever spring, most commonly formed using selective etching of single-crystal Si.

As the sample is scanned under the tip, the tip-cantilever assembly deflects in response to forces exerted by the sample. The cantilever assembly and typical schematic are shown in Fig. 6-45 [175]. A displacement sensor then measures those deflections, usually from the back of the cantilever. Displacement sensors include electron tunneling devices, optical interferometry, optical-beam deflection, and capacitance, resistance, and piezoelectric sensors.

The SFM can be operated in air, liquid, and vacuum. Strengths include the ability to analyze nonconductive samples (polymers, biological samples, insulators), operate in air or liquid environments, attain atomic resolution; image static and dynamic features; ease of operation; and the availability of multiple scanning method on a single instrument. The limitations of SFM include artifacts in imaging resulting from forces; difficulties in operation in ultrahigh vacuum; difficulty in providing atomic resolutions; difficulty in preparing cantilevers; and difficulty in interpreting images.

6.5.3 Ballistic Electron-Emission Microscopy (BEEM)

The BEEM provides nanoscale microscopy, allowing the characterization of interface properties with nanometer spatial resolution and the energy spectroscopy of carrier transport [176–179]. It uses an STM tip as an injector of ballistic electrons into a sample heterostructure. In general, the sample consists of at least two layers separated by an interface of interest. Ballistic electron-emission microscopy operates as a multielectrode system, with electrical contact to each layer of the sample structure. As a tip-sample bias voltage is applied, electrons tunnel across the vacuum gap and enter the sample as hot

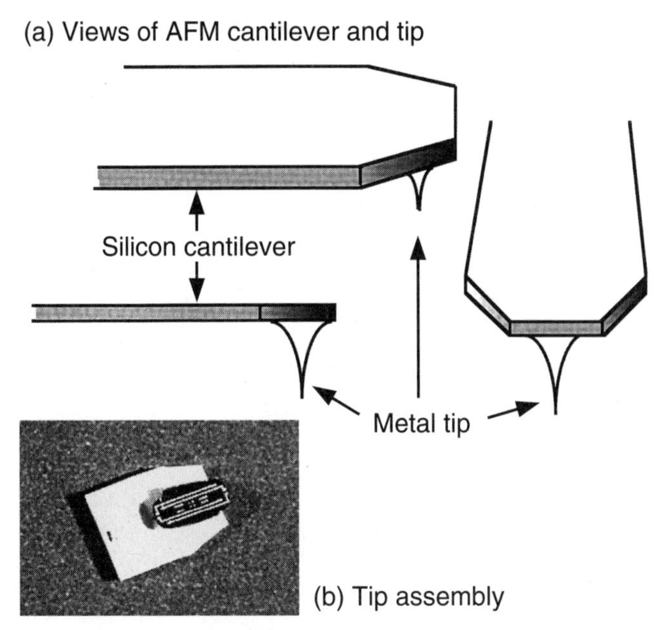

FIGURE 6-45 Schematic representation of cantelever for scanning force instrument.

carriers. Because characteristic attenuation lengths are typically hundreds of angstroms, many of these hot electrons may propagate through the base layer and reach the interface before scattering. Ballistic electron-emission microscopy provides information on subsurface regions of semiconductors and has been used to evaluate Schottky barriers and other junctions [180].

If conservation laws restricting total energy and momentum parallel to the interface are satisfied, these electrons may cross the interface and be measured as current in the collector layer. In an n-type semiconductor, the band bending accelerates the collected carriers away from the interface and prevents their leakage back into the base. By varying the voltage between tip and base, the energy distribution of the hot electrons may be controlled, and a spectroscopy of interface carrier transport may be performed.

For tunnel voltages less than the interface barrier height, none of the injected electrons have total energy equal to or greater than the barrier height, and the measured collector current is zero. As the voltage is increased to values in excess of the barrier (i.e., $eV > E_f + q\phi_b$), some of the hot electrons cross the interface into the semiconductor conduction band, and a collector current is observed. This location of the threshold in the spectrum defines the interface barrier height. The magnitude of the current above threshold and the threshold spectrum shape also yield important information on interface transport. Information on conduction-band structure and valence band structure at the interface is readily attainable [181–183].

6.5.4 Nanoscale Electrical and Optical Characterization

The ability to laterally confine an electron field provided by the STM (and AFM) opens possibilities to perform electro-optical characterization with nanometer spatial resolutions [184–187]. This led to the development and investigation of a host of measurement techniques having spatial resolutions in the range of 10 to 1000 nm and spanning the range from compositional determinations through electro-optical characterizations [184–192]. Based on macroscale electron-beam techniques, the STM is used to provide nanoscale EBIC (NEBIC), electroluminescence (NEL), and cathodoluminescence (NCL). Nanoscale EBIC, illustrated schematically in Fig. 6-46, is an analogue of the conventional (bulk) methodology [179]. The transverse electron beam-induced current is measured through the sample when the STM probe is scanned across it. The energy of the incoming electron beam is, of course, much lower than that used for conventional EBIC (tens of volts compared to thousands of volts). The generation volume (which dictates the spatial resolution for the information gathered) is not calculated through the normal formulations because of the very low energy of the incoming electron beam. Nanoscale EBIC has been applied to evaluate junctions and defects in various semiconductors. An NEBIC evaluation of a grain boundary in $CuInSe_2$ is presented in Fig. 6-47. The effect of oxygen incorporation is indicated with the change in the NEBIC response (lowering of the carrier collection in the region), and has been interpreted as enhanced *p*-type doping of the region. Nanoscale CL (NCL) has been used to identify the levels in these oxygen-processed grain boundaries. Figure 6-48 presents NCL data [182], showing the evolution of the Se vacancy and oxygen at Se site signals as a function of grain-boundary treatment.

The development of the near-field optical microscope (NSOM) has provided for microscopies and spectroscopies with spatial resolutions beyond those predicted by classical wavelength limitations [193–196]. This near-field evanescent field effect has been incorporated with the enhanced spatial resolution of the STM or AFM. Such configuration is

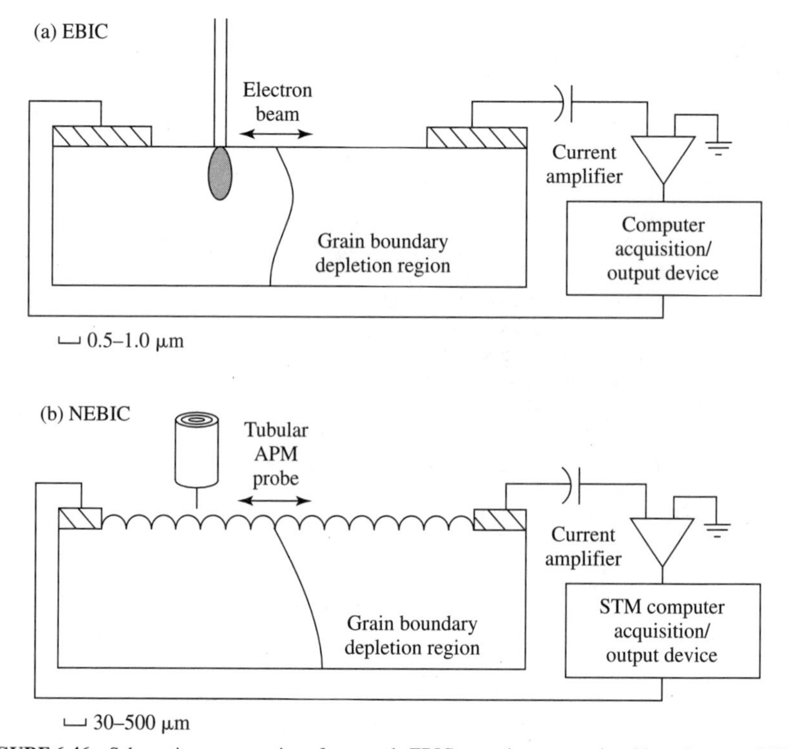

FIGURE 6-46 Schematic representation of nanoscale EBIC operation, comparing this to the normal EBIC configuration.

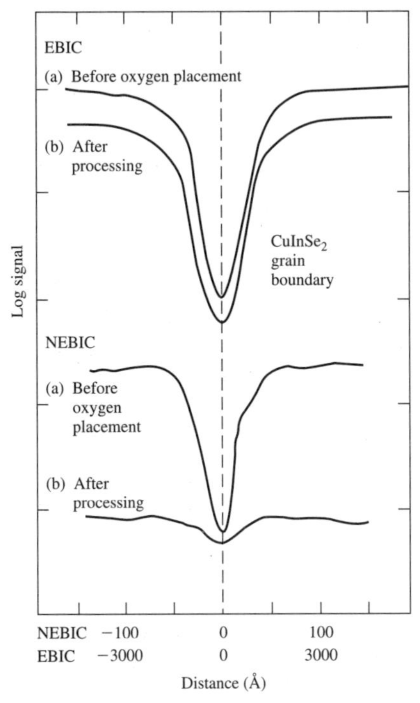

FIGURE 6-47 Nanoscale EBIC scans across grain boundary in $CuInSe_2$, comparing resolution to macroscale EBIC performed on same area.

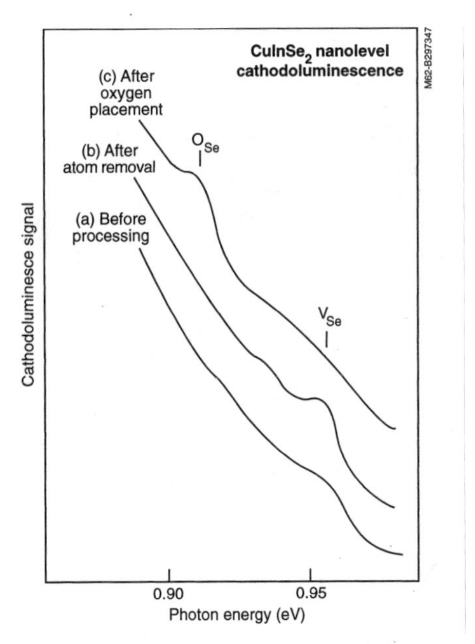

FIGURE 6-48 Nanoscale cathodoluminescence (NCL) measurement of same grain boundary represented in Fig. 6-47.

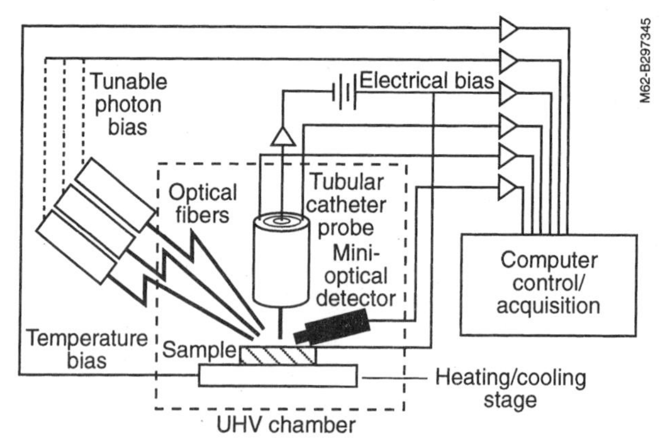

FIGURE 6-49 Schematic representation of STM using photon excitation for atom imaging and atom manipulation.

presented in Fig. 6-49, in which the STM/AFM uses photon energy sources for optical input into the thin film surface. The pulsed-laser signal is introduced using specially prepared optical fibers, analogous to those used in an NSOM. The tapered and metal-clad optical fibers introduce the pulsed and tunable laser signal onto the sample surface, configured in the near-field operational mode. Nanoscale PL (NPL) can be performed with spatial resolutions around 100-nm. The effect of this operation is shown in Fig. 6-50*a,b,* which shows the detected signal (again, via an optical fiber) from the far-field operation

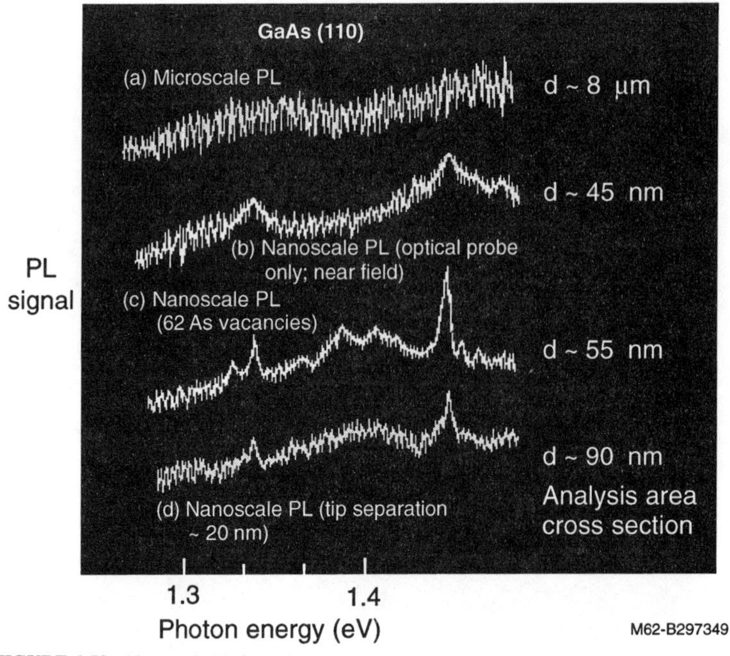

FIGURE 6-50 Nanoscale PL from GaAs surface, using the configuration of Fig. 6-49.

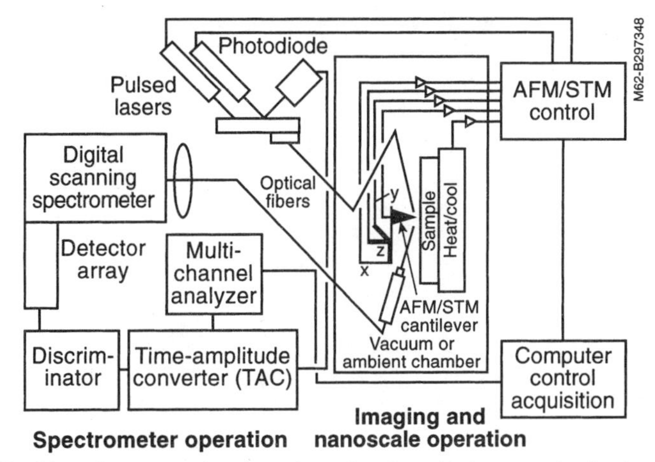

FIGURE 6-51 Schematic representation of experimental configuration for nanoscale minority-carrier lifetime spectroscopy using single photon-counting method.

and the near-field case for the thin-GaAs sample. The evolution of several peaks are apparent. The spatial resolution can be enhanced further when the STM tip is used to provide an electric field, which interacts with the optical signal to enhance the NPL output (Fig. 6-50*c*). The effect of this electric field is further shown in Fig. 6-50*d*, in which the STM tip is drawn slightly away from the surface. This nanoscale characterization technique is used to identify electronic defect levels (As and Ga vacancies) for the first time in $In_xGa_{1-x}As$ thin layers. The power has been the ability to image the defects and characterize them with the same instrument.

The experimental apparatus of Fig. 6-51 can also be used to evaluate the minority-carrier properties [197]. Using the single-photon PL decay method described earlier, nanoscale minority-carrier-lifetime spectroscopy can be realized. Figure 6-52 provides

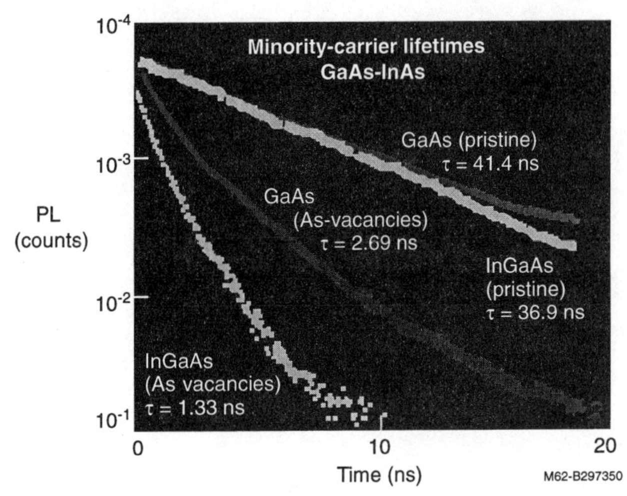

FIGURE 6-52 Minority carrier lifetime data for GaAs and InGaAs samples, showing the decay in the signal and the extraction of the lifetime.

such data for GaAs and InGaAs samples, showing the decay in the signal and the extraction of the lifetime. These data are used to show the effects of defect generation in the surface region on the surface lifetime of these carriers. The technique also allows for mapping of these parameters (PL response and minority-carrier lifetime) over nano-regions. Such data are presented in Fig. 6-53 for two regions of a GaAs film grown on InP, with purposely generated As defects in two distinct regions. The electronic levels associated with the As shallow levels from the NPL and the minority-carrier lifetime from the nanoscale minority-carrier spectrometer are clearly delineated.

The area of nanoscale characterization, using these evolving proximal probe techniques, is one of the more important and exciting areas of research and development.

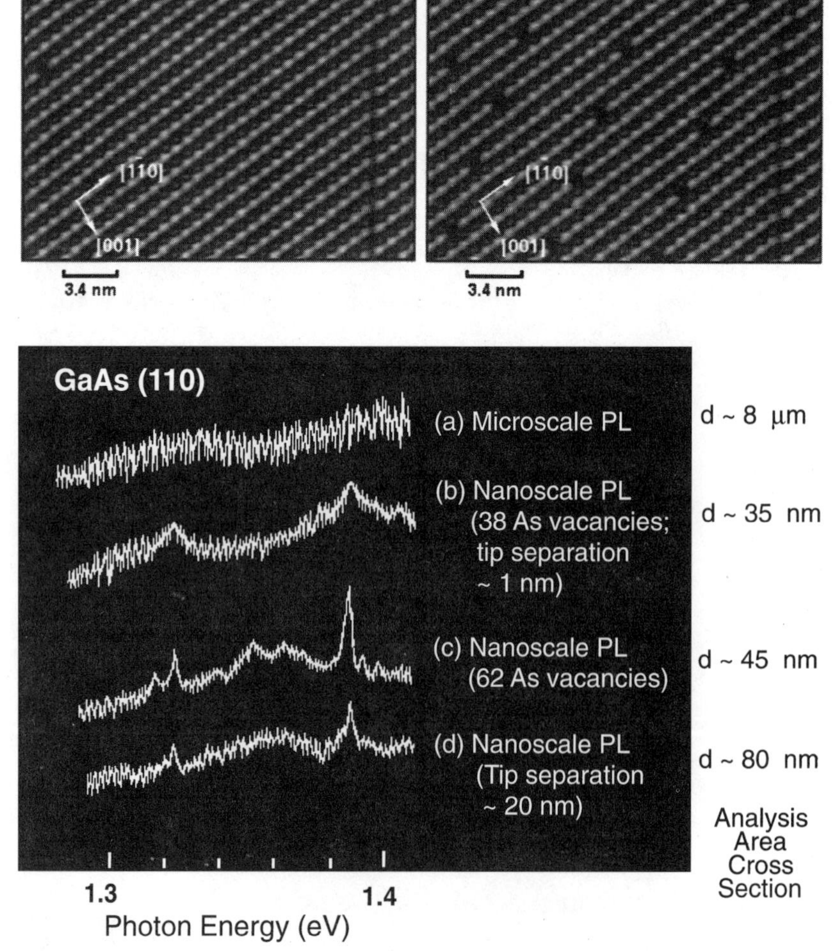

FIGURE 6-53 STM images of GaAs surface, showing pristine *(a)* and As-vacancy engineered regions using the proximal probe.

It is an area that will lead and guide researchers to develop new devices, evolve atomically engineered materials, and optimize interfaces. The range of these techniques has been expanding, and now covers from composition through electro-optical characterization.

6.6 FUTURE CHARACTERIZATION DIRECTIONS AND NEEDS

Almost every characterization technique that has been applied to bulk semiconductor analysis has had either direct or modified use for thin-film investigations. Thin films have special measurement constraints because of the larger influence of surfaces, the internal defects and chemistry, and the details and controlling influence of the microstructure. As the requirements for thin-film devices have evolved, so have the many characterization methods that serve the development of these technologies. The demand is for higher spatial resolution, higher accuracy and precision, better reproducibility, automation and in-situ analysis, ease of operation, and determination of properties that are fundamental to the optimization of semiconductor materials and devices. Measurements and characterization will continue to respond to the demands of thin film research and industry. The contributions have been substantial, and these contributions should be expected to continue in direct response to the evolution of electronics and materials science.

REFERENCES

1. C. R. Brundle, C. A. Evans, Jr., and S. Wilson, (eds.), *Encyclopedia of Materials Characterization,* Butterworth, Heinemann, Boston, 1992.
2. A. Ulman, (ed.), *Characterization of Organic Thin Films,* Butterworth, Heinemann, Boston, 1992.
3. G. W. Ewing, (ed.), *Analytical Instrumentation Handbook,* Marcel Dekker, New York, 1990.
4. B. G. Yacobi, D. B. Holt, and L. L. Kazmerski, *Microanalysis of Solids,* Plenum Press, New York, 1994.
5. J. T. Waber, (ed.), *Characterization and Behavior of Materials with Submicron Dimensions,* World Scientific, Singapore, 1985.
6. L. C. Feldman and J. W. Mayer, *Fundamentals of Surface and Thin Film Analysis,* North-Holland, New York, 1986.
7. J. J. Hren, J. I. Goldstein, and D. C. Joy, *Introduction to Analytical Electron Microscopy,* Plenum Press, London, 1979.
8. E. Lifshin, (ed.), *Characterization of Materials,* VCH Pub., New York, 1992.
9. K. N. Tu and R. Rosenberg, (eds.), *Analytical Techniques for Thin Films,* Academic Press, New York, 1988.
10. J. Sibilia, *A Guide to Materials Characterization and Chemical Analysis,* VCH Pub., New York, 1988.
11. P. J. Grundy and G. A. Jones, *Electron Microscopy in the Study of Materials,* Edward Arnold, London, 1976.
12. L. I. Maissel and R. Glang, *Handbook of Thin Film Technology,* McGraw-Hill, New York, 1970.
13. D. K. Schroeder, *Semiconductor Material and Device Characterization,* John Wiley & Sons, New York, 1990.

14. K. L. Chopra, *Thin Film Phenomena,* McGraw-Hill, New York, 1969.
15. R. N. Hall, *Solid State Electron,* vol. 3, 1961, p. 320.
16. E. H. Sondheimer, *Phys Rev,* vol. 80, 1950, p. 401.
17. E. H. Putley, *The Hall Effect and Related Phenomena,* Butterworth, London, 1960.
18. L. J. van der Pauw, *Philips Research Reports,* vol. 13, 1958, p. 1.
19. See, for example, S. T. Thornton and A. Rex, *Modern Physics,* W. B. Saunders, New York, 1993, pp. 79–81.
20. J. Jerhot and V. Snejdar, *Thin Solid Films,* vol. 52, 1978, p. 379.
21. M. V. Garcia-Cueca, J. L. Morenza, and J. M. Codina, *J Appl Phys,* vol. 58, 1985, p. 1080.
22. C. H. Seager and T. G. Castner, *J Appl Phys,* vol. 49, 1978, p. 3879.
23. H. P. Strunk, J. H. Werner, B. Fortin, and O. Bonnaud, *Polycrystalline Semiconductors III,* Scitec, Switzerland, 1994.
24. C. Herring, *J Appl Phys,* vol. 11, 1960, p. 1939.
25. J. Y. W. Seto, *J Appl Phys,* vol. 46, 1975, p. 5247.
26. D. B. Holt and R. Ogden, *Solid-State Electron,* vol. 19, 1976, p. 37.
27. K. Seiler, *J Appl Phys,* vol. 54, 1983, p. R1.
28. H. J. Leamy, *J Appl Phys,* vol. 53, 1982, p. R51.
29. L. L. Kazmerski, in *Advances in Solar Energy,* vol. 3, K. W. Böer, (ed.), Plenum, New York, 1986, pp. 60–67.
30. O. von Roos, *Appl Phys Lett,* vol. 35, 1978, p. 408.
31. L. L. Kazmerski, *Vacuum,* vol. 43, 1992, p. 1011.
32. R. Cohn and Caledonia, 1970.
33. R. H. Geis, in Ref. 1, pp. 129–130.
34. T. E. Everhardt and P. H. Hoff, *J Appl Phys,* vol. 42, 1971, p. 5837.
35. C. Donolato, in *Polycrystalline Semiconductors,* Springer-Verlag, Berlin, 1985.
36. D. Sawyer and H. K. Kessler, *IEEE Trans Electron Dev,* vol. ED-27, 1980, p. 864; also, E. L. Miller, A. Shumka, S. S. Chern, *Proc 15th IEEE Photovoltaic Spec Conf,* IEEE, New York, 1981, pp. 1126–1133.
37. I. L. Eisgruber, R. J. Matson, J. R. Sites, and K. A. Emery, *Proc IEEE Photovoltaic Spec Conf and World Conference on Photovoltaic Energy Conversion,* IEEE, New York, 1995, pp. 283–286; also, H. Takakura, K. Fujimoto, K. Okuda, C. Coluzza, and Y. Hamakawa, *Jpn J App Phys,* vol. 5J, 1983, p. 569.
38. J. I. Goldstein, in Ref. 94, pp. 83–120.
39. R. J. Matson, K. A. Emery, I. L. Eisgruber, and L. L. Kazmerski, *Proc 12th European Photovoltaic Solar Energy Conf,* Kluwer Scientific, The Netherlands, 1994, pp. 1222–1225.
40. See, for example, *ASTM Annual Book of Standards,* vol. 10.05, ASTM, New York, 1985, pp. 39–55; also, ASTM Publ. F28-75, 1975.
41. R. L. Mattis and A. J. Baroody, Jr., *NBS Tech Note,* NBS, Gaithersburg, MD, 1972, pp. 736–42.
42. J. M. Borrego, R. J. Gutmann, N. Jensen, and O. Paz, *Solid State Electron,* vol. 30, 1987, p. 195.
43. M. C. Chen, *J Appl Phys,* vol. 64, 1988, p. 945.
44. G. Beck and M. Kunst, *Rev Sci Instrum,* vol. 57, 1986, p. 197.
45. R. K. Ahrenkiel, in *Semiconductors and Semimetals,* vol. 39, R. K. Ahrenkiel and M. S. Lundstrom, (eds.), Academic Press, New York, 1993, pp. 39–150.
46. D. K. Schroder, in *Semiconductor Material and Device Characterization,* John Wiley & Sons, New York, 1990, pp. 359–447.

47. R. K. Ahrenkiel, in *Current Topics in Photovoltaics,* T. J. Coutts and J. D. Meakin, (eds.), Academic Press, London, 1988.

48. R. K. Ahrenkiel, B. M. Keyes, and D. J. Dunlavy, *J Appl Phys,* vol. 70, 1991, p. 225; also, R. K. Ahrenkiel, B. M. Keyes, D. J. Dunlavy, and L. L. Kazmerski, *Proc 10th EC Photovoltaic Energy Conf,* Kluwer Academic Publ, the Netherlands, 1992, pp. 533–536.

49. J. W. Orton and P. Blood, *The Electrical Characterization of Semiconductors: Measurement of Minority-Carrier Properties,* Academic Press, London, 1990.

50. R. Z. Bachrach, *Rev Sci Instrum,* vol. 43, 1972, p. 734; also, J. N. Demas, in *Excited State Lifetime Measurements,* Academic Press, New York, 1983.

51. R. K. Ahrenkiel, *Solar Cells,* vol. 16, 1986, p. 549.

52. E. Yablonovitch, *Solid St Electron,* vol. 35, 1992, p. 261.

53. R. K. Ahrenkiel (in press).

54. Miller, Lang, and Kimerling, *Annual Rev Materials Science,* Annual Rev., Inc., New York, 1977, pp. 377–448.

55. R. K. Ahrenkiel, *Solar Cells,* vol. 16, 1986, p. 549.

56. S. D. Brotherton and P. Bradley, *J Appl Phys,* vol. 53, 1982, p. 1543.

57. R. Brunwin, B. Hamilton, P. Jordan, and A. R. Peaker, *Electron Lett,* vol. 15, 1979, p. 349.

58. H. J. Leamy, G. E. Pike, and C. H. Seager, *Grain Boundaries in Semiconductors,* vol. 5, Mat. Res. Soc., Pittsburgh, 1982.

59. H. Moller, *Solar Cells,* vol. 31, 1991, p. 77.

60. J. Narayan and T. Y. Tan, *Defects in Semiconductors,* vol. 2, Mat. Res. Soc., Pittsburgh, 1981.

61. J. H. Werner and H. P. Strunk, (ed.), *Polycrystalline Semiconductors II,* Springer-Verlag, Berlin, 1991.

62. H. C. de Graaff, M. Huybers, and J. G. de Groot, *Solid-State Electron,* vol. 25, 1982, p. 67.

63. M. Peisl and A. W. Wieder, *IEEE Trans Electron Dev,* ED-30, 1983, p. 1792.

64. S. Hirae, M. Hirose, and Osaka, *J Appl Phys,* vol. 51, 1980, 1043.

65. C. H. Seager and G. E. Pike, *Appl Phys Lett,* vol. 35, 1979, p. 709.

66. J. Werner, W. Jantsch, and H. J. Queisser, *Solid-State Commun,* vol. 42, 1983, p. 415.

67. J. Werner, J. Jantsch, K. J. Frohner, and H. J. Queisser, *Mat Res Soc Proc,* vol. 5, 1982, p. 99.

68. E. H. Nicollian and J. R. Brews, *MOS Physics and Technology,* Wiley Interscience, New York, 1982.

69. D. E. Newbury, D. C. Joy, P. Echlin, C. E. Fiori, and J. I. Goldstein, *Advanced Scanning Electron Microscopy and X-Ray Microanalysis,* Plenum, New York, 1986.

70. D. B. Holt and D. C. Joy, (eds.), *SEM Microcharacterization of Semiconductors,* Academic Press, New York, 1989.

71. O. Johari, (ed.), *Scanning Electron Microscopy,* Series 1975–1996, SEM Inc., O'Hare, IL.

72. G. W. Lorimer, M. H. Jacobs, and P. Doig, (eds.), *Quantitative Microanalysis with High Spatial Resolution,* The Metals Soc., London, 1981.

73. P. R. Buseck, M. M. Cowley, and L. Eyring, (eds.), *High Resolution Transmission Electron Microscopy and Associated Techniques,* Oxford University Press, New York, 1988.

74. J. J. Hren, J. I. Goldstein, and D. C. Joy, (eds.), *Introduction to Analytical Electron Microscopy,* Plenum Press, New York, 1979.

75. D. B. Williams, *Practical Analytical Electron Microscopy in Materials Science,* Philips Electronic Instruments, Mahway, NJ, 1984.

76. C. E. Lyman, H. G. Stenger, and J. R. Michael, *Ultramicroscopy,* vol. 22, 1987, p. 129.

77. V. P. Dravid et al., *Phil. Mag.,* vol. A 61, 1990, p. 417.

78. G. Cliff and G. W. Lorimer, *J. Microscopy,* vol. 103, 1975, p. 203.
79. J. Mansfield, *Convergent Beam Electron Diffraction of Alloy Phases,* Hilger, Bristol, UK, 1984.
80. A. V. Crewe, J. Wall, and J. Langmore, *Science,* vol. 168, 1970, p. 1338.
81. W. Katz and P. Williams, (eds.), *Applied Materials Characterization,* vol. 48, Mat. Research Soc., Pittsburgh, 1986.
82. D. B. Williams, *Practical Analytical Electron Microscopy in Materials Science,* Philips Electronic Instrum., New Jersey, 1984.
83. N. M. Johnson, S. G. Bishop, and G. D. Watkins, *Microscopic Identification of Electronic Defects in Semiconductors,* vol. 46, Mat. Res. Soc., Pittsburgh, 1985.
84. J. C. Bravman, R. M. Anderson, and M. L. McDonald, (eds.), *Specimen Preparation for Transmission Electron Microscopy of Materials, MRS Proc.,* vol. 115, Mat Res Soc, Pittsburgh, 1988.
85. M. L. Anderson, (ed.), *Specimen Preparation for Transmission Electron Microscopy II, MRS Proc.,* vol. 199, Mat. Res. Soc., Pittsburgh, 1990.
86. K. C. Thompson-Russell and J. W. Edington, *Electron Microscopy Specimen Preparation Techniques in Materials Science,* Monograph No. 5, Philips Technical Library, Eindhoven and Delaware, The Netherlands, 1977.
87. L. Reimer, *Scanning Electron Microscopy,* Springer-Verlag, Berlin, 1985.
88. D. B. Holt and D. C. Joy, *SEM Microcharacterization of Semiconductors,* Academic Press, London, 1989.
89. J. C. Russ, *Computer Assisted Microscopy,* Plenum Press, New York, 1990.
90. W. Krakow, D. Smith, and L. W. Hobbs, *Electron Microscopy of Materials,* vol. 31, Mat. Res. Soc., Pittsburgh, 1984.
91. J. I. Goldstein, D. E. Newbury, P. Echlin, D. C. Joy, C. Fiori, and E. Lifshin, *Scanning Microscopy and X-Ray Microanalysis,* Plenum Press, New York, 1981.
92. S. J. B. Reed, *Electron Microprobe Analysis,* Cambridge University Press, London, 1975.
93. D. C. Joy, A. D. Romig, and J. I. Goldstein, (eds.), *Principles of Analytical Electron Microscopy,* Plenum Press, New York, 1984.
94. J. Goldstein, D. E. Newbury, P. Echlin, D. Joy, C. Fiori, and E. Lifshin, (eds.), *Scanning Electron Microscopy and X-ray Microanalysis,* Plenum Press, New York, 1981, pp. 205–273.
95. J. D. Geller, *Scanning Electron Microscopy/1977,* O. Johari, (ed.), SEM, Inc., O'Hare, IL, 1977, pp. 281–288.
96. J. J. Hren, J. I. Goldstein, and D. C. Joy, (eds.), *Introduction to Analytical Electron Microscopy,* Plenum Press, New York, 1979.
97. See, for example, L. L. Kazmerski, in K. Boer, (ed.), *Advances in Solar Energy,* Plenum Press, New York, 1986, pp. 1–123.
98. J. Goldstein, in Ref. 94, pp. 95–118 and 33–338; also, Ref. 28, p. R51.
99. P. E. Russell and C. R. Herrington, *Scanning Electron Microscopy/1982,* O. Johari, (ed.), SEM, Inc., O'Hare, IL, 1982, pp. 1077–1082.
100. S. J. B. Reed, in Ref. 92, pp. 132–152.
101. L. L. Kazmerski, in *Advances in Solar Energy,* vol. 3, K. W. Boer, (ed.), Plenum Press, New York, 1986, pp. 5–8.
102. V. D. Scott and G. Love, (eds.), *Quantitative Electron-Probe Microanalysis,* John Wiley & Sons, New York, 1983.
103. G. W. Lorimer, M. H. Jacobs, and P. Doig, (eds.), *Quantitative Microanalysis with High Spatial Resolution,* Plenum Press, New York, 1979.
104. A. W. Czanderna, (ed.), *Methods of Surface Analysis,* Elsevier, New York, 1975.
105. G. E. McGuire, *Auger Electron Spectroscopy Reference Manual,* Plenum Press, New York, 1979.

106. G. E. Muilenburg, (ed.), *Handbook of x-ray Photoelectron Spectroscopy,* Perkin-Elmer, Eden Prairie, MN, 1979.
107. C. R. Brundle and A. J. Baker, (eds.), *Electron Spectroscopy: Theory, Techniques and Applications,* vol. I-X, Academic Press, New York, 1979–present.
108. J. W. Robinson, *Handbook of Spectroscopy,* CRC Press, Boca Raton, FL, 1981.
109. T. Sekine et al., *Handbook of Auger Electron Spectroscopy,* JEOL Ltd., Tokyo, 1982.
110. D. Briggs and M. P. Seah, (eds.), *Practical Surface Analysis By Auger and x-ray Photoelectron Spectroscopy,* Wiley, Chichester, 1983.
111. L. C. Feldman and J. W. Mayer, *Fundamentals of Surface and Thin Film Analysis,* North Holland, New York, 1986.
112. M. Thompson, M. D. Baker, A. Christie, and J. F. Tyson, *Auger Electron Spectroscopy,* Wiley-Interscience, New York, 1986.
113. A. Beninghoven, F. G. Rüdenauer, and H. W. Werner, *Secondary Ion Mass Spectrometry,* John Wiley & Sons, New York, 1987.
114. K. N. Tu and R. Rosenberg, (eds.), *Analytical Techniques for Thin Films,* Academic Press, New York, 1988.
115. J. M. Walls, *Methods of Surface Analysis,* Cambridge University Press, London, 1990.
116. A. W. Czanderna, (ed.), *Crucial Role of Surface Analysis in Studying Surfaces,* Florida Atlantic University, Boca Raton, FL, 1984.
117. J. C. Riviere, *Surface Analysis Techniques,* Oxford University Press, London, 1990.
118. J. F. Watts, *An Introduction to Surface Analysis by Electron Spectroscopy,* Oxford University Press, London, 1990.
119. J. F. Ferguson, *Auger Microprobe Analysis,* Cambridge University Press, London, 1989.
120. A. W. Czanderna and D. M. Hercules, (eds.), *Ion Spectroscopies for Surface Analysis,* Plenum Press, New York, 1991.
121. D. Briggs, A. Brown, and J. C. Vickerman, *Handbook of Static Secondary Ion Mass Spectrometry,* John Wiley & Sons, New York, 1989.
122. R. G. Wilson, F. A. Stevie, and C. W. Magee, *Secondary Ion Mass Spectrometry,* John Wiley & Sons, New York, 1989.
123. F. Fiermans, J. Vennick, and W. Kedeyser, (eds.), *Electron and Ion Spectroscopy of Solids,* Plenum Press, New York, 1978.
124. W. M. Bullis, D. G. Seiler, and A. C. Diebold, (eds.), *Semiconductor Characterization: Present Status and Future Needs,* AIP, New York, 1996.
125. J. K. G. Panitz, D. Sharp, and C. R. Hills, *J Vac Sci Technol A,* vol. 3, 1985, p. 1; also, G. K. Wehner, in Ref. 104; Also, R. A. Kant, S. M. Meyers, and S. T. Picraux, *J Appl Phys,* vol. 50, 1979, p. 214.
126. J. A. McHugh, in Ref. 80, pp. 223–278.
127. J. M. Morabito and R. K. Lewis, in Ref. 80, pp. 159–222.
128. L. L. Kazmerski et al., *J Vac Sci Technol A,* vol. 5, 1987, p. 2814.
129. F. A. Stevie, P. M. Kohora, D. S. Simons, and P. Chi, *J Vac Sci Technol A,* vol. 6, 1988, p. 76.
130. R. M. Bradley and J. M. E. Harper, *J Vac Sci Technol A,* vol. 6, 1988, p. 2390.
131. R. G. Wilson, R. A. Stevie, and C. W. Magee, *Secondary Ion Mass Spectrometry: A Practical Handbook for Depth Profiling and Bulk Impurity Analysis,* John Wiley & Sons, New York, 1989.
132. J. R. Bird and J. S. Williams, (eds.), *Ion Beams for Materials Analysis,* Academic Press, New York, 1988.
133. O. Auciello and R. Kelly, (eds.), *Ion Bombardment Modification of Surfaces,* Elsevier, Amsterdam, 1984.
134. A. Zalar, *Thin Solid Films,* vol. 124, 1983, p. 223.

135. F. A. Stevie, P. M. Kahora, D. S. Simons, and P. Chi, *J Vac Sci Technol A,* vol. 6, 1988, p. 76.
136. M. P. Seah, in *Methods of Surface Analysis,* J. M. Walls, (ed.), Cambridge University Press, Cambridge, 1989.
137. H. Ibach and D. L. Mills, *Electron Energy Loss Spectroscopy and Surface Vibrations,* Academic Press, New York, 1982; also, W. H. Weinberg, *Methods of Experimental Physics,* vol. 22, 1985, p. 23.
138. J. F. Gempere, D. Delafosse, and J. P. Contour, *Chem Phys Lett,* vol. 33, 1975, p. 95.
139. D. K. Biegelsen, G. Rozgonhyi, and C. Shank, *Energy Beam-Solid Interactions and Transient Thermal Processing,* vol. 35, Mat. Res. Soc., Pittsburgh, 1985; also, J. C. C. Fan, vol. 23, 1984.
140. C. R. Brundle, in Ref. 1, p. 322.
141. H. H. Madden and G. Ertl, *Surf Sci,* vol. 5, 1973, p. 211.
142. W. Reuter and K. Wittmaack, *Appl Surf Sci,* vol. 5, 1971, p. 525.
143. J. P. Coad, H. E. Bishop, and J. C. Riviere, *Surf Sci,* vol. 21, 1970, p. 253.
144. R. Noufi, R. J. Matson, R. C. Powell, and C. R. Herrington, *Sol Cells,* vol. 16, 1986, p. 279; also, T. P. Massopust, P. J. Ireland, L. L. Kazmerski, and K. J. Bachmann, *J Vac Sci Technol A,* vol. 2, 1984, p. 1123.
145. D. Finello and H. L. Marcus, in D. Briggs, (ed.), *Handbook of x-ray and Photoelectron Spectroscopy,* Heyden and Son, London, 1977, pp. 138–139.
146. L. L. Kazmerski, P. J. Ireland, and P. Sheldon, *J Vac Sci Technol,* vol. 17, 1978, p. 1353.
147. P. Swift, D. Shuttleworth, and M. P. Seah, in Ref. 110, pp. 437–444.
148. K. Siegbahn, *ESCA: Atomic, Molecular, and Solid State Structure Studied by Means of Electron Spectroscopy,* vol. 20, series IV, Nova Acta Regime Soc. Scien., Upsala, 1967; Also, *ESCA Applied to Free Molecules,* North Holland.
149. E.-E. Koch, *Handbook on Synchrotron Radiation,* vol. 1b, North Holland, New York, 1983.
150. C. D. Wagner et al., *Handbook of X-Ray Photoelectron Spectroscopy,* Perkin-Elmer, Physical Electronics, Eden Prairie, MN, 1979; also, D. Briggs, Ref. 100, pp. 381–392.
151. A. Benninghoven, *Ann Phys,* vol. 15, 1958, p. 549; *Z Phys,* vol. 230, 1970, p. 402; *Surf Sci,* vol. 28, 1971, p. 541; also, A. Benninghoven, and E. Loebach, *J Rad Chem,* vol. 12, 1972, p. 95.
152. D. Briggs, *Polymer,* vol. 25, 1984, p. 1379; also, J. R. Tesmer and M. Nastasi, (eds.), *Handbook of Modern Ion Beam Materials Analysis,* MRS, Pittsburgh, 1996.
153. G. K. Wehner, Ref. 104, pp. 5–38; also, J. A. McHugh, Ref. 104, pp. 223–278.
154. W. K. Chu, J. W. Mayer, and M. A. Nicolet, *Backscattering Spectrometry,* Academic Press, New York, 1978. Also, *Thin Solid Films,* vol. 17, 1973, p. 1.
155. H. Niehus and E. Bauer, *Surface Sci,* vol. 47, 1975, p. 222; also, G. R. Sparrow, *Relative Sensitivities for ISS,* Adv. R&D, St. Paul, MN, 1976.
156. R. E. Reason, in H. W. Baker, ed., *Modern Workshop Technology,* Cleaver-Hume Press, Bristol, U.K., 1960.
157. N. Schwartz and R. Brown, Trans. *8th AVS National Symposium and 2nd Internl Congr Vac Sci Technol,* Pergamon Press, New York, 1961, pp. 836–837.
158. E. T. K. Chow, *J Vac Sci Technol,* vol. 2, 1965, p. 203.
159. S. Tolansky, *An Introduction to Interferometry,* Logmans, Green & Col, Ltd., London, 1955; also, *Surface Microtopgraphy,* Interscience Publ., London, 1960.
160. J. G. Gottling and W. S. Nicol, *J Opt Soc Am,* vol. 56, 1966, p. 1227.
161. N. A. Burnham and R. J. Colton, *J Vac Sci Technol A,* vol. 7, 1989, p. 2906.
162. S. Gauthier and C. Joachim, (eds.), *Scanning Probe Microscopy: Beyond the Images,* Les Editions de Phys., France, 1991.
163. G. Binnig, H. Rohrer, Ch. Gerber, and E. Wiebel, *Phys Rev Lett,* vol. 50, 1983, p. 120.

164. R. J. Behm, N. Garcia, and H. Rohrer, (eds.), *Scanning Tunneling Microscopy and Related Methods,* Kluwer Academic Publ., Dordrecht, 1990.
165. Ph. Avouris, (ed.), *Atomic and Nanometer-Scale Modification of Materials: Fundamentals and Applications,* Kluwer Academic Pub., Dordrecht, 1993.
166. D. A. Bonnell, (ed.), *Scanning Tunneling Microscopy: Theory and Practice,* VCH Publ., New York, 1991.
167. S. N. Magonov, *Surface Analysis with STM and AFM,* VCH Publ., New York, 1996.
168. D. W. Pohl and D. Courjon, (eds.), *Near Field Optics,* Kluwer Scientific Publ., The Netherlands, 1993.
169. S. Amelinckx, D. VanDyck, J. Van Landuyt, and G. Van Tendeloo, (eds.), *Handbook of Microscopy,* VCH Publ., New York, 1996.
170. J. Tersoff and D. R. Hamann, *Phys Rev Lett,* vol. 50, 1983, p. 1998.
171. P. K. Hansma and J. Tersoff, *J Appl Phys,* vol. 61, 1987, p. 15.
172. D. Whitehouse, *Handbook of Surface Metrology,* IOP Publ, Bristol, 1996.
173. D. Rugar and P. Hansma, *Physics Today,* vol. 43, 1992, p. 23.

CHAPTER 7
DIAMOND FILMS

Bradley A. Fox
Electronic Materials Center
Kobe Steel USA Inc.

7.1 INTRODUCTION

Diamond thin films manufactured by chemical vapor deposition (CVD) (see Chapter 1) represent an emerging technology that offers tremendous potential for numerous commercial applications. By 1993, around 2000 cumulative papers and patents on diamond CVD films have been published since its initial discovery in the 1950s [1]. The interest in diamond CVD films in the middle to late 1980s was sparked by several advances in diamond deposition. More-detailed historical perspectives have been given in other reviews [2–4]. As a summary, Eversole first achieved vapor-phase deposition of diamonds in 1953 [5]; however, it was not reported until 1962 [6]. Diamond film deposition was confirmed by Angus and coworkers [7,8]. This initial achievement of diamond CVD film was around the same time as the development of high-pressure, high-temperature (HPHT) synthesis of diamonds [9,10]. Unfortunately, the CVD technique of flowing methane over a diamond grit at elevated temperatures produced low diamond-deposition rates and significant co-deposition of graphite. In addition, these first experiments were not initially published. The difficulties of CVD diamond film growth and the ability to synthesize HPHT diamond in competing technology limited the development of CVD diamond technology until new deposition breakthroughs were achieved.

Several developments in the 1970s and early 1980s demonstrated that the growth rate could be increased, co-deposition of graphite could be suppressed, and non-diamond substrates could be used. The growth rate was increased through the cyclic introduction of atomic hydrogen (H) [11]. Atomic hydrogen was also used to permit nucleation of diamond on non-diamond substrates [12–15]. Additional improvements were made to this technique [16], and the growth rate was improved to 0.1 to 1 μm/hr on non-diamond substrates. In addition, several different reactor configurations for diamond deposition were developed. The improved growth rate, deposition on non-diamond substrates, reduction in the co-deposition of graphite, and the development of different deposition techniques demonstrated the potential viability of CVD diamond films.

The truly exceptional properties of diamond are the motivation behind the tremendous research effort in CVD diamond films. Most people are familiar with the hardness and optical brilliance of natural diamond stones, but there are many other properties of diamond that may be used to solve the materials problems of tomorrow. Some of the properties of diamond are summarized in Table 7-1. It is important to realize that hardness, saturated electron velocity, thermal conductivity, and stiffness are all the highest values of any known material. Combine these properties with its chemical inertness, low coefficient of expansion, low coefficient of friction, radiation hardness, and optical transparency, and it is easy to understand the current interest in diamond.

Diamond films grown using CVD are currently finding use as windows, heat sinks, tool coatings, and speaker diaphragms. Even more exciting are the development projects such as radiation detectors, field emitters for flat panel displays, pressure sensors, and active electronics. The potential advantage of CVD diamonds over natural or HPHT diamonds is their capacity for thin film deposition over large areas with controlled morphology and dopant concentration. One study [17] projects the CVD diamond market to be $1.1 billion by the year 2000, growing to approximately $16 billion by 2020 [17]. Electronics production is projected to account for 75% of the CVD diamond market. Production of tool coatings, optical coatings, and thermal coatings are projected to account for 8%, 6%, and 6% respectively. The remaining 5% of the market would be accounted for by several miscellaneous applications. In this book, electronic, optical, and thermal applications of CVD diamond will be addressed because they are the most relevant. Applications of CVD diamond films for use in tools have been reviewed by others [18].

This chapter contains three primary sections. The first is about the nucleation and growth of diamond films; the second is about the characterization and properties of diamond films; and the final section contains a summary of some potential applications of diamond films. The breadth of this subject prevents a complete discussion of the significant work in this fascinating field from being presented here. Selected topics were chosen that should be of most interest to the reader. Other areas are extensively covered in other sources, as noted in the references.

TABLE 7-1 Diamond Properties

Property	Value
Crystal structure	Cubic Fd3m
Lattice parameter	0.356725 nm
Atom density	1.77×10^{27} cm^{-3}
Hardness (Knoop)	5700-10,400 kg/mm^2
Young's modulus	10.5×10^{11} N/m^2
Anisotropy	1.21
Coefficient of thermal expansion	0.8×10^{-6}
Thermal conductivity	20 W/cm/°K
Specific heat	6.195 J/°K·mol
Debye temperature	1890
Refractive index	2.42
Bandgap	5.45 eV
Resistivity (IIa)	$>10^{14}$ Ω·cm
Hole effective mass (Hall)	0.75
Electron effective mass	0.57
Dielectric constant	5.7
Hole mobility	2000 cm^2/V·s
Electron mobility	2000 cm^2/V·s
Electron saturated current velocity	2.7×10^7 cm/s
Hole saturated current velocity	1×10^7 cm/s
Electric breakdown field	1×10^7 V/cm

7.2 NUCLEATION AND GROWTH

7.2.1 Thermodynamics and Kinetics

One of the fundamental questions about diamond CVD is why diamond forms when graphite is thermodynamically stable at atmospheric and lower pressures. Initially, it was thought that diamond would not nucleate because it is metastable at atmospheric pressure [19]. The thermodynamic preference for graphite at atmospheric pressure ($\approx 10^{-4}$ GPa) is demonstrated through the phase diagram shown in Fig. 7-1 [20]. Bridgeman was one of the first to postulate that growth of metastable phases of carbon was possible [21]. At 298°K and 1 atm, the free energy of graphite is only 2.9 kJ/mol (0.03 eV/atom) less than diamond [5]. Because the thermal energy at room temperature (given by the product of the temperature and Boltzmann's constant) is also ~0.03 eV/atom, the energy difference between the two polytypes of carbon (graphite or diamond) is not substantial. It is the energy barrier between the two polytypes that inhibits phase transformations from one polytype to the other. The activation energy for graphitization of the {110} diamond surface, where two carbon bonds must be broken, was 728 ± 50 kJ/mol or ≈250 times the free energy difference [22]. On the {111} surface, where three carbon bonds must be broken, the activation energy for graphitization is 1060 kJ/mol [23]. Fortunately, the presence of atomic hydrogen during growth overcomes these difficulties and allows the formation of diamond. This mechanism will be discussed in the Section 7.2.3. A schematic overview of a possible sequence of events

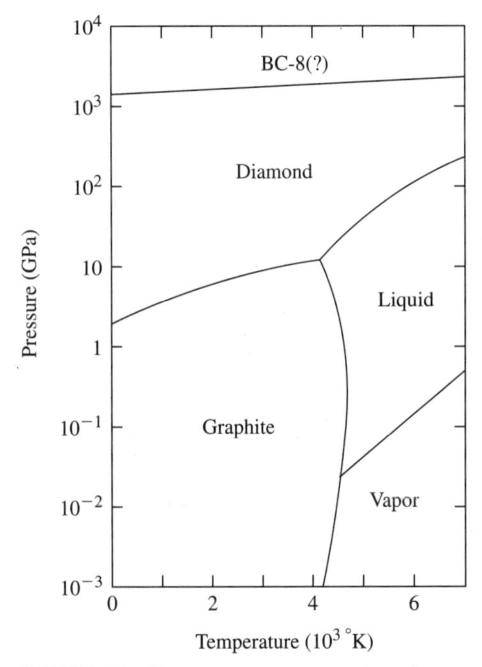

FIGURE 7-1 The pressure-temperature phase diagram for carbon. *(From H. Kanda and T. Sekine, G. Davies (ed.), in* Properties and Growth of Diamond, *INSPEC, Exeter, England, 1994, p. 404. Reprinted with permission. Copyright 1994 INSPEC.)*

for the nucleation and growth of diamond from the vapor phase on a nonreacting surface is shown in Fig. 7-2 [24]. The growth sequence is not understood and this atomistic representation is overly simplified; however, this diagram helps to visualize how diamond film growth may occur.

7.2.2 Nucleation

The mechanism of diamond nucleation is not clear. Several plausible mechanisms have been proposed although a detailed discussion of the different possibilities will not be presented. Rather, only one possible mechanism will be given to illustrate how diamond nuclei may be formed. One mechanism for the nucleation of diamond on a non-diamond substrate is through the formation of a graphite-like or a sp^2 bonded intermediate [25]. This mechanism has been supported by x-ray photoelectron spectroscopy measurements during diamond nucleation on platinum (Pt) [26] and by other techniques [5]. This mechanism suggests that the initial nuclei may be graphitic in nature, but transform into diamond nuclei because the H-passivated diamond nuclei are energetically more stable. When the nuclei reach a critical size, the barrier energy for a phase transformation into graphite inhibits the diamond from converting into the thermodynamically stable graphite. The morphology of diamond nuclei may be spherical clusters of diamond microcrystals, cubes, cubo-octahedrons, or flat hexagon plates [5]. For some substrate materials, the nuclei may be epitaxially oriented to the substrate. This will be discussed in more detail in Section 7.2.5.

Diamond film deposition occurs on many different materials; however, the chemical and physical nature of the substrate can affect the nucleation rate and subsequent film morphology. The substrates for diamond deposition are usually one of the following three types [24]:

1. Substrates with little or no solubility of carbon or reaction with carbon
2. Substrates with high carbon solubility
3. Substrates that form strong carbides

The incubation time for diamond nucleation is shortest on metals that can achieve a rapid supersaturation of carbon on the surface [24,27]. This means that materials

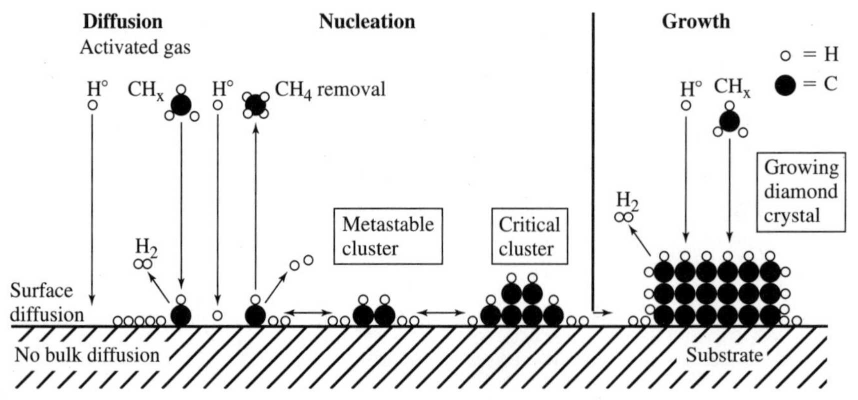

FIGURE 7-2 A schematic diagram for the growth mechanism of diamond. (*From B. Lux and R. Haubner,* Electrochemical Society Meeting Spring Meeting, *Electrochemical Society, Washington, DC, 1991, p. 314.)*

with a high carbon solubility often inhibit the supersaturation of the surface and are not ideal for the deposition of diamond. Carbide formers can allow high-nucleation densities, but under certain conditions also delay nucleation of diamond because the formation of the carbide competes with the nucleation of the diamond by acting as a carbon sink [24,27].

It is worthwhile to discuss several of the most technologically important substrates. Diamond film deposition on silcon (Si) was first demonstrated in 1966 [28]. Diamond nucleation on Si often results in a low nucleation density unless the surface is specially prepared, as discussed below. Diamond deposition on insulating substrates, such as Si_3N_4 and α-Al_2O_3 [29], has also been demonstrated. In addition to materials that facilitate diamond nucleation, it is also important to identify materials that inhibit the nucleation of diamond. This is valuable for selective area deposition of diamond. Two important materials the inhibit the deposition of diamond are amorphous Si [30] and SiO_2. Both thermally grown SiO_2 [31] and sputtered SiO_2 [32,33] are used to inhibit the nucleation of diamond and selectively deposit diamond on patterned substrates.

Although nucleation of diamond on non-diamond substrates is possible, the nucleation density is often too low, 10^3 to 10^4 cm^{-2}, to obtain a continuous film within a reasonable deposition time. To increase the nucleation density, polishing, ion implantation, or bias-enhanced nucleation are often used [34]. Polishing of substrates is generally performed with diamond grit and allows nucleation densities of 10^7 to 10^8 cm^{-2} [34–36]. It has been postulated that residual diamond polishing compound serves as a nucleation site for diamond growth [37]. This nucleation of diamond on the residual diamond grit was confirmed under certain conditions through high-resolution electron microscopy [38, 39]. Abrasion with non-diamond materials, such as cubic boron nitride (cBN) [29], SiC [40], and stainless steel [41], also produced high nucleation densities; therefore, damage must also play a role in increasing the nucleation density. In addition to mechanical polishing, ultrasonic vibration with diamond powders increased nucleation density 10^3 to 10^5 times over non-scratched surfaces [42]. Biasing of the substrate during the initial stages of diamond deposition produced enhanced nucleation, as high as 10^{10} cm^{-2} [34]. Biasing may also assist in the formation of aligned nuclei on non-diamond substrates [43,44]. Interestingly, if biasing is maintained throughout growth, the diamond film quality is degraded [34].

7.2.3 Chemistry

The deposition of diamond film depends on carbon and hydrogen where the presence of oxygen is often used to facilitate or improve diamond growth. The carbon source is typically methane or acetylene; however, solid sources as well as many other carbon sources are also used. Diamond growth from numerous carbon compounds, including alcohols and ethers, has been demonstrated [45]. Several researchers have investigated specific combinations of reactant species. These have included studies of $CH_4 \cdot H_2 \cdot H_2O$ [46], $CO \cdot H_2$ [47], $CH_4 \cdot CO \cdot H_2$ [48], and $CO \cdot H_2 \cdot Ar$ [49]. The best carbon precursor to diamond growth is still under dispute; however, a ternary-phase diagram of carbon, hydrogen and oxygen developed by Bachmann et al. [50] demonstrates that the gas chemistry that favors diamond growth has a C:O ratio of 1:1 in the presence of hydrogen. This diagram is shown in Fig. 7.3 [50]. For C:O ratios more than 1:1 growth of non-diamond carbon is prevalent. For C:O ratios less than 1:1, no growth is observed. This diagram appears to depend only on the relative concentrations of C, H, and O and not on the reactant species.

The effect of atomic hydrogen in the deposition of diamond remains an area of active research. There are many hypotheses [5,51–55] and some inconsistencies exist [56,57]. Some of the possible roles for atomic hydrogen in diamond nucleation and growth include:

1. Generation of specific gas-phase species that promote the nucleation and growth of diamond

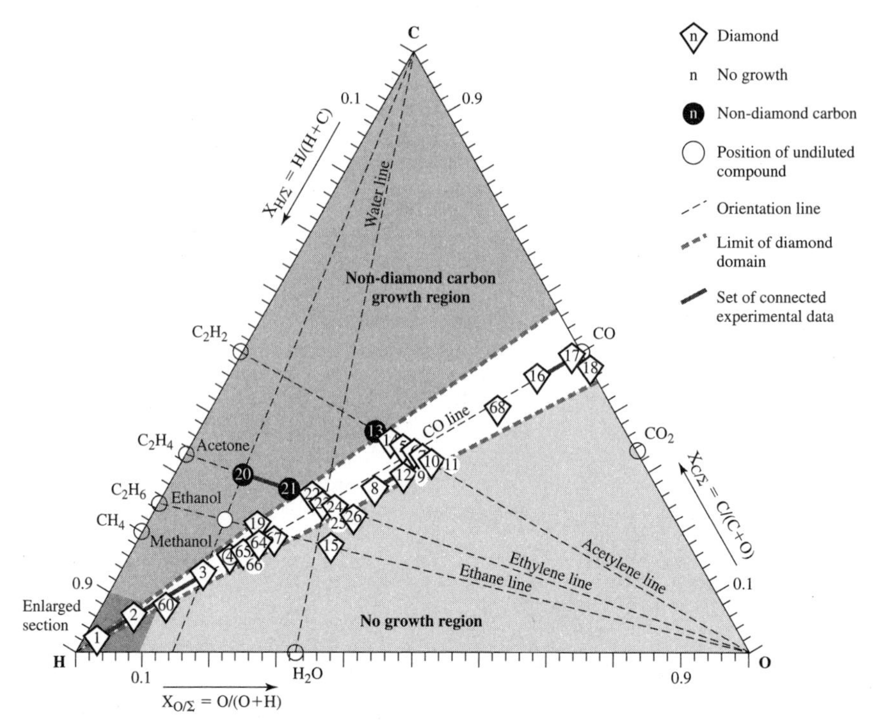

FIGURE 7-3 A tertiary C–H–O phase diagram for the growth of diamond. *(From P. K. Bachmann, D. Leers, and H. Lydtin,* Dia Rel Mat *vol. 1, 1991, p. 1. Reprinted with permission. Copyright 1991 Elsevier Science S.A.)*

2. Hydrogenation of unsaturated carbon bonds on the growth surface to promote sp^3 bonding
3. Reduction in the critical nucleus size for diamond relative to graphite
4. Preferential etching of non-diamond-bonded carbon
5. Stabilization of the diamond surface.

Although deposition of diamond does not require oxygen, its presence does provide more flexibility in the choice of growth conditions. As indicated by Bachmann's diagram, the C : O ratio is typically 1 : 1 for diamond growth. Oxygen was first used as a reactant for diamond growth by Hirose and Terasawa [45]. The introduction of O_2 produced low-defect-density diamond with up to 10% acetone at 800 torr and a growth rate of up to 10 μm/hr. Without oxygen, the maximum acetone concentration for diamond growth was 1% at only 100 torr. With the addition of oxygen non-diamond carbon deposition is suppressed and the growth rate can be several times higher than with methane and hydrogen mixtures. Additionally, the deposition temperature may be as low as 450°C. The presence of oxygen augments the role of atomic hydrogen [58].

It is difficult to generalize the process parameters for CVD diamond because it has been deposited through a diverse number of techniques. However, some typical values for diamond CVD will be summarized. The pressure is typically 10 torr to atmospheric pressure and the substrate temperature is usually 700°C to 1000°C. There is some means to form atomic hydrogen, where the growth rate is increased as the temperature of the gas phase increases [50]. The gas chemistry consists of various sources of carbon and hydrogen and sometimes oxygen. The carbon concentration is often <1% and the balance of the gas is often hydrogen. When oxygen is used, it is often in low concentrations, <1%, while maintaining a C:O ratio of 1. A map of the growth morphology of diamond as a function of methane concentration and substrate temperature is shown in Fig. 7-4 [59].

7.2.4 Reactors

Diamond growth by CVD typically requires atomic hydrogen and a carbon source at temperatures between 500°C and 1200°C (above or below leads to diamond-like carbon (DLC) or graphite). Diamond growth at temperatures as low as 300°C has also been reported [60,61]. The growth chemistry allows various reactor configurations to be used for diamond growth. Numerous authors have reviewed the different reactor configurations [62–65]. The more common deposition techniques are microwave plasma, including electron cyclotron resonance (ECR), hot filament, combustion flame, direct current (dc) discharge, and dc arc jet. A summary of the various deposition techniques and ranges of processing conditions are shown in Table 7-2.

It is not clear whether any one technique is superior to the other. The issues that distinguish the reactors are deposition area, growth rate, quality, uniformity, and reproducibility. It may be that thick film applications >100 μm may prefer a high-growth-rate reactor. Active electronics may require high-quality thin films where a slower growth rate is tolerable, but uniformity and reproducibility are more important. Several commercial diamond deposition systems are now available. These are primarily microwave plasma and arc jet deposition systems.

Microwave Plasma and ECR. Commercial microwave generators are available at 2.45 GHz and 915 MHz frequencies and use a plasma environment to generate atomic hydrogen. The initial use of microwave plasmas was by the National Institute for Research in Inorganic Materials in Japan [66]. A schematic diagram of a microwave plasma reactor

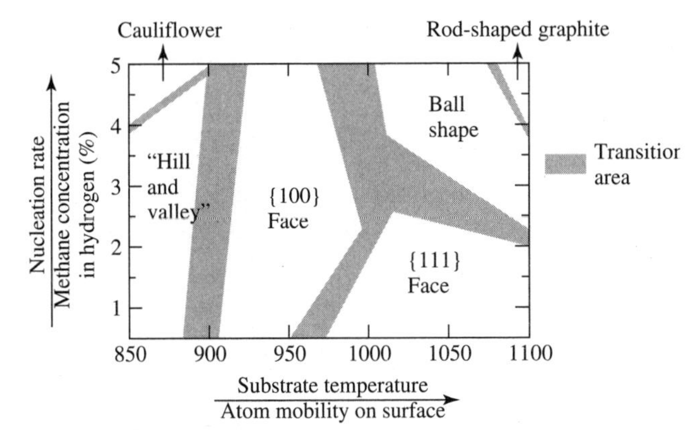

FIGURE 7-4 A correlation of the growth morphology to the methane concentration and substrate temperature. *(From W. Zhu, A. R. Badzian, and R. F. Messier,* Proc SPIE Diamond Optics III, *Bellingham WA, SPIE, 1990, p. 187. Reprinted with permission. Copyright 1990 Society of Photo-Optical Instrumentation Engineers.)*

TABLE 7-2 Summary of Diamond Deposition Growth Techniques

Technique	Temp (°C)	Pressure (torr)	Flow (slm)	Growth Rate (μm/hr)
Microwave	800-1000	5-100	0.1-1	0.1-5
ECR	400-600	0.1-50	0.1-1	0.08-0.1
Hot filament	700-1000	10-100	0.1-1	1-10
dc discharge	>600	5-760	0.1-1	20-250
dc arc jet		500-760	>10	80-1000
Combustion flame	800-1000	760	1-2	140
RF torch	700-1200	500-760	>10	60-180
RF plasma	800-1000	1-30	0.1-1	0.5-3

is shown in Fig. 7-5 [64]. The substrate is typically immersed in the plasma; however, remote or downstream plasma deposition has been reported [67]. The diamond quality degraded, as measured by electrical conductivity, when the sample was moved downstream of the plasma [68]. Because of the higher quality of samples immersed in the plasma, this technique is more prevalent. When the sample is immersed in the plasma it is heated by the excited plasma species and the sample temperature and the microwave power are coupled. For more independent control of the microwave power and sample temperature, secondary heating or cooling is sometimes used.

The introduction of microwaves into the deposition chamber occurs by several different methods. The microwaves may be directed either parallel to the substrate, often termed "side launch," or normal to the substrate, often termed "end launch." The chamber pressure is limited to the pressure and power where a plasma will be maintained. Typically, a hydrogen ambient is used and the pressure range is 5 to 100 torr. Plasma deposition systems with microwave power up to 75 kW have been developed for microwave plasma CVD. A magneto-microwave plasma deposition system has also been used where Helmholtz-type coils were placed around the cylindrical waveguide and a magnetic field was used to control the deposition area [69]. Although the magnetic field (875 G) satisfied the ECR conditions, the deposition pressure for good diamond deposition was 4 to 50 torr which prevented complete electron gyrations because of the limited mean free path of the electrons. For true ECR con-

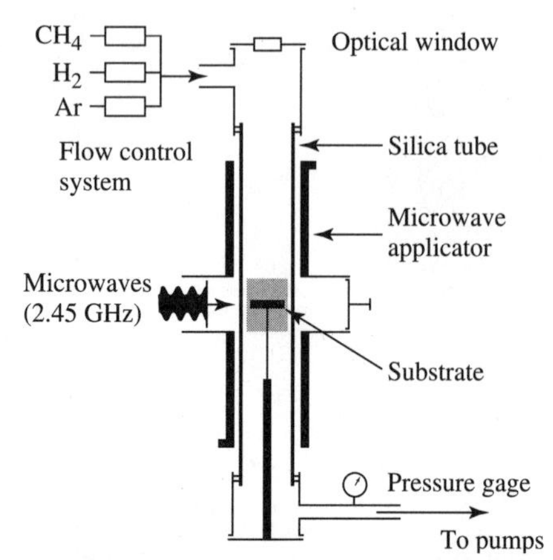

FIGURE 7-5 A schematic diagram for a microwave CVD diamond-growth reactor. (*From W. Zhu et al.,* Proc of the IEEE, *vol. 79, 1991, p. 621. Reprinted with permission. Copyright 1991 IEEE.)*

ditions in this reactor, the pressure must be below 10 mtorr. Although the technique is not a true ECR deposition, it does show promise for large area (7 to 8 cm plasma), low-temperature deposition. Unfortunately, no diamond film has been grown under true ECR conditions where Raman spectroscopy has confirmed it to be predominately sp^3-bonded [62].

The primary advantage of microwave plasma reactors is its reproducibility. Additionally, the availability of commercial microwave reactors has made this technique extremely popular. Microwave reactors are able to generate a high density of high-energy electrons in the plasma while the gas and the substrate remains relatively cool [70]. Independently controlled substrate heating or cooling assists in separating the microwave power from the substrate temperature. One disadvantage of the microwave plasma technique is the difficulty in distributing a uniform plasma density over a large area. Because the plasma density determines the local gas chemistry, variations are likely to cause nonuniform diamond deposition. This nonuniformity may either be in film thickness or quality. Also, the growth rates are typically 0.1 to 5 μm/hr; therefore, thick films require long deposition times.

Hot Filament. The hot filament or thermally activated deposition uses an extremely hot (>2000°C) filament to form atomic hydrogen and other activated species. This was one of the first methods used for diamond CVD [16,71]. A substrate is placed in close proximity to the filament for diamond deposition. In this technique, it is also helpful to use external heating and cooling to control the temperature of the substrate. A schematic diagram of a hot-filament reactor is shown in Fig. 7-6 [64]. In this technique, the filament thermally cracks the

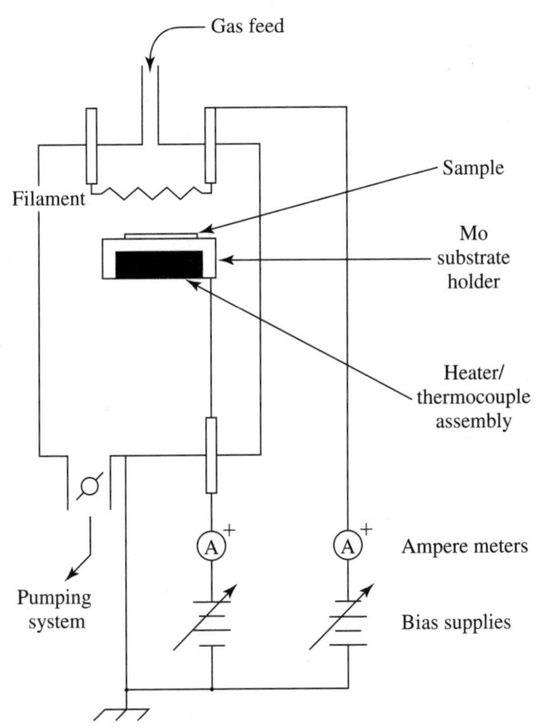

FIGURE 7-6 A schematic diagram for a hot filament CVD diamond-growth reactor. *(From W. Zhu et al.,* Proc of the IEEE, *vol. 79, 1991, p. 621. Reprinted with permission. Copyright 1991 IEEE.)*

hydrogen into atomic hydrogen and dissociates the methane or other carbon source gas. The substrate can be positively biased to induce electron bombardment on the growth surface, which enhances the dissociation of gases at the surface. This may enhance both the nucleation density and the growth rate [40,72].

The advantage of this technique is its simplicity. It really only needs a vacuum chamber, a filament, a dc power supply, and a gas supply to deposit diamond. The disadvantages are uniformity, purity, reproducibility, and filament embrittlement. The uniformity problem with this technique is caused by a filament acting as a line source of activated gas species for deposition. This difficulty may be overcome with a network of filaments that produce a planar front of activated gas species. The filament also serves as a source of impurities because the filament material is often incorporated into the diamond film [73]. The reproducibility of this technique is more difficult because the filament often changes with time because of carburization [74]. The resistance of the filament was shown to vary with carburization time; therefore, the input power to the filament must be modified with time to maintain a constant deposition environment. Additionally, carburization of the filament consumes some carbon; therefore, the gas chemistry is also changing with time during this technique. The filament is often made of tungsten (W) and other materials that have a high solubility for carbon which embrittles the filament. This embrittlement makes the filament prone to catastrophic mechanical failure. Carburization generates a volume expansion of the filament that leads to cracking and reduces the lifetime of the filament [75]. Requirements for good filament materials were discussed by Anthony [76]. Tungsten, tantalum (Ta), Pt [77] and rhenium (Re) [73] have been used as filaments. Carbon filaments were also tried, but did not lead to the growth of diamond. It was thought that the atomic hydrogen reacted with the carbon filament to form hydrocarbons and was not transported to the substrate to enable diamond deposition [78].

Combustion Flame. The combustion flame technique for the deposition of diamond was first reported by Hirose [79]. The technique involves burning hydrocarbons at atmospheric pressure, just as in a welding or cutting torch [80]. The source gases are typically an acetylene-oxygen combination, but ethane, methane, propane, ethylene, methanol, and ethanol have also been used with oxygen. The gases are expelled through a nozzle and burned while the substrate for diamond growth is placed in the flame as shown in Fig. 7-7 [64]. The high gas pressure and temperature is thought to generate atomic hydrogen similar to other diamond deposition processes. Alternately, because the role of atomic oxygen is similar to the role of atomic hydrogen, this may also count for the high rates of good quality diamond deposition with this method [61,81]. The key to this technique is maintaining temperature control. The flame is often 3000°C and the substrate is typically maintained around 800°C to 1200°C. This requires precise control of the position of the substrate relative to the flame tip and a cooled substrate.

The combustion flame in the fuel-rich mode exhibits three distinct regions: a primary combustion zone, an intermediate zone (feather), and outer zone (secondary combustion zone). The secondary combustion zone in a fuel-rich flame occurs because of the diffusion of oxygen from the surrounding atmosphere and the unburned fuel from the primary combustion zone. The substrate is typically placed in the intermediate or feather of the combustion flame. In this region, there is atomic hydrogen produced from the primary combustion zone. A fuel-rich mixture is used so that there will be unburned hydrocarbons in the intermediate zone for diamond growth. One drawback of this technique is that appreciable amounts of nitrogen (N) that have diffused from the atmosphere are present in the feather region [82]. However, recent growth in an enclosed chamber may alleviate this problem [83].

dc Discharge. In this process, dc discharge-assisted deposition of diamond is attained by applying a high dc voltage and high current density between two parallel elec-

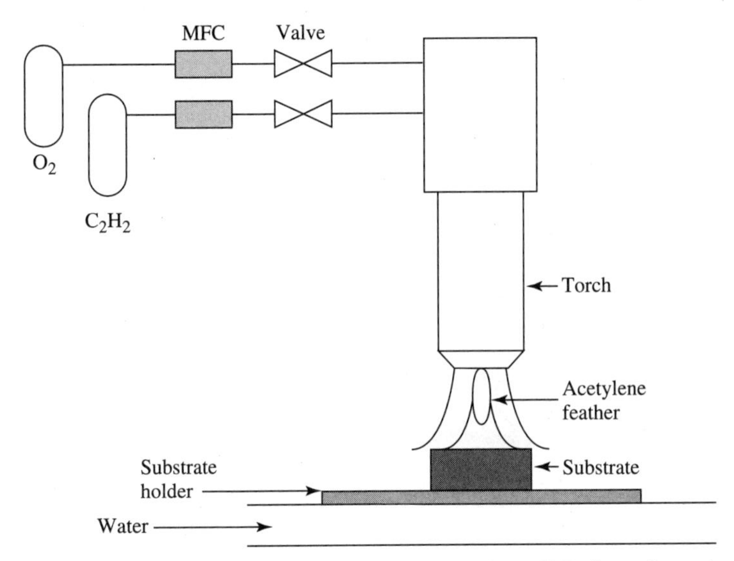

FIGURE 7-7 A schematic diagram for a combustion flame CVD diamond growth reactor. *(From W. Zhu et al.,* Proc of the IEEE, *vol. 79, 1991, p. 621. Reprinted with permission. Copyright 1991 IEEE.)*

trodes at ≈200 torr. The substrate has been placed on both the anode electrode [29] and on the cathode electrode [84]. With a 1 kV applied voltage and a current density of 4 A/cm^2, a nucleation density of 10^8 cm^{-2} was obtained on Si and Al_2O_3 where the growth rates were ≈20 μm/hr [29]. Under these conditions, the dc discharge was believed to be between a glow and an arc and produced a high gas temperature to excite the reactant species. The electron bombardment of this technique heats the substrate to ≈800°C; therefore if a lower deposition temperature is desired, the substrate must be cooled [75]. Numerous variations of this technique have been developed [64].

dc Arc Jet. One significant modification of the dc discharge technique that has gained considerable attention is the dc arc jet (plasma jet). It generates gas temperatures in excess of 4000°C which enhances gas dissociation. It is related to the dc discharge reactor; however, in this technique the gases are heated and decomposed by an arch discharge between two cylindrically symmetric electrodes as shown in Fig. 7-8 [64]. The heating and decomposition of the gases create a rapid expansion and the gases are forced out an orifice in a jet 5 to 10 cm long toward a cooled substrate. Using an atmospheric dc discharge plasma jet, growth rates of 930 μm/hr have been obtained [85].

7.2.5 Diamond Film Morphology

It is difficult to summarize diamond film morphology because there are so many different possible growth techniques and substrates that have been used for diamond deposition. Therefore, specific examples of typical morphologies will be discussed; however, this brief treatment does not cover all of the possible variations. The film morphology of homoepitaxial, heteroepitaxial, and polycrystalline films will be discussed. In

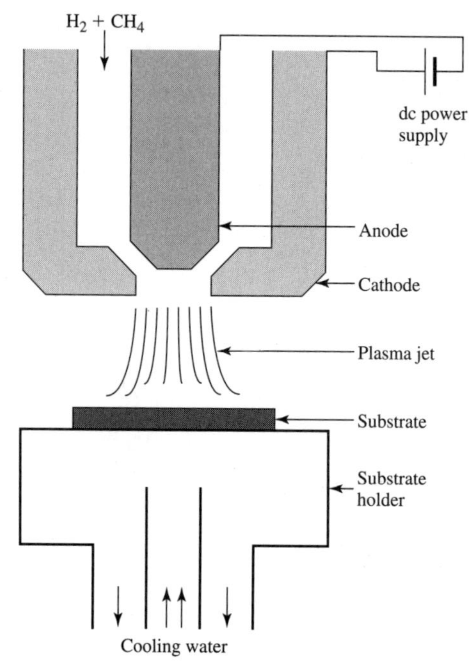

FIGURE 7-8 A schematic diagram for a dc plasma jet diamond-growth reactor. *(From W. Zhu et al.,* Proc of the IEEE, *vol. 79, 1991, p. 621. Reprinted with permission. Copyright 1991 IEEE.)*

Fig. 7-9, [86] the morphology of a polycrystalline, highly oriented, and homoepitaxial diamond film is shown. Additionally, typical defects observed in CVD diamond will be discussed.

Homoepitaxial Growth. Homoepitaxial diamond growth on diamond was the first type of deposition used to demonstrate the feasibility of CVD diamond growth [6]. In that experiment the increase in weight of diamond particles was used to verify the deposition of diamond. Today, homoepitaxial growth on diamond is often used for precise control of the doping concentration for electronic device structures. The CVD of boron-doped, *p*-type diamond was first demonstrated on diamond particles [87,88] and later on diamond substrates where the semiconducting nature of the boron-doped epitaxial film was verified by Hall effect measurements [89].

The orientation of the homoepitaxial film strongly influences the growth rate, impurity incorporation, and surface morphology. In a study of the simultaneous deposition of {100}, {110} and {111} oriented homoepitaxial diamond films, the growth rate of {110} films was faster than of the {100} or the {111} films which exhibited similar growth rates [90]. The incorporation of boron (B) was similar on the {111} and {110} planes, but both incorporated more boron than on the {100} planes. In a study of {110} and {100} boron doped films, similar trends in the growth rate and boron concentration were observed and the crystalinity of the {110} films was inferior to the {100} films [91]. It is generally believed that the {100} surface offers the best quality CVD diamond [90,92].

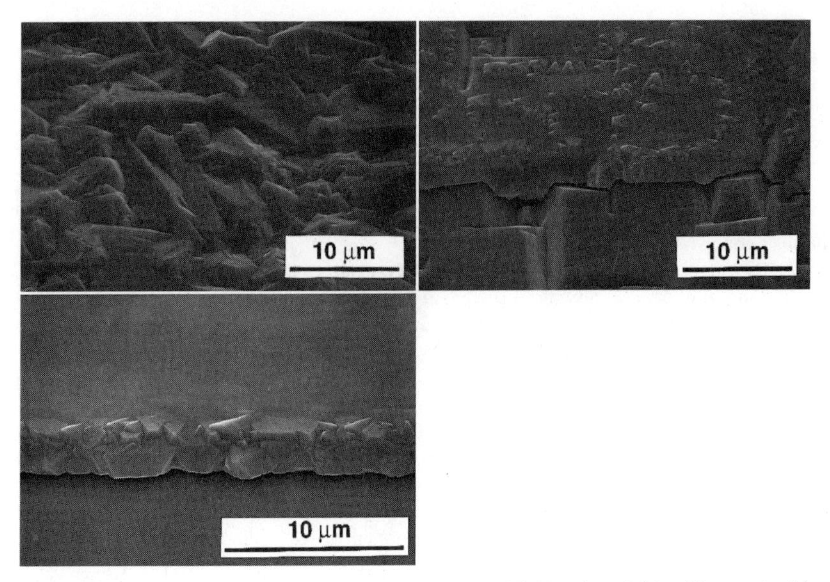

FIGURE 7-9 The film morphology of a polycrystalline. *(a)* Highly oriented *(b)* and homoepitaxial diamond film *(c)*. *(From B. A. Fox et al.,* Dia Rel Mat, *1994, p. 382. Reprinted with permission. Copyright 1994 Elsevier Science S.A.)*

Although it is desirable to use epitaxial growth for high-quality diamond-film deposition, the availability of high-quality, large-area single crystal substrates currently limits widespread application of homoepitaxial deposition. The size of natural diamonds is typically <10 mm, and the price and availability of these crystals prohibits any large-scale use of these diamonds. Similarly, synthetic diamonds are typically limited to <6 mm in diameter. In addition to the size limitations, the defect density of existing single-crystal diamond is inferior to single-crystal Si or GaAs. A high defect density of twins, stacking faults, and dislocations is prevalent in natural diamond crystals where the dislocation density in IIa diamonds is $10^8/cm^2$ [93]. The full width at half maximum (FWHM) of the Bragg angle which is an inverse measure of a crystal quality, by x-ray diffraction has been as large as 2.8°. Synthetic diamonds have been grown where the FWHM was 12 to 17 arcsec [94]. Therefore high-quality diamonds, as measured by x-ray diffraction, are attainable but have not been produced in commercial volumes.

Heteroepitaxial Growth. The ability to grow monocrystalline diamond films on a non-diamond substrate is desired because large-area, high-quality, single-crystal diamond substrates are not available. Unfortunately, the characteristics of diamond that generate its exceptional properties also make it difficult to achieve heteroepitaxial growth of diamond. To understand why heteroepitaxial diamond growth is difficult, a brief summary of the epitaxial growth mechanisms will be discussed. Deposition typically occurs by one of three growth modes [95]. The first is monolayer-by-monolayer growth, which is often referred to as Frank-van der Merwe growth or two-dimensional growth [96, 97]. Alternately, island growth may occur, which is often referred to as Volmer-Weber growth or three-dimensional growth [98]. The third growth mode is a combination of the island growth and monolayer-by-monolayer growth and is often referred to as Stranski-Krastanov growth. In this growth mode, monolayer-by-monolayer growth occurs first, then island growth develops. The growth mode that develops depends on the relative surface free energy. Monolayer-by-monolayer growth is desired for growth of large area films. For monolayer-by-monolayer growth to occur, the relative magnitude of the surface free energy of the overgrowth γ_O, the substrate γ_S, and the interface γ_I satisfy the following expression [99]

$$\gamma_O - \gamma_S < \gamma_I \tag{1}$$

In this expression, the surface free energy of the interface incorporates the elastic strain energy because of difference between the lattice constant of the overgrowth and the substrate, and the misfit energy of any atom disregistry that occurs at the interface [100].

There are several difficulties with diamond heteroepitaxy. One is that the surface free energy of diamond is extremely high, 6 J/m^2 on the {111} plane [101] and 5.3 to 9.2 J/m^2 on other low-index planes [102]. This is evident in its high surface tension and nonwettability. As a result, it is difficult for Eq. 1 to be satisfied so that heteroepitaxial films may be made by monolayer-by-monolayer growth. A second reason is that the lattice constant of diamond is only 0.3567 nm, which is small for the diamond crystal structure. As a comparison, Si has a lattice constant of 0.5646 nm [103], which is 60% larger than the lattice constant of diamond. The magnitude of the surface free energy of diamond and the lattice misfit inhibit heteroepitaxial growth of a monocrystalline film. A third difficulty for diamond heteroepitaxy is that diamond does not form a continuous solid solution with other elemental semiconductors. This prevents the use of compositional grading to decrease the lattice mismatch to nondiamond substrates. A fourth difficulty for diamond epitaxy is the covalent nature of the bonds. For metal epitaxy, if there is not a direct lattice match, it is often possible to select a crystal face of the substrate that has an atom spacing similar to that of the epitaxial film. Because the bonds are metallic and nondirectional, epitaxy results. For covalent semiconductors, with localized and directional bonds, similar atomic spacing may not be sufficient to obtain epitaxy. It is also necessary to have a one-to-one correspondence of dangling bonds on either side of the interface [104]. However, surface reconstruction effects may change the bonding at the surface in the case of nonmatching covalent substrates and accommodate the difference in dangling bonds [104].

Even with the difficulties mentioned above, heteroepitaxial diamond films have been grown on cubic boron nitride cBN [105,106]. The crystal structure of cBN is zinc blende. It is similar to a diamond structure except that the two-atom basis consists of different atoms in the zinc blende structure rather than the same atoms as in the diamond structure. The surface energy of cBN is 5.4 J/m^2, which is similar to that of diamond [105]. The lattice constant is 0.36 nm [105] and is similar to that of diamond; therefore, the elastic strain energy should not be large. Unfortunately, large-area single crystals of cBN have not been made [107]; therefore, cBN does not offer significant commercial opportunities for heteroepitaxial deposition of diamond.

Although the discussion above suggests that the potential for diamond heteroepitaxy is limited, there is still some promise. Cluster-binding calculations suggest that growth of diamond on non-diamond substrates may be possible under certain conditions [108]. Modeling and experiments of potential epitaxial orientations for diamond on β SiC have been investigated [109]. Additionally, as with the formation of diamond, the surface energy model is only a thermodynamic model whereas kinetics and other factors may also play a role. When the deposition or supersaturation rate exceeds the rate at which the surface atoms may occupy their equilibrium positions, monolayer-by-monolayer growth may occur even if it is not energetically favored [110].

Although heteroepitaxial films have only been demonstrated on cBN, heteroepitaxial nuclei have been grown on various substrates. The following epitaxial relationships have been observed for individual nuclei:

- Diamond on graphite [111]

$$(0001)_G \parallel (111)_D \quad \text{and} \quad \langle 11\,\bar{2}0\rangle_G \parallel \langle 1\,\bar{1}\,0\rangle_D$$

- Diamond on BeO [112]

$$(0001)_{\text{BeO}} \parallel (111)_D \quad \text{and} \quad \langle 11\,\bar{2}0\rangle_{\text{BeO}} \parallel \langle 1\,\bar{1}\,0\rangle_D$$

- Diamond on β-SiC [43]

$$\{100\}_{\beta\text{-SiC}} \parallel \{100\}_D$$

- Diamond on α silicon carbide [113]

$$(0001)_{\text{SiC}} \parallel (111)_D \quad \text{and} \quad \langle 11\,\bar{2}0\rangle_{\text{SiC}} \parallel \langle 1\,\bar{1}\,0\rangle_D$$

- Diamond on Ni [114]

$$\{100\}_{\text{Ni}} \parallel \{100\}_D$$

Additionally, if the nucleation density is high and the individual nuclei are grown large enough that they coalesce, continuous films of highly oriented nuclei separated by low-angle grain boundaries have been demonstrated.

In addition to nucleation of heteroepitaxial nuclei on non-diamond substrates, the preferential growth of these nuclei is also important for highly oriented diamond films. Because the nucleation process may only yield 50% oriented nuclei [43,65], the growth conditions must be controlled to promote growth of the oriented particles to form a continuous film. Control of the specific growth directions is possible by taking advantage of the van der Drift texture evolution theory [115]. This theory describes how nuclei with the direction of fastest growth normally oriented to the substrate surface will dominate the film because their grains will grow most rapidly. According to this model, the shape of a nuclei is governed by the relative growth rates of the different crystallographic planes through the relative growth rate parameter α. The relative growth parameter is proportional to the ratio of growth velocity of the $\{100\}$ plane v_{100} to the growth velocity on the $\{111\}$ plane v_{111}

$$\alpha = \sqrt{3}\frac{v_{100}}{v_{111}} \tag{2}$$

The relative growth rate parameter varies from 1 which produces a cube with $\{100\}$ faces to 3 which produces an octahedron with $\{111\}$ faces [116]. Between 1 and 3, various cubo-octahedra exist as shown in Fig. 7-10 [117]. The relative growth rates of individual planes depend on the processing parameters. An example of how the

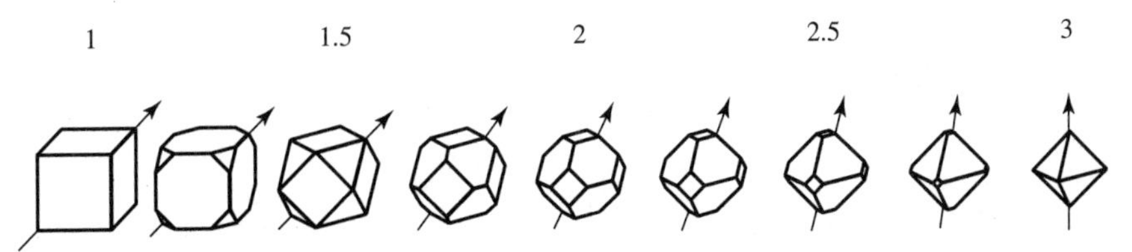

FIGURE 7-10 The preferred morphology as a function of the growth parameter where the arrows indicate the direction of fastest growth. *(From C. Wild et al.,* Dia Rel Mater, *vol. 3, 1994, p. 373. Reprinted with permission. Copyright 1994 Elsevier Science S.A.)*

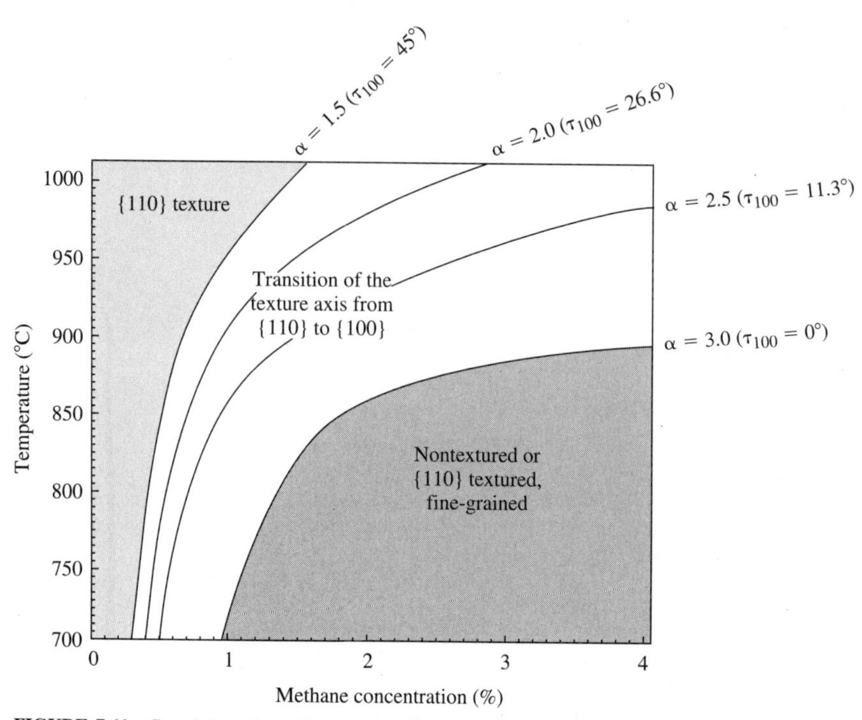

FIGURE 7-11 Correlation of the film texture with the methane concentration and temperatue. *(From P. Koidl, C. Wild, and N. Herres,* Proc NIRIM International Symposium on Advanced Materials, *Tsukuba, Japan, 1994. Reprinted with permission. Copyright 1994 International Communications Specialists Inc.)*

growth temperature and methane concentration modify the relative growth rate is shown in Fig. 7-11 [118].

Polycrystalline Growth. Unless the growth conditions and substrate are specifically selected, most diamond deposition on nondiamond substrates is polycrystalline. The morphology of individual grains within a polycrystalline diamond film may vary considerably, from cauliflower to highly faceted grains. The diamond film morphology depends strongly on the process parameters [64]. The temperature is one critical parameter which controls the growth rate [119] and crystal habit [16]. Additionally, the gas composition affects the film growth. Although the film morphology depends on the specific reactor design and gas chemistry, one example of the relationship between gas chemistry, temperature, and film morphology is shown in Fig. 7-4 [59]. The most common morphology of continuous films arises from films with a ⟨110⟩ fiber texture. Fiber texture means the grains are oriented with a preferential crystal direction normal to the substrate along the growth axis, but there is no orientational relationship about the growth axis. A fiber texture is almost inevitable for thick films resulting from the van der Drift evolutionary selection. In a ⟨110⟩ fiber texture film, the individual nuclei are {111} faceted and the films consists of columnar structures where the columns are in a ⟨110⟩ direction. An example of the morphology of a typical polycrystalline diamond film with a ⟨110⟩ fiber texture is shown is Fig. 7-12 [120]. X-ray texture analysis was used to show that typical polycrystalline diamond films with {111} facets exhibit ⟨110⟩

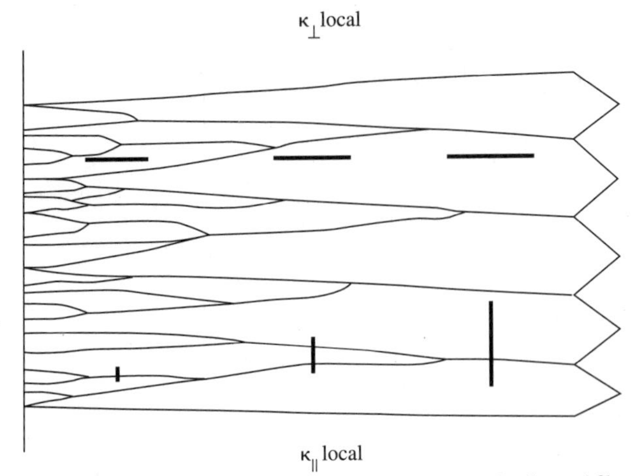

FIGURE 7-12 A schematic diagram of the cross section of a diamond film where the substrate is on the left and the growth surface is on the right. *(From J. E. Graebner et al.,* Nature, *vol. 359, 1992, p. 401. Reprinted with permission. Copyright 1992 Macmilliam Magazines Limited.)*

fiber texture [121]. Columnar growth yields anisotropic properties, such as the thermal conductivity, because the grain size increased as the film thickness increased [120]. The mobility has also increased as the thickness increased, and this was attributed to film quality and not grain size. [122]

7.2.6 Structural Defects

Transmission electron microscopy has revealed that the most commonly observed structural defects in CVD diamond are {111} twins followed by dislocations and {111} stacking faults [123,124]. The defect density has been correlated to the deposition technique and growth conditions [125]. Perfect *a*/2<110> dislocations and *a*/6<112> partial dislocations have been imaged in CVD diamond [126]. Additionally, {111} micro-twins are prevalent on {111} oriented grains, but are not as common on {100} oriented grains [127,128]. The crystallographic variation in defect formation means that diamond films with cubic habits typically have fewer defects than crystals with octahedral habits. Cross section transmission electron microscopy (TEM) of CVD diamond particles indicate that planar defects often originate from a point. This suggests that the defects are generated at nucleation, and that nucleation occurs at a single point near the substrate/diamond interface [125,129].

Twins not only alter the internal structure of the diamond film, but are often responsible for changes in the morphology of the film [118,130,131]. Because twins occur on {111} planes and have a mirror symmetry, dramatic alteration of the surface occurs. Two examples of the effect of twins on the diamond film morphology are the five-fold twin [124] and the cubo-octahedral twin with re-entrant {100} faces [132]. Lower and higher order twins with $\sigma = 3$, 9, 27 and 81 have been observed [133,134]. A correlation of the relative growth rates α of the {111} and {100} faces, from Eq. 2, and the diamond twin morphology has been observed [117,135]. Three different types of twins were defined [117]:

1. $T_{\{100\}}$ twin plane for {100} facet where the twin is inclined to {100}
2. $T_{\{111\},p}$ twin plane for {111} facet where the twin is parallel to {111}
3. $T_{\{111\},i}$ twin plane for {111} facet where the twin is inclined to {111}

Inclined twins are sometimes referred to as penetration twins whereas parallel twins are referred to as contact twins. For the $T_{\{100\}}$ twin, if $\alpha > 2$, the twin disappears, and if $\alpha < 2$, the twin increases in size. For the $T_{\{111\},p}$ twin because it is parallel to the growth surface, it propagates laterally, and affects the adjacent facets. If $\alpha > 2$, the twin propagates to {111} facets to form a $T_{\{111\},i}$ twin and for $\alpha < 2$, the twin propagates to {100} facets for form a $T_{\{100\}}$ twin. For a $T_{\{111\},i}$ twin, if $\alpha > 1.5$, the twin grows larger and if $\alpha < 1.5$ the twin vanishes [117]. As a summary, to minimize the presence of twins, for {111} facets α should be <1.5 and for {100} facets α should be >2. These two conditions are mutually exclusive and suggest that twin free deposition is unlikely for multifaceted polycrystalline growth [117].

7.2.7 Impurities

Intentional and unintentional noncarbon elements are sometimes incorporated into diamond during growth. Most elements have been observed in natural diamond [136]. The ability of diamond to incorporate impurities during formation in nature suggests that incorporation during CVD is also likely. To discuss the impurities in diamond it is important to understand the classification system used for diamond. This classification system was developed based on natural diamond, but is relevant to synthetic diamond as well.

Diamonds were initially classified by their nitrogen content, which is often determined by infrared or ultraviolet absorption spectroscopy [137]. Diamonds with appreciable nitrogen, typically 100 to 1000 ppm, are referred to as type I. Diamond without substantial nitrogen (<100 ppm) are referred to as type II. The type I diamonds are further subdivided according to the manner in which the nitrogen is incorporated. If the nitrogen is aggregated, it is referred to as type Ia, and if a single nitrogen is substitutional, it is referred to as type Ib. The aggregate type Ia diamonds are further subdivided on the basis of the structure of the nitrogen aggregate, referred to as A or B. The A aggregate consists of an nitrogen pair on adjacent substitutional sites [138] and is named type IaA. In contrast, the B aggregate consists of four nitrogens and a vacancy [139] and is named type IaB. Often in nature, both aggregates are observed and the diamonds are classified as type Ia/A/B. For type II diamond, two subclassifications are commonly used. Type IIa contains low nitrogen and has a high electrical resistivity ($\approx 10^{16}$ Ωcm), while type IIb diamond contains boron and is semiconducting.

Nitrogen is readily incorporated into the diamond lattice because of its negative formation energy [140]. Nitrogen may be incorporated in many forms. In addition to a single nitrogen and the A and B aggregates, other nitrogen-base complexes have also been observed in diamond. It is believed that nitrogen is initially incorporated into natural [141] and synthetic [142] diamond in substitutional positions. In natural diamond annealing for $\approx 3 \times 10^9$ years at 1200°K to 1600°K produces nitrogen aggregates. The A center [141,143] and B center [141] have also been observed through high-temperature annealing. Because CVD diamonds are formed in a short time, nitrogen incorporation is likely to be substitutional. However, lengthy high-temperature growth or sample processing may allow nitrogen aggregation. Although not as commonly observed, nitrogen interstitials have also been reported [144].

Because of the large concentrations of hydrogen used in most CVD diamond-growth techniques, an abundance of hydrogen is available for incorporation into the film. In fact, whereas for natural and HPHT diamond nitrogen is thought to be the dominant defect, for CVD diamond hydrogen is dominant over nitrogen as the most important impurity [145]. In a detailed investigation of hydrogen by infrared spectroscopy in CVD diamond, it was observed that hydrogen is incorporated into noncrystalline and defective regions of the polycrystalline film, mainly at grain boundaries and dislocations [146]. Seven different C–H stretching vibrations in the 2800 to 3100 cm^{-1} spectral range in polycrystalline diamond have been identified and summarized. Six of these vibrations are related to amorphous hydrogenated carbon and one to hydrogenated diamond. The relative abundance of the hydrogenated diamond peak is $<10\%$ of the amorphous peaks and suggests that hydrogen is primarily incorporated into the defective regions of polycrystalline diamond films [146]. In homoepitaxial diamond films, grain boundaries are not present, but the hydrogenated diamond peak remains; therefore, hydrogen is also incorporated into the lattice and not simply at grain boundaries [147]. It has been proposed that the hydrogenated diamond peak is caused by hydrogen on a substitutional site where it forms a sp^3 bond with a neighboring C atom [148]. Hydrogen in CVD diamond is so important that it has been proposed that a new classification of IIc diamond be used [147].

Although oxygen is a common reactant species used to improve CVD diamond growth, its incorporation and effect on diamond properties are not well understood. Additionally, diamond growth with oxygen has been shown to exhibit suppressed boron and silicon peaks, as well as broad luminescence peaks at 1.9 and 2.4 eV [149–152]. Based on diffusion experiments of a CrO_3-treated diamond surface in a hydrogen plasma, a broad oxygen-induced cathodoluminescent peak was observed at 4.64 eV [153]; however, this feature has not been observed in CVD diamond. The influence of oxygen on the properties of CVD diamond requires additional investigation.

Silicon is another impurity that is commonly observed in CVD diamond. Ion implantation experiments of Si into diamond have demonstrated that Si is related to the 1.681 eV (734.7 nm) defect [154,155]. The quadratic dependence of the luminescence on the implantation dose suggested that the center consisted of two Si atoms. The required annealing above 600°C suggests that a vacancy is either in the center, or the formation of the center is enhanced by vacancy diffusion [156]. This defect has also been observed in CVD diamond [156,157]. Because of the proximity of the defect to the 1.673 eV neutral vacancy defect, there was some initial debate whether this defect was caused by the neutral vacancy (GR1) or Si. Detailed cathodoluminescence has shown the phonon replica of the 1.681 eV peak is 0.064 eV, and the GR1 phonon replica is 0.036 eV. Because the phonon replica of the cathodoluminescence peak in the CVD diamond was 0.065 eV, the peak was conclusively demonstrated to be caused by the Si and not the isolated vacancy [158]. The source of the silicon is thought to be attributed to the etching of amorphous silica walls of the reactor or etching of the Si substrate [158,159].

Boron is often intentionally added to diamond film deposition because it makes it a *p*-type semiconductor. Boron sources have included gas sources, such as diborane [91], or solid sources like a BN substrate holder [160]. As noted in the section on homoepitaxial growth, the boron incorporation depends on the growth orientation. The impurity incorporation can be understood best by the use of a diagram generated for impurity incorporation in diamond in different directions. It was developed for synthetic HPHT diamond [161], but is probably applicable to CVD diamond [162]. This diagram is shown in Fig. 7-13 [162]. Boron is most readily incorporated on the {111} plane, then the {110} plane, and finally the least boron incorporation occurs on the {100} plane.

7.2.8 Additional Growth Techniques

The focus of this section has been on CVD of diamond films. There are other techniques for the growth of diamond or near-diamond materials that warrant a brief discussion. In

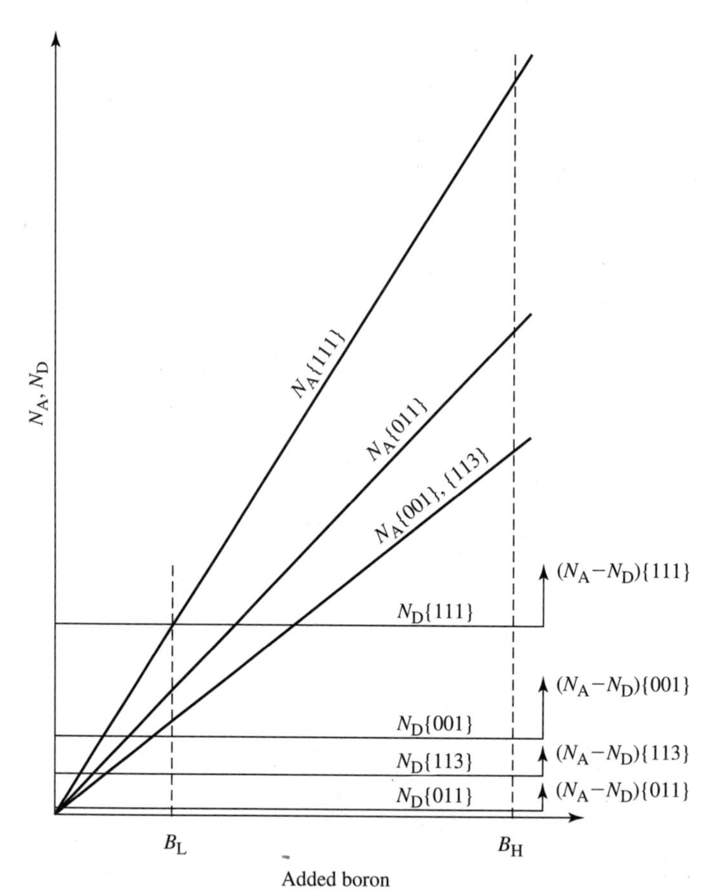

FIGURE 7-13 A diagram of the relative incorporation rates of boron and nitrogen on different crystallographic planes. (*From A. T. Collins,* Phil Trans R Soc London A, *vol. 342, 1993, p. 233. Reprinted with permission. Copyright 1993 The Royal Society.*)

addition to CVD diamond, synthetic diamond has also been made through HPHT synthesis or through dynamic shock wave deposition. Diamond films consist of sp^3-bonded carbon; however, other bonding configurations are possible for carbon. Graphite, an alternate C polytype with a hexagonal structure, is sp^2 bonded. Additionally, sp^1 hybridization is also possible. Carbon films with different carbon hybridization may be deposited by numerous chemical and physical vapor deposition techniques and yield a wide variety of films with differing properties.

High-pressure, high-temperature growth of diamond is a bulk-growth technique and has been reviewed in detail by others [163,164]. Primarily, two alternative approaches to diamond growth may be used. In one approach, carbon is directly converted into diamond. To achieve this, the pressure and temperature must be high enough to make diamond the thermodynamically stable polytype of carbon. For direct conversion of carbon into diamond, 13 GPa and 3300°K are often required [165]. In this technique, lonsdalite, sometimes incorrectly referred to as hexagonal diamond, may form [165]. Lonsdalite has a wurtzite structure that is hexagonal where the carbon is sp^3 bonded; however, it has *ababab* packing instead of the *abcabc* packing that is observed for diamond. The formation of hexagonal lonsdalite depends on the pressure and temperature and often occurs at a lower temperature than is required for the formation of cubic diamond.

The second approach to form bulk diamond involves the use of molten metal catalysts that allow less severe temperatures and pressures to be used than are required for the direct conversion synthesis. The conditions required for this technique are 1825°K to 1875°K and 5 to 6 GPa [164]. In this technique, carbon transport occurs across a temperature gradient in a molten metal from the source charge to the diamond seed crystal. The dissolved carbon crystallizes at the diamond seed crystals which are held at a lower temperature. The carbon source charge may be either graphite or diamond grit and the molten are typically Ni, Co, or Fe.

Another synthesis process for diamond involves dynamic shock waves. In this technique, extremely high pressures and temperatures are generated by an explosive charge. The first successful shock wave synthesis obtained pressures of 300 kbar for 1 μs [166]. In this experiment, graphite was transformed into particles the size of a micrometer. This technique is not suitable for the production of large bulk diamond but is suitable for producing diamond grit for fine polishing processes [167].

In addition to the synthesis of diamond, there are many near-diamond materials that may be formed [168–170]. These carbon films are often differentiated by the concentration of hydrogen in them and are grouped into two classifications. Diamond-like carbon (DLC) films are typically hydrogenated carbon films without any long-range structure. However, they may contain a microcrystalline phase. The films are called diamond-like because many of their properties resemble those of diamond, particularly, the high hardness and low coefficient of friction. These films contain a mixture of sp^3, sp^2, and sp^1 hybridized carbon. Hydrogen is often incorporated in high percentages from 10% [171] to 60% [172]. There is a correlation of the hydrogen incorporation and the carbon coordination. It is thought that the hydrogen passivates dangling bonds in the amorphous structure [170]. Hydrogen is often necessary for a wide optical bandgap and high electrical resistivity. Additionally, its presence inhibits the carbon from transforming into the graphite phase.

The second classification of carbon films is amorphous diamond. Amorphous diamond films are hydrogen-free, with the hydrogen concentration is below 10%, generally 0.5% to 5% [173]. The amorphous diamond films are primarily sp^3 bonded, where the fraction of sp^3 bonding may be as high as 85% [174]. This means that there is considerable C–C sp^3 bonding, in contrast to DLC which is predominantly C–H sp^3 bonded.

There are a wide variety of deposition techniques for DLC and amorphous diamond films. Diamond-like carbon films have been deposited by dc plasma, RF plasma-assisted CVD, sputtering, and ion-beam deposition, arc discharge, and laser ablation [169]. In these deposition techniques, carbon and hydrogen are present; the temperature at deposition is typically <325°C, and some sort of ion bombardment facilitates the formation of the DLC film. For amorphous carbon films, the various deposition techniques include magnetron sputtering [175], ion-beam sputtering [176], laser ablation [177], and vacuum-arc deposition [178]. In these techniques, the carbon species that arrives at the surface either has a high energy or is bombarded with energy when on the surface. Because these films contain low hydrogen concentrations, little or no hydrogen is intentionally present during deposition.

7.3 PROPERTIES OF DIAMOND

Numerous techniques have been used to characterize and understand the properties of diamond films. All of the techniques cannot be discussed here, but several primary characterization methods will be presented. The specific properties discussed include electrical, optical, and thermal. These were selected because they correlate to the potential applications of CVD diamond discussed in Section 7.4. Because of the limited amount of information on CVD diamond, and because the results of natural and HPHT diamond are relevant, characterization of all of these will be discussed.

7.3.1 Electrical Properties

Diamond is comprised of covalently bonded carbon atoms in a diamond cubic crystal structure. A band diagram for diamond is shown in Fig. 7-14 [179]. Diamond has an indirect bandgap of 5.45 eV with its conduction band minimum occurring in the $X\ \langle 100 \rangle$ direction [180]. The temperature dependence of the bandgap has been modeled with temperature and correlates well with measured values [181]. At 700°K, the bandgap is still 5.34 eV. The wide bandgap makes undoped diamond a good insulator. Additionally, it may be doped to make it semiconducting. A good understanding of the electrical properties of diamond is hampered by the structural defects and impurities that exist within natural and synthetic diamond. Many properties are limited by the defects within the diamond. Further improvements in the diamond quality will provide a better understanding of the fundamental properties of diamond.

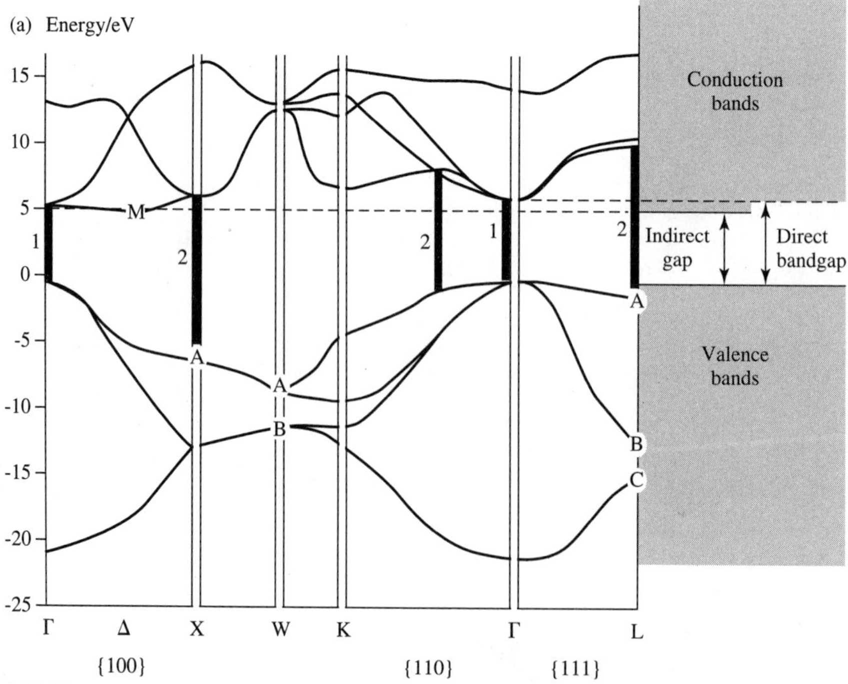

FIGURE 7-14 The band diagram for diamond. *(From F. P. Bundy, H. M. Strong, and R. H. Wentorf Jr in P. L. Walker and P. A. Thrower (eds.),* Chemistry and Physics of Carbon, *Marcel Dekker, New York, 1973, p. 213. Reprinted with permission. Copyright 1979 Academic Press LTD.)*

Undoped Diamond. The large bandgap of diamond provides a negligible concentration of intrinsic charge carriers in diamond. For example, at room temperature, the calculated intrinsic carrier concentration is $\approx 1 \times 10^{-27}$ cm^{-3}. A plot of the calculated intrinsic carrier concentration of diamond versus inverse temperature is shown in Fig. 7-15 [182]. As a comparison, the intrinsic carrier concentration for Si is also plotted. The wider bandgap of diamond produces substantially fewer intrinsic carriers than in Si. As a matter of perspective, a diamond the size of the earth would only have ~1 electron-hole pair at room temperature. Therefore, intrinsic conduction in diamond is unlikely. Experimentally, the resistivity of undoped diamond at room temperature is $>10^{16}$ Ω·cm and the activation energy is ≈1.4 eV [183]. Because an intrinsic conduction mechanism would have an activation energy of half the bandgap, or ≈2.7 eV, the conductivity is not dominated by intrinsic conduction. Even at 1000 K, the intrinsic carrier concentration is only 10^6 cm^{-3}. Because electronic devices are typically doped $>10^{17}$ cm^{-3}, intrinsic conduction will not degrade the operation of diamond electronics at the temperatures of interest. Additionally, the large bandgap makes undoped diamond a suitable insulator for isolation of electrical devices.

Although diamond is fundamentally a good insulator, as-deposited CVD diamond film often has a resistivity of $\approx 10^6$ Ω·cm [184]. Subsequent annealing of the film increases its resistivity to $\approx 10^{13}$ Ω·cm. When the annealed CVD diamond film was subjected to a hydrogen plasma, the resistivity was reduced to $\approx 10^6$ Ω·cm, similar to the as-deposited diamond. These results suggest that the near-surface regions of the as-deposited diamond is influenced by atomic hydrogen, which allows higher conductivity. In another experiment, natural diamond was placed in an hydrogen plasma to produce a low resistivity and the sample was annealed at different temperatures to observe the change in resistivity with

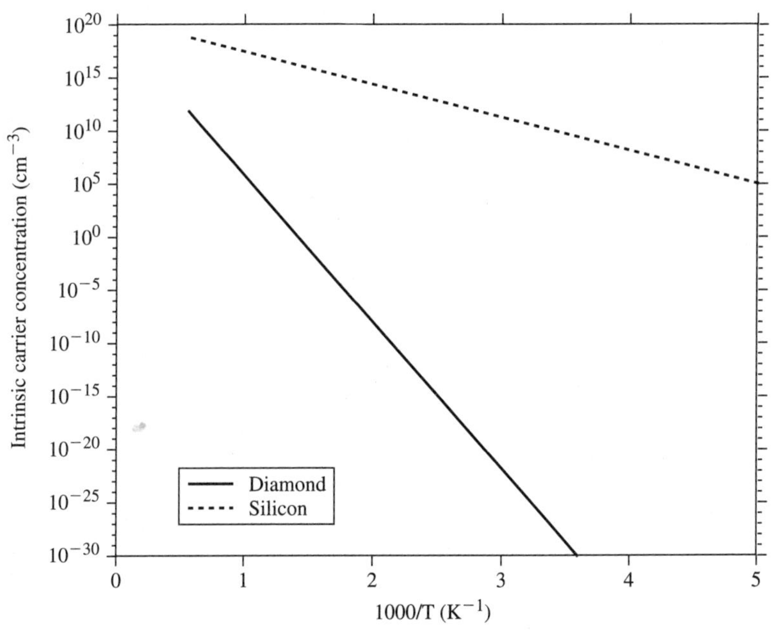

FIGURE 7-15 The intrinsic carrier concentration as a function of temperature for silicon and diamond.

annealing [185]. Heat treatment at 780°C for 2 hours in a nitrogen ambient produced a resistivity of $\approx 10^{14}$ Ω·cm. The removal of the hydrogen effect was also demonstrated by exposure of the as-deposited diamond surface to an oxygen plasma [186] or by simply cooling the as-grown sample in an oxygen-containing ambient [187].

Semiconducting diamond. The potential use of diamond as a semiconductor for electronic devices is one of the reasons that CVD of diamond has generated a tremendous amount of interest. Custers [188] was the first to realize that diamond was a semiconductor. Boron was shown to be an acceptor in natural [189,190] synthetic [191], and CVD diamond [89,192,193]. Boron lies 0.37 eV above the valence band [190]. Unfortunately, there have not been any definitive confirmations of *n*-type dopants in diamond that have a measurable conductivity at room temperature. Damage has been shown to yield *n*-type behavior and phosphorus has been shown to be incorporated into the lattice, but successful demonstration of reproducible, temperature-stable, low resistivity, and controllable *n*-type conduction still remains to be proven.

Doping in most semiconductors may be accomplished through in situ doping during deposition, diffusion, or ion implantation. Although all of these means are possible in diamond, there are some difficulties with diffusion doping and ion implantation into diamond that make these techniques more challenging. Diffusion doping of diamond is difficult because of the low diffusivity of most elements in diamond [194]. The diffusion of boron into diamond was achieved by a 30s, 1400°C anneal [195]. However, the penetration depth of boron was only 50 nm and the diffusivity of the B was estimated to be 10^{-12} cm²/s at 1400°C. This diffusion coefficient is 100 times smaller than the diffusion coefficient of B in Si at this temperature. Ion-implantation doping of diamond is also more difficult because of residual damage to the diamond [196–198]. If the damage exceeds a critical threshold, then a subsequent anneal will produce graphite and not diamond. Even if the diamond is not changed into graphite, residual damage often remains. For diamond, solid state recrystallization, as commonly observed for Si is not an option.

Dopants. To make full use of diamond as a semiconductor it is desirable to have both *p*- and *n*-type diamond. A summary of the dopants and vacancy level for diamond are shown in Fig. 7-16 [145]. Although some early investigations suggested that aluminum (Al) was the acceptor in natural diamond, the active acceptor in diamond was determined to be B [190,191]. The availability of a donor in diamond that will produce *n*-type material has yet to be identified.

Nitrogen is thought to be a donor in diamond. The isolated nitrogen and A aggregate nitrogen have donor-like properties at 1.7 and 4.0 eV respectively, while the B aggregate does not appear to behave like a donor [199]. An absorption threshold at 2.0 eV was attributed to substitutional nitrogen [200]. This level is too deep to contribute significant electrons to the conduction band. Although the substitutional nitrogen donor level in diamond is deep, electrons may be transferred to the conduction band by thermal and optical excitation [201]. Thermal resistance measurements of both natural and synthetic type Ib diamond indicated an activation energy of 1.6 to 1.7 eV [201,202]. The sign of the thermoelectric coefficient of the charge carriers in the synthetic type Ib diamond indicated that the carriers were electrons [202]. For the A aggregate (a pair of substitutional nitrogen atoms) a photoconductivity threshold was measured at 4.05 eV [203]. This is also too deep to contribute significant concentrations of electrons to the conduction band. Photo-Hall measurements were also used to verify that nitorgen acts as a donor. Photo-Hall measurements on a type IIa natural diamond indicated that the charge carriers were electrons with a room temperature mobility of 2800 cm²/V·s [203]. Previous photo-Hall

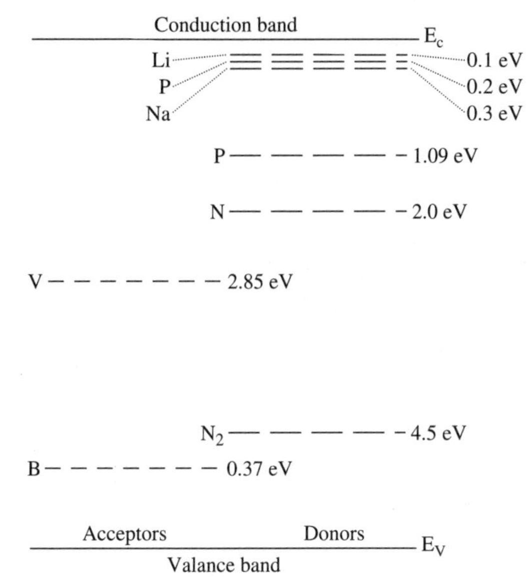

FIGURE 7-16 The defect levels in diamond.

measurements also indicated that electrons were generated; however, the measured mobilities were lower [204,205]. The difference in measured values was attributed to the use of type I diamonds in the experiments that recorded lower values for the electron mobility [203].

Phosphorus (P), sodium (Na) and lithium (Li) are also potential n-type dopants. Phosphorus-doped polycrystalline diamond films were deposited on n-type Si [206, 207]. Although Hall-effect measurement indicated *n*-type conduction, the possibility of conduction through the substrate was likely. Additionally, the conductivity was not a function of the dopant concentration which indicates P might not be controlling the conduction in these films. It is important to note that although the P/C ratio in the gas phase was $\approx$10,000 ppm, the solid phase P/C ratio was $<$1 ppm. This data supports the theory that incorporation of phosphorus into diamond is low. Additional evidence has been provided on the difficulty of phosphorus incorporation into polycrystalline and homoepitaxial diamond [208]. The phosphorus incorporation into polycrystalline diamond films was approximately 10 times that in homoepitaxial diamond films and suggests that phosphorus was incorporated into the grain boundaries. Additionally, the presence of P produced a microcrystalline grain morphology instead of the well-faceted diamond grains that were obtained without phosphorus. Ion-implantation of Li and Na has been used to increase the conductivity of diamond [209]. Unfortunately, the conduction mechanism was shown to be variable range hopping between implant sites, and not valence-band conduction.

Although definitive *n*-type diamond has not been obtained without external excitation, theoretical analysis suggests that there are some potential candiates to produce *n*-type material. Phosphorus is a substitutional impurity that has the potential to donate an electron. The activation energy of phosphorus is estimated to be 1.09 eV [210] or 0.2 eV [211]. It is unclear which value is more appropriate for phosphorus. Irrespective of the activation energy, there are some other complications if phosphorus is used as a dopant. It has a high energy of segregation inside diamond; therefore, its incorporation is expected to be low [211]. In addition to substitutional impurities, interstitial lithium has been proposed as a donor with an activation energy of 0.1 eV. However, it also has a high energy of segregation and its incorporation is expected to be low [211]. An additional problem with lithium is that it is a fast diffuser in diamond. At room temperature, it will diffuse more than 1 μm in one year. This leads to deactivation of the lithium through trapping at a defect site or loss of control of the extent of the *n*-type dopant region. Sodium is also a potential *n*-type impurity in diamond that sits on an interstitial site [212]. Sodium is also thought to have a low incorporation into diamond. Sodium should be immobile at moderate temperatures and should be more stable as a dopant. These theories suggest some potential for *n*-type diamond, but verification of these theories requires experimental investigation.

Mobility. The mobility of diamond is assessed by three primary techniques. These include Hall-effect measurements, photo-induced conductivity measurements, and time-of-flight measurements. Hall-effect measurements of natural, single-crystal diamonds have shown a hole mobility as high as 2010 cm²/V·s in selected high-quality diamonds. Other researchers have reported similar, but somewhat lower values [213–215]. Time-of-flight measurement of the hole mobility in high-resistivity natural diamond was 2100 cm²/V·s, similar to that obtained by Hall-effect measurements [216,217]. Photoinduced conductivity measurements in type IIa single crystal diamond have a

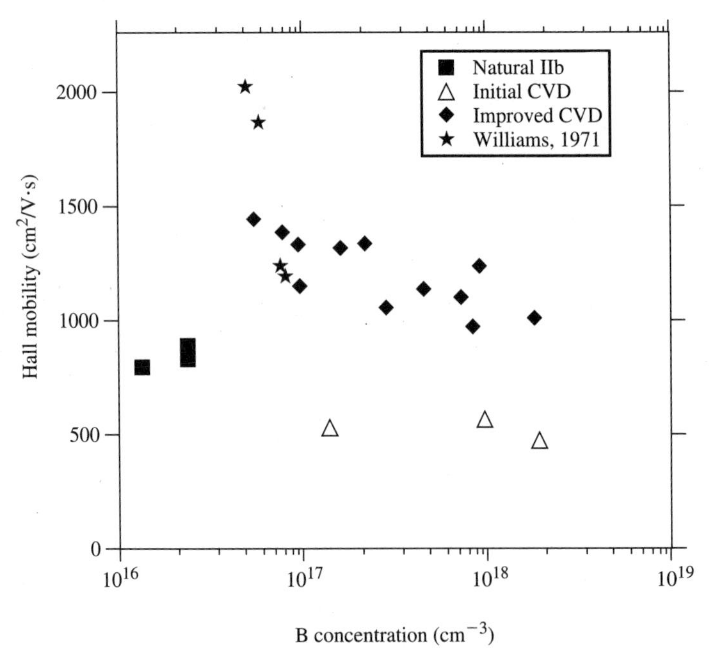

FIGURE 7-17 The room temperature mobility as a function of boron concentration. (*From B. A. Fox, et al.,* Dia Rel Mat, *vol. 4, 1995, p. 622. Reprinted with permission. Copyright 1995 Elsevier Science S.A.)*

room temperature hole mobility of ≈1400 $cm^2/V \cdot s$ [218]. It is unclear if the values for the mobility mentioned above are indicative of the upper limit. Natural diamonds are known to contain defects that may limit the ultimate mobility of holes in diamond. A summary plot of the room temperature Hall mobility of natural type IIb diamond and homoepitaxial diamond films versus B concentration is shown in Fig. 7-17 [219].

The temperature-dependent mobility of natural and two CVD diamond films is shown in Fig. 7-18 [220]. The scattering mechanisms for charge carriers in diamond are not well-understood. This is because of both the lack of a defect-free diamond to obtain the ideal mobility, and the lack of fundamental understanding of the critical parameters for theoretical analysis. In the absence of defects such as vacancies, dislocations, stacking faults, twins, and grain boundaries, the mobility in elemental semiconductors is often limited by impurity scattering and lattice scattering. The mobility is determined by the lowest mobility through a sum of the individual mechanisms based on Mathiessen's rule.

Impurity scattering may occur by either ionized [221] or neutral impurities [222] and is typically more important at lower temperatures. Ionized impurity scattering follows an approximate $T^{1.5}$ dependence [221] although the scattering caused by neutral impurities is temperature-independent [222]. Unfortunately, a calculation of the mobility limited by ionized impurities was too large to explain the low-temperature diamond mobility data [223]. Additionally, neutral impurity scattering was estimated at ≈7000 $cm^2/V \cdot s$, but this was also too high to explain the observed mobility [223]. Experimental data have shown a linear dependence of mobility on N_D for natural

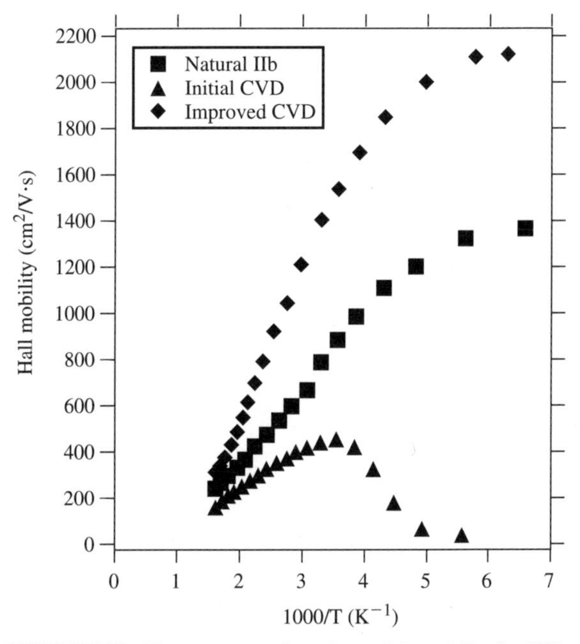

FIGURE 7-18 The temperature dependence of the mobility for CVD and natural diamond samples. *(From D. M. Malta, et al.,* Second International High-Temperature Electronics Conference, *1994, p. P-75. Reprinted with permission. Copyright 1994 High Temperature Electronics Conference.)*

diamonds and suggests that the mobility is limited by the ionized impurities [224]. The best fit to the experimental data is obtained through an empirical, temperature-independent ionized-impurity factor [224]. This contradicts the theory that ionized impurities have a $T^{1.5}$ temperature dependence. Clearly, the low-temperature mobility behavior is complicated and not yet understood.

At higher temperatures, lattice scattering is typically dominant and occurs by acoustic-phonon or nonpolar optical-phonon scattering. Acoustic-phonon scattering has an ideal temperature dependence of $T^{-1.5}$ [225]. The temperature dependence of nonpolar optical-phonon scattering cannot be reduced to a simple temperature dependence, but typically has a stronger temperature dependence than acoustic-phonon scattering [226]. The observed high-temperature dependence of the mobility is $T^{-2.2}$ to $T^{-3.1}$ for natural and CVD diamond [91,160,214,227]; therefore, acoustic-phonon scattering cannot account for the temperature dependence of the mobility. The large slope in elevated-temperature regime is likely caused by the combination of optical and acoustic-phonon scattering [227]. The transition temperature between acoustic-phonon-dominated and optical-phonon-dominated scattering is estimated to be 400°K [217].

The maximum hole mobility in B-doped CVD diamond measured by the Hall effect is 1590 $cm^2/V \cdot s$ [228]. This compares favorably with the highest hole mobility of 2010 $cm^2/V \cdot s$ measured in a natural diamond [227]. As discussed earlier, it is unclear if the mobility of 2010 $cm^2/V \cdot s$ measured in a natural diamond [227] represents a limit to the mobility in a defect-free diamond. Additional improvements in diamond growth are required to answer this question. For electrons, the mobility maximum is much less understood. A value of 2000 $cm^2/V \cdot s$ is typically used, but measurements by photo-Hall [203–205,213], time-of-flight [216,229], and photo-induced conductivity [218] yield a variety of values.

Carrier Concentration. The carrier concentration p in diamond for a single acceptor level in the presence of compensating donors may be approximated by [230]

$$\frac{p(N_D + p)}{N_A - N_D - p} = \frac{N_V}{g} \exp\left(-\frac{E_a}{k_B T}\right) \tag{3}$$

where N_D = donor concentration
N_A = acceptor concentration
N_V = density of states of the valence band
g = degeneracy of the valence band
E_a = activation energy of the acceptor
k_B = Boltzmann's constant
T = absolute temperature

This equation does make some simplifying assumptions [162]; however, this equation is shown to correlate well to the measured carrier concentration [227]. An example of the fit of this expression to the temperature-dependent carrier concentration is shown in Fig. 7-19 [220].

The temperature dependence of the carrier concentration is important for electronic device applications. Typically, the carrier concentration of a material may be defined by three regions, as shown in Fig. 7-20 for both Si and diamond doped $\approx 10^{17}$ cm^{-3} [182].

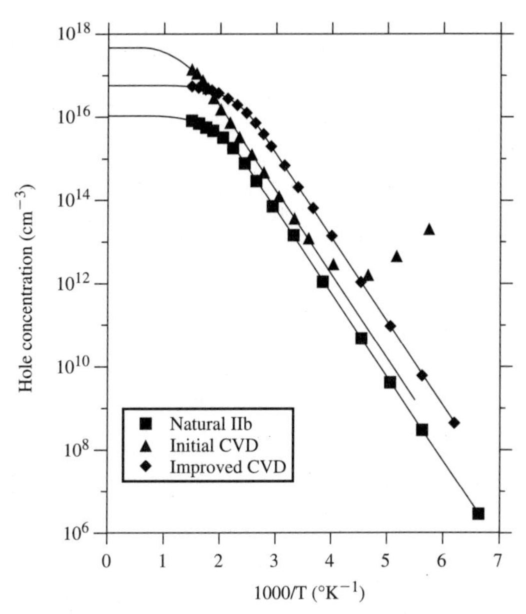

FIGURE 7-19 The temperature dependence of the carrier concentration for CVD and natural diamond samples. *(From D. M. Malta, et al.,* Second International High-Temperature Electronics Conference, *1994, p. P-75. Reprinted with permission. Copyright 1994 High Temperature Electronics Conference.)*

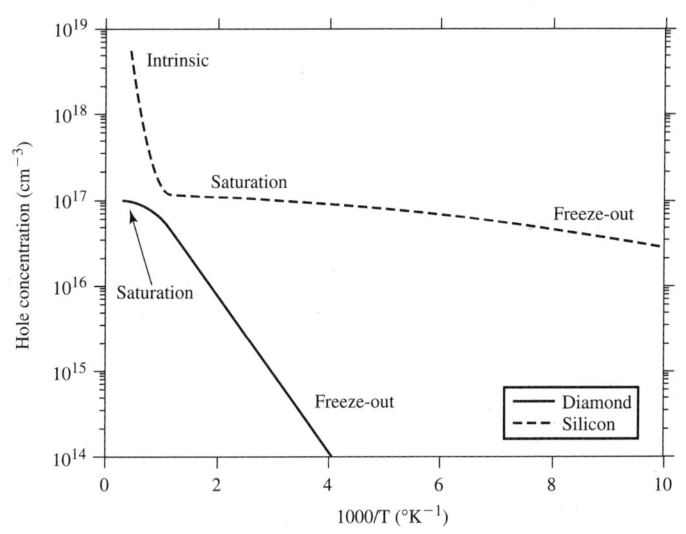

FIGURE 7-20 The carrier concentration as a function of temperature for silicon and diamond doped $\sim 10^{17}$ cm^{-3}.

In the freeze-out range, the carrier concentration is thermally activated from an impurity level. A saturation range occurs when the impurities in the material are fully ionized and the carrier concentration is relatively stable. In the intrinsic range, the carrier concentration is determined by thermal generation of electron-hole pairs that exceeds the impurity concentration in the material. For boron-doped natural diamond, the acceptor concentration is $\approx 10^{16}$ cm^{-3} and the donor concentration is $\approx 10^{15}$ cm^{-3}. At room temperature diamond is in the freeze-out regime. The hole concentration is approximately 10^{13} cm^{-3} and is only 0.1% of the acceptor concentration [227]. This differs substantially from Si which is in the saturation range at room temperature. A comparison of the temperature dependence of the carrier concentration of Si and diamond indicates that Si exhibits all three temperature regimes, although diamond only exhibits a freeze-out and saturation regime. Diamond does not exhibit an intrinsic regime up to the temperature at which it transforms to graphite because of its 5.45 eV bandgap.

The effect of the compensating species for diamond is different than the effect for Si because diamond is in the freeze-out regime and Si is in the saturation regime. The role of compensation can be demonstrated through the data shown in Fig. 7-21 [231]. The carrier concentration is approximately 0.01% of the boron concentration. The detrimental effects of compensation may further be demonstrated through Fig. 7-22 [182]. The uncompensated acceptor concentration, N_A-N_D, is held constant at 1×10^{16} cm^{-3} while the compensation is varied. The room temperature carrier concentration varies by more than two orders of magnitude.

Activation Energy. The most widely accepted value for the activation energy of boron is 0.3685 eV [199,227]. Other values for diamond are in the 0.29 to 0.40 eV range [214,223,227,232]. This variation is likely to be caused by both experimental error and the fact that the activation energy varies with dopant concentration. Because ionization is not complete, differences in the activation energy are evident through changes in the temperature-dependent carrier concentration and conductivity of the material. It is unclear whether the activation energy is determined by the acceptor concen-

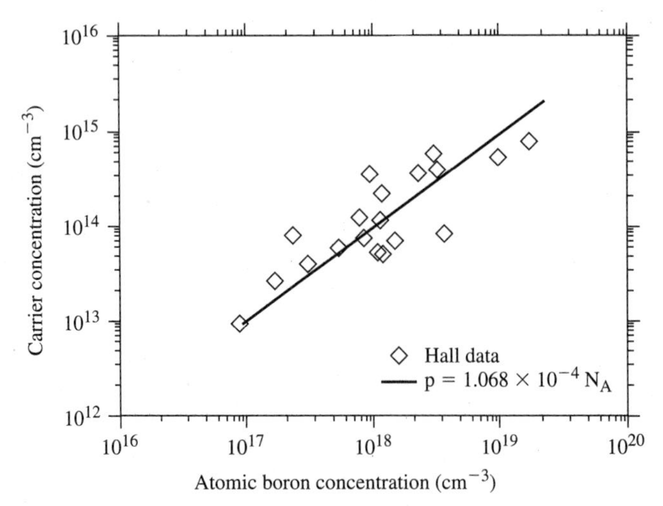

FIGURE 7-21 The carrier concentration as a function of the boron concentration for homoepitaxial diamond films.

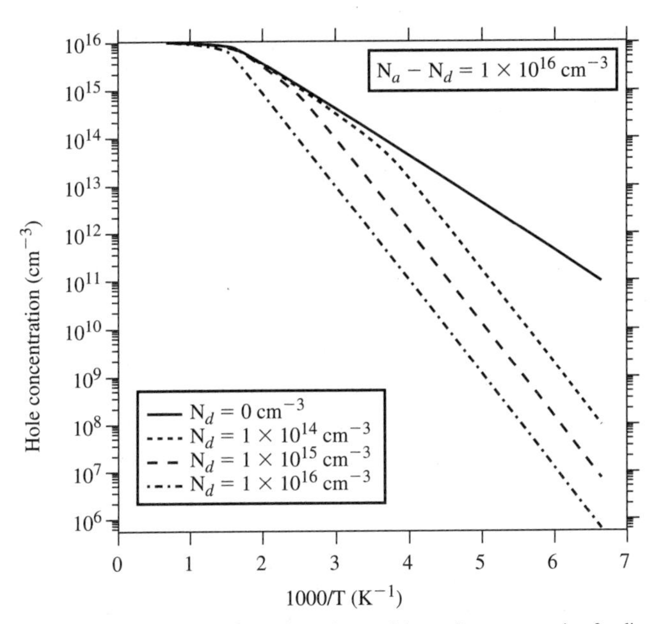

FIGURE 7-22 The temperature dependence of the carrier concentration for diamond films with different concentrations of compensation.

tration or the donor concentration, or some combination of them in *P*-type material. The general expression is

$$E_A = E_A^o - \alpha N^{1/3} \tag{4}$$

where *N* may be defined in several different ways. It has been defined as the majority-dopant concentration, the minority-dopant concentration, the difference in the majority- and minority-dopant concentration, or the ionized-impurity concentration [233,234]. One expression for diamond where *N* is the concentration of ionized impurities was used to calculate a value of 8.6×10^{-8} eV/cm for α [235]. Using this expression, it can be calculated that the activation energy becomes negligible when the ionized impurity concentration is $\sim 8 \times 10^{19}$ cm^{-3}. Other experimental evidence for lowering of the activation energy has been correlated to the carrier concentration instead of ionized-impurity concentration; however, the effect is the same [236].

Nonvalence Band Conduction. General conduction is often described through the relation [237,238]

$$\sigma = \sigma_1 \exp\left(\frac{-E_1}{k_B T}\right) + \sigma_2 \exp\left(\frac{-E_2}{k_B T}\right) + \sigma_3 \exp\left(\frac{-E_3}{k_B T}\right) \tag{5}$$

where σ_i is the conductivity coefficient and E_i is the activation energy. Each term is related to a different conductivity mechanism where σ_1 is related to valance-band conduction, σ_2 is related to impurity-band conduction, and σ_3 is related to nearest neighbor-hopping conduction. Typically, $\sigma_1 < \sigma_2 < \sigma_3$ and $E_1 > E_2 > E_3$, so valence-band conduction dominates at the highest temperature, impurity-band conduction dominates at intermediate temperatures, and hopping conduction dominates at low temperatures. Other investigators have shown that at lower temperature, variable range-hopping [239,240] may also occur with a temperature dependence of

$$\sigma = \sigma_4 \exp\left(-\frac{T_o}{T}\right)^n \tag{6}$$

where $n = 0.25$. Nonvalence-band conduction has been observed in both synthetic [241] and CVD diamond at low temperatures [160].

One method to differentiate nearest neighbor and variable range-hopping is through characterization of samples with the different compensation ratios. In one experiment, compensation was introduced via 0.7 MeV electron irradiation [242]. A single activation energy was not able to be assigned to the different compensation ratios; however, a plot of $\ln \sigma$ versus $T^{-1/4}$ produced straight lines. This suggests that variable range-hopping was the active conduction mechanism. Additional support was provided by the $T^{3/4}$ dependence of the activation energy, which is expected for variable range-hopping. For synthetic diamond, below 100°K to 150°K variable range-hopping is thought to be the dominant conduction mechanism [233].

Another important conduction mechanism is the Mott metal-insulator transition [239]. For p-type material, above the transition concentration $N_{A,M}$ metallic conduction occurs. This transition occurs at

$$N_{A.M.}^{1/3}\, a_H \approx 0.2 \tag{7}$$

where a_H is the Bohr radius [239]. For diamond, where the Bohr radius is 0.325 nm, the Mott transition is predicted to occur at $\approx 2 \times 10^{20}$ cm^{-3}. This transition has been observed at $\approx 10^{20}$ to 10^{21} cm^{-3} in reasonable agreement with the theory [243,244].

Electron Emission. The wide bandgap of diamond produces another interesting effect. The conduction band approaches the vacuum level. The energy difference between the vacuum level and the conduction band is known as the electron affinity, while the work function is the energy difference between the Fermi level and the vacuum level. Both quantities are used in evaluation of electron emission. In most metals the work function is 4 to 5 eV. In diamond, there is some evidence that the conduction band may exceed the vacuum level, yielding a negative electron affinity [245]. For example, capacitance-voltage measurements determined that the electron affinity of diamond was -0.7 eV for the {111} plane [246]. Ultraviolet photoelectron spectroscopy results have also suggested that the conduction band lies above the vacuum level [247]. Exposure of natural type IIb {111} diamond to argon (Ar) and hydrogen plasmas suggests that the nature of the surface may alter the electron affinity of diamond [249]. Using photoemission spectroscopy, a negative electron affinity exists for a hydrogen-terminated surface, although surfaces exposed to an argon plasma had a positive electron affinity of 1.0 eV. A comparison of the work function for diamond, Si molybdenum (Mo), and cesium (Cs) is shown in Fig. 7-23 [250].

The emission characteristics of diamond are not well understood, but some initial investigations have been performed. The emission of an Si tip and a diamond-coated Si tip were compared [251]. The current from the diamond-coated tip was ≈ 10 times greater than the current from the Si tip. Additionally, the emission current of the diamond-coated Si tip was more stable than the emission current from the Si tip. Diamond-coated molybdenum broad-area emission has also been investigated [252]. The turn-on field to generate current was 3 to 5 MV/m, although molybedenum samples have a turn-on field of

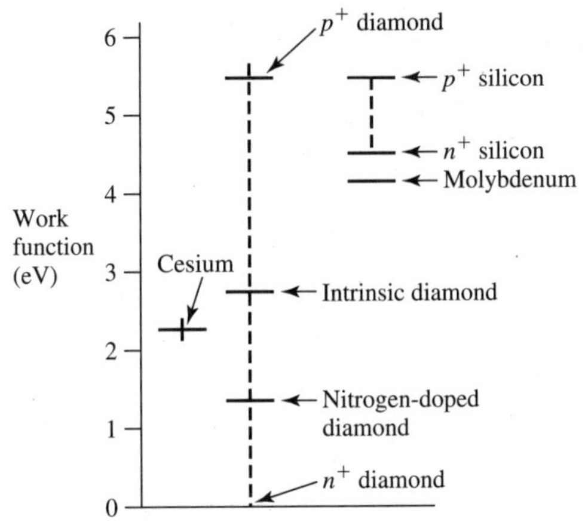

FIGURE 7-23 The work function of diamond relative to some other materials. *(From K. V. Ravi,* Mat Sci, Eng, *vol. 19, 1993, p. 203. Reprinted with permission. Copyright 1993 Elsevier Science S.A.)*

>10 MV/m. Additionally, the diamond emission had a more uniform spatial distribution than the molybdenum emitter. These initial investigations indicate that the electron emission from diamond is viable and warrants continued study.

7.3.2 Optical Properties

Optical characterization is one of the simplest and most powerful assessment techniques for the quality of diamond films. For example, infrared spectroscopy is used to distinguish between the different types of diamond. Additionally, Raman spectroscopy is a simple technique to qualify the amount of sp^3 and sp^2 bonding. The luminescence of diamond under different excitation means also aids in the characterization of diamond, including cathodoluminescence, photoluminescence, and electroluminescence. Finally, optical transmission is directly related to the performance of diamond in various window applications discussed later in Section 7.4.2.

The transmission of light through diamond is one of its attractive features. The refractive index of diamond is

$$n^2 - 1 = \frac{4.3356\lambda^2}{\lambda^2 - (0.1060)^2} + \frac{0.3306\lambda^2}{\lambda^2 - (0.1750)^2} \tag{8}$$

where λ is the wavelength in micrometers. This expression is valid from 0.225 μm (UV) to the infrared region [253]. The refractive index may be used to calculate the reflection coefficient R. In air, this expression is

$$R = \left(\frac{n-1}{n+1}\right)^2 \tag{9}$$

For 1 μm electromagnetic radiation, the refractive index is 2.4 and the reflection coefficient is 17% [254]. The low refractive index of a diamond leads to high surface reflection.

The transmission through diamond depends both on the reflection coefficient and the thickness and is given by

$$T = \frac{[(1 - R)^2 \exp(-\mu t)]}{[1 - R^2 \exp(-2\mu t)]} \tag{10}$$

where μ is the absorption coefficient. A plot of the absorption coefficient of diamond is shown in Fig. 7-24 [255]. Using Eq. 10 and the absorption coefficient, for a 100 μm thick film, the transmission is >60% above 230 nm, except near the absorption edge and the two- and three-phonon regions. In the infrared regions, the transmission is typically 70% [255].

Raman Spectroscopy. Raman spectroscopy uses the scattering of light within a material to characterize its structure. An incident laser beam is scattered inside a crystal because of lattice vibrations or phonons. The light may be elastically scattered (Rayleigh scattering) where the frequency has the same frequency as the incident light, or inelastically scattered where the frequency has been shifted by an amount equal to the vibrational frequency of the lattice. There are two type of inelastic scattering. Normal Raman (or Stokes) scattering occurs when energy from the incident light beam generates a phonon. Anti-Stokes scattering occurs when the incident light annihilates an existing lattice phonon. Because the vibrational nature of phonons is dependent on the atomic bonding, Raman scattering may be used to characterize the nature of the chemical bonds.

Raman spectroscopy is probably the most widely used technique to verify the presence of diamond. Seto, et al. [256,257] were the first to realize that Raman spectroscopy might be more useful than x-ray or electron diffraction in assessing the structure of CVD carbon deposits. This is because the wavelength shift of sp^3 bonded diamond differs from that of sp^2 bonded graphite. In this technique, the film is probed with photons from a laser beam, typically the 514 nm line from an Ar-ion laser. The Raman frequency is the maximum frequency phonons may propagate through a diamond crystal. Diamond exhibits a sharp Raman peak at 1332 cm^{-1} [258] because of the 165 meV longitudinal optical phonon. Features associated with sp^3 bonding occur below 1332 cm^{-1} whereas features associated with sp^2 bonding occur above 1332 cm^{-1}, because the bond between sp^2 carbon is stronger than the bond between sp^3 bonded carbon [259]. Therefore, graphite (band G) or non-diamond (disorder band D) exhibits broad peaks near 1355 cm^{-1} [260] and near 1550 cm^{-1},

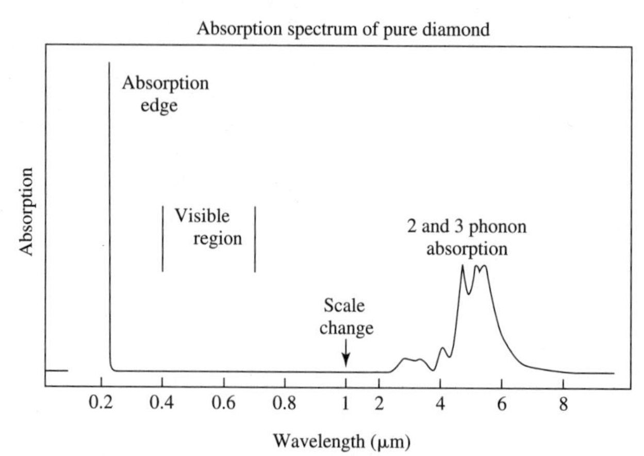

FIGURE 7-24 The absorption spectra of diamond. *(From A. T. Collins,* Physica B, *vol. 185, 1993, p. 284. Reprinted with permission. Copyright 1993 Elsevier Science B.V.)*

TABLE 7-3 Raman Spectra

Peak position (cm^{-1})	Description
1140	Amorphous sp^3 carbon
1284	^{13}C fundamental peak
1315-1325	Lonsdalite (hexagonal carbon)
1332	^{12}C fundamental peak
1355	Microcrystalline graphite
1500	Amorphous sp^2 carbon
1580	Graphite

respectively [261]. The ability to distinguish diamond from non-diamond depositions is the reason for the popularity of this technique. A summary of the Raman peak positions and the structural feature associated with each peak is given in Table 7-3.

Raman spectroscopy has also been used to assess the quality and structure of diamond films. An example of the various Raman spectra for natural diamond, high-quality CVD diamond, mixed graphite and diamond, and graphite is shown in Fig. 7-25 [262]. In addition to assessing whether diamond is present, the relative peak height of the 1332 cm^{-1}

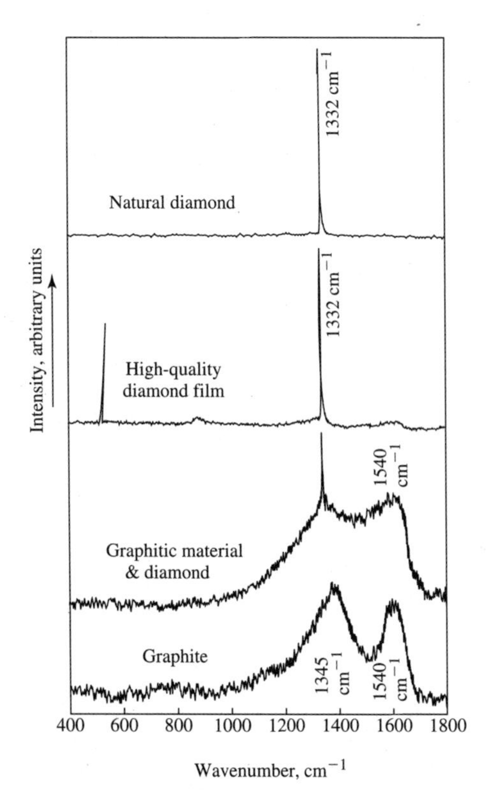

FIGURE 7-25 A comparison of the Raman spectra for different types of carbon samples. *(From P. K. Bachmann and R. Messier,* Chem Eng News, *1989, p. 24.)*

diamond peak to the ≈1550 cm^{-1} non-diamond peak is often used to determine the quality of the deposition [263]. The cross section of graphite (1580 cm^{-1}) is 50 times that of diamond (1332 cm^{-1}); therefore, Raman scattering enhances the graphitic component within a primarily diamond sample [264]. Additionally, graphite is absorbing whereas diamond is transparent in the visible range. These differences make quantifying the amount of graphite present in diamond difficult [259]. Although there are some questions about the quantitative usefullness of this ratio [57], it is still commonly used as a quality assessment.

In addition to the ratio technique, the full width at half maximum (FWHM) of the first order diamond Raman peak has also been used as a quality assessment. The FWHM of the 1332 cm^{-1} peak has been correlated to thermal conductivity [265]. Additionally, a correlation was observed between the dislocation density and the FWHM of the Raman line for both ^{13}C and ^{12}C epitaxial diamond films. Increases in the dislocation density of 5 to 10 times produced broadening of the Raman peak by 20% to 30% [266,267]. Another study has shown a correlation of the Raman FWHM to the collection distance of a radiation detector [268]. In many cases a narrow FWHM of the Raman 1332 cm^{-1} peak is a necessary condition for high quality, but not sufficient to ensure high quality.

Raman characterization is used to assess other properties of diamond as well. Grain orientation may be characterized through polarized Raman spectroscopy. The orientation relationships of monocrystalline diamond deposits on cBN [269] and oriented growth on Si [44, 270] are verified through polarized Raman spectroscopy. Additionally, changes in the peak position are correlated with film stress. The Raman peak may shift, positively or negatively, because of internal stress. Shifts in the range of 4 to 13 cm^{-1} for diamond films deposited on alumina, tungsten carbide, and Si have been observed [271]. The peak shift was correlated to the internal stress through the factor 2.9 to 3.6 cm^{-1} GPa^{-1} [272].

The position of the Raman peak is also related to the isotopic purity of the diamond. The presence of the heavier ^{13}C shifts the Raman frequency to lower frequencies than the ^{12}C diamond. The shift is based on the square root of the atomic mass $(^{12}/_{13})^{0.5}$; therefore, the Raman line occurs at 1280 cm^{-1} for ^{13}C instead of 1332 cm^{-1} for ^{12}C [273,274]. The isotopic frequency shift of the Raman peak allows the nature of some defects to be established. For example, a C–N Raman peak occurs at 1344 cm^{-1}. A change from ^{14}N to ^{15}N did not shift the peak; however, a change from ^{12}C to ^{13}C shifted the peak by $(^{12}/_{13})^{0.5}$ to 1292 cm^{-1} [275]. Through isotopic investigations, Raman was used to establish that in this defect, the nitrogen is stationary and the carbon vibrates.

Absorption Spectroscopy. The absorption or transmission of ultraviolet (UV), visible (VIS), or infrared (IR) electromagnetic radiation provides information about the defects and bonding within diamond. The infrared absorption of diamond is probably the most thoroughly studied of these three techniques. One advantage of the IR spectra is that it provides quantative information about the defect concentration within a sample. Infrared absorption is observed if there is a change in a dipole moment. Therefore, not all transitions involving vibrational states may be observed by this technique. A summary of the IR transitions is given in Table 7-4. More detailed overviews of this characterization technique are available [276–279].

The common absorption feature shown in the IR spectra in Fig. 7-26 [145] is the two-phonon absorption that occurs between approximately 1500 and 2700 cm^{-1}. The IR spectra of the type IIa diamond represents the absorption resulting from the diamond lattice and is not generated by defects. In a monovalent material, radiation is not absorbed with the production of one phonon because no dipole moment is produced. The introduction of impurities changes the crystal symmetry and allows dipole moments to be generated by one-phonon absorption. The most common impurities in diamond are nitrogen, boron and hydrogen. The effects of the impurities on the IR absorption are evident in Fig. 7-26. The type Ia and Ib spectra are for diamonds that contain nitrogen. The type IIb diamond contains boron and the type IIc diamond contains hydrogen.

TABLE 7-4 Absorption Spectra

Peak position (eV)	Peak position (nm)	Peak position (cm^{-1})	Description
0.140	8850	1130	Substitutional isolated N
0.146	8511	1175	B nitrogen aggregate
0.148	8354	1197	A nitrogen aggregate
0.159	7800	1282	B nitrogen aggregate
0.159	7800	1282	A nitrogen aggregate
0.112–0.165	11,111–7519	900–1330	Substitutional boron
0.170	7300	1370	Platelets: originally attributed to nitrogen; now thought to be a stacking fault
0.180	6897	1450	C–H bending
0.207–0.310	6000–4000	1667–2500	Two phonon absorption
0.338	3561	2808	Substitutional boron
0.359–0.384	3488–3226	2900–3100	$C–H_n$ symmetric and antisymmetric stretching
0.4584	2703	3700	Three-phonon absorption

Nitrogen has a dramatic effect on many of the properties of diamond. Substitutional nitrogen atoms increase the lattice spacing because the effective volume of a single substitutional nitrogen atom is 1.4 ± 0.06 times that of the carbon atom it replaces [280]. The absorption characteristics of diamond may be used to determine the local bonding structure of incorporated nitrogen. Nitrogen degrades the octahedral symmetry of the diamond lattice (centrosymmetric) which becomes tetrahedral (noncentrosymmetric). As a result, one-phonon absorption occurs in the range of 1000 to 1400 cm^{-1}. Each nitrogen defect type has a unique absorption characteristic. The IR absorption spectra of a type Ia and Ib is shown in Fig. 7-26. For the type Ia diamond, the peaks resulting from A aggregates, B aggregates, and platelets (*p*) are shown. Platelets are planar aggregates that are parallel to the {100}. The platelets have a thickness of approximately one-third the unit cell length and may be several micrometers width [281,282]. Although originally thought to be attributed to nitrogen, it is now believed that the platelets are related to stacking faults [283].

Detailed investigations of the nitrogen defects in diamond have been performed. The IR absorption of the A aggregate occurs with a maximum at 1282 cm^{-1} (0.159 eV) and a subsidiary peak at 1212 cm^{-1} (0.150 eV) [138]. The concentration of nitrogen is 17.5 times the absorption coefficient (in cm^{-1}) at 1282 cm^{-1} [284]. The IR absorption of the B aggregate consists of a sharp peak at the Raman frequency of 1282 cm^{-1} (0.159 eV), a broader maximum at 1175 cm^{-1} (0.146 eV), a shoulder at 1096 cm^{-1} (0.138 eV), and a subsidiary peak at 1010 cm^{-1} (0.125 eV). The concentration for nitrogen in a B aggregate is 103.8 times the absorption coefficient (in cm^{-1}) at 1282 cm^{-1} [139]. It was the two groups of peaks in the IR spectra that led researchers to identify two defects labeled A and B [285] which were later established as A and B aggregates. Because the peaks for A and B aggregates in diamond overlap, the contribution from each component must be resolved. This was done through characterization of a series of samples with different amounts of A and B aggregates [286]. The third common form of nitrogen is substitutional where the IR absorption occurs at 1130 cm^{-1} and a sharp peak at 1344 cm^{-1} from a local mode vibration. The concentration of substitutional nitrogen is 22.0 times the absorption coefficient (in cm^{-1}) at 1130 cm^{-1} [287]. Nitrogen interstitials have also been observed and have an absorption at 1450 cm^{-1} [144].

Boron and hydrogen also are important impurities in diamond. For boron, the prominent absorption often occurs at 2462 cm^{-1} (0.305 eV), 2802 cm^{-1} (0.347 eV), and 2930 cm^{-1} (0.363 eV). Additionally, approximately 20 absorption lines have been

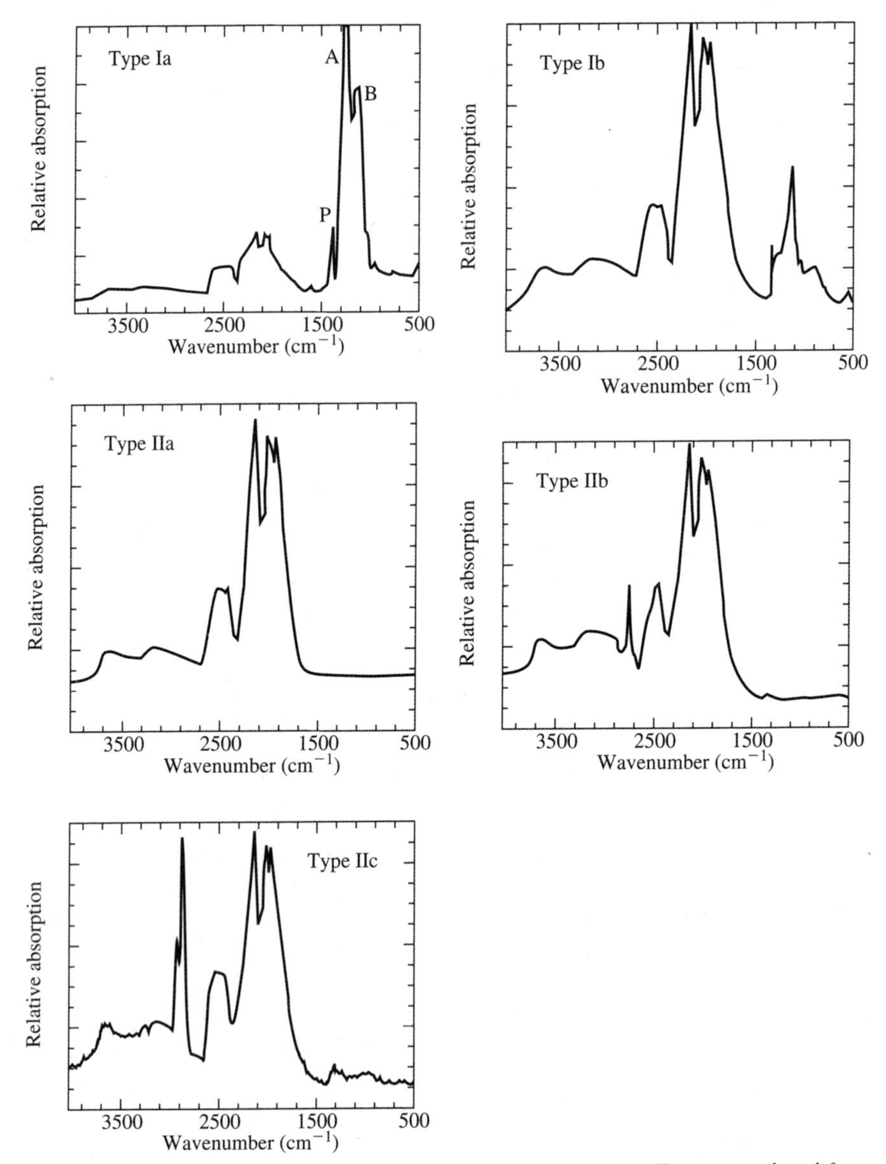

FIGURE 7-26 The IR spectra for type Ia, Ib, IIa, IIb and IIc samples. *(Figure was adapted from W. J. P. van Enckevort, in K. E. Spear and J. P. Dismukes (eds.),* Synthetic Diamond, *John Wiley & Sons, New York, 1994, p. 307.)*

identified between 2600 and 3000 cm^{-1} [288, 289]. The boron concentration has been correlated to the IR absorption by two means. The concentration is 1.2 times the absorption coefficient in cm^{-1} at 1282 cm^{-1} [278]. Alternately, the integrated absorption between 0.325 and 0.360 eV has been correlated to the boron concentration [190]. For hydrogen,

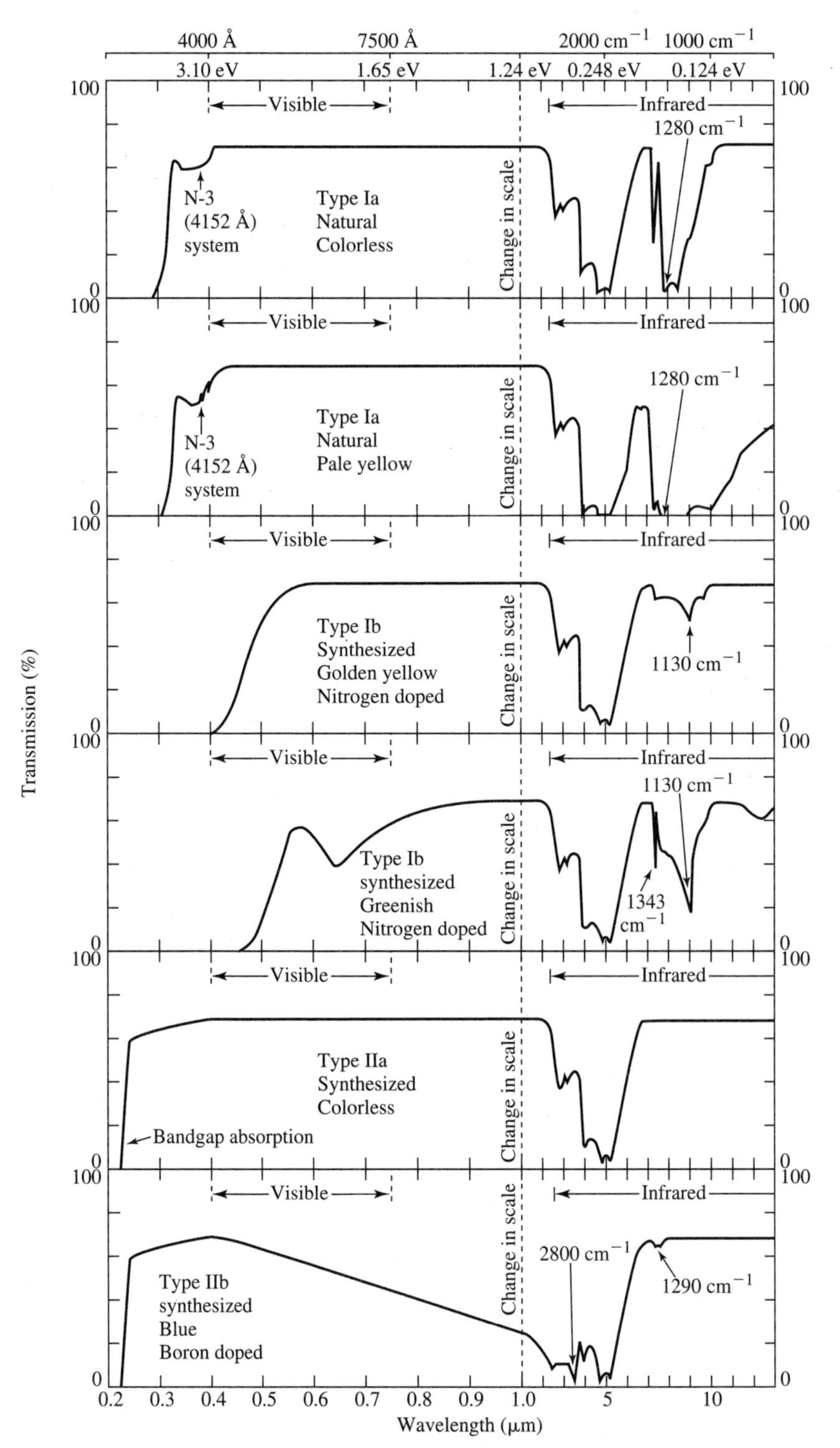

FIGURE 7-27 The transmission spectra for type Ia, Ib, IIa, and IIb diamond samples. (*Figure was adapted from J. I. Pankove and C.-H. Qui in* Synthetic Diamond, *1994, p. 307. Reprinted with permission from R.M. Chrenko.)*

C–H bands occur at 2800 to 3100 cm^{-1} [146]. The hydrogen concentration has been estimated as 1000 times the absorption coefficient in cm^{-1} at ≈2840 cm^{-1} [147].

The UV and VIS spectra are also used to characterize diamond. The transmission spectra over the entire energy range of bandgap of diamond is shown in Fig. 7-27 for type Ia, Ib, IIa, and IIb diamond [290]. The distinct differences in the transmission spectra make the absorption spectra a useful technique for characterizing diamond.

Luminescence Spectroscopy and Imaging. In addition to the absorption spectra, the luminescence spectra observed when excited states relax is beneficial in assessing the properties of diamond. Cathodoluminescence, where states are excited with an electron beam, is probably the most commonly used luminescence technique. It may be used for both imaging and spectroscopic investigation. Photoluminescence is generated when the sample is exposed to photons and electroluminescence may be investigated when the excitation caused by an applied voltage is sufficient to generate light.

Many of the defects investigated by their luminescence form vibronic bands that generate zero-phonon lines. In addition to the zero-phonon lines, phonon side bands are also generated on the low energy side of the zero-phonon line. These two aspects of the luminescence are used to evaluate the properties and behavior of a defect. The nomenclature for most of the zero-phonon lines was developed in the 1950s [291]. Since that time other zero-phonon lines have been observed. Unfortunately, not all have been named according to this system. In this nomenclature *N* stands for defects in natural diamond, *R* for those caused by radiation (*GR* is general radiation) and *H* is for those observed after heat treatment. Extensive investigation of the defects in natural diamond has been performed. In Table 7-5, the peak position and description for several common defects in diamond are summarized.

For cathodoluminescence (*CL*) measurements, the wavelength of the emitted photon is related to the energy difference of the transition that generated the light; therefore CL provides information about the defect levels within a material. The CL spectrum typically consists of sharp peaks and broad bands. The sharp peaks are indicative of discrete transitions between energy levels whereas broad bands indicate variable transition energies. Cathodoluminescence is a near-surface technique in which the depth of the luminescence is ≈0.007 to $0.014V^{1.825}$ [μm] [292]. Typical acceleration voltages are 10 to 50 keV and correspond to emission depths of 0.5 μm to 17.7 μm. More than 100 luminescence cen-

TABLE 7-5 Luminescence Spectra

Peak position (eV)	Peak position (nm)	Name	Description
4.582	270		Related to carbon interstitial
3.188	388.8		Carbon interstitial and single nitrogen atom
2.985	415	N3	N_3-V
2.85	435	Blue band A	Broad band—dislocations
2.33	572		Doublet line—nitrogen related
2.2	563	Green band A	Broad band—thought to be related to boron
2.156	574.9		Single N and V
1.945	637.2		Vacancy at a single substitutional nitrogen atom (absorption, no CL)
1.682	737		Silicon
1.673	741	GR1	Neutral vacancy

ters in diamond have been observed and summarized [277,278]. The emission intensity of one defect may vary with the concentration of other defects and the efficiency of individual defects differs; therefore, CL does not provide an quantitative determination of the defect density.

As previously mentioned, the cathodoluminescence spectra typically consist of broadband and sharp-zero phonon lines and their sidebands. Two broad peaks are often observed and are referred to as green and blue band A because they occur in the green and blue wavelengths, respectively. Originally, the broad bands were thought to be attributed to donor-acceptor transitions where the energy was determined by the separation between the donor and the acceptor [232]. More recently, these transitions have been associated with dislocations [293]. A comparison of the CL spectra and x-ray topographic images in a synthetic diamond revealed a strong correlation of the CL intensity of the 2.85 eV band (blue) to the dislocation density [294]. The 2.2 eV band (green) is thought to be attributed to a boron related center [294]. The origin of these luminescence bands is still under investigation.

The sharp peaks are often associated with impurity atoms or aggregates. As expected, nitrogen containing defects generate numerous CL peaks. Some of the most common nitrogen peaks occur at 2.156, 2.33, 2.807, and 3.188 eV. The 2.156 eV (574.9 nm) defect is thought to consist of nitrogen and a vacancy [295]. Near 2.33 eV, a doublet line occurs which is thought to be related to nitrogen [294]. Separate lines at 2.807 and 3.188 eV are thought to be caused by a single N atom and a C interstitial [296]. Another defect that is believed to be related to a C interstitial occurs at 4.582 eV [297]. A final defect of interest that is often observed in CVD diamond occurs at 1.681 eV. As mentioned in Section 7.2.7, this defect occurs near the 1.673 eV neutral vacancy (*GR*1) defect, but cathodoluminescence was able to uniquely identify Si as the source of this peak through the spacing of the phonon sideband [158].

The photoluminescence and electroluminescence spectra are similar to the cathodoluminescence spectra for most defects. Therefore, these techniques will not be discussed in detail. Laser light is often used for photoluminescence. The luminescence occurs from defects with lower energy than the incident photons. This feature allows selective excitation of a defect by controlling the incident laser energy. The intense beam can excite lower concentrations of defects than investigated through absorption and this makes it an important technique as the quality of the diamond improves. Although not as heavily investigated, electroluminescence generated by an applied voltage is also used to characterize diamond. Yellow-green luminescence was observed for Schottky barrier point contacts probed at 300°C to 750°C [298]. Boron-doped diamond produces electroluminescence similar to the cathodoluminescence. Broad peaks in the blue (2.8 to 2.9 eV) and green (2.3 to 2.4 eV) regions were observed [293]. Insulating diamond layers with p^+ contacts in a p^+-i-p^+ configuration have also been used to generate blue, orange, and green light from the A band, 575 nm zero-phonon line, and the H3 center, respectively [299]. Correlation of the electroluminescence and cathodoluminescence was observed.

Photoconductivity. The intrinsic and extrinsic photoconductivity of insulating and semiconducting diamond has been investigated using 0.35 to >6 eV light [203,300,301]. In this technique, diamond is exposed to a pulsed high-laser beam and the resulting current pulse is recorded. The intrinsic photoconductivity occurs when the excitation energy exceeds the bandgap, although extrinsic photoconductivity occurs when the excitation energy is less that of the bandgap.

Investigation of the boron defect in semiconducting diamond revealed a phenomena attributed to photothermal ionization. The photoconductivity maximum occurred at energies less that of the ionization energy. The maxima also corresponded to the maxima in the absorption spectra [302]. The optical energy required to produce conductivity was less than that of the ionization energy because of a two-stage process where the optical energy caused a transition to an excited state and the thermal energy produced a hole in the valence band that allowed conductivity [303].

The substitutional and A-aggregate nitrogen centers have also been investigated through photoconductivity and photo-Hall experiments. These experiments were discussed previously in Section 7.3.1. Photoconductivity of type Ib diamond was used to determined a 1.7 eV ionization energy for substitutional nitrogen [202,203]. The photoconductivity threshold for the A aggregate in diamond was 4.05 eV [203,304,305].

For intrinsic photoconductivity, the sample is excited with above bandgap light. Two approaches were used in these experiments. In one experiment, the transient photoconductivity is investigated because of a pulsed excitation of the sample. Alternately, an integrated investigation of pulses may be used to correlate to continuous excitation of the sample. This analysis is typically performed on a lateral device and the top 2 μm of the sample is probed [306]. The collection distance d is defined as

$$d = \mu E t = Vt \tag{11}$$

where E = electric field

μ = mobility

t = lifetime (determined from the photoconductivity measurements)

The transient photoconductivity measurements allow the mobility and lifetime to be determined independently. The mobility obtained from this measurements is the lifetime-weighted sum of the electron and hole mobilities [218]. At high excitation density, the interaction of electrons and holes limits the mobility and causes a decrease in mobility with increased carrier density. At low excitation density, the mobility saturates at a mobility which is independent of excitation density. The mobility as a function of carrier density is

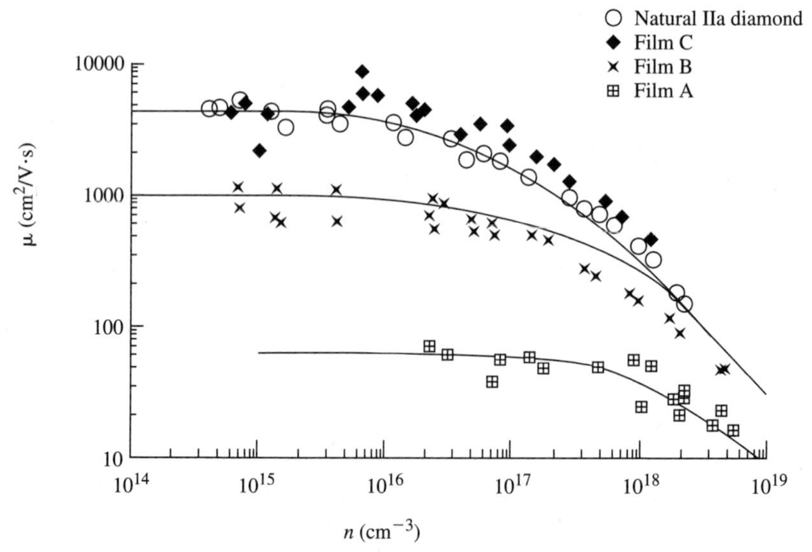

FIGURE 7-28 The mobility as a function of the concentration of excited carriers in a photo-induced conductivity experiment. *(From L. S. Pan, et al.,* Dia Rel Mat, *vol. 2, 1993, p. 1012. Reprinted with permission. Copyright 1993, Elsevier Science S.A.)*

shown in Fig. 7-28 [307]. Carriers were speculated to recombine at dislocations and at nitrogen sites within the material in the type IIa diamond. In the CVD diamond, the decay was faster than the system response; therefore, the maximum lifetime was 70 ps. Substantial differences were observed in the photoconductivity on the growth side and on the substrate side of the samples. The mobility differed by a factor of 50 at low excitation densities and both the mobility and the decay time increased as sample thickness increased [307]. The collection distance increased linearly with increasing sample thickness [308]. Recent measurements of high-quality CVD diamond recorded combined electron-hole pair mobilities of 4000 $cm^2/V \cdot s$, similar to that of the best natural type IIa diamond [309]. The lifetimes of these CVD films was 150 ps whereas the lifetime for type IIa diamond was 300 ps. The collection distance of the CVD diamond was 15 μm and was half of that for the natural type IIa diamond. However, these results indicate a strong improvement in the quality of CVD diamond. It was speculated that the difference in lifetime may be attributed to the higher defect density in CVD diamond, but does not suggest a strong influence of the grain boundaries. This leads to optimism that CVD diamond will be a viable radiation detector.

7.3.3 Thermal Properties

The thermal conductivity describes the ease of thermal transport in a material. Diamond has an extremely high thermal conductivity. Often diamond is mentioned in heat sink applications; however, it is important to note that diamond has high thermal conductivity, but not a high heat capacity. Thermal conductivity describes the transfer of heat through a material whereas heat capacity describes how much heat is required to raise its temperature by 1°K. Therefore, diamond is more appropriately described as a heat spreader while other materials must be used as the heat sink. The exceptional thermal conductivity of diamond compared to other semiconductors and to copper (Cu) is shown in Fig. 7-29 [310].

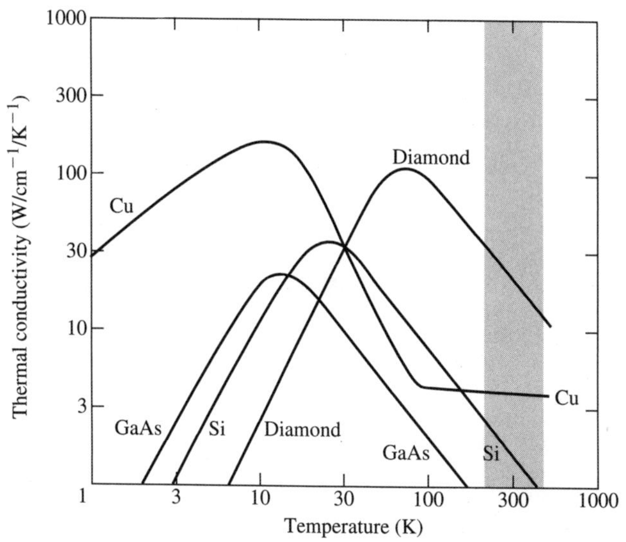

FIGURE 7-29 The thermal conductivity as a function of temperature for different materials. *(From M. W. Geis, N. N. Efremow, and D. D. Rathman,* J Vac Sci Technol A, *vol. 6, 1994, p. 1953. Reprinted with permission. Copyright 1988 American Institute of Physics.)*

Thermal conductivity is the random transfer of thermal energy in a material. Detailed theoretical descriptions of the thermal conductivity may be found in several texts [311,312]. A brief overview of thermal conductivity will be discussed. The thermal conductivity coefficient relates the heat flow in a material to its thermal gradient. In solids, heat is typically conducted through electrons or phonons. Metals have a high concentration of electrons and their thermal conductivity is dominated by electrons. Therefore, metals which are good thermal conductors are also good electrical conductors. For nonmetallic materials where electrons are not available for thermal conduction, thermal transport typically occurs through phonons. Some applications require both electrical isolation and thermal conduction. Materials in which thermal conduction is dominated by phonons are ideally suited for these applications because they allow for thermal conductivity without electrical conductivity. This is especially true for diamond where the 5.45 eV bandgap produces negligible thermally generated electron-hole pairs even at elevated temperatures. Therefore diamond is a good electrical insulator while allowing heat conduction through phonons.

An important material property for thermal conductivity is the Debye temperature (θ). The Debye temperature is the characteristic temperature which describes the actions of crystal vibrations. For diamond, the value depends on the technique used to estimate it. From a high temperature estimation the value is 1890°K, whereas for a low temperature estimation the value is 2340°K [313]. By either estimate, the Debye temperature of diamond is high and corresponds to a tight lattice and strong bonds where phonon flow through the material is easily accomplished [314] and the intrinsic phonon-phonon scattering is weak at a room temperature [283]. The maximum of the thermal conductivity occurs at approximately 1/30 of Debye temperature; therefore, the maximum of the thermal conductivity of diamond occurs at approximately 60 to 80°K. The highest thermal conductivity reported for natural diamond is 175 W/cm/°K at 65°K [315]. The maximum in the thermal conductivity indicates a change in the dominant scattering mechanism.

Phonon-Scattering Mechanisms. The thermal conductivity depends on the phonon group velocity, the phonon relaxation time, and the spectral specific heat [283]. In nonmetallic crystals the dominant phonon-scattering types are intrinsic phonon-phonon processes, phonon-impurity scattering, and boundary scattering. Specifically, vacancies, isotopes, impurities, double and triple carbon bonds, dislocations, stacking faults, grain boundaries, surface, or other phonons may all increase the scattering and decrease the thermal conductivity [316]. These separate scattering rates all add according to Mathiessen's rule.

Intrinsic phonon-phonon scattering occurs because of an anharmonic coupling of lattice waves which may be of two forms. There are normal processes (those which conserve crystal momentum), and umklapp processes (those that involve a nonzero reciprocal lattice vector). Only the umklapp processes contribute directly to the thermal resistance. Numerous estimates of phonon-phonon scattering have been developed. Using a simple model, the intrinsic phono-phonon limited thermal conductivity was shown to depend on 1/T [317].

Impurities and phonons also interact to scatter the phonons. A general expression for the phonon relaxation time in the presence of dilute concentrations of impurities suggests that the thermal conductivity depends on both the atomic mass difference and the difference in the elastic constants of the host and the impurity [283]. For isotopic impurities, the elastic contstants are similar and phonon scattering is caused by the difference in mass [318]. Because neither the mass nor the elastic constants are strongly temperature-dependent, this scattering mechanism does not exhibit a strong temperature dependence, but acts to reduce the total mobility through Mathiessen's rule.

Boundaries also scatter phonon flow within a material. The boundary may be the physical extent of the sample or defect boundaries [319]. Boundary scattering dominates at low temperatures where phonon-phonon and phonon-impurity scattering are weak. This expression is both temperature- and frequency-independent and suggests that the thermal conductivity should follow the temperature dependence of the specific heat which varies as T^3. This model assumes a perfect interface and this may not be ideal because of interface roughness or slight transmission thorough the boundary.

The temperature dependence of the thermal conductivity may be characterized by two different regimes. At low temperatures, boundaries dominate and the thermal conductivity follows a temperature dependence of approximately T^3. Above approximately $\theta/30$, phonon-phonon interactions dominate and the thermal conductivity follows a T^{-1} temperature dependence [320]. Phonon-defect interactions produce reductions in the thermal conductivity that depend on the nature of the defect.

Thermal Conductivity of CVD Diamond. The initial attempts to measure the thermal conductivity of CVD diamond did not produce spectacular results. The early values for the thermal conductivity of CVD diamond were 10 W/cm/°K [321] to 11 W/cm/°K [40]. These values are ≈50% of the values for type IIa diamond. Although disappointing, these low values were not surprising based on the wide variation of the thermal conductivity in natural diamond, where the thermal conductivity of natural type IIa diamonds is superior to type Ia diamonds. The nitrogen content of these two types of diamonds differ and a correlation of the thermal conductivity and optical absorption features caused by nitrogen has been made [322–324]. Therefore, it is likely that impurities and defects limited the thermal conductivity in the initial CVD diamond samples.

A correlation between the thermal conductivity and methane concentration was found on microwave CVD diamond films measured by IR tomography [321,325]. The methane gas concentration varied from 0.1% to 3% methane and the best thermal conductivity of 10 W/cm/°K was found for the films grown with 0.1% methane. For the low-methane concentrations, a strong 1333 cm^{-1} Raman line was also observed. As the methane concentration increased, the strength of the 133 cm^{-1} line decreased and a broad feature at 1500 cm^{-1} appeared. The 1500 cm^{-1} line is attributed to sp^2-bonded carbon and indicates that the thermal conductivity is strongly influenced by the bonding within the CVD diamond film. Additionally, it was shown that the full width at half maximum (FWHM) of the 1332 cm^{-1} Raman line correlated to the thermal conductivity [265]. As the FWHM increased, the thermal conductivity decreased. For a FWHM of 4.6 cm^{-1}, the thermal conductivity was 11.3 W/cm/°K whereas for a FWHM of 12.2 cm^{-1}, the thermal conductivity was only 3.6 W/cm/°K. The thermal conductivity decreased almost linearly between these two extremes. Although there is a correlation between the Raman characteristics and the thermal conductivity, the mechanisms and sensitivities for Raman and thermal conductivity differ; therefore, there is not a one-to-one correspondence between these values [283].

Recent improvements in the diamond CVD process have resulted in improvements in thermal conductivity. Several groups have deposited CVD diamond that has had thermal conductivity similar to natural diamond [265,321]. The films with the highest thermal conductivity were grown with the slowest growth rates [326] and with the most atomic hydrogen or oxygen [316]. These growth conditions improve the quality of the diamond by reducing the sp^2 content of the CVD diamond. Another important observation was that the thermal conductivity improves with increased thickness. The thermal conductivity was 21 W/cm/°K at the top surface of a 350 μm-thick film but only 7 W/cm/°K at the substrate interface. The columnar structure also generates an anisotropy of the thermal conductivity. The thermal conductivity perpendicular to

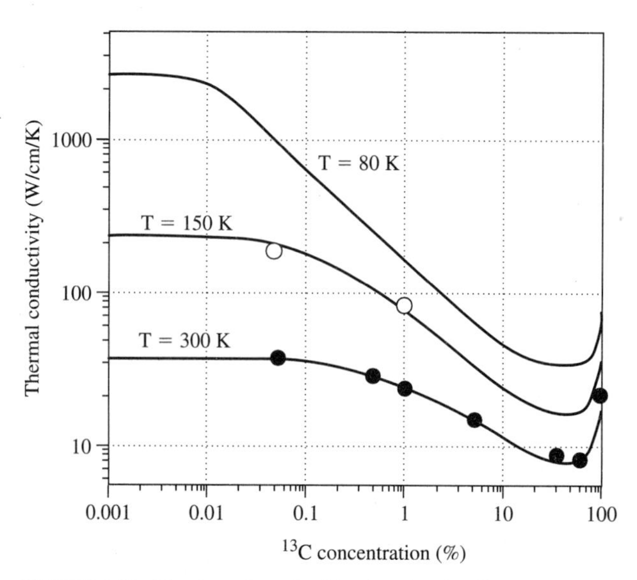

FIGURE 7-30 The thermal conductivity as a function ^{13}C concentration for different temperatures. (*From L. Wei et al.,* Phys Rev Lett, *vol. 70, 1990, p. 1104. Reprinted with permission. Copyright 1993 The American Physical Society.)*

the growth plane was 50% higher than the thermal conductivity parallel to the growth plane [120].

Isotopic Purity Effects on the Thermal Conductivity. Because thermal conduction in diamond is related to phonons, a change in the isotopic concentration of diamond alters its thermal conductivity. Estimates for the increase in thermal conductivity range from 5% [327] to 50% [328] for isotopic purification. The thermal conductivity of type IIa diamond is limited by the presence of $\approx$1.1% ^{13}C in natural diamond [329]. Isotopic impurities are randomly distributed throughout the crystal and the mass difference leads to scattering of phonons and a loss of thermal conductivity [316].

Isotopic deposition of ^{13}C on natural diamond [330] was demonstrated in the early years of CVD. An initial investigation of isotopically enhanced CVD diamond did not demonstrate improvement of thermal conductivity from the presence of scattering centers (impurities, crystalline defects, and grain boundaries) that limited the thermal conductivity. These scattering centers dominated the thermal conductivity and did not all allow the effect of isotopic purity to be observed where a thermal conductivity of only 6 W/cm/°K was measured [331]. Recent improvements in CVD diamond thermal conductivity approaching 20 W/cm/°K at room temperature warranted repeating this experiment because this value of the thermal conductivity approached that of natural diamond. The improvement in thermal conductivity through isotopic enrichment was verified for CVD diamond where the isotopic concentration was 0.055% ^{13}C [332]. The in-plane thermal conductivity was 21.8 W/cm/°K whereas the perpendicular thermal conductivity was 26 W/cm/°K. Both are highest values reported for CVD diamond. The perpendicular thermal conductivity of CVD diamond is slightly higher than the highest reported thermal conductivity for natural diamond.

An interesting combination of CVD and HPHT diamond synthesis was used to grow the highest thermal conductivity diamond to date. Isotopically enriched HPHT diamond was formed using a source charge containing isotopically enriched CVD diamond. Diamond was used as the source charge to maintain constant pressure during synthesis. The volume reduction for graphite to diamond at 50,000 atm and 1500°K is ≈30%. This large volume change would alter the cell pressure during synthesis and would lead to a degradation in properties of the HPHT diamond. Two HPHT diamonds with ^{13}C concentrations of 0.5% and 0.07% were prepared. The 0.07% ^{13}C specimen had a room-temperature thermal conductivity of 33.2 W/cm/°K, a more than 50% improvement over natural diamond [333]. Theoretical and measured thermal conductivity as a function of isotopic purity is shown in Fig. 7-30 [334].

More recently, using the same process, additional improvements in the thermal conductivity have been observed. A thermal conductivity of 410 W/cm/°K was measured at 104°K [334]. There was not a maximum at 104°K, but this was the lowest temperature where the thermal conductivity was measured. Using a thermal conductivity model by Calloway [335], the thermal conductivity at room temperature should not be improved for samples isotopically purer than 99.9% ^{12}C [334]. At 80°K, the thermal conductivity improves in samples containing up to 99.999% ^{12}C, which is the current practical limit for methane isotopic purification [334].

7.4 APPLICATIONS

Diamond CVD is expected to enable numerous applications [336]. Thorough reviews of these applications have been reported [250,314,337–339]. The intent of this section is to highlight some of the most promising applications related to electronics, optics, and thermal management. Not discussed are the costs associated with CVD diamond manufacturing [1] or the potential markets for CVD diamond [17], because these topics have already been reviewed.

7.4.1 Electronic Applications

The potential electrical applications of diamond are varied. Three basic electrical applications will be discussed: a thermistor that relies on the temperature-dependent conductivity to measure temperature; a field-effect transistor which takes advantage of the high-mobility, saturated current velocity, and breakdown field of diamond; and electronic emission devices that take advantage of the electron affinity of diamond. For all of these devices, other properties such as the thermal conductivity, chemical and radiation inertness, mechanical strength, and optical transparency can all contribute to the overall performance of the device.

Thermistors. Boron-doped diamond has been proposed as a good material for thermistor applications. A thermistor is a temperature-sensitive resistor. The 0.37 eV depth of the boron acceptor gives diamond a negative coefficient of resistance; that is, the resistance decreases as the temperature increases. Additionally, diamond has a low specific heat and high thermal conductivity; therefore, diamond thermistors should have shorter response times than more traditional metal oxide thermistors or platinum resistance temperature detectors (RTD). Another motivation for fabricating thermistors from semiconducting diamond is that the performance of metal oxide thermistors degrades above 350°C to 400°C. Platinum RTDs are an alternative, but offer a low sensitivity at elevated temperatures and may be cost-prohibitive. Thermistor applications are governed by the following diamond properties: temperature operating range; ability to operate in hostile environments; small dimensions and weight; high resistance (contact and lead resistance are negligible); high sensitivity; and short response times [340,341].

The first report of a diamond thermistor was on natural diamond in which cylindrical rod of diamond 4 mm × 0.5 mm was used [342]. Later thermistors fabricated from synthetic HPHT diamond determined a minimum recorded temperature change of 10^{-3}°K [340]. This indicates the precision of temperature measurement with diamond thermistors because of their high temperature sensitivity. Thin film CVD diamond is more flexible for thermistor fabrication and design because the thickness and geometry may be readily defined and the dopant concentration may be controlled through in situ doping. An investigation of the effect of B concentration on the performance of rectangular and serpentine thermistor structures has been performed [343]. The temperature response of the thermistors with 0.1 ppm B_2H_6 or 1 ppm B_2H_6 in the gas phase is shown in Fig. 7-31 [344]. Polycrystalline boron doped diamond film on Si_3N_4 substrates with a 1.5 μm undoped diamond passivation layer were investigated. The boron concentration determined the sensitivity of the device. For the serpentine structure, poor sensitivity was observed at 1 ppm diborane where there was only a 100× change in the resistivity over 500°K range. For 0.1 ppm diborane, a better sensitivity was observed and was 10^4× over a 500°K temperature range. Unfortunately, the serpentine geometry produced a room temperature resistance of 10^{11} Ω which is substantially greater than the 1 MΩ resistance desired for practical applications. The room-temperature resistance may be changed through a geometry change; therefore, a 0.1 ppm diborane film with a rectangular geometry was used to produce a

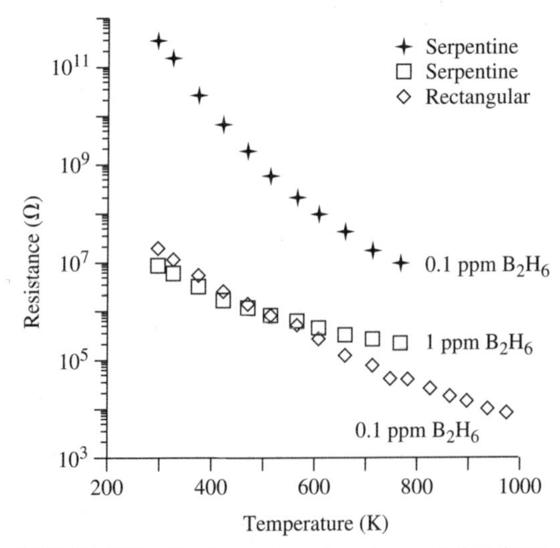

FIGURE 7-31 The temperature dependence of thermistors with different boron concentrations and geometry. (*From B. A. Fox, et al.,* Second International High-Temperature Electronics, *1994, p. vi-3. Reprinted with permission. Copyright 1994 High-Temperature Electronics Conference.)*

thermistor with a lower room-temperature resistance of approximately 10 MΩ and good temperature sensitivity. Thermistors were shown to operate from 25°C to 700°C over multiple temperature cycles. The β value, which indicates the temperature sensitivity of this device, was 5500°K and was better than typical metal oxide thermistors which have β values of 2000°K to 4000°K. There was some concern over elevated temperature operation in oxygen containing ambients. Unpassivated devices had a change in resistance that was shown to be caused by oxidation of the diamond and the Ti/Au metalization [343].

To protect the diamond surface from the oxidizing ambient, some samples were passivated with sputtered Si_3N_4 or SiO_2 and tested from 300°C to 700°C. An unpassivated (100) sample had increased resistance with temperature cycling. The primary failure mode was thought to be contact oxidation; however, hexagonal platelets on the surface indicated that surface graphitization of the diamond occurred. For a silicon nitride passivated thermistor, the resistance deviation was $<1\%$ up to 20 cycles, but after 25 cycles, some deviation in the resistance was observed. Examination of the thermistor by scanning electron microscopy revealed cracks in the passivation that were not present before 20 cycles and these may have allowed contact oxidation. For the silicon dioxide-passivated thermistors, the performance was worse. After only 4 cycles, the SiO_2 passivation buckled and cracked. This was thought to be caused by the difference in thermal expansion of the SiO_2 and the diamond. Although passivation with silicon nitride does appear to offer some increased protection from oxidizing ambients, significant improvements are required for long-lifetime, high-temperature devices [345].

Field-Effect Transistors. The wide bandgap, high thermal conductivity, high breakdown voltage, and high saturated carrier velocity have made diamond a promising material for electronic applications. Figures-of-merit have been developed to compare the performance of different materials. The Johnson and the Keyes figures-of-merit demonstrate the applicability of diamond to high-temperature, high-frequency, and high-power applications [346]. Additionally, other figures-of-merit have been evaluated and also demonstrate the superiority of diamond [347,348]. The correlation of these figures-of-merit to actual device performance has been questioned [162,224,349]. One of the problems with understanding diamond or other wide-bandgap semiconductor electronics is that it is often assumed that the equations and understanding developed for Si or gallium arsenide are directly transferrable to these materials. In a re-evaluation of diamond as a high-frequency device wherein the incomplete ionization of *p*-type material was considered, the prospects of boron-doped diamond relative to GaAs and SiC were not as favorable [350]. Despite these theoretical studies, several proof-of-concept devices have been fabricated that warrant further study.

The development of diamond field-effect transistors (FETs) has progressed rapidly. Early FETs exhibited only modulation, whereas more recent devices exhibited saturation, pinch-off, and transconductance as high as 1.3 mS/mm [351]. Additionally, amplifiers [352] and digital logic circuits [353] have been fabricated from diamond. Numerous diamond field-effect transistors have been made from diamond. The primary differences in them is the morphology of the diamond, the technique for boron incorporation, and the gate structure. Diamond transistors have been fabricated from polycrystalline, highly oriented, and single crystal diamond. Boron has been incorporated by ion implantation, diffusion, and in situ B doping. The gate structures for the FETs have been Schottky metal semiconductor contracts, insulating diamond, or a dielectric such as SiO_2. For this chapter, the diamond FETs have been classified according to gate structure and the performance of various devices are reviewed.

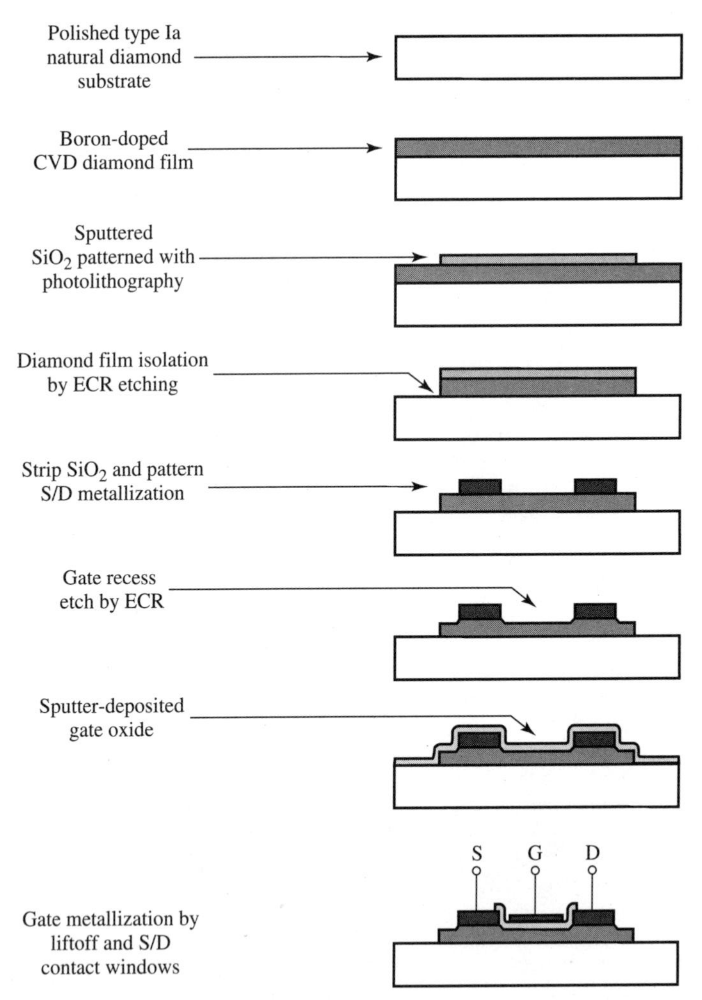

FIGURE 7-32 A fabrication sequence and structure for a diamond MOSFET. (*From S. A. Grot, G. S. Gildenblat, and A. R. Badzian,* IEEE Electron Dev Lett, *vol. 13, 1992, p. 462. Reprinted with permission. Copyright 1992 IEEE.)*

A typical structure and fabrication sequence for a field-effect transistor is shown in Fig. 7-32 [354].

Metal-Semiconductor-Field-Effect Transistor (MESFET). Schottky gate metal diamond field-effect transistors have been made by several laboratories. However, there have been fewer reports of this structure than others mentioed below, probably because of the difficulty in repeatedly obtaining rectifying contacts to CVD diamond. Additionally, the series resistance has limited the source-to-drain currents. One MESFET configuration consisting of an active region formed in a natural type IIa diamond via diffusion of boron. [195]. At 1400°C, for 30 to 60 s, boron diffused ≈50 nm into the natural diamond to form a *p*-channel with a boron concentration of 3×10^{19} cm^{-3}. The operation of this device showed a transconductance 0.7 μS/mm and pinchoff was observed at 5 V; however, saturation of the current was not observed. The source-to-drain current was $\approx$1 μA and was thought to be limited by compensation and the impedance of the ohmic contacts. Another MESFET device consisted of an in situ boron-doped homoepitaxial film [355]. An undoped synthetic {100} diamond was used as the substrate and the boron-doped film was grown in a microwave CVD reactor where B_2H_6 was used as the doping source. The homoepitaxial film was $\approx$2 μm thick and had a carrier concentration of approximately 10^{15} cm^{-3}. The series resistance of the device was high and resulted in a low source-to-drain current that was only $\approx$1 nA. It was also speculated that an insulating interfacial layer existed between the A1 gate and the diamond film and slight modulation of the current was observed. Although these devices verified the operation of diamond MESFETs, difficulties with rectifying contacts in the MESFET structures led to the investigation of other devices which did not rely on rectifying gate contacts.

Metal Undoped Diamond-Semiconductor Field-Effect Transistor (MiSFET). A MiSFET was fabricated on an in situ-doped homoepitaxial diamond film that was selectively deposited with a tungsten mask so that the undoped layer was under the gate only [356]. The boron-doped film thickness was 0.8 μm and the B concentration was estimated to be 1×10^{18} cm^{-3} with a mobility of approximately 100 $cm^2/V\cdot s$. The transconductance of the device was 0.5 μS at 26°C with a tendency toward saturation at room temperature. The operating temperature of the device was limited by an Al contact and the need for a lower resistance ohmic contact was expressed. In addition to the homoepitaxial MiSFET, polycrystalline devices have also been fabricated [357]. A 30 nm in situ B-doped layer with a B concentration of approximately 5×10^{18} cm^{-3} was selectively deposited on an undoped polycrystalline diamond layer on a silicon nitride substrate. A 0.6 μm undoped diamond film was selectively deposited over the gate. At room temperature, the transconductance was 43 pS/mm and there was a tendency for saturation of the drain current al $\approx$1 nA with an applied gate voltage of 8 V. Modulation was also observed at 400°C; however, significant gate leakage occurred at high temperatures.

Metal-Oxide-Semiconductor-Field-Effect Transistor (MOSFET). Single crystal MOSFET devices have been fabricated through both in situ boron doping of homoepitaxial films and ion implantation of boron into natural diamond. Two in situ-doped devices were mesa-isolated recessed-gate MOSFET's [354]. The transconductance of one device was 87 μS/mm whereas another had a transconductance of 60 μs/mm. Both devices had incomplete pinchoff that was thought to be caused by a parallel conduction path. Saturation of the current was observed, but excessive series resistance was caused by the impedance of source and drain contacts. Gate insulator leakage current limited the high-temperature operation of the device to 350°C [354]. A single-crystal MOSFET has also been made by ion-implantation of B. A triple implant of B may be used to produce a uniformly doped film [358]. In this device, a triple implant produced a 210 nm-thick doped layer. The properties of the film were measured by Hall effect where the room-temperature carrier concentration was 5×10^{15} cm^{-3} and the mobility was 30 $cm^2/V\cdot s$. The device had well-defined saturation and complete pinchoff. The transconductance per unit gate width was 3.9 μS/mm. The gate current increased as the temperature increased and eventually led to gate failure above 400°K Another ion-implanted device was fabricated by the same researchers [359]. The transconductance of the device was 48 μS/mm when operated in the enhancement mode and was an order of magnitude higher than the previous ion-implanted devices. Two devices were connected in a driver-load configuration where a dc voltage gain of approximately 2 was observed.

Previous limitations of diamond FET devices included high source-to-drain resistances, high impedance contacts, high gate-leakage currents, and parallel conduction paths that prevented pinchoff. Improvements in the operation of homoepitaxial diamond

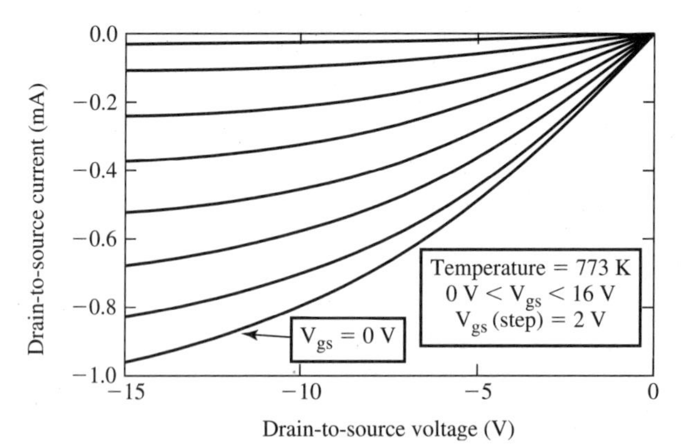

FIGURE 7-33 The current voltage characteristics of a diamond MOSFET operating at 500°C. *(From D. L. Dreifus, et al.,* Second International High-Temperature Electronics Conference, *1994, p. VI-35. Reprinted with permission. Copyright 1994 High-Temperature Electronics Conference.)*

MOSFETs have recently been demonstrated [351]. The transistor characteristics of a diamond MOSFET operating at 500°C is shown in Fig. 7-33 where pinchoff and saturation were observed [352]. Boron-doped homoepitaxial diamond films were grown on {100} type IIa natural diamond substrates. These devices were tested to 550°C before gate leakage prevented device operation. Some MOSFETs had source-to-drain currents as large as 9.7 mA and a transconductance of 1.3 mS/mm. These improvements have also been demonstrated for MOSFETs fabricated on highly oriented diamond films [270].

The improved device performance allowed diamond MOSFETs to be connected in a common source amplifier configuration [352]. The gain characteristics were measured at

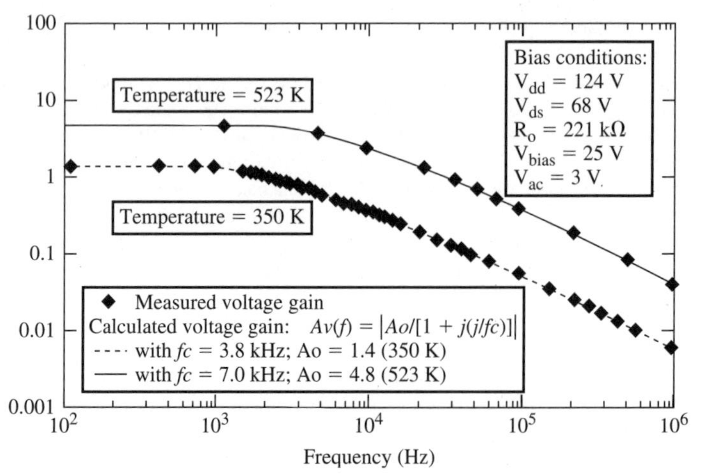

FIGURE 7-34 The voltage gain as a function of frequency for a diamond ac amplifier. *(From D. L. Dreifus, et al.,* Second International High-Temperature Electronics Conference, *1994, p. VI-35. Reprinted with permission. Copyright 1994 High-Temperature Electronics Conference.)*

75°C and 250°C for frequencies between 20 Hz and 1 MHz. The frequency response is shown in Fig. 7-34 [352]. The gain was ≈5 at 250°C and began to roll off at 5 kHz at a rate of 20 dB/decade. The roll off was caused by the resistance-capacitance time constant of the device. Two MOSFETs were also combined into a logic circuit and are shown in Fig. 7-35 [353]. Both NAND and NOR structures were fabricated and demonstrated operation at 300°C, 350°C, and 400°C.

Other Transistor Devices. In addition to field-effect transistors, other diamond electronic devices have been fabricated. High-temperature point-contact transistors have been made on a HPHT diamond shaped in the form of a truncated pyramid [360]. The top and bottom surfaces were {100} and the sides were {111}. Capacitance-voltage measurements suggested an uncompensated acceptor concentration of approximately 10^{16} cm^{-3}. The base contact was made to a {111} face and the emitter and collector Schottky

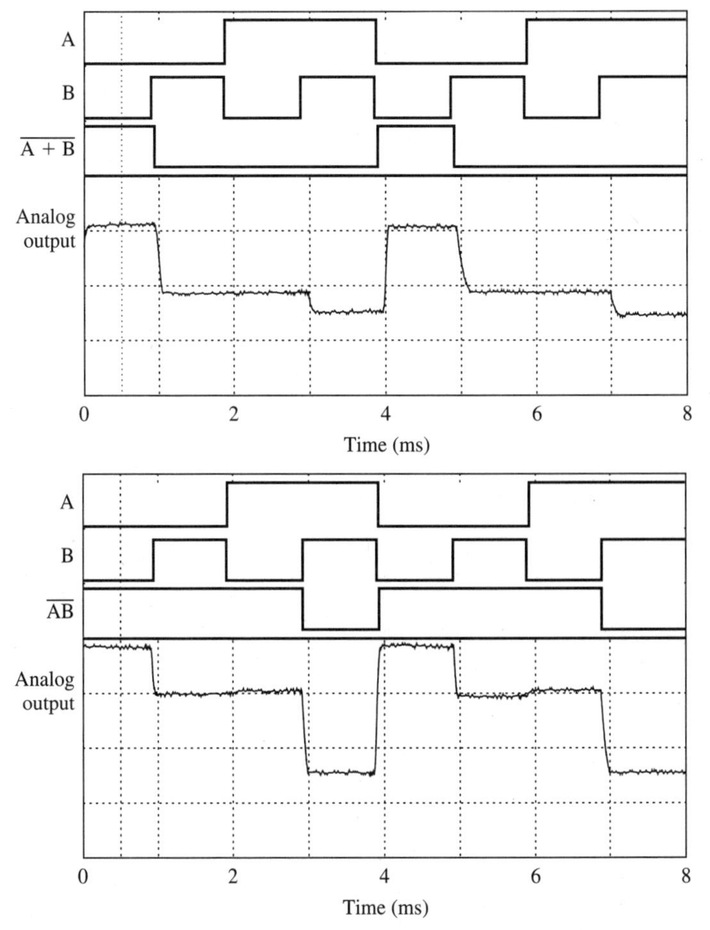

FIGURE 7-35 The operation of *(a)* NAND and *(b)* NOR diamond logic devices. *(From J. S. Holmes, A. J. Tessmer, and D. L. Dreifus,* Second International High-Temperature Electronics Conference, *1994, p. VI-35. Reprinted with permission. Copyright 1994 High-Temperature Electronics Conference.)*

contacts were made by positioning the probes on the {100} surface. Typically, the probes were within 10 to 20 μm of each other. At room temperature the small signal current gain was 2 to 25 and the small signal power gain was 6 to 35. At 510°C, the small signal current gain was 1.6 to 0.5 while the small signal power gain was 4.5 to 1.3.

Although the difficulties associated with *n*-type doping have been discussed, a bipolar junction transistor has been fabricated by producing a damage region through carbon implantation that behaves like an *n*-type region [361]. In this device a type IIb *p*-type diamond was implanted with carbon ions to produce *n*-type collector and emitter regions. The *n*-type implant was estimated to be 0.3 μm deep. The switching voltage was ≈20 V at room temperature, but was reduced to 2 V at 250°C. The high switching voltage at room temperature, was attributed to the higher-resistance electrical contact to the base. The current amplification was only 0.11 because of the nonideal geometry of the device. In an idealized device it was estimated that the amplification would be 733.

In another device configuration a *p-i-p* injection transistor has been fabricated. Ion implantation was used to define the *p* regions in an undoped natural diamond substrate. The Hall hole mobility of injected holes was ≈700 $cm^2/V{\cdot}s$, even after a 1600°C anneal. In this device, 200 mA of current was observed at a gate voltage of −10 V, with a source-to-drain bias of 100 V [362].

Electron Emission. Another promising area use of diamond is through electron emission. Diamond as an electron emitter may offer promise as a cold cathode electron source [363] for use in flat-panel vacuum fluorescent displays [364], or high-speed radiation-hardened vacuum microelectronic devices and circuits [365].

Theoretical considerations of electron emitters suggest that the desired material would be one with a low work function. In addition to low temperature emission, there are numerous benefits to an emission source with a low work function. Emission sources often have a small radius of curvature to enhance the electric field. The emission is sensitive to changes in the radius of curvature, but a low work function device would be less sensitive to geometrical effects. The turn-on voltage to initiate electron emission is reduced in a

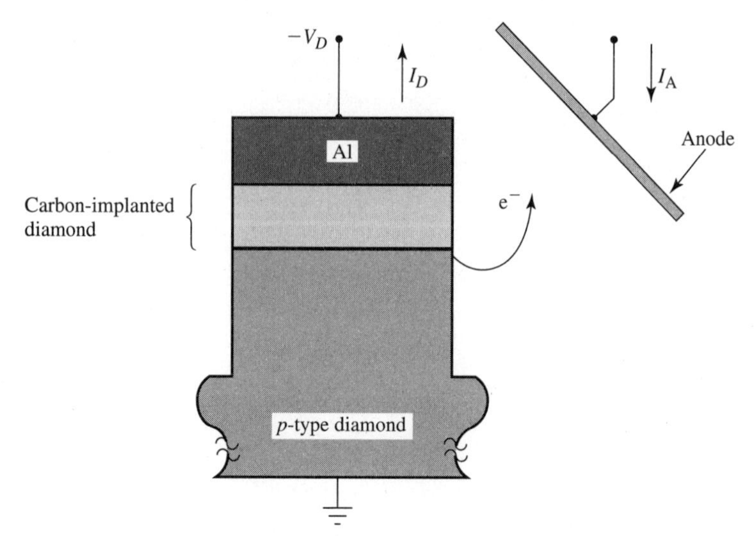

FIGURE 7-36 A schematic diagram of a diamond cold cathode. (*From M. W. Geis, et al.,* Applications of Diamond and Related Materials, *Y. Tzang, M. Yoshikawa, and M. Murakawa (eds.), 1991, p. 309. Reprinted with permission. Copyright 1991 Elsevier Science S.A.)*

low-work-function material and is more abrupt; therefore, voltage variations do not alter the emission current as significantly as in higher-work-function emitters with a more gradual turn-on. A final benefit of a low-work function material is that the saturation emission current is higher in low-work-function devices. Based on these benefits, the low work function of diamond may make it ideally suited for these applications.

Several structures have been investigated for electron emission from diamond. A schematic diagram of a potential field-emitter structure is shown in Fig. 7-36 [366]. This device includes a *p-n* junction. Other configurations for diamond emitters have been fabricated. Diamond was used as a tip for electron emission and these tips were made into an array as shown in Fig. 7-37 [367]. Diamond was also used as a coating for other materials. Both broad-area coatings of 15 mm molybdenum substrates and [252] coating of a Si tip [251] were used.

The potential of diamond as a material for a cold cathode has been investigated. The resilience of diamond emission when exposed to oxygen at 10^{-2} torr was demonstrated and the device has the highest emission efficiency of any non-cesiated cathode [366]. In addition to a negative or low work function, field-enhanced electron emission may be generated by the formation of a diamond tip with a small radius of curvatuve [368]. A diamond field-emitter array of pyramidal tips was fabricated by CVD deposition [367]. A 1 mm diameter tungsten wire was placed 100 μm from a diamond emitter. At an applied voltage of 6 kV, a current of 100 μA was observed and the current followed the Fowler-Nordheim relationship.

FIGURE 7-37 An array of diamond emitter tips. *(From K. Okano, et al.,* Appl Phys Lett, *vol. 64, 1994, p. 2742. Reprinted with permission. Copyright 1994 American Institute of Physics.)*

7.4.2 Optical Applications

The wide transmission range of diamond makes its suitable for many window applications. Its chemical inertness, mechanical strength, and thermal conductivity allow diamond to be used for applications where other materials would fail. The existence of optical centers in diamond also make light emission devices viable.

Windows, Mirrors, and Monochrometers. The high refractive index of diamond leads to only 60% to 70% transmission over most of the spectral range. To reduce the reflection a coating may be applied to diamond to match the refractive index. One problem with the approach of using a coating on the diamond is that the protective properties of diamond are lost. As an alternate application, diamond may be used as the protective coating of a window material to take advantage of some of its protective properties. Because of its low refractive index, diamond may be used as a film coating for other materials and reduce the surface reflection. For example, the surface reflection of Si at 0.6 nm is 30%, but a quarterwave coating of diamond reduces this to only 6.6% [23]. The transmittance of diamond-coated Si is $>90\%$ for wavelengths from 0.1 to 2.5 nm with negligible transmittance difference from 15 to 300°K.

One proposed application for diamond windows is for high-energy free-electron lasers (FEL) or other high-power laser systems [369]. Diamond is desirable for this application because conventional materials absorb too much laser energy and the optical properties become nonlinear as the energy levels increase. The absorption coefficient of highly pure diamond near 1 μm is expected to be $<10^{-4}$ cm^{-1}. This combined with the high thermal conductivity at 77°K, makes diamond an excellent candidate for free-electron laser windows. In this application, diamond would be coated with an antireflective layer. The high power of the free-electron laser locally heats the diamond window, which is cooled at the edges. A thermal gradient is induced across the sample. The change in refractive index with temperature is known as the thermo-optic coefficient and is 10^{-5}/°K for diamond [370]. The change in refractive index of the window generates optical distortion of the laser beam. A distortion of $<\lambda/4$ is required. For a diamond window, the laser beam power is limited to 750 kW at 1 μm. This power level exceeds the power handling capability of present day laser windows [371].

Diamond may also be used for mirrors and monochrometers in a FEL. If diamond is used as a mirror, a traditional 2500 m FEL requires only 15 m and a 60 m FEL requires only 2 m. The length difference is caused by higher power densities that may be handled by diamond mirrors; therefore, the separation distance does not need to be as large [369]. The improved transmissivity of diamond in normal-incidence monochrometers for free-electron lasers provides superior performance. The diamond monochrometers may be thicker because of the improved transmissivity and lower structure factor. The thicker window of diamond with a high thermal conductivity provides greater heat dissipation than with conventional materials [372]. A radially water-cooled diamond window sustained heat loads of 20 W/mm^2 at the European Synchronous Radiation Facility (ESRF) in Grenoble, France, without measurable radiation damage [373].

In a related application, third-generation synchrotron radiation sources now under construction require crystal monochrometers that both diffract the x-rays and yield sufficient brightness [374]. For this application the Bragg slope plane errors and dilatation from the x-ray beam must be small. It is desired that these distortions are smaller than the Darwin angle width of the Bragg reflection and the opening angle of the photon beam. In the second-generation synchrotron sources, traditional Si and germanium (Ge) have already reached their limits. The third-generation sources are expected to have 100 times the difficulties and Si and Ge monochrometers may limit the beam brightness. The primary problem is absorption of the x-ray beam by the monochrometer, which heats the monochrometer and causes distortion. The monochrometer must be cooled and this generates a thermal gradient across the monochrometer that broadens the Bragg angle. The thermal gradient generates a curvature in the monochrometer caused by the difference in thermal expansion. The thermal gradient depends on the incident power, thermal conductivity, and coefficient of thermal expansion of the window. For second generation NSLS X25 wiggler beam line that operates at 8 keV, the incident power is 2 W/mm^2. For (111) Si and a 4 mm beam, the slope error is only 24 arcsec whereas the beam opening is 40 arcsec and Si is a suitable monochrometer material. For the third-generation advanced-photon source (APS), the undulating beam line may have a power density as high as 200 W/mm^2. For a Si monochrometer, the integrated slope error is 1200 arcsec and makes Si unsuitable for third-generation synchrotron monochrometers. The high thermal conductivity and low thermal expansion of diamond make it suitable for third-generation synchroton sources. A figure-of-merit for monochrometer materials is the ratio of the thermal conductivity to the coefficient of thermal expansion. For diamond, this figure-of-merit is 50 times that of Si. With a diamond monochrometer in the APS source, the integrated slope error would only be 24 arcsec and it is suitable for this application.

Diamond is an excellent window material for IR radiation because of its good transmission over a wide spectral range. Even in the 2.5 to 6 μm two-phonon regime, the maximum absorption coefficient is only 12 cm^{-1}. Therefore, for windows with thicknesses below 0.25 mm, this absorption has little effect on the IR transmission [254]. A review of diamond windows in infrared optics systems is discussed by Ditchburn [375]. Both diamond windows or diamond-coated windows of ZnS, ZnSe, or Si are being used for optical applications [250]. The ZnS and ZnSe IR detector materials are often used in military systems; however, they are relatively soft with a hardness of only 210 to 240 kg/mm^2. The low hardness leads to a susceptibility to damage through use that degrades their performance. There are difficulties in depositing diamond directly on ZnS or ZnSe because of the presence of atomic H required for growth [250]. As a result, thin diamond films have been mechanically bonded to ZnS [376]. The difficulty of depositing diamond directly the window material is not a problem for Si IR detectors [377]. The improved performance of diamond-coated Si over uncoated Si detectors in water-drop impact tests has been verified [250].

Another potential application for diamond IR window is for in situ monitoring in the plastics industry. The chemical inertness as well as the transparency of diamond make it suited for his application. It is often desirable to measure the chemical composition during manufacturing; for example, the measurement of carbonyl groups in the manufacturing of polyethylene. Real-time monitoring of the process provides immediate indication of potential degradation in the process. Sapphire has been used in this sort of application in the near IR, but diamond transmits mid-IR wavelengths better than sapphire. The mid-IR wavelength is often required for organic groups [378].

Another application for diamond in the IR regime is in the domes of high-speed missiles. There are numerous optical, elastic, and mechanical properties required for use in high-speed missiles. Key properties are the transmissivity at particular wavelengths, heating caused by the air friction at supersonic speeds, and the rupture stress from a pressure differential. The optical transmission of the window is limited by the signal-to-noise ratio where the noise is caused by gray body emission as the dome heats up from air friction. For medium wavelength heat seekers (3 to 5 μm) in the two-phonon absorption region, the absorption coefficient must be <0.02 cm^{-1} to produce a signal-to-noise ratio <3 dB. The absorption coefficient in the 3 to 5 μm range is ≈ 12 cm^{-1}; therefore, diamond is not a suitable window material in this wavelength regime. For long wavelength heat seekers (8 to 12 μm) where one-phonon absorption occurs, the absorption is substantially less than the two-phonon absorption, and CVD diamond is a suitable window material in this wavelength regime [371]. The heating of the dome for a missile traveling at Mach 6 was calculated to have a stagnation temperature of 1860°K. At this temperature, diamond transforms to graphite in the presence of oxygen. Therefore, for these high speeds, a protective coating may be required for diamond. Alternately, if the speed were limited to Mach 4, the stagnation temperature is low enough that the diamond does not transform into graphite. The most common failure mechanism in conventional missile domes is thermal shock resulting from the instant temperature changes during launch. For diamond, the high allowable stress and the low thermal expansion lead to excellent thermal shock resistance. A 1 mm-thick missile dome should be able to tolerate the pressure differential required where the safety factor is 4.

Diamond has also been used as x-ray windows and as a mask membrane for x-ray lithography. Diamond is suited for this application because it has a low atomic number, high transmissivity, and superior mechanical strength [250]. Common x-ray windows are made of beryllium (Be) that are 8 or 12.5 μm thick. Be windows cannot be used to detect x-rays from elements with atomic numbers less than Na, which has an atomic number of 7 [250]. Additionally, the Be windows with thicknesses less than 8 μm are difficult to fabricate without pinholes or other flaws. Therefore because the applications requires leak-tight seals with 1 atm pressure differentials, thinner Be windows are not probable. Diamond windows as thin as 0.5 to 1.0 μm diamond can be made and still maitain a leak-tight seal. As a result, a diamond window may be used to analyze the x-ray spectra of elements with atomic numbers as low as that of boron, which has an atomic number of 5 [250]. A comparison of performance of beryllium and diamond windows is shown in Fig. 7-38 [250].

Another potential application for diamond related to x-rays is for the manufacture of x-ray masks for x-ray lithography [379–383]. Typical materials for x-ray masks include Silicon, silicon nitride, silicon carbide, and boron nitride. Some of the existing problems with these materials are the dimensional tolerance of the boron nitride and silicon nitride because of the presence of dissolved hydrogen in the films and the opaque nature of Si [250]. The most important properties for this application are the modulus of elasticity

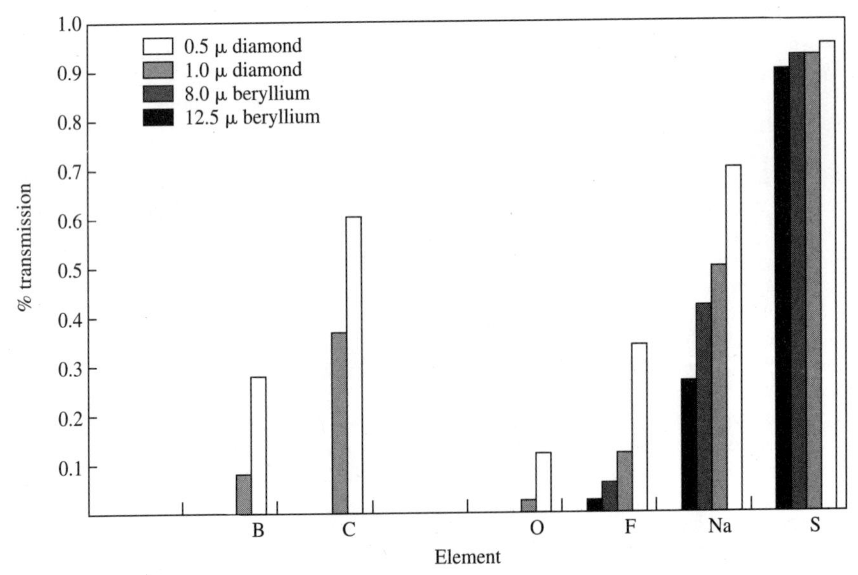

FIGURE 7-38 A comparison of the x-ray transmission of diamond and beryllium. *(From K. V. Ravi,* Diamond Materials, Electrochemical Society Symposium, *1991, p. 301. Reprinted with permission. Copyright 1991 Electrochemical Society.)*

and resistance to radiation damage. Diamond is ideal in both of these properties because it is transparent to x-ray radiation, dimensionally stable under x-ray fluxes, and has a high modulus of elasticity for minimal distortion [250]. One requirement for this application is that the deposited-diamond film is under tensile stress to prevent buckling of the film.

Diamond Photoemission Devices. Light-emitting diodes have been fabricated in several colors [299]. Blue (A-band electroluminescence), orange (575 nm zero-phonon line-nitrogen-related center) or green emission (H3 center) have all been observed depending on the nature of the diamond substrate. The highest brightness was observed for the green emission. The structures consisted of p^+-i-p^+ or M-i-p^+ structures where the light is emitted from the i regions when the deep traps have been filled. Other structures have also been used to generate light. A double-insulated electroluminescent structure was fabricated with CVD diamond that emitted blue light (430 nm) [384]. A Schottky junction to a boron-doped homoepitaxial layer emitted light at 530 nm [385].

Diamond lasers have also been demonstrated. Because diamond is an indirect bandgap material, the bandgap recombination of 227 nm electromagnetic radiation is not plausible [386]. However, lasing action was observed in diamond because of the H3 color center (carbon vacancy and two N atoms) [387]. The H3 centers may be formed through electron irradiation and annealing and have demonstrated good stability. Under an excitation with a wavelength of 490 nm, laser light emission was observed at 530 nm with a critical inversion density of 4.2×10^{12} cm^{-3}. Other centers such as the N3 center did not have lasing action [387]. The prospects and requirements for potential lasing centers have recently been reviewed [388].

Diamond has also been used as a Q-switch. In this application, short pulses of high-intensity light are generated. The GR1 center is a neutral vacancy with zero-phonon lines at 1.665 and 1.673 eV, with a decay time of 2.5 ns at 70°K and 1 ns at 300°K [389]. These centers were produced in natural diamond by a 10^{18} cm^{-2} dose of 4 MeV electrons. The photochemical stability of the irradiated diamond was good and no change in the optical transmission was observed after 2000 pulses of 25 ns duration and 500 to 700 MW/cm^2 intensity. The Q-switch for a laser resonator was obtained where the 4 to 5 μm pulse of the ruby laser was converted into 1 to 3 pulses of 0.3 to 0.5 μs duration and intensity that was 10 to 15 times that of the free generation of the ruby laser. Antireflective coatings improved the efficiency of the Q-switch and pulses were obtained with a duration of <1 ns. Similar effects were measured for the H3 center with an absorption band at 450 to 510 nm, which may be used with argon-ion lasers [390].

Although diamond was shown to have photoemission properties, it is not ideally suited for these applications [224]. The indirect bandgap of diamond makes these transitions inefficient. An additional difficulty is that the high resistance of diamond makes minority-carrier injection difficult. Although defects may be introduced into diamond to produce light emission, other materials are probably better suited for these sorts of applications.

Radiation and Particle Detectors. Metal-diamond-metal structures have been used to detect ionizing radiation. The radiation impinges on the undoped diamond to generate free carriers. Any free carriers generated in the diamond produce a current that may be used to detect the presence of the ionizing radiation. For diamond, any electromagnetic radiation with an energy >5.5 eV is sufficient to create carriers. This includes, UV-rays, x-rays, and γ rays. Additionally, high-energy particles such as α particles, electrons, neutrons, pions, or other exotic particles may be detected. Radiation detectors typically operate in two different modes. They may operate in either pulse-counting or current modes. In the pulse-counting mode, individual events are recorded where either the entire energy spectra may be recorded through spectroscopic techniques or simply the number of events may be counted. In the current mode of operation, the current is proportional to the intensity of the radiation field. Carrier generation is roughly proportional to the intensity of absorbed radiation divided by the average energy to form an electron-hole pair ($\approx 3\ E_g$).

Currently, the most common detector is a *p-n* junction made of Si. The *p-n* junction is required to reduce thermally generated currents that limit the applied electric field and also produce noise. Diamond does not require a *p-n* junction because of its high resistivity and low intrinsic-carrier concentration. This also enables high-temperature operation. Its low dielectric constant reduces device capacitance. The large bandgap reduces the sensitivity to visible radiation, eliminating the need for light-tight enclosures. The high saturated velocity of carriers enables light count-rate operation of these devices. Additionally the radiation hardness and chemical inertness make diamond suitable for these detector applications [306].

For high-energy particle colliders, detectors that can operate at high event rates and severe radiation environments are required. The main detector criteria are a charge collection time of 1 ns (dictated by event rate) and charge generation of approximately 10^4 charges per ionizing particle (dictated by charge-sensitive preamplifier). As a result, the detector should be 200 μm thick. Because the collection distance must exceed the detector thickness to prevent detrimental space charge buildup, the collection distance must be 200 μm. Additionally, the mobility must be 2000 $cm^2 V \cdot s$ and the carrier lifetime 1 ns in an electric field of 10^4 V/cm. Current state-of-the-art techniques for CVD diamond show a collection distance of 20 μm; therefore, improvements in the material are required for this application to be realized [307]. Initial proof-of-concept detectors have been fabricated and tested with the 20 μm collection distance CVD diamond and demonstrated the detection of single minimum ionizing particles [306].

7.4.3 Thermal Applications

The high thermal conductivity of diamond makes it a good heat spreader for discrete devices, power modules, microprocessor modules, multichip modules, and other devices or components that require heat dissipation. Because diamond is a good electrical insulator, it is suitable for applications where both heat conduction and electrical insulation are required. Heat spreaders that rely on electronic heat conduction, or metals, are also good electrical conductors and are not able to provide suitable electrical isolation. As mentioned in the section on thermal conductivity, diamond has a high thermal conductivity, but not a high heat capacity; therefore, it is a heat spreader rather than a heat sink. A schematic diagram of a conventional thermal management structure using aluminum oxide and a diamond heat spreader is shown in Fig. 7-39 [391].

A good design of the heat spreader is required to maximize its heat dissipation capability. Detailed thermal conductivity modeling was developed for diamond heat spreaders [392–395]. Experience yields a rule-of-thumb that the heat spreader thickness should be approximately one-third to one-half the lateral dimension. If it is too thick, the heat spreader performance is not significantly improved because of the additional thermal resistance produced by the excess thickness. If it is too thin, there is insufficient material to spread the heat laterally away from the source and the heat flow is all one-dimensional flow, which is more inefficient than two-dimensional heat flow. Typical lateral dimensions for heat spreaders are $\approx$5 times the heat source device dimension for centrally mounted chips, and $\approx$2.5 times the device dimension for edge mounted ships [314]. These rules-of-thumb assume an isotropic thermal conductivity. Therefore, modifications are required for CVD diamond because it has an anisotropic thermal conductivity as a result of the fiber texture. Because of the high cost of diamond, the total volume (thickness and diameter) must be evaluated to determine the best cost/performance relationship. An increased area heat spreader with decreased thickness had equal performance to a thicker,

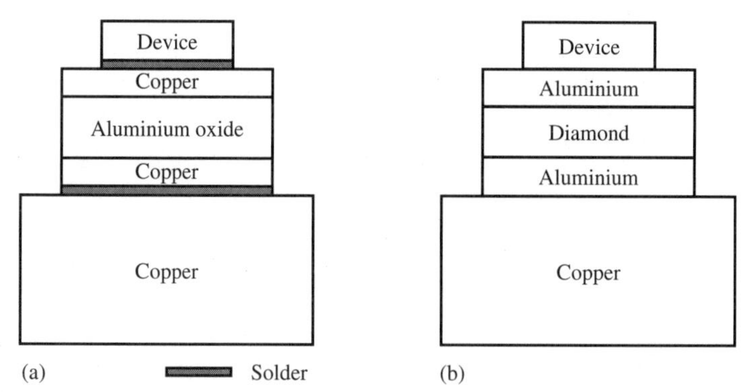

FIGURE 7-39 A schematic diagram of *(a)* a conventional heat spreader and a *(b)* a diamond heat spreader. *(From B. Fiegl, et al.,* Dia Rel Mat, *vol. 3, 1994, p. 936. Reprinted with permission. Copyright 1994 Elsevier Science S.A.)*

smaller area film; however, the volume of the larger area heat spreader was about half that of the thicker one and is more cost effective [396].

Laser diode submounts account for the largest usage of diamond heat sinks. In the late 1980s, several hundred thousand submounts per year were purchased. The market size was estimated to be millions per year by 1995. Initial use for these devices began around 1970 [397–399]. Low-power commercial laser diodes typically do not require heat sinks; however, higher-power lasers generate sufficient heat that the junction temperature rises. This significantly degrades the device performance or reliability unless the heat is dissipated. One of the areas where high-power lasers are required is for optical repeaters in long-distance fiber optics. Sufficient heat dissipation is particularly required for submarine cables where lifetime is critical. Optical pumping of other lasers is also becoming a prominent application of laser diodes and also requires heat dissipation.

Diamond heat spreaders have been used for a multitude of different laser diode materials. These include InP [400], GaAs [401], AlGaAs [402] and, InGaAsP [403]. In an early demonstration of the benefits of a diamond heat spreader, continuous operation of a GaAs laser diode at 200°K was obtained by the use of a natural diamond heat spreader. The use of diamond increased the operating temperature 50°K over that of a Cu heat spreader. The limiting thermal resistance in this device is the heat dissipation through the GaAs and not the diamond [401]. More recent results of an AlGaAs laser diode on a diamond submount demonstrated 100 W of output power at room temperature. This was a record output power obtained from this type of laser and the thermal resistance of the heat dissipation was reduced by 50% over a Cu heat sink by the presence of the diamond submount [402]. Other researchers found the maximum light output of laser diode arrays on diamond heat sinks was increased by 10% over Cu heat sinks [404]. The improved performance and availability of CVD diamond lead AT&T to announce the first commercial use of CVD in diamond submounts for laser applications [405]. Continued work on heat spreaders for InP laser diodes has shown that metalization of diamond to InP laser is critical [400].

Other device applications also require heat dissipation. Microwave diodes were the first devices to use diamond heat sinks [406]. Impact ionization avalanche transit time (IMPATT) diodes operate in the 1 to 100 GHz frequency range. The avalanche mechanism generates heat and the frequency of operation is sensitive to thermal fluctuations; therefore, heat generated must be dissipated and diamond heat sinks were required. In semiconductor power modules, increased switching power density and miniaturization has led to the need for improved power dissipation. Typical power losses in these devices are 200 W/cm^2. Aluminum oxide, with a thermal conductivity of only 0.25 W/cm/°K is typically used, but is a poor heat conductor compared to most metals and vastly inferior to diamond. In addition to thermal conduction, these modules must provide electrical isolation. For the CVD diamond heat spreader, the electrical resistivity was 10^9 to 10^{13} Ω·cm [391]. In addition to the heat dissipation, the heat spreader must also be able to have a high dielectric strength (5000 V for at least 1 min). Although no hard breakdown field was observed, the breakdown field was estimated at 10^5 to 10^6 V/cm; therefore, a minimum diamond thickness of only 100 μm was required for proper electrical insulation. Because of the improved electrical and thermal properties of diamond, the thermal resistance in the diamond power module was 35% less than that in the aluminum oxide module.

Another area where heat dissipation is required is for multichip modules (MCMs), which are discussed in Chapter 9. Multichip modules are a semiconductor package that places multiple chips onto a single substrate instead of packaging devices individually. MCMs can achieve higher performance by combining multiple state-of-the-art chips within a single module. Also, smaller devices on MCMs give higher yields than large area, complex chips; and the package size may be dramatically reduced. The primary technologies for MCMs include organic laminated structures, multilayer ceramic structures, and thin film interconnects on a ceramic substrate such as diamond, alumina, or Si. In one application where a high density of interconnections is desired, 90% of the area is occupied by chips. The close spacing of chips with higher and higher powers produces significant thermal management problems. Diamond allows improved packing density of chips and modules over current alumina and Si substrates [407]. Diamond MCMs have been used for mounting 2 microprocessor chips, static random-access memory (SRAM), and address and data buffers on the same MCM for a computer motherboard [408]. The total heat required to be dissipated is 243 W where the junction tempeature cannot exceed 85°C and the ambient temperature is 40°C in air. The diamond MCM lowers the junction temperature 10°C to 15°C. In another potential MCM application for an image processor, it is desirable to have 8 MCMs/board with a microprocessor, logic application-specific integrated circuit (ASIC), and dynamic randam-access memory (DRAM) on each MCM [408]. Additionally, each board must be able to be mounted with a 8.6 mm center-to-center board limitation. Because diamond is an improved thermal conductor, a thinner diamond substrate may be used than with other ceramic substrate materials and allow closer packing of the boards.

7.5 SUMMARY

An overview of the deposition technology for CVD diamond films, the properties of diamond films, and the potential applications for diamond films has been presented. The area of diamond CVD is a relatively new one; but it is growing rapidly because of the great promise of this unique material. There are still numerous technical challenges to overcome before the projected $16 billion industry in the year 2020 may be realized. The primary technical hurdles are: increasing the deposition area; increasing the growth rate; identifying or developing a large-area heteroepitaxial or homoepitaxial substrate for single-crystal film deposition; and reducing the defects. Also, for electronic applications a suitable n-type dopant should be identified to allow a bipolar device to be realized. Although there are technical hurdles to be overcome, the unique properties of diamond combined with the ability to control the film through CVD makes it a promising technology for numerous electrical, optical, and thermal applications.

ACKNOWLEDGEMENT

The author would like to thank Kobe Steel USA Inc., Electronic Material Center, for the assistance and support that it provided during the preparation of this work. Additionally, critical review of this work was provided by Drs. J. T. Glass, D. L. Dreifus, and L. S. Plano. The author would also like to acknowledge his collegues at Kobe Steel USA Inc., Electronic Material Center, Kobe Steel Europe, Ltd., and Kobe Steel, Ltd., Electronics Research Laboratory, for their contributions to this work.

REFERENCES

1. J. V. Busch, and J. P. Dismukes, in K. E. Spear and J. P. Dismukes (eds.), *Synthetic Diamond,* John Wiley & Sons, New York, 1994, p. 581.
2. N. Setaka, in K. E. Spear and J. P. Dismukes (eds.), *Synthetic Diamond,* John Wiley & Sons, New York, 1994, p. 57.
3. D. V. Fedoseev, in K. E. Spear and J. P. Dismukes (eds.), *Synthetic Diamond,* John Wiley & Sons, New York, 1994, p. 41.
4. J. C. Angus, in K. E. Spear and J. P. Dismukes (eds.), *Synthetic Diamond,* John Wiley & Sons, New York, 1994, p. 21.
5. J. C. Angus et al., *Phil Trans R Soc London A* vol. 342, 1993, p 195.
6. W. G. Eversole, *Synthesis of Diamond,* U. S. Patents 3030187 and 3030188, 1962.
7. J. C. Angus, H. A. Will, and W. S. Stanko, *J Appl Phys* vol. 39, 1968, p 2915.
8. J. C. Angus, *Air Force Cambridge Research Lab Report,* AFCRL-66-107; AD-63-705, 1966.
9. F. P. Bundy et al., *Nature* vol. 176, 1955, p. 51.
10. H. Liander and E. Lundblad, *Ark. Kemi* vol. 16, 1960, p. 139.
11. J. C. Angus et al. *Sin Almazy* vol. 3, 1971, p. 38.
12. B. V. Deryagin et al., *Dkol Akad Nauk SSSR* vol. 231, 1976, p. 333.
13. D. V. Fedoseev et al, *Izv Akad Nauk SSSR Ser Khim* vol. 6, 1978, p. 1252.
14. B. V. Deryagin et al., *Sov Phys-Cryst* vol. 21, 1976, p. 239.
15. B. V. Deryagin and D. V. Fedoseev, *Russian Chem Rev* vol. 39, 1970, p. 783.
16. S. Matsumoto et al., *J Mater Sci* vol. 17, 1982, p. 3106.
17. C. J. Russel, in K. E. Spear and J. P. Dismukes (eds.), *Synthetic Diamond,* John Wiley & Sons, New York, 1994.
18. B. Lux and R. Haubner, in R. F. Davis (ed.), *Diamond Films and Coatings,* Noyes Publications, Park Ridge, NJ, 1993, p. 183.
19. A. Neuhaus, *Angew Chem* vol. 66, 1954, p. 525.
20. H. Kanda and T. Sekine, in G. Davies (ed.), *Properties and Growth of Diamond* INSPEC, Exeter, England, 1994, p. 404.
21. P. W. Bridgeman, *Sci Am* vol. 193, 1955, p. 42.
22. G. Davies and T. Evans, *Proc R Soc London* vol. A328, 1972, p. 413.
23. M. N. Yoder, in R. F. Davis (ed.), *Diamond Films and Coatings,* Noyes Publications, Park Ridge, NJ, 1993, p. 1.
24. B. Lux and R. Haubner, *Electrochemical Society Meeting Spring Meeting,* Electrochemical Society, Washington, DC, 1991, p. 314.

25. J. C. Angus, R. W. Hoffman, and P. H. Schmidt, *Science and Technology of New Diamond,* KTK/Terra, Tokyo, Japan, 1988, p. 9.
26. D. N. Belton and S. J. Schmieg, *Surf Sci* vol. 233, 1990, p. 131.
27. P. O. Joffeau, R. Haubner, and B. Lux, *MRS Spring Meeting,* Materials Research Society, Pittsburg, PA, 1988, p. 15.
28. K. Wasa and S. Hayakawa, Japanese Patent Application, 51-84840, 1966.
29. K. Suzuki et al., *Appl Phys Lett* vol. 50, 1987, p. 728.
30. K. Kobashi et al., *Proc 1st Int Symposium on Diamond and Diamond-like Films,* The Electrochemical Society, Los Angeles, 1989, p. 139.
31. R. Ramesham et al., *New Diamond Science and Technology,* Materials Research Society, Washington, DC, 1990, p. 943.
32. S. A. Grot et al., *Appl Phys Lett* vol. 58, 1991, p. 1542.
33. T. Inoue et al., *J Appl Phys* vol. 67, 1990, p. 7329.
34. B. R. Stoner et al., *Phys Rev B* vol. 45, 1992, p. 11067.
35. L. Ferrari et al., *Appl Surf Sci* vol. 56–58, 1992, p. 100.
36. H. Maeda et al., *J Mater Sci* vol. 28, 1993, p. 129.
37. P. K. Bachmann et al., *Diamond and Diamond-Like Materials Synthesis,* Materials Research Society, Pittsburgh, 1988, p. 99.
38. S. Iijima, Y. Aikawa, and K. Baba, *J Mater Res* vol. 6, 1991, p. 1491.
39. S. Iijima, Y. Aikawa, and K. Baba, *Appl Phys Lett* vol. 57, 1990, p. 2646.
40. A. Sawabe and T. Inuzuka, *Thin Solid Films* vol. 137, 1986, p. 89.
41. C.-P. Chang et al., *J Appl Phys* vol. 63, 1988, p. 1744.
42. K Higuchi and S. Noda, *Dia Rel Mat* vol. 1, 1992, p. 220.
43. B. R. Stoner and J. T. Glass, *Appl Phys Lett* vol. 60, 1992, p. 698.
44. X. Jiang et al., *Appl Phys Lett* vol. 62, 1993, p. 3438.
45. Y. Hirose and Y. Terasawa, *Jpn J Appl Phys* vol. 25, 1986, p. L519.
46. Y. Saito et al., *J Mater Sci* vol. 23, 1988, p. 842.
47. T. Ito et al., *Proceedings of the 1st International Conference on Science and Technology of New Diamond,* Tokyo, Japan, 1988, p. 107.
48. C. Chen et al., *Proceedings of the Surf Eng Int Conf in Japan,* 1989, p. 317.
49. Y. Muranaka, H. Yamashita, and H. Miyadera, *J Vac Sci Technol* vol. 9, 1990, p. 76.
50. P. K. Bachmann, D. Leers, and H. Lydtin, *Dia Rel Mat* vol. 1, 1991, p. 1.
51. T. R. Anthony, "Metastable Synthesis of Diamond," in R. Freer (ed.), *The Physics and Chemistry of Carbides, Nitrides, and Borides,* conference proceedings, 1990, p. 133.
52. M. Frenklach, *J Appl Phys* vol. 65, 1989, p. 5142.
53. J. C. Angus et al., *MRS Bull* October, 1989, p. 38.
54. W. Banholzer, *Surf and Coatings Tech* vol. 53, 1992, p. 1.
55. W. L. Hsu, *J Vac Sci* vol. 6, 1988, p. 1803.
56. W. Piekarczyk, *J of Crystal Growth* vol. 119, 1992, p. 345.
57. W. A. Yarbrough and R. Messier, *Science* vol. 247, 1990, p. 688.
58. S. J. Harris and A. M. Weiner, *Appl Phys Lett* vol. 55, 1989, p. 2179.
59. W. Zhu, A. R. Badzian, and R. F. Messier, *Proc SPIE Diamond Optics III* SPIE, Bellingham, WA, 1990, p. 187.
60. R. A. Rudder et al., *Appl Phys Lett* vol. 60, 1992, p. 329.
61. A. Inspektor et al., *Surf Coat Technol* vol. 39/40, 1989, p. 211.

62. L. S. Plano, in L. S. Pan and D. R. Kania (eds.), *Electronic Properties and Applications,* Kluwer Academic Publishers, Boston, 1995, p. 61.
63. W. A. Yarbrough, A. Inspektor, and R. Messier, *Mat Sci Forums* vol. 52 & 53, 1989, p. 151.
64. W. Zhu et al., *Proc of the IEEE* vol. 79, 1991, p. 621.
65. C.-P. Klages, *Appl Phys A* vol. 56, 1993, p. 513.
66. M. Kamo et al., *J of Cryst Growth* vol. 62, 1983, p. 642.
67. D. J. Pickrel et al., *Appl Phys Lett* vol. 56, 1990, p. 2010.
68. B. R. Stoner et al., *J Electron Mat* vol. 21, 1992, p. 629.
69. H. Kawarada, K. S. Mar, and A. Hiraki, *Jpn J Appl Phys* vol 26, 1987, p. L1032.
70. W. L. Hsu, *J Appl Phys* vol. 72, 1992, p. 3102.
71. S. Matsumoto et al., *Jpn J Appl Phys* vol. 21, 1982, p. L183.
72. A. Sawabe and T. Inuzuka, *Appl Phys Lett* vol. 46, 1985, p. 146.
73. F. Jansen, I. Chen, and M. A. Machonkin, *J Appl Phys* vol. 66, 1989, p. 5749.
74. T. D. Moustakas, *Solid State Ionics,* vol. 32/33, 1989, p. 861.
75. T. D. Moustakas, in K. E. Spear and J. P. Dismukes (eds.), *Synthetic Diamond,* John Wiley & Sons, New York, 1994, p. 145.
76. T. R. Anthony and W. F. Banholzer, in P. K. Bachmann and A. Matthews (eds.), *Diamond, Diamond-like, and Related Coatings, 1991,* Elsevier, Amsterdam, 1991, p. 717.
77. B. Singh et al., *Appl Phys Lett* vol. 52, 1988, p. 451.
78. T. R. Anthony, *MRS Fall Meeting,* Boston, MA, 1987.
79. Y. Hirose, *1st Int Conf on New Diamond Science and Technology,* Tokyo, 1988.
80. Y. Hirose and M. Mitsuizumi, *New Diamond* vol. 4, 1988, p. 34 (in Japanese).
81. J. A. Mucha, D. L. Flamm, and D. E. Ibbotson, *J Appl Phys* vol. 65, 1989, p. 3448.
82. Y. Matsui et al., *Jpn J Appl Phys* vol. 28, 1989, p. 1718.
83. N. G. Glumac and D. G. Goodwin, *Thin Solid Films* vol. 212, 1992, p. 122.
84. M. Peters et al., *SPIE Conference on Innovative Science and Technology,* Los Angeles, 1988.
85. N. Ohtake et al., *First International Symposium on Diamond and Diamond-like Films,* The Electrochemical Society, 1989, p. 93.
86. B. A. Fox et al., *Dia Rel Mat* 1994, p. 382.
87. D. J. Poferl, N. C. Gardner, and J. C. Angus, *J Appl Phys* vol. 44, 1972, p. 1428.
88. J. C. Angus, et al., *International Conference on Applications of Synthetic Diamonds in Industry,* Kiev, 1971, presentation.
89. A. E. Aleksenko et al., *Sov Phys Dokl* vol. 22, 1977, p. 166.
90. G. Janssen et al., *Dia and Rel Mat* vol. 1, 1992, p. 789.
91. N. Fujimori, H. Nakahata, and T. Imai, *Jpn J of Appl Phys* vol. 29, 1990, p. 824.
92. K. A. Snail and L. M. Hanssen, *J Crystal Growth* vol. 112, 1992, p. 651.
93. D. P. Malta et al., *J Mater Res* vol. 8, 1993, p. 1217.
94. D. R. Black, H. E. Burdette, and W. Banholzer, *Dia Rel Mat* vol. 2, 1993, p. 121.
95. E. Bauer, *Z Kristallogr* vol. 110, 1958, p. 423.
96. F. C. Frank and J. H. van der Merwe, *Proc Roy Soc London A* vol. 198, 1949, p. 216.
97. F. C. Frank and J. H. van der Merwe, *Proc Roy Soc London Ser A* vol. 198, 1949, p. 205.
98. M. Volmer and Z. Weber, *Phys Chem* vol. 119, 1926, p. 277.
99. J. H. van der Merwe and E. Bauer, *Phys Rev B* vol. 39, 1989, p. 3632.
100. M. W. H. Braun et al., *J Appl Phys* vol. 69, 1991, p. 2679.

101. Y. H. Lee et al., *Appl Phys Lett* vol. 57, 1990, p. 1916.
102. J. E. Field (ed.), in *The Properties of Diamond,* Academic Press, London, 1979, p. 281.
103. S. M. Sze, *Physics of Semiconductor Devices,* John Wiley & Sons, New York, 1981.
104. W. S. Verwoerd, *Surf Sci* vol. 304, 1994, p. 24.
105. S. Koizumi et al., *Appl Phys Lett* vol. 57, 1990, p. 563.
106. M Yoshikawa et al., *Appl Phys Lett* vol. 57, 1990, p. 428.
107. W. A. Yarbrough, *J Vac Sci Technol A* vol. 9, 1991, p. 1145.
108. B. R. Stoner and J. T. Glass, *Appl Phys Lett* 1994, submitted.
109. W. Zhu et al., *Phys Rev B* vol. 47, 1993, p. 6529.
110. J. H. van der Merwe, J. Wolterdorf, and W. A. Jesser, *Mat Sci Eng* vol. 81, 1986, p. 1.
111. Z. Li et al., *J Appl Phys* vol. 73, 1993, p. 711.
112. A. Argoitia et al., "BeO/Diamond Heteroepitaxy," from Suzuki et al., *Appl Phys Lett* vol. 64, 1994, p. 557.
113. T. Suzuki, M. Yagi, and K. Shibuki, *Apply Phys Lett* vol. 64, 1994, p. 557.
114. Y. Sano et al., *Phil Trans R Soc Lond A* vol. 342, 1993, p. 225.
115. A. van der Drift, *Philips Res Rep* vol. 22, 1967, p. 267.
116. R. E. Clausing et al., *Materials Research Society Symposium International Proceedings NDST-2,* Materials Research Society, Pittsburg, PA, 1991, p. 575.
117. C. Wild et al., *Dia Rel Mater* vol. 3, 1994, p. 373.
118. P. Koidl, C. Wild, and N. Herres, *Proc NIRIM International Symposium on Advanced Materials* Tsukuba, Japan, 1994.
119. B. V. Spitzyn, *J Crystal Growth* vol. 99, 1990, p. 1162.
120. J. E. Graebner et al., *Nature* vol. 359, 1992, p. 401.
121. C. Wild, N. Herres, and P. Koidl, *J Appl Phys* vol. 68, 1990, p. 973.
122. L. S. Pan et al., *Science* vol. 255, 1992, p. 830.
123. B. E. Williams, H. S. Kong, and J. T. Glass, *J Mater Res* vol. 5, 1990, p. 801.
124. B. E. Williams and J. T. Glass, *J Mater Res* vol. 4, 1989, p. 373.
125. G.-H. M. Ma et al., *Dia Rel Mat* vol. 1, 1991, p. 25.
126. J. Narayan, *J Mater Res* vol. 5, 1990, p. 2414.
127. Z. Wang et al., *Applications of Diamond Films and Related Materials,* 1991, p. 489.
128. A. R. Badzian et al., *New Diamond Science and Technology,* Materials Research Society, Washington, DC, 1991, p. 549.
129. G.-H. M. Ma, et al., *J Mater Res* vol. 5, 1990, p. 2367.
130. M. P. Everson et al., *J Appl Phys* vol. 75, 1994, p. 169.
131. J. C. Angus et al., *J Mater Res* vol. 7, 1992, p. 3001.
132. K. Hirabayashi and N. I. Kurihara, *Jpn J Appl Phys* vol. 30, 1991, p. L49.
133. W. Luyten, G. Van Tendeloo, and S. Amelinckx, *Phil Mag A* vol. 66, 1992, p. 899.
134. D. Shechtman et al., *J Mater Res* vol. 8, 1993, p. 473.
135. M. A. Tamor and M. P. Everson, *J Mater Res* vol. 9, 1994, p. 1839.
136. J. P. F. Sellschop, in J. E. Field (ed.), *The Properties of Diamond,* Academic Press, New York, 1979, p. 107.
137. W. Kaiser and W. L. Bond, *Phys Rev* vol. 115, 1959, p. 857.
138. G. Davies, *J Phys C,* vol. 9, 1976, p. L537.
139. J. H. N. Loubser and J. A. van Wyk, *Diamond Conference,* Reading, 1981, unpublished.

140. S. A. Kajihara, A. Antonelli, and J. Bernholc, *Physica B* vol. 185, 1993, p. 144.
141. T. Evans and Z. Qi, *Proc R Soc London Ser A* vol. 381, 1982, p. 159.
142. R. M. Chrenko, H. M. Strong, and R. E. Tuft, *Philos Mag* vol. 23, 1971, p. 313.
143. R. M. Chrenko, R. E. Tuft, and H. M. Strong, *Nature* vol. 270, 1977, p. 141.
144. G. S. Woods, *Philos Mag B* vol. 50, 1984, p. 673.
145. W. J. P. van Enckevort, in K. E. Spear and J. P. Dismukes (eds.), *Synthetic Diamond,* John Wiley & Sons, New York, 1994, p. 307.
146. B. Dischler et al., *Physica B,* vol. 185, 1993, p. 217.
147. G. Janssen et al., *Surf and Coatings Tech* vol. 47, 1991, p. 113.
148. K. Baba, Y. Aikawa, and N. Shohata, *J Appl Phys* vol. 69, 1991, p. 7313.
149. A. Deneuville et al., *Dia Rel Mater* vol. 2, 1993, p. 737.
150. J. Ruan, W. J. Choyke, and K. Kobashi, *Appl Phys Lett* vol. 62, 1993, vol. 1379.
151. J. Ruan, K. Kobashi, and W. J. Choyke, *Appl Phys Lett* vol. 60, 1992, p. 3138.
152. J. Ruan, K. Kobashi, and W. J. Choyke, *Appl Phys Lett* vol. 60, 1992, p. 1884.
153. Y. Mori et al., *Appl Phys Lett* vol. 1, 1992, p. 47.
154. A. M. Zaitsev, A. A. Gippius, and V. S. Vavilov, *Kratkie Soobscheniya po Fizike* (in Russian), vol. 10, 1981, p. 20.
155. V. S. Vavilov et al., *Sov Phys-Semicond (USA)* vol. 14, 1980, p. 1078.
156. C. D. Clark and C. B. Dickerson, *Surface Coatings Technol* vol. 47, 1991, p. 336.
157. A. T. Collins et al., *Phys Rev Lett* vol. 65, 1990, p. 891.
158. A. T. Collins, M. Kamo, and Y. Sato, *J Mater Res* vol. 5, 1990, p. 2507.
159. R. J. Graham, T. D. Moustakas, and M. M. Disko, *J Appl Phys* vol. 69, 1991, p. 3212.
160. E. P. Visser et al., *J Phys Condens Matter* vol. 4, 1992, p. 7365.
161. R. C. Burns, V. Cvetkovic, and C. N. Dodge, *J Crystal Growth* vol. 104, 1990, p. 257.
162. A. T. Collins, *Phil Trans R Soc London A* vol. 342, 1993, p. 233.
163. F. P. Bundy, H. M. Strong, and R. H. Wentorf Jr, in P. L. Walker and P. A. Thrower (eds.), *Chemistry and Physics of Carbon,* Marcel Dekker, New York, 1973, p. 213.
164. R. C. Burns and G. J. Davies, in K. E. Spear and J. P. Dismukes (eds.), *Synthetic Diamond,* John Wiley & Sons, New York, 1994, p. 395.
165. F. P. Bundy, *J Chem Phys* vol. 38, 1963, p. 631.
166. P. S. DeCarli and J. C. Jamieson, *Science* vol. 133, 1961, p. 1821.
167. O. R. Bergmann, N. F. Bailey, and H. B. Coverly, *Metallography* vol. 15, 1982, p. 121.
168. J. Robertson, *Adv Phys* vol. 35, 1986, p. 317.
169. A. Grill and B. S. Meyerson, in K. E. Spear and J. P. Dismukes (eds.), *Synthetic Diamond,* John Wiley & Sons, New York, 1994, p. 91.
170. H. Tsai and D. B. Bogy, *J Vac Sci Tech A* vol. 5, 1987, p. 3287.
171. P. V. Koeppe et al., *J Vac Scis Technol A* vol. 3, 1985, p. 2327.
172. S. Kaplan, F. Jansen and M. Machonkin, *Appl Phys Lett* vol. 47, 1985, p. 750.
173. J. J. Cuomo et al., *J Vac Sci Technol* vol. A 9, 1991, p. 2210.
174. C. J. Morath et al., *Appl Phys* vol. 76, 1994, p. 2636.
175. J. J. Cuomo et al., *Appl Phys Lett* vol. 58, 1991, p. 466.
176. C. Weissmantel et al., *Thin Solid Films* vol. 96, 1982, p. 31.
177. C. L. Marquardt, R. T. Williams, and D. J. Nagel, *Materials Research Society Symposium,* 1985, p. 325.

178. I. I. Askenov and V. E. Strelnitskij, *Surf Coating Technol* vol. 47, 1991, p. 98.
179. A. M. Stoneham, in J. E. Field (ed.), *The Properties of Diamond,* Academic Press, London, 1979, p. 185.
180. G. S. Painter, D. E. Ellis, and A. R. Lubinsky, *Phys Rev B* vol. 4, 1971, p. 3610.
181. K. P. O'Donnell and X. Chen, *Appl Phys Lett* vol. 58, 1991, p. 2925.
182. B. A. Fox, unpublished.
183. S. F. Adams et al., *Second International Conference on Diamond and Diamond Materials,* Electrochemical Society, Washington, DC, 1991, p. 471.
184. M. I. Landstrass and K. V. Ravi, *Appl Phys Lett* vol. 55, 1989, p. 975.
185. M. I. Landstrass and K. V. Ravi, *Appl Phys Lett* vol. 55, 1989, p. 1391.
186. H. Nakahata, I. Takahiro, and F. Naoji, *179th Meeting of the Electrochemical Society,* Washington, DC, 1991, p. 487.
187. Y. Mori et al., *Jpn J Appl Phys* vol. 31, 1992, p. L1191.
188. J. F. H. Custers, *Physica* vol. 18, 1952, p. 489.
189. E. C. Lightowlers and A. T. Collins, *J Phys D Appl Phys* vol. 9, 1976, p. 951.
190. A. T. Collins and A. W. S. Williams, *J Phys C Solid St Phys* vol. 4, 1971, p. 1789.
191. R. M. Chrenko, *Phys Rev B* vol. 7, 1973, p. 4560.
192. K. Okano et al., *Jpn J of Appl Phys* vol. 28, 1989, p. 1066.
193. J. Mort et al., *Appl Phys Lett* vol. 55, 1989, p. 1121.
194. M. E. Baginski, T. A. Baginski, and J. L. Davidson, *J Electrochem Soc* vol. 137, 1990, p. 2984.
195. W. Tsai et al., *IEEE Elect Dev Lett* vol. 12, 1991, p. 157.
196. J. F. Prins, in J. E. Field (ed.), *The Properties of Natural and Synthetic Diamond* Academic Press, New York, 1992, p. 301.
197. J. F. Prins, *Phys Rev B* vol. 38, 1988, p. 5576.
198. V. S. Vavilov et al., *Sov. Phys-Semiconductors* vol. 4, 1970, p. 12.
199. A. T. Collins and E. C. Lightowlers, in J. E. Field (ed.), *The Properties of Diamond,* Academic Press, London, 1979, p. 79.
200. H. B. Dyer et al., *Philos Mag* vol. 11, 1965, p. 763.
201. R. G. Farrer, *Solid State Commun* vol. 7, 1969, p. 685.
202. L. A. Vermeulen and R. G. Farrer, *Diamond Research 1975 (Suppl Ind Diam Rev),* 1975, p. 18.
203. P. Denham, E. C. Lightowlers, and P. J. Dean, *Phys Rev* vol. 161, 1967, p. 762.
204. C. C. Klick and R. J. Maurer, *Phys Rev* vol. 81, 1951, p. 124.
205. A. G. Redfield, *Phys Rev* vol. 94, 1954, p. 526.
206. K. Okano et al., *Dia and Rel Mat* vol. 3, 1993, p. 35.
207. K. Okano et al., *Appl Phys A* vol. 51, 1990, p. 344.
208. J. R. Flemish et al., *Dia Rel Mater* vol. 3, 1994, p. 672.
209. S. Prawer et al., *Appl Phys Lett* vol. 63, 1993, p. 2502.
210. K. Jackson, M. R. Pederson and J. G. Harrison, *Phys Rev B* vol. 41, 1990, p. 12641.
211. S. A. Kajihari et al., *Phys Rev Lett* vol. 66, 1991, p. 2010.
212. S. A. Kajihara, A. Antonelli, and J. Bernholc, *Diamond, Silicon Carbide and Related Wide Bandgap Semiconductors,* Materials Research Society, 1990, p. 315.
213. E. A. Konorova and S. A. Shevchenko, *Sov Phys-Semicond* vol. 1, 1967, p. 299.
214. P. T. Wedepohl, *Proc Phys Soc* vol. 70B, 1957, p. 177.
215. J. G. Austin and R. Wolfe, *Proc Phys Soc V* vol. 69, 1956, 329.
216. C. Canali et al., *Nucl Inst and Methods* vol. 160, 1979, p. 73.

217. L. Reggiani et al., *Solid State Comm* vol. 30, 1979, p. 333.
218. L. S. Pan et al., *J Appl Phys* vol. 73, 1993, p. 2888.
219. B. A. Fox et al., *Dia Rel Mater* submitted for publication.
220. D. M. Malta et al., *2nd High Temperature Electronics Conference,* Charlotte, NC, 1994, p. P-75.
221. E. Conwell and V. F. Weisskopf, *Phys Rev* vol. 77, 1950, p. 388.
222. C. Erginsoy, *Phys Rev* vol. 79, 1950, p. 1013.
223. R. T. Bate and R. K. Willardson, *Proc Phys Soc (London)* vol. 74, 1959, p. 363.
224. A. T. Collins, *Semicond Sci Technol* vol. 4, 1989, p. 605.
225. J. Bardeen and W. Shockley, *Phys Rev* vol. 80, 1950, p. 72.
226. K. Seeger, *Semiconductor Physics,* Springer-Verlag, New York, 1985.
227. A. W. S. Williams, *Electrical Transport Measurements in Natural and Synthetic Semiconducting Diamond,* University of London, Ph.D. Thesis, 1971.
228. D. L. Dreifus et al., unpublished results, 1994.
229. F. Nava et al., *Solid State Comm* vol. 33, 1980, p. 475.
230. J. S. Blakemore, *Semiconductor Statistics,* Dover Publications, Inc., New York, 1987.
231. L. P. Plano, D. M. Malta, and B. A. Fox, *Electrochemical Society Conference,* Honolulu, Hawaii, 1993.
232. P. J. Dean, *Phys Rev* vol. 139, 1965, p. A588.
233. P. P. Debye and E. M. Conwell, *Phys Rev* vol. 93, 1954, p. 693.
234. J. Monecke et al., *Phys Stat Sol* vol. 103, 1981, p. 269.
235. J. C. Bourgoin, J. Krynicki, and B. Blanchard, *Phys Stat Sol (a)* vol. 52, 1979, p. 293.
236. K. Nishimura, K. Das, and J. T. Glass, *J Appl Phys* vol. 69, 1991, p. 3142.
237. N. F. Mott and W. D. Twose, *Adv Phys* vol. 10, 1961, p. 107.
238. E. A. Davis and W. D. Compton, *Phys Rev* vol. 140, 1965, p. A2183.
239. N. F. Mott and E. A. Davis, *Phil Mag* vol. 17, 1968, p. 1269.
240. N. F. Mott, *Philos Mag* vol. 19, 1969, p. 835.
241. A. W. S. Williams, E. C. Lightowlers, and A. T. Collins, *J Phys C Solid St Phys* vol. 3, 1970, p. 1727.
242. B. Massarani, J. C. Bourgoin, and R. M. Chrenko, *Phys Rev B* vol. 17, 1978, p. 1758.
243. M. Werner et al., *Appl Phys Lett* vol. 64, 1994, p. 595.
244. A. S. Vishnevskii et al., *Sov Phys Semicond* vol. 15, 1981, p. 659.
245. F. J. Himpsel et al., *Phys Rev B* vol. 20, 1979, p. 624.
246. M. Geis, J. Gregory and B. Pate, *IEEE Trans Electron Dev* vol. 38, 1991, p. 619.
247. N. Eimori et al., *Diamond Materials,* Electrochemical Society, Honolulu, HI, 1993, p. 934.
248. R. J. Nemanich et al., *Physica B,* vol. 185, (1993), p. 528.
249. J. van der Weide and R. J. Nemanich, *Appl Phys Lett* vol. 62, 1993, p. 1878.
250. K. V. Ravi, in K. E. Spear and J. P. Dismukes (eds.), *Synthetic Diamond,* John Wiley & Sons, New York, 1994, p. 533.
251. J. Liu et al., *Appl Phys Lett* submitted for publication (TBP).
252. N. S. Xu, R. V. Latham, and Y. Tzeng, *2nd International Conference on the Applications of Diamond Films and Related Materials,* MYU, Tokyo, 1993, p. 779.
253. F. Peter, *Z Phys* vol. 15, 1923, p. 358.
254. M. Seal and W. J. P. van Enckevort, *Diamond Optics,* SPIE, 1988, p. 144.
255. A. T. Collins, *Physica B,* vol. 185, 1993, p. 284.
256. Y. Sato et al., NIRIM Report Number 39, 1984.

257. Y. Sato et al., *44th Jap Appl Phys Soc Fall Meeting,* 1984.
258. R. A. Solin and A. K. Ramdas, *Phys Rev B* vol. 1, 1970, p. 1687.
259. W. Zhu, H.-S. Kong, and J. T. Glass, in R. F. Davis (ed.), *Diamond Films and Coatings,* Noyes Publications, Park Ridge, NJ, 1993, p. 244.
260. R. J. Nemanich and S. A. Solin, *Phys Rev B* vol. 20, 1979, p. 392.
261. F. Tuinstra and J. L. Koenig, *J Chem Phys* vol. 53, 1970, p. 1126.
262. P. K. Bachmann and R. Messier, *C&EN,* May 15, 1989, p. 24.
263. R. W. Prior et al., *Diamond Optics II,* Society of Photo-Optical and Instrumentation Engineers, 1989, p. 68.
264. N. Wada, and A. Solin, *Physica B* vol 105, 1981, p. 353.
265. J. A. Herb et al., *Proceedings of the First International Symposium on Diamond and Diamond-Like Films,* The Electrochemical Society, Los Angeles, 1989, p. 366.
266. M. Mitsuhashi et al., *Appl Surf Sci* vol. 60–61, 1992, p. 565.
267. M. Mitsuhashi, S. Karasawa, and S. Ohya, *Thin Solid Films* vol. 228, 1993, p. 76.
268. S. Zhao, *Characterization of the Electrical Properties of Polycrystalline Diamond Films,* Ohio State University, Ph.D., 1994.
269. M. Yoshikawa et al., *Appl Phys Lett* vol. 58, 1991, p. 1387.
270. B. R. Stoner et al., *Diamond-Film Semiconductor,* Society of Photo-Optical Instrument Engineers, Los Angeles, 1994, p. 2.
271. D. S. Knight and W. B. White, *J Mater Res* vol. 4, 1989, p. 385.
272. B. J. Parsons, *Proc R Soc London* vol. A352, 1977, p. 397.
273. M. Kamo, H. Yurimoto, and Y. Sato, *Appl Surf Sci* vol. 33/34, 1988, p. 553.
274. R. M. Chrenko, *J Appl Phys* vol. 63, 1988, p. 5873.
275. A. T. Collins and G. S. Woods, *Phil Mag* vol. B46, 1982, p. 77.
276. C. D. Clark, E. W. J. Mitchell, and B. J. Parsons in J. E. Field (ed.), *The Properties of Diamond,* Academic Press, London, 1979, p. 23.
277. C. D. Clark, A. T. Collins, and G. S. Woods, in J. E. Field (ed.), *The Properties of Natural and Synthetic Diamond,* Academic Press, New York, 1992, p. 35.
278. G. Davies, "The Optical Properties of Diamond," in P. L. Walker and P. A. Thrower (eds.), *Chemistry and Physics of Carbon,* vol. 13, Marcel Dekker, New York, 1977, p. 1.
279. J. Walker, *Rep Prog Phys* vol. 42, 1979, p. 1605.
280. A. R. Lang, M. Moore, and J. C. Walmsley, in J. E. Field (ed.), *The Properties of Natural and Synthetic Diamond,* Academic Press, New York, 1992, p. 215.
281. P. Humble, J. K. Mackenzie, and A. Olsen, *Philos Mag A* vol. 54, 1985, p. L49.
282. A. R. Lang, *Proc Phys Soc* vol. 84, 1964, p. 871.
283. D. T. Morelli, *Chem and Phys of Carbon* vol. 24 1994, p. 45.
284. G. S. Woods et al., *J Phys Chem* vol. 51, 1990, p. 1191.
285. G. B. B. M. Sutherland, D. E. Blackwell, and W. G. Simeral, *Nature* vol. 174, 1954, p. 901.
286. G. Davies, *Diamond Research* vol. 21, 1972, p. 1.
287. G. S. Woods, J. A. van Wyk, and A. T. Collins, *Phil Mag* vol. B62, 1990, p. 589.
288. S. D. Smith and W. Taylor, *Proc Phys Soc* vol. 79, 1962, p. 1142.
289. J. J. Charette, *Physica* vol. 27, 1961, p. 1061.
290. J. I. Pankove and C.-H. Qui, in K. E. Spear and J. P. Dismukes (eds.), *Synthetic Diamond,* John Wiley & Sons, New York, 1994, p. 401.
291. C. D., Clark, R. W. Ditchburn, and H. B. Dyer, *Proc Roy Soc* vol. A234, 1956, p. 363.
292. G. Davies, in J. E. Field (ed.), *The Properties of Diamond,* Academic Press, London, 1979, p. 165.

293. H. Kawarada et al., *J Appl Phys* vol. 67, 1990, p. 983.
294. L. H. Robins and D. R. Black, *J Mater Res* vol. 9, 1994, p. 1298.
295. A. T. Collins and S. C. Lawson, *J Phys Condens Matter* vol. 1, 1989, p. 6929.
296. A. T. Collins and G. S. Woods, *J Phys C* vol. 20, 1987, P. L797.
297. A. T. Collins et al., *J Phys* vol. C21, 1988, p. 1363.
298. V. S. Tatarinov, Y. S. Mukhachev, and W. A. Parifianovich, *Sov Phys Semicond* vol. 13, 1979, p. 956.
299. B. Burchard et al., *Dia Rel Mater* vol. 3, 1994, p. 947.
300. A. T. Collins, E. C. Lighowlers, and P. J. Dean, *Phys Rev* vol. 183, 1969, p. 725.
301. L. Pan et al., *Appl Phys Lett* vol. 57, 1990, p. 623.
302. A. T. Collins and E. C. Lightowlers, *Phys Rev* vol. 171, 1968, p. 843.
303. A. T. Colins, in G. Davies (ed.), *Properties and Growth of Diamond* INSPEC, Exeter, England, 1994, p. 265.
304. E. A. Konorova, L. A. Sorokina, and S. A. Shevchenko, *Sov Phys-Solid State* vol. 7, 1965, p. 876.
305. L. A. Vermeulen and F. R. N. Nabarro, *Phil Trans R Soc A* vol. 262, 1967, p. 251.
306. D. R. Kania et al., *Dia Rel Mater* vol. 2, 1993, p. 1012.
307. L. S. Pan et al., *Dia and Rel Mat* vol. 2, 1993, p. 820.
308. M. A. Plano et al., *Appl Phys Lett* vol. 64, 1994, p. 193.
309. M. A. Plano et al., *Science* vol. 260, 1993, p. 1310.
310. M. W. Geis, N. N. Efremow, and D. D. Rathman, *J Vac Sci Technol A* vol. 6, 1988, p. 1953.
311. R. Berman, *Thermal Conduction in Solids,* Clarendon, Oxford, 1976.
312. C. Kittel, *Introduction to Solid State Physics,* John Wiley & Sons, New York, 1986.
313. B. R. Pamplin and L. I. Berger, in R. C. Weast (ed.), *Handbook of Chemistry and Physics,* vol. 70, 1989, p. E-106.
314. M. Seal, *Phil Trans R Soc. Lond A* vol. 342, 1993, p. 313.
315. G. A. Slack, *J Phys Chem Solids* vol. 34, 1973, p. 321.
316. T. R. Anthony, *Phil Trans R Soc Lond A* vol. 342, 1993, p. 245.
317. G. Leibfried and Schlomann, *Nachr Acad Wiss Gottingen IIa4* 1954, p. 71.
318. P. G. Klemens, in R. P. Tye (ed.), *Thermal Conductivity,* Academic Press, New York, 1969, p. 1.
319. H. B. Casimir, *Physica* vol. 5, 1938, p. 495.
320. R. Berman, *Contemp Phys* vol. 14, 1973, p. 101.
321. A. Ono et al., *Jpn J Appl Phys* vol. 25, 1986, p. L803.
322. M. Martinez, Oxford University, D. Phil., 1976.
323. A. J. Schorr, *Proc Industrial Diamond Conference,* Industrial Diamond Association of America, Moorestown, NJ, 1969.
324. E. A. Burgenmeister, *Physica* vol. 93B, 1978, p. 165.
325. A. Nishikawa et al., *Proceedings of the 1st International Symposium on Diamond and Diamond-like Films,* Electrochemical Society, Los Angeles, 1989, p. 524.
326. J. E. Graebner et al., *Appl Phys Lett* vol. 60, 1992, p. 1576.
327. G. A. Slack et al., *J Phys Chem Solids* vol. 48, 1987, p. 641.
328. R. Berman and M. Martinez, *Diamond Research,* Supplement to Industrial Diamond Review, 1976, p. 7.
329. R. Berman, E. L. Foster, and J. M. Ziman, *Proc R. Soc London* vol. A237, 1956, p. 344.
330. Y. Sato et al., *J Surface Sci Soc Jpn* vol. 1, 1980, p. 60.
331. T. R. Anthony et al., *J Appl Phys* vol. 69, 1991, p. 8122.

332. J. E. Graebner, T. M. Hartnett, and R. P. Miller, *Appl Phys Lett* vol. 64, 1994, p. 2549.

333. T. R. Anthony et al., *Phys Rev B* vol. 42, 1990, p. 1104.

334. L. Wei et al., *Phys Rev Lett* vol. 70, 1993, p. 3764.

335. J. Calloway, *Phys Rev* vol. 113, 1959, p. 1046.

336. M. N. Yoder in K. E. Spear and J. P. Dismukes (eds.), *Synthetic Diamond,* John Wiley & Sons, New York, 1994, p. 3.

337. D. L. Dreifus, in L. S. Pan and D. R. Kania (eds.), *Electronic Properties and Applications,* Kluwer Academic Publishers, Boston, 1995, p. 371.

338. M. Seal *Diamond and Related Materials* vol. I, 1992, p. 1075.

339. M. Seal, in J. E. Field (ed.) *The Properties of Natural and Synthetic Diamond,* Academic Press, New York, 1992, p. 607.

340. L. F. Vereschchagin et al., *Sov Phys Semicond* vol. 8, 1974, p. 1581.

341. M. Werner, V. Schlichting, and E. Obermeier, *Dia Rel Mat* vol. 1, 1992, p. 669.

342. G. B. Rodgers and F. A. Raal, *Rev Scientific Instrum* vol. 31, 1960, p. 663.

343. J. P. Bade et al., *Dia Rel Mat* vol. 2, 1993, p. 816.

344. B. A. Fox et al., *Second International High Temperature Electronics Conference,* Charlotte, NC, 1994, p. VI-3.

345. S. R. Sahaida et al., *Applied Diamond Conference,* MYU, Tokyo, Japan, 1993, p. 371.

346. R. F. Davis et al., *Mater Sci Engr* vol. B1, 1988, p. 77.

347. K. Shenai, R. S. Scott, and B. J. Baliga, *IEEE Trans on Elect Devices* vol. 36, 1989, p. 1811.

348. R. J. Trew, J.-B. Yan, and P. M. Mock, *Proc of IEEE* vol. 79, 1991, p. 598.

349. A. T. Collins, *Mater Sci Eng* vol. B11, 1992, p. 257.

350. M. Shin, R. J. Trew, and G. L. Bilbro, *2nd High Temperature Electronics Conference,* Charlotte, NC, 1994, p. IV-15.

351. D. L. Dreifus et al., *MRS Spring Meeting,* Materials Research Society, San Francisco, 1994.

352. D. L. Dreifus et al., *Second International High Temperature Electronics Conferences,* Charlotte, NC, 1994, p. VI-29.

353. J. S. Holmes, A. J. Tessmer, and D. L. Dreifus, *2nd High Temperature Electronics Conference,* Charlotte, NC, 1994, p. VI-35.

354. S. A. Grot, G. S. Gildenblat, and A. R. Badzian, *IEEE Electron Dev Lett* vol. 13, 1992, p. 462.

355. H. Shiomi, Y. Mishibayashi, and N. Fujimori, *Jpn J of Appl Phys* vol. 28, 1989, p. L2153.

356. N. Fujimori and Y. Nishibayashi, *Dia Rel Mater* vol. 1, 1992, p. 665.

357. K. Kobashi et al., *2nd International Conference on the Applications of Diamond Films and Related Materials,* NYU, Tokyo, Japan, 1993, p. 35.

358. C. R. Zeiss et al., *IEEE Elect Dev Lett* vol. 12, 1991, p. 602.

359. J. R. Zeidler et al., *Dia Rel Mater* vol. 2, 1993, p. 1341.

360. M. W. Geis et al., *IEEE Electron Dev Lett* vol. 8, 1987, p. 341.

361. J. F. Prins, *Appl Phys Lett* vol. 41, 1982, p. 950.

362. A. Denisenko et al., *Mat Sci & Engr* vol. B11, 1992, p. 273.

363. J. K. Cochran, K. J. Lee, and D. N. Hill, *J Mater Res* vol. 3, 1988, p. 67.

364. G Lasbrunie and R. Meyer, *Display Tech Appl* vol. 8, 1987, p. 37.

365. R. Greene, H. Gray, and G. Campisi, *IEDM Tech Dig conference proceedings,* 1985, p. 172.

366. M. W. Geis et al., Y. Tzeng et al., (eds.), *Applications of diamond films and related materials,* Elsevier, Amsterdam, 1991, p. 309.
367. K. Okano et al., *Appl Phys Lett* vol. 64, 1994, p. 2742.
368. C. Wang et al., *Electron Lett* vol. 27, 1991, p. 1459.
369. J. R. Seitz, *US Patent 3,895,313,* Laser Systems with Diamond Optical Elements, 1975.
370. J. Fontanella, et al., *Appl Opt* vol. 16, 1977, p. 2949.
371. C. A. Klein, *Dia and Rel Mat* vol. 2, 1993, p. 1024.
372. J. D. Stephenson, *Phys Stat Sol (a)* vol. 141, 1994, p. K83.
373. G. Grübel, *ESRF Beamline Handbook,* ESRF, Grenoble, France, 1993, p. 41.
374. L. E. Berman et al., *Nucl Ins and Meth in Phys Res* vol. A329, 1993, p. 555.
375. R. W. Ditchburn, *Opt Acta* vol. 29, 1982, p. 355.
376. R. E. Witkowski, J. P. McHugh, and W. D. Partlow, *Proceedings of the Fourth Electromagnetic Windows Symposium,* Office of Naval Technology, Arlington, VA, 1991, p. 233.
377. A. L. Lin, M. G. Stapelbroek, and K. V. Ravi, *Proceedings of the First International Symposium on Diamond and Diamond-like Films,* Electrochemical Society, Los Angeles, 1989, p. 261.
378. L. B. Kilham and M. W. Le Blon. US Patent 4,910,403, 1990.
379. J. A. Herb et al., *Sensors and Actuators* vol. A21-23, 1990, p. 982.
380. J. J. Cuomo et al., *Applications of Diamond Films and Related Materials,* Elsevier, Amsterdam, 1991, p. 169.
381. B. Löchel, *Microelectron/Engng* vol. 17, 1992, p. 175.
382. H. Windesham and G. F. Epps, *J Appl Phys* vol. 68, 1990, p. 5665.
383. H. Windesham et al., *New Diamond Science and Technology,* Materials Research Society, Pittsburg, PA, 1991, p. 791.
384. Y. Taniguchi, *New Diamond,* conference proceedings, 1990, p. 104.
385. Y Nishibayashi et al., *Proc of the 36th Meeting of the Soc Jpn Appl Phys* 1989, p. 481.
386. C. D. Clark, P. J. Dean, and P. V. Harris, *Proc Roy Soc Lond A* vol. 277, 1964, p. 312.
387. S. C. Rand and L. G. DeShazer, *Optics Ltrs* vol. 10, 1985, p. 418.
388. S. C. Rand, in G. Davies (ed.), *Properties and Growth of Diamond,* INSPEC, Exeter, England, 1994, p. 235.
389. G. Davies et al., *J Phys C* vol. 20, 1987, p. L13.
390. V. P. Mironov, E. F. Martinovich, and V. A. Brigorov, *Dia Rel Mater* vol. 3, 1994, p. 936.
391. B. Fiegl et al., *Dia Rel Mater* vol. 3, 1994, p. 658.
392. J. V. Beck, A. M. Osman, and G. Lu, *J Heat Transfer* vol. 115, 1993, p. 51.
393. J. Doting and J. Molenaar, *Proc 4th SEMI-THERM,* IEEE, Piscataway, NJ, 1988,
394. P. Hui and H. S. Tan, *J Appl Phys* vol. 75, 1994, p. 748.
395. J. Molenaar and G. W. M. Staarink, *Proc First Eur Symp on Mathematics in Industry,* Teubner, Stuttgart, 1985, p. 113.
396. G. Lu and E. F. Borchelt, *Photonics Spectra,* vol. 27, 1993, p. 88.
397. J. E. Ripper et al., *Appl Phys Lett* vol. 18, 1971, p. 155.
398. I. Hayashi et al., *Appl Phys Let* vol. 17, 1970, p. 109.
399. J. C. Dyment, J. E. Ripper, and T. H. Zachos, *J Appl Phys* vol. 40, 1969, p. 1802.
400. A. Katz et al., *J Appl Phys* vol. 75, 1994, p. 563.
401. J. C. Dyment and L. D'Asaro, *Appl Phys Lett* vol. 11, 1967, p. 292.
402. M. Sakamoto et al., *Electron Lett* vol. 8, 1992, p. 197.

403. K. Kurihara, K. Sasaki, and M. Kawarada, *Fujitsu Scientific and Technical Journal* vol. 25, 1989, p. 44.

404. G. Zuo et al., *New Diamond Science and Technology,* Materials Research Society, Washington DC, 1991, p. 893.

405. C. T. Troy, *Photonics Spectra,* vol. 26, 1992, p. 28.

406. C. B. Swan, *Proc IEEE,* vol. 55, 1967, p. 451.

407. W. Banholzer and C. L. Spiro, *Dia Films and Tech* vol. 1, 1991, p. 115.

408. G. Lu and K. Blakken, Gorham Conference, Orlando, 1994.

CHAPTER 8
THIN FILM OPTICAL MATERIALS

Angus Macleod
Thin Film Center Inc.

8.1 INTRODUCTION

Because thin film optical materials are used in optical coatings, it is impossible to discuss their properties adequately without discussing the properties of the coatings they form. This chapter, therefore, includes significant information on optical coatings.

Optical coatings are used to modify the optical properties of surfaces, that is, interfaces between optical media. Virtually all optical instruments consist, at least in part, of a series of interfaces that separate optical media of different kinds. These interfaces are usually smooth (on a scale of a fraction of a wavelength) and part of the light incident on them is reflected and part refracted. Typical applications of these effects are mirrors, where the light is largely reflected, and lenses where the refraction of the light is most important. The intrinsic properties of simple interfaces are rarely satisfactory, and are often quite unsatisfactory. Their properties must be altered if the optical instrument is to operate correctly. Optical coatings accomplish this.

The simplest types of optical coatings are the thin metal layer that transforms a transparent substrate into a front-surface mirror, and the single layer of magnesium fluoride that reduces the reflectance of a glass surface from around 4.25% to approximately 1.25%. Enormous numbers of such coatings are produced annually. More complicated are hot mirrors that transmit the luminous portion of a beam of light and reflect the heat; cold mirrors that reflect luminous light and transmit heat; narrowband filters, longwave pass and shortwave pass filters; beam splitters and combiners; polarizers; notch filters; color separation filters; multiplexers and demultiplexers; laser mirrors; phase retarders; and the list goes on and on.

These components all have in common that they are constructed from one or a series of thin films of materials supported on a substrate. They are of such regularity and precision of thickness that interference effects are combined with their intrinsic optical properties to achieve the desired performance. The word "thin" in the context of optical coatings can really be taken to mean that interference is supported.

In the early part of the nineteenth century, telescope mirrors were made from speculum metal because this enabled a polished surface to reflect well. The metal was not very stable, however, and could not hold a good figure for more than a few hours. It was normal practice to work the surface of such a mirror immediately before viewing. A great advance was made when it was realized that the high reflectance could be assured by a thin layer of silver deposited over a substrate material chosen for its stability rather than its reflectance. The common modern front-surface mirror is of this type. The metal is usually aluminum and the substrate glass.

This combination of coating and substrate properties into the final component is typical of optical components. The substrate supports the coating and gives a stability to its shape, but the coating in turn modifies the properties of the substrate surface in a desired way. The modification may be purely optical but more often today the mechanical and environmental properties must also be improved. For example, the antireflection coating on a plastic eyeglass lens is intended not only to reduce the reflection loss, but also to ensure a particular color and to improve the resistance to abrasion, moisture, and other environmental influences.

The materials used for optical coatings are usually either insulators, metals, or semiconductors. Their uses can be classified in two principal ways, either as transparent layers supporting interference, or as opaque, highly reflecting films, referred to simply as dielectric or metallic. Thus semiconductors that are used in their region of transparency beyond the intrinsic edge are called dielectric.

The following discussion is limited to media with linear response to electromagnetic waves.

8.2 PROPAGATION OF LIGHT THROUGH MEDIA

Light is a high-frequency form of electromagnetic radiation and is therefore characterized by oscillating magnetic and electric fields. It interacts with materials principally through the electrons. Although an electric field exerts a force on a stationary electron, a magnetic field can influence the electron only when it is moving. For the interaction to be significant, the speed of the electron must be a sizable fraction of the speed of light. At the high frequencies of light waves, the electron, oscillating under the influence of the electric field, never reaches a high enough velocity. Thus any force exerted by the magnetic field is vanishingly small. The interaction between light and the medium through which it propagates therefore results from the electric field. Thus amplitude, polarization, phase retardation, and similar terms in optics conventionally relate to the electric field of the wave. We discuss plane harmonic waves where the vectors E, H, and $\hat{s}$, the electric and magnetic fields, and the unit vector in the direction of propagation form a right-handed set in that order. E and H are functions of time and distance along $\hat{s}$, the direction of propagation, only. We adopt the complex form of the harmonic wave with electric field

$$\mathrm{E}e^{i(\omega t - \kappa z)} \tag{1}$$

where this indicates a plane wave propagating along the positive direction of the z-axis and contains the sign convention for the phase factor. The exponent $i(\omega t - \kappa z)$ could also have been written $i(\kappa z - \omega t)$ but it is important that a consistent choice be made. ω is the angular frequency given by $2\pi/\tau$ with τ the period of the wave, κ is the wavenumber given by $2\pi/\lambda$ where λ is the spatial period, or wavelength, of the wave. E is the complex electric field amplitude, which contains the relative phase of the wave. E can be written as $|\mathrm{E}|\exp(i\phi)$.

The interaction of the wave with the medium is such that it slows down to a velocity that is characteristic of the medium. It is normal to put this property of the medium in the form of a dimensionless quantity.

$$\text{Refractive index} = n = \frac{\text{Velocity of light in free space}}{\text{Velocity in the medium}} \tag{2}$$

When a wave passes from free space to a medium where the refractive index differs from unity the frequency remains constant while the wavelength changes. Unfortunately we normally use the wavelength to characterize the wave rather than frequency. To avoid problems of a changing wavelength we adopt the free space wavelength and therefore alter the expression for κ to $2\pi n/\lambda$ where λ is now the free space wavelength.

The expression for the wave is now

$$\mathrm{E}\exp\left[i\left(\omega t - \frac{2\pi nz}{\lambda}\right)\right] \tag{3}$$

Absorption in a medium implies that the wave will decay exponentially as it propagates. This behavior can be included in Eq. 3 by adding an imaginary term to the refractive index n.

$$n \longrightarrow (n - ik) \tag{4}$$

where k is known as the extinction coefficient. The quantity $(n - ik)$ is usually called the complex refractive index, sometimes just refractive index. Sometimes n is the real part of refractive index, or often just refractive index. The symbol N may be used to indicate $(n - ik)$. The terminology may be confusing but the meaning is usually clear from the context.

Eq. 3 now becomes

$$\mathrm{E}\exp\left[i\left(\omega t - \frac{2\pi(n - ik)z}{\lambda}\right)\right] = \mathrm{E}\exp\left(-\frac{2\pi kz}{\lambda}\right)\exp\left[i\left(\omega t - \frac{2\pi nz}{\lambda}\right)\right] \tag{5}$$

So far we have said little about the magnetic field. The magnetic and electric fields of a harmonic wave are related through a parameter of the medium, y, known as the characteristic admittance.

$$\frac{H}{E} = y \tag{6}$$

If we are dealing with complex amplitudes that include relative phase, this expression reduces to

$$\frac{\mathrm{H}}{\mathrm{E}} = y \tag{7}$$

where H is the magnetic field amplitude and y and $(n - ik)$ are indirectly related through the permittivity and permeability of the medium. At optical and still higher frequencies the relative permeability can always be assumed as unity. Then the relationship between y and $(n - ik)$ becomes a direct one. If Υ is the admittance of free space, (1/377 siemens), then

$$y = (n - ik)\Upsilon \tag{8}$$

This allows a further simplification. Because we deal with relative quantities in most of what we do we can change the units of characteristic admittance to units of the admittance of free space, Υ. We need to be careful to change back to SI units for y whenever the absolute values of the quantities are required.

$$y = n - ik = N \text{ free space units} \tag{9}$$

and the same number can be used for both y and $(n - ik)$.

Unfortunately the validity of Eq. 9 causes confusion about the nature of the parameters and frequently leads to an incorrect use that, although it gives a correct numerical answer, is invalid at lower frequencies. The confusion has been compounded over the years by the many different systems of units.

An important observable quantity is the rate of transport of energy by a beam of light. In the complex wave representation the irradiance, (that is, the mean rate of energy flow in the direction of the wave per unit area) is given by the complex form of the Poynting vector.

$$I = \frac{1}{2}\mathrm{Re}(\mathrm{EH}^*) \tag{10}$$

where the relative phases of the electric and magnetic fields must be included in the complex amplitudes E and H. The SI symbol for irradiance is E but because this will inevitably lead to confusion with E, the electric field, we will use I instead. An older term for irradiance is intensity but in SI units intensity is a different quantity. The units of irradiance are watt·m^{-2}.

Eq. 10 may also be written as

$$I = \frac{1}{2}\mathrm{Re}(y)\mathrm{EE}^* \tag{11}$$

If we include the decaying exponential in Eq. 5 in the electric field amplitude we find the irradiance in an absorbing medium to be

$$I = \frac{1}{2}\mathrm{Re}(y)\mathrm{EE}^* \exp\left(-\frac{4\pi kz}{\lambda}\right) = I_0\exp(-\alpha z) \tag{12}$$

where α is the absorption coefficient. The relationship between extinction coefficient and absorption coefficient is therefore

$$\alpha = \frac{4\pi k}{\lambda} \tag{13}$$

Watch the units when evaluating this expression. α is traditionally given in cm^{-1} whereas λ is in nm. α must be very large for k to become important and a good first approximation is to treat k as zero in the transparent regions of dielectric films.

The major optical properties of materials that make them useful in thin film coatings are explained very well by classical models. For dielectric materials the medium is assumed to consist of an array of polarizable objects that acquire a dipole moment in the electric field of the wave. At high frequencies this dipole moment is a consequence of the movement of an electron, although at lower frequencies it may be caused by relative movement of two oppositely charged ions. In either case the charged particles are bound so that they cannot move through the medium but oscillate around their mean positions. Except at the resonances, the extinction coefficient is small whereas the real part of refractive index can be quite large. In a metal or semiconductor, free charges are able to move through the material and are responsible for an increased extinction coefficient and usually rather lower real part of refractive index. Therefore, the two principal models are, bound charges and free charges. In fact real materials usually have a combination of these effects with one predominating, and there are other phenomena not normally significant in coatings.

The calculation of the optical constants of the materials [1,2] is through the susceptibility, X, given by

$$X = \frac{P}{\epsilon_0 E} \tag{14}$$

where P is the polarization. At optical frequencies

$$X = N^2 - 1 = (n - ik)^2 - 1 \tag{15}$$

If the polarizability of a single particle (atom or molecule) is α and there are N of them per unit volume, then a simple expression for P would be

$$P = \mathrm{N}\alpha E \tag{16}$$

but this neglects the interactions between the polarized particles themselves. When this is taken into account the expression becomes

$$P = \mathrm{N}\alpha\left(E + \frac{P}{3\epsilon_0}\right) \quad \text{giving} \tag{17}$$

$$\frac{X}{3 + X} = \frac{\mathrm{N}\alpha}{3\epsilon_0} \tag{18}$$

If now the particle j has a bound electron with charge e with mass m, natural frequency ω_j and damping factor γ_j with oscillator strength f_j then

$$\frac{N^2 - 1}{N^2 + 2} = \frac{\mathrm{N}e^2}{3m\epsilon_0}\sum_j \frac{f_j}{\omega_j^2 - \omega^2 + i\gamma_j\omega} \quad \text{and} \tag{19}$$

$$N^2 = (n - ik)^2 = (n^2 - k^2) - i^2nk \tag{20}$$

Only in the regions where the frequency is close to a resonance is the extinction coefficient significant unless γ is very large. Most dielectric optical materials useful in thin film optical coatings have one pronounced electronic resonance at short wavelengths, then a transparent region followed by a further resonance at longer wavelengths in the infrared region that is caused by relative movement of the atoms of the molecules, particularly strong when the molecules are ionic. In the transparent region the refractive index falls slowly toward longer wavelengths in what is termed normal dispersion.

In metals many electrons are substantially free with no resonant frequency but with some damping, and these free electrons effectively prevent the interaction between the polarized particles that led to the correction term in Eq. 17 so that Eq. 16 applies. Then we can write

$$N^2 - 1 = \frac{\mathrm{N}e^2}{m\epsilon_0}\left(\frac{f_f}{i\gamma_f\omega - \omega^2} + \sum_j \frac{f_j}{\omega_j^2 - \omega^2 + i\gamma_j\omega}\right) \tag{21}$$

The bound electrons are important primarily in semiconductors and poor conductors. In a good metal the relationship reduces to

$$N^2 - 1 = \frac{\omega_p^2}{i\dfrac{\omega}{\tau} - \omega^2} \tag{22}$$

We have replaced the damping constant γ by τ^{-1} where τ is called the relaxation time, and we have introduced the plasma frequency ω_p where

$$\omega_p = \frac{Ne^2}{m\epsilon_0} \tag{23}$$

Then we can show

$$n^2 - k^2 = 1 - \frac{\omega_p^2}{\omega^2 + \tau^{-2}} \qquad \text{and} \tag{24a}$$

$$2nk = \frac{\omega_p^2}{\omega\tau(\omega^2 + \tau^{-2})} \tag{24b}$$

The properties of the metals are therefore determined by the relaxation time τ which is consistent with a frequency in the infrared region and the plasma frequency that is usually in the ultraviolet region. The higher ω_p and the longer τ the better are the optical properties of the metal. Generally metals reflect strongly at frequencies below the plasma frequency and transmit poorly at frequencies above.

A detailed account of the optical properties of a wide range of selected materials has been provided by Palik [3,4].

8.3 BEHAVIOR OF LIGHT AT INTERFACES

In the development of the theory of optical thin films there are many decisions to be made concerning which direction will be called positive, which phase will be considered as leading, and so on. In almost all cases one decision is as good as another, but when the decision has been made it must be rigidly adhered to, otherwise disaster will occur. These matters are conventions rather than fundamentals and those we use here are the usual ones in thin film optics.

We have already introduced a convention in the previous chapter. This is that the phase factor of the complex wave shall be written as

$$\exp[i(\omega t - \kappa z)] \qquad \text{and not} \tag{25}$$

$$\exp[i(\kappa z - \omega t)] \tag{26}$$

An immediate consequence of this is that we write $(n - ik)$ and not $(n + ik)$ for the complex refractive index of an absorbing medium.

We now consider what happens at an interface between two media.

The model that we use of an ideal interface is a flat, featureless, abrupt discontinuity between two different media. It can be represented as a plane with no thickness. Although real interfaces may sometimes be different than this model, it is nevertheless a very good model in the majority of cases. Light in the form of a plane wave that is incident at such an interface is split into two plane waves, one of which is reflected and the other refracted. The plane of incidence is defined as the plane that contains the direction of the incident wave and the normal to the surface. The angle of incidence is the angle that the direction of incidence makes with the normal. The direction of the reflected wave is contained within the plane of incidence and the angle of reflection. This means that the angle between the reflected direction and the normal is equal to the angle of incidence, but on the other side of the normal. The refracted wave is also contained within the plane of

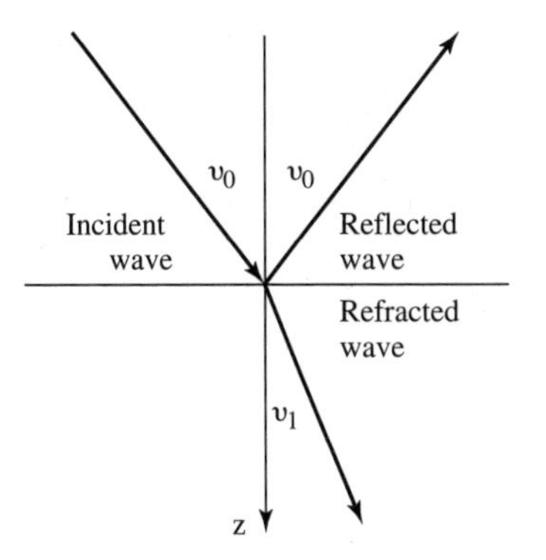

FIGURE 8-1 Shows the directions of the incident, reflected, and refracted waves at a boundary between two media labeled *0* and *1*.

incidence but the angle between the ray direction and the normal is related to the angle of incidence by Snell's law.

The boundary conditions are such that the total electric and magnetic fields parallel to the boundary are continuous across it. The directions of the waves are then derived from the condition that the components of their velocities along the direction of the surface must all be equal, otherwise the boundary conditions cannot hold for all *x, y,* and *t.* This implies the equality of the two angles marked υ_0 in Fig. 8-1, in which the angles of incidence and reflection are equal. The equality of the components of incident- and refracted-beam velocities gives Snell's law, which for absorption-free media is written

$$n_0 \sin\upsilon_0 = n_1 \sin\upsilon_1 \tag{27}$$

The complex form of Snell's law, where $(n - ik)$ is used, is difficult to interpret because the angles become complex. An alternative approach is better and will be considered shortly.

When the magnitudes of the components at the boundary are considered then the polarization of the waves must be included. Fortunately the two simple cases of electric field in the plane of incidence, known as *p*-polarization, and electric field normal to the plane of incidence, known as *s*-polarization, can be completely decoupled and considered separately. We must, however, set up a sign convention for the directions of the fields. Unfortunately, although there is no disagreement in the literature about the normal incidence convention, there are two opposite conventions that are frequently used for oblique incidence. The convention shown in Fig. 8-2 is the one most often used in thin film optics. It is chosen so that as the angle of incidence decreases toward normal incidence, both conventions tend to be the same at normal incidence. The other common convention inverts the positive direction for *p*-polarization in the reflected wave and is used especially in ellipsometry. Here it is helpful if the reference axes to describe the elliptically polarized

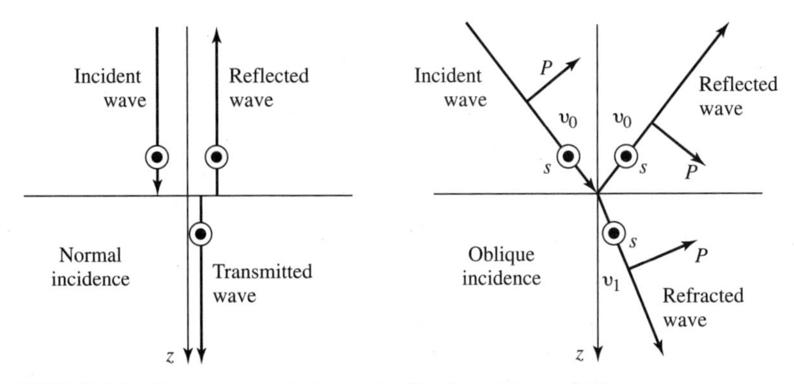

FIGURE 8-2 The conventions for the positive directions of electric field.

light in both incident and reflected beams can coincide with the positive directions of the electric vector.

The boundary conditions, that the electric and magnetic field components are continuous across the interface, permit us to calculate the fraction of the incident light that is reflected and transmitted. At normal incidence this is fairly straightforward and the results can be expressed as two ratios, ρ and τ respectively, given by

$$\rho = \frac{E_{refl}}{E_{inc}} = \frac{(y_0 - y_1)}{(y_0 + y_1)} \quad \text{and} \tag{28}$$

$$\tau = \frac{E_{trans}}{E_{inc}} = \frac{2y_0}{(y_0 + y_1)} \tag{29}$$

ρ and τ are known as the amplitude reflection and transmission coefficients, respectively. At normal incidence they are equivalent to what are known as the Fresnel coefficients because of the equivalence of the numerical values of y and $(n - ik)$, but this is not the case at oblique incidence because of the way in which the coefficients will shortly be defined. Note that the definitions apply to the amplitudes actually at the boundary, either just inside the incident medium or just inside the emergent medium as appropriate. Any loss in propagation through either of the two media is not included. The phase difference between the waves is included in Eq. 28 and Eq. 29 and is measured at the interface. This is especially important in the case of the reflected beam. A phase difference between two counter-propagating beams must always be referred to a reference plane because the phase difference varies with position, passing through 360° with each halfwave of travel.

The irradiances of the waves are important and their ratios are known as reflectance and transmittance. When the incident medium is absorbing it is impossible to separate the incident and reflected irradiances because there is an interchange of energy between them, so we are unable to calculate their ratio. (The problem is an old one and very well understood. For further details see for example [5].) The limitation of an absorption-free incident medium is not an important restriction because the absorption coefficient has to be very large for k to be important. Then it is virtually impossible to make any kind of measurement of reflectance or transmittance. With this restriction of real y_0, reflectance and transmittance are given by

$$R = \frac{I_{refl}}{I_{inc}} = \rho\rho^* = \frac{(y_0 - y_1)(y_0 - y_1)^*}{(y_0 + y_1)(y_0 + y_1)^*} \tag{30}$$

$$T = \frac{I_{trans}}{I_{inc}} = \frac{\mathrm{Re}\,(y_1)}{y_0}\tau\tau^* = \frac{4y_0\mathrm{Re}\,(y_1)}{(y_0 + y_1)(y_0 + y_1)^*} \tag{31}$$

At oblique incidence the polarization must be taken into account and either the magnetic vector (s-polarization) or the electric vector (p-polarization) must be resolved along the interface. Furthermore, it is important that the reflectance and transmittance should add to unity because there is no possibility of absorption at an interface of zero thickness. The refraction of the transmitted beam implies that the sum of the irradiances of the transmitted and reflected beams cannot add to that of the incident beam because unit area normal to the beams will not subtend equal areas at the interface. Thus at oblique incidence the reflectance and transmittance is defined with respect to the components of irradiance normal to the interface. The electric and magnetic fields involved in the calculation of these normal irradiances are parallel to the interface. Thus it makes sense to use these components in the oblique-incidence amplitude coefficients and this is the usual procedure in thin film optics. These are not the same as the Fresnel coefficients because they use the full amplitudes.

Given the sign convention in Fig. 8-2, there is a straightforward procedure to calculate the various quantities. The admittances, y, should be replaced by the tilted admittances, η, where η is given by

$$\eta_s = y \cos\upsilon \tag{32}$$

$$\eta_p = \frac{y}{\cos\upsilon} \tag{33}$$

and υ is given by Snell's law,

$$n_0 \sin\upsilon_0 = (n - ik)\sin\upsilon \tag{34}$$

κ in the phase factor should be replaced by an oblique incidence form

$$\kappa = \frac{2\pi(n - ik)\cos\upsilon}{\lambda} \tag{35}$$

where the index $(n - ik)$ has taken the same form as η_s. Then

$$\rho = \frac{(\eta_0 - \eta_1)}{(\eta_0 + \eta_1)} \tag{36}$$

$$\tau = \frac{2\eta_0}{(\eta_0 + \eta_1)} \tag{37}$$

$$R = \frac{(\eta_0 - \eta_1)(\eta_0 - \eta_1)^*}{(\eta_0 + \eta_1)(\eta_0 + \eta_1)^*} \tag{38}$$

$$T = \frac{4\eta_0 \operatorname{Re}(\eta_1)}{(\eta_0 + \eta_1)(\eta_0 + \eta_1)^*} \tag{39}$$

For absorbing media, Eqs. 32 to 34 are correct but not in a very good form. A better set of expressions for η is

$$\eta_s = \sqrt{n^2 - k^2 - n_0^2 \sin^2 \upsilon_0 - 2ink} \quad \text{fourth quadrant} \tag{40}$$

$$\eta_p = \frac{(n - ik)^2}{\eta_s} \tag{41}$$

These are absolutely consistent with the earlier expressions but are more convenient for calculation.

A good example of the use of Eqs. 40 and 41 is shown by the phenomenon of total internal reflection. Here a beam of light is incident internally on the hypotenuse of a prism of dielectric material, as shown in Fig 8-3.

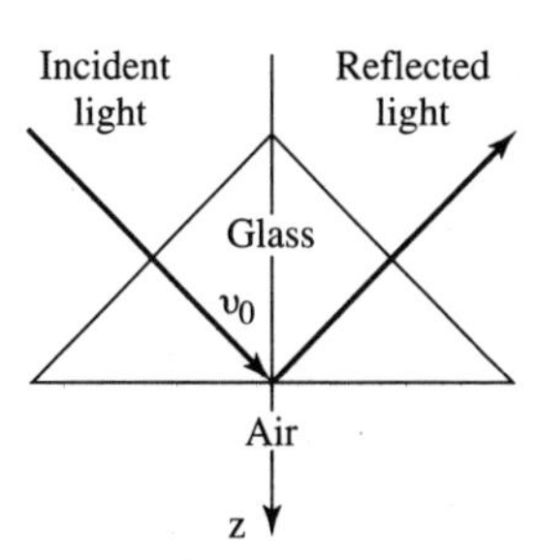

FIGURE 8-3 Light incident on the base of a prism beyond the critical angle.

Here k is zero but υ_0 is large enough that $n_0 \sin\upsilon_0 > n$. Then, from Eqs. 40 and 41

$$\eta_s = -i\sqrt{n_0^2 \sin^2\upsilon_0 - n^2} \tag{42}$$

and

$$\eta_p = \frac{in^2}{\sqrt{n_0^2 \sin^2 \upsilon_0 - n^2}} \tag{43}$$

In either case the admittance is purely imaginary and so the reflectance is unity. This is known as total internal reflectance. That angle of incidence where $n_0 \sin\upsilon_0 = n$ is known as the critical angle and the total internal reflectance exists for all greater angles of incidence. The component of the wave propagating along the z-axis in the emergent medium is given by Eqs. 35 and 42 as

$$\mathrm{E}\exp\left[\frac{2\pi\left(\sqrt{n_0^2 \sin^2 \upsilon_0 - n^2}\right)z}{\lambda}\right]\exp(i\omega t) \tag{44}$$

where the polarization may be either s or p. This no longer has the form of a progressive wave. Rather it is an oscillation that is decaying along the z-axis without change of phase. Because the magnetic field will be given by the admittance times (δ), the electric field and the admittance are purely imaginary, and the Poynting vector and therefore the irradiance is zero. This phenomenon is known as an evanescent wave.

8.4 INTERFERENCE IN THIN FILMS

When light is incident on a surface that is coated with a thin film, part of the light is reflected and part is transmitted. Inside the film light is reflected backward and forward between the two interfaces. By definition, if the film is thin then all the beams have a consistent phase relationship and combine in a process of interference. In a thick film there is no interference and the beams combine incoherently, that is, their irradiances are simply added. The interference varies the proportions of the resultant reflected and transmitted beams, which are functions of the wavelength and of the properties of the film, particularly the thickness. The deliberate control of the reflectance and transmittance by the choice of film system deposited over the surface is the objective of thin film coatings.

We deal with linear systems and in such systems the combination of two light beams involves the addition of the magnetic and the electric fields of the rays. This is known sometimes as the principle of linear superposition. Two beams of light propagating in the same direction with identical frequencies and polarization and with a constant phase difference can be represented as a complex amplitude that contains the relative phase information together with a phase factor that is identical to that for each beam. The resultant beam is then

$$(\mathrm{E}_1 + \mathrm{E}_2)\exp[i(\omega t - \kappa z)] \tag{45}$$

The magnetic field amplitude of the wave is

$$(\mathrm{H}_1 + \mathrm{H}_2) = y(\mathrm{E}_1 + \mathrm{E}_2) \tag{46}$$

$$\mathrm{E}_1 = |\mathrm{E}_1|\exp(i\phi_1), \text{ etc.} \tag{47}$$

and so the irradiance of the wave is

$$\frac{1}{2}\mathrm{Re}\left[(\mathrm{E}_1 + \mathrm{E}_2) \cdot (\mathrm{H}_1 + \mathrm{H}_2)^*\right]$$

$$= \frac{1}{2}\text{Re}(y)\text{E}_1\text{E}_1{}^* + \frac{1}{2}\text{Re}(y)\text{E}_2\text{E}_2{}^* + \text{Re}(y)|\text{E}_1| \cdot |\text{E}_2|\cos\phi \tag{48}$$

$$= I_1 + I_2 + 2\sqrt{I_1}\sqrt{I_2}\cos\phi$$

where ϕ is the phase difference between the two waves. The first two terms are the irradiances of the two beams individually. The third term is the interference term. When $\cos\phi$ is positive and unity the interference is said to be constructive and when $\cos\phi$ negative and unity the interference is destructive. It may appear that in constructive interference Eq. 48 includes something for nothing because the maximum irradiance is $(\sqrt{I_1} + \sqrt{I_2})^2$. The necessary energy, however, is always extracted from somewhere else in the system. In the case of interference in thin films, the superposition of the interfering beams results from partial reflection or transmission at different interfaces. A net transmitted beam always accompanies a net reflected beam. One draws any necessary extra energy from the other so that the conservation of energy is always satisfied.

Multiple beam interference involves more than two beams, usually an infinite number of gradually falling irradiance and regular phase difference, and is an elaboration of two-beam interference. Thin film interference is of the multiple-beam class, and although it is traditionally associated with interferometric etalons like the Fabry-Perot (a device similar to the single-cavity filter of Fig. 8-12 but consisting of two reflectors with air space between them), in fact Poisson and Fresnel derived multiple-beam expressions for thin films much earlier [6].

The boundary conditions are that the tangential electric and magnetic fields are continuous across the boundary. We know the total electric and magnetic fields at the rear boundary of the film. They are simply the fields associated with the emergent wave in the substrate. We now discuss the inside of the film and the multiple beams that exist there. Because the process is linear, it does not matter in what order we add the various beams. We therefore add all the beams traveling in the direction of incidence, making one resultant positive-going wave, and we separately add all the beams going in the opposite direction into one negative-going wave (Fig. 8-4). When we have these waves we can use them to transfer the total electric and magnetic fields at the rear interface to give the total electric and magnetic fields at the front interface. We need to know the phase changes that happen to the waves as they traverse the film, and from the expressions that we have for the progressive harmonic waves we know that the phase difference is a lag and given by

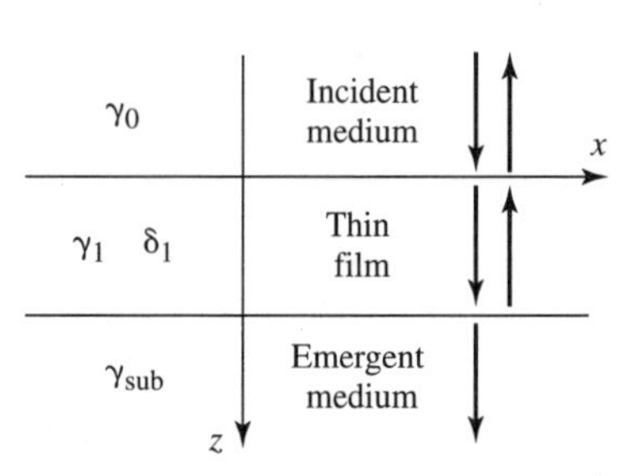

FIGURE 8-4 The arrangement of resultant beams in a single thin film on a substrate that is also the emergent medium. The origin of the coordinate system is the intersection of the z-axis with the upper surface of the thin film.

$$\delta_1 = \frac{2\pi(n_1 - ik_1)d_1}{\lambda} \tag{49}$$

where we allow for the possibility of absorption by including $k_1 \cdot \delta$ is known as the phase thickness of the layer. In most cases we will effectively have zero extinction coefficient.

Because the process is linear we can normalize it in any way we like. We force the tangential electric field at the rear interface to be unity, then the corresponding magnetic field is y_{sub}. Then if we denote the tangential electric and magnetic fields at the front surface by B and C respectively, they are given by the expression

$$\begin{bmatrix} B \\ C \end{bmatrix} = \begin{bmatrix} \cos\delta_1 & \dfrac{i\sin\delta_1}{y_1} \\ iy_1\sin\delta_1 & \cos\delta_1 \end{bmatrix} \cdot \begin{bmatrix} 1 \\ y_{sub} \end{bmatrix} \tag{50}$$

and the 2 × 2 matrix is known as the characteristic matrix of the thin film. Eq. 50 can readily be extended to a series of these films. We number them from the incident medium to the substrate as 1 to q. The expression is

$$\begin{bmatrix} B \\ C \end{bmatrix} = \prod_{j=1}^{q} \left\{ \begin{bmatrix} \cos\delta_j & \dfrac{i\sin\delta_j}{y_j} \\ iy_j\sin\delta_j & \cos\delta_j \end{bmatrix} \right\} \cdot \begin{bmatrix} 1 \\ y_{sub} \end{bmatrix} \tag{51}$$

$$\text{where} \quad \delta_j = \frac{2\pi(n_j - ik_j)d_j}{\lambda} \tag{52}$$

Eq. 51 can readily be extended to oblique incidence by replacing y_j by the η_j appropriate to the particular plane of polarization and δ_j by

$$\delta_j = \frac{2\pi d_j(n_j - ik_j)\cos\upsilon_j}{\lambda} = \frac{2\pi d_j\sqrt{(n_j^2 - k_j^2 - n_0^2\sin^2\upsilon_0 - 2in_jk_j)}}{\lambda} \tag{53}$$

the fourth quadrant solution for the square root being taken.

The reflectance and transmittance are then given from the values of B and C by

$$\rho = \frac{(y_0B - C)}{(y_0B + C)}$$

$$R = \rho\rho^* = \frac{(y_0B - C)(y_0B - C)^*}{(y_0B + C)(y_0B + C)^*}$$

$$\tau = \frac{2y_0}{(y_0B + C)} \tag{54}$$

$$T = \frac{4y_0\mathrm{Re}(y_{sub})}{(y_0B + C)(y_0B + C)^*}$$

Note that the expression for ρ compares the amplitudes at the front surface of the multilayer whereas the expression for τ compares the emergent ray at the rear interface with the incident ray at the front interface. This is important when phase differences are extracted from the expressions. For oblique incidence, the η and η_{sub} for the appropriate plane of polarization should be used.

Because B and C represent electric and magnetic field, their ratio represents an admittance, that of the multilayer on the substrate and measured at the front interface. As far as reflectance is concerned the system can be replaced by a simple surface with admittance given by

$$Y = \frac{C}{B} \tag{55}$$

The multilayer can be considered an admittance transformer that transforms the admittance of the substrate, y_{sub} into Y. This is a particularly powerful way of dealing with the properties of a multilayer, particularly a dielectric one.

The calculation of the properties of a thin film system is accomplished by application of the above expressions. Particularly when dielectric materials are used and there is no absorption, it is convenient to combine n and d into an optical thickness. Dimensionless quantities are always preferred, and this can be converted into a dimensionless quantity by dividing by a reference wavelength λ_0. The phase thickness can then be written

$$\delta = 2\pi\left(\frac{nd}{\lambda_0}\right) \cdot \left(\frac{\lambda_0}{\lambda}\right) = 2\pi\left(\frac{nd}{\lambda_0}\right) \cdot g \tag{56}$$

where $g = (\lambda_0/\lambda)$ is a dimensionless quantity related to frequency and very useful in design.

When $\delta = \pi/2$ the film is a quarterwave thick and the matrix takes a very simple form. It can readily be shown that the admittance of the substrate is transformed according to the rule

$$y_{\text{sub}} \rightarrow \frac{y_1^2}{y_{\text{sub}}} \tag{57}$$

This is known as the quarterwave rule. A double application of the quarterwave rule gives the effect of adding a halfwave of dielectric material to the substrate.

$$y_{\text{sub}} \rightarrow \frac{y_1^2}{(y_1^2/y_{\text{sub}})} = y_{\text{sub}} \tag{58}$$

Because they have no affect on the substrate, halfwave layers are sometimes called absentee layers.

The quarter and halfwave rules give straightforward ways of calculating the properties of multilayers of such thicknesses at the appropriate wavelengths. For other wavelengths the full expressions can be used.

Dielectric materials at oblique incidence have phase factors that are proportional to $\cos\upsilon$. Thus the phase thickness becomes significantly less as the coating is tilted. The admittances vary in their response according to the polarization of the light but they too have the same $\cos\upsilon$ factor. Thus the characteristics of dielectric coatings move toward shorter wavelengths as the angle in incidence increases. The effect on the admittances is to move the admittances of high- and low-admittance materials closer together for p-polarization and further apart for s-polarization. This causes the s-polarized performance to become more pronounced but the p-polarized performance to become weaker.

Metals, on the other hand, are characterized by high k and small n. At oblique incidence the tilted performance depends on the expression

$$\sqrt{n^2 - k^2 - n_0^2 \sin^2\upsilon_0 - 2ink} \tag{59}$$

$n_0^2\sin^2\upsilon_0$ is small compared with k^2 and so may be neglected when Eq. 59 becomes just $(n - ik)$. Metals, therefore, are barely affected by tilting. Unfortunately the incident medium is dielectric and has the normal correction, so the combination of the dielectric incident medium and metal film in a reflector does imply slight changes. The most serious change is a difference in phase shift between light that is s-polarized and light that is p-polarized. This difference, referred to as a phase retardation, can become very significant and cause considerable changes in the ellipticity of the polarization.

Another very interesting phenomenon is the surface plasmon resonance. Total internal reflectance was discussed in the previous section. The wave that penetrates into the emergent medium is evanescent. It decays exponentially and carries no energy. The wave in a metal is of the form

$$\mathrm{E}\exp\left[i\left(\omega t-\frac{2\pi(n-ik)z}{\lambda}\right)\right]=\mathrm{E}\exp\left(-\frac{2\pi kz}{\lambda}\right)\exp\left[i\left(\omega t-\frac{2\pi nz}{\lambda}\right)\right] \tag{60}$$

Because n is very small in a high-performance metal, the wave can almost be written as

$$\mathrm{E}\exp\left(-\frac{2\pi kz}{\lambda}\right)\exp(i\omega t) \tag{61}$$

that is, an evanescent wave. Therefore it is probably not surprising that a surface wave can exist on a metal that has an evanescent tail into both the metal and the surrounding medium so that the wave is held on the surface. This type of surface wave is known as a surface plasma wave or a surface plasmon. The velocity of this wave is rather less than the speed of light in free space and, in fact, it corresponds to the component of velocity along the surface of a wave that is totally internally reflected at just beyond the critical angle. In this way it is possible to couple energy into the surface wave, in the same way that light can be coupled into a waveguide. The conditions are such that the wave must be p-polarized. (The electron oscillations are longitudinal). The metal film is deposited in a given thickness on the outer surface of the hypotenuse of a prism and the coupling is accompanied by a steep and narrow drop in the internal reflectance of the prism hypotenuse for p-polarized light. There is nothing mysterious about this resonance. It is predicted by the normal thin film calculation techniques. The higher the optical performance of the metal, the narrower is the resonance. As the thickness of the metal departs from the ideal value the depth of the resonance is reduced. The ideal thickness will actually yield a reflectance of zero at the resonance angle.

The resonance can be used to derive the optical constants of the metal. There are three features of the resonance that can be readily measured; the depth, the width, and the angular position. There are three parameters that must be measured for a metal film at any wavelength. These are thickness d, refractive index n, and extinction coefficient k. It is necessary that the thickness of the metal be such that a resonance is obtained. In fact, there are two possible solutions for the parameters except when the reflectance in the center of the resonance is zero, which makes the solutions degenerate and coincident. To distinguish between the various solutions it is necessary only to identify the correct value of d. This can be done either by making measurements at two wavelengths when the value of d that is correct will be found at both, or by biasing the resonance sufficiently far from zero reflectance that the two possible values are forced to be so far apart that the correct one is unmistakable.

The resonance is a very sensitive function of the conditions at the outer surface of the metal. As dielectric material is added, the resonance moves toward greater angles of incidence. This phenomenon can be used in sensitive measurements.

8.5 MULTILAYER OPTICAL COATINGS

Because of the simplicity of the quarterwave rule, and because the quarterwave layer yields an interference extremum, quarterwaves are often used in optical coatings. The normal shorthand notation for designs is based on quarterwaves. In this notation, a quarterwave layer is represented by a capital letter. H is usually used for a layer of high admittance and L for one of low admittance, but if more than two materials are involved in the design then the choice of additional letters is quite arbitrary. For example, A, B or M, N may be used. HH or $2H$ represent a halfwave of high admittance, 0.5 L an eighth of a wave of low admittance. $(HL)^3L$ is an example of a still shorter way of writing $HLHLHLL$. The design is written from left to right with the substrate at one side and the incident medium at the other, with no agreed convention about whether or not the substrate should be on the right. Many people begin with the substrate on the left because the layers then appear in the order to be deposited. Others prefer the incident medium at the left because that agrees with the usual convention in optics. In this chapter we place the incident medium on the left and the substrate on the right.

The quarterwave rule shows us that the addition of a quarterwave film of admittance given by

$$y_1 = \sqrt{y_0 y_{sub}} \qquad (62)$$

will act as a perfect antireflection coating because the reflectance of the combination is zero. For a glass surface in air as the incident medium, the required admittance is $\sqrt{1.00 \cdot 1.52} = 1.23$. Such a low refractive index is not possible in a form that is strong enough to withstand the environmental attack to which an antireflection coating is subjected. Magnesium fluoride of admittance 1.38 is normally used. The performance over the visible region of a magnesium fluoride coating is shown in Fig. 8-5.

The reflectance at the minimum is around 1.25%, a usual improvement on the 4.25% reflectance of an uncoated surface. The single-layer coating is probably the most common dielectric coating made. It has the enormous advantage that for even large errors in the deposited film thickness, the reflectance of the coated surface can never exceed that of the uncoated surface. That is not the case in the vast majority of coatings.

An improvement over a limited range of wavelength can be achieved by a two-layer system consisting of a thin high-admittance layer next to the substrate, followed by a thick low-admittance layer. The thin high-admittance layer raises the reflectance of the substrate to the point where it becomes possible to antireflect it exactly. The phase shift moves into the third quadrant so that the low admittance layer must now be thicker than a quarterwave. The characteristic of such a layer is shown in Fig. 8-6. It is much narrower than that of the single layer coating and for that reason is usually known as a V-coat.

Halfwave layers of dielectric material are absentee layers at λ_0 and therefore can be inserted anywhere in a design with no effect on the reflectance or transmittance at that wavelength. The halfwave does affect the performance elsewhere when it is no longer a halfwave and in some cases can improve it. Full details of the way in which the insertion point and the admittance of the layer should be best chosen [7] are beyond the scope of

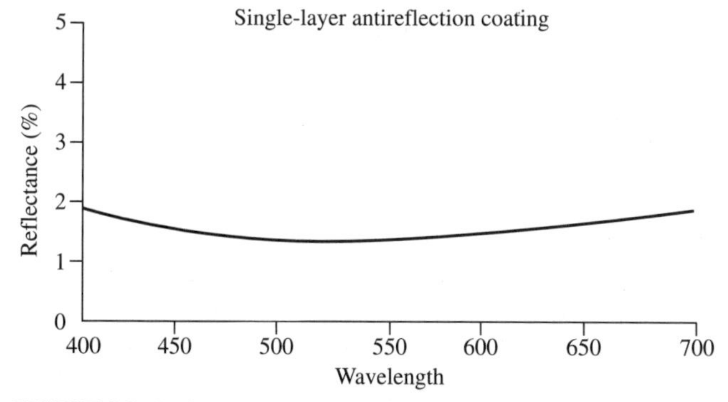

FIGURE 8-5 Performance curve of a typical antireflection coating consisting of a single quarterwave layer of magnesium fluoride on glass. The reference wavelength is 510 nm.

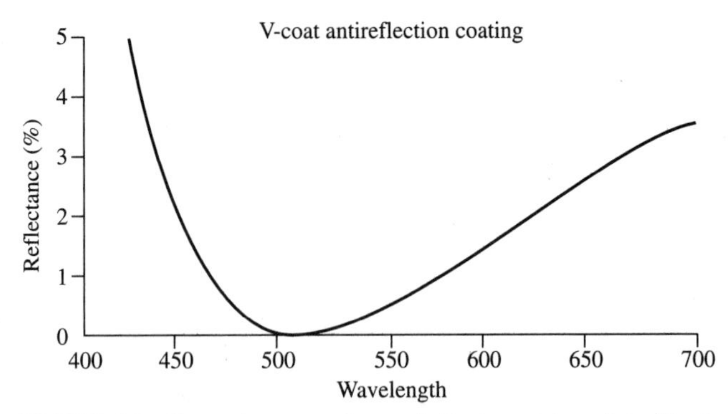

FIGURE 8-6 The performance of an antireflection coating of the V-coat type. The coating has design: Air | 1.31 L 0.28 H || Glass with $\lambda_0 = 510$ nm, $y_H = 2.35$, $y_L = 1.46$, $y_{\text{Air}} = 1.00$ and $y_{\text{Glass}} = 1.52$.

this chapter. Briefly, the layer should be inserted at a point where the admittance within the coating is real and the layer characteristic admittance should be either higher or lower than the admittance of the insertion point. In the classic V-coat the insertion point is one quarterwave away from the incident medium within the low-index layer and the admittance of the halfwave should be high. This leads to a design: Air|$LHHL'H'$|Glass where L' and H' indicate layers that are different from quarterwaves. Here H' corresponds to the appropriate layer in the V-coat design whereas L' is that part of the original low-admittance layer that is greater than a quarterwave. A small adjustment by refinement permits a slightly better performance and results in the curve shown in Fig. 8-7. This new design uses only two materials. It is often used, either in this form or slightly changed, in high-performance antireflection coatings for the visible region. There are various other methods of deriving the design [8]. Silicon oxide is frequently used instead of magnesium fluoride to give additional environmental resistance.

There are many different methods that are used in the design of antireflection coatings. They tend to be particularly suitable for automatic computer refinement, which is a very

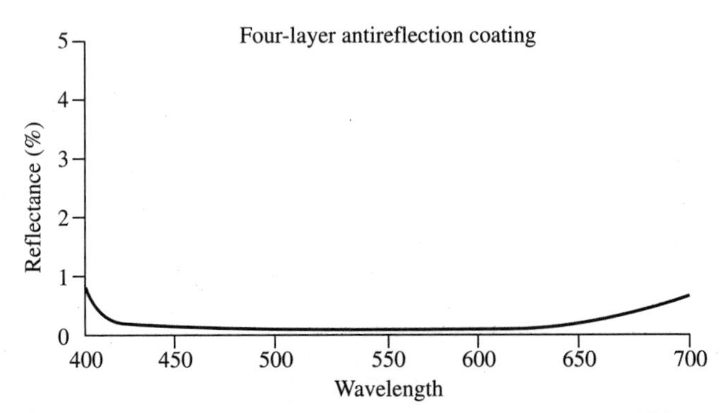

FIGURE 8-7 The four-layer design based on the V-coat with additional flattening halfwave layer. The design is: Air | 1.01 L 2.14 H 0.334 L 0.259 H | Glass with $\lambda_0 = 510$ nm. The layers are L, magnesium fluoride, and H, zirconium dioxide. *(Courtesy of Thin Film Center Inc.)*

popular technique for design of antireflection coatings. Figure 8-8 shows a broadband antireflection coating designed using automatic computer methods.

The simplest type of high-reflectance coating is a metal layer, and metals like aluminum (Al), gold (Au), and rhodium (Rh) are frequently used. The most common coating of this type is Al, which is used for the ultraviolet, visible, and infrared regions. It is sometimes uncoated, but more frequently has a single dielectric overcoat to improve the abrasion resistance. Dielectric overcoats always reduce reflectance, except for a very small range of thicknesses that are just less than a halfwave. The lowest possible index should be chosen for the overcoating layer because this gives less reduction in reflectance. Silica is a good choice for Al reflectors. Gold has excellent infrared reflectance although poor in the visible region, especially green and blue. It is sometimes coated but overcoats do not always adhere well, so it is often unprotected. Rhodium is used where tough, hard, corrosion-resistant coatings are required. Its reflectance is lower than Al and it is very expensive. It has a very fine microstructure so it has the additional advantage of low scattering. Metal layers all have some absorption and can never achieve the very-high reflectances possible with purely dielectric multilayers.

A single quarterwave layer changes the admittance of a substrate from y_{sub} to y^2/y_{sub} and so a succession of quarterwaves will give an admittance of the form

$$Y = \frac{y_1^2\, y_3^2\, y_5^2 \ldots}{y_2^2\, y_4^2\, y_6^2 \ldots y_{sub}} \tag{63}$$

with y_{sub} in the denominator, as shown, or numerator, depending on an odd or even number of layers, respectively. Thus if there are alternate high- and low-index layers the admittance can be made either very high or very low, in each case leading to high reflectance. Under normal circumstances, the highest reflectance is obtained with high-index layers next to the substrate and outermost. If there are x high index layers and $x - 1$ low index, then the reflectance is given by

$$R = \left[\frac{y_0 - \dfrac{y_H^{2x}}{y_L^{2(x-1)}\, y_{sub}}}{y_0 + \dfrac{y_H^{2x}}{y_L^{2(x-1)}\, y_{sub}}} \right]^2 \tag{64}$$

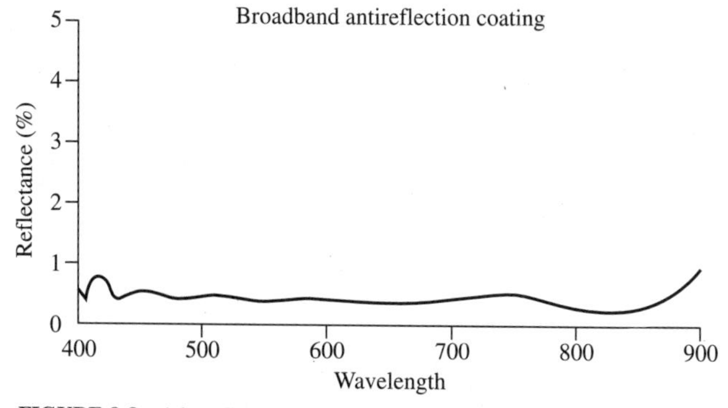

FIGURE 8-8 A broadband antireflection coating designed by automatic computer refinement. It consists of fourteen layers of silica and titania and is designed for a glass substrate. *(Courtesy of Thin Film Center Inc.)*

With lossless layers, the possible reflectance is limited by the number of layers that can be deposited. The higher this number, the higher the reflectance. In practice, the ultimate reflectance that can be achieved depends on the residual losses, absorption, and scattering. The best results reported to date are total losses of the order of 0.0001% [9]. These results were obtained in the red region by the ion-beam sputtering of silica and tantala layers.

The high reflectance associated with the quarterwave stack is obtained only over a limited region. Outside that region the interference is no longer completely constructive. The reflectance falls and transmittance rises, although there are fringes of higher reflectance that decay with distance from the high-reflectance region. Because of the cyclic nature of interference phenomena the high-reflectance peak is repeated at regular intervals of frequency. A typical quarterwave stack characteristic is shown in Fig. 8-9.

The width of the high reflectance zone is given by $2\Delta g$ where $g = \lambda_0/\lambda$ and

$$\Delta g = \frac{2}{\pi} \cdot \arcsin\left\{\frac{y_H - y_L}{y_H + y_L}\right\} \tag{65}$$

The principal features of the quarterwave stack are thus the regions of high reflectance and low transmittance around frequencies or wavelengths for which the layer optical thicknesses are odd integral numbers of quarterwaves, separated by regions of relatively high transmittance with appreciable ripple. By itself, the quarterwave stack is used as an efficient reflector with very-low losses. The quarterwave stack is also used as a subunit of many different types of optical coating.

The characteristic of an edge filter is a sharp transition between a transmitting region and a reflecting or absorbing region. The sharp transition is known as the edge. Filters in which the transmission band is situated at wavelengths longer than the edge are known as long-wave-pass filters. Those where it is shorter are known as short-wave-pass filters. In the dichoric beam splitter the rejected band is reflected rather than absorbed so that two spectral regions are separated.

The simplest type of edge filter is an absorption filter. There are many colored glasses and dyed-gelatin filters which can be used as long-wave-pass filters for the near ultraviolet, visible, and infrared regions. Most materials have sharp onsets of absorption because of electronic transitions in the shorter wavelength region. Semiconductors adsorb in the infrared and visible, and dielectrics have adsorption edges in the ultraviolet. Short-wave-pass edges also exist but they are usually much less sharp and the choice is more limited

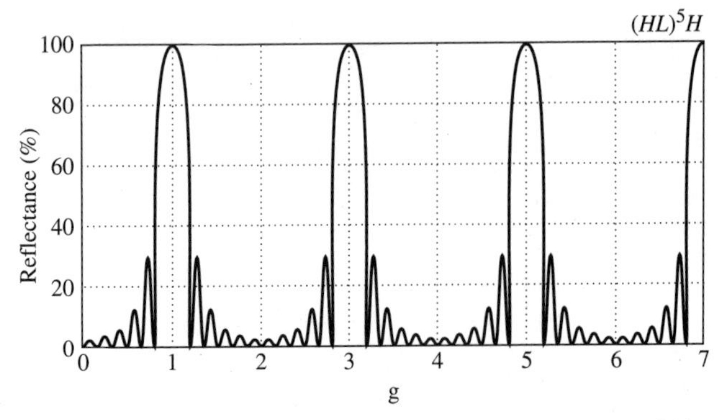

FIGURE 8-9 A typical quarterwave stack characteristic. The design of this coating is Air | $(HL)^5H$ | Glass with $y_H = 2.35$, $y_L = 1.35$ and $y_{glass} = 1.52$. The characteristic is plotted in terms of $g(= \lambda_0/\lambda)$.

than for long-wave-pass edges. If a natural-edge filter exists, it is often the best solution. If not, then thin film interference coatings are used.

The basic edge filter is the quarterwave stack. Because it is characterized by regions of high reflectance separated by regions of transmittance, it can be used either as long-wave-pass or short-wave-pass by choosing the position of the stop bands. The principal deficiency of the simple quarterwave stack is the prominent ripple in the pass bands. The ripple can be considerably reduced by altering the thicknesses of a few of the outermost layers so that the multilayer is matched to the surrounding media. The simplest arrangement replaces the outermost quarterwave layers with eighthwave layers. For the usual materials in the visible region the eightwave layers should be of high admittance for a long-wave-pass filter and of low admittance for a short-wave-pass filter. A long-wave-pass filter of this kind is shown in Fig. 8-10.

Two common types of edge filter, probably more correctly described as dichroic beam splitters, are heat-reflecting filters and cold mirrors. The heat-reflecting filter is a short-wave-pass filter that reflects the near infrared and transmits the visible region, whereas the cold mirror reflects the visible and transmits the near infrared. Because the coating used in Fig. 8-10 does not have quite wide enough high-reflectance zone for a cold mirror, usually two stacks of this type of coating are used. One is deposited over the other and displaced in wavelength to cover completely the 400 to 700 nm region.

A bandstop filter transmits on either side of a spectral region that is rejected (the stop-band). Bandstop filters, (sometimes also known as minus filters), are essentially quarter-wave stacks, which are designed to have low ripple on either side of the high-reflectance zone, instead of on just one side as in edge filters. The elimination of the ripple is, however, a more difficult task than in the edge filter. There are effective analytical techniques, but the most common approach uses computer refinement to design sets of matching layers on either side of the basic quarterwave stack, (Fig. 8-11). The width of the rejection zone can be reduced by the use of materials with closer admittances or by detuning the layers of the stack so that they are no longer quarterwaves. This technique was used in Fig. 8-11.

Narrowband filters are also based on the quarterwave stack. Consider a design

$$HLHLHLHHLHLHLH \tag{66}$$

This consists of two quarterwave stacks placed one over the other. The central layer, *HH*, is a halfwave and therefore contributes nothing to the reflectance because it is an absentee

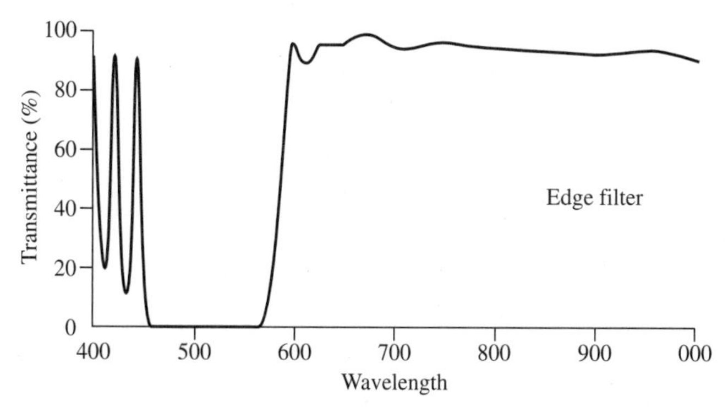

FIGURE 8-10 A longwave pass filter for the visible region of design Air | $(0.5\,H\,L\,0.5\,H)^{11}$ | Glass. *H,* titania *L,* silica $\lambda_0 = 510$ nm.

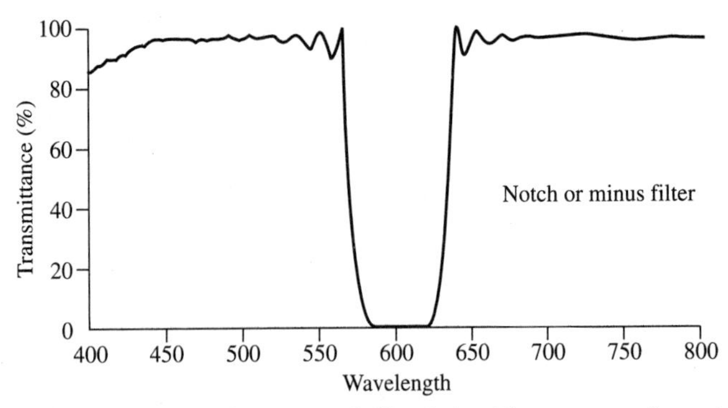

FIGURE 8-11 A band-stop or notch filter designed by computer refinement from a starting design of a 60-layer quarterwave stack of zinc sulfide and cryolite.

layer. If it were to be removed then there would be a halfwave of low admittance in the center and it too would be an absentee that could be removed with no change in reflectance. We can see that the entire structure is an absentee system and it therefore has no effect on the reflectance of whatever structure it partly forms. The absentee character depends on a process of interference where many of the rays have quite large path differences so that the condition varies rapidly with wavelength. The reflectance characteristic of the quarterwave stacks is quickly established. The characteristic of the coating is that of a nar-

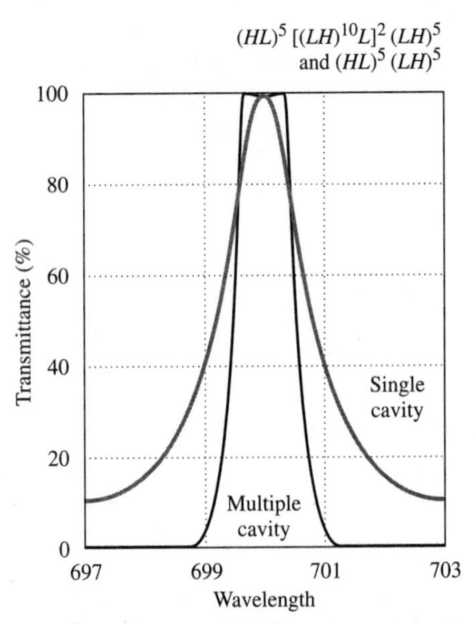

FIGURE 8-12 Narrowband filter characteristics of the single-cavity and triple-cavity type showing the much more rectangular profile of the multiple-cavity design. The materials used for the design are *H,* zinc sulfide and *L,* cryolite. The layer sequence is shown in the formulae at the top of the diagram.

row band of high transmittance surrounded by high reflectance, that is, a narrowband transmission filter. The central halfwave layer is usually referred to as a cavity layer and the entire structure is known as a single-cavity filter. A multiple-cavity filter can be constructed by coupling several single-cavity structures together to give forms like

$$HLHLH\overline{LHH}LHLHLHLHLHLH\overline{LHH}LHLHLH$$

or

$$HLHLH\overline{LHH}LHLHLHLHLHLH\overline{LHH}LHLHLHLHLHLH\overline{LHH}LHLHLH \tag{67}$$

where the bars indicate the cavity layers. Examples of a single- and a multiple-cavity filter are shown in Fig. 8-12.

The dielectric filters have the characteristic of the quarterwave stacks from which they are constructed. The high reflectance of the rejection regions is limited in extent. Metal layers are rather different. The refractive index and optical admittance of a metal layer is of the form $(n - ik)$ where n is small and k is large. As the wavelength increases, k increases so that the properties of the metal are strengthened at longer wavelengths rather than weakened as are the dielectrics. A metal layer can be antireflected by a system consisting of a reflector and a phase matching layer, as shown in Fig. 8-13.

A metal layer can be antireflected on each side. If the metal is not so thick that the transmittance is vanishingly small, the antireflection coating will correspond to a pass-band and the remainder of the characteristic will represent high reflectance. The width of the region over which antireflection is achieved depends very much on the structure of the reflector. The greater the number of layers involved in the reflector the narrower the antireflection characteristic and the greater the thickness of metal that can be used. Figures 8-14 and 8-15 show a simple three-layer broad-band system that corresponds to a heat-reflecting filter and a narrow-band filter that can be used for suppressing the long-wave transmittance of the filters of Fig. 8-12.

The higher-order reflectance bands of the quarterwave stack are sometimes a problem. The reason for their existence is that the various beams that make up the interference effects that create the primary bands will combine constructively again when the phase shift of each has been increased by an integral number of full waves. This will be the case if all layer's optical thicknesses are increased by an integral number of halfwaves. Thus there will be high-reflectance bands at values of g of 3, 5, 7, 9 and so on. These bands could be suppressed if the individual beams could be suppressed. To accomplish this it is

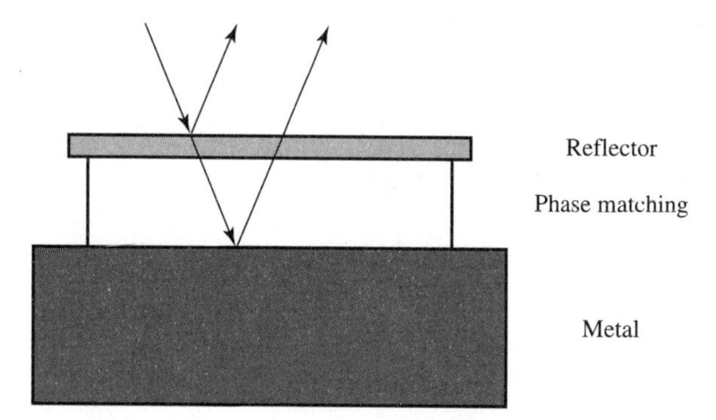

FIGURE 8-13 The outer reflector can be arranged to ensure that the two beams shown are of equal amplitude. Then the phase matching layer thickness can be chosen to make the two beams interfere destructively and the net reflectance is zero.

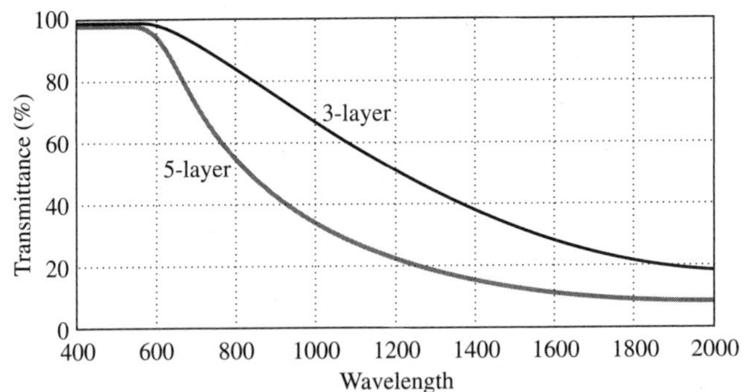

FIGURE 8-14 A heat-reflecting filter based on silver layers. The incident and emergent media are glass and the structures are titania | silver | titania for the three-layer structure and additional outer quarterwaves of magnesium fluoride for the five-layer structure. The extra quarterwaves increase the reflectance of the matching assemblies and permit greater thickness of silver, hence the improved performance.

necessary that the various interfaces that produce these beams be antireflected. Of course, this antireflection coating should not be effective at the primary reflectance band, or else the characteristic will disappear there as well as at the other wavelengths. The ideal coating for this purpose is an inhomogeneous layer, that is a layer with an admittance that varies smoothly from that of the substrate or emergent medium to that of the incident medium. Such a coating with its characteristic curve is shown in Fig. 8-16.

This coating gives virtually perfect antireflecting properties for wavelengths that are long enough to assure slightly greater optical thickness than a halfwave. If this coating is inserted between the high-admittance and low-admittance quarterwaves of the quarterwave stack, adjusting the thicknesses so that the period remains a halfwave, then the antireflecting properties can be assured at $g \geq 2$ although at $g = 1$ the antireflecting properties will be poor. The first order peak of reflectance, therefore, remains high, although not quite as high as before, whereas the higher order peaks are completely suppressed. The

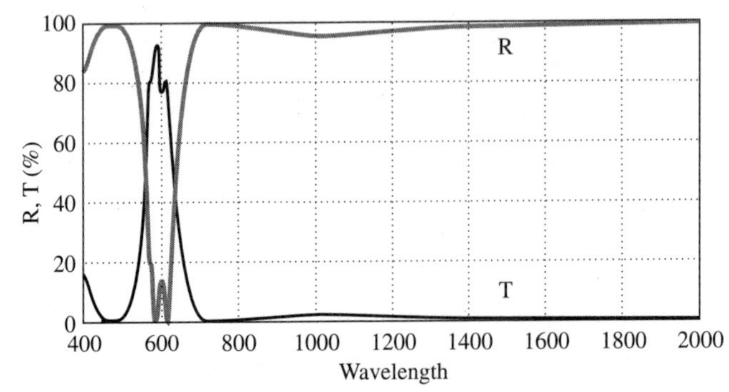

FIGURE 8-15 A narrow-band filter based on the use of a silver layer with antireflection coatings. The antireflection coatings consist of a five-layer system, four quarterwaves with a phase-matching layer. The complete design is Glass | *HLHLH'* Ag *H'LHLH* | Glass with *H*, zinc sulfide and *L*, cryolite.

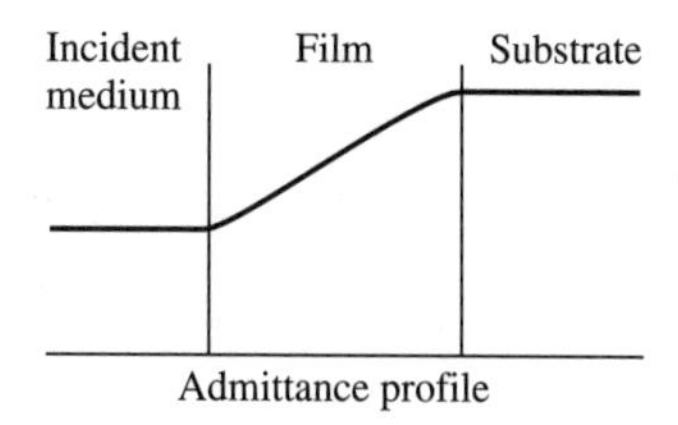

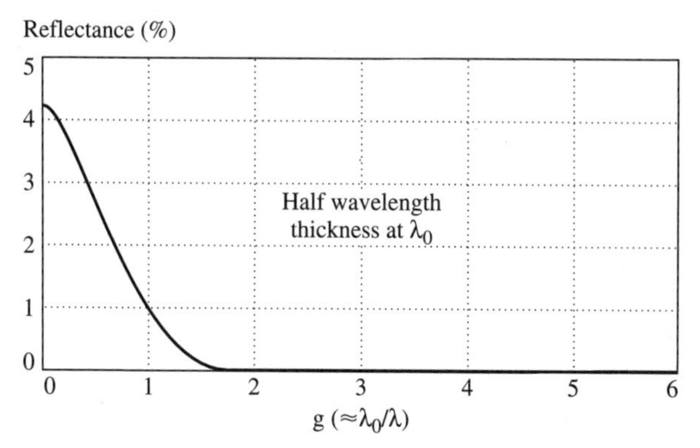

FIGURE 8-16 An inhomogeneous antireflection coating based on a fifth order polynomial solution where both the admittance and the gradient of admittance are continuous functions. (The profile shown in the upper sketch is not an exact representation.)

profile of the admittance variation is oscillatory and quite similar to a sine or cosine wave. (There is discussion about the best profile but in any practical case dispersion quickly alters it as the wavelength changes.) This type of filter is called rugate and a typical characteristic of a rugate filter is shown in Fig. 8-17.

A principal use of a rugate filter is the blocking of a narrow wavelength band such as the output of a laser or spectral lamp, but the potential of the technique is much greater.

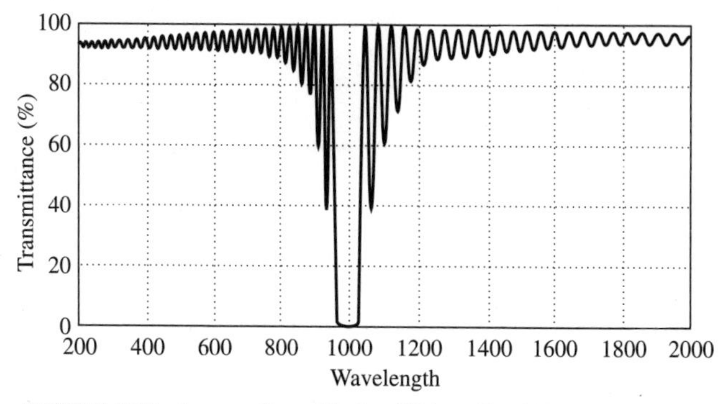

FIGURE 8-17 A rugate filter with sinusoidal profile of admittance and index. Note the absence of any higher-order peaks at 500, 333, 250, and 200 nm. The profile of this particular rugate is the exponential of a sine function.

The continuous nature of the admittance profile permits the use of an approximate solution that involves a Fourier transform of the logarithm of the admittance function and a relatively simple function of performance. This relationship can be inverted readily and so the theoretical construction of a design may be analytically performed rather than iteratively performed. The simple filter of Fig. 8-17 rejects a single wavelength but the characteristic of the filter and the corresponding admittance/index variation through the structure can be more complicated. Except for a few special applications the technique is not yet used by industry, but the uses are growing and there is research literature available on the topic [10,11].

Design techniques for thin-film optical coatings are well covered in a number of books [7,12–15].

8.6 PREPARATION AND FORMATION OF OPTICAL THIN FILMS

There are many processes [16–18] that are used for the deposition of optical coatings. The most common are still physical vapor deposition (PVD) techniques, although there is increasing interest in chemical vapor deposition (CVD) processes. Even liquid deposition processes are used in special applications. For regular coating production, however, vacuum evaporation is still the most common.

In vacuum evaporation (as discussed in Chapter 1), the material to be deposited is vaporized and the vapor then condenses on the substrates that are, even if heated, at temperatures below the solidification point of the evaporant. The process is carried out in vacuum. Although this process is called "physical" there is in fact a great deal of chemistry involved. The vapor frequently is a decomposed version of the source material. And in some cases, to restore the source composition to the film, additional reacting gases must be added to the system, a process usually termed reactive evaporation.

The production of the vapor is by simple heating when the name given to the process is evaporation. Small furnaces, either directly heated refractory metal crucibles, called boats because of their shape, or water-cooled crucibles with contents that are bombarded by energetic (7 to 8 kV at around 1 A) beams of electrons, electron-beam sources, are used as sources of vapor. The vaporized material travels in a direct line to the substrate where it freezes in position, and the deposited thickness follows roughly the same laws as those of illumination. To ensure greater degrees of uniformity of deposit the substrates are often moved in various ways over the sources. A high degree of uniformity can be achieved by a circular motion with two separated axes of rotation, one fast and the other slow. Materials like the refractory oxides are known to decompose on evaporation and to lose oxygen. To restore the level of oxygen in the deposited film, additional oxygen is added to the coating chamber and reacts with the growing film in the reactive evaporation process. Occasionally the added oxygen may be activated in some way, usually by passing it through an electrical discharge.

An alternative technique of vapor production is known as sputtering (discussed in Chapter 1). Here the source material is bombarded by a beam of ions of some hundreds or sometimes thousands of electron volt energy. The collision process with the source material, the target, produces a collision cascade in the material that culminates in the ejection of single molecules of source material at relatively high energies compared with that of thermally evaporated molecules. The two main classifications of sputtering depend on the way in which the production of ions is carried out. In the earlier processes this was by means of an electrical discharge in the coating chamber itself. Usually there are two electrodes, a cathode and an anode. The substrates to be coated are held at or near the anode and the cathode is the material target. If the parts are metallic a dc discharge is sufficient; if they are dielectric then radio-frequency discharges are used. This type of two-electrode sputtering is called diode sputtering.

A reactive gas can be introduced, oxygen or nitrogen being most common, when the process is called reactive sputtering. The pressure in the coating chamber must be reasonably high for this process to be successful, otherwise there is difficulty in maintaining the discharge that generates the ions. This affects the mechanical properties of the coating. A reduction in pressure is possible if the electrons in the discharge can be made to execute longer paths. This is achieved by adding magnetic fields to the electric ones to force a cycloidal motion of the electrons. Because this is the same technique used in the original magnetron microwave sources, the technique is called magnetron sputtering. The magnetic field is derived usually from permanent magnets that are built into the source that now also contains both anode and cathode, the anode being in the form of an outer structure and the cathode, a flat plate within it. Such structures are often made long and narrow so that they are suited to the linear motion of substrates in continuous in-line coating processes. Again there are dc and R_f magnetron processes.

A simpler technique generates the ions by a discharge in a separate ion source. A neutralized beam of ions is then extracted to perform the sputtering. The discharge chamber is quite small and mounted often within the main deposition chamber and an electric field extracts the ions, sometimes through a perforated grid. Such separate ion sources are much more controllable than a discharge in the main chamber, so the films achieved are of very-high quality. The process, known as ion-beam sputtering, is slower than diode or magnetron sputtering and is principally used for very-high-quality coatings such as interferometer mirrors. Not all materials are suitable for sputtering. In particular the fluorides present considerable difficulties because of preferential sputtering of fluorine atoms. The film is then fluorine-deficient and optically absorbing. The fluorine vacancies can be filled with oxygen. There is usually some oxygen in the residual atmosphere of the deposition chamber which removes the absorption, at least at longer wavelengths, but the film becomes an oxyfluoride with altered (usually raised) index of refraction and frequently degraded environmental resistance.

A major problem with thermally evaporated films is that their packing density is low because they freeze very rapidly on the substrates. This low packing density makes the films mechanically weak and also permits the adsorption of large amounts of material, especially water vapor when the coating is exposed to the atmosphere, so that the films are not very stable. The low packing density also has the effect of making the residual stress in the coatings tensile. Because thin films are quite weak in the tension this is a further problem that often causes tensile cracking of the films. Addition of energy to the condensation process, usually in the form of bombardment by ions, can improve the packing density considerably. The increased packing density in turn improves the stability of the films in other ways. Films have smaller temperature-induced shifts and resist decomposition at higher temperatures. The increased reactivity of the bombarding ions permits the deposition of compounds, such as nitrides, that are difficult or impossible by normal vacuum evaporation.

The sputtering process is itself an energetic process and the films that are produced by it are of higher packing density and have better adhesion than thermally evaporated films. They also tend to have compressive stress. In low-voltage ion-plating, a high-current beam of low-voltage electrons is directed into the region above the hearth in an electron-beam source. This results in a very-high degree of ionization of evaporant material. The electrical circuit from ion gun to electron-beam source has a complete return path and it is completely isolated from the rest of the structure. The substrate carrier is also electrically isolated. There are many mobile electrons and the isolated substrates acquire a charge that is negative with respect to the electron-beam source. This attracts the positive ions from the source so that they arrive at the film surface with additional momentum that is transferred to the film and compacts it.

Laser evaporation, or laser ablation (discussed in Chapter 1), is another process that has a much higher energy associated with arriving atoms and molecules at the film surface. Here the vapor is produced thermally by directing the beam from a pulsed laser onto the target. This produces a very-high local temperature and the vapor molecules have much higher energy than in conventional thermal evaporation. The laser is external to the chamber and there must be a window for entry of the beam. A mirror usually directs the beam onto the target. Because the mirror is in the line-of-sight from the target it is coated along with the substrates, and there are various techniques for limiting the effects of the deposition, including rotating the mirror gradually from behind a mask. The process tends to be used for coatings that are intentionally inhomogeneous. In such cases the design can often be represented as a succession of very thin homogeneous layers of either of two materials. Each laser pulse can be arranged to deposit a basic unit of the appropriate material. The normally severe control and monitoring problems that accompany inhomogeneous layer deposition are eased considerably. Pulsed laser deposition (PLD) is becoming accepted as a process used in semiconductor device manufacturing, and so may also be applied to the optical-coating method.

Ion-assisted deposition is a process with the advantage that it is easy to implement in conventional equipment. It consists of thermal evaporation using bombardment of the growing film with a beam of energetic ions. All that is required to put it into operation in a conventional evaporation system is the addition of an ion source. The most common types of ion sources for this purpose are broad-beam sources, often with extraction grids. The beam of ions is neutralized outside the discharge chamber by adding electrons, often from a hot filament immersed in the beam to avoid space charge limitation. There are other processes, but the principal ones are sputtering, ion plating, and ion-assisted deposition.

The major benefit of these energetic processes is an increase in film-packing density. The improvements are achieved at comparatively low substrate temperatures which helps with the difficult coating of plastic substrates. It has been theoretically demonstrated by computer modeling that the major effects are caused by the additional momentum of the molecules, either supplied by collisions with the incoming energetic ions, or derived from the additional kinetic energy of the evaporant [19]. Experimental evidence exists that shows correlation of the effects with momentum rather than energy of the bombarding ions [20]. Advantages of these processes are the increased solidity of the films, making them more bulklike and hence increasing their strength; the improved adhesion resulting from a mixing of materials at the interfaces between layers; and reduction of the sometimes quite high tensile stress in the layers. The increase in packing density also reduces the moisture sensitivity and can often eliminate it altogether. Ion-beam processes have been reviewed [21,22]. Fulton [23] has recently written a very useful account of ion-assisted deposition particularly for volume manufacturing.

Other processes that are currently in use include CVD, a variety of plasma-assisted deposition processes, and some liquid processes mostly based on sol gel techniques.

In CVD, the growth of the film results from the production of the film material actually at or very near the film surface by a chemical reaction that is induced in the material, making up the surrounding vapor. Frequently this reaction is induced by the high temperature of the substrate itself, and this is the classic form of CVD. The flow of the reacting components over the substrate surface is an important feature of the process and the uniformity of the deposit is closely related to this. The high temperature necessary for the substrates is a difficulty associated with this process. An alternative technique for driving the reaction is by an electrical discharge. Such processes are known by a variety of names, but plasma-enhanced CVD is quite common, as discussed in Chapter 1. A similar process uses the discharge to polymerize a monomer in the vapor phase and is also known by a variety of names including plasma polymerization.

Another problem in most of these processes is connected with the reaction itself. If it is too fast a porous, sooty deposit results that is useless for optical purposes. The reaction must, therefore, be quite slow. This limits the possible range of processes and materials. The most effective way around this problem is by pulsing the process so that although the reaction may be very fast, only small amounts of film material are deposited with each pulse. Even the thermal CVD may be pulsed by the way in which the reactants are injected into the carrier gas, but the pulsing of the plasma processes is more common. This will usually be a radio-frequency plasma and the pulsing can be very rapid as long as it permits each small increment of film to reach equilibrium before arrival of the next. The CVD process has the advantage that the films are not quenched as rapidly as the films in most PVD processes, so they usually have better microstructure.

Sol gel processes (discussed in more detail in Chapter 1) [24] involve the deposition from solution of a metal organic gel that is hydrolyzed by the addition of water. Subsequent heat treatment converts this into a tough, dense oxide layer. Titania and silica can be deposited in this way. Recently it was discovered that films deposited in this way, with virtually no heat treatment so that they were porous and weak, had an unprecedentedly high laser-damage threshold. Sol gel coatings are now often used for the antireflection of components used in very big lasers. They have an exceptionally high laser-damage threshold and can also be readily stripped off the substrate, which can then be recoated without the need for repolishing. Because the large fusion lasers are in exceptionally well-controlled clean conditions, the lack of strength and the sensitivity to contamination is of lesser importance.

The usual process for optical-coating production is thermal evaporation. In that process the microstructure of the thin films tends to be columnar, with the columns roughly normal to the coated surface, although the orientation does depend on the conditions during deposition. Most optical thin films therefore exhibit form birefringence with the optic axis normal to the surface. Because the optical coatings themselves have pronounced anisotropic behavior at oblique incidence, the smaller-form birefringence is largely unnoticed in practice. The columnar structure, unfortunately, has other less-benign implications. A major problem is the pore structure that accompanies the columns. The columns are loosely packed so that there are pore-shaped voids between them and the packing density is less than unity. The intrinsic stress in the films is tensile because of the bonds that are stretched across the voids. At very low values of relative humidity the internal surface of the films is coated with a thin film of moisture. But at higher values the pores actually fill with liquid water through a process known as capillary condensation. The water has a refractive index of 1.33 compared with the 1.00 of the empty void, and so the refractive index of the film rises. This rise in refractive index is accompanied by a corresponding change in optical thickness because the physical thickness does not change. The characteristic of the coating in turn moves to a longer wavelength. This is a source of major instability in optical coatings. Because the moisture can desorb under certain conditions, especially when the system is heated, the shifts can be considerable. Desorption processes are not always exactly the reverse of adsorption ones, and so there is often considerable hysteresis. Moisture also reduces surface energy by as much as an order of magnitude in high-energy surfaces such as those of the refractory oxides. This reduction in surface energy contributes to a reduction in abrasion resistance and adhesion. Stress levels also tend to become less tensile but there is still usually considerable strain energy in the films. The strain energy is frequently still high enough to be capable of driving an adhesion failure should one start to develop [7].

Almost all of this is negative in character. It is clear that an increase in packing density would be of great value, and this is achieved by the energetic processes. The benefit of these processes comes from almost purely mechanical effects. Most of the processes involve bombardment by energetic ions, or it may be that the bombardment is by energetic evaporant itself. The momentum that is transferred from the arriving particles to the film being bombarded may pack the material more tightly, squeezing out the voids. The increased packing moves the stress in the film to compressive rather than tensile. However, in the ion-assisted deposition process where the bombardment can be sensitively controlled, the stress can be arranged to be very low and either neutral or slightly tensile, as well as compressive. Because films are weak in tension but strong in compression, the compressively stressed state gives much tougher films, provided that the stress is not so high that the films fail in shear. Because the voids are squeezed out the moisture sensitivity is greatly reduced and may disappear entirely. A further unexpected advantage is a great reduction in temperature sensitivity because the coefficients of change of optical thickness with temperature are greatly reduced. The bombarding ions can be arranged to react with the material of the film. For example, nitrides and oxynitrides can be produced by ion-assisted deposition. They are very difficult to produce in other ways. Disadvantages associated with the processes are few. Differential sputtering of material from the films and implantation of a small amount of the bombarding species are two disadvantages. The price is small considering the large improvement in other properties.

8.7 CHARACTERIZATION OF OPTICAL THIN FILMS

Many different forms of characterization are necessary for optical thin films [25,26]. Techniques like electron microscopy, electron diffraction, x-ray diffraction, Rutherford backscattering spectroscopy, atomic force microscopy, and so on, are common to virtually all thin films and are discussed in Chapter 6. Here we are concerned with the optical characterization. Like many other forms of characterization, it involves the measurement of certain aspects of optical behavior that permit subsequent prediction of the behavior of similar thin films. The process is usually described as the measurement of this or that parameter of the thin film. In reality it is one in which the parameters of a model are adjusted until the model reproduces the measured behavior of the film that is being used for the parameter extraction. The model is then used for predictions and is therefore critical to the success of the process. Films that behave like the model are called well-behaved. Films that conform poorly to the model are said to be anomalous. Figure 8-18 shows typical film refractive index extracted from transmittance behavior, Fig. 8-19.

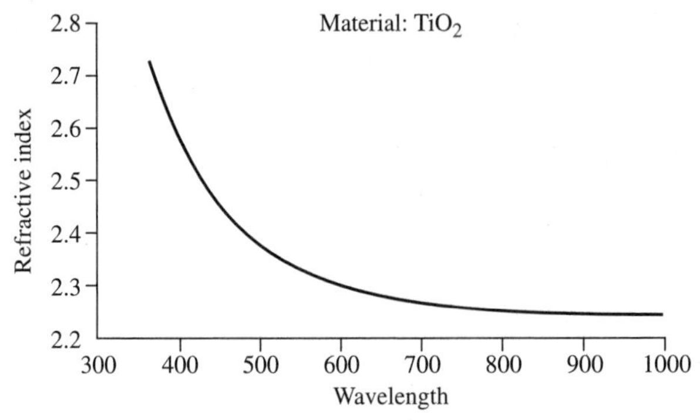

FIGURE 8-18 The refractive index of a thin film of TiO_2. *(Courtesy of Thin Film Center Inc.)*

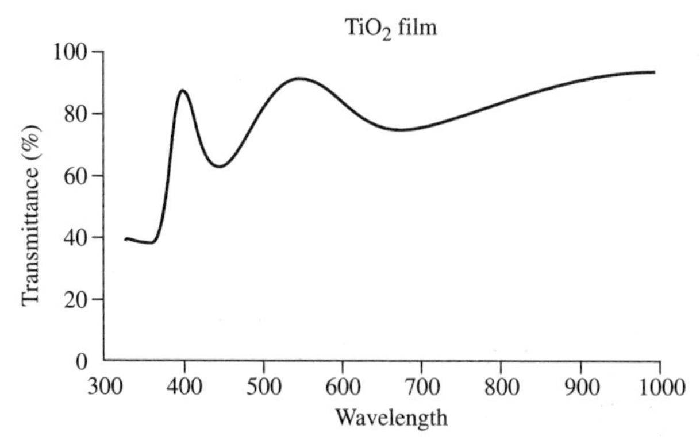

FIGURE 8-19 Fringes in transmission exhibited by a film of TiO_2 on a substrate of glass. *(Courtesy of Thin Film Center Inc.)*

The model that is almost always used for the interpretation of thin film behavior is a parallel-sided slab of isotropic material characterized by a refractive index n, an extinction coefficient k and a thickness d. The optical admittance of the film is numerically equal to the refractive index. The model may be elaborated by including an element of inhomogeneity in the index n. This may be represented by Δn, where the extreme values of the index are $n \pm \Delta n$; or by n_{outer} and n_{inner}, the index at the outer and inner surfaces of the film. Only rarely is inhomogeneity in the extinction coefficient included. The behavior of films is not a sensitive function of the form of the variation of index through the film, provided it is slow. Usually a linear variation is assumed. Another complication that is sometimes used is the representation of the film by a bilayer, each element of the bilayer being homogeneous and isotropic. This second model has been much used in interpreting the behavior of films where a transition region exists between film and substrate or film and surrounding medium, and the representation of such a transition region is sometimes called a *couche de passage*. In all cases n, k and Δn are functions of λ.

Given the optical constants and thicknesses of any series of thin films on a substrate, the calculation of the optical properties is straightforward. The inverse problem, that of calculating the optical constants and thicknesses of even a single thin film, given the measured optical properties, is much more difficult. There is no general analytical solution to the problem of inverting the equations. The traditional methods of measuring optical constants, therefore, rely on special limiting cases that have straightforward solutions. More recent techniques rely more on computer iteration for reduction of the measurements [7,27].

Perhaps the simplest case of all is represented by a quarterwave of material on a substrate, which is both lossless and dispersionless. That is, k is zero and n is constant. The reflectance is given by

$$R = \left(\frac{1 - n_f^2/n_{\text{sub}}}{1 + n_f^2/n_{\text{sub}}} \right)^2 \tag{68}$$

where n_f is the index of the film, n_{sub} that of the substrate, and the incident medium is assumed to have an index of unity. Then n_f is given by

$$n_f = n_{\text{sub}}^{1/2} \left(\frac{1 \pm R^{1/2}}{1 \pm R^{1/2}} \right)^{1/2} \tag{69}$$

where the refractive index of the substrate, n_{sub}, must be known. The measurement of reflectance must be reasonably accurate. For instance, if the refractive index is around 2.3, with a substrate of glass, then the reflectance should be measured to around one third of a percent (absolute ΔR of 0.003) for a refractive index measurement accurate in the second decimal place. The technique can be adapted to give results in the presence of slight dispersion.

If absorption is present changes in technique are necessary. When absorption is heavy the extinction coefficient can be calculated from

$$\frac{1-R}{T} = \exp\left(\frac{4\pi k_f d_f}{\lambda}\right) \tag{70}$$

This is just a simple comparison of what emerges from the film compared with what enters, with the assumption that interference is not important. If something better is required we can write

$$\psi = \frac{T}{1-R} = \frac{\mathrm{Re}(n_{sub})}{\mathrm{Re}(BC^*)} \tag{71}$$

where ψ is a quantity known as potential transmittance. Equation 71 is readily derived from Eq. 54. If we write $\beta = \dfrac{2\pi k_f d_f}{\lambda}$ then we can show that near the extrema,

$$\psi = \frac{1}{1 + \left(\dfrac{n_s}{n_f} + \dfrac{n_f}{n_s}\right)\beta} \quad \text{so that} \tag{72}$$

$$k_f = \frac{\lambda}{2\pi d_f[(n_{sub}/n_f) + (n_f/n_{sub})]} \cdot \frac{1-R-T}{T} \tag{73}$$

In the methods discussed so far we assume that the thickness of the film is unknown, except as it can be deduced from the measurements of reflectance and transmittance, and the extrema are the principal indicators of film thickness. However, it is possible to accurately measure film thickness in other ways, such as multiple-beam interferometry, electron microscopy, or by using a stylus step-measuring instrument. When there is an independent, accurate measure of physical thickness, the problem of calculating the optical constants is much simpler.

The most frequently used technique of this type was devised by Hadley [28]. Because two optical constants, n_f and k_f, are involved at each wavelength, two parameters must be measured. These can most conveniently be R and T. In the ideal form of the technique, if a value of n_f is assumed, then by trial and error one value of k_f can be found. This, together with the known geometrical thickness and the assumed n_f, yields the correct measured value of R and then a second value of k_f that similarly yields the correct value of T. A different value of n_f will give two further values of k_f and so on. We can plot two curves of k_f against n_f, one corresponding to the T-values and the other to the R-values. Where they intersect we have the correct values of n_f and k_f for the film. The angle of intersection of the curves gives an indication of the precision of the result. Hadley's original work was a book on the reflectance and transmittance values of a complete range of films as a function of the ratio of geometrical thickness to wavelength, with n_f and k_f as parameters. The method can now be readily programmed complete with precision estimates. This method can be applied to any thickness of film, not just at the extrema, although maximum precision is achieved near optical thicknesses of odd quarterwaves. However, at halfwave optical thicknesses it is unable to yield any results. As with many other techniques, it has multiple solutions when the films are thick. In practice a range of wavelengths is used, which adds an element of redundancy and helps to eliminate some of the less-probable solutions. It is not useful with films where there is perceptible inhomogeneity except near the odd quarterwave thicknesses.

Hadley's method involves simple iteration and does not require any powerful computing facilities. It does require the additional measurement of film thickness, which is different than the measurements of R and T and can lead to problems unless the measurement is very accurate. A different approach was developed by Pelletier and his colleagues [29], which retains the measurement of R and T, but measures in addition R' the reflectance of the film from the reverse side, and uses the fact that the geometrical thickness of the film does not vary with wavelength. Therefore if information over a spectral region is used, there will be sufficient redundancy to permit an accurate estimate of geometrical thickness. Initially only R and T are used. When the thickness has been determined, a refining computer method determines accurate values of the optical constants over the whole wavelength region. For dielectric layers used in optical coatings, k_f will usually be small, often negligible over at least part of the region. A preliminary calculation involving an approximate value of n_f is able to yield a value for geometrical thickness, which in most cases is sufficiently accurate for the subsequent determination of the optical constants. Finally R' is involved with the other two measurements in a method of successive approximations to minimize a figure of merit consisting of a weighted sum of the squares of the differences between measured T, R, and R' and the calculated values of the same quantities using the assumed values of n_f and k_f. The method has also been adapted to include the effect of inhomogeneity Δn.

A straightforward technique is described by Manifacier, Gasiot, and Fillard [30] and has been elaborated by Swanepoel [31,32]. Provided the absorption in a thin film is small, the transmittance at the quarterwave and halfwave points is a fairly simple function of n_f, k_f, and d. The transmittances at these points for one single film can only be measured for different wavelengths because the film is certainly not a halfwave at a wavelength for which it is a quarterwave. Manifacier, Gasiot, and Fillard avoided this problem by drawing two envelope curves around the transmittance characteristic for the film. These envelope curves are then supposed to mark the loci of quarterwave and halfwave points, assuming that the thicknesses of the film were to vary by a small amount. At each wavelength point this gives two values of transmittance corresponding to the two envelopes. These correspond to transmittances of a film thickness equal to an integral number of halfwave or of an odd number of quarterwaves at that particular wavelength. If we indicate these transmittances by T_{max} and T_{min} respectively for a film of high index on a substrate of lower index, then we can write

$$\alpha = \frac{C_1[1 - (T_{max}/T_{min})^{1/2}]}{C_1[1 + (T_{max}/T_{min})^{1/2}]} \quad \text{where} \tag{74}$$

$$\alpha = \exp(-4\pi\, k_f\, d_f/\lambda)$$

$$4\pi\, k_f\, d_f/\lambda = m\pi \text{ (quarter or halfwave thickness)}$$

$$C_1 = (n_f + n_0)(n_{sub} + n_f)$$

$$C_1 = (n_f - n_0)(n_{sub} - n_f) \tag{75}$$

$$T_{max} = 16 n_0\, n_{sub} n_f^2 \alpha/(C_1 + C_2\alpha)^2$$

$$T_{max} = 16 n_0\, n_{sub} n_f^2 \alpha/(C_1 - C_2\alpha)^2$$

Then from Eqs. 74 and 75, if we define N as

$$N = \frac{n_0^2 + n_{\text{sub}}^2}{2} + 2n_0 n_s \frac{T_{\max} - T_{\min}}{T_{\max} T_{\min}} \tag{76}$$

n_f is given by

$$n_f = \left[N + (N^2 - n_0^2 n_s^2)^{1/2} \right]^{1/2} \tag{77}$$

After n_f has been determined, Eq. 74 can be used to find a value for α. The thickness d can then be found from the wavelengths corresponding to the various extrema and the extinction coefficient k_f from the values of d and α. The method has the advantage of explicit expressions for the various quantities, which makes it easily implemented on machines as small as programmable calculators. Unfortunately, as with many of the other techniques, the results can suffer from appreciable errors in the presence of inhomogeneity. It is possible to extend the method to deal with inhomogeneity provided reflectance measurements are included.

Computers provide the advantage that we no longer need to devise methods of optical-constant measurement with the principal objective of ease of calculation. Instead, methods can be chosen simply on the basis of precision of results, regardless of the complexity of the analytical techniques that are required. This is the approach advocated by Hansen [33], who developed a reflectance attachment that makes it possible to measure the reflectance of a thin film for virtually any angle of incidence and plane of polarization. The particular measurements are chosen to suit each individual film.

For rapid, straightforward measurement of refractive index, a method provided by Abelès [34] is useful. The reflectance for p-polarization is the same for substrate and film at an angle of incidence that depends only on the indices of film and incident medium and not at all on either substrate index or film thickness. However, layers that are a halfwave thick at the appropriate angle of incidence and wavelength will give a reflectance equal to that of the uncoated substrate regardless of index. Snell's law and the expressions for p-admittances give

$$\tan \upsilon_0 = n_f / n_0 \tag{78}$$

The measurement of index reduces to the measurement of the angle υ_0 at which the reflectances are equal. Heavens [28] shows that the greatest accuracy of measurement is obtained when the layer is an odd number of quarterwaves thick at the appropriate angle of incidence. This is because there is the greatest difference in the reflectances of the coated and uncoated substrate for a given angular misalignment from the ideal. It is possible to achieve an accuracy of around 0.002 in refractive index if the film and substrate indices are within 0.3 of each other, but not equal. Hacskaylo [35] developed an improved method based on the Abelès technique. It involves incident light that is plane-polarized with the plane of polarization almost but not quite parallel to the plane of incidence. The reflected light is passed through an analyzer and the analyzer angle, for which the reflected light from the uncoated substrate and from the film-coated substrate are equal, is plotted against the angle of incidence. A very sharp zero at the angle satisfying the Abelès condition is obtained, which permits accuracy of 0.0002 to 0.0006 in the measurement of indices in the range 1.2 to 2.3. It is not necessary for the film index to be close to the substrate index.

For inhomogeneous films, provided that the variation of index throughout the film is either a smooth increase or decrease, (so that there are no extrema within the film) we use a very simple technique to determine the difference in behavior at the quarterwave and halfwave points. We assume that the film is absorption-free and that its properties can be calculated by a multiple-beam approach, which considers the amplitude reflection and transmission coefficients at the boundary only. The characteristic matrix for the layer is then given by [36]

$$\begin{bmatrix} (y_b/y_a)^{1/2}\cos\delta & \dfrac{i\sin\delta}{(y_a\,y_b)^{1/2}} \\ i(y_a\,y_b)^{1/2}\sin\delta & (y_a/y_b)^{1/2}\cos\delta \end{bmatrix} \tag{79}$$

Now we consider cases where the layer is either an odd number of quarterwaves or an integral number of halfwaves. We apply the expression in Eq. 79 in the normal way and find the well-known relations

$$R = \left(\frac{y_0 - y_a\,y_b/y_{\text{sub}}}{y_0 + y_a\,y_b/y_{sub}}\right)^2 \quad \text{for a quarterwave, and} \tag{80}$$

$$R = \left(\frac{y_0 - y_a\,y_{\text{sub}}/y_b}{y_0 + y_a\,y_{\text{sub}}/y_b}\right)^2 \quad \text{for a halfwave} \tag{81}$$

The expression for a quarterwave layer is indistinguishable from that of a homogeneous layer of admittance $(y_a y_b)^{1/2}$ and so it is impossible to detect the presence of inhomogeneity from the quarterwave result. The halfwave expression is quite different. Here the layer is no longer an absentee layer and cannot therefore be represented by an equivalent homogeneous layer. The shifting of the reflectance of the halfwave points from the level of the uncoated substrate in absorption-free layers is a sure sign of inhomogeneity and can be used to measure it.

The Hadley method of deriving the optical constants takes no account of inhomogeneity. Therefore any inhomogeneity introduces errors. The Marseille method, however, includes halfwave points and has sufficient information to accommodate inhomogeneity. The matrix expression is a good approximation when the inhomogeneity is not too large and when the admittances y_a and y_b are significantly different from those of substrate and incident medium. To avoid any difficulties resulting from the model, the Marseille group actually uses a model for the layer consisting of ten homogeneous sublayers with linearly varying values of n but identical values of k and thickness d. The halfwave points still give the principal information on the degree of inhomogeneity. They are also affected by the extinction coefficient k and this must also be taken into account. One halfwave point within the region of measurement can be used to measure inhomogeneity that is assumed constant over the rest of the region. Several halfwave points can yield values of inhomogeneity that can be fitted to a Cauchy expression, that is an expression of the form

$$\frac{\Delta n}{n} = A + \frac{B}{\lambda^2} + \frac{C}{\lambda^4} \tag{82}$$

The envelope method has also been extended [27] to deal with inhomogeneous films using the inhomogeneous matrix expression for the calculations. The extinction coefficient k, as in the Marseille method, is assumed constant through the film.

The surface plasmon resonance mentioned earlier is used as a tool for the measurement of the optical properties of metals. The method is particularly attractive because its three parameters, width, angular position, and depth correspond roughly to the three parameters n, k, and d of the metal. There are two solutions that merge into one when the reflectance at the minimum is zero. The thicknesses associated with the two solutions are quite distinct and can be used to determine the correct one, or measurements may be made at two different wavelengths when the correct solutions can be detected by their consistent thickness.

The disadvantages of the technique are that only metals that yield a good resonance are suitable and that the thickness should be such that a reasonable resonance is obtained. A film that has a good resonance and that is well-characterized can be used as a substrate for a very thin metal or dielectric film. The characteristics of the displaced resonance can then be used to extract the optical parameters of the very thin film. This technique has been used in many applications from electrochemistry to biochemistry in the characterization and monitoring of very thin films of all kinds.

Ellipsometry [37] is a technique that measures the ratio of the amplitude coefficients for *p*- and *s*-polarized light at oblique incidence.

$$\frac{\rho_p}{\rho_s} = \left|\frac{\rho_p}{\rho_s}\right| e^{i(\phi_p - \phi_s)} \tag{83}$$

This can be written in terms of two quantities, Ψ and Δ

$$\tan\psi = \left|\frac{\rho_p}{\rho_s}\right| \quad \text{and} \tag{84}$$

$$\Delta = (\phi_p - \phi_s) \tag{85}$$

Note that the convention normal in ellipsometry for ϕ_p is 180° different from the normal thin film convention used in this chapter.

Both ψ and Δ are relative quantities that can be extracted from measurements of the orientation and ellipticity of the polarization of light before and after reflection at oblique incidence from a surface, provided that the beam contains both *s*- and *p*-polarized light. The *s*-polarized reflected beam is effectively used as a reference for the *p*-polarized beam and vice versa so that absolute calibration is not required. The instrument is known as an ellipsometer. There are many different arrangements, some completely automatic and others requiring critical adjustment by the operator. In one of the simplest arrangements linearly polarized light is incident on a quarterwave plate set at 45° to the plane of incidence. The resulting beam is incident on the surface under measurement and then the reflected beam passes through an analyzer. The polarizer and analyzer are then rotated until complete extinction is achieved. The two measurements, the angle of rotation of analyzer and polarizer, are sufficient to permit derivation of ψ and Δ.

For a simple surface that is characterized by optical constants $(n - ik)$ the two quantities ψ and Δ are sufficient to derive both optical constants. A surface covered by a thin film has more parameters, n and k for the substrate and n, k, and d for the film. Clearly there are insufficient measurements for the complete extraction of the parameters. Additional measurements over both a range of wavelengths and a range of angles of incidence can lead to the complete extraction in a process called variable-angle spectral ellipsometry.

Ellipsometry is an exceedingly sensitive and useful technique. The measurements, however, are somewhat remote from the actual parameters that are to be extracted and complete reliance must be placed on their integrity. It is like a process of navigation by dead reckoning. Extreme care is needed in setting up the instrument so that the various angles are completely accurate.

Some values of optical constants corresponding to films of common materials are presented in Table 8-1. Because of differences in preparation conditions, film thickness, microstructure, purity, and even the model used in reducing the measurements, the optical constants of films of identical materials may vary enormously from each other and from the measured bulk values. The figures in the table are, therefore, indications only. No table of values can represent accurately the optical constants of films of particular materials in every case.

TABLE 8-1 Optical constants of films of common materials

	550 nm		1.5 μm		2.5μm	
Material	n	k	n	k	n	k
Ag	0.055	3.320	0.305	10.615	0.833	17.98
Al	0.834	6.033	2.038	12.690	3.250	19.975
Al_2O_3	1.671	0.000	1.651	0.000	1.631	0.000
Au	0.331	2.324	0.360	10.400	0.820	17.300
BaF_2	1.480	0.000	1.480	0.000	1.480	0.000
Bi_2O_3	1.910	0.000	1.910	0.000	1.910	0.000
C	1.838	0.443	2.095	0.085	2.134	0.025
Cr	3.116	4.423	4.259	4.784	3.820	7.638
Cu	0.670	2.863	0.680	10.300	1.278	16.163
Fe	2.891	3.352	3.612	5.471	4.149	8.076
Gd_2O_3	1.925	0.000	1.915	0.000	1.915	0.000
Ge	3.950	1.975	4.850	0.225	4.500	0.000
HfO_2	1.995	0.000	1.971	0.000	1.971	0.000
ITO	2.050	0.014	1.597	0.329	1.474	1.062
LaF_3	1.576	0.000	1.576	0.000	1.576	0.000
MgF_2	1.383	0.000	1.348	0.000	1.312	0.000
Na_3A1F_6	1.35	0.000	1.35	0.000	1.35	0.000
Ni	1.868	3.322	3.319	6.648	4.042	9.680
PbTe	1.751	2.899	4.517	0.358	5.814	0.220
Pd	1.641	3.845	2.905	8.120	4.106	11.461
Pt	2.131	3.715	5.143	7.003	3.775	7.880
Rh	1.967	5.016	3.636	10.031	4.218	16.216
Sb_2O_3	2.037	0.000	1.8740	0.000	1.849	0.000
Si	4.400	0.630	3.635	0.000	3.530	0.000
SiO	2.002	0.0255	1.855	0.000	1.830	0.000
SiO_2	1.460	0.000	1.445	0.000	1.430	0.000
Ta_2O_5	2.140	0.000	2.100	0.000	2.100	0.000
Ti	2.544	3.341	3.680	4.535	4.570	5.390
TiO_2	2.318	0.000	2.250	0.000	2.250	0.000
W	3.236	2.487	2.705	4.850	1.675	8.345
Y_2O_3	1.788	0.000	1.773	0.000	1.773	0.000
ZnS	2.356	0.000	2.234	0.000	2.200	0.000
ZrO_2	2.057	0.000	2.035	0.000	2.035	0.000

Source Thin Film Center Inc.
Note: thin film constants depend on preparation conditions, microstructure, etc., and so these values should be considered as indications only.

8.8 APPLICATION OF OPTICAL THIN FILMS

It is impossible to give a comprehensive list of the applications of optical coatings because there are so many of them. Any optical instrument with a surface that directs or redirects the light will certainly make use of a coating on the surface. We will therefore make a few general statements about applications and consider just a few of them briefly.

A large field of application of optical coatings is in the reduction of glare from video-display units. The source of illumination in these units is usually a phosphor that is stimulated to emit visible radiation by bombarding it with a beam of energetic electrons. The phosphor is coated on the inside of a thick glass plate that is a vacuum window. The outer surface is not normally antireflection-coated and so it reflects an image of the lighting system in the room. Because the window is usually dished, partly to support the atmospheric pressure and partly to ease the requirements on the scanning electron beam, it acts as a convex mirror. Therefore it is very difficult for the viewer to escape specularly reflected glare from one or another part of the screen. An antiglare filter consists of a glass or plastic plate that is fixed over the display unit. The plate is slightly absorbing. Because the glare is derived from light outside the unit, the glare light is attenuated both on its journey into the unit and on its journey out, whereas the signal is attenuated only once. Figure 8-20 may make this clear.

If the glare is denoted by the ratio of the undesired light to the desired light, then a glare without the filter of G will become

$$\text{Glare} = G \cdot \frac{T_f^2}{T_f} = G \cdot T_f \tag{86}$$

This, however, assumes that the brilliance of the display is unchanged, so that the signal light is attenuated by its single passage through the filter. In practice, the brilliance is increased so that the perceived brightness remains at its original level. Then the reduction in glare is even greater.

$$\text{Glare} = G \cdot T_f^2 \tag{87}$$

An internal transmittance of 50%, for example, will then give a four-fold glare reduction. This is a substantial improvement but there is a problem. The surfaces of the filter itself are sources of glare and an antireflection coating is required. The outer surface is the most important and requires a coating that has good performance over the entire visible spectrum. The inner surface is not as important because although it does contribute to the glare, its contribution is attenuated by the filter along with that of the screen and is not always coated, especially if the cost must be kept low.

A similar application occurs in picture frames where the double surface of the glass reflects about 8.5% of the incident light. Because the picture in the frame is not self-luminous, the glare cannot be reduced by decreasing the transmittance of the glass. Both surfaces must be antireflected with a good broadband multilayer coating. An alternative approach reduces the specular reflectance of the surface by making it slightly scattering.

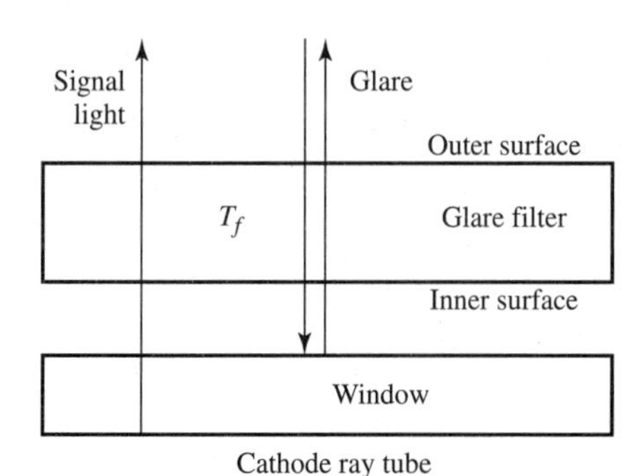

FIGURE 8-20 Arrangement of an antiglare filter in front of a visual display unit. The internal transmittance of the filter is T_f.

The glass must be kept sufficiently close to the picture to avoid any blurring so it is not suitable for most other purposes.

Narrowband filters are used in the improvement of signal-to-noise ratio in instruments where a spectral line, usually from a laser, is used in measurement. The measurement might be of distance, reflectance, direction, or the level of a chemical or pollutant and so on. The transmission characteristic of a single-cavity narrowband filter is roughly that the integrated transmittance as a function of frequency between the 50% transmission points of successive orders is equal to the integrated transmission between the 50% points of a single-transmission peak. This might immediately appear to disqualify the single-cavity filter altogether. In fact, this is virtually the only type of application for which it is most suitable. The line to be observed will normally be derived from a laser or low-pressure discharge lamp and will be very narrow compared with the filter bandwidth. (The line width should not be greater than one third of the width of the filter.) This implies that the filter will pass the line without reduction. If the interfering light is broadband without narrow spectral lines, the amount of background light that is transmitted will be purely a function of the bandwidth of the filter. The narrower the filter is, the smaller the amount of background light. We can assume a uniform distribution of energy in terms of frequency and take the interfering light as that amount between orders.

The transmittance of a single-cavity filter at the peak wavelength is high and the uniformity of transmission is good. Therefore such filters are very suitable for this application. In the visible and near infrared regions the filter is usually backed up by additional blocking with a metal-dielectric filter and an absorption glass filter, so that the transmittance in the rejection region is further reduced. An alternative application of a narrowband filter is as a replacement for a monochromator where the source is essentially white and the filter selects a narrow band of wavelengths to simulate a monochromatic source. Here the single-cavity filter is quite unsuitable and a multiple-cavity filter is required.

The characteristic of a filter changes with the angle of incidence and this affects its usefulness. For small tilts the principal effect is a shift toward shorter wavelengths, but for large tilts significant differences between performance for *p*- and *s*-polarization appear. There may be considerable distortion of the characteristic from that at normal incidence. The changes may be used to advantage in polarization-sensitive coatings. The alternative of achieving similar performance for both *s*- and *p*-polarizations at oblique incidence is very difficult and not usually possible unless the range of wavelengths or angle of incidence is restricted. For narrowband filters the acceptance cone for incident light is limited by the acceptable degradation of the performance. For tilts in collimated light small enough to permit the use of second-order approximations for sine and cosine, the filter characteristic can be assumed to be displaced toward shorter wavelengths by an amount

$$\Delta\lambda = \frac{\upsilon_0^2 \lambda_p}{2\tilde{n}^2} \tag{88}$$

where υ_0 is the angle of incidence in air measured in radians and $\tilde{n}$ is the effective index of the filter that lies between the upper and lower indices of the material of the films and depends on the particular design. For filters in the visible region 1.5 is a reasonable value to assume if the correct value is not known. If the maximum and minimum angles of incidence in the acceptance cone of the filter are used to calculate the maximum and minimum displacement of the filter characteristic, then if the halfwidth of the filter is given by W_0 and the two wavelength shifts are $\Delta\lambda_1$ and $\Delta\lambda_2$, the effective reduction in peak transmittance of a single-cavity filter is approximately

$$\frac{(\Delta\lambda_1 - \Delta\lambda_2)^2}{3W_0^2} \tag{89}$$

and the increased halfwidth

$$\sqrt{W_0^2 + (\Delta\lambda_1 - \Delta\lambda_2)^2} \tag{90}$$

For a total shift of approximately one third of the halfwidth the peak transmittance reduction will be around 4% whereas the halfwidth will be increased by 5%. These are usually acceptable figures. For multiple-cavity filters the performance degradation is less so that Eqs. 89 and 90 represent worst-case estimates. For larger tilts, greater than roughly 15° in air, there is increasingly significant polarization splitting that makes redesign of the filters necessary for most applications.

The shift with tilt can be used to tune the filter over a small spectral region. The tuning is always toward shorter wavelengths.

There are many applications, both accidental and intentional, where strongly colored interference fringes are used in an artistic way. These fringes are usually produced by the growth of an oxide film over a metal and the growth may be induced by heating the metal in air or by anodic oxidation. This works well with some metals but not with others. Aluminum, for example, is almost completely devoid of any colored effects when its oxide film grows. The thermal blueing of steel, however, is well-known and has been practiced for centuries.

For a ten to one ratio of maximum to minimum reflectance in the fringes, we need amplitudes for the two interfering beams that have a ratio smaller than two to one, or individual reflectances in a ratio smaller than four to one (Fig. 8-21). Therefore the metal should have a reflectance that is not more than four times greater than the reflectance of its oxide. For example, an oxide of index 2.0 has a reflectance of just over 11%, implying that the metal should have a reflectance of not greater than 44%. Thus iron, steel, titanium, and copper all have intense colors but silver and aluminum show virtually no colors at all. Titanium is especially popular with artists. Its oxide is of rather higher index than 2.0 and its reflectance is quite low so the fringes are very pronounced.

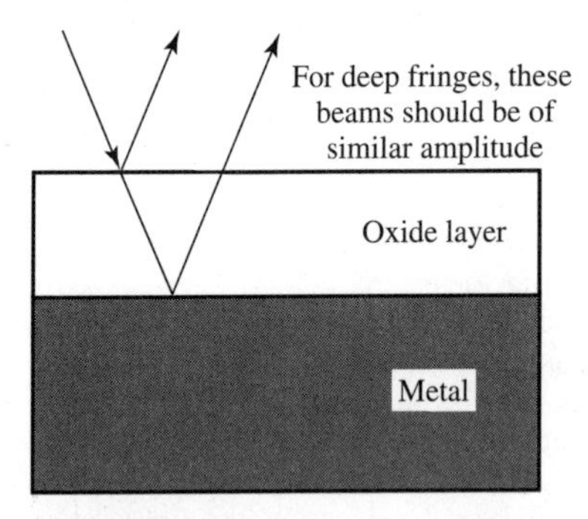

FIGURE 8-21 For pronounced interference fringes where the minimum is near zero, the beams reflected from the front and rear surfaces of the oxide film over a metal should have roughly equal amplitudes.

Metals with higher reflectance can be antireflected if the outer reflectance is increased. This can be achieved by the addition of a thin metal layer. An interposed dielectric layer acts as a phase-matching layer and adjustment of its thickness varies the wavelength corresponding to minimum reflectance (Fig. 8-22). The outer metal layer may be further protected by a dielectric overcoat. Such coatings may be used purely for decorative purposes but also have been used as low-emittance absorbers in solar energy applications. There the minimum reflectance (and maximum absorptance) corresponds to the emission from the sun, whereas the reflectance increase at longer wavelengths reduces the thermal emittance so that the radiation loss from the hot absorber is minimized. Similar coatings may be used in optical data storage. The irradiation of the coating with a high-energy light pulse to write on it causes local heating that damages the coating so that the reflectance locally increases. The phase matching layer may be thermoplastic so that a hole can easily be melted in it.

The surface plasmon resonance has been used in many applications as a sensitive detector of changes at the outer surface of the metal. The study of tarnishing processes, contamination, and biological systems are just some of the applications [38–41]. A great ad-

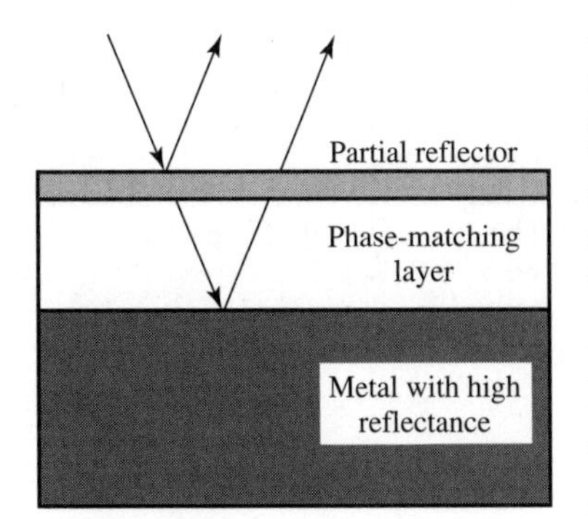

FIGURE 8-22 A metal with high reflectance can be antireflected if the outer reflectance of Figure 8-21 is increased by means of an additional partial reflector, usually a thin metal layer, and suitable adjustment of the dielectric-phase matching layer.

vantage of the method is that the measurement takes place at the side remote from the metal surface where the changes are taking place. The measurements are also straightforward and the extraction of the parameters of the film from the measurements of the resonance is well-conditioned.

In optoelectronics there are several applications related to optical fibers and semiconductor lasers. In the traditional semiconductor laser a junction region is bounded by two parallel facets arranged vertically with respect to the junction region. The two facets form the reflectors of the laser cavity and stimulated emission of recombination radiation in the junction region gives the necessary gain. For the vast majority of such lasers the natural reflectance of the facets is normally acceptable and the facets are simply coated with a halfwave absentee passivating layer. However, there are, some applications where an antireflection coating must be placed on one of the facets so that the device may be mounted in an external cavity. Such antireflection coatings present considerable problems that are not completely resolved. One difficulty is that the junction region is not a single material but is layered like a waveguide, complete with cladding, and the irradiance is distributed over both guide and cladding. A simple coating covers both guide and cladding and has a different performance for each. If the coating is optimized for one area it will not be optimum for the other. We can assume that the admittances involved lie roughly in the region 3.2 to 3.5. A single-layer antireflection coating matched to the geometric mean, 3.347, would give a residual reflectance for each extreme value of 0.05%. This means that reflectances as low as 0.1% are reasonably attainable with a simple coating.

Unfortunately there are requirements for reflectances still lower by at least an order of magnitude. For such coatings the distribution of irradiance in the laser mode must be taken into account, and it has been theoretically shown that reflectances lower than 0.001% are possible [42]. The antireflection of an optical fiber has some of the same characteristics. The mode is contained not only in the core but also in the cladding and a single coating covers both. For exceedingly low reflectances the mode distribution must be taken into account in the coating design. Such coatings require exceedingly tight control of thickness and refractive index.

A vertical-cavity surface-emitting laser (VCSEL) is a multilayer coating with one or more active layers. The earlier structures resembled quarterwave stacks and the buildup and emission of light was through a process known as distributed feedback. The theory of such active devices is beyond the scope of this chapter, but we can understand some of the principal features by thinking of them as simple thin film coatings. The characteristic curve of a quarterwave stack is shown in Fig. 8-9. The distribution of electric field through a quarterwave stack illuminated at the central wavelength ($g = 1$) has the form of a standing wave, increasing exponentially in amplitude toward the front surface. Such a distribution can be maintained only by a strong incident irradiance. It is so completely unlike the field distribution required for a laser that a distributed feedback laser based on such a quarterwave stack would not emit at the central wavelength. However, the field distribution corresponding to wavelengths coinciding with the sharp steep dips on either side of the high-reflectance zone show a low field at either end of the stack rising to a high value in the center. This corresponds very well to the distribution required for a laser. Such structures, therefore, can strongly emit at either of these wavelengths. They are usually composed of alloys of indium, gallium, arsenic, and phosphorus, with admittances around 3.2 to 3.5, and so the admittance contrast is low and the high-reflectance zone is narrow. Therefore the two possible wavelengths for lasing are close together and this creates problems for such devices. A better arrangement that is usually used is to fabricate a structure that is essentially a single-cavity interference filter. This has a natural field distribution at the single, central wavelength that is immediately suitable for lasing.

An excellent review of current problems and solutions in electro-optical devices has been written by Brown [43].

REFERENCES

1. A.M. Portis, *Electromagnetic Fields: Sources and Media,* John Wiley & Sons, New York, 1979.
2. O.S. Heavens and R.W. Ditchburn, *Insight into Optics,* John Wiley & Sons, Chichester, West Sussex, England, 1991.
3. E.D. Palik (ed.), *Handbook of Optical Constants of Solids,* Academic Press, Orlando, Florida, 1985.
4. E.D. Palik, *Handbook of Optical Constants of Solids II,* Academic Press, San Diego and London, 1991.
5. H.A. Macleod, *Antireflection coatings on absorbing substrates,* Society of Vacuum Coaters, 38th Annual Technical Conference, Chicago, 1995, pp. 172–175.
6. Z. Knittl, "Fresnel historique et actuel," *Optica Acta* vol 25, 1978, pp. 167–173.
7. H.A. Macleod, *Thin film optical filters,* Adam Hilger, Bristol, 1986.
8. A.J. Vermeulen, "Some phenomena connected with the optical monitoring of thin film deposition and their application to optical coatings," *Optica Acta* vol 18, 1971, pp. 531–538.
9. R. Lalezari et al., "Measurement of ultralow losses in dielectric mirrors," Optical Society of America, Topical Meeting on Optical Interference Coatings, Tucson, AZ, 1992, pp. 331–333.
10. J.A. Dobrowolski and D. Lowe, "Optical thin film synthesis program based on the use of Fourier transforms," *Appl Opt* vol 17, 1978, pp. 3039–3050.
11. B.G. Bovard, "Rugate filter theory: an overview," *Appl Opt* vol 32, 1993, pp. 5427–5442.
12. A. Thelen, *Design of Optical Interference Coatings,* McGraw-Hill, New York, 1988.
13. S.A. Furman and A.V. Tikhonravov, *Basics of Optics of Multilayer Systems,* Editions Frontières, Gif-sur-Yvette, France, 1992.
14. Z. Knittl, *Optics of Thin Films,* John Wiley & Sons and SNTL, London, New York, Sydney, Toronto, and Prague, 1976.
15. J.D. Rancourt, *Optical Thin Films: Users' Handbook,* Macmillan Publishing Company, New York, 1987.
16. J.L. Vossen and W. Kern, *Thin Film Processes,* Academic Press, New York, San Francisco, and London, 1978.
17. J.L. Vossen and W. Kern, *Thin Film Processes II,* Academic Press, San Diego, 1991.
18. L. Holland, *Vacuum Deposition of Thin Films,* Chapman and Hall, London, 1956.
19. K.-H. Müller, "Model for ion-assisted thin film densification," *J Appl Phys* vol 59, 1986, pp. 2803–2807.
20. J.D. Targove, L.J. Lingg, and H.A. Macleod, "Verification of momentum transfer as the dominant densifying mechanism in ion-assisted deposition," Optical Society of America, Optical Interference Coatings, Tucson, AZ, 1988, pp. 268–271.

21. P.J. Martin and R.P. Netterfield, "Optical films produced by ion-based techniques," in E. Wolf, (ed.) *Progress in Optics,* vol 23, Elsevier Science Publishers BV, 1986, pp. 115–182.
22. U.J. Gibson, "Ion-beam processing of optical thin films," in M.H. Francombe and J.L. Vossen, (eds.) *Physics of Thin Films,* vol 13, Academic Press, 1987, pp. 109–150.
23. M.L. Fulton, "Applications of ion-assisted deposition using a gridless end-Hall ion source for volume manufacturing of thin film optical filters," *Proc SPIE* vol 2253, 1994, pp. 374–393.
24. I.M. Thomas, "Sol-gel coatings for high-power laser optics: past, present, and future," *Proc SPIE* vol 2114, 1993, pp. 232–243.
25. H.K. Pulker, "Characterization of optical thin films," *Appl Opt* vol 18, 1979, pp. 1969–1977.
26. H.K. Pulker, *Coatings on Glass,* Elsevier, Amsterdam, 1984.
27. D. P. Arndt et al., "Multiple determination of the optical constants of thin film coating materials," *Appl Opt* vol 23, 1984, pp. 3571–3596.
28. O. S. Heavens, "Measurement of optical constants of thin films," in G. Hass and R. E. Thun, (eds.), *Physics of Thin Films,* vol 2, Academic Press, New York and London, 1964, pp. 193–238.
29. E. Pelletier, P. Roche, and B. Vidal, "Détermination automatique des constantes optiques et de l'épaisseur de couches minces: application aux couches diélectriques," *Nouv Rev d'Optique* vol 7, 1976, pp. 353–362.
30. J.C. Manifacier, J. Gasiot, and J.P. Fillard, "A simple method for the determination of the optical constants n, k, and the thickness of a weakly absorbing thin film." *J Phys E* vol 9, 1976, pp. 1002–1004.
31. R. Swanepoel, "Determination of the thickness and optical constants of amorphous silicon," *J Phys E* vol 16, 1983, pp. 1214–1222.
32. D. Minkov and R. Swanepoel, "Computerization of the optical characterization of a thin dielectric film," *Opt Eng* vol 32, 1993, pp. 3333–3337.
33. W. Hansen, "Optical characterization of thin films: theory," *J Opt Soc Am* vol 63, 1973, pp. 793–802.
34. F. Abelès, "La détermination de l'indice et de l'épaisseur des couches minces transparentes," *J Phys Rad* vol 11, 1950, pp. 310–314.
35. M. Hacskaylo, "Determination of the refractive index of thin dielectric films," *J Opt Soc Am* vol 54, 1964, pp. 198–203.
36. F. Abelès, "Recherches sur la propagation des ondes électromagnetiques sinusoidales dans les milieus stratifiés," *Ann Phys 12th series,* vol 5, 1950, pp. 706–784.
37. R.M.A. Azzam, "Ellipsometry," in M. Bass, (ed.), *Handbook of Optics,* vol 2, McGraw Hill, New York, 1995, pp. 27.1–27.27.
38. F. Abelès and T. Lopez-Rios, "Ellipsometry with surface plasmons for the investigation of superficial modifications of solid plasmas," in *Polaritons,* proceedings of the First Taormina Research Conference on the Structure of Matter, 1972, Taormina, Italy, E. Burstein and F.D. Martini, (eds.), Pergamon Press, New York, 1974, pp. 241–246.
39. W. Fink, H.v.d. Piepen, and W. Schneider, "Optical surface plasmon resonance as a measurement tool," *Japanese J Appl Phys* vol 14, 1974, pp. 419–423.
40. B. Rothenhäusler and W. Knoll, "Surface-plasmon microscopy," *Nature,* vol 332, 1988, pp. 615–617.
41. Z. Salamon et al., "Conformational changes in rhodopsin probed by surface plasmon resonance spectroscopy," *Biochemistry,* vol 33, 1994, pp. 13706–13711.
42. T. Saitoh, T. Mukai, and O. Mikami, "Theoretical analysis and fabrication of antireflection coatings on laser-diode facets," *J Lightwave Technol* vol LT-3, 1985, pp. 288–293.
43. R.G.W. Brown, "Thin film coatings for optoelectronic devices," in F. R. Flory (ed.), *Thin Films for Optical Systems,* Optical Engineering, vol 49, Marcel Dekker, New York, 1995, pp. 551–574.

CHAPTER 9
THIN FILM PACKAGING AND INTERCONNECT

Dr. Philip Garrou
Dow Chemical

9.1 INTRODUCTION

Since the 1960s most integrated circuits (ICs) have been attached to a lead frame and either encapsulated in a non-hermetic plastic package or sealed into a hermetic ceramic package. These individually packaged chips (single-chip packages) are then interconnected by copper (Cu) traces on a glass-reinforced epoxy laminate printed wiring board (PWB). Although these technologies had served the industry well through the years, it became clear during the mid-1980s that the performance advances inherent in future generations of advanced ICs (large-scale integrated circuits [LS], very large-scale integrated circuits [VLSI] and ultralarge-scale integrated circuits [ULSI]) could not be fully exploited in systems using such conventional packaging and interconnect. For example, at system frequencies of 100 MHz or greater, severe loss of signal or severe signal distortion are observed when traditional packaging and interconnect are used. [1,2]. It has been reported by NTT that a silicon-bipolar-based switching system containing devices with intrinsic speed of 2.2 Gbit/s is slowed when interconnected in single-chip packages on an impedance-controlled PWB to 1.0 Gbit/s [3].

In the past thin film technologies have been relegated to IC fabrication and microwave hybrid applications.

As discussed in the previous chapters of this book, thin film processing utilizes deposition and patterning techniques normally associated with the fabrication of integrated circuits. Metal deposition is carried out by techniques such as evaporation, sputtering, and chemical vapor deposition (CVD). Metal patterning is performed by lithography and dry etching. The dielectrics used to separate conductor layers have traditionally been CVD oxide or spin-on glass (SOG) materials. Such techniques produce very fine features and very high performance, but this performance has a cost penalty when compared to technologies used in the PWB industry. Techniques normally confined to PWB processing (such as electroplating and wet etching) use much less costly equipment. Furthermore, they can be performed on much larger areas (albeit on much larger features), resulting in lower cost per fabricated unit area. Glass-reinforced polymeric resins have also traditionally been used instead of ceramics and/or inorganic dielectrics.

Over the last decade, the IC packaging industry has melded these processes together to create the low-cost, high-density thin film technology that is needed for the future.

9.1.1 High-Density Packaging

The more-recently developed high-density technologies take the best of both thin film IC technologies and PWB technologies. The supporting base for this high-density evolution has been "multichip module" (MCM) technology. Although no universally accepted definition exists, MCM technology involves the interconnecting of bare die on high-density interconnect structures. Figure 9-1 depicts (to scale) the interconnection of four single-chip packages on a printed wiring board compared with the interconnection of the same four bare chips in an MCM package.

Hybrid, laminate, and IC practitioners all developed variations on their technologies to fabricate MCM structures. Initial attempts by industry groups to label such technologies using acronyms such as MCM-C (ceramic), MCM-L (laminate), and MCM-D (deposited thin film dielectric) have been useful; however, they quickly became obsolete as technology advanced into hybridized forms of all three, as shown in Fig. 9-2 [4].

This new technology has evolved to meet different market requirements. Thin film technology can, in fact, be used on a cost-performance basis to package very high input/output (I/O) count single chips; to package either a few chips or to package

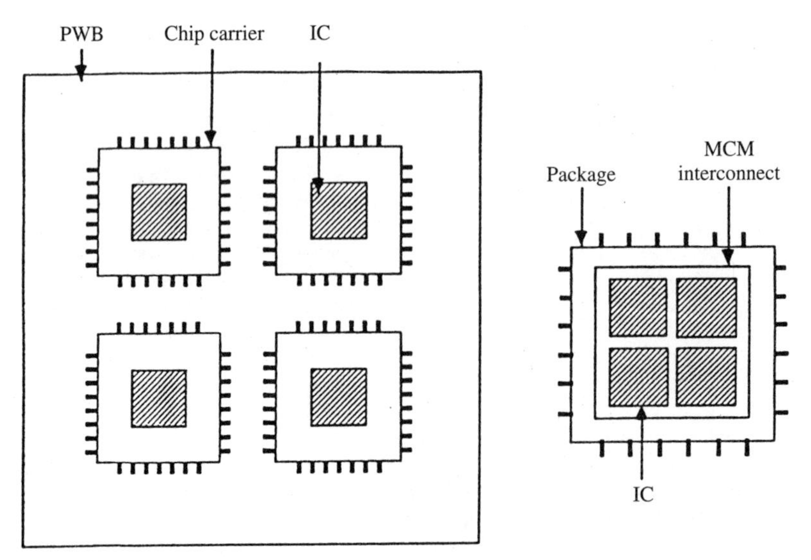

FIGURE 9-1 Interconnection in single-chip versus multichip packaging.

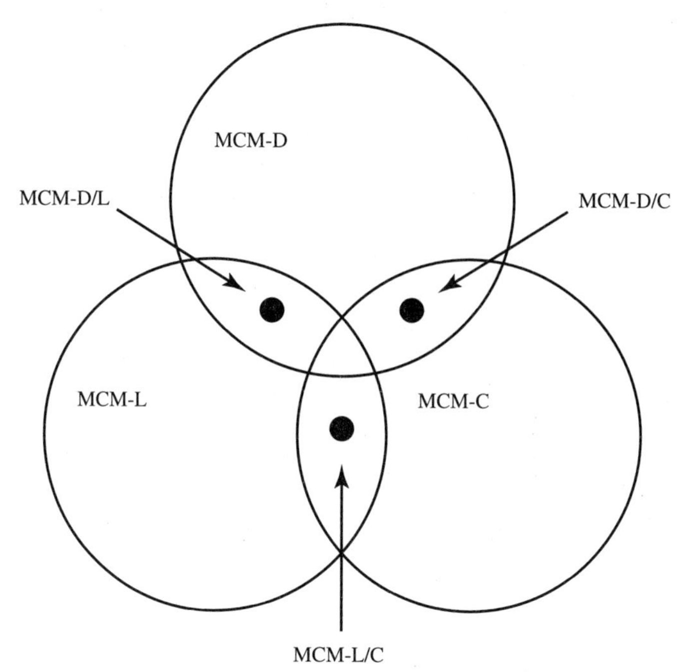

FIGURE 9-2 MCM packaging variations.

>20 chips; and to deliver highly complex military and avionic systems within a single module.

9.2 IC AND SYSTEM DRIVERS

9.2.1 IC Advances

Semiconductor advances are directly related to a continuing decrease in feature size. Smaller feature sizes provide increased gate density, increased number of gates per chip, and increased clock rates. With reduced feature size, devices exhibit reduced parasitics, which allows faster switching. The shorter gate-to-gate distances reduce interconnect delays. Such benefits are offset by an increased number of I/O and an increase in the power that is dissipated by the chip.

As shown in Fig. 9-3, chip integration has increased 35% to 50% per year over the last two decades [5]. The 1 billion-transistor DRAM (dynamic random access memory) is expected early in the next decade. A projection of IC technology (see Fig. 9-4) past the year 2000 was recently developed by the Semiconductor Industry Association. The average die size for DRAM and application specific IC (ASIC) devices is projected to increase to about 1000 mm on a side during the next decade, to accommodate the proposed >1 billion transistors and the 3000+ I/Os. If the industry achieves anything near these levels of integration, the needed I/O for central processor unit (CPU) and other logic chips will require major changes in semiconductor packaging.

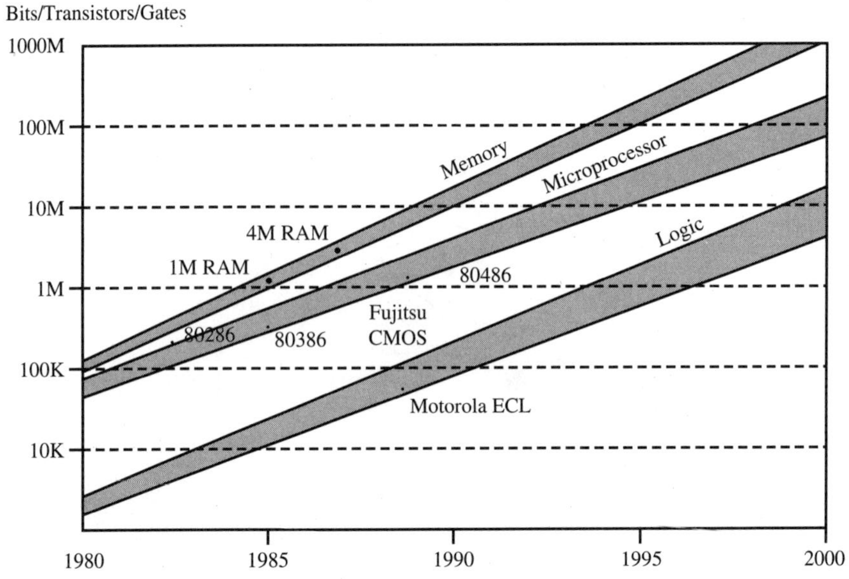

FIGURE 9-3 MOS integration versus time.

With transistor feature size decreasing by one-half every six years (causing I/O requirements to double every three years), the industry is facing severe I/O interconnection constraints. Many VLSI die are already limited by the pitch of the peripheral bond pads (typically 125 μm). Tighter pad pitch on high-I/O die would require expensive high-precision packaging equipment. It appears inevitable that the expected 300 to 500 I/O VLSI die will be interconnected by area (grid) array techniques. These high-I/O area array attachments will require very high-interconnect density signal redistribution layers. Such interconnect technology should deliver shorter signal paths, lower capacitive loads, and reduced circuit noise, as discussed in section 9.3.

9.2.2 System Drivers

The thin film MCM package offers several performance advantages over traditional single-chip packaging. For example, Fig. 9-5 reveals that circuit accessibility is improved dramatically by placing the chips closer together. For a large functional partition with multiple chip crossings, a high circuit accessibility will result in a faster cycle time [2].

9.2.3 Size and Weight Reduction

Figure 9-6 depicts the transformation of a Hughes system from dual inline package (DIP) packaged chips interconnected on 28-PWB boards to equivalent systems packaged by surface mount technology (SMT) and high-density multilayer interconnect (MCM) [6]. The gains in volume, and therefore weight, are apparent.

Similar results have been reported for telecommunication applications [7].

THE NEW SIA ROADMAP

	1995	1998	2001	2004	2007	2010
Minimum Feature Size (microns)						
	0.35	0.25	0.18	0.13	0.10	0.07
DRAM Bits/Chip						
	64M	256M	1G	4G	16G	64G
Max. No. of Input/Output Contacts						
Chip	900	1350	2000	2600	3600	4800
Package	750	1100	1700	2200	3000	4000
Package Cost (cents/pin)						
	1.4	1.3	1.1	1.0	0.9	0.8
μP On-Chip Clock Frequency (MHz)						
	150-300	200-450	300-600	400-800	500-1000	625-1100
Chip Size (mm^2)						
DRAM	190	280	420	640	960	1400
μP	250	300	360	430	520	620
ASIC	450	660	750	900	1100	1400
μP On-Chip Wiring Levels						
	4-5	5	5-6	6	6-7	7-8
Minimum No. of Logic Mask Levels						
	18	20	20	22	22	24
Wafer Diameter (mm)						
	200	200	300	300	400	400
Desktop μP Chip Voltage						
	3.3	2.5	1.8	1.5	1.2	0.9
High Performance μP Power Dissipation (W)						
	80	100	120	140	160	180

Source: SIA

FIGURE 9-4 SIA IC roadmap.

9.2.4 Comparison of Wiring Capability

Table 9-1 shows a comparison of current and expected wiring capabilities of the different MCM classifications [8]. For example, eight layers of PWB with six lines per channel (of 100 mil) is capable of 480 in/in^2 (200 cm/cm^2) of wiring. This is increased significantly if thin film redistribution is incorporated (MCM-LD). MCM-D offers the highest

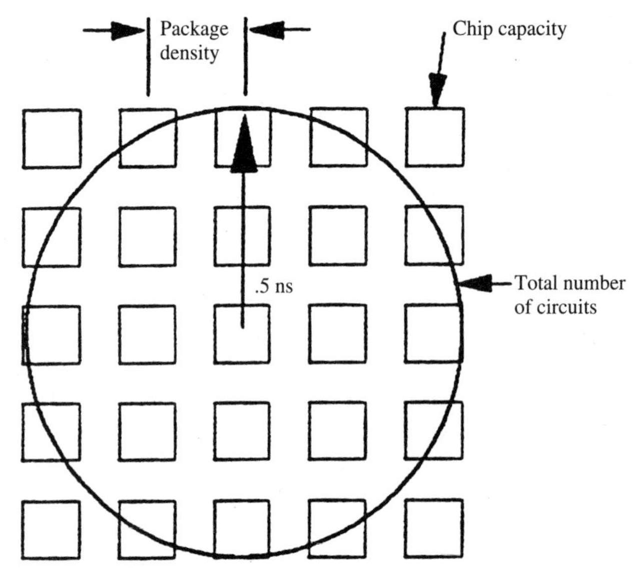

FIGURE 9-5 Improved circuit accessibility in an MCM [2].

interconnect density per layer, but MCM-C offers the overall highest wiring capability because 40 or more layers can be fabricated.

It is clear that commercial applications, whether consumer, computer, or telecommunication, all require the same type of advances: they need to be faster, smaller, lighter, and more reliable. All this is expected, of course, at an equal or (preferably) lower cost. Only technologies that meet these criteria will be implemented into production in the late 1990s.

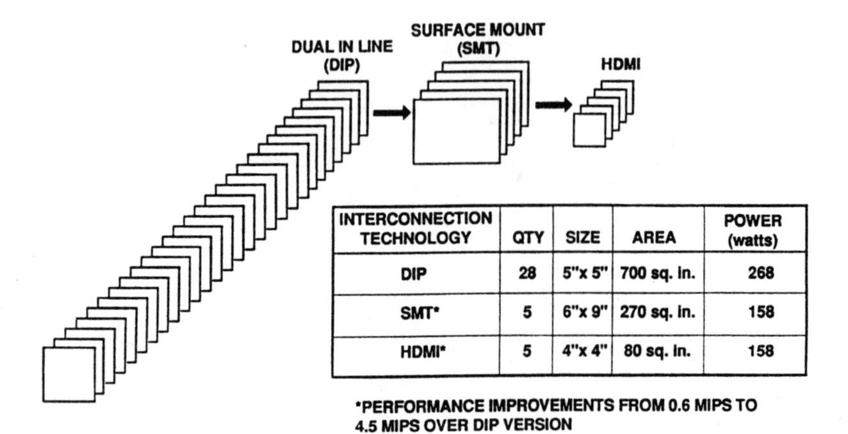

INTERCONNECTION TECHNOLOGY	QTY	SIZE	AREA	POWER (watts)
DIP	28	5"x 5"	700 sq. in.	268
SMT*	5	6"x 9"	270 sq. in.	158
HDMI*	5	4"x 4"	80 sq. in.	158

*PERFORMANCE IMPROVEMENTS FROM 0.6 MIPS TO 4.5 MIPS OVER DIP VERSION

FIGURE 9-6 Advantages of higher-density packaging [6].

TABLE 9-1 Wiring capability of typical MCM technologies [8]

	Typical wiring capability (in/in²)	Typical signal planes	Total wiring capability (in/in²)
MCM-L	(Std.) 30	4	120
		8	240
	(Extended) 60	4	240
		8	480
	"LD" thin film 250	2	500
		4	1000
MCM-C	(Std.) 50	10	500
		40	2000
	(Extended) 100	10	1000
		40	4000
MCM-D	(Std.) 300	1	300
		2	600
	(Extended) 1000	1	1000
		2	2000

9.3 ELECTRICAL CHARACTERISTICS OF THIN FILM CONNECTIONS

The electrical characteristics of a high-density thin film interconnect are dependent on the feature geometries and materials of construction that are chosen.

9.3.1 Conductor Resistivity

The dc resistance of an interconnect line is given by Eq. 1:

$$R = \frac{\rho L}{wt} \tag{1}$$

where R = resistivity of the trace
ρ = conductor resistivity
L = trace length
t = conductor thickness
w = width

Thus a thin narrow line typical of thin film interconnect will have a higher resistance. This will be offset, however, by the shorter conductor trace lengths inherent to such constructions. (The resistivities of typical conductor metallizations are shown later in Table 9-5).

9.3.2 Skin Effect

At high frequencies current tends to flow more toward the surface of the conductor trace closest to the ground plane. This is known as the skin effect. For example, the skin depth for Cu at 1 GHz is about 2 μm. If the thickness of a Cu trace were thinner than 2 μm for a signal propagating at 1 GHz, severe signal degradation would occur. This is also an issue if highly resistive metals such as chromium (Cr) or titanium/tungsten (Ti/W) are used as a barrier/adhesion layer for a signal trace (see section 9.4.9), because the percentage of the signal propagating in the barrier layer will be skewed from the ratio of the thickness of the barrier layer to the conductor metallization, the skewing being worse as the frequency increases [9,10].

9.3.3 Propagation Velocity and Signal Loss

The propagation velocity (V) of a signal in a thin film interconnect is given by Eq. 2:

$$V = \frac{c}{\sqrt{\epsilon_r}} \tag{2}$$

where c is the speed of light and ϵ_r is the dielectric constant of the insulator layer. Therefore, the lower the dielectric constant of the insulating layer, the faster the signal will propagate [11]. (Dielectric constants for thin film dielectric materials are shown later in Table 9-9, section 9.4.)

Dissipation factor (ϵ'') and loss tangent (ϵ_r/ϵ'') are also important electrical parameters, especially at high frequencies. Low values are indicative of minimal conversion of electrical energy to heat and little overall signal loss—both being important factors. The problem of signal attenuation is particularly important in the interconnection of low-power complementary metal oxide semiconductor (CMOS) devices when signal loss cannot be tolerated.

The dielectric strength/breakdown voltage defines the applied voltage that causes current to flow through the insulator. (The materials listed in Table 9-9 all have values in the range 10^6 V/cm.)

9.3.4 Crosstalk and Impedance

High-speed conductor paths must be thought of as transmission lines, not simple resistive/capacitive (RC) elements.

When the time delay of a signal is greater than the rise time of the pulse, full signal voltage won't be achieved by the time the pulse reaches its destination. Reflection will occur, therefore, causing noise and signal ringing. To maximize signal transfer and minimize reflections, one must control impedance. Impedance must be matched to device I/O and must remain stable over the entire length of the trace to ensure good signal-to-noise ratio. Impedance is controlled by the thickness and dielectric constant of the insulator and by the width and spacing of the signal lines.

While the conductor cross section determines the dc resistance, the spacing between the conductors determines the cross talk. The noise induced by cross talk on unswitched lines is particularly troublesome in emitter-coupled logic (ECL) circuits, in which a number of lines are switched simultaneously with very short rise times. Acceptable cross talk can be obtained by placing the ground plane closer to the line than the line-to-line separation. However, close spacing between ground plane and line increases capacitance. Higher capacitance negatively affects the characteristic impedance (Z), as expressed in Eq. 3:

$$Z = \sqrt{\frac{L}{C}} \tag{3}$$

where L and C are the inductance and capacitance per unit length, respectively. Capacitance is defined by Eq. 4:

$$C = \epsilon_r \frac{W}{t} \tag{4}$$

where ϵ_r = dielectric constant
W = interconnect width
t = dielectric thickness

Geometries that produce high characteristic impedance have low capacitance, which minimizes delay time and minimizes the power required to charge a conductor line. Interconnect capacitance reduces the maximum clock rate for interconnects that are not terminated transmission lines.

Sullivan and co-workers noted that significant capacitance exists on VLSI devices. It can be removed by interconnecting chips on thin film MCM interconnects versus interconnecting single-chip packages on a PWB, as shown in Table 9-2 [12]. This reduction in capacitance would only occur for nets interconnecting on the MCM.

This is exemplified in a comparison study reported by Gustafsson and co-workers [13]. Table 9-3 compares the capacitive load for a processor application using tape automated bonding (TAB) connected circuits for the processor chip and memories on a high-density thin film substrate compared with a standard solution using pin grid array (PGA) packages. The increased packaging density for the thin film solution gives a lower capacitive load, which means that the power dissipation decreases from 16 to 10 W. They note that if the drivers on the CMOS chips had been designed to be used in a thin film packaging solution, the power dissipation would have been reduced even further.

TABLE 9-2 Capacitance of CMOS I/O pads in PWB and thin film MCM environments [12]

	PWB	Thin film MCM
Bond pad	1.0 pF	0.1 pF
ESD diodes	1.0 pF	0.0 pF
Driver capacitance	1.5 pF	0.2 pF
Total	3.5 pF	0.3 pF

TABLE 9-3 Comparison of processor application using thin film MCM versus single chip package/PWB solutions [13]

	PGA + SMT on PWB	TAB + WB on thin film
Layout size	200 cm^2	50 cm^2
Critical conductor propagation delay	2.6 ns	0.8 ns
Critical conductor capacitive load	67 pF	42 pF
Estimated power dissipation at 33 MHz	16 W	10 W
Package + interconnect contribution to cycle time at 33 MHZ	8%	2.5%

9.4 MATERIALS OF CONSTRUCTION

9.4.1 Carrier Substrates

The interconnect carrier substrate should offer a flat and highly polished surface to build upon, because the subsequent thin film fabrication requires the use of lithographic techniques.

Irregularities in a surface may be thought of in the three categories, as depicted in Fig. 9-7; namely, roughness, waviness, and flatness. Surface roughness (h) is reported as an average value, either as root mean square (RMS) or centerline average (CLA). By taking an average, one must keep in mind that the average value can mask a single deep peak or valley(s) that could cause discontinuities in deposited thin films.

Lack of adequate flatness (h3) will interfere with thermal contact for heating and cooling and will distort patterns during exposure.

The substrate should be inert to the process chemicals, gas atmospheres, and temperatures used during the fabrication of the interconnect. Modulus is important because the substrate must be strong enough to withstand handling, thermal cycling, and thermal shock.

The carrier substrates should meet certain CTE constraints because it must interface with both silicon chips and the next level of packaging, which is very likely a PWB. Thermal conductivity is also important because a mechanism to conduct the heat away from the closely spaced, heat-generating semiconductor chips is required.

Four classes of materials are being used today: silicon wafers, metals, ceramics, and laminates. Each carrier material has a unique set of properties in terms of the required surface roughness, flatness, mechanical and thermal stability, and/or compatibility with subsequent thin film processing. Table 9-4 presents some relevant electrical, physical, and mechanical properties of substrate materials now being used or being considered for use.

Silicon. As a substrate material, silicon has many advantages [14]. It has an exact coefficient of thermal expansion (CTE) match to the silicon chips mounted on it. The thermal conductivity of silicon is 10 times higher than conventional alumina. Silicon wafers are extremely flat, smooth, and dimensionally stable. They allow the fabrication of integral decoupling capacitors in the substrate itself [15].

Low-cost, highly polished wafers are compatible with automated IC processing equipment. When silicon wafers are used as the substrate, the power, ground, and signal layers are built on top of the substrate. The resulting circuitized substrate must be placed into a package for connection to the next level of the system.

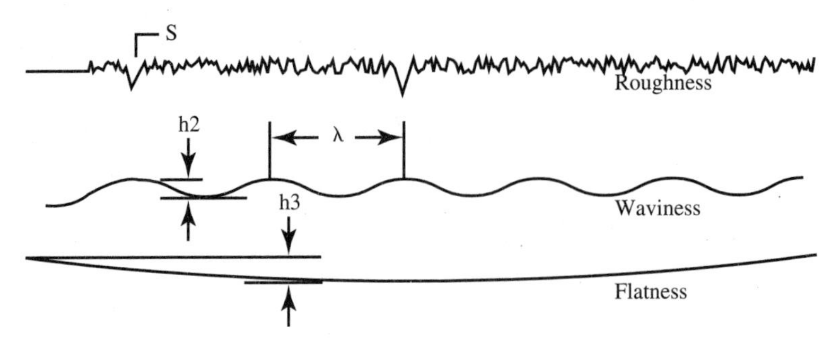

FIGURE 9-7 Characteristics of surface finish.

TABLE 9-4 Properties of MCM carrier materials

Substrate	CTE (ppm/°C)	TC (W/m·°K)	ϵ_r	Flex mod (kpsi)
Si	2.7-3.5	150	11.7	2
Al	22-24	238		
Cu	17	400		
Cu/Mo/Cu	5.1	166(x,y) 151 (z)		
$Al_2 0_3$ (96%)	6.0-7.7	20	9	30-50
$Al_2 0_3$ (99.6)	6.5	27	10	30-40
Mullite	4-5	4-7	5.5-6.5	18-40
BeO	8	250	6.7	20
AlN	4.1	175-200+	8.5	40-50

Problems associated with the use of silicon wafers include poor flexural strength, a relatively high dielectric constant, and excessive warping during processing caused by the stresses imposed by the dielectric and metallization layers. In general, warpage three times that of the same thickness ceramic substrate can be expected when a dielectric/metal thin film structure is deposited on it because of the lower Young's modulus of the silicon carrier [19].

Metallic Substrates. Metallic substrates optimize the thermal and mechanical properties while minimizing substrate raw material and processing cost. Metallic substrates can easily be manufactured in 400 mm^2 sizes or cut to the shape of a silicon wafer to be accepted on IC processing equipment. Metallic carriers have excellent thermal properties, but can react with common processing chemicals such as the common wet etchants used to define conductor patterns.

DEC, and later Micro Module Systems (MMS), have commercialized technology based on aluminum (Al) carriers [20,21].

Ceramic Carrier Technology. Many ceramic materials have been examined or are under consideration as carriers, including alumina-based ceramics, berillia, and aluminum nitride, as shown in Table 9-4. Ceramics can be used in two distinct ways: (1) as a mechanical support upon which the power, ground, and signal layers are fabricated by thin film techniques; or (2) as a co-fired ceramic structure that contains power and ground distribution planes and some of the signal layers, and upon which thin film signal layers and/or chip signal redistribution layers are fabricated.

Ceramic processing can be categorized in several ways, including thick film versus thin film and high-temperature versus low-temperature co-fire. Thick film processes, such as doctor blading of dielectric and conductor layers, and co-firing processes, such as lamination of individual green sheets, use materials with coarse granularity that are not easily adapted to thin film processing [19]. Thin film processes require much smoother surface finishes to meet the requirements of photolithography.

High-temperature co-fired parts require high-temperature firing because of their high ceramic content. This necessitates the use of refractory metallizations that are inherently more resistive. Low-temperature co-fired parts have higher glass content, and, thus, they can be fired at lower temperatures, with higher conductivity metallizations.

To apply thin film to metallized ceramic substrates, the carrier must present a flat, smooth surface. The topographies of ceramic carriers are rough because of distortions that occur during sintering, inherent ceramic porosity, and the protruding top-surface

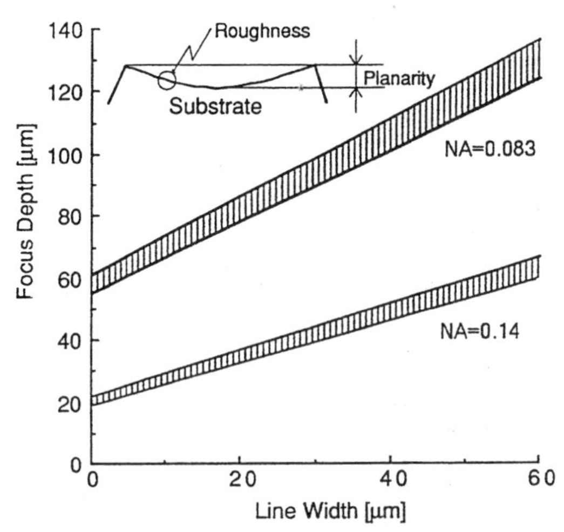

FIGURE 9-8 Relationship between linewidth, focus depth, and substrate roughness [22].

metallurgy. This may exceed the photolithographic depth of focus. The relationship between line width and projection-aligner focus depth (with different numerical apertures) is shown in Fig. 9-8. To define patterns on rough surfaces, a larger focus depth is required. This in turn inhibits fine line definition. As a rule of thumb, the substrate should have a roughness of less than 1000 Å and should have a flatness of 10 μm or less to be suitable for thin film lithography [21].

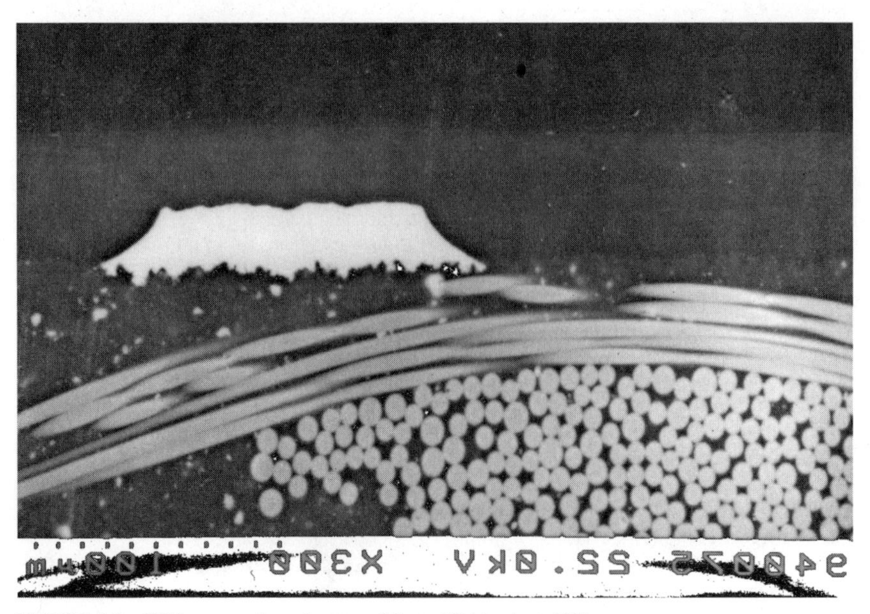

FIGURE 9-9 SEM cross section of patterned Cu on FR4 laminate [22].

The lapping and polishing needed for the desired post-fired surface finish has a large role in determining the cost of the co-fired carrier. In many cases, the rougher finish of the co-fired ceramic versus silicon carriers can be circumvented by an initial polymer planarization coat before the deposition of thin film metallization [6].

Laminate Substrates. The copper-clad glass-reinforced laminates most commonly used in the PWB industry have much rougher surface finishes than conventional MCM-D substrates such as silicon or polished ceramic. This roughness is the result of two principal factors; the use of woven glass fabric in the laminate and the imprint of the copper tooth in the laminate during lamination. Typical surface roughnesses of glass-reinforced laminate after copper etch is shown in Fig. 9-9 [22].

More recently, laminates have been developed that are based on non-woven oriented glass fibers or chopped random fibers, which exhibit considerably less surface topography than that observed with woven-glass reinforced laminates.

Aramid paper-reinforced laminates such as TL-01 (commercialized by Tejin) have TCE values of 6 ppm and a high softening point (Tg) of 194°C [23]. Such Aramid-based laminates exhibit outstanding dimensional stability and a smooth surface, as shown in Fig. 9-10. It is formed by impregnating a non-woven aramid fabric with

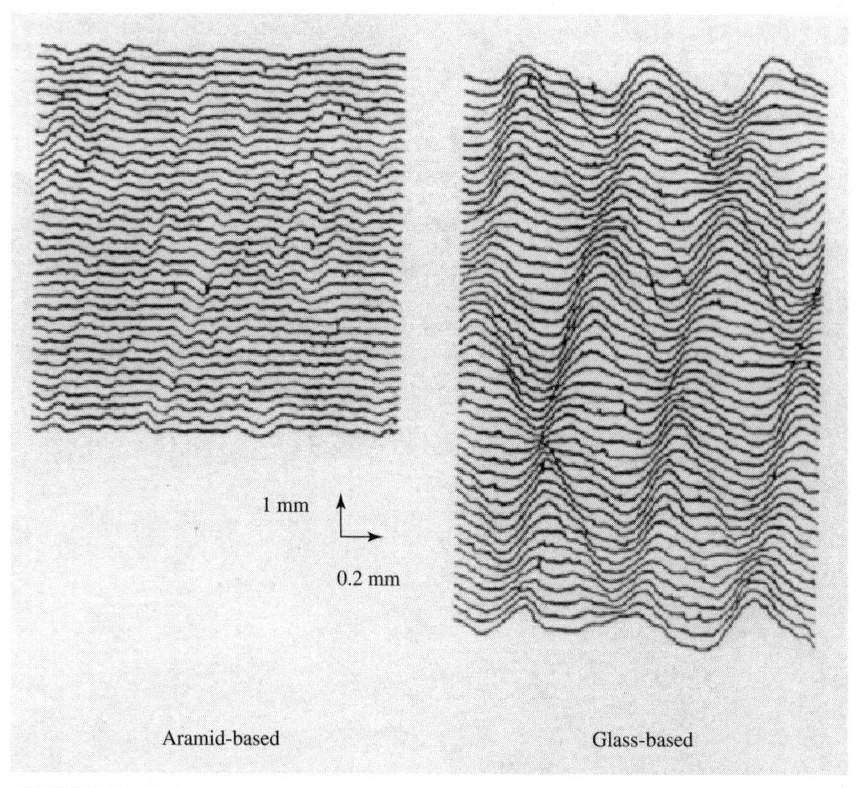

FIGURE 9-10 Surface roughness comparison between aramid and glass weave reinforced laminates.

epoxy resins. A similar material, known as Thermount™, has been developed by DuPont [24].

Additionally finer-tooth copper is also being used to satisfy the need for thinner foils with tighter control of thickness. In addition, the thicker copper foils (17- to 35-μm thick) used on laminates result in significant topographical variation.

The degree of warpage of a PWB base substrate is dependent on several factors, such as the overall thickness of the base substrate, the construction (whether balanced or imbalanced), and the distribution of the copper at various layers in the board construction. Warpage values of approximately 5.0 to 7.5 μm per mm are common. In order to compensate for this high degree of warpage, contact lithography is most commonly used in the PWB industry, with heavy glass panels used to sandwich the substrate and artwork together, thereby providing intimate contact between the mask and substrate. Additionally, because automated in-line processing equipment is used extensively throughout the PWB fabrication process with roller driven feeds to assist in transfer of panels through the in-line processing cells, these levels of warpage can generally be accommodated. It should be noted, however, that deposited dielectrics will, in general, increase substrate warpage when using one sided constructions.

When a laminate interconnection medium is used as the base substrate, deposited dielectrics are selected that are curable at temperatures at or below the Tg of the laminate system used (see section 9.4.3, Polymers).

9.4.2 Metallization

Interconnect Metallization. Table 9-5 presents the resistivity (ρ), CTE, and thermal conductivity (TC) of the metals typically used to fabricate thin film packaging. The specific conductor material chosen will depend on the design, the electrical requirements, and the process that is chosen to fabricate the MCM.

Aluminum is being used by many commercial organizations for thin film interconnect (see Table 9-15, section 9.5). It is a low-cost material, has adequate conductivity, and can be deposited and patterned by typical IC techniques. It is also relatively resistant to further oxidation. It can be sputtered or evaporated, but it cannot be electroplated.

Copper offers significant conductivity improvements over aluminum, and it is more resistant to electromigration. Copper can be deposited by sputtering, evaporation, electroplating, and electroless plating. Copper rapidly oxidizes, and its oxides have poor adhesion

TABLE 9-5 Physical properties of metals used in the fabrication of thin film structures

Metal	ρ ($\mu\cdot\Omega\cdot$cm)	CTE (10^{-6}/°C)	TC (W/cm·°C)
Cu	1.67	18	~400
Al	2.6-4.3	23	240
Au	2.3	14	300
Ni	6.9	13	90
W	5.5	45	200
Cr	13-20	63	66
Ti	55	90	22
Pt	10.5	90	70
Pd	11	110	70

to interfacing polymer dielectric insulators. A barrier/adhesion layer of Cr or Ti is required when interfacing polyimide (PI) dielectrics whereas this is not necessary with benzocyclobutene (BCBs) (see section 9.4.4).

Gold (Au) is used in thin film structures to minimize via contact resistance problems caused by oxidation of Cu and Al. Gold can be deposited by sputtering, evaporation, electroplating, or electroless plating. Although its resistivity is excellent, its cost is high. Adhesion to Au is usually poor, so it requires an adhesion layer (50 to 200 nm of Ti or Ti/W are typical).

Pad-level metallization not only must be conductive and show good adhesion to the dielectric, but also must be compatible with the chosen assembly process. Pad-level metal is usually finished in Au because of its corrosion resistance. It normally requires adhesion/barrier layers such as nickel (Ni), Cr, or Ti/W. One to three μm of gold is needed for wire bonding or TAB, whereas only a flash is required for a soldering surface.

Barrier Metallization/Adhesion Layers. Barrier metallization such as Cr or Ti is required for structures containing Cu conductor and polyimide dielectrics to achieve good adhesion between the Cu and the PI and to prevent chemical attack on the copper by the PI—more specifically, the polyamic acid precursor (see section 9.4.3, Polymers). The barrier layer also serves to protect the underlying copper during subsequent oxygen RIE.

Chromium is known to be a strong oxygen attractor; thus it is necessary to reduce the Cr thickness in order to reduce interface contact resistance. A layer of at least 100 Å is needed to get continuous film coverage.

During any etching step, the Cu/Pl nodules or deposits impede clearing of the vias and, thus, the formation of low resistance contacts. Cr/Cu/Cr are subtractively etched in separate Cr and Cu etching solutions. In this process, the side walls of the Cu are left bare, (not covered with Cr), therefore the copper is still subject to interaction with the PI or to oxidation.

A short ion beam etching step can also be employed to remove upper layer of Cr (a few hundred Å) before via formation [25].

Metal Deposition Techniques. Metallization for thin film packaging is deposited by thin film vacuum processes, such as evaporation or sputtering, or by wet processes, such as electrolytic or electroless plating.

Film thickness uniformity is the key to repeatable manufacturing of thin film conductor patterns. In general, vacuum deposition processes result in more uniform films than electroplating processes.

Evaporation. Metal deposition using evaporation involves the thermal heating of the metal in a vacuum and the subsequent condensation of the metal onto a substrate. Conventional thermal evaporation is a low-energy process. Impact onto the substrate is mild; therefore, adhesion to smooth surfaces is usually poor. Since the deposition is a direct line-of-sight process, the metal is deposited with an angle of incidence near 90°. This is advantageous for lift-off processing, as will be discussed following. For common evaporation sources such as E-beam guns, deposition occurs in an upward vertical direction, requiring that substrates be suspended in a tooling fixture. Evaporators usually require manual loading and are not easily adapted to automated wafer handling.

Sputtering. Metal deposition by sputtering involves bombardment of a metal target source by energetic particles, typically argon ions (Ar^+) in a plasma. Metal atoms are dislodged from the target and deposited onto a substrate. Substrate heating can be a problem because of short target-to-substrate distances, but adhesion to a variety of substrates is inherently better. Direct current magnetrons are common sputtering sources that reduce the problem of substrate heating by magnetic confinement of electrons in the plasma.

Sputtering usually suffers from low deposition rates and poor target utilization because of uneven flux distribution, which is important for gold deposition. Sputtering is adaptable to automation. It has been reported that stress in sputtered metals can be controlled by controlling the process variables [26,27]. Sputtering tends to be a conformal process that results in very good step coverage of features.

Contamination must be removed from the substrate surface to ensure metal adhesion at the interface. Ion milling of RF sputtering is used to remove several hundred Å of thickness from the surface of the substrate before coating. This process removes organic contaminants and, in some cases, is thought to cause surface roughening, which aids in adhesion.

Electroplating. Electroplating involves the selective deposition of ions from a plating solution. An electrical potential is applied to a previously deposited seed layer to initiate the electrochemical reaction. The substrate acts as the anode, and metal ions are reduced and deposited on exposed regions of the seed layer (usually through a patterned photoresist mask). Electroplating techniques require tight process control. The bath must be continually monitored to maintain reproducible plating conditions. Impurities in the plating baths result in poor adhesion, pitting, stress, cracking, and other undesirable attributes. Plating rate depends on bath flow conditions, which affect the concentration of metallic ions near the substrate surface. Many factors can influence the plating uniformity, including current density, bath agitation, anode/cathode spacing, and rack design. Motorola has reported that a thickness variation of less than 10% can be achieved over 16-inch^2 substrates [28].

Non-uniform deposition rates can present a processing problem. Variations in current density caused by different dimensions on the pattern being plated will result in greater deposition on finer features.

The main advantage of electroplating is a high deposition rate, typically 1 μm per minute.

Since Cr, Cu, Au and Ni can be electroplated, surface pad metallurgy such as Cu-Ni-Au can be created by plate up technology. Thin gold (1500 Å) is typically used for soldered components, while thicker gold (>1 μm) is needed for bonded interconnects.

Electroless Plating. Unlike electroplating, which requires a continuous metallic base held at a given current or voltage, electroless plating requires only a catalytic seed layer without electrical contacts. The metal is built up by selective deposition onto the catalytic seed layer, through the reduction of metallic ions in aqueous solution.

Electroless plating solutions for Cu, Au, and Ni are available. Nickel provides a high deposition rate, but it has low conductivity. Gold typically exhibits a slow deposition rate of approximately <2 μm/hr.

The electroless Cu deposition reaction is based on the reduction of Cu ions in an alkaline solution of formaldehyde and hydroxyl ions. The overall reaction requires a solution pH in the range of 11.5 to 13 at a plating temperature between 40°C and 70°C. PIs are not chemically stable in such highly alkaline solutions; thus, bulk stability and interface stability problems have been reported [29]. Surfaces need to be activated (usually with a palladium salt) to initiate selective deposition.

Metal Stress. Total metal film stress is comprised of thermal and intrinsic stress. Thermal stress results from the thermal coefficient of expansion mismatch between the deposited film and the substrate. The level of intrinsic stress and whether it is tensile or compressive stress is influenced by system geometry and operating conditions as much as by process factors [26]. Stress increases as the coating gets thicker.

Windischmann has pointed out that ion bombardment during sputter deposition can alter microstructure, resulting in coating density changes and, under specific conditions, reversal from tensile to compressive stress [27]. Metal stress can be addressed through process parameters such as gas type and pressure, temperature, substrate composition, and deposition rate.

Metal Definition Techniques

Subtractive Definition. In subtractive definition, a blanket layer of metal is deposited by one of the techniques described previously. The metal layer is then masked and exposed areas are removed by wet chemical etching. Because wet etching is isotropic, the technique works best for patterns where the width of the feature is several times the thickness. When common metal etchants that are high in acid concentration are used for subtractive etching, the fine lines created for the interconnect are severely undercut and might be completely removed. Less aggressive etchants such as ammonium persulfate are reported to result in approximately 4% reduction in line width for 25-μm lines [28]. For high aspect ratio features, anisotropic dry etching such as RIE is necessary.

Because of the isotropic nature of subtractive etching, the resist mask is sized smaller than the desired metal feature dimensions to compensate for the undercutting during etching. Subtractive etching of aluminum has less undercut than copper, resulting in better line profile. Differences in center-to-edge dimensions across a substrate are usually compensated for by overetching. This becomes more important as the substrate size increases.

The primary defect type observed during subtractive etching of copper and aluminum is electrical shorts caused by resist flaws and contamination.

Additive Electroplating. Additive electroplating is feasible for both copper and gold wiring.

In a conventional conformal via electroplating process, a blanket layer of metal is deposited as the plating base. The metal base is patterned using a photoresist and the desired features are plated up in the exposed regions of the photoresist. Electrical isolation of the features is achieved by subsequent wet etching of the plating seed layer, as depicted in Fig. 9-11.

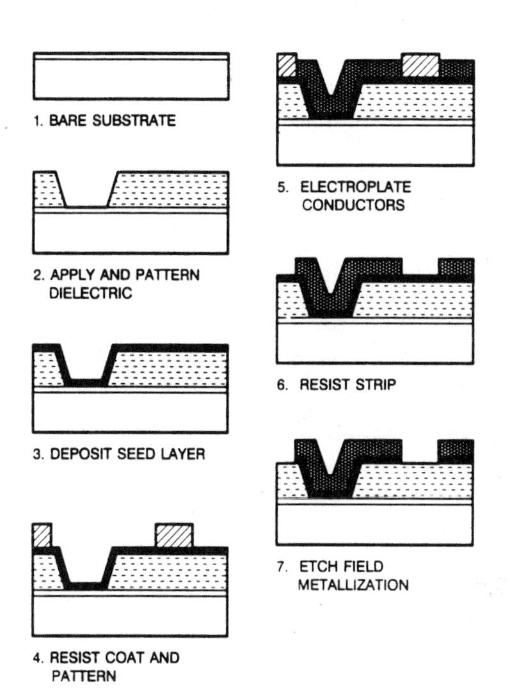

FIGURE 9-11 Process flow for conformal via additive electroplating process.

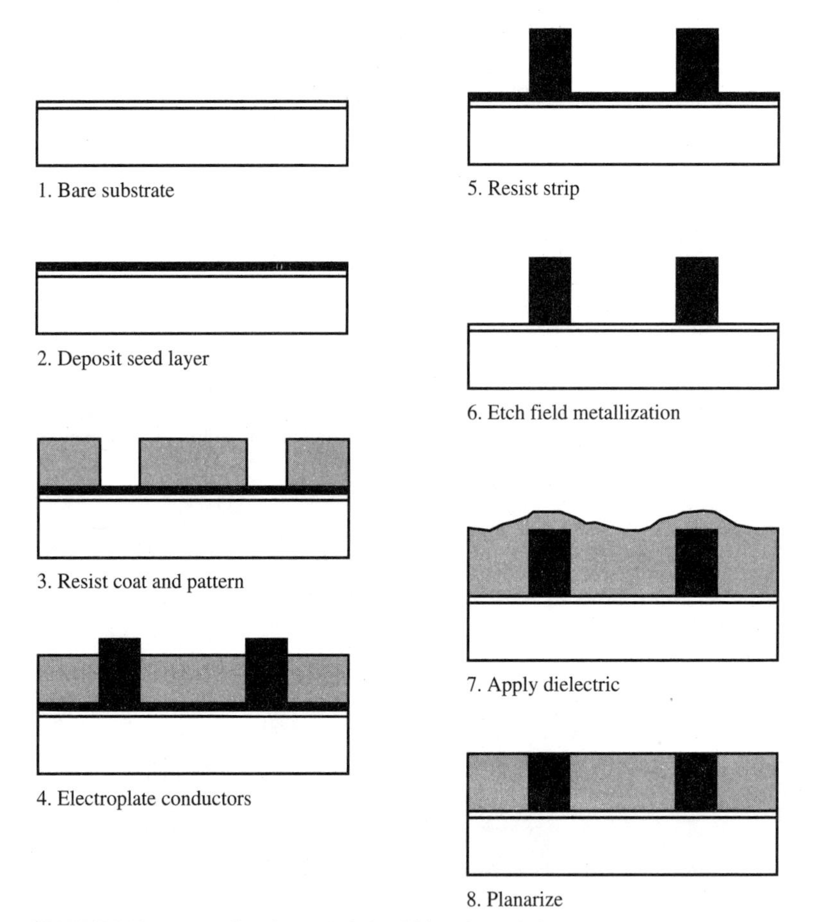

FIGURE 9-12 Process flow for stacked via additive electroplating process.

In a conventional stacked via process, a seed is similarly deposited and features are patterned in the resist and plated up. The resist is subsequently stripped, the seed metallization etched away and the dielectric is applied and mechanically planarized to expose the "stud" tops as depicted in Fig. 9-12.

Additive electroplating allows high aspect ratio wiring because feature tolerance is determined by photoresist lithography tolerance. Thickness uniformity is determined by uniformity of the electroplating process.

Lift-Off Processing. Lift-off processing involves evaporation of metal through a defined mask [30,31]. The mask is fabricated such that the metal deposited on top of the mask and the metal deposited through the mask openings onto the base substrate are discontinuous as shown in Fig. 9-13. The mask is subsequently dissolved, which lifts off the metal deposited on the mask surface. The process gives better feature definition, but is more complex and expensive than other metal definition processes.

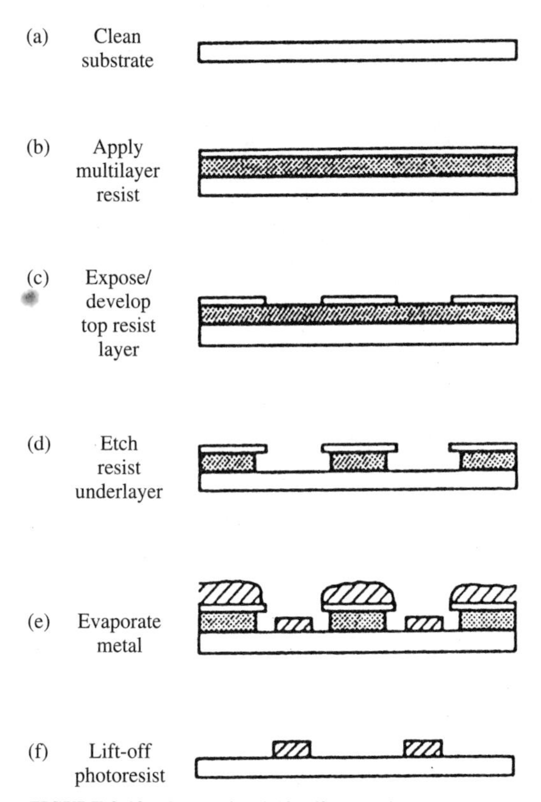

FIGURE 9-13 Conventional lift-off processing.

9.4.3 Thin Film Dielectrics

Two types of thin film dielectrics are currently being used for thin film packaging: polymeric and inorganic. Although ICs have traditionally used SiO_2 as the dielectric layer, thin film packaging solutions have evolved to a much greater extent with the use of polymeric materials. The impetus for this has been the capability of polymers to form thicker layers (with much lower stress); the speed and cost of deposition versus typical CVD SiO_2 techniques; and the better planarization (SiO_2 is conformal) and lower dielectric constant offered by the polymers. On the other hand, the SiO_2 offers much better thermal conductivity, thermal stability, and chemical inertness.

Inorganics. SiO_2 is the only inorganic to receive any real commercial attention. Thick SiO_2 layers are difficult to deposit crack-free because of the normally high stress that CVD techniques impart. Patented technology has been described for the controlled compressive stress physically enhanced chemical vapor deposition (PECVD) deposition of $>10\ \mu m$ coatings [32,33]. It is claimed that SiO_2-based modules are more rugged, offer a harder surface for wirebonding, have a higher tolerance for high-temperature IC rework, and are more reliable under thermal cycling and thermal shock.

Polymers. The widespread use of polymers as insulating layers in microelectronic packaging is relatively recent [34]. NEC was the first to use a PI dielectric in their SX-2 supercomputer (in the late 1980s) [35].

The suitability of a specific polymeric material is highly dependent on the process chosen to fabricate the structure and the intended application of the module. Commercially available materials are usually categorized by the via formation technique that they are compatible with, such as dry etching, wet etching, or photosensitive techniques.

Polyimides (PIs). Polyimides are a class of polymers containing the imide functional unit. PIs are classified by the method of via formation that they support, which includes dry etchable, wet etchable, and photosensitive materials.

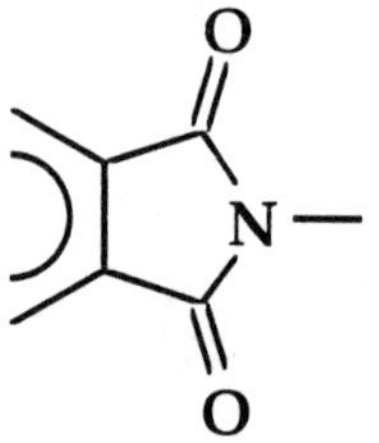

The imide functional unit

Commercial materials used in thin film packaging are derived from the reaction of an aromatic dianhydride with an aromatic diamine [36, 37]. Most products are supplied as soluble polyamic acid (PAA) intermediates, which evolve water upon curing to yield the final PI structure. Some are supplied as preimidized materials. The traditional solvent has been N-methylpryolidone (NMP). In order to obtain good film quality, high-molecular weight PAAs are needed, resulting in low-solubility, high-viscosity systems. PAA solutions are thermally unstable and must be refrigerated. The curing reaction typically requires a temperature of 350°C to 450°C.

Low CTE polyimides. Most PIs that are spin-coated show unequal *x-y* and *z* CTE values. PIs having rigid backbones, such as biphenyl dianhydride/p-phenylenediamine (BPDA/PDA), reveal very high modulus and low biaxial (in-plane) CTE values [38,39]. These low stress PIs are highly anisotropic, revealing *z*-axis CTES $\gg$100 ppm [40].

Polyamic esters. Recent IBM studies identified polyamic esters as PI materials that can overcome some of the undesirable characteristics of polyamic acids such as high curing temperatures, reactivity with copper metallizations, and low planarization capabilities [41]. Because the production of these materials require significantly more processing steps, it is expected that they will be more expensive.

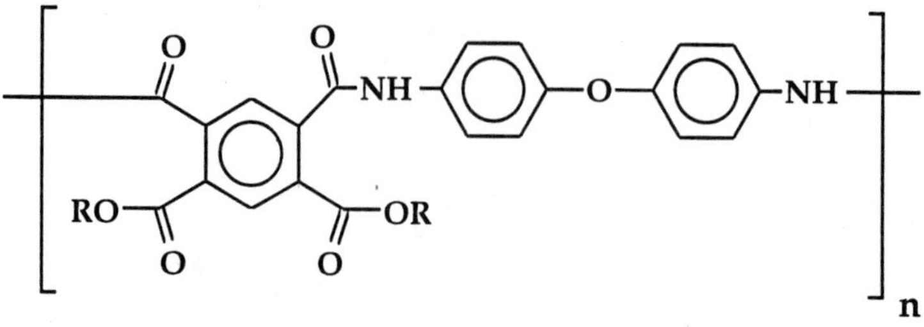

Polyamic ester

More detail on the chemistry of polyimides is available in reference 42.

TABLE 9-6 Pyralin™ PI chemistries

	Chemistry type		
	I	III	V (low CTE)
Standard	PI 2545	PI 2525	PI 2611
Photosensitive		PI 2722	PI 2730
Wet etch			WE 1111

Commercially available polyimides

Pyralin™. The Pyralin series of PI products from DuPont is based on several amine/dianhydride chemistries, as shown in Table 9-6 [43].

PIQ™. PIQ polyimides (commercialized by Hitachi Chemical) have structures exemplified by PIQ-L100. These structures show increased thermal stability as a result of the polyisoindoloquinazolinedione structure [44].

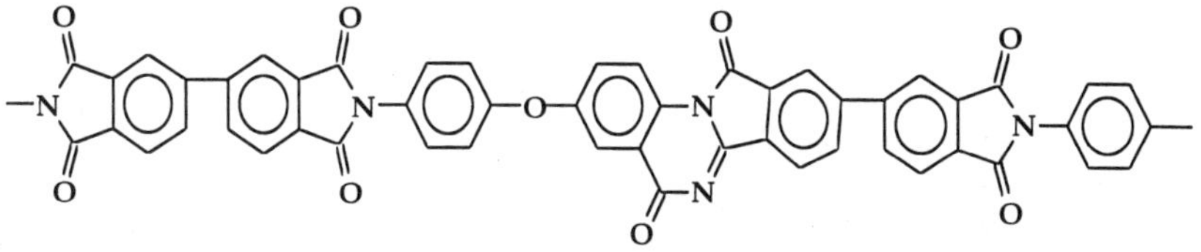

PIQ-L100, a PIQ polymide

Probamide™. Polyimides commercialized by Olin/Ciba Geigy (OCG) under the trade name Probamide™ encompass several structural varieties. The 400 series is based on BTDA and ortho alkyl substituted diamines. These PIs are inherently photosensitive as the result of the presence of the benzophenone moiety. These materials are not as oxidatively stable as other PI materials [45]. The 500 series is based on BPDA/PPD chemistry, and the 600 series is based on polyamic acid esters.

Ultradel™. The Ultradel™ 4000 series is a wet-processable, partially fluorinated PI [46].

Photoneece™. Photoneece™ PIs typified by photosensitive UR 3800 have been developed by Toray based on PMDA/ODA chemistry and an ionically attached photosensitive crosslinker. Such materials have received extensive commercial attention in Japan.

Benzocyclobutenes (BCBs)

Cyclotene™. Benzocyclobutenes (BCBs) commercialized by Dow Chemical under the tradename Cyclotene® are a family of thermoset resins [47,48]. The structure of siloxy containing DVS-BCB is shown in the following figure. The resins are supplied partially polymerized (B-staged) to obtain appropriate viscoelastic handling properties. BCB solutions require no refrigeration, and they are stable at room temperature. Benzocyclobutenes have totally different chemical structures when compared to PIs; therefore, they have significantly different physical, electrical, and processing characteristics. Current commercial materials polymerize thermally without the evolution of water or other by-products. Curing is typically carried out at 210°C to 250°C—several hundred degrees lower than is possible for polyimides.

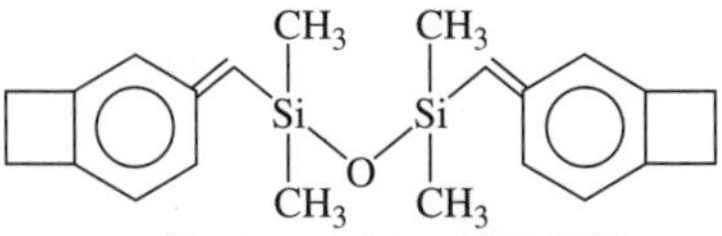

Siloxy containing DVS-BCB

9.4.4 Dielectric Coating

Spin Coating. Organic dielectrics are typically applied by spin coating. The spin coating technique is depicted in Fig. 9-14. The dielectric is dispensed onto the wafer (Fig. 9-14*a*); spread across the wafer at about 500 rpm (Fig. 9-14*b*); spun at higher speed (2000 to 4000 rpm) to achieve a uniform coating of desired thickness (Fig. 9-14*c*); and the edge bead is removed using a backside wash cycle that causes solvent to curl back over the lip of the wafer and wash off the bead created by surface tension at the edge of the wafer (Fig. 9-14*d*). It is important to remove organic contaminants from the substrate surface before film deposition because spin-on polymers tend to pull away from these areas, leaving pin-holes.

The applied polymer dielectric must produce the desired thickness in the fewest coating applications possible, exhibit thickness uniformity across the wafer, and produce a smooth, planar, pinhole-free film.

Cured film thickness is a function of both spin speed and spin time. Nonuniformity across the substrate can result if the spin speed is either too slow or too fast [49].

A typical PI spin curve is shown in Fig. 9-15. Notice the shrinkage that occurs during cure. This requires that a layer about 30% to 50% thicker than would be needed electrically be deposited to compensate for this shrinkage.

IBM reported that PI coating uniformity across a ceramic substrate is typically 8% [25]. Coating uniformity for BCB has been reported as $<2\%$ for the dry etchable DVS material [47]. Variations in dielectric thickness can affect subsequent dry etching; certain areas may overetch while other areas might not clear. A 10% overetch during processing (to make up for nonuniform/non-planar dielectric layers) is typical for reported PI process sequences.

Spin coating of polymer dielectrics usually involves the application of an adhesion promoter to the substrate surface, followed by the dielectric itself. Some polyimides are self-priming (contain adhesion promoter in the dielectric precursor solution).

Spray Coating. There is very little published literature on spray coating of dielectric layers. The thickness, uniformity, and texture of the sprayed surface are influenced by the viscoelastic properties of the polymer, the concentration of the polymer in solvent, and the critical surface tension at the interface [50]. A schematic of a spray coating operation is shown in Fig. 9-16. AT&T PolyHIC substrates are manufactured using a spray coating process (see section 9.5.1).

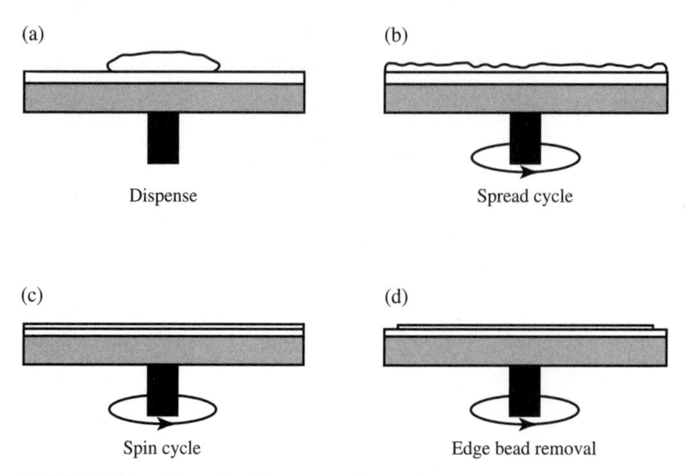

FIGURE 9-14 Schematic of the spin coating process.

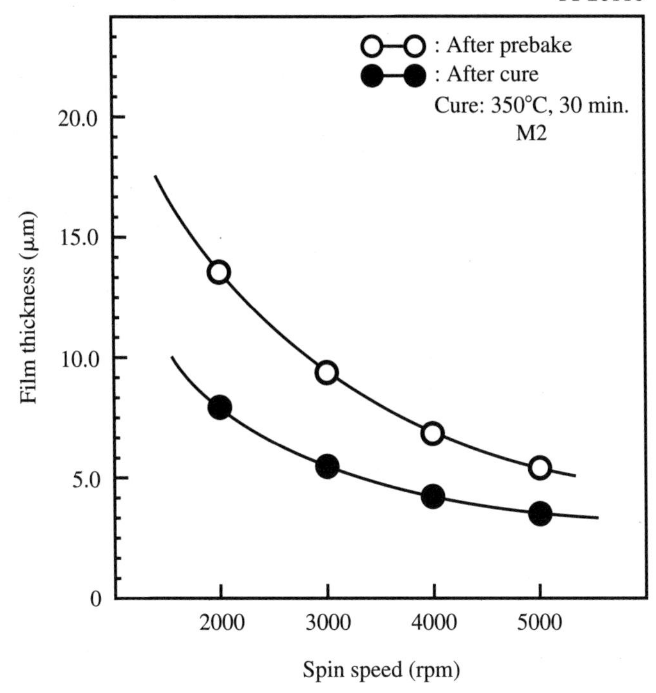

FIGURE 9-15 Spin curve for PI-2611.

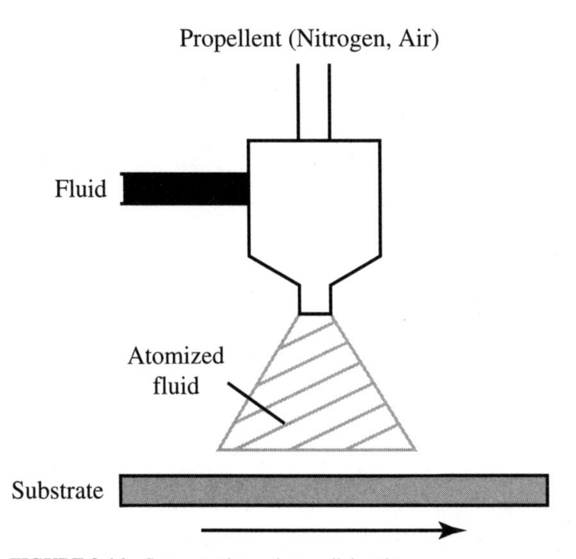

FIGURE 9-16 Spray coating polymer dielectrics.

Other Coating Techniques. As the emphasis in thin film package manufacturing shifts to lowering costs, more emphasis is being placed on alternative coating technologies. Some of these are discussed in section 9.7.2

9.4.5 Dielectric Curing

PI Curing. During polyamic acid-based PI curing, NMP solvent is lost and water is released. Inadequate curing can have a pronounced effect on the properties of the final PI film. IR monitoring of films undergoing cure has demonstrated that the rate of reaction is dependent on both temperature and thickness [51].

BCB Curing. Extensive kinetic studies have revealed a curing activation energy (Ea) of 36 ± 0.9 Kcal/mol [52]. No volatiles are evolved upon cure, thus shrinkage is low (<5%). Curing must be carried out in an inert (nitrogen) atmosphere (<100 ppm O_2 recommended). Curing can take place in a box oven, on a nitrogen-purged hot plate, or in a controlled-atmosphere belt furnace. Processing to partial (approximately 70% to 80%) cure is recommended to achieve the best layer-to-layer adhesion [53]. Full cure can be obtained on a time scale of minutes to hours by controlling the curing temperature. In a typical convection oven, soft curing is achieved in 30 minutes at 210°C and a full hard cure is obtained in about 1 hour at 250°C. Curing is best monitored by IR spectroscopy [54].

Rapid Thermal Curing. Traditionally, polymer dielectric curing has been carried out in a box oven configuration in a batch mode. The heat up, hold, and cool down cycles reported by most dielectric manufacturers can occupy 4 to 12 hours of processing time and create logistical work flow and work-in-progress problems. In addition, an oven loaded with hundreds of wafers puts a lot of product at risk, especially when the product is near completion. To circumvent such problems, rapid thermal curing (RTC) was developed [55].

Rapid thermal curing of BCB in a belt furnace decreases typical cure times from 4 to 5 hours to less than 15 minutes with no significant effect on stress, adhesion, or planarization [55].

Hot plate RTC has also been described [56]. Kohl and co-workers extended this concept in their investigation of the curing of various PI and BCB materials using RTC hot plate processing [57]. They concluded that RTC is a viable process, resulting in polymer films that have equivalent mechanical and electrical properties when compared to standard furnace curing for BCB and preimidized PIs. Polyamic acid-based PIs were shown to be incompatible with RTC techniques because of inferior properties developed with RTC compared with standard furnace curing. Higher dielectric constants and much higher stresses were observed.

9.4.6 Dielectric Via Formation

Via formation in the dielectric layer is necessary to interconnect the signal, ground, power, and pad layers. The main techniques of via formation include: dry etching, wet etching, use of photosensitive dielectrics, and laser ablation.

Laser ablation can be further defined as scanning laser ablation (SLA) or projection laser ablation. These techniques are discussed later. Irrespective of the technique used to create them, good vias are synonymous with low-resistance vias. The key to fabrication of low resistance vias—by any technique—is minimizing the amount of residue to ensure a clean metal-to-metal contact area.

Myszka has defined general via resolution limits for photosensitive, RIE, and SLA versus PI dielectric thickness as shown in Table 9-7 [28].

Stacked Stud Vias. As discussed in 9.4.2, metal features can be defined in photoresist and plated up. The features are coated with PI and the dielectric is cured. Excess PI is then removed by chemical/mechanical polishing of the structure. The uniformity of the PI removal during the planarization process determines the uniformity of the dielectric thickness. In subsequent layers, the vias may be built directly on top of the previous layer because of the flatness uniformity that is achieved.

Conformal Vias. If subtractive processing is being used, vias cannot be vertically stacked unless they are filled by plating them up. Instead, they must be staggered or staircased in a so-called conformal via process. In such processing, blanket metal deposition conformally follows the side-wall angle created by the dielectric via etching process and therefore is filled in by the next layer of dielectric. The next layer of vias must therefore be offset or staircased. When using a highly planarizing material, the vias may be fabricated in folded staircase technology (i.e., layer three may be fabricated directly on top of layer one.)

Dry Etching. Both plasma and reactive-ion etching take advantage of the presence of plasma-activated ionic species to remove material. The difference between the two etching processes is that plasma etching uses chemical reactions to remove material, while reactive-ion etching employs both chemical reactions and ion bombardment.

The final shape profile and size of the etched via is determined by the masking technique. Sloped sidewall vias are obtained with plasma etching, whereas tailored slopes or vertical sidewalls are more easily attainable with reactive-ion etching.

Dry Etch "Soft" Masking. In order to use a soft mask such as a photoresist to pattern a polymer dielectric there must be a measurable difference between the resist and dielectric etch rates.

BCB shows poor resist selectivity to most photoresists in oxygen/fluorine plasmas. A commercial resist has been identified that results in a 3:2 BCB/resist selectivity at an etch rate of 0.6 μm/min when etched in a 1:4 O_2 to CF_4 plasma [58].

A thick positive photoresist is usually used to mask cured PI. The thickness of the photoresist must be substantially thicker than the PI because the etch rate of photoresists is about the same as that of PIs.

Dry Etch Hard Masking. If the thickness of the dielectric precludes the use of a photoresist mask, a variety of metal or inorganic (SiO_2) masks can be used [59]. Hard mask deposition at or above the Tg of the dielectric has resulted in cracking of the mask and the underlying dielectric layers [59]. Typical mask thicknesses used for RIE processing are in the range of 3000 to 5000 Å [28]. For scanning laser ablation processes, 2 to 4 μm thick metal layers are required to withstand the fluences typically used for scanning laser

TABLE 9-7 Typical resolution limits for via generation approaches in PI [28]

	Dielectric thickness		
	10 μm	15 μm	20 μm
Photo	25	50	50
RIE	5	10	10
SLA	5	10	10

ablation (discussed in the following text). Depositing, patterning, and etching a hard mask is a costly and time-consuming process.

Etch Gas Compositions. Although PIs can be etched in straight oxygen plasma, many practitioners have preferred to etch in CF_4/O_2 or SF_6/O_2 in order to obtain better control of the sidewall slope for subsequent metallization and approximately five times higher etch rates [60].

As a result of its silicon content, DVS-BCB cannot be etched in pure O_2; instead it requires fluorinated etchant gasses. The process has been studied in detail [47 and ref. therein]. Etch rates >1 μm per minute have been reported. Formation of surface SiO_2 has been detected at some O_2/CF_4 gas compositions [58].

Wet Etching. Wet etching of PIs is carried out by developing areas of the partially imidized films in aqueous alkaline solution. Process control requires tight specification on the delivered degree of imidization and the thickness of the coatings. The usually highly isotropic nature of the wet etch process results in shallow wall profiles. Others have reported that wet etching of PI is very sensitive to temperature variations during preimidization. Using 12 μm cured coatings, the via-forming process is reported to be limited to vias greater than 50 μm in thickness [61].

Photosensitive. The creation of vias in a photosensitive polymer simply entails exposure and development of the polymer before polymer curing. For a negative working photo polymer, the exposed areas are cross-linked and thus are made insoluble, while the areas that are not exposed are washed away during the development. Detail on the processing of photo dielectrics are given in the following text.

Laser Processing. Laser ablation has been shown to be a viable technique for via formation. The mechanism of ablation is the thermally induced breaking of polymer chains and the subsequent volatilization of the chain fragments. Materials that strongly absorb at wavelengths of radiation emitted by excimer lasers (193 nm for Arf, 248 nm for KrF, and 308 nm for XeCl) show the highest ablation efficiency. The primary variables in the ablation process are the fluence, the number of pulses, and the focus of the image on the surface.

Scanning Laser Ablation. Scanning laser ablation involves the rastering of a laser across a hard mask-patterned dielectric layer to form vias. Mask undercutting, an issue with RIE via-generation processes, does not occur for laser processing. The wall angles are also much closer to vertical.

Excimer SLA of a variety of dielectrics has been reported [62].

Projection Laser Ablation. Projection laser ablation involves the use of a discrete mask to project the via ablation pattern onto the substrate surface in a manner analogous to semiconductor photolithographic steppers.

The glass ceramic/Pl thin film packages used in the IBM ES 9000 (see section 9.6) are thought to be the first commercial application of projection laser ablation [63].

9.4.7 Photosensitive Dielectrics

Photosensitive dielectrics can be divided into two types: negative-working or positive-working. In positive-working materials, the area irradiated undergoes photolytic cleavage and is removed upon development. No positive-working materials are commercially available at present, although it is an area of intense research.

Negative-working materials contain photoreactive groups that lead to photocross-linking with adjacent polymer chains when exposed to UV light. This leads to a solu-

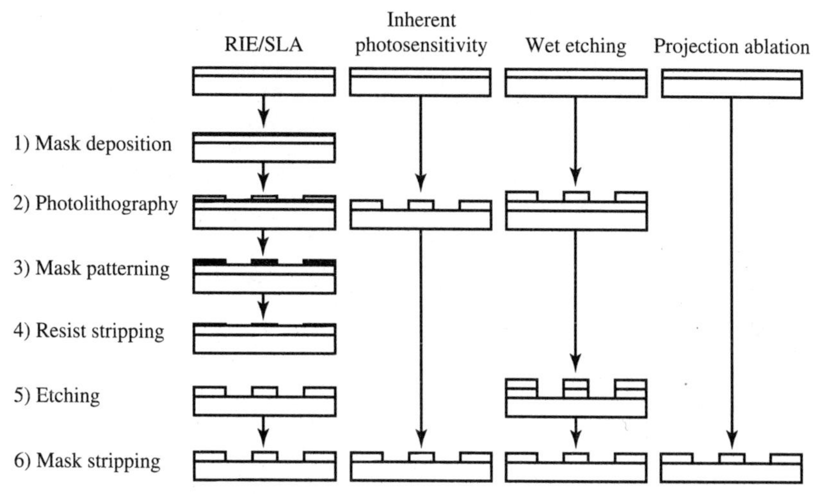

FIGURE 9-17 Comparison of via generation alternatives.

bility differential between the exposed and unexposed portions of the film and, thus, development like a negative photoresist. The three common frequencies of exposure are 365 nm (I line), 405 nm (h line) and 436 nm (g line). Negative-working materials swell because of absorption of organic solvent during development, resulting in limited resolution capabilities.

Although photosensitive materials are inherently more expensive to manufacture, they are generally used to lower overall processing costs. Several processing steps can be omitted when using photoimagible polymers or projection laser ablation as shown in Fig. 9-17 [49].

Photosensitive Polyimides (PSPIs). Commercial materials are limited to negative working materials. Negative working photosensitive polyimides can be further categorized by the source of photoactivity. PIs that are not inherently photosensitive are modified by covalent attachment or ionic attachment of photosensitive groups to the parent PI molecules. These are referred to as ester-type and salt-type PSPIs, respectively.

Mechanical Properties. The properties of ester-based and salt-based PSPIs for the same diamine/diacid combinations are shown in Table 9-8 [64]. While the ester-based materials show better resolution in thicker layers, the salt-based materials reveal vastly superior mechanical properties.

TABLE 9-8 Comparison of properties for cured PSPI films [64]

PSPI type	Elongation (%)	Tensile strength (Mpa)	Peel strength (g/cm)*
ester	<1	<10	0
salt	10	116	260

*Adhesion to Si water after 20 hr of PCT.

Storage, Application, and Prebake. Negative-working PSPIs require refrigeration and have a finite shelf life. Viscosity tends to increase with time at elevated temperatures (RT), causing thickness variations upon application and exposure variations because the concentration of cross-linker changes with time.

Prebake is used to optimize the solvent content of the PSPI film. Too much solvent can result in delamination; too little solvent can result in cracking.

Exposure. Linewidth and via definition are usually dependent on precise control of the UV exposure and the thickness of the dielectric.

Exposure can be performed with either projection steppers or scanners or with printers in the hard contact, soft contact, or proximity mode. Contact exposure will produce steeper wall profiles and better resolution than projection exposure because of the reduction in focus effects [65]. The use of contact exposure in production is limited due to mask cleaning considerations.

The most important factor in obtaining the desired feature profile is the exposure process. In general the mask dimension defines the bottom PSPI dimension, while the top PSPI dimension will change (blow out) on development and cure. The feature profile for most PSPIs can be modified by varying the exposure wavelength and/or dose. Increasing the dose will result in steeper wall profiles. An increase in dose has been reported to increase the amount of residue (scum) in the vias [66].

Develop and Post-develop Processes. Developing can be carried out in three ways: (1) dip tank immersion, which is usually enhanced by stirring or ultrasonics, (2) puddle developing and rinsing, usually on an automated track coater, or (3) spray developing in spray development systems.

In order to decrease via blowout upon shrinkage, several post-exposure stabilization methods for PSPIs have been suggested, such as blanket exposure, which enhances the cross-linking before thermal cure [67].

Curing and Descum. Severe film shrinkage occurs when negative working PSPIs are cured (up to 60%, depending on the specific chemistry of the material and the thickness of the layer) [63]. During curing, the crosslinkers and any remaining unreacted photogroups are volatilized along with the water that is generated during the curing process. This film shrinkage has been reported to produce cusping or crowning at via edges [66–68].

Shrinkage not only affects the film thickness but also the developed feature wall profile. Fig. 9-18 shows the effect of shrinkage on isolated, semi-dense, and dense features. Note the top via dimension is significantly smaller for the semi-isolated feature [67].

Although the goal of the development process is to remove all unexposed material, most current processes require a so-called "descum" process to remove the thin polymer residue that is left behind and to ensure good electrical contact between metal layers. This process is usually carried out with an oxygen plasma source such as a Matrix asher or barrel etcher.

Niwa reports that the PSPI materials and non-photo polyamic acid materials reveal insoluble residues formed in the via holes during development caused by Cu dissolution into the PI [69].

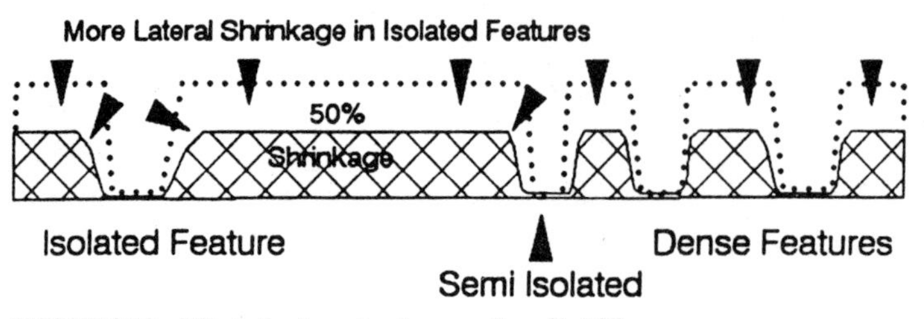

FIGURE 9-18 Effect of pattern density on wall profile [67].

The following process sequence has been suggested by Kotzias for the fabrication of Al/PSPI layers [66].

- Etch metallization and remove photoresist
- Dehydration bake
- Oxygen plasma descum (for adhesion)
- Application of PSPI and hotplate prebake
- Projection exposure
- Develop using track spray coater
- Final cure
- Plasma descum
- RF backsputter and deposition of next PI layer

Commercially Available Materials. Representative commercially available PSPI materials are listed in Table 9-9. The use of PSPIs has been extensively detailed by, among others, NTT [70], NEC [71], Boeing [72], and AT&T [66].

Photosensitive Benzocyclobutene (PBCB). Photosensitive BCB formulations that generate approximately 5 and 10 μm-thick final cured films are commercially available as the

TABLE 9-9 Properties of thin film polymer dielectrics

Vendor	Product designation	Photo	ϵ_r	Tg (°C)	CTE (ppm)	Tensile modulus (GPa)	Tensile strength (MPa)	Elongation (%)
Amoco	Ultradel 4212	—	2.9	295	50	2.8	101	30
	7501	+	2.8	400	24	1.3	122	70
Ciba-Geigy	Probimide 400	+	3.0	350	39		140	56
	500	—	3.0	>350	6-7	11.6	444	28
	600	—	3.0	>350	23	2.7	194	100
Cemota	Syntorg IP-2000	—	2.9	350	55	—	119	4
Dow	Cyclotene 3022	—	2.65	>350*[1]	52	2.9	85	9
	Cyclotene 4026	+	2.65	>350*	52	2.9	85	9
Dupont	Pyralin 2540	—	3.5	400	26	1.3	160	60
	2611	—	2.9 (x,y)	350	5	6.6	600	60
	WE-1111	—		385	19	4.4	300	55
	2730	+	2.9 (x,y)	350	15	—	170	—
Hitachi	PIQ-13	–	3.4	290	50		116	10
	L100	–	3.2	360	3		320	22
	PL-2135	+	3.3	270	40	3.6	124	10
Toray	Photoneece UR-3800	+	3.3	280	45	3.4	145	30
	UR-5100	+	—	—	25	—	200	20

Cyclotene™ 4000 series. Unlike the dry etchable Cyclotenes, the PSBCBs must be refrigerated and have a finite shelf life [73,74]. The process sequence for photo BCB is shown in Fig. 9-19 [75].

The use of PBCB has been detailed by NEC [11,76], Siemens [7], and MMS [77].

Cyanate Ester. AT&T uses a proprietary cyanate ester resin called PHP 92 in their Poly-HIC technology [78]. The formulation consists of a blend of triazine resin (40% to 65%), acrylated acrylonitrile-butadiene rubber, acrylated epoxy, and other minor components. These photoactivated materials cure at <300°C. These formulated products are manufactured for internal use and are not available commercially.

Epoxies. When a laminate interconnection medium is used as the base substrate, deposited dielectrics are selected that are curable at temperatures at or below the Tg of the laminate system. Consequently, the laminate-based photovia processes that have been described in the literature to date have used inherently photodefineable dielectric materials with curing temperatures less than 250°C. Epoxy-based systems such as Ciba-Geigy's Probimer 52 [79] and Shipley's FP-9111 [80] have been reported. BCB can also be rapidly cured at temperatures as low as 210°C, making it compatible with most high-temperature laminates.

General properties for all the aforementioned dielectrics are compared in Table 9.9.

9.4.8 Polymer Physical Properties

Mechanical Properties. The mechanical properties of a polymer are determined in part from stress-strain curves such as the one shown in Fig. 9-20. Stress is the force per unit area that is being exerted on the material. Strain is the change in dimension that occurs in response to the stress. Point 1 in Fig. 9-20 is the yield point and is a measure of the strength of the material and its resistance to deformation. Point 2 in Fig. 9-20 is the break point. Between point 1 and point 2 is a region where the material undergoes plastic deformation.

The ratio of stress to strain is the elastic modulus. The greater the elastic modulus, the greater the force needed to deform the polymer. Modulus is thus a measure of the stiffness or hardness of the polymer.

The ratio of tensile stress to tensile strain is called Young's modulus:

$$E = \sigma/\epsilon \tag{5}$$

where E = Young's modulus
σ = tensile stress
ϵ = tensile strain

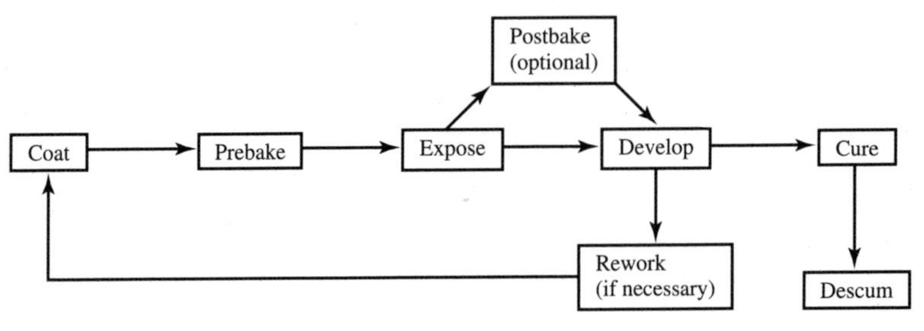

FIGURE 9-19 Processing sequence for PBCB.

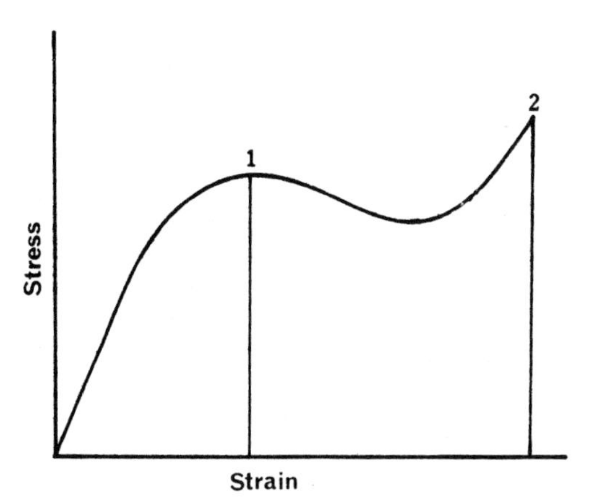

FIGURE 9-20 Typical polymer stress strain curve.

Hard, brittle polymers have high moduli of elasticity, no well-defined yield point, and low strain at break. Hard, strong polymers have high moduli of elasticity with yield points and moderate stress at break. Hard, tough polymers have high moduli of elasticity with yield points and high stress and strain at break.

Another important mechanical property is the glass transition temperature (Tg). This is the temperature at which the polymer will deform in response to an external load. Above its Tg, a thermoset polymer takes on the properties of a cross-linked rubber, it will deform but it cannot flow. If processing of a thin film structure is to occur at a temperature above the Tg of the polymer dielectric, the mechanical integrity of the structure may be compromised.

Water Absorption. Water absorption impacts both the electrical properties and the processibility of MCM dielectrics. Absorption of water, with a dielectric constant of 78, raises the dielectric constant (ϵ_r) of an insulating polymer and thus affects circuit performance. Uncontrolled outgassing of moisture from underlying layers of dielectric during subsequent high temperature processing can result in blistering and delamination of the thin film structures being fabricated [49,81,82]. Bakeout cycles are routinely employed during the fabrication of MCM's with polymers that exhibit significant moisture absorption. Gridded ground planes are used to allow moisture to escape [83].

Jensen has noted that the ϵ_r of PMDA/ODA-based PI increases linearly from 3.1 at 0% RH to 4.1 at 100% RH [84]. He concluded that the electrical design tolerances must accommodate the resultant variations in electrical characteristics or the MCM must be hermetically sealed to control the humidity level.

Moisture uptake is specific to the polymer in question and the conditions of the experiment. Data has been reported for exposure of varying thicknesses of polymer film, for varying times, at varying RH. Values for a variety of materials are given (with experimental conditions where available) in Table 9-10.

Planarization. Planarization by polymeric materials recently has been reviewed by Shiltz [31].

Relief existing on a surface before photolithography will affect the resolution and quality of subsequent conductor traces. One of the main functions of the insulator layer is to planarize the underlying topography and, thus, provide a flat focal plane for the next layer. Lack of topographic planarity can lead to subsequent non-uniform metal thickness (poor

TABLE 9-10 Water uptake by polymer dielectrics under varying conditions

Polymer	Condition	Water uptake (%)
PI 2555	1 hr @ 50% RH	2-3
PI 2611	2 hr water boil	0.4
PIQ 13	Unknown	2.3
PIQ I100	2 hr water boil	0.5
IP 200	Unknown	0.8
Prob 400	100% RH	5.0
Prob 500	50% RH	0.74
UR 3800	24 hr immersion @ RT	1.1
Cyclotene 3022	24 hr water boil	0.2
Cyclotene 4026	24 hr water boil	0.3

step coverage). These thinned and weakened areas are susceptible to cracking [85]. The degree of planarization (DOP) for a polymer dielectric is described in Fig. 9-21 [86].

Thin film coatings of BCB exhibit excellent planarization properties. The degree of planarization is dependent on the film thickness versus the feature thickness, the number of coatings, the feature width, height, and spacing [87]. Typical planarization versus feature width is shown in Fig. 9-22 for single coatings of varying thicknesses.

In conformal via structures, BCB dielectrics not only planarize underlying conductor layers, but also planarize the conformal vias so that the next orthogonal layer of conductor can be routed directly over the vias. Under identical conditions, BCB revealed 65% to 70% planarization of via features versus 10% to 15% for typical polyimide dielectrics [88]. Photosensitive BCB exhibits slightly less planarization, as reported by NEC [76].

Senturia and co-workers have shown that the DOP obtained by fully cured PI films is the composite drying and curing [89]. During drying, loss of planarization occurs from shrinkage when the polymer can no longer flow. For polyimides, imidization and its associated water loss that occur during curing cause further shrinkage, which adds to planarization loss. Shrinkage is reported to be as high as 50% or more depending on the polyimide. Molecular weight also plays a role in planarization; the lower the MW, the better the planarization. DOP for PIs reportedly can be improved by optimizing spin conditions [90].

Planarization for a variety of dielectric materials is given in Table 9-11.

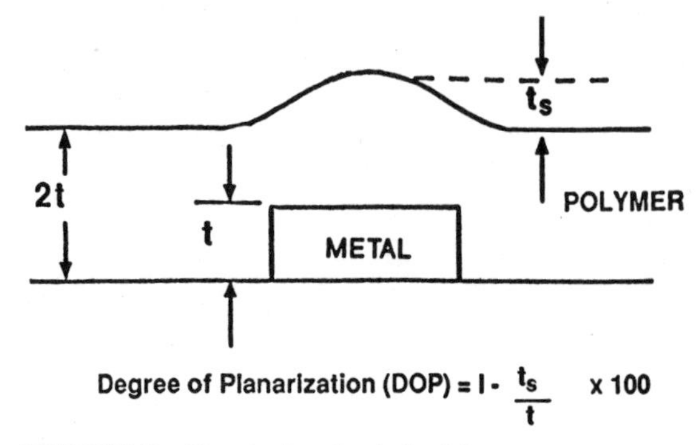

FIGURE 9-21 Planarization of an isolated feature.

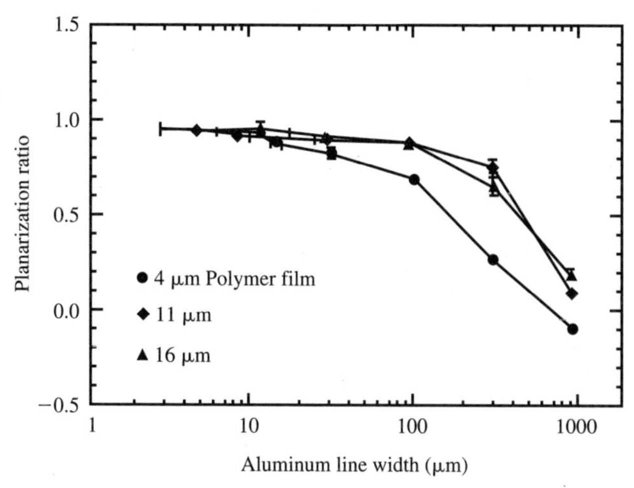

FIGURE 9-22 DVS-BCB planarization over 4-μm thick isolated aluminum lines [87].

Chemical Resistance. Polymer dielectrics must be resistant to the processing and cleaning chemicals used in the fabrication of the structures. The usual manifestations of poor chemical resistance is swelling, cracking, or crazing (microcacking) of the polymer layer. Thermosets are generally much more resistant than thermoplastic materials. Most of the materials listed in Table 9-9 are reported to be resistant to normal processing chemicals [82].

PIs are not stable to strongly basic solutions which causes problems with typical electroless plating solutions [29]. BCB films are not stable in the presence of strongly oxidizing acid solutions.

Preimidized PIs are soluble before curing and after solvent removal. As such, solvent resistance is poor unless they are taken to an elevated temperature to induce cross-linking of the linear molecules. Poor solvent resistance is manifested by crazing of the polymer surface.

Thermal Stability. A material is thermally stable at a given temperature when its properties at that temperature do not change. Thermal stability for an MCM polymer dielectric needs to be in excess of the temperatures experienced during subsequent processing and/or

TABLE 9-11 Planarization of typical thin film polymer dielectrics

Polymer	Planarization
PI 2525	30
PI 2611	18-28
PI 2722	28
IP 200	25
PIQ 13	25
PIQ L100	25-28
Cyclotene 3022	90-95
Cyclotene 4026	70-90

repair steps. In general, the dielectric will be subjected to temperatures less than 350°C during chip attach or other assembly processes.

Fully aromatic PIs are stable in air and nitrogen at 400°C. The Hitachi PIQ materials were specially designed to produce the most thermally stable materials available. DVS-BCB stability is limited to 350°C under nitrogen.

Polyimides that are not fully aromatic, (such as the preimidized materials, and BCBs are inherently more susceptible to oxidation because of the presence of benzylic or aliphatic functionality in the cured films. The use of antioxidants increases the stability of DVS-BCB in air high temperatures [91].

Stress. Stress is a reliability issue because excessive stress buildup can exceed the adhesive strength of an interface, resulting in delamination, or exceed the tensile strength of the polymer, resulting in cohesive fracture of the dielectric layers. Intrinsic stress is built into a film because of the shrinkage and contraction that occur after a film has adhered to the underlying surface. Thermal stress is generated after cure as the polymer cools down (from its Tg), because of the mismatch in CTE between the polymer film and the substrate as described by Eq. 9.6.

$$\Delta\sigma_f = \frac{E_f(\alpha_s - \alpha_f)/\Delta T}{1 - \nu_f} \qquad (6)$$

where σ_f = film stress
E_f = Young's modulus
α_s = CTE substrate
α_f = CTE film
ΔT = temperature difference
ν_f = film Poisson ratio

The biaxial stress (σ_f) is generally determined from substrate warpage measurements. Stress is related to substrate warpage according to Equation 9.7.

$$\Delta\sigma_f = \frac{E_s t_s^2}{6(1 - \nu_s)t_f K} \qquad (7)$$

where E_s = Young's modulus of substrate
t_s = substrate thickness
t_f = film thickness
K = substrate curvature ($1/R_u - 1/R_c$ where R is the radius of curvature for coated and uncoated samples, respectively)
ν_s = Poisson's ratio for the substrate

The most pronounced effects of stress are observed at stress concentrators such as the corners or edges of patterned features [47,48]. It is for this reason that round vias are favored over square vias and sloped via bottoms are favored over sharp undercutting via bottoms.

Stresses that are generated by polymer dielectrics can lead to fabrication problems for substrate warpage, especially when using silicon substrates [6,47]. If wafer bow becomes too severe, vacuum fixtures cannot hold the wafers down flat enough to allow sharp focusing over the entire wafer diameter. Problems have also been encountered in automated cassette-to-cassette handling equipment as a result of wafer sticking when the warpage is too severe. Warpage problems increase with substrate dimensions and the total number of dielectric layers. Ceramic substrates, with higher moduli, are more resistant to such bowing. Tensile stresses are also imparted by the metallization layers.

TABLE 9-12 Stress as a function of relative humidity [92]

	0% RH	50% RH	100% RH
PI-2525	41	31	18
PIQ-L110	8	5	0
PIQ-2611	4	0	−5

Polymer stresses for a series of PIs have been correlated with relative humidity [92]. As shown in Table 9-12, humidity has a tendency to swell the polyimide layers and thus somewhat relax the stress. This effect of RH may account for some of the differences in published stress data. Published stress values for a variety of dielectric materials are given in Table 9-13.

Adhesion. Polymer adhesion is needed at four interfaces: polymer to substrate, polymer to metal, metal to polymer, and polymer to polymer. Adhesion is closely related to other issues such as water absorption and stress. Temperature and humidity are known to severely degrade adhesion. Adhesive failure usually manifests itself as a delamination between interfaces.

Adhesion is very sensitive to surface preparation. Cleaning of the surface before polymer deposition removes reactive layers such as oxides and/or particulate contamination. Surface treatment with plasma or ion beam techniques or adhesion promoters are used to induce better adhesion.

Adhesion promoters are formulated to chemically react with both the substrate surface and the deposited film. PI [93] and BCB adhesion promoters are usually amino silanes (APS). The adhesion promoter is either included in the dielectric formulation (self priming PIs) or applied as a separate solution before the dielectric (BCB and some PIs).

It is reported that the adhesion of deposited Cu onto PI can be increased 25-fold by oxygen RIE of the PI surface [94,95].

Thin Film Thermal Management. Reliability at all levels of device packaging, thin film packaging included, is directly related to operating temperature. Higher temperatures are known to accelerate various device-failure mechanisms such as corrosion and electromigration [96]. Device degradation occurs at an exponentially faster rate as the chip operating temperature increases (Arrhenius function). For example, for activation energies of 0.8 to 1.0 eV, a 25°C increase in operating temperature increases the failure rates five to six times.

TABLE 9-13 Thin film polymer stress values

Polymer	Stress (Mpa)
PI 2525	40-54
PI 2611	4-10
PIQ 13	28-30
PIQ L100	4-5
Prob 400	55
Prob 500	5
UR 3800	45
Cyclotene 3022	24-38
Cyclotene 4026	28-30

Thus, the system designer must consider thermal management early in the design cycle. Materials, material interfaces, and overall strategies for heat removal should be chosen carefully to avoid premature system failures.

Definition of Terms. Heat can be removed from a body by *conduction* and *convection*. Conduction describes the removal of heat through a material that has a given intrinsic resistance to heat flow (thermal conductivity). Convection describes the process by which heat is transferred from a hotter to a cooler region by a moving medium such as a gas or a liquid. Where possible, air cooling is preferred over liquid cooling because of its simplicity and lower cost. The velocity of the air is usually limited in electronic systems by noise tolerance. (Noise is created by the generated turbulence.)

Thermal resistance (θ_{AB}) is defined as the temperature difference between points A and B divided by the power that is dissipated between those two points. The most commonly used thermal resistance expressions in packaging are θ_{JC} (thermal resistance from IC junction to case) and θ_{CA} (thermal resistance from the case to ambient) which add up to give θ_{JA} (thermal resistance from junction to ambient). In normal practice, heat is transferred from junction to case by conduction and from case to ambient by convection (see Eq. 9.8).

$$\theta_{JA} = \theta_{JC} + \theta_{CA} \tag{8}$$

The measured thermal resistance (θ_{JA}) may thus be defined as shown in Eq. 9.7:

$$\theta_{JA} = T_j - T_a/P_c \tag{9}$$

where T_j = device junction temperature
T_a = ambient temperature
P_c = chip power dissipation

The thermal resistance of a package is a function of many variables such as die size, die thickness, die attach material, and package material. The thermal resistance of the package material is proportional to the thickness of the material and inversely proportional to its thermal conductivity. The thermal resistance at an interface, however, is a function of the roughness and flatness of the substrate making contact and the pressure being applied to the contacting surfaces. In many applications, the use of highly thermally conductive materials will be nullified by poor surface finish, poor physical contact, or poor choice of die attach material for the chosen application. We refer the reader to several excellent references to further examine the complexity of thermal design and management [97,98].

Polymers have significantly lower thermal conductivities than most other materials used in microelectronics processing, including metals, ceramics, and glasses. This is the result of their low density when compared to these other materials.

A listing of the thermal conductivity of common electronic materials is given in Table 9-14.

The configuration shown in cross section in Fig. 9-23 is typical of a thin film packaging structure. The heat is conducted through die attach (50 μm), thin film (40 μm), substrate (1 mm alumina), and thermal grease (<25 μm) to the heat sink and then to the environment by air convection.

Figure 9-24 compares strategies for thermal management. Thermal vias are routed like normal interconnect vias, but their sole purpose is to carry heat away from under the die areas. Thermal walls are created as a cutout in the thin film structure and the chips are mounted directly to the support substrate. Thermal vias limit the amount of interconnect that can be routed under the die (shown as 75% to 20% of theoretical maximum in Fig. 9-24). Thermal wells obviously do not allow for routing under the die [99].

TABLE 9-14 Thermal conductivity of microelectronic packaging materials

Material	Approximate thermal conductivity (W/m·°K)
Copper	400
AlN	200
Aluminum	200
Silicon wafer	100
95/5 solder	40
Alumina ceramic	20
Die attach (Ag filled epoxy)	4
LTCC (glassy alumina ceramic)	2
Epoxy molding compound (SiO_2-filled)	1
Thermal grease	1
BCB or PI dielectric	0.2

9.4.9 The Metal-Polymer Interface

The Copper-Dielectric Interface. Copper easily corrodes at Cu-PI interfaces, and the oxide produced has poor adhesion [100]. The mechanical and electrical properties of PI can degrade when cured in the presence of copper [101,102]. Copper has been shown to diffuse into PI by oxidation at the Cu-PI interface, migration of copper ions into the PI matrix, and finally reduction of the ionic copper to small nodules of copper metal by the matrix. The incorporation of copper reduces the breakdown voltage and increases the dielectric constant of the PI. Thin diffusion-barrier layers of passivating metals stop this diffusion of metal into the PI [103 and references therein].

As copper is converted to its oxides in the PI-driven corrosion process, the resistance of the copper interconnects increases because copper oxides have higher resistivity.

Niwa and associates have shown that Cu (II) diffuses approximately 8 μm from a copper-PSPI interface after a one-hour prebake at 80°C [69]. This copper-rich scum is

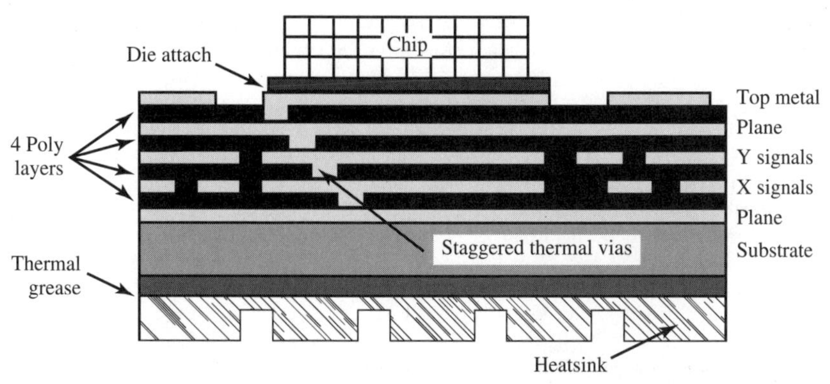

FIGURE 9-23 Cross section of typical thin film interconnect structure [99].

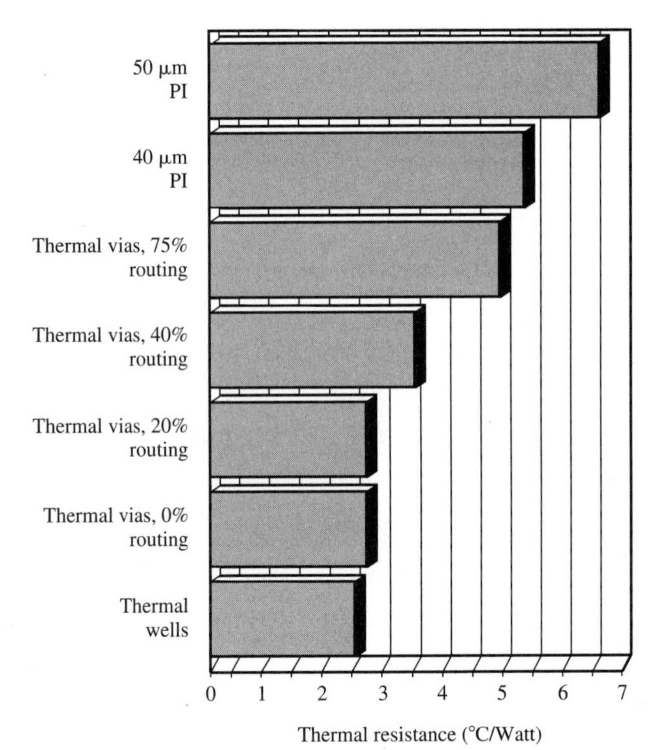

FIGURE 9-24 Relative thermal performance for thermal vias and thermal wells [99].

difficult to remove with a typical oxygen descum process. During any etching step, the Cu-PI nodules or deposits impede clearing of the vias and thus the formation of low-resistance contacts.

Cech and co-workers have shown that the corrosion rate for unprotected lines is eight times higher than for unprotected lines. Resistance increases after 85% RH/85°C testing were four to five times lower for fully encased lines than for 60% encased (top and bottom only) lines [104].

Rao has used transmission electron microscopy (TEM) to examine PI-Cr and PI-Cu interfaces on a completed MCM substrate that has been subjected to 120°C, 85% RH, and 40 V bias for 856 hours [105]. While the Cr-PI interfaces appear clean, Cu nodules, the largest of which is 25 nm, can be observed out to 100 nm from the interface.

Skin effects for high-frequency transmission are expected to be adversely affected by these highly resistive barrier layers. The degradation is expected to be most severe for thin conductor lines with relatively thick passivation/barrier layers [25].

Adema and associates have shown that sputtered Si_3N_4 and PECVD SiO_xN_y behave as diffusion barriers between Cu-PI interfaces, which result in improved performance at high frequencies over unprotected transmission lines [106]. Copper was shown to diffuse through sputtered SiO_2 whereas BCB was shown to be impervious to copper migration, as expected.

The Copper-BCB Interface. Copper diffusion is not observed when BCB is deposited directly on Cu metallization [106,107] even after 1500 hours at 85% RH/85°C. NEC has examined direct BCB-Cu-BCB interfaces in MIL 883 testing and seen no corrosion or other failures [76].

9.4.10 Reliability

Many material stability issues are of concern for a thin film package, especially those manufactured with a polymer dielectric. Although there are no specific MIL STD tests for thin film packages per se, most vendors have been performing MIL 883 tests on their packages. The 883B tests performed by most computer manufacturers exercise the parts between −55°C and 125°C. The telecommunication industry, driven by Bellcore standards, has preferred MIL 883C, which requires −65°C to +150°C testing. Some manufacturers request HAST testing (121°C/2 atm) instead of more typical 85% RH/85°C testing.

9.5 THIN FILM PROCESSES AND APPLICATIONS

Since the initial references to thin film interconnect first started appearing in the early 1980s, a plethora of process sequences have been strung together to fabricate test structures and demonstration vehicles. Some of those which have been commercial, are currently commercial, or have been taken to preproduction are discussed in the following text.

9.5.1 AT&T

PolyHIC. AT&T PolyHIC (Polymer Hybrid Integrated Circuit) technology for their switching and transmission equipment has been in production in their Merrimack Valley facility since 1987 [108]. Their standard technology uses copper conductors and TaN resistors on a thin film alumina substrate. The cyanate ester (CE)-based dielectric is described in section 9.4.3.

Process flow is shown in Fig. 9-25 [109]. Laser-drilled vias are generated in the alumina substrate for designs requiring double-sided structures. The first level of metal is created by sputtering a Ti adhesion layer and plating-up copper metal. TiN resistors are fabricated and laser trimmed. The dielectric is applied by spray deposition, exposed, developed, and cured. Conformal, staggered, 100-μm square vias are normally developed in the nominal 50 μm dielectric thickness. A second level can be created by repeating this process sequence.

Line widths and spaces range from 50 to 100 μm. Top-level metallurgy is Ti/Cu/Pd/Ni/Au. The Pd passivates the Ti/Cu interface. The Ni acts as a barrier layer between the Cu and the Au. The gold is used to permit soldering and wire bonding of devices.

AVP. The initial AT&T silicon-on-silicon process was known as the advanced VLSI process (AVP process) [15]. It was based on nickel-passivated Cu metallization. More recently, an Al/PI process was developed that reportedly requires much simpler processing and is more economical [110,111].

Capacitors, TaSi resistors, and discrete bipolar npn transistors are incorporated into the active substrate. The aluminum is deposited by sputtering, and it is photolithographically patterned and defined by wet chemical etching. Typical conductor dimensions are 10 to

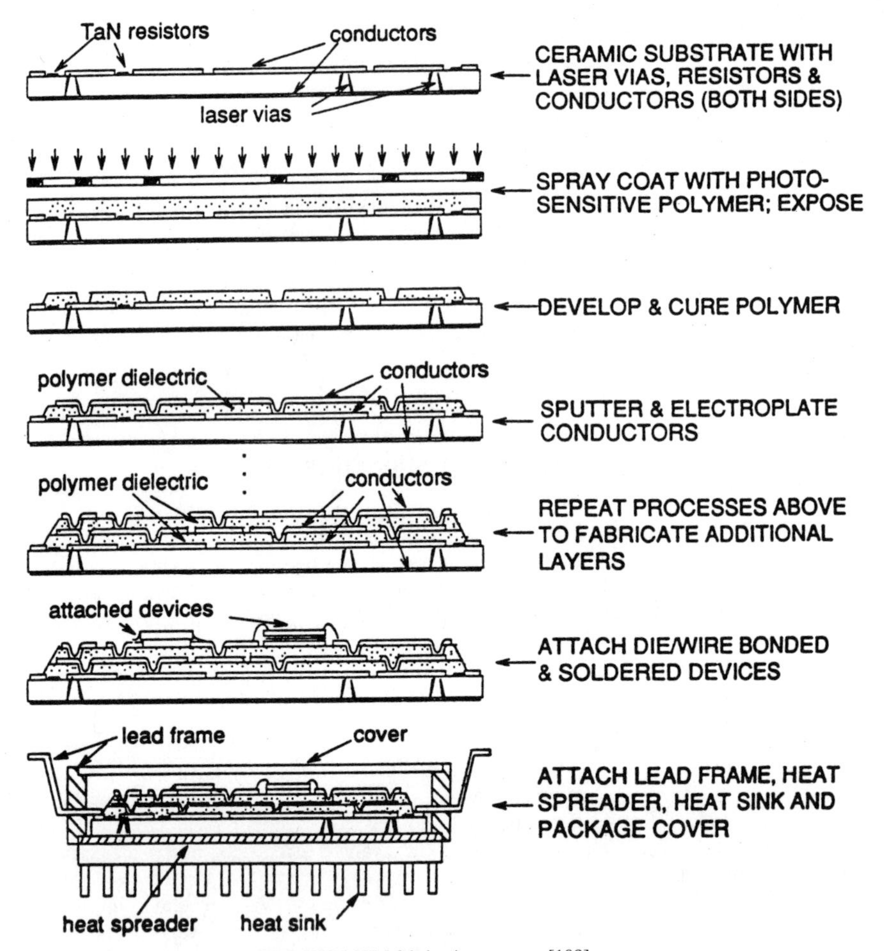

FIGURE 9-25 Schematic of PolyHIC MCM fabrication process [109].

40-μm wide and 1.5 to 2.5-μm thick. PSPI is 3-μm thick. A cross-section of the process is shown in Fig. 9-26.

9.5.2 IBM

IBM has been developing and implementing multilayer thin film technology since the early 1980s [12]. Cu/PI and Al/PI thin film structures have been implemented on silicon, alumina, and glass ceramic carriers. Several of these technologies are detailed in the following text.

Single Chip Packages. Two layers of Cr/Cu/Cr wiring isolated by PI dielectric are fabricated on alumina ceramic and used to interconnect chip I/Os to the package pins. Vias are defined by wet etching. The process flow is shown in Fig. 9-27. Such packages are used for high-performance logic and memory devices [112].

chip
wire-bond pad
polyimide 1
aluminum
polyimide 2
solder bump
solder wettable metal
polyimide 2
aluminum 2
polyimide 1
aluminum 1
nitride
SiO2
TaSi
p+
n+
p
n
p+
capacitor
resistor
npn transistor
p

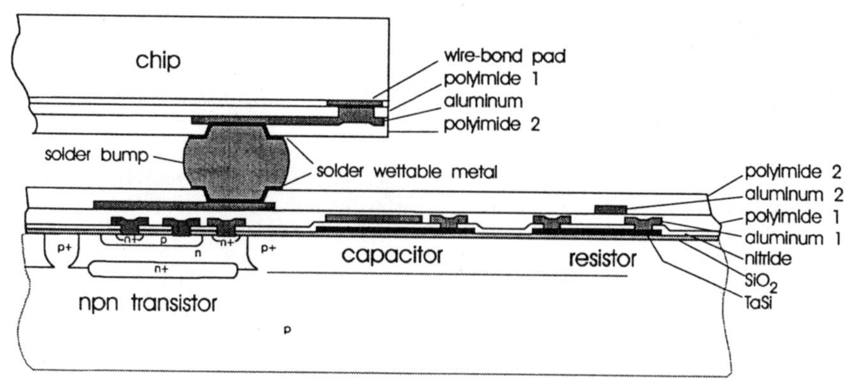

FIGURE 9-26 Cross section of AT&T Si-on-Si process [110].

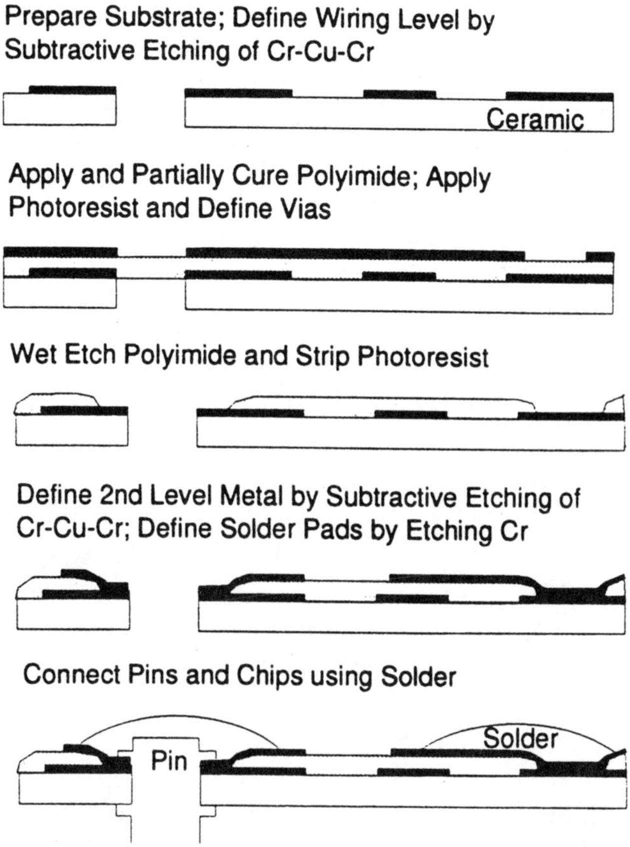

Prepare Substrate; Define Wiring Level by Subtractive Etching of Cr-Cu-Cr
Ceramic
Apply and Partially Cure Polyimide; Apply Photoresist and Define Vias
Wet Etch Polyimide and Strip Photoresist
Define 2nd Level Metal by Subtractive Etching of Cr-Cu-Cr; Define Solder Pads by Etching Cr
Connect Pins and Chips using Solder
Solder
Pin

FIGURE 9-27 Process flow for IBM thin film single chip packages [112].

Thin Film Redistribution on Cofired Ceramic. Thin film technology was developed for signal redistribution on systems such as the ES9000 [113]. The thin film structure for the ES9000 processor utilize a planarized stud process for high wiring density, Cu/PI thin film layers, projection laser ablation for via formation, chemical mechanical polishing for planarization, and laser CVD for thin film repair [114].

Thin Film Al/PI on Silicon. Al/PI thin film technologies were developed in IBM to package CMOS chips in MCMs [83,115,116]. The thin film structure is defined by a conformal via process. The PI is patterned by RIE through photoresist. The wiring is patterned by subtractive etching of aluminum. The process and materials were chosen to be compatible with a CMOS manufacturing line [83]. Aluminum (4% to 9% Cu) conductors are 3 μm thick and typically 13 μm wide on 25 μm pitch. Minimum wiring dimensions are 8 μm. The dielectric is 3 to 6 μm PI, vias are dry etched. Via size is 8 μm for wiring, 50 μm for power and ground. The top surface metallization is evaporated Cr/Cu/Au for standard IBM C4 joining technology.

9.5.3 Hughes

The Hughes HDMI-1 process is based on dry-etch, low-CTE PI and a Al interconnect. Substrates have been fabricated on 150-mm silicon or ceramic (alumina or aluminum nitride) wafers.

HDMI-1 consists of power, ground, signal 1, signal 2, and chip attach metal layers, which are sputtered in 5 μm Al and wet etched. The top Al layer is sputtered with a thin Ti/W layer and a thin Au layer, and then 5 μm of Au are plated-up to form the pad layer. Vias are plasma etched after standard lithography is used to define the pattern. Vias are typically staggered to avoid the buildup of non-planarity because the vias are not filled. A cross section of the substrate is shown in Fig. 9-28 [16].

9.5.4 Micro Module Systems (MMS)

The MMS process retains many of the processing steps of the initial DEC process for the VAX 9000 while substantially reducing the manufacturing costs inherent in such a large

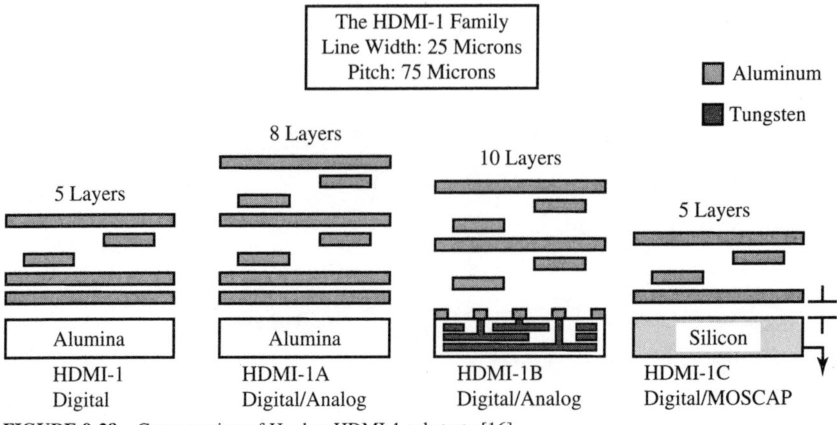

FIGURE 9-28 Cross section of Hughes HDMI-1 substrate [16].

CPU system. Aluminum wafers (150 mm) are used as carriers. They are coated on the backside and edges with dielectric to protect the Al during processing. Chromium is used as an adhesion/barrier layer, and copper signal lines are electroplated-up. Fluorinated oxygen plasma is used to etch the vias. The aluminum substrate is used as one of the power distribution layers as well as one electrode of a decoupling capacitor. Typically, 15-μm lines and 50- to 70-μm line pitches are designed [17]. More recently, a Cu/BCB low-cost process has been described [77].

9.5.5 n-Chip SiCB™

n-Chip (a division of Flextronics) fabricates thin film interconnect on a silicon substrate. The n-Chip technology uses proprietary plasma-enhanced CVD to deposit undoped SiO_2 dielectric ($\epsilon_r = 3.75$). SiO_2 is deposited by a high-rate PECVD process. Control of stress during the deposition allows layers as thick as 10.5 μm [117–119]. The technologies are shown in Table 9-15 [119]. A cross section of the nC2000 is shown in Fig. 9-29. Integral decoupling capacitors are common to all structures. Resistor layers and electroplated copper are used in the 2000 and 3000 series for higher performance and higher frequency operation.

9.5.6 GE/Ti HDI

The GE high-density thin film interconnect process (HDI) is a chips first approach [120–122]. In the nominal process, bare chips are inserted top down in cavities that have been created in a ceramic substrate and the interconnect structure is formed over the chips and substrate. The interconnect is formed by lamination of a thin PI sheet over the substrate to form the first dielectric layer. Via holes are made through the film by laser ablation down to the I/O pads. Connections to the chip pads and the interconnect traces are formed by a 4 μm Ti/Cu/Ti sputter/plate up process. The metal interconnect pitch is typically 4 mil. Multiple layers are formed by repeating the PI film lamination, via formation, and metallization processes. Typically two orthogonal signal layers as well as power and ground layers are processed.

9.5.7 IMC

The IMC Swedish process is based on 125-mm silicon wafers. Interconnects are made in four layers with BCB as dielectric. All metal layers consist of sputtered aluminum. Thin film resistors and capacitors are fabricated in a TiN layer. Die are wirebonded to the substrate [123].

9.5.8 Thomson

The Thomson MCM process manufactures copper/PI structures on 100 to 150-mm^2 ceramic substrates [124]. Conductor widths range from 20 to 75 μm, with corresponding pitch and via diameters from 50 to 200 μm and 10 to 60 μm, respectively. The typical five-layer structure consists of power, ground, x, y, and pad layer. Both Al and Au wire bonding are used [125].

9.5.9 IBM Japan

The addition of thin film redistribution layers to laminate in order to interconnect high-I/O area array chips was first publicly disclosed by IBM-Japan (126,127), who call it their

TABLE 9-15 Comparison of merchant thin film MCM technologies

	AT&T PolyHIC	IBM MLTF	Hughes HDMI-1	MMS	n-Chip nC1000	TI/GE	IBM Yasu	Kyocera	IMC	Thomson
Conductor	Cu	Cu	Al	Cu	Al	Cu	Cu	Cu	Al	Cu
Adhesion/barrier	—	Cr	—	Cr	—	Ti	—	Cr	—	Cr
Thickness (μm)	8	5	5	2-8	2	4	—	3, 5	3	5
Width (μm)	50-100	13-25	25	20	15	40	75-100	30	20	20-75
Pitch (μm)	125-225	25-100	75	75	—	100	—	76.2	50	50-200
Pad level	Ti/Pd/Cu/Ni/Au	Cr/Cu/Ni/Au	TiW/Ni/Au	Cu/Ni/Au	Al	TiW/Au	—	Ni/Au	Al, Ni/Au	Cu/Au
Substrate	Alumina	Ceramic or silicon	Silicon, alumina, or AlN	Aluminum	Silicon	Ceramic or laminate	Laminate	Ceramic	Silicon	Ceramic
Dielectric	CE	PI	PI 2611	PI, BCB	SiO_2	Kapton	Epoxy	PI	BCB	PI
Photosensitive	+	–	–	–	N/A	–	+	+/–	+/–	+
Thickness (μm)	50	5.5-12.5	5-10	3.5-12	7	35	40	20	7	5-15
Via dia (μm)	100	10-20	35	30-50		25	125	30	30	15-60
Via type	Staggered conformal		Conformal	Conformal	Conformal	Conformal staggered	Conformal	Conformal	Conformal	Conformal
Chip attach	WB	Al, Au WB; TAB	WB	WB, FC	—	Direct contact	FC	WB, FC; TAB	Al and Au; WB	Al, Au; WB

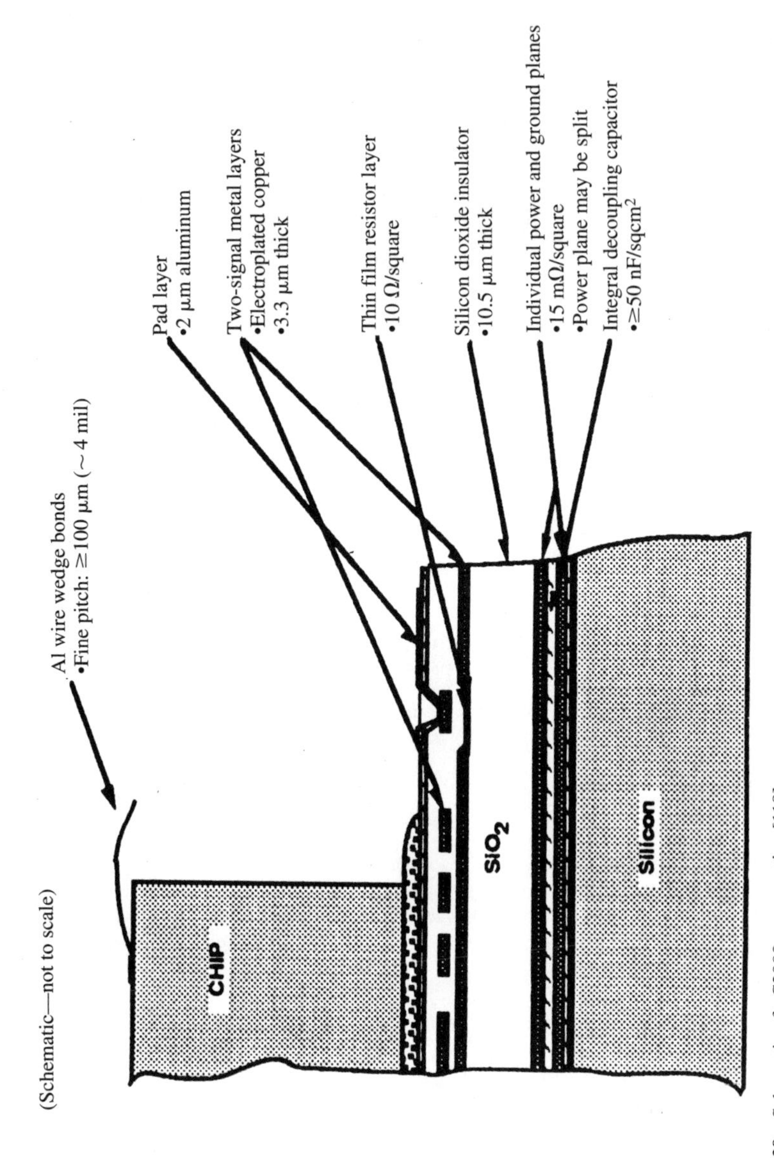

FIGURE 9-29 Schematic of nC2000 cross-section [119].

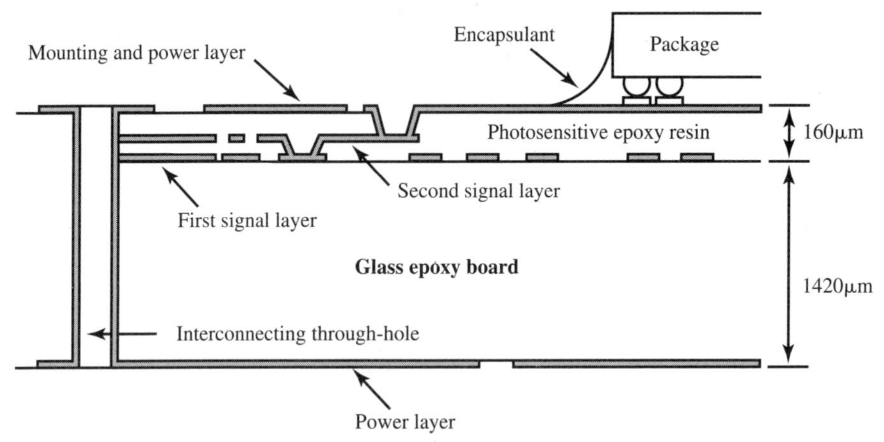

FIGURE 9-30 Cross section of the IBM Yasu SLC technology.

surface laminar circuit (SLC) technology. They use a modified epoxy photo resist (Ciba Geigy Probimer 52) to create 75-μm lines and 100-μm vias. The dielectric is applied by curtain coating to a full PWB panel substrate. A cross section of this structure is shown in Fig. 9-30. The key technology for implementation of such structures is the under-chip epoxy buffer layer that redistributed the stress created by the CTE differential between epoxy board and the area array direct chip attach (DCA).

9.5.10 Kyocera

Kyocera has divulged very little about their thin film MCM packages [128]. The substrate tends to be co-fired multilayer mullite. A special blend of PI has been developed to give reliable adhesion after high-temperature storage and pressure cooker testing. Typical ground rules are given in Table 9-15.

9.5.11 NTK

Kambe and co-workers have described a process based on a co-fired alumina substrate [129]. TaN thin film termination resistors are formed on the top layer of the co-fired ceramic, then five Cr/Cu/Cr metal layers and five PSPI layers are deposited. Plated-up post vias are used. Line widths of 25-μm and 50-μm vias on 150-μm pitch are typical.

A comparison of many of these technologies is shown in Table 9-15.

9.6 SELECTED THIN FILM APPLICATION AREAS

Applications for thin film packaging have traditionally been performance driven, dominated by high clock rate, I/O-intensive, interconnect rich architectures such as tightly coupled multiprocessors, crossbar switches and multiple bus architectures. However, it has become increasingly clear that thin film technology will have to deliver system cost savings as well as performance benefits in order to expand market acceptance. As long as thin film

MCM technology stays focused on low-volume, high-performance applications, costs will remain relatively high.

9.6.1 Mainframe Computers

Thin film MCM technology has been implemented by mainframe manufacturers in systems such as the IBM 390/ES9000 (130), the NEC SX3 (131), the Hitachi M880 (132) and the DEC VAX-9000 (17).

9.6.2 Workstations

n-Chip SPARC Module. n-Chip has produced a 40-MHz SPARC CPU module that has been commercialized in the UNIX-based workstation made by Tadpole [133]. The small footprint of the module was important in this laptop application, as was the 20% to 30% reduction in power consumption afforded by the low-capacitance interconnect. The 90-MHz hyper SPARC module is manufactured using n-Chip's nC1000 technology (see 9.5.5).

The 4-6 chips are more manufacturable than a monolithic version, yet they are partitioned in such a way that they do not reduce overall system performance.

This module, shown in Fig. 9-31, is being used in desktop workstations, deskside server applications, parallel computers, and notebook applications.

FIGURE 9-31 Hypersparc module fabricated by SiCB™ technology [133].

NEC RISC-3000 Prototype. Figure 9-32 shows a prototype of a high-density RISC 3000 module fabricated with Cu/BCB technology. Good reliability was obtained without barrier/adhesion layer metals such as Ti, Cr, or Ti/W. The direct Cu/BCB interface resulted in a 30% reduction in signal transmission speed versus the Cr/Pd Au being used in contact with PI [76].

9.6.3 Military and Space

Each succeeding generation of military and space products tends to have higher performance and smaller form factor than its predecessor. Performance has traditionally come from increasing device integration. Size and weight requirements can restrict the use of standard SMT single-chip packages mounted on PWBs and backplanes. The most common approach to resolving this dichotomy is to minimize the ratio of device package area to active silicon area by MCM approaches. It is for this reason that military and space applications have driven the leading edge of MCM technology over the last decade. The newest trend is to get even higher performance in yet smaller volumes by three dimensional packaging of the MCMs.

While some of the military and space vendors have their own substrate fabrication and MCM assembly, most have relied on outside vendors for one or both of these activities. The following descriptions are not comprehensive, but rather attempt to give examples of some of these activities.

Hughes. Using their HDMI technology, Hughes has developed numerous military MCM applications. An example is the HAC-32™ processing element by Hughes data processors [134]. It is a 20-MIPS general-purpose processing element based on a 32-bit JIAWG processor, an Intel I960™ MX RISC microprocessor, and 194-meg SRAMS in a 2″ × 4″ footprint. Physically the HAC-32 is about one-tenth the size

FIGURE 9-32 Prototype of R-3000 module [76].

and weight of the comparable elements if individually packaged. The HAC-32 supports a range of military applications, including the F-22 Advanced Tactical Fighter and Navy electronic warfare systems, such as those in the F/A-18, F-14, A-6, and Harrier aircraft.

TI. TI has developed several thin film MCM technologies. Their work in the military arena can be exemplified by the Aladdin program, which seeks to package mainframe processor performance in a soup can-sized volume [135]. The resulting 2-GFLOPS processor occupies a cylindrical volume of 4.5″ diameter and 3.1″ in height. The technology used to fabricate these parts was silicon-on-silicon, Al/PI technology. The six layers of metal and five layers of PI were processed on 125-mm wafers. Ti/W was used as an adhesion layer for the 8-μm PI layers. Twenty 5-μm vias were dry etched on 50-μm pitch.

After the substrate is populated with capacitors, ICs, and three-dimensional memory, it is mounted into its housing using gold compression bonds from the substrate pads to the housing.

Honeywell. Honeywell's technology is exemplified by the Advanced Spaceborn Computer Module (ASCM) program, which is a general purpose space-qualified processor based on 1750A processor design. There are three types of MCMs in the ASCM system. The generic VLSI spaceborn computer (GVSC) module contains three CPUs and two floating point processors (FPPs). The memory MCM contains nine 8K × eight SRAMs and a memory line driver. The third MCM contains cache memory (six 8K × eight SRAM die and a cache controller chip). The system has been fabricated in co-fire MCM-C and in MCM-D technology [136].

9.6.4 Telecommunication

Large-capacity, high-speed switching and transmission systems will require MCM packaging technologies to implement signal switching [137]. Saving space as a result of the elimination of multiple single-chip packages and passive components can also be of great value in mobile telecommunications [123].

AT&T. Fig. 9-33 depicts a timing generator which operates above 300 MHz. The circuit contains seven chips, over 100 thin film resistors, and 30 capacitors. The circuit operates with controlled impedance lines, tightly controlled delays, and terminating resistors [133].

NTT. NTT has described thin film technology for asynchronous transfer mode (ATM) switching [139].

The main uses of ATM switching systems will be high-speed communication, including voice, video, and data. Future ATM switching systems will be constructed of many ATM switching elements with high-speed interconnects between them. Thin film MCM technology is expected to allow the fabrication of entire ATM switching elements within a single package [139].

Figure 9-34 shows a cross section of an ATM switch using thin film technology. For a 4 × 4 ATM shared memory switch, the MCM contains four BiCMOS LSIs, 16 bipolar LSIs for conversion between TTL and ECL, two 8 : 1 multiplexor bipolar SSIs, and two 1 : 8 demultiplexor LSIs. The MCM has six layers of Cu/PI thin film, two of which are meshed power supply layers. To increase the yield of the MCM ATM switches, TAB technology is used to allow pretesting of the LSI devices.

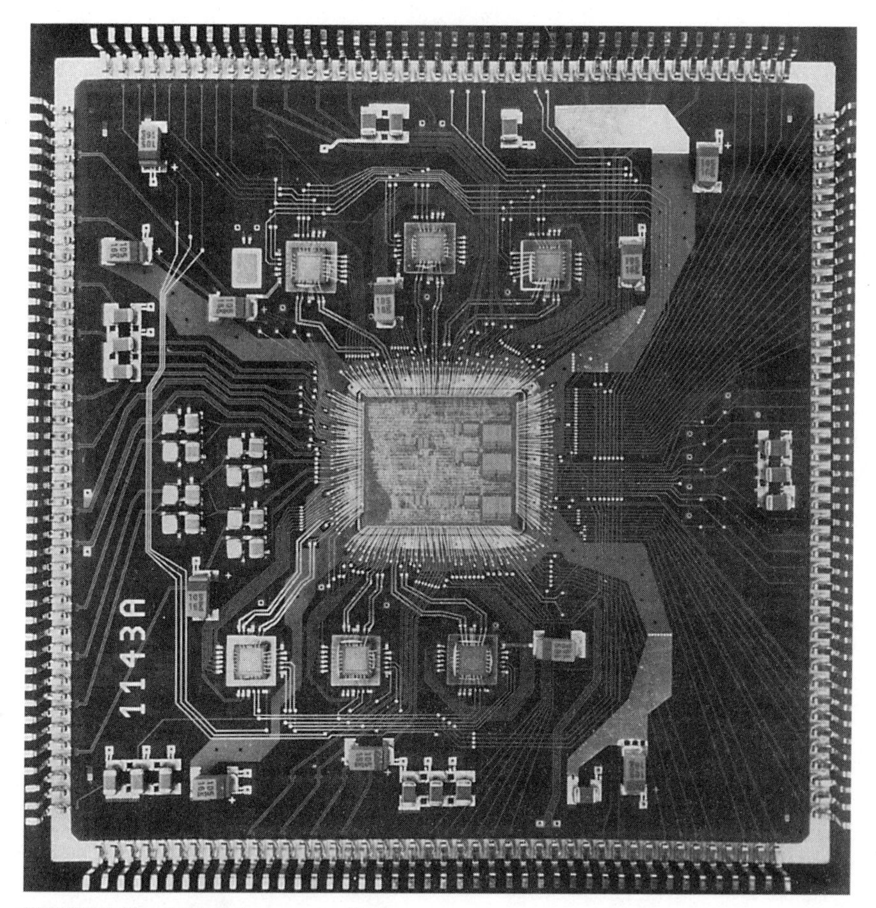

FIGURE 9-33 AT&T timing generator [138].

9.6.5 Consumer

The approach in considering consumer applications has been to consider the cost savings that can be achieved at the system level by using thin film technology to replace existing components [140,141]. The lower parasitic loading results in less power dissipation and thus, at the system level reduced power supply costs, less demand for cooling, and denser packaging of boards in cabinets. The smaller footprint of the thin film MCM can reduce the size and, thus, cost of motherboards, and it appears most importantly of all, allows one to incorporate large numbers of passive components directly into the substrate [141,142].

AT&T. AT&T has described the examination of silicon-based MCM for their consumer telephone business [142,143]. Key to the low cost is the fabrication of the needed passive components directly on the carrier, and the adoption of existing SMT assembly infrastructure to the chip mounting on the carrier wafers. The leadless, bare die are furnished with solder wettable flip chip pads in place of conventional wire-bond pads. The active

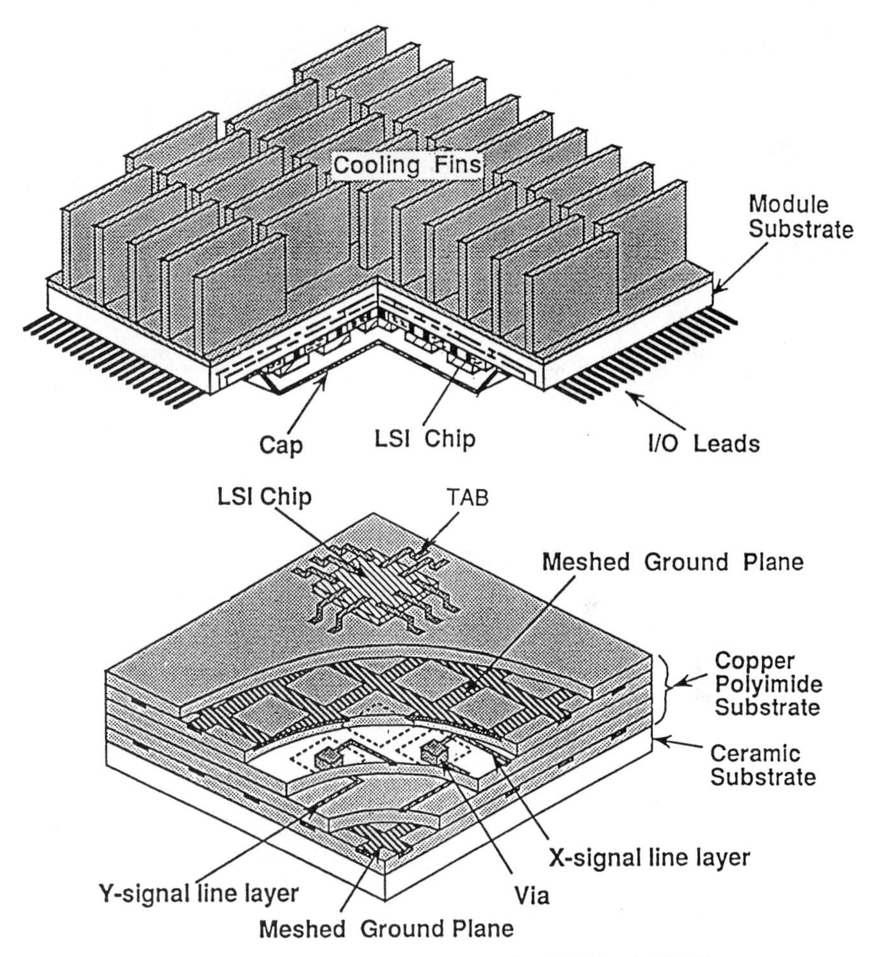

FIGURE 9-34 Cross section and functional block diagram of NTT ATM switch [139].

MCM interconnect is a three level, 10-μm interconnect with 119 resistors, eight capacitors, seven diodes, and 152 transistors [142].

Because each MCM interconnect substrate is only 9.5 mm × 6.65 mm in size, this permits patterning of 148 per 125 mm wafer. Figure 9-35 shows the populated MCMs on the silicon carrier. It has been noted that the cost savings from mounting and stocking the passive components is greater than the cost of the components themselves.

9.7 THIN FILM POLYMER IC APPLICATIONS

Polymer dielectrics are beginning to get a lot of attention in IC applications. As was shown in the SIA roadmap in Fig. 9-4, the number of IC wiring levels on a chip is expected to increase to seven or eight by the year 2010.

FIGURE 9-35 Populated MCM wafer [142].

Polymers have been used as chip overcoats (stress buffer) and multilevel insulation since the 1980s, when PIs were introduced for these applications [144].

9.7.1 Stress Buffer for Si Devices

Large die such as Si DRAMs packaged in plastic molding compound are subject to significant stress. The stress can cause cracking of the silicon or the protective inorganic passivation (usually Si_3N_4) and/or metal displacement. Early practitioners used wet-etchable PIs as a stress buffer layer, which required that features be defined in a layer of photoresist followed by wet etching of the exposed PI to open the pad areas, dry etching of the passivation, and subsequent stripping of the photoresist. More recently, photosensitive PI and BCB have been used to cut down the number of process steps and improve on the wet etch PI resolution and side wall profile. The increased resolution is needed to open the fuse links that are being built into today's chips to allow rerouting/repair. These fuse link windows are typically 10 microns wide and projected to get smaller.

The current goal is to develop a one-mask process in which the photodielectric is exposed and developed and the primary passivation (Si_3N_4) dry etched to open the pads and fuse links without photoresist protection [145]. The goal is for minimal amount of dielectric to be removed when it is exposed to the fluorinated Si_3N_4 etch gas.

9.7.2 Silicon Inner Layer Dielectric (ILD)

The combination of aluminum metallization, tungsten via fill (plugs), and silica dielectric (deposited as conformal CVD SiO_2 or spin-on-glass materials) have led the IC industry to 0.25-μm feature manufacturing technology. It is clear, however, that progress towards the goal of <0.10-μm features past the year 2000 is hindered by both the current dielectric and metallization [5].

As features get smaller, so does the Al interconnect cross section and, thus, the resistance of the line increases. Interconnect lengths are not scaling relative to the other on-chip geometries since chips are actually getting larger with time and some traces must still carry signals across the full dimensions of the chip. This increase in resistance will result in increasing RC time delays [146]. It is felt that this can be overcome by lower-resistance copper interconnect and lower ϵ (dielectric constant) dielectrics. In fact, the technology roadmap calls out for planarizing dielectrics with $\epsilon < 3.0$ and low-resistance conductors as high-priority requirements for 0.18-μm technology to evolve by 2001.

Some will replace the dielectric first, thus continuing to require greater than 425°C stability of the polymer dielectric to meet Al anneal requirements. Others will replace both the dielectric and the Al as they move below 0.25-micron geometries.

PIs were first proposed for interlayer dielectric applications in the 1980s [147]. Selective chips have been produced by IBM and Hitachi through the 1980s with PI dielectric. In general, however, the industry standardized on aluminum metallization and silicon dioxide dielectric.

Despite obstacles [148], there have been many reports on Cu/PI ULSI technology over the last five years [149, 150] that use barrier metal protection of the Cu (top, bottom, and sides) and chemical mechanical polishing to achieve planarization. Recently, processes based on Cu/BCB have also been reported [151–153]. It is suggested that by 1998 polymers will begin to replace spin-on-glass dielectrics in high-end applications.

GaAs ILD. Figure 9-36 shows an SEM of a multilevel (four metal layer) GaAs chip fabricated by the Triquint QED2 process where the BCB dielectric has been etched away to reveal the multilevel metal [154]. This four-level metal structure could not be fabricated with traditional air bridge technology.

Bumping and On-Chip I/O Redistribution. Area array bonding technology delivers superior electrical performance, reliability, and reduced footprint [155,156]. Tight peripheral bond pad pitches caused by some of today's microprocessors and ASIC chips require mating with a high-density interconnect, such as the MCMs described above, or alternatively can be designed into the chip to produce area array I/O, or chips produced with peripheral I/O can be redistributed on chip-to-area array of bumps with looser pitch. Redistribution is shown in Fig. 9-37 in cross section. Such redistribution requires thin film polymers and techniques similar to those described above for MCMs.

Figure 9-38 shows the rewiring of an 8.8-mm, 328-pin chip with a peripheral pad pitch of 4 mil (100 μm) [156]. The array pitch is 350 μm. BCB photo dielectric was used for both the dielectric and as an insulator to allow rewiring and bump formation.

FIGURE 9-36 Triquint QED2 multilevel metal GaAs process [154].

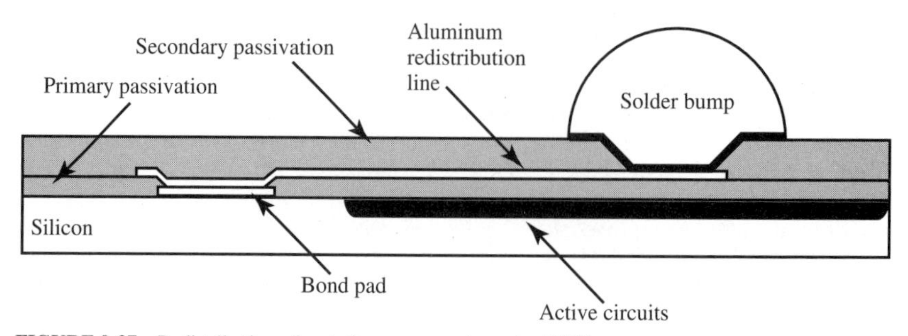

FIGURE 9-37 Redistribution of pads for area array bumping [155].

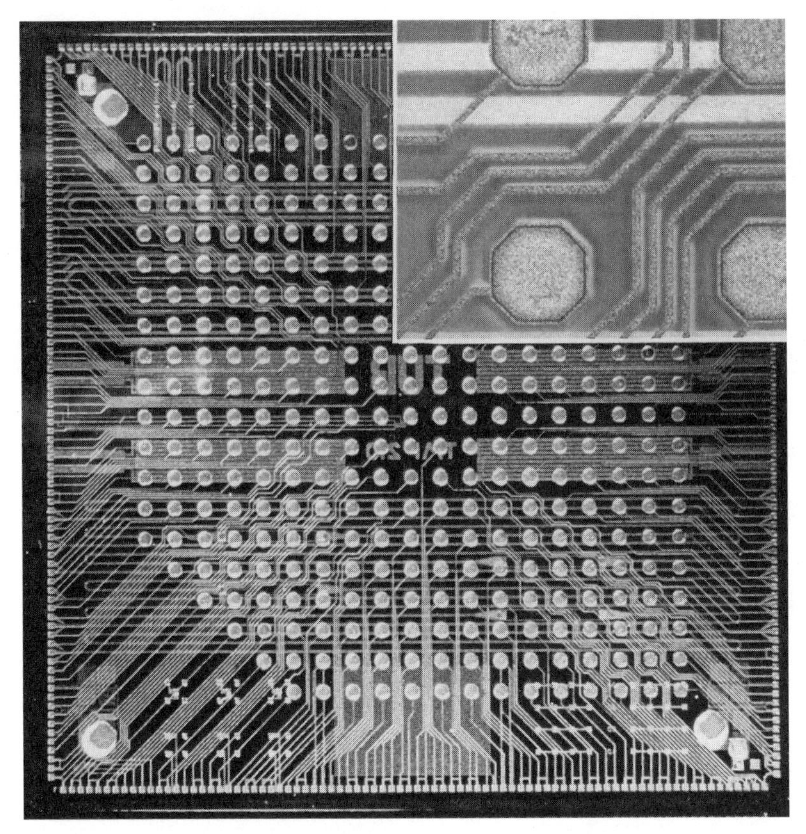

FIGURE 9-38 Area array thin film redistribution on a 328-pin IC [156].

9.8 OPTICAL INTERCONNECT

The introduction of new types of interactive multimedia services puts increased demand on the existing telecommunications infrastructure. It has been suggested that the required increase in transmission capacity and bandwidth will be impossible to achieve without the use of optical transmission in switching, transport, and access system equipment [157]. Low-cost polymeric optical waveguide technologies are projected to offer breakthrough optical solutions in these areas [157,158]. The core and cladding of these optical waveguides are deposited and defined by traditional thin film processing techniques. Figure 9-39 shows GaAs chips intracting over thin film polymer optical interconnect.

In addition, electrical busses in electronic systems are currently unable to keep pace with on-chip clock speeds. Optical backplanes based on passive, multimode polymer waveguides offer the promise of maintaining on-chip clock speeds across chip, package, board, and cabinet boundaries [159].

The incentive for using polymer optical waveguides for short ($<$100 cm) optical interconnects (PWB and backplane applications) is that the polymer waveguide technology has the potential to become more cost effective when compared to alternative glass or polymer fiber technologies [159,160].

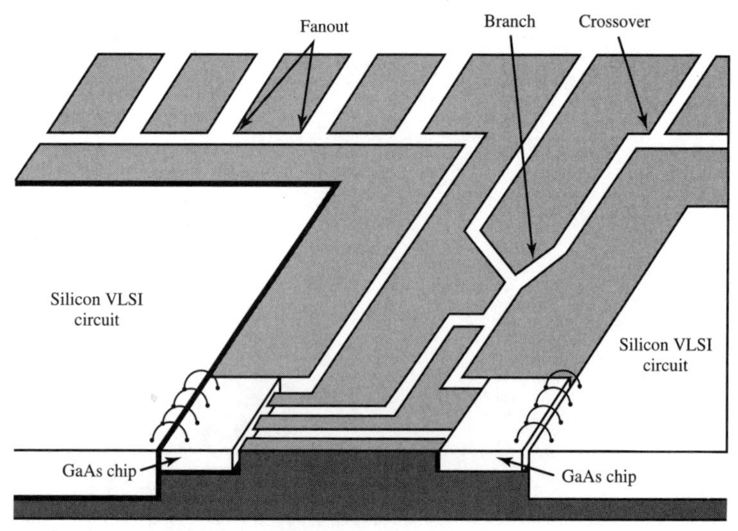

FIGURE 9-39 Thin film optical interconnect.

9.9 FLAT PANEL DISPLAYS

Flat panel displays for lap top computers (and soon televisions) has become a multi-billion dollar industry. There are both passive and active matrix technologies [161]. The cross section of a passive-matrix flat panel display is shown in Fig. 9-40. PI is universally used as the alignment layer. PI is also being used as the matrix for some of the color filter pigments [162]. The glass barrier layers (to prevent sodium migration) are currently SiO_2,

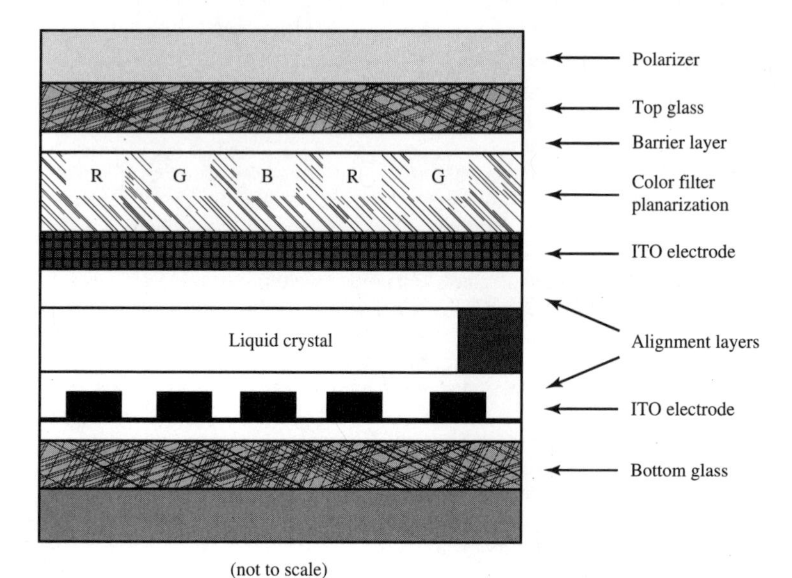

FIGURE 9-40 LCD cross-section.

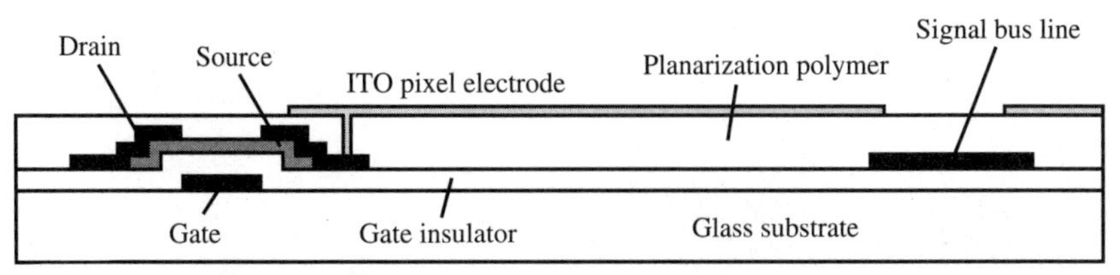

FIGURE 9-41 Cross section view of a high aperature ratio TFT LCD [163].

and the color filter planarization layer is usually acrylic. BCB is being examined as a planarization layer for high-aperature ratio TFT LCDs, as shown in Fig. 9-41 [163].

9.10 SENSORS AND MICROMACHINING

Polymeric thin films have also gained acceptance in many sensor [164] and micromachining [165] applications.

9.11 RECENT ADVANCES IN THIN FILM PROCESSING

Although many avionic, telecommunication (data and video), HDTV, and computer workstation thin film applications are at the advanced prototype stage and are poised for introduction in the late 1990s and the years following, reduction of thin film packaging costs is essential. It is generally accepted that cost will be the main driving force behind widespread implementation of MCM technology. However, as long as MCM technology remains focused on low-volume, high-performance applications, costs are expected to remain high [166].

As the IC industry learned early on, layout of multiple dies on one substrate is the path to better economics. MCM module size has a significant impact on the number of parts that can be produced on a given size substrate (Si, ceramic, or laminate). Since substrates are processed individually, throughput is clearly a linear function of modules per substrate. As such, costs will be a strong function of substrate area utilization. Therefore, it is very important to minimize the size of the part and/or modify its shape during the design phase to reduce the ultimate cost of the MCM module.

There are two basic methods of increasing throughput: (1) fabrication of very small MCM structures, as exemplified by the recent work of AT&T [142,143] in which 9.5 mm $\times$ 6.65 mm substrates are fabricated on a 125-mm wafer; or (2) fabrication of larger MCM substrates on large-area carriers, large-area processing (LAP). This latter concept, first proposed by Motorola/Dow [167] and implemented by Z-Systems and IMC [168], is depicted in Fig. 9-42. It was proposed that modified flat panel display equipment could be used to fabricate such thin film MCM substrates on such large formats [169,170].

Several recent papers have examines the feasibility of large area processing (LAP) for thin film interconnect fabrication [169–171]. Many of the unit operations are based upon equipment and techniques used by the flat panel display (FPD) industry. There are several important deviations of the proposed MCM LAP processing from current FPD manufacturing, including substrate options and potential coating techniques. For LAP thin film fabrication, potential substrates include laminates, glass, and other rigid materials.

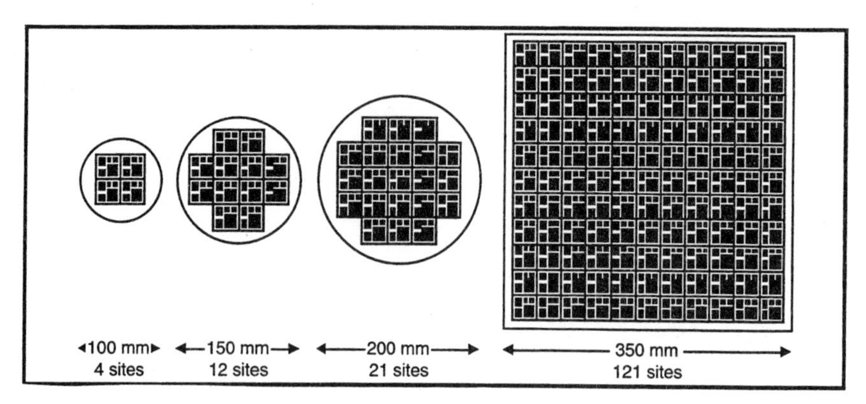

FIGURE 9-42 Increased throughput from LAP.

FIGURE 9-43 Comparison of 150 mm and 400 mm (LAP) thin film processing [171].

9.11.1 Deposition Techniques for LAP

New deposition techniques are required for LAP processing to replace the spin coating typically associated with IC thin film fabrication. This operation introduces the largest change from traditional semiconductor processing. It is assumed that spin coaters will be replaced by high efficiency coaters such as Meniscus, Extrusion, Patch, or Curtain. Recall, curtain coating is the method used by IBM Yasu in their SLC type of LAP processing on laminate (see section 9.5.9). Coating techniques such as extrusion coating and meniscus coating result in greater than 90% efficiency in coating application versus the typical approximately 5% efficiency for spin coating.

The Envision® thin film cost model has been used to determine the economic impact of large-area processing techniques on MCM manufacturing costs [172]. Such large-panel, LCD-like processing is projected to reduce MCM-D substrate costs approximately 10-fold.

Thin film substrates manufactured on a 150-mm line are compared to the same substrates manufactured on the Dow/MMS 400-mm LAP line in Fig. 9-43 [171].

REFERENCES

1. W. Pence and J. Krusius, "The Fundamental limits for Electronic Packaging and Interconnect," *IEEE Trans CHMT,* vol. 10, 1987, p. 176.
2. E. Davidson "The Coming of Age of MCM Packaging Technology," *Proceed Int Conf MCMs,* Denver, 1992, p. 103.
3. N. Yamanaka et al., "Multichip 1.8-Gb/s High Speed Space Division Switching Module," *IEEE Proceed ECTC,* 1990, p. 562.
4. P. E. Garrou, "MCM Technologies: A Global Perspective," *Proceed Japan Ceramics Soc Meeting,* 1994, p. 1.
5. B. McClean (ed.), *Status 1997—A Report on the Integrated Circuit Industry,* Integrated Circuit Engineering Corp., Tempe, AZ, 1997.
6. R. Himmel, J. Licari, "Fabrication of Large-Area, Thin-Film Multilayer Substrtes," *Proceed ISHM,* 1989, p. 454.
7. W. Radliket et al., "MCM-D Technology for a Communication Application," *Proceed Int Conf MCMs,* Denver, 1994, p. 402.
8. L. Buda, R. Gedney, and T. Kelley, "High Density Packaging Future Outlook," *Proceed ECTC,* 1992, p. 36.
9. L. T. Hwang et al., *Proceed SPIE,* Boston, 1990.
10. B. Gilbert and W. Walters, "Design Guidelines for Digital MCMs Operating at High System Clock Rates," *Int J Microcircuits and Electronic Packaging,* 1992, p. 171.
11. T. Shimoto et al., "High Density Multilayer Structure using BCB Dielectric," *Proceed 7th Int Microelectronics Conf,* 1992, p. 325.
12. P. Sullivan, T. Gabara, and R. Frye, "Low Power Considerations in MCMs," *Proceed ISHM,* 1993, p. 577.
13. K. Gustafsson, G. Flodman, and M. Bakszt, "Multichip Modules for Telecom Applications," *Proceed IEPS,* 1990, p. 16.
14. J. Hagge, "Ultra-reliable Packaging For Silicon-on-Silicon WSI," *IEEE Trans CHMT,* vol. 12, 1989, p. 170.
15. C. Bartlett, J. Segelken, and N. Teneketges, "Multichip Packaging Design for VLSI Based Systems," *Proceed ECC,* 1987, p. 518.
16. K. P. Shambrook and P. Trask, "High Density Multichip Interconnect," *Proceed ECTC,* 1989, p. 656.

17. S. Abbasi, "The Technology and Manufacture of the VAX 9000 Multichip Unit," in Doane and Franzione (eds.), *MCM Technology: The Basics,* Van Nostrand Reinhold, New York, 1993.
18. S. Mok, "Volume Implementation of a MCM-D Based Cache SRAM product for Workstation and PC Application," *Proceed Int Conf MCMs,* 1994, p. 320.
19. R. Tummala, in R. Tummala and E. Rymaszewski (eds.), *Microelectronic Packaging Handbook,* Van Nostrand Reinhold, New York, 1989.
20. K. Takeda, Y. Naritsukada, and M. Harada, "High Density Module with Small Packaged LSIs for Mainframe Computers," *Proceed Int Conf MCMs,* Denver, 1993, p. 260.
21. G. B. Leung and S. A. Sands, "A Thin Film on MLC Application," *Proceed ECTC,* 1991, p. 10.
22. A. J. Strandjord, R. H. Heistand, J. N. Bremmer, P. E. Garrou, and T. G. Tessier, "A Photosensitive BCB on Laminate Technology (MCM-LD)," *Proc 44rd ECTC,* 1994, p. 374.
23. H. Hirakawa et al., "A New Aramid/Epoxy Laminate for Advanced SMT," *IEEE Trans CHMT,* vol. 13, 1990, p. 570.
24. J. Diekman and M. Mirhej, "Non-woven Aramid Papers: A New PWB Reinforcement Technology," *Proceed IEPS,* 1990, p. 123.
25. J. Paraszczak et al., "Fabrication and Performance Studies of Multilayer Polymer/Metal Interconnect Structures for Packaging Applications," *Proceed ECTC,* 1991, p. 362.
26. D. Darrow and S. Vilmer-Bagen, "Comparative Analysis of Thin Film Metallization Methodologies for High Density Multilayer Hybrids," *Proceed Int Conf MCMs,* Denver, 1992, p. 56.
27. H. Windischmann, "Intrinsic Stress in Sputtered Thin Films," *J Vac Sci Technol,* A9, 1991, p. 4.
28. E. Myszka et al., "The Development of A Multilayer Thin Film Cu/PI Process for MCM-D Substrates," *Int Conf MCMs,* Denver, 1992, p. 1.
29. P. Kersten, G. Schammler, H. Reichl, "An Additive Approach to Multichip Modules, Using Electroless Metallization Processes," *Int J Microcircuits and Electronic Packaging,* vol. 17, 1994, p. 73.
30. A. Schiltz et al., "Lift-off Techniques for Multichip Modules," *Proceed NEPCON (West),* 1990, p. 975.
31. A. Schlitz, "A Review of Planar Techniques for Multichip Modules," *IEEE Trans CHMT,* vol 15, 1992, p. 236.
32. T. Horton, "MCM Driving Forces, Applications, and Future Directions," *Proceed NEPCON (West),* 1991, p. 487.
33. P. Marella and D. Tuckerman, "Prequalification of n-Chip SiCB Technology," *Proceed Int Conf MCM,* Denver, 1992, p. 376.
34. A. Evans and J. Hagge, "Advanced Packaging Concepts—Microelectronics Multiple Chip Modules Utilizing Silicon Substrates," *Proceed 1st Int SAMPE Elect Conf,* 1987, p. 37.
35. H. Murano and T. Watari, "Packaging Technology for the NEC-SX3 Supercomputer," *IEEE Trans CHMT,* vol. 15, 1992, p. 411.
36. G. Bower and L. Frost, *J Polym Sci A,* vol. 1, 1963, p. 3135.
37. L. Frost and I. Keese, *J Appl Polym Sci,* vol. 8, 1964, p. 1039.
38. S. Numata et al., "Thermal Expansion Behavior of Various Aromatic Polyimides," *J Appl Polym Science,* vol. 31, 1986, p. 101.
39. S. Numata et al., "Chemical Structures and Properties of Low Expansion Polyimides," *Proceed 2nd Int Symp Polyimides,* 1985, p. 164.
40. D. Boese et al., *App Phys Letters,* vol. 59, 1991, p. 1043.
41. W. Volksen et al., "Polyamic Alkyl Esters: Versatile Polyimide Precursors for Improved Dielectric Coatings," *Proceed ECTC,* 1991, p. 572.
42. C. Feger and C. Feger, "Selection Criteria for Multichip Module Dielectrics," in D. Doane and P. Franzone (eds.), *Multichip Module Technologies and Alternatives,* Van Nostrand Reinhold, New York, 1993.

43. B. T. Merriman et al., "New Low CTE Polyimides for Inorganic Substrates," *Proceed ECTC,* 1989, p. 155.
44. A. Saiki et al., "Development of Polyimide Isoindoloquinazolinedione in Multilevel Interconnections for Large-Scale Integration (LSI)," in E. Feit and C. W. Wilkens (eds.), *Polymeric Materials for Electronic Applications,* ACS Symp Series #184, 1982, p. 123.
45. J. Pfeifer and O. Rohde, "Direct Photoimaging of Fully Imidized Solvent-Soluble Polyimides," *Proceed 2nd Int Conf on Polyimides,* 1985, p. 130.
46. H. Neuhaus, "A High Resolution, Anisotropic Wet Patterning Process Technology for MCM Production," *Proceed Int Conf MCMs,* 1992, p. 256.
47. D. Burdeaux, P. Townsend, J. Carr and P. E. Garrou, "Benzocyclobutene (BCB) Dielectrics for the Fabrication of High Density Thin Film Multichip Modules," *J Electronic Materials,* vol. 19, 1990, p. 1357.
48. R. H. Heistand et al., "Advances in MCM Fabrication with Benzocyclobutene Dielectric," *Proceed ISHM,* 1991, p. 96.
49. C. Chao et al., "Multilayer Thin Film Substrates for Multichip Packages," *IEEE Trans CHMT,* vol. 12, 1989, p. 180.
50. P. Rickeri, J. Stephanie, and P. Slota, "Evaluation of Photosensitive Polyimides for Packaging Applications," *Proc ECC,* 1987, p. 220.
51. C. A. Pryde, *J Polymer Sci,* part A, vol. 27, 1989, p. 711.
52. T. Stokich et al., "Real Time FT-IR Studies of the Reaction Kinetics for the Polymerization of DVS-BCB," *MRS Symp Proceed,* vol. 222, 1991, p. 103.
53. R. Heistand et al., "Advances in MCM Fabrication With Benzocyclobutene Dielectric," *J Microcircuits and Microelectronic Packaging,* vol. 15, 1992, p. 183.
54. C. Mohler et al., "Micro ATR as a Probe of BCB Layers for MCM-LD Applications," *Proceed MRS Symp,* vol. 323, 1993, p. 295.
55. P. E. Garrou et al., "Rapid Thermal Curing of BCB Dielectric," *IEEE Trans CHMT,* vol. 15, 1993, p. 46.
56. M. G. Dibbs et al., "Cure Management of BCB Dielectric for Electronic Applications," *Int SAMPE Elect Conf,* vol. 6, 1992, p. 1.
57. T. C. Hodge et al., "Rapid Thermal Curing of Polymer Interlayer Dielectrics," *Int J Microcircuits and Electronic Packaging,* vol. 17, 1994, p. 10.
58. B. Rogers et al., "Soft Mask for Via Patterning in Benzocyclobutene," *Proceed ISHM,* 1993, p. 187.
59. T. G. Tessier et al., "Process Considerations in Fabricating Thin Film Multichip Modules," *Proceed IEPS,* 1989, p. 294.
60. T. G. Tessier et al., "Via Processing Options for MCM-D Fabrication: Excimer Laser Ablation versus Reactive Ion Etching," *Proceed ECTC,* 1991, p. 827.
61. G. Schammler et al., "Comparison of the Metallization of Chemically and Laser-Etched Structures in BPDA-PDA Polyimide," *IEEE Trans CHMT,* vol. 16, 1993, p. 720.
62. T. G. Tessier and G. Chandler, "Compatibility of Common MCM-D Dielectrics with Scanning Laser Ablation Via Generation Processes," *Proceed ECTC,* 1992, p. 763.
63. H. Ahne et al., "PI Patterns Made Directly from Photopolymers," in K. L. Mittel (ed.), *Polyimides,* Plenum Press, New York, 1984, p. 933.
64. J. Kojima et al., "Photosensitive Polyimide for IC Devices," *Proceed ECTC,* 1989, p. 920.
65. E. Perfecto et al., "Factors that Influence Photosensitive PI Lithographic Performance," *Proceed Int Conf MCMs,* Denver, 1993, p. 40.
66. B. Kotzias and D. Murray, "Review of PSPI process for Manufacturing Multichip Modules," *Int J Microcircuits and Elect Pkging,* vol. 16, 1993, p. 339.
67. E. Perfecto et. al., "Engineering PSPIs for MCM-D Applications," *Int J Microcircuits and Elect Pkging,* vol. 16, 1993, p. 319.

68. R. Hubbard and G. Lehman-Lamer, "Very High Speed Multilayer Interconnect Using Photosensitive PI," *Proceed ISHM,* 1988, p. 374.

69. K. Niwa et al., "Patterning Techniques of PSPI on Copper," *Proceed Int Microelect Conf,* Yokohama, 1992, p. 51.

70. S. Sasaki, T. Kon, T. Ohsaki, and T. Yasuda, "A New Multichip Module Using a Copper/PI Multilayer Substrate," *IEEE CHMT,* vol. 12, 1989, p. 658.

71. T. Watari and H. Murano, "Packaging Technology for the NEC SX Supercomputer," *IEEE Trans CHMT,* vol. 8, 1985, p. 462.

72. K. K. Chakravorty et al., *IEEE Trans CHMT,* vol. 13, 1990, p. 200.

73. E. Rutter et al., "A Photodefinable BCB Resin for Thin Film Microelectronic Applications," *Proceed Int MCM Conf,* 1992, p. 394.

74. E. Moyer et al., "Photodefinable BCB Formulations for Thin Film Microelectronics Applications, Part II," *Proceed IEPS,* 1992, p. 37.

75. A. Strandjord et al., "Process Optimization and Systems Integration of a Cu photo BCB MCM-D," *Int J Microcircuits and Microelectronic Packaging,* 1996, p. 260.

76. T. Shimoto, K. Matsui, and K. Utsumi, "Cu/PhotoBCB Thin Film Multilayer Technology for High Performance Multichip Modules," *Proceed Int Conf MCMs,* Denver, 1994, p. 115.

77. M. Skinner et al., "Twinstar Dual Pentium Processor Module," *J Microcircuits and Microelectronic Packaging,* 1996, p. 358.

78. E. Sweetman, "Characteristics and Performance of PHP-92: AT&T's Triazine-Based Dielectric for Polyhic MCMs," *Proceed Int Conf MCMs,* Denver, 1992, p. 401.

79. Y. Tsukada, "Design and Performance of SLC-MCM," *Proceed NEPCON (West),* p. 521, 1994.

80. P. Knudsen and R. Brainard, "A Photoimangeable Dielectric for Sequential PWB Fabrication," *Proc Surface Mount Int,* 1993, p. 351.

81. G. Clatterbaugh and H. Charles, "The Application of Photosensitive Polyimide Dielectrics in Thin Film Multilayer Hybrid Circuit Structures," *Proceed ISHM,* 1988, p. 320.

82. T. G. Tessier et al., "Polymer Dielectric Options for Thin Film Packaging Applications," *Proceed ECC,* 1989, p. 127.

83. A. Kimura et al., "Si-on-Si Technology: Process and Characteristics of Si Carrier Substrate," *Proceed Int Microelect Conf,* Yokohama, 1992, p. 337.

84. R. Jensen et al., "Characteristics of Polyimide Material for Use in Hermetic Packaging," *Proceed VHSIC Pkging Conf,* Houston, 1987, p. 193.

85. J. McDonald et al., "Technique for the Fabrication of Wafer Scale Interconnections in Multichip Packages," *IEEE Trans CHMT,* vol. 12, 1989, p. 195.

86. L. B. Rothman, "Properties of Thin Polyimide Films," *J Electrochem Soc, Solid State Science and Tech,* 1980, p. 2216.

87. T. Stokich et al., "Planarization with Cyclotene 3022 (BCB) Polymer Coatings," *Material Research Soc Symp Proceed,* vol. 308, 1994, p. 517.

88. T. Tessier, "A Comparison of Common MCM-D Dielectric Material Performance," *Proceed 6th SAMPE Electronics Conf,* 1992, p. 347.

89. D. Day et al., "Polyimide Planarization in Integrated Circuits," in K. L. Mittal (ed.), *Polyimides,* Plenum Press, 1984, p. 767.

90. C. C. Chao and W. Wang, in K. L. Mittal (ed.), *Polyimides,* Plenum Press, 1984, p. 783.

91. T. Stokich et al., "Thermal and Oxidative Stability of Polymer Thin Films Made from DVS-BCB," *MRS,* Anaheim, 1991.

92. J. Pan and S. Poon, "Film Stress in High Density Thin Film Interconnect," *MRS Symp Proceed,* vol. 154, 1989, p. 27.

93. J. Greenblatt, C. Araps, and H. Anderson, "Aminosilane Polyimide Interactions and Their Implications in Adhesion," in K. L. Mittal (ed.), *Polyimides,* Plenum Press, 1984, p. 573.

94. H. Sotou, "PIQ PI for Multichip Module Applications," *Proceed NEPCON (West),* 1989, p. 921.
95. K. W. Paik and A. Ruoff, *J Adhesion Science Tech,* part A, vol. 6, 1988, p. 1004.
96. R. Hannemann, "Electronic Systems Thermal Design for Reliability," *IEEE Trans Reliability,* R-26, 1977.
97. M. Mahalingham, "Thermal Management in Semiconductor Device Packaging," *Proceed IEEE,* vol. 73, 1985.
98. I. Turlik and R. Darveaux, "Thermal Management," in P. Garrou and I. Turlik (eds.), *Multichip Module Handbook,* McGraw-Hill, 1997, Chap. 13.
99. C. Ryan, A. Keeley, "Correlation of Modeled Thermal Resistance and Experimental Measurements . . . ," *Proceed IEPS,* 1991, p. 447.
100. S. Chambers and K. Chakravorty, "Oxidation at the PI/Cu Interface," *J Vacuum Science Tech,* vol. A6, 1988, p. 3008.
101. M. Burrell et al., "Study of the Enhanced Oxidative Degradation of Polymer Films at Polymer/Copper Oxide Interfaces," *J. Vaccuum Science Tech,* vol. A, 1988, p. 2893.
102. D. Shih et al., "A Study of the Chemical and Physical Interaction Between Copper and Polyimide," *J Vaccuum Science Tech,* vol. A7, 1989, 1402.
103. P. E. Garrou, "Polymer Dielectrics for Multichip Module Packaging," *Proceed IEEE,* vol. 80, 1992, p. 1942.
104. J. Cech, F. Burnett and C. Chien, "Reliability of Passivated Copper Multichip Module Structures Embedded in PI," *IEEE Trans CHMT,* vol. 16, 1993, p. 752.
105. S. Rao, "TEM Examination of MCM-D Substrate with Copper Conductors," *Proceed Int Conf MCMs,* Denver, 1992, p. 286.
106. G. Adema et al., "Passivation Schemes for Cu/PI Thin Film Interconnections Used in MCMs," *Proceed ECTC,* 1992, p. 776.
107. R. Heistand et al., "Advances in MCM Fabrication with BCB Dielectric," *Int J Microcircuits and Microelect Packaging,* vol. 15, 1992, p. 183.
108. C. Shiflett et al., "High Density Multilayer Hybrid Circuits Made with Polymer Insulating Layers (Polyhics)," *Proceed ISHM,* 1986, p. 481.
109. A. Shah, E. Sweetman, and C. Hoppes, "A Review of AT&T's PolyHIC MCM Technology," *Proceed NEPCON (West),* 1991, p. 850.
110. M. Lau et al., "A Versatile, IC Process Compatible MCM-D for High Performance and Low Cost Applications," *Proceed Int Conf MCMs,* Denver, 1993, p. 107.
111. R. Day et al., "A Silicon-on-Silicon Multichip Technology with Integrated Bipolar Components in the Substrate," *IEEE MCM Conf,* Santa Cruz, 1994, p. 64.
112. K. Prasad and E. Perfecto, "Multilevel Thin Film Packaging: Applications and Processes for High Performance Systems," *IEEE Trans CHMT,* part B, vol. 17, 1994, p. 38.
113. T. Redmond et al., "Polyimide Copper Thin Film Redistribution on Glass Ceramic/Copper Multilevel Structures," *Proceed ECTC,* 1991, p. 689.
114. E. Perfecto et al., "Multi-level Thin Film Packaging Technology at IBM," *Proceed Int MCM Conf,* Denver, 1993, p. 474.
115. M. Bregman et al., "A Thin Film MCM for Workstation Applications," *Proceed ECTC,* 1992, p. 968.
116. A. Kimura et al., "Fabrication and Characteristics of Silicon Carrier Substrates for Silicon-on-Silicon Packaging," *Proceed Int Conf MCMs,* Denver, 1992, p. 23.
117. B. McWilliams, "Comparison of Multichip Interconnect Technologies," *Proceed IEPS,* 1991, p. 63.
118. D. Pierson, S. Drobac, and D. Parry, "An 80-Mhz MIPS R6000 CPU Using Multichip Module Technology," *Proceed NEPCON (West),* 1993, p. 954.
119. J. Denman, "n-Chips Silicon Circuit Board Technology," *Proceed NEPCON (West),* 1994, p. 2039.

120. R. Fillion et al., "Non-Digital Extensions of Embedded Chip MCM Technology," *Proceed Int Conf MCMs,* Denver, 1994, p. 464.

121. R. Fillion et al., "Development of a Plastic Encapsulated Power Multichip Technology for High Volume, Low Cost Commercial Electronics," *Proceed ISHM,* 1994, p. 84.

122. R. Fillion, "Non-Digital Extensions of an Embedded Chip MCM Technology," *Proceed Int Conf MCMs,* Denver, 1994, p. 464.

123. J. Strandberg et al., "High Reliability 4-layer MCM-D Structure with BCB as Dielectric," *Proceed Int MCM Conf,* Denver, 1995, p. 223.

124. J. Droguet, "A Pilot Line for MCM Substrates," *Proceed 8th ISHM Europe,* Rotterdam, 1991, p. 280.

125. J. Droguet, "An MCM-D Foundry From Copper/PI TFML Bare Substrates to MCMs," *Proceed Int Conf MCMs,* Denver, 1994, p. 384.

126. Y. Tsukada, Y. Maeda, and K. Yamanaka, "A Novel Solution For MCM-L Utilizing Surface Laminar Circuit and Flip Chip Attach Technology," *Proceed ICMCM,* Denver, 1993, p. 252.

127. Y. Tsukada et al., "Surface Laminar Circuit and Flip Chip Attach Packaging," *Proceed IMC,* Yokohama, 1992, p. 252.

128. J. Tanaka et al., "A Multichip Module PI on Mullite Ceramic," *Proceed IEPS,* 1990, p. 421.

129. R. Kambe et al., "Cu/PI Multilayer Substrate for High-Speed Signal Transmission," *Proceed ECTC,* 1991, p. 14.

130. R. Tummala et al., "Packaging Technology For IBM's Latest Mainframe Computers (S/390/ES9000)," *Proceed ECTC,* 1991, p. 682.

131. D. Akihiro et al., "Packaging Technology for the NEC SX-3/SX-X Supercomputer," *Proceed ECTC,* 1990, p. 525.

132. T. Inoue et al., "Microcarrier for LSI Chip Used in the HITAC M-880 Processor Group," *IEEE Trans CHMT,* vol. 15, 1992, p. 7.

133. D. Tuckerman et al., "A High-Performance Second Generation Sparc MCM," *Proceed Int Conf MCMs,* Denver, 1994, p. 314.

134. R. Brown and R. Boone, "A Comparison of MCM Packaging Technologies for a 32-Bit High-Performance Processor," *Proceed ISHM,* 1993, p. 562.

135. R. E. Terrill et al., "Aladdin, Lessons Learned," *Proceed Int MCM Conf,* Denver, 1995, p. 7.

136. W. Jacobsen et al., "Application of MCM-C and MCM-D for Spaceborn Data Processors," *Proceed Int Conf MCMs,* Denver, 1993, p. 525.

137. T. Ohsaki, "Electronic Packaging in the 1990s—A Perspective From Asia," *IEEE Trans CHMT,* vol. 14, 1991, p. 254.

138. N. Teneketges et al., "A High-Performance MCM-Based Spaceborn Processor," *Proceed Int Conf MCM,* Denver, 1992, p. 183.

139. Y. Doi et al., "An ATM Switch Hardware Technology Using Multichip Packaging," *IEEE Trans CHMT,* vol. 16, 1993, p. 60.

140. J. Noren and P. Brofman, "Interconnection Considerations for a Hybrid MCM," *Proceed Int Conf MCMs,* Denver, 1994, p. 299.

141. R. Frye et al., "Silicon-on-Silicon MCMs with Integrated Passive Components," *Proceed IEEE MCM Conf,* Santa Cruz, CA, 1992, p. 155.

142. D. Cokely and C. Strittmatter, "Redefining the Economics of MCM Applications," *Proceed Int Conf MCMs,* Denver, 1994, p. 306.

143. T. Dudderar et al., "AT&T μSMT Assembly: A New Technology For Large-Volume Fabrication of Cost Effective Flip-Chip MCMs," *Proceed Int Conf MCMs,* Denver, 1994, p. 266.

144. A. Wilson, "Use of Polyimides in VLSI Fabrication," in K. L. Mitel (ed.), *Polyimides,* 1984, Plenum Press, p. 715.

145. A. Strandjord et al., "Photosensitive BCB for Stress Buffer and Passivation Applications: A One-Mask Manufacturing Process," *Proceed ECTC,* San Jose, CA, 1996, p. 1260.

146. S. Bothra et al., "Analysis of the Effects of Scaling on Interconnect Delay in ULSI Circuits," *IEEE Trans Electrion Devices,* vol. 40, 1993, p. 591.

147. G. Samuelson, "Polyimide for Multilevel VLSI," in Feit and Wilkins (eds.), *Polymers for Electronic Applications,"* Amer Chem Soc Symp Series, vol. 184, 1982, p. 93.

148. D. Gardner et al., "Encapsulated Copper Interconnection Devices Using Sidewall Barriers," *Proceed VMIC Conf,* 1991, p. 99.

149. C. Kaanta et al., "Dual Damascene: A ULSI Wiring Technology," *Proceed VMIC,* 1991, p. 144.

150. S. Roehl et al., "High-Density Damascene Wiring and Borderless Contacts for 64 M DRAM," *Proceed VMIC,* 1992, p. 22.

151. S. Bothra, M. Kellam, and P. Garrou, "Feasibility of BCB as an Interlevel Dielectric in Integrated Circuits," *J Electronic Materials,* vol. 23, 1994, p. 819.

152. Y. Hayashi et al., "A New Two-Step Metal-CMP Technique for High-Performance Multilevel Interconnects Featured by Al and Cu in Low ϵ Organic Film," *IEEE 1995 Symposium on VLSI Technology-Digest of Technical Papers,* 1995, p. 88.

153. C. Case, A. Kornblit, and J. Sapjeta, "Evaluation of Cyclotene 5021 as a Low-Dielectric Constant ILD," *Proceed VMIC Conf,* 1996, p. 63.

154. E. Finchem et al., "A Multilevel High-Density Interconnect Process Designed and Developed for Manufacturability," *U.S. Conf on GaAs Manufacturing Technology (Mantech) Proceed,* 1994, p. 163.

155. D. Mis et al., "Flip Chip Production Experience," *Proceed ISHM,* 1996, p. 291.

156. J. Simon, M. Topper, H. Reichl, and G. Chimel, "A Comparison of Flip Chip Technology with Chip Size Packages," *Proceed IEPS,* 1995, p. 665.

157. M. Robertsson et al., "Optical Interconnects in Packaging for Telecom Appl," *Proceed 10th European Microelectronics Conf,* 1995, p. 581.

158. G. Palmkog et al., "Low-Cost Single Mode Optical Passive Coupler Devices with an MT-Interface Based on Polymeric Waveguides in BCB," in press.

159. J. Bristow et al., "Optical Polymer Waveguide Interconnect Technology," *Proceed ISHM,* 1996, p. 138.

160. Y. Liu et al., "Optioelectronic Packaging and Polymer Waveguides for MCM and Board-Level Optical Interconnect Applications," *Proceed 45th ECTC,* Las Vegas, 1995, p. 185.

161. W. O'Mara, "Active Matrix Flat Panel Displays," *Solid State Technology,* vol. 34, 1991 (Dec), p. 65.

162. W. Latham and D. Hawley, "Color Filters from Dyed Polyimides," *Solid State Technology,* vol. 31, 1988 (May), p. 223.

163. J. Lan et al., "Electrical and Optical Properties of Low-Dielectric Constant Planarization Polymer for High-Aperature Ratio a-Si:H TFT LCDs," *Proceed Material Research Soc Symp,* Spring, 1997, in press.

164. G. Harsanyi, "Polymeric Films in Microelectronic Sensors," *Proceed ISHM,* 1996, p. 191.

165. M. Allen, "PI Processes for the Fabrication of Thick Electroplated Structures," *Proceed 7th Int Conf on Solid State Sensors and Actuators,* 1993, p. 60.

166. R. Frye and A. Shah, "Targeting Low-Cost, High-Volume MCM Applications," *Proceed Int MCM Conf,* Denver, 1993, p. 12.

167. T. Tessier and P. Garrou, "Overview of MCM Technologies: MCM-D" *Proceed ISHM,* 1992, p. 235.

168. C. Hodges et al., "Large Panel Format MCM Substrates," *Proceed Int MCM Conf,* Denver, 1993, p. 125.

169. A. Strandjord, P. E. Garrou, R. Heistand, and T. Tessier, "MCM-LD: Large Area Processing Using PS-BCB," *IEEE Trans CPMT,* part B, vol. 18, 1995, p. 269.

170. G. White et al., "Large Format Fabrication—A Practical Approach to Low-Cost MCM-D," *Proceed Int MCM Conf,* Denver, 1994, p. 88.
171. E. Chieh et al., "Development of Large Area Processing for Thin Film Substrates," *Proceed Int MCM Conf,* Denver, 1997, p. 234.
172. D. Frye, M. Skinner, R. Heistand, P. E. Garrou, and T. Tessier, "Cost Implications of Large Area MCM Processing," *Proceed Int MCM Conf,* Denver, 1994, p. 69.

CHAPTER 10
THIN FILM FOR MICROWAVE HYBRIDS

Richard Brown
Richard Brown Associates, Inc.

10.1 INTRODUCTION

Microwave electronics are found in military applications such as radar; commercial uses such as satellite communications, microwave relay links, and personal communications; and in science. Examples are radio astronomy and spectroscopy. Microwave electronics may be accomplished either by monolithic microwave integrated circuits (MMICs) or hybrid microwave integrated circuits (MICs). MMICs have reasonable cost, high performance, small size, and reproducibility. MICs, on the other hand, are still required to provide small-volume packages with passive-support electronics. Combinations of active and passive devices often provide the optimum solution by offering economical design, reliability, and flexibility. The advantages of solid-state devices can only be achieved if the supporting elements and components can be accurately reproduced in similar size. Thus thin film technology is used almost exclusively to make the components necessary to meet performance requirements.

This chapter will present the design information necessary to understand the materials and processing requirements of microwave hybrids. Information on materials technology including substrates, dielectrics, conductors, and resistive films follow. The various fabrication techniques for thin film circuits are discussed in light of their applicability for microwave circuitry. Selected packaging requirements for high-speed applications are also reviewed.

Any discussion of microwaves should mention the contributions of Faraday, Maxwell, and Hertz. Faraday fostered the field concept, and Maxwell discovered that light, by its very nature, is electromagnetic wave. He developed a mathematical theory of electrostatics and magnetostatics in terms of fields, extending the theory to include Faraday's laws of electromagnetic frequency. The essence of the classical Maxwellian equations is the theoretical prediction of the existence of electromagnetic waves, and was the starting point for the concept of an electromagnetic spectrum that extends from dc to γ rays. Hertz confirmed in 1888 Maxwell's theories of electromagnetic propagation. Indeed, Hertz may be considered the first microwave engineer, and possibly the only one for at least a generation. In 1893, Hertz conducted a series of experiments proving beyond doubt the existence of wave motion [1]. Although the propagation of electromagnetic (EM) waves through metal pipe was used by German investigators before the turn of the century, and scientific interest continued in this new field, it was almost 40 years before microwaves were commercially applied. In the 1930s, a growing need for short- and long-distance communication technology rapidly led to the development of radar. The subsequent move to higher frequencies led to increasingly smaller components, the first steps toward miniaturization. In the 1960s, the use of chromium (Cr) doping of gallium arsenide (GaAs) by Cronin and Haisty to achieve high-resistivity substrate growth [2] and the development of solid state GaAs devices by Mead [3] had profound implications in the microwave industry by truly miniaturizing microwave circuitry.

Materials and process requirements for microwave hybrids are far more stringent than those for low-frequency circuits. Conductor and substrate properties are of particular importance because they substantially affect circuit design and performance. Many of the advantages of solid-state devices can only be achieved if the MIC circuitry is of comparable

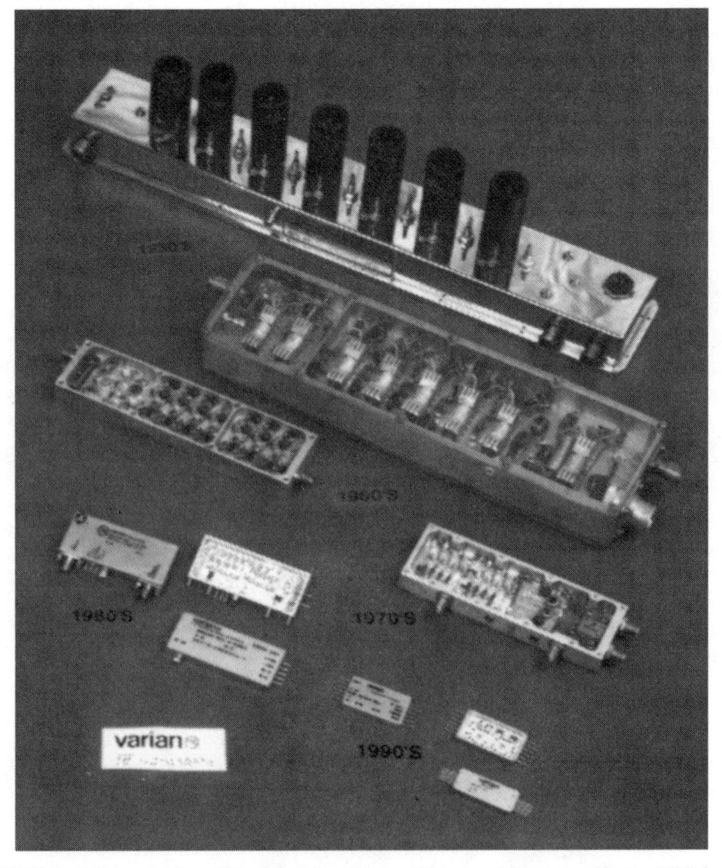

FIGURE 10-1 Miniaturization of logarithmic amplifier. *(Photograph courtesy of Varian RF Subsystems, Beverly, MA. The assistance of Dr. David Stevenson is greatly appreciated.)*

size. Hybrid technologists have been steadily improving their ability to produce sophisticated, smaller, low-cost circuits, while simultaneously improving reliability. Continuing efforts in size and cost reduction are shown in the design of the logarithmic amplifier shown in Fig. 10-1. The original two-pound amplifier, with a volume of 140 in^3, has been systematically shrunk over the last four decades to only 0.1 in^3 and a weight of 0.5 oz [4]. This remarkable reduction in size and weight has been accomplished with a corresponding increase in performance and reliability.

10.2 PLANAR TRANSMISSION STRUCTURES

In the early 1950s, new waveguides were developed that would allow the microwave industry to catch up with the miniaturization taking place in the electronics industry. Therefore, development of innovative and complex circuits using waveguides, coaxial cables, and two-wire transmission systems was severely hindered by the high cost and other constraints of these systems. These new types of transmission lines, shown in Fig. 10-2 as planar, are used in conventional printed-circuit techniques. These milestone developments provided designers with new degrees of design freedom.

The most commonly used type of transmission line is illustrated in Fig. 10-2*a*, the microstrip, which consists of a dielectric substrate with a narrow conductor on one surface, and a ground plane on the other side. The coplanar waveguide is formed by moving the microstrip ground plane to the upper surface of the dielectric and splitting it so that the resultant ground electrodes run adjacent and parallel to the signal conductor, Fig. 10-2*b*. The stripline, Fig. 10-2*c* utilizes a center conductor buried within the dielectric and sandwiched

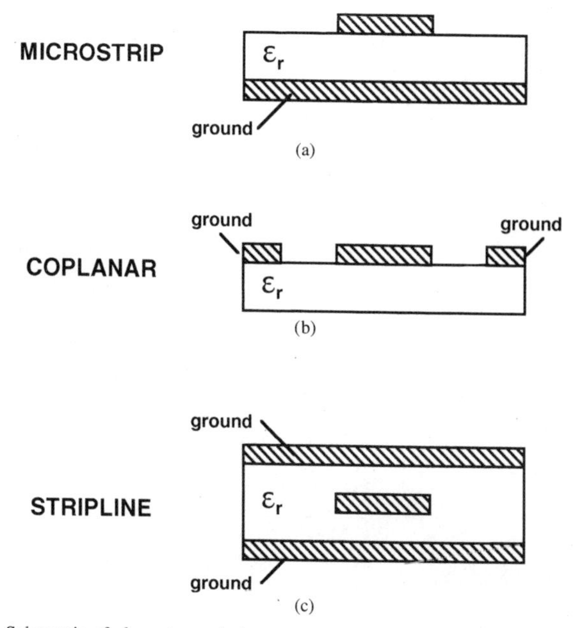

FIGURE 10-2 Schematic of planar transmission structures.

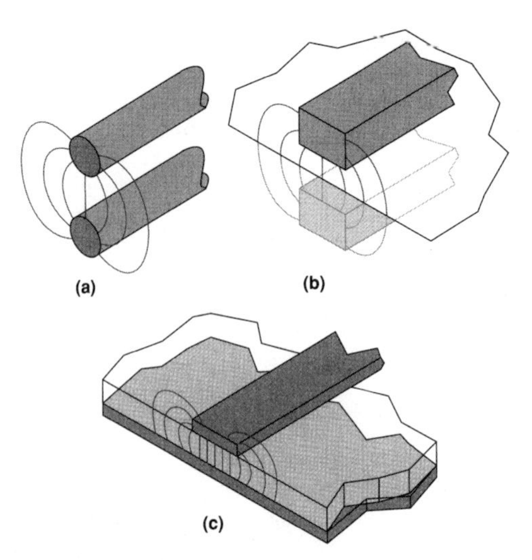

FIGURE 10-3 Evolution of microstrip transmission line. (*a*) Two-wire conductor; (*b*) single wire above ground with image; (*c*) microstrip.

between two ground planes. The planar transmission systems on which microwave-printed circuit techniques are based, can be thought of as a progression of either the coaxial or parallel-line transmission systems.

Parallel-wire transmission line can be converted to a planar configuration by transforming the round wires to rectangular strips separated by a dielectric. The evolution of this concept, shown in Fig. 10-3, was pioneered by the Federal Communications Research Laboratories in 1952 [5] as Microstrip. The field in microstrip line is predominantly confined to the area beneath the conductor strip and the ground plane. However, as an unshielded system it is more difficult to confine all the energy in the vicinity of the strip, and the dielectric medium plays an even more important role. This chapter shows how the properties of the dielectric affect the design and performance of these planar transmission circuits.

It was not until late 1969 that work by Wen [6] produced the coplanar waveguide (CPW) configuration. This planar configuration, shown in Fig. 10-4, is formed by moving the microstrip ground plane to the upper surface of the substrate, and splitting it so that the resultant ground electrodes run adjacent and parallel to the signal strip, forming a unipla-

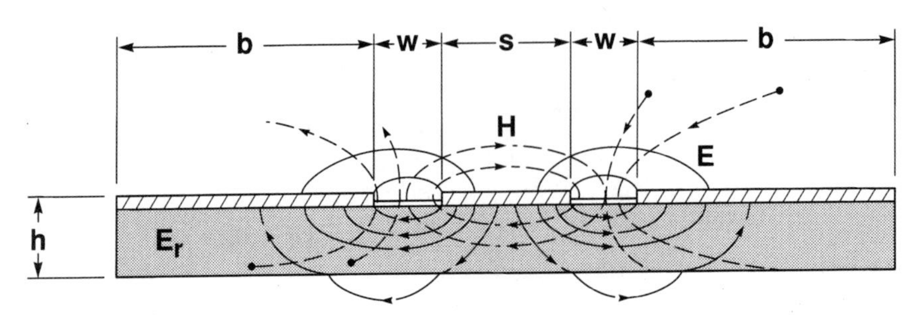

FIGURE 10-4 Coplanar waveguide. (Note the ground plane has moved to the top surface.)

nar transmission structure. The electric field of the CPW, shown in Fig. 10-4, is comprised almost exclusively of fringing fields. In early notational convention, the dimension s represented the width of the signal conductor, and w the gap between conductor and ground plane. CPW may be thought of as two slot lines joined at the center of the signal line along its length. Many authors assign values of a and b to each side of the slot line so that

$$a = s/2 \tag{1}$$

$$b = s/2 + w \tag{2}$$

The a, b notations have been used for ease of comparison.

Round coaxial line can be transformed so that the center and outer conductors are square or rectangular, and the side walls extended to infinity, Fig. 10-5. The resultant flat transmission system, first reported by Barrett and Barnes [7] in 1951, has a form factor amenable to printed circuit technology, yet maintains the advantages of the co-axial system. The entire field is essentially confined to a region in the immediate vicinity of the center conductor, and is uniformly distributed within the dielectric, directly between the center conductor and the outer ground planes. This technology was refined by Sanders Associates (Tri-plate) and Airborne Instrument Laboratories (Strip-Line). Stripline has become the generic term. Table 10-1 compares the three waveguide structures.

The starting point for the design of any microwave hybrid module begins with an understanding of the material properties and fabrication methods used. Currently, substrates for hybrid circuitry are available with a wide variety of properties, as opposed to MMICs, where the dielectric constant is fixed. Materials had been a key drawback to MIC development in the 1950s, but the following decade had a period of rapid growth from new and

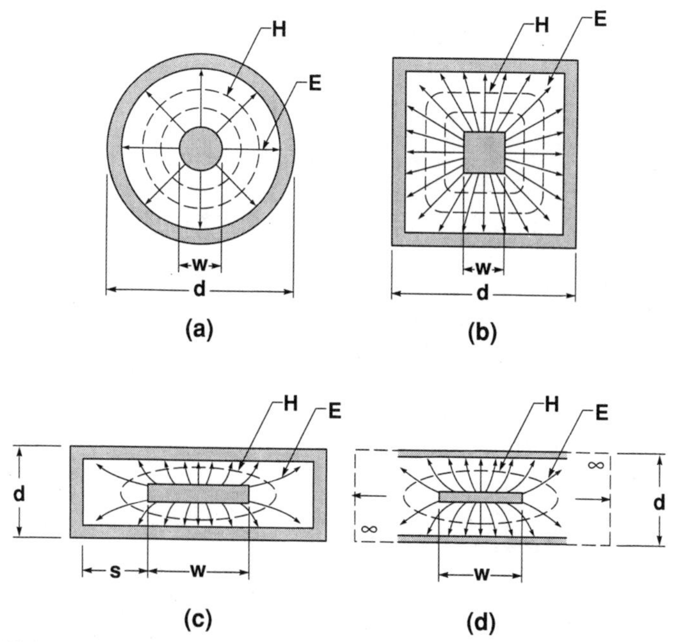

FIGURE 10-5 Evolution of a flat strip transmission line. (*a*) Coaxial line; (*b*) square line; (*c*) rectangular line; (*d*) stripline.

TABLE 10-1 Comparison of MIC waveguide structures

Characteristic	Microstrip	Coplanar waveguide	Stripline
Dielectric constant	10	10	2.8
Effective dielectric constant	6.5	5.5	2.8
Practical impedance range (Ω)	20-125	40-110	5-150
Power handling capability	High	Medium	Low
Radiation loss	Low	Very low	Very low
Unloaded Q	Medium	Low	High
Dispersion	Medium	Very low	Low
Component assembly			
Series configuration	Easy	Easy	Easy
Shunt configuration	Difficult	Easy	Moderate
Fabrication difficulties	Narrow lines and spaces, access to ground	Narrow lines and spaces	Double-sided etching Temperature limitations
Relative circuit size	Smallest	Larger than microstrip	Largest

improved electronic materials. During the 1960s, material suppliers introduced new microwave substrate materials for stripline and microstrip applications. The cross-linked and glass-reinforced polystyrenes and epoxy-based materials were supplanted by woven or randomly fiber-filled polytetrafluoroethane (PTFE). High density, fine-grained ceramics began to appear. High-dielectric-constant ceramics ($\epsilon_r > 10$) were developed with a wide range of dielectric properties, and materials with higher thermal conductivity also emerged. Improvements in polishing these ceramics allowed for the metallization of reproducible thin film circuits. The MIC industry also adapted technologies from industries ranging from the semiconductor to printed wiring board, refining them for batch, low-cost processes. All of these developments stimulated the growth of microwave hybrids in a wide spectrum of applications.

MIC requirements on materials and processes are far more stringent than those for low-frequency circuits. Conductor and substrate properties are of particular importance. Furthermore, many of the advantages of solid-state devices can only be achieved if the MIC circuitry is of comparable size. Toward this end, hybrid technologists have been steadily improving their ability to batch produce sophisticated, smaller, low-cost microcircuits.

10.3 TRANSMISSION LINE PARAMETERS

10.3.1 Dielectric Properties

The way an electric or a magnetic field propagates inside a medium depends on the dielectric, magnetic, and resistive properties of the individual materials. In dielectric materials where charges are rigidly fixed and cannot move under an applied field, the permittivity (ϵ) is a qualitative measure of the ease with which the dielectric can be polarized to form dipoles to cancel an applied electric field, in effect a measure of the dielectric medium's ability to store energy.

$$\epsilon = \epsilon_o \epsilon_r \quad (3)$$

where in free space $\epsilon_o = 8.854 \times 10^{-12}$ F/m

The permittivities of other media are expressed as the relative dielectric constant ϵ_r, where

$$\epsilon_r = \frac{\epsilon}{\epsilon_o} \quad (4)$$

Air, where $\epsilon = \epsilon_o$, has a dielectric constant of 1. The values of other media are tabulated in their appropriate sections.

Similarly, the permeability and the relative permeability are defined by

$$\mu = \mu_o \mu_r \quad (5)$$

where $\mu_o = 4.0\pi \times 10^{-7}$ H/m, and

$$\mu_r = \frac{\mu}{\mu_o} \quad (6)$$

μ_r is typically 1 for nonmagnetic materials.

10.3.2 Wavelength

The term *microwave frequencies,* as used here, commonly refers to electromagnetic radiation from about 1 to 100 GHz. Figure 10-6 indicates the locations of the frequencies of interest along with their associated free-space wavelengths. The free-space wavelength (λ_o) may be readily calculated from the basic relationship between the propagation velocity of light and the frequency, Eq. 7, where

$$\lambda = \frac{c}{f} \quad (7)$$

where c is the velocity of light in free space (in m/s) and f is the frequency in hertz, and

$$c = \frac{1}{\sqrt{\mu_o \epsilon_o}} \sim 3 \times 10^8 \text{ m/s} \quad (8)$$

An FM signal is about 100 (MHz), or 100 million Hz. At 1 GHz the frequency is 1 billion Hz, and at 16 GHz, a common microwave operating frequency for satellite communications, 16 billion (16×10^9) Hz.

The propagation or phase velocity of light, v_p, in a medium is

$$v_p = \frac{1}{\sqrt{\mu\epsilon}} = \frac{1}{\sqrt{\mu_o \mu_r \epsilon_o \epsilon_r}} = \frac{c}{\sqrt{\mu_r \epsilon_r}} \quad (9)$$

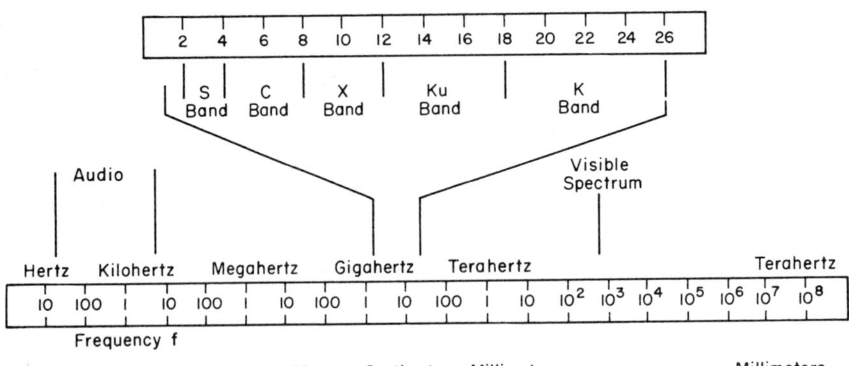

FIGURE 10-6 Electromagnetic spectrum.

For nonmagnetic materials, $\mu_r = 1$, and Eq. 9 reduces to

$$v_p = \frac{c}{\sqrt{\epsilon_r}} \tag{10}$$

The wavelength in a homogeneous medium, or the guide length (λ_g) is defined as

$$\lambda_g = \frac{c}{f\sqrt{\mu_r \epsilon_r}} \tag{11}$$

Again for a homogeneous, nonmagnetic medium, $\mu_r = 1$ and Eq. 11 reduces to

$$\lambda_g = \frac{c}{f\sqrt{\epsilon_r}} \tag{12}$$

The effective dielectric constant (ϵ_{eff}) takes into account the inhomogeneous dielectric nature of transmission structures with mixed dielectrics, such as microstrip and coplanar structures. It accounts for the differences in electrical environment above and in the substrate by considering the relative substrate dielectric ϵ_r, and the air above ($\epsilon_r = 1$).

The guide length in a nonhomogeneous medium is given by

$$\lambda_g = \frac{c}{f\sqrt{\epsilon_{\text{eff}}}} \tag{13}$$

Size reduction in distributed circuits may then be attained using high-dielectric-constant materials. The guide length, λ_g, is also reduced as the frequency increases.

10.3.3 Field Concepts

What makes life relatively easy at the lower frequencies, but difficult at microwave frequencies, is the size of the signal wavelength relative to the signal processing circuitry.

At low frequencies, the wavelengths are very large. At microwave frequencies, the wavelengths are relatively small. For example, at 10 GHz the free-space wavelength is only 3 cm (slightly over an inch) and comparable to the circuit size. The wavelengths at microwave frequencies are of the same order of magnitude as the dimensions of the circuit devices, and the signal propagation time between circuit parts is comparable to the period of the sinusoidal currents. As such, conventional circuit concepts of voltage and current must be replaced by field concepts. In conventional circuitry, the potential difference (V) between two points (y) usually means the line integral of the electric field strength (E):

$$V = \int E \cdot dy \tag{14}$$

taken at an instantaneous time along a trace connecting the two points. This idea is viable only if Eq. 14 is independent of the trace. This is indeed the case when the wavelength is large compared to the trace length. If, however, the wavelength is small compared to the trace length, the line integral depends upon the trace, and therefore the significance of the term "voltage" is lost. Electric and magnetic fields then, must be substituted for concepts of voltage and current.

Maxwell's field equations and their solutions are beyond the scope of this chapter. It will suffice to point out that as a result of his generalization, a close correspondence between circuit and field concepts can be established. For example, the power at high frequency is given by:

$$P = E \cdot H \tag{15}$$

Similarly, corresponding to the formula for power in conventional circuitry:

$$P = V \cdot I \tag{16}$$

The three parameters which are of most importance to transmission lines are the characteristic impedance (Z_o), the line wavelength (λ_g), and the attenuation constant (α). These parameters are very sensitive to conductor shape and the conductivity of the metallization, the discontinuities (surface flaws), and the nature of the dielectric medium.

10.3.4 Impedance

In free space, a propagating sinusoidal electric field generates a magnetic field of equal amplitude so their ratios are always the same, as illustrated in Fig. 10-7. Although free space has no mechanical resistance to motion, it will impede any electromagnetic wave propagation. The ratio of the electric field strength (E) to the magnetic field strength (H) is known as the characteristic or wave impedance (η_o) of the medium. In Standard International (SI) units, E is expressed in volts/meter, and H is expressed in amperes/meter. Z_o indicates the characteristic impedance in ohms.

The electric and magnetic field strengths are related to the permittivity and the permeability of the medium, respectively. The characteristic impedance of free space (η_o) may be expressed as their ratio,

$$\eta_o = \sqrt{\frac{\mu_o}{\epsilon_o}} = 120\pi\Omega = 377\ \Omega \tag{17}$$

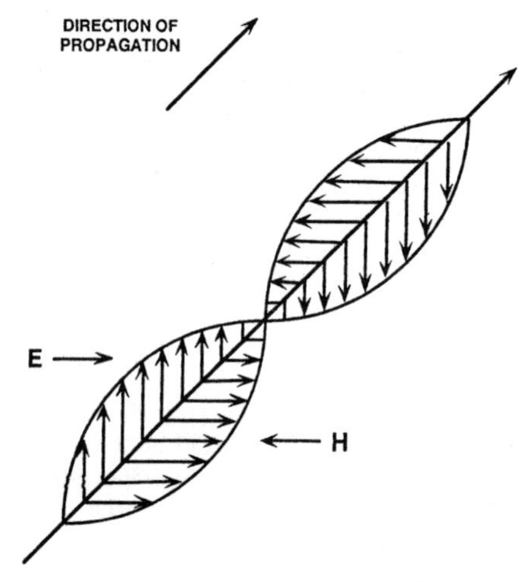

FIGURE 10-7 Schematic of electromagnetic wave showing mutually perpendicular E and H fields.

However, transmission lines are comprised of metals and operate in a dielectric media with properties very different than free space. An electromagnetic wave may travel unimpeded in free space, but the situation is different inside the substrate. The impedance of these lines depend on the geometry and properties of the media in which they are located. A schematic of the distributed parameters of a transmission line is shown in Fig. 10-8.

Lumped elements (Δy) represent infinitesimal lengths of the physical transmission line. The inductance (L), the resistance (R), the capacitance (C) and the conductance (G), are respectively all per unit length. In a planar transmission line configuration, L and R represent the series inductance and the resistance of the metal conductor, while C represents the

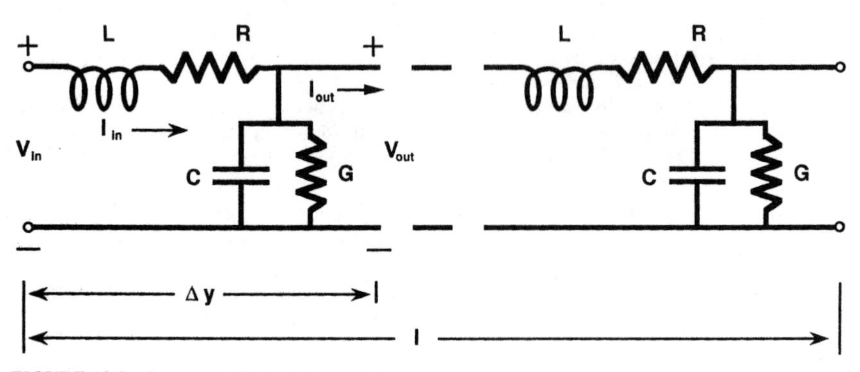

FIGURE 10-8 Schematic of transmission line.

shunt capacitance to ground from the metal conductor, and G represents the dielectric leakage resistance. The impedance for the structure shown in Fig. 10-8 is given by the ratio of the voltage to the current at point y, and time t, as,

$$\frac{v(y,t)}{i(y,t)} = Z_o = \sqrt{\frac{R + j\omega L}{G + j\omega C}} \tag{18}$$

If the series resistance of the transmission line (R) and the dielectric losses (G) are very small, they may be ignored. Equation 18 then reduces to

$$Z_o = \sqrt{\frac{L}{C}} \tag{19}$$

where L and C are the inductance and the capacitance per unit length, respectively. A constant ratio of these two parameters maintains a constant characteristic impedance. This basic relationship explains the fact that as the substrate dielectric constant changes, and thus C, the inductance L, of the conductor must also change if the characteristic impedance is to remain constant. Graphically, this is illustrated in Fig. 10-9, where root capacitance per unit length and impedance of a 0.025 cm wide line on a 0.5 mm substrate ($\epsilon = 50$) [8] are plotted against percentage of ground coverage. Changes in ground coverage directly affect line shunt capacitance, and without altering the line inductance, reduce the line impedance. The effect of changing the inductance-capacitance ratio on impedance is greatest at lower values, because from Eq. 19, the impedance is proportional to the reciprocal of the square root of the capacitance.

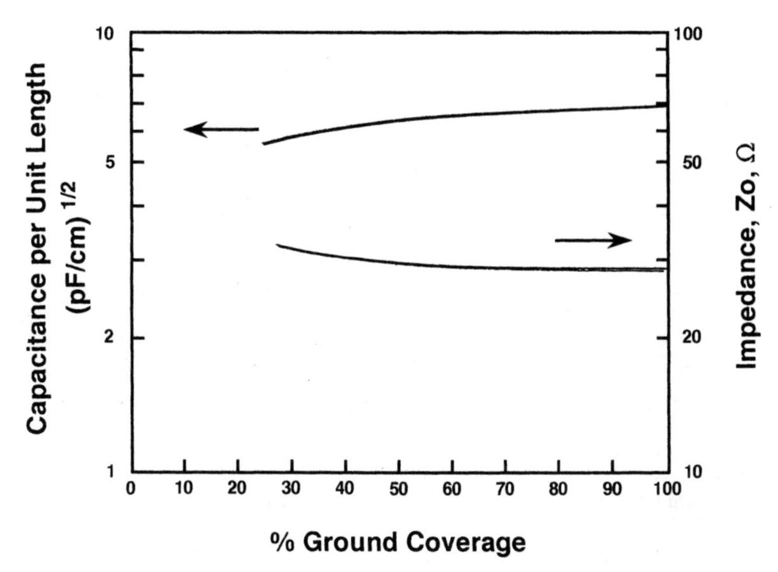

FIGURE 10-9 Capacitance per unit length and characteristic impedance of a microstrip line versus percent ground coverage.

10.4 MICROSTRIP LINE

The microstrip transmission line is one of the most widely used transmission media at microwave frequencies. Its open-ended configuration facilitates the integration and the mounting of discrete monolithic components for ease in production, assembly, tuning, and repair. A basic microstrip circuit consists of a dielectric substrate of thickness h and a relative dielectric constant ϵ_r, coated on one side with a patterned line of width w and thickness t, and a metal ground plane on the other side. From Fig. 10-3, some of the electric field is in the air and the remaining fraction (q) is in the substrate.

The impedance of a microstrip is given by

$$Z_o = \frac{377h}{\sqrt{\epsilon_r} w[1 + 1.735\epsilon_r - 0.0724\frac{w}{h} - 0.836]} \tag{20}$$

The microstrip design problem is one of finding the proper values of w for values of h (w/h ratio) and the effective dielectric constant (ϵ_{eff}), all of which are interdependent. The effective dielectric constant is made up of the dielectric constant of air ($\epsilon_r = 1$), in addition to the dielectric constant of the substrate in excess of that of air ($\epsilon_r - 1$), times the filling fraction (q). The filling fraction, (a measure of the amount of field in the substrate) is a quantity vital to the determination of ϵ_{eff}. Graphic methods were once used [9], but CAD is now routinely used to determine both q and ϵ_{eff}.

$$\epsilon_{eff} = 1 + q(\epsilon_r - 1) \tag{21}$$

This expression holds true for all media discussed in this chapter.

The inductance decreases with increasing cross section, and because of the field distribution, the line width is particularly important. For a 50-Ω line, this effect is shown in Fig. 10-10 which plots the ratio of the line width (w) to the substrate thickness (h) as a function of substrate dielectric constant ϵ_r. Various substrate materials are annotated at appropriate ϵ_r values. A 50-Ω line on fused silica, $\epsilon_r = 3.88$ is twice the line width on comparably thick alumina substrate, $\epsilon_r = 9.9$, illustrating the reason a higher dielectric constant material can result in miniaturization.

For example, many useful curves are obtained when line widths on 0.025 in substrates with different dielectric constants are plotted against the characteristic impedance, as shown in Fig. 10-11. For a 50-Ω line, the line width is 0.025 inches for $\epsilon_r = 10$. For the same impedance, the line broadens to 0.080 inches when the value of $\epsilon_r = 2$. Although easily processed, space may not permit such wide lines, negating the use of the lower dielectric constant material. Not shown on Fig. 10-11, a material with $\epsilon_r = 50$ requires only a 0.0025 in wide line, as shown in Fig. 10-12. Widths of this magnitude demand high substrate surface quality, and special processing may be needed, particularly for some ferrites and titanates with porous bodies. It is readily apparent from Figs. 10-11 and 10-12 that for a given impedance, the w/h ratio decreases with increasing substrate dielectric constant. Accordingly, where minimum substrate area is the dominant factor, the line width w can be reduced by either using thinner substrates or high-dielectric-constant substrates. Unfortunately, as seen later, other substrate and practical processing considerations may negate either approach. Conversely, where wide lines are permissible or desirable, lower-dielectric-constant materials provided by polymer-dielectric or polymer-based substrates can be used.

Figure 10-13 plots the effective dielectric constant, ϵ_{eff} against the impedance varying the dielectric constant. This figure illustrates the effect of line width widening on

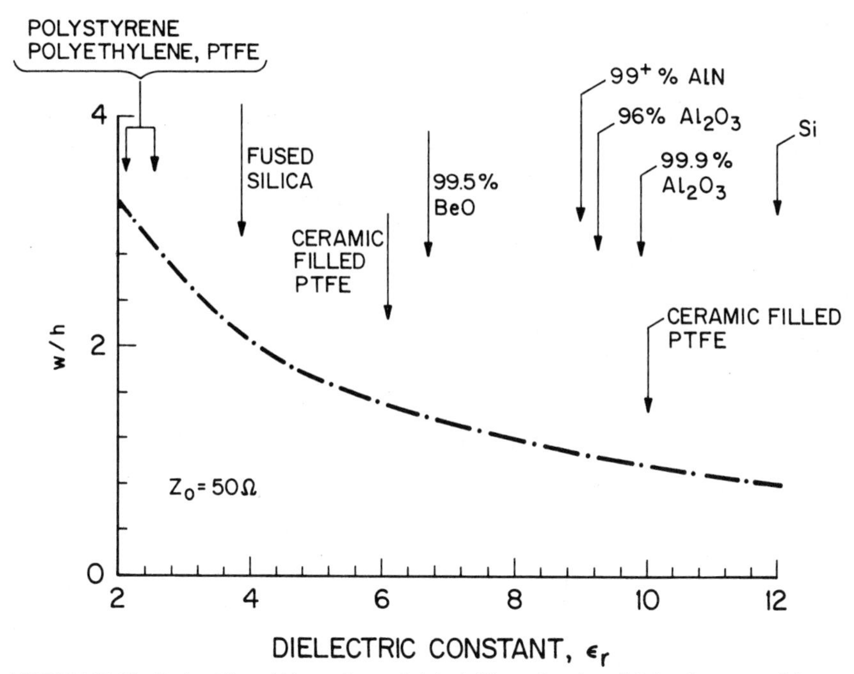

FIGURE 10-10 Ratio of line width to substrate height (w/h) as a function of dielectric constant (ϵ_r).

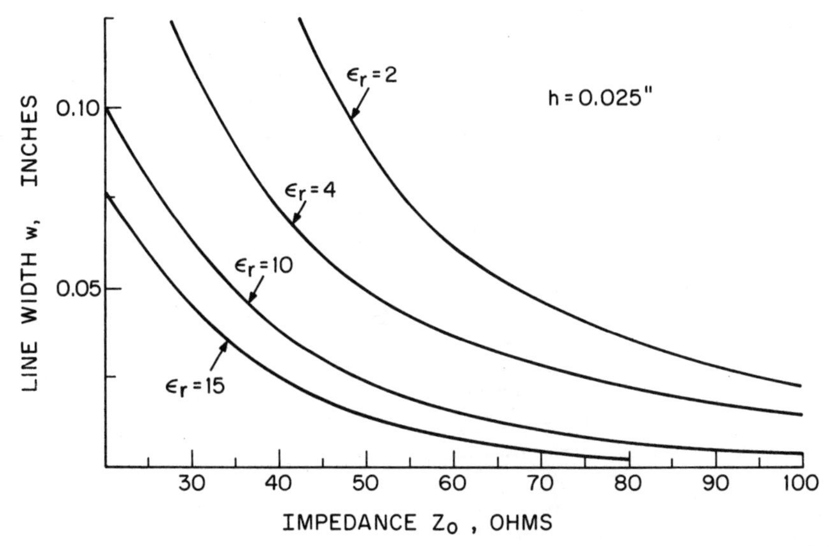

FIGURE 10-11 Effect of substrate dielectric constant on conductor line width and characteristic impedance.

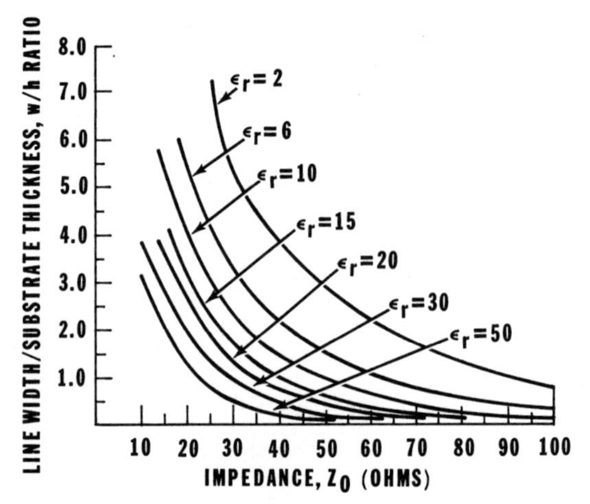

FIGURE 10-12 Effect of dielectric constant and characteristic impedance on microstrip *w*/*h* ratio.

ϵ_{eff}. The effect is most pronounced in high-ϵ_r dielectrics because the fields are more constrained in dielectrics of lower ϵ_r materials. However, the line-width changes with low ϵ_r materials are minimal with changes in impedance, because line widths are wide to begin with. Figure 10-14 illustrates how using higher dielectric constant materials also reduces the guide length (λ_g) of a 50-Ω line for a given frequency of operation, achieving additional size reduction. Figure 10-11 showed that as the frequency increases for a

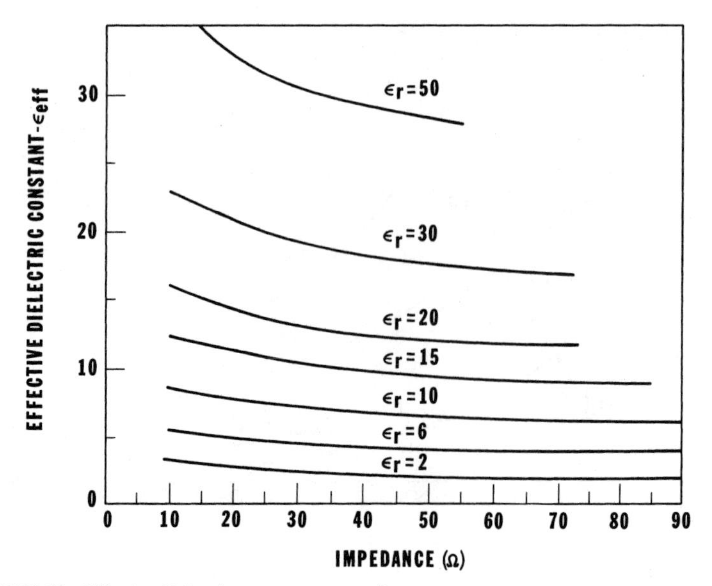

FIGURE 10-13 Effective dielectric constant versus *w*/*h*.

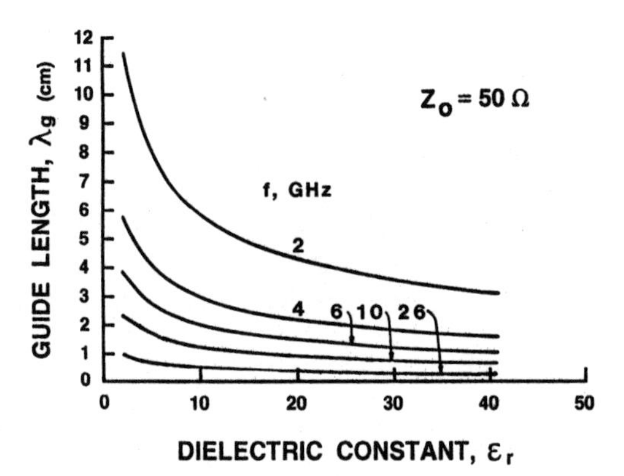

FIGURE 10-14 Guide length versus λ_g for various frequencies.

circuit on a substrate of a given dielectric constant, more of the field penetrates the substrate, a situation akin to widening the conductor line. As a consequence, the resultant increase in the effective dielectric constant also serves to reduce the guide length. However, because $\sqrt{\epsilon_{\text{eff}}}$ is used, at a given frequency the guide length changes less with increasing ϵ_r.

It should be noted that for a given frequency, line widths can only be changed by changing the substrate thickness for single-crystal materials such as for gallium arsenide, because these materials have only one dielectric constant. On the other hand, hybrids with a variety of dielectric constants available have much broader design options. The use of high-dielectric substrates can provide tuning circuitry of equivalent or smaller size compared to active devices.

10.5 COPLANAR TRANSMISSION LINE

Coplanar waveguide (CPW) circuits offer a number of advantages over traditional microstrip circuits. Their widespread application has been limited by the absence of good models and algorithms. In contrast to microstrip, where the field is predominantly in the substrate, in CPW the field is more evenly distributed between the substrate and the air. The basic problem in coplanar design is relating impedance to the *a*/*b* ratio and the conductor thickness (*t*) (Fig. 10-4). Because fields are distributed more or less equally between the substrate and another medium, usually air, the effective dielectric constant can be given as

$$\epsilon_{\text{eff}} = \frac{\epsilon_r + 1}{2} \tag{22}$$

The impedance in coplanar waveguide is primarily a function of *a* and *b*, as defined in Eqs. 1 and 2. Over the frequency range for which CPW is nondispersive, the characteristic impedance is given by

$$Z_o = \frac{377}{4\sqrt{\epsilon_{\text{eff}}}} \left(\frac{K'(k)}{K(k)} \right) \tag{23}$$

where $K'(k)$ is the complete elliptical integral of the first kind, and $K(k)$ is its complementary function, defined by

$$K'(k) = K(k') \tag{24}$$

and

$$k' = \sqrt{1 - k^2} \tag{25}$$

and

$$k = \frac{s}{s + 2w} = \frac{a}{b} \tag{26}$$

For computation of the complete elliptical integrals (K/K'), the reader is directed to Hilberg [10]

For $0.7 < k < 1$, K/K' may be approximated by

$$\frac{1}{\pi}\ln\left[\frac{2(1 + \sqrt{k})}{1 - \sqrt{k}}\right] \tag{27}$$

For $0 < k < 0.7$, K/K' may be approximated by

$$\frac{\pi}{\ln\left[\frac{2(1 + \sqrt{k'})}{1 - \sqrt{k'}}\right]} \tag{28}$$

Using Eqs. 27 and 28, Stegens [11] plotted the impedance as a function of a/b, Fig. 10-15, using ϵ_r as a variable, assuming metal thickness $(t) = 0$ and the normalized substrate thickness $(h/b) = \infty$.

CPW does have a small amount of parallel-plate capacitance between signal track and ground plane because of sidewall metal thickness on both sides of the gap. Bachert [12] corrected for thickness effects by calculating new shape factor and ϵ_{eff}. His results were replotted as impedance versus the ratio of a/b, Fig. 10-16. The asymptotic value of 49 Ω (rather than 50 Ω) is caused by slightly overestimating the capacitance in the thickness correction equations. Nevertheless, at very low a/b ratios, significant impedance errors may arise. Giani and Naldi [13] showed that very small h/b ratios result in sharply higher impedance values than those calculated by $h/b = \infty$.

10.6 STRIPLINE TRANSMISSION LINE

The fields in stripline structures are confined entirely within the dielectric. As such, $\epsilon_{\text{eff}} \sim \epsilon_r$. The characteristic impedance for structures where the strip width (w) is wide enough so that fringing fields do not interact, (that is, $w/b \geq 0.35$) is given by

$$Z_o\sqrt{\epsilon} = \frac{94.15}{\left(\frac{w/b}{1 - t/b} + \frac{C_f'}{0.0885\epsilon_r}\right)} \tag{29a}$$

$$C_f' = \frac{0.0885\epsilon}{\pi}\left[\frac{2}{1 - t/b}\ln\left(\frac{1}{1 - t/b} + 1\right) - \left(\frac{1}{1 - t/b} - 1\right)\ln\left(\frac{1}{(1 - t/b)^2} - 1\right)\right](\text{pf/cm}) \tag{29b}$$

General curves from Howe [14] for the characteristic impedance of dielectrically loaded stripline are presented in Fig. 10-17. These curves take into account the conductor width (w), the thickness (t), and the dielectric thickness (b). For most stripline applications, these curves are sufficient to provide a useful approximation to the parameters.

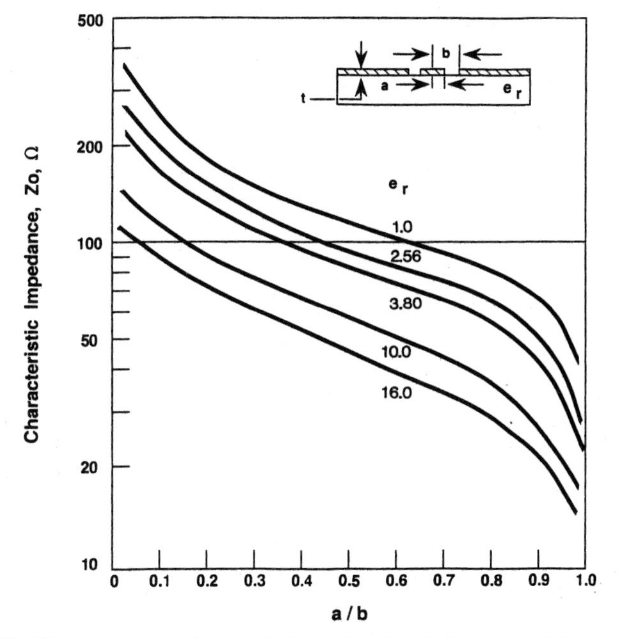

FIGURE 10-15 Characteristic impedance of CPW as a function of a/b, ϵ_r as a variable.

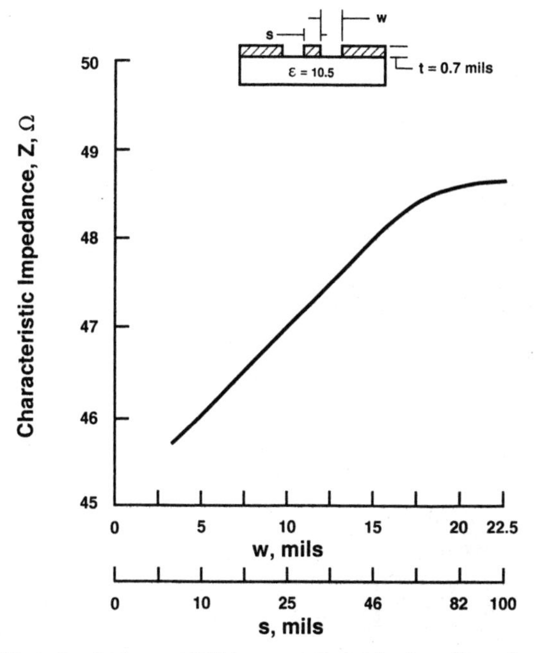

FIGURE 10-16 Effect of neglecting metal thickness on calculated values of impedance.

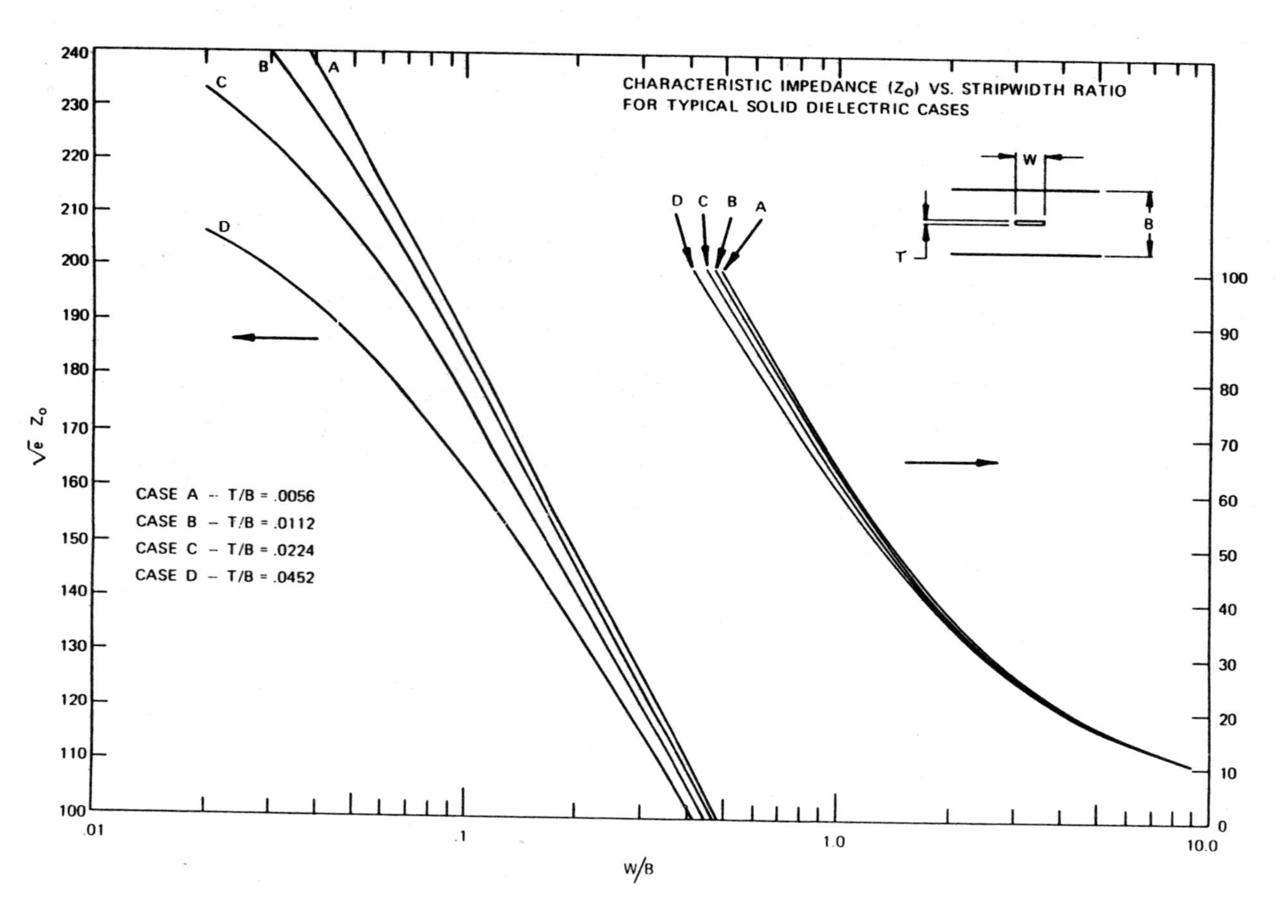

FIGURE 10-17 Characteristic impedance of stripline using commonly available *t/b* ratios.

10.7 PROPAGATION VELOCITY

Inside the transmission line media, inductive and capacitive parasitics along the transmission line combine to slow down the propagating electromagnetic wave. If the inductance and the capacitance values are known, the propagation velocity or phase velocity (v_p) may be expressed by

$$v_p = \frac{1}{\sqrt{LC}} \tag{30}$$

More commonly, v_p is expressed in terms of ϵ_{eff} by

$$v_p = \frac{c}{\sqrt{\epsilon_{\text{eff}}}} \tag{31}$$

Figure 10-18 plots the normalized values of v_p for all three commonly used transmission structures, $Z_o = 50\ \Omega$. Although ϵ_{eff} was used for the calculations, the data are plotted against the dielectric constant ϵ_r. The effect of increasing the non-air fraction of the dielectric on reducing v_p is clearly evident. Propagation within the stripline, where $\epsilon_r \neq 1$, is slower than on coplanar waveguide, where air accounts for about one half of the dielectric.

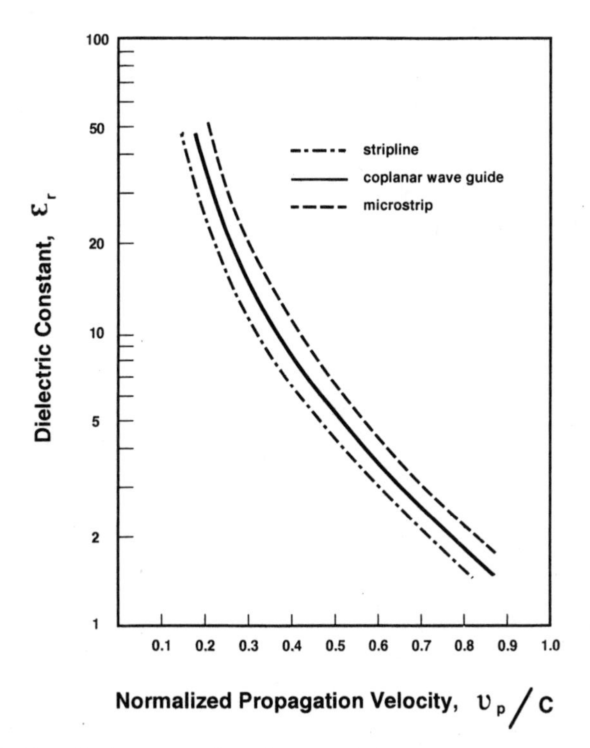

FIGURE 10-18 Normalized v_p versus ϵ_r for various transmission structures.

10.8 ANISOTROPY

Substrates used for hybrid microelectronics are rarely isotropic. Ignoring anisotropy introduces errors during circuit design, which may be significant. Thus the electrical parameters of interconnection traces on anisotropic substrates must be considered to secure proper circuit performance.

Anisotropy in the dielectric arises from three major sources:

1. Crystallographic orientation in single crystals such as sapphire.
2. Matrix contributions in polycrystalline materials because of preferred orientation in processing. Tape-produced alumina is a prime example.
3. Fillers in organic dielectrics used for clad substrates. An example of this is woven glass mat in clad PTFE substrates.

Dielectric anisotropy produces additional fringing capacitance because field perturbation results from a dielectric distortion. This additional capacitance impacts transmission line impedance, particularly for narrow lines and structures using edge coupling such as filters and couplers, where a uniform dielectric constant between adjacent lines is critical.

In single crystals such as sapphire (α-alumina), anisotropy is well defined, $\epsilon_r = 9.34$ for the *a*-axis and 11.55 for the *c*-axis. High-density polycrystalline alumina composed of many grains of sapphire, if perfectly oriented, would yield a fired body with an ϵ_r of 10.08 [15]. However, tape-process substrates have nonrandom, or preferred orientation. Values of 10.22 to 10.80 were reported for 99.96% alumina, suggesting preferred orientation with the *c*-axis perpendicular to the substrate surface. This condition develops during the casting operation and is accentuated during high-temperature sintering.

Polymer composites show larger variations in anisotropy. According to Traut [16], in composites the degree of anisotropy increases with increasing

- Differences in properties between filler or fiber and polymer matrix
- Degree of filler or fiber orientation
- Volume fraction of filler or fiber
- Filler diameter and length
- Orientation on nonspherical fillers
- Unevenness of layered filler distribution through laminate thickness.

Filler type, (woven or random, ceramic, glass, or combination) also contribute to composite anisotropy. Figure 10-19 plots the effect of increased dielectric constant (more filler) on the anisotropy ratio ϵ_{xy}/ϵ_z for both woven and random-fiber-filled PTFE. Increasing filler percentage results in increased anisotropy, with randomly filled substrates having less.

Szentkuti [17] plotted the effects of dielectric anisotropy on the effective dielectric constant (ϵ_{eff}) on sapphire, Epsilam-10,* and isotropic materials of comparable dielectric properties as a function of *w*/*h*, Fig. 10-20. With narrow lines, there was a 3% increase in $(\epsilon_{\text{eff}})^{1/2}$ with sapphire over the isotropic materials, and with the filled PTFE, a 9% increase in $(\epsilon_{\text{eff}})^{1/2}$. There is, however, very little difference with large *w*/*h* ratios. This is explained by the fact that the fringing fields are larger for narrow strips, and it is only the horizontal component (ϵ_x) of the fringing field that is affected by the dielectric constant in the *x* direction.

*Trade Mark—Arlon, Inc., Bear, DE.

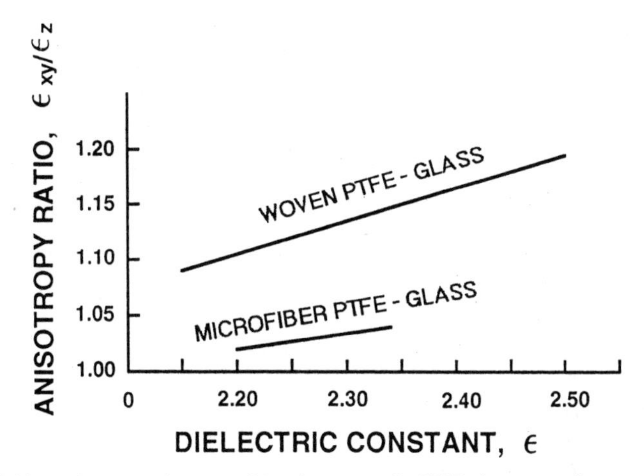

FIGURE 10-19 Anisotropy ratio versus dielectric constant for PTFE-glass composites.

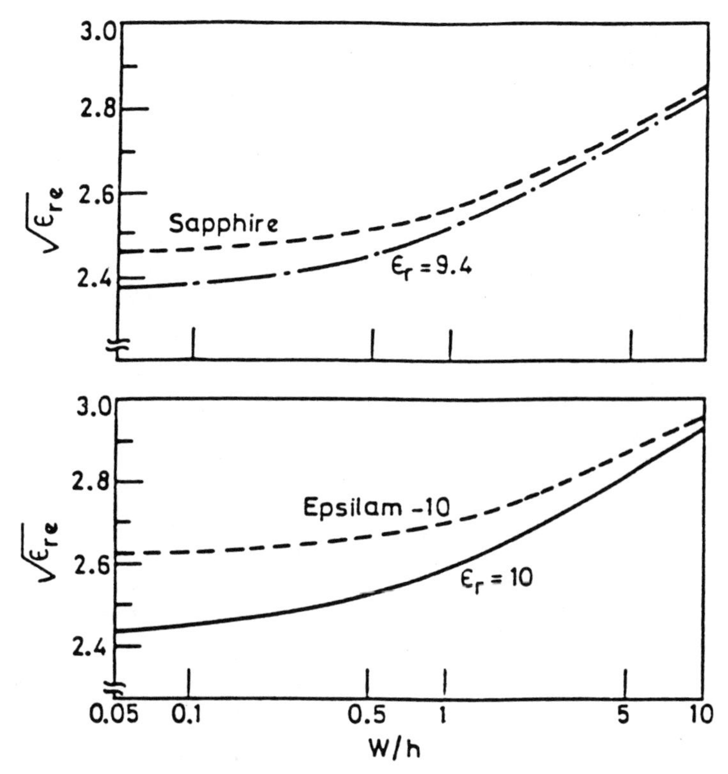

FIGURE 10-20 Effect of dielectric anisotropy on effective dielectric constants of microstrip lines on sapphire ($\epsilon_x = \epsilon_z = 11.6$, $\epsilon_y = 9.9$) and Epsilam-10 substrates.

10.9 LOSSES

The major losses (α) in any transmission line are the sum of the dielectric loss (α_d), the conductor loss (α_c), and the radiation loss (α_r). Radiation losses frequently may be ignored, or minimized by proper shielding, so that the total loss may be expressed by

$$\alpha = \alpha_d + \alpha_c \tag{32}$$

The circuit quality factor is expressed by

$$\frac{1}{Q} = \frac{1}{Q_d} + \frac{1}{Q_c} \tag{33}$$

where Q = total quality factor,
Q_d = dielectric quality factor,
Q_c = conductor quality factor.

10.9.1 Dielectric Losses

A good dielectric is one which withstands high potentials without appreciable conduction. Under an applied electric field, energy in the form of an electric charge is stored by the dielectric material. Many of these charge carriers are naturally polarized dipoles, and realign themselves by rotation in the direction of the applied field. In an alternating field, the total recoverable energy depends on the ability of charge carriers to reorient themselves as the polarity of the field changes. As a result of this rotation, part of the electrical energy is converted into heat and is lost. The effectiveness of reversibility depends on the time available.

Energy losses are important in high-frequency designs, not only because they represent a lack of efficiency, but also because energy losses change the impedance of the circuit. The lost energy in a dielectric may be characterized by its loss tangent, tan δ. The loss tangent (or dissipation factor), the ratio of ϵ'', the lost energy (out-of-phase component), to ϵ', the stored energy (in-phase component),

$$\tan\delta = \frac{\epsilon''}{\epsilon'} \tag{34}$$

is schematically depicted in Fig. 10-21.

A perfect vacuum is the only dielectric medium from which the stored energy can be totally recovered. In nonpolar materials such as polyethylene, the field response is limited to ionic and electronic displacement so that within the microwave frequencies, the dielectric losses are very low. Highly polar cross-linked materials such as polyimides have higher losses. High-density ceramics such as aluminum oxide also have low losses because they elastically respond to electronic movement. In general, tan δ is small, typically ranging from 10^{-2} to 10^{-4}.

The quality factor (Q_d) of a dielectric, the ratio of the energy stored/energy lost is related to tan δ, by

$$\frac{1}{Q_d} = \tan\delta \tag{35}$$

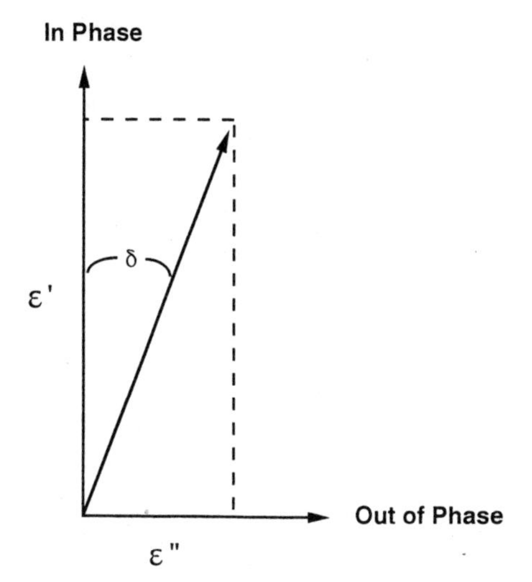

FIGURE 10-21 Schematic of loss tangent.

10.9.2 Conductor Losses

Skin Depth. The formation of eddy currents is one consequence of Maxwell's fourth equation.

$$\nabla \times E = -\frac{\partial B}{\partial t} \tag{36}$$

where E = electric field intensity
$B = \mu H$ = magnetic flux intensity
t = metal thickness

Magnetic fields penetrate most materials, so when alternating current flows, the changing magnetic field generated by the oscillating current creates an opposing current. This inductive current itself creates an opposing magnetic field to cancel any changes. The resistivity and permeability of the metal determine the degree of opposition. Because most metals have a permeability of one, only the resistivity usually needs consideration. Also as the frequency increases, so does the rate of change of the magnetic field, and correspondingly, the magnitude of opposition. With increasing frequency, the field and thus the current is concentrated closer to the conductor surface, as shown schematically in Fig. 10-22. Skin depth is defined as the distance from the metal surface beyond which the current density falls below $1/e$ (about 37%) of its original magnitude.

Mathematically, the skin depth (δ) is expressed as

$$\delta(\text{cm}) = \sqrt{\frac{\rho}{\pi f \mu}} \tag{37}$$

where ρ = resistivity of the metal in ohm·cm $\times 10^{-6}$
f = frequency in Hertz
μ_o = permeability, 1.26×10^{-8} H/cm

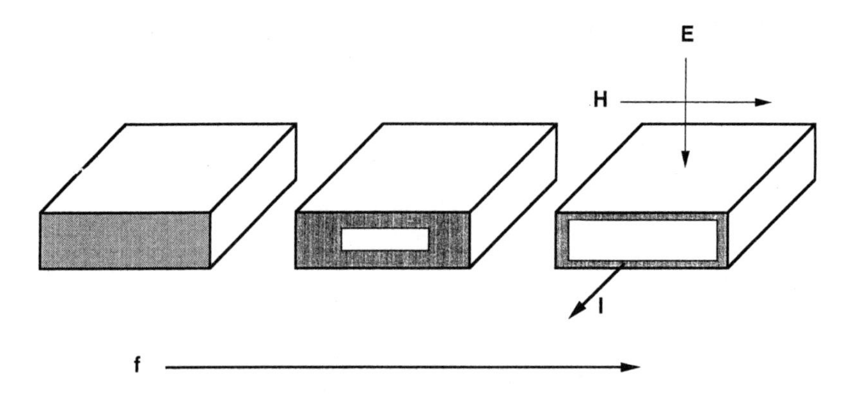

FIGURE 10-22 Schematic of conductor cross section showing effect of frequency on skin depth.

It becomes readily apparent that conductor conductivity, the surfaces, and the interfaces become critical in determining the circuit performance. The skin depth for various conductors as a function of frequency is shown in Fig. 10-23. The low value for nickel is because of its permeability.

For most circuitry, a conductor thickness equal to three skin depths is sufficient to carry the current. This is illustrated by a plot of attenuation versus conductor thickness, normal-

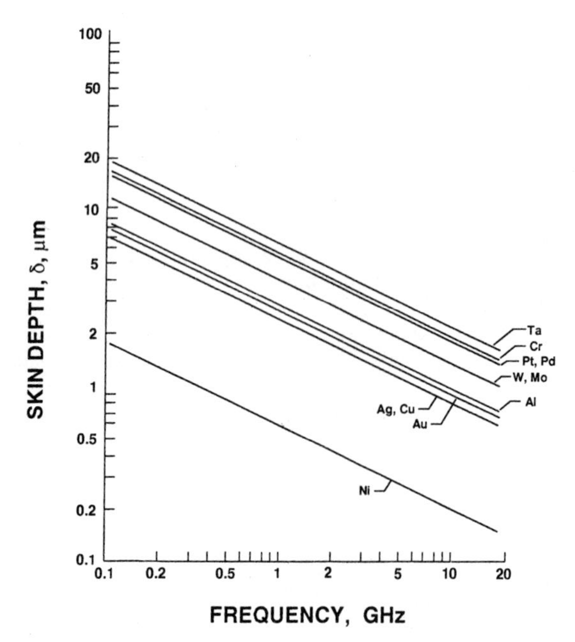

FIGURE 10-23 Skin depth as a function of frequency.

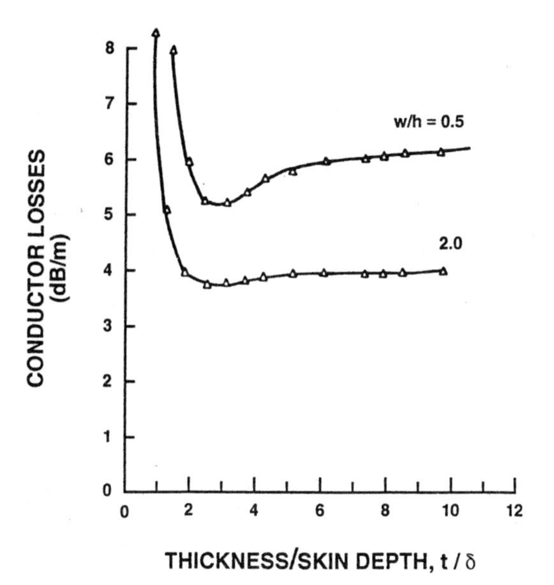

FIGURE 10-24 Conductor attenuation as a function of thickness to skin depth ratio. $\epsilon = 11$, $h = 0.020$ inches (0.508 mm).

ized to skin depth (t/δ) in Fig. 10-24 [18]. Horton's data also illustrates the effect of line width on attenuation. Wider lines with less inductance have lower losses. It will be shown later that certain components such as coils and interdigited capacitors require conductors with thicknesses of approximately ten skin depths.

Attenuation. Conductor losses (α_c) arise primarily from

- Current flow distribution
- Conductor resistance
- Surface roughness

Figure 10-25 compares the attenuation per unit length of several transmission line structures. It is clear that stripline and microstrip line have higher losses per unit length. However, conductor lengths are relatively short; as a result, the overall losses are lower.

10.10 MICROSTRIP TRANSMISSION LINE

For microstrip, the conductive loss α_c (nepers/unit length) can be written as [19,20]

$$\alpha_c = \frac{R}{2Z_o} = \frac{R_{gp} + R_{st}}{2Z_o}$$
$$R_{st} \cong \frac{R_s}{w} \tag{38}$$

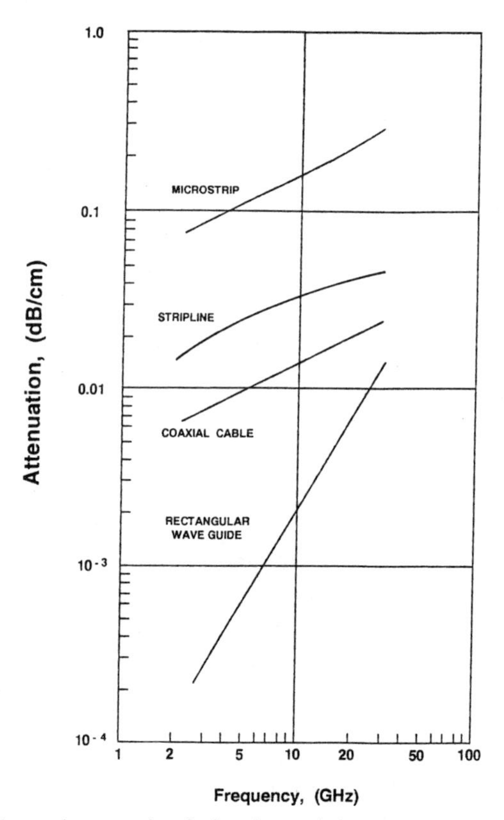

FIGURE 10-25 Attenuation properties of selected transmission structures versus frequency.

where α_c = conductive loss
R_s = line resistance
R_{st} = strip conductor resistance
R_{gp} = ground-plane resistance
Z_o = impedance

and

$$R_{gp} \cong \frac{R_s}{2h + w}$$
$$Z_o = \text{function}\left[f\left(\frac{w}{h}\right)\right] \tag{39}$$

From Eq. 39, α_c is inversely proportional to substrate thickness (h) and decreases with increasing w/h ratio. This is expected because wider trace widths of the same thickness have less inductance, thus lower loss. For CPW, the conductor loss per unit length for a fixed value of $2b$, varying $2a$ at 10 GHz, is plotted in Fig. 10-26 [11]. The very low losses over a wide range of center conductor width is attributable to relatively low parasitics, because field distribution is almost independent of spacing between center track and ground.

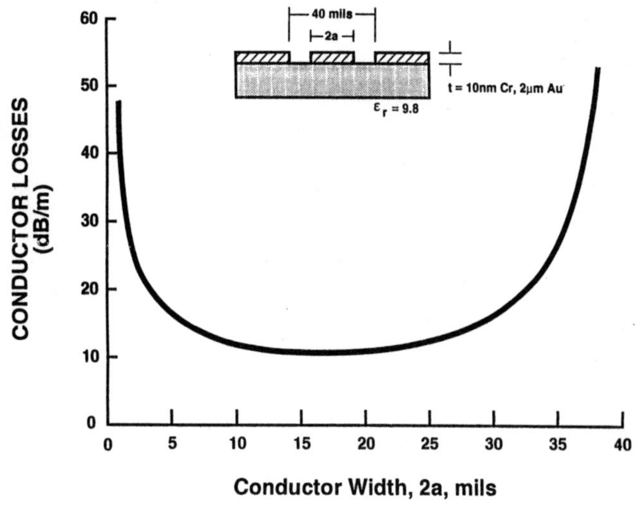

FIGURE 10-26 Conductor loss per unit length as a function of center strip width.

Nevertheless, Kulke et al. [21] have shown that at high frequency, high losses can occur when narrow lines and spacing are used to maintain a constant 50 Ω impedance, as shown in Fig. 10-27. Although losses may be significant, such designs may be advantageous because large reductions in circuit size can be realized. (Note the *w* and *s* nomenclature on the inset drawing in Fig. 10-27.)

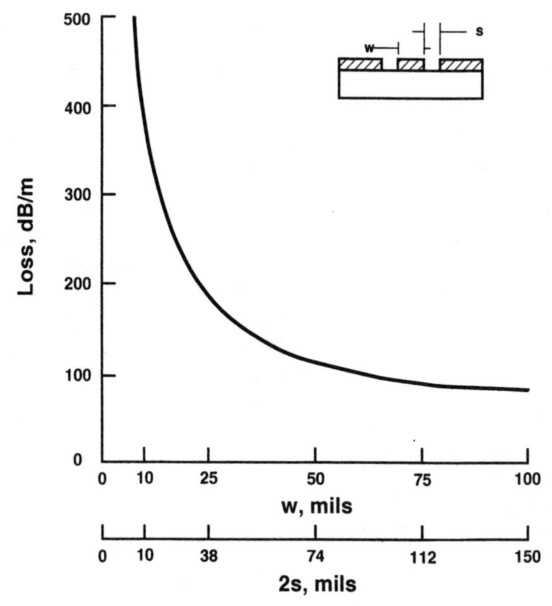

FIGURE 10-27 Line losses in *dB/m* of CPW as a function of line width. *S* values to maintain $Z_o = 50\ \Omega$ are included.

10.11 Q MEASUREMENTS

Unfortunately, predicted ohmic losses cannot take into account large current densities at conductor edges or variations in sheet resistivity from processing. Q measurements of resonators offer a convenient and simple method of determining losses resulting from geometry, processing, dielectric, and other system contributions. A general definition of the Q of a resonant circuit is

$$Q = \frac{\text{Time average energy stored in the circuit}}{\text{Circuit energy loss per second}} \tag{40}$$

A useful expression from Collin [22] for determining unloaded Q is

$$Q = \frac{f_o}{f_2 - f_1} \tag{41}$$

where f_o = resonator center frequency
f_1 = lower -3 dB response frequency
f_2 = upper -3 dB response frequency.

For thick, high-quality films, theoretical Q values may be as high as 500. As a rule of thumb, many workers in the field prefer to use 60% to 75% of the theoretical unloaded Q to determine actual conductor losses. Conductor loss per unit guide length may be derived from the following expression:

$$\alpha_c = \frac{23.7}{\lambda g \times Q} \tag{42}$$

As is expected, α_c decreases with increasing Q. For maximum Q, conductors should have as high a conductivity as possible.

10.12 SURFACE ROUGHNESS

The two metals used almost exclusively as conductors for microwave hybrids are copper (Cu) and gold (Au), $\sigma = 5.8 \times 10^7$, and 4.1×10^7 S/m, respectively. This conductivity is sufficient to produce low-loss transmission lines. Most hybrid substrate surfaces, as opposed to polished, semiconductor surfaces, have some degree of surface roughness. The major contribution to attenuation is this substrate surface roughness. Surface texture increases the line length, promotes discontinuities, and causes localized current crowding, all contributing to an increase in attenuation.

The effect of surface roughness on surface resistivity has been discussed by a number of authors. Morgan [23] calculated the increase in conductor loss on waveguides resulting from grooves of various geometries transverse to the current flow. Lending [24], Sobol [25] and Benson [26] also investigated the increase in surface resistance from surface roughness. Bhasin et al. [27] looked at attenuation on glass-filled PTFE surfaces. Hammerstadt and Bekkada [28] provide a simple formula for calculating the excess loss from surface roughness approximately Morgan's curves.

$$\frac{\alpha_c}{\alpha_{co}} = \left[1 + \frac{2}{\pi} \arctan 1.4\left(\frac{\Delta}{\delta}\right)^2\right] \tag{43}$$

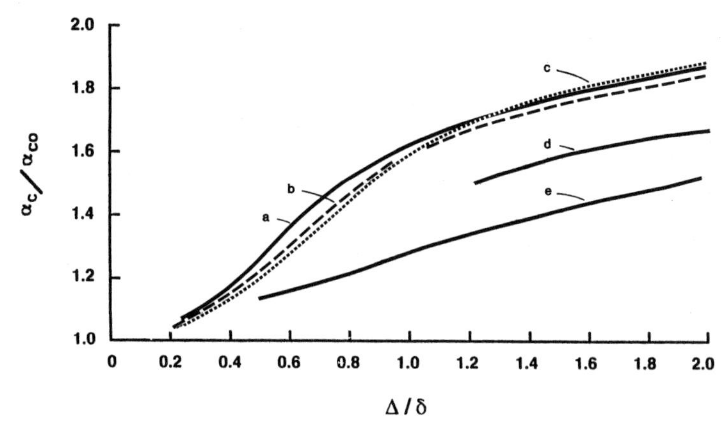

FIGURE 10-28 Excess conductor loss from surface roughness; (*a*) ref. 25, (*b*) ref. 23, (*c*) ref. 28, (*d*) ref. 26, (*e*) ref. 24.

where Δ = surface roughness in microns
δ = skin depth in microns
α_c = losses from roughness
α_{co} = losses on a smooth surface at zero-frequency

The increase in resistance from surface roughness is plotted in Fig. 10-28. The frequency dependence of the increase in loss is taken into account by the term Δ/δ. These plots show that when $\Delta/\delta > 0.4$, additional losses in excess of 10% result.

10.13 HIGH-RESISTANCE ADHESION LAYERS

In general, conductors such as Cu and Au do not adhere well to the substrate. It is usually necessary to use a thin metal layer of high resistance as an adhesion layer between the conductor and substrate. Table 10-2 groups conductors into three categories: (1) good electrical conductors with poor substrate adhesion, (2) poor electrical conductors with good substrate adhesion, and (3) moderate conductors with fair adhesion. Group 1 metals serve as the main current-carrying conductor. Thin layers of group 2 metals serve as adhesion layers for group 1 metals.

TABLE 10-2 Characteristics of thin film conductors

Group	Conductor	Conductivity	Adherence to dielectric
1	Ag, Cu, Au	High	Poor
2	Cr, Ta, Ti, TaN Ti-W, Ni-Cr	Low	Excellent
3	W, Mo	Moderate	Good

TABLE 10-3 Microwave metalization systems

Adherence	Seed	Plated	Barrier	Plated
Cr	Cu	Cu	Ni	Au
Cr	Cu	Cu	Pd	Au
Cr	Au	Au		
Mo	Au	Au		
Ti	Pd	Au		
Ti	Au	Au		
Ti:W	Au	Au		
Ti	Cu	Cu	Ni	Au
Ta	NiCr			Au
Ta/Ti			Pd	Au
Ta_2O/Ti	Cu	Cu	Ni	Au
Ta_2ON/Cr	Au	Au		

Table 10-3 lists some common thin film multilayer systems. Barrier layers are usually needed to prevent diffusion between two dissimilar conductors, particularly during high-temperature operations such as resistor stabilization and wire bonding.

The effect of high-resistivity adhesion layers on conductor loss may be considerable if the thickness of the layer(s) is not very small compared to the skin depth. Caulton

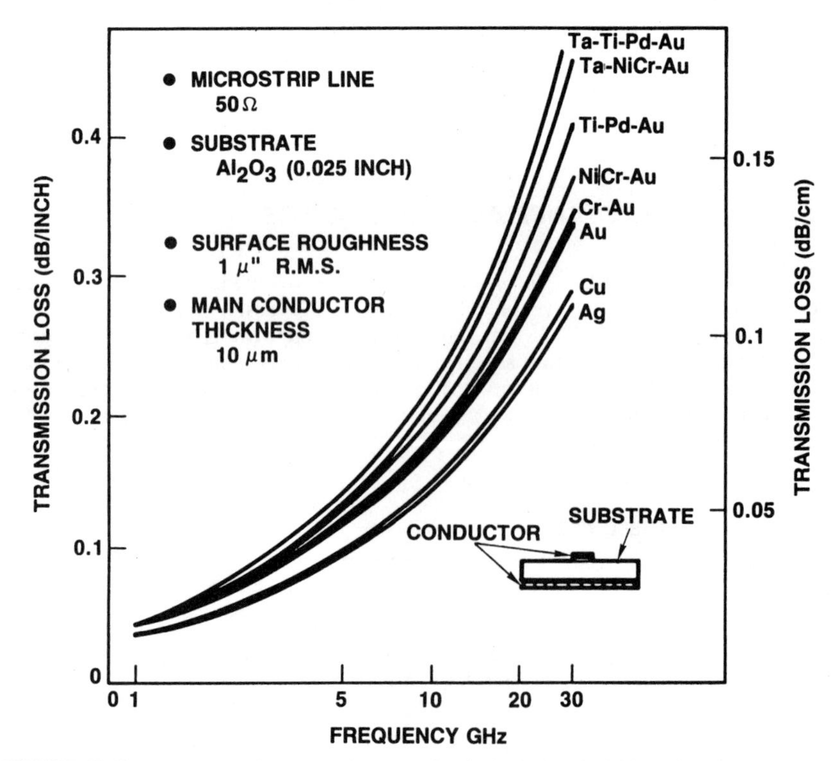

FIGURE 10-29 Transmission loss versus frequency for single, dual, and triple metal conductors.

et al. [19] examined the effects of adhesion layers with and without barrier layers. They found that low losses could be maintained if the adhesion layer was 40 to 50 nm. They also showed significant increases in loss with increasing frequency. The same effect was found when a barrier layer was used. From 30 to 40 mm of titanium with 200 to 400 nm of palladium result in losses up to 10%, compared to pure Au, at 10 GHz and 30% to 50% at 50 GHz. Similar results were reported by van der Peet [29] who also found increases in α_c when the conductor layer was thin or narrow. Gunshinam et al. [30] in Fig. 10-29 summarize the transmission loss versus frequency for single, dual, and triple-metal conductor systems.

10.14 GROUND PLANE EFFECTS

The contribution of the ground-plane resistance (R_{gp}) to the overall conductor losses cannot be ignored. Using a value of resistivity slightly higher than pure Au compared to a perfect conductor, Fig. 10-30 illustrates the effect of ground-plane resistivity (ρ) on conductor attenuation (α_c) as a function of w/h.

Ground-plane continuity is important to maintain line impedance. Openings in the ground plane change the shunt capacitance (C) in that area, and the characteristic impedance. By fabricating the ground plane in mesh form, and changing the mesh pitch, Kasuya et al. [31] were able to demonstrate an increase in impedance with corresponding increase in percentage of ground opening, as shown in Fig. 10-31.

A summary set of curves relating the conductor and the dielectric losses as a function of frequency is shown in Fig. 10-32 [32].

10.15 SUBSTRATES

Substrate properties such as surface finish, and the fabrication processes such as metallization and definition, which impact the accuracy of line width and gap width, ultimately determine the circuit performance. For optimum performance, the microwave substrates should have the following attributes:

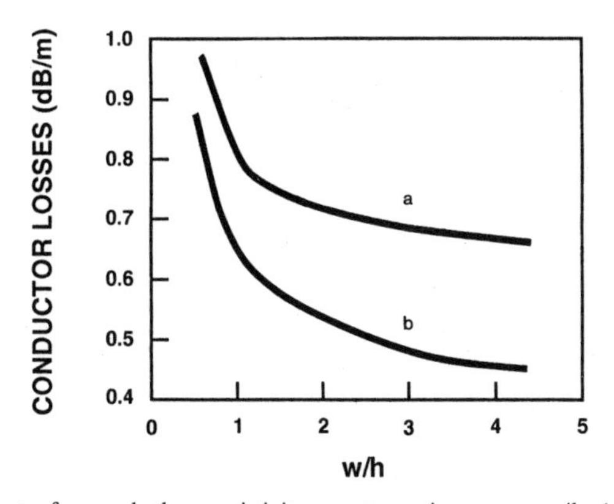

FIGURE 10-30 Effect of ground plane resistivity on attenuation versus w/h; (a) $\rho = 3.0\ \mu\cdot\Omega\cdot$cm, ($b$) $\rho = 0, f = 1$ GHz.

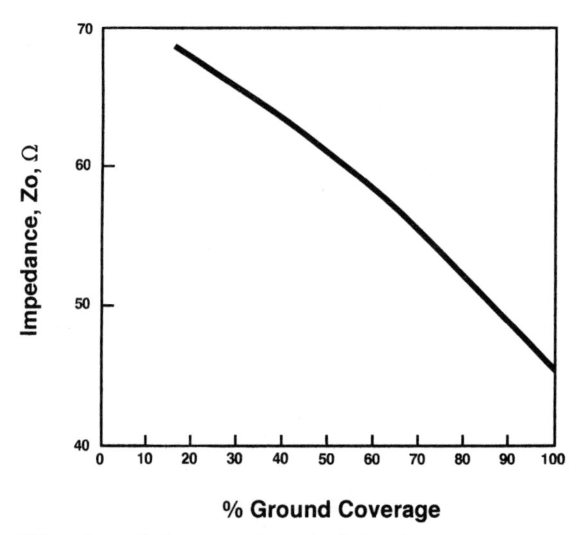

FIGURE 10-31 Effect of ground plane coverage on circuit impedance.

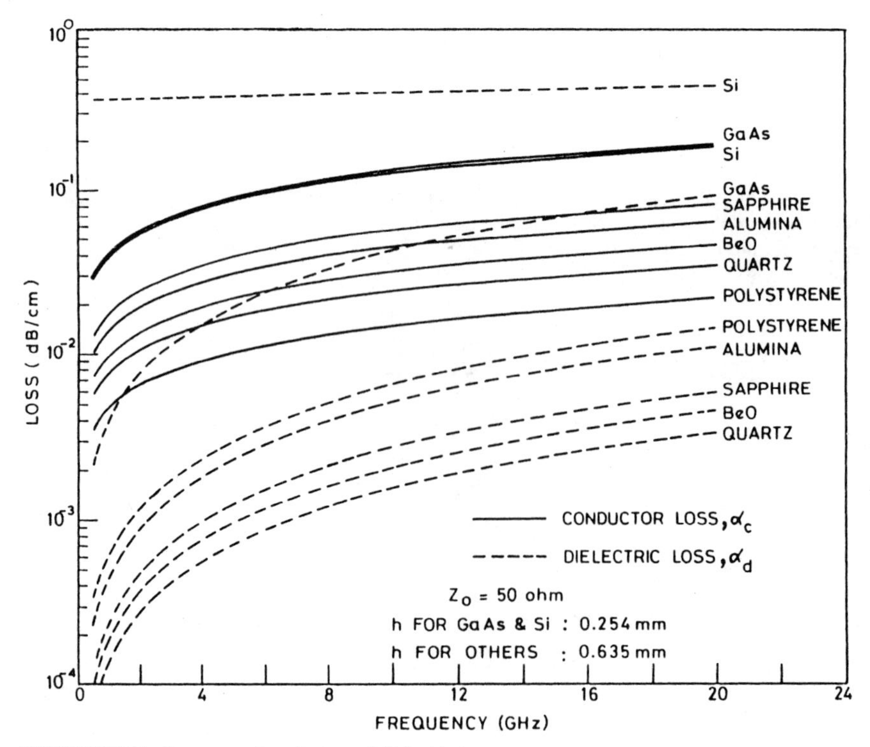

FIGURE 10-32 Summary of conductor and dielectric losses.

1. Low loss tangent to reduce dielectric loss.
2. A uniform, isotropic dielectric constant to minimize, for example, impedance changes within the circuit.
3. Uniformly consistent dielectric constant within a manufactured batch, and from batch to batch, to allow for MIC production without compensation of circuit design.
4. A smooth surface finish to minimize conductor ohmic losses.
5. High thermal conductivity for power circuits.
6. Good thermal expansion coefficient compatibility with components and package requirements.
7. High chemical resistance.

Moreover, the proper choice of a particular substrate material depends on factors such as:

1. Is the substrate cost justifiable for the particular application?
2. What technology is to be used?
3. What frequency and temperature ranges are involved?
4. Are the substrates available with sufficient area to realize the circuit design?
5. If necessary, can tight physical dimensions be maintained either initially or through secondary processing?

Substrates used for microwave circuits are "hard," usually a polycrystalline ceramic, or "soft," a metal-clad polymer-composite. Their important characteristics are listed in Table 10-4.

10.15.1 Glass

Glasses are produced by melting crushed glass (cullet) in a tank-type furnace and withdrawing the molten glass in sheet form. Glass is the lowest in cost of all materials used for thin film substrates. On the other hand, glasses have low mechanical strength and poor

TABLE 10-4 Properties of hard substrates

Material	Surface roughness polished (μin)	Tan δ 1E −4 (1 MHz)	Dielectric constant	Thermal conductivity (W/m·°K)	Thermal expansion (ppm/°C)
Sapphire	$\ll 1$	11	9.3/11.7*	40	8.3*
Alumina 99.5%	<1	1-2	10	30	7.1
Alumina 96%	1-3	6	9	28	7
Beryllia 99.5%	1-4	4	6.6	250	8.5
Aluminum nitride	1-2	7-20	9-10	170	4.5
Silicon carbide	1-5	500	40	270	3.7
Cordierite	1-3	4	4-6	5	3-6
Fused silica	$\ll 1$	<1	3.8	1	0.5
Glass (alkali-free)	$\ll 1$	10-20	5	1	4
Glass (sealing)	$\ll 1$	60	4.1	1	3.2

*Depending on orientation.

thermal conductivity, creating problems in handling and thermal-compression bonding. For microwave applications where superior chemical and electrical properties are necessary, two types of glasses, fused silica and solder glass, are the only practical glass substrates.

An extremely smooth surface, low dielectric constant, and low loss tangent make fused silica an ideal choice for microwave hybrid integrated circuits. Thin film circuits defined on fused silica are characterized by precisely measured fine lines and the almost total absence of yield problems resulting from surface defects. Thus this substrate finds widespread application for millimeter-wave circuits.

Fused silica may be produced in two ways. In the first, ground quartz is first melted by flame fusion. A long, thick disc of silica is made by slowly pulling, while rotating, on the molten pool of SiO_2. This boule (typically 72 in in diameter and 26 in thick) is later sliced and polished to thickness. This technique is similar to the thermal flame fusion process for making silicon. The water content, which directly affects the adherence of deposited films and other properties, ranges in the 100 to 200 ppm range.

Using an oxyhydrogen flame, SiO_2 may also be made by hydrolyzing silicon tetrachloride.

$$2H_2O + SiCl_4 \rightarrow SiO_2 + 4HCl \tag{44}$$

A disc, 5 in to 6 in thick and 30 in to 50 in in diameter, is grown. Again, subsequent cutting and polishing creates substrates, 0.010 in to 0.020 in thick. However, the water content, typically 5 to 10 times higher ($\sim$1000 ppm) than the material made by flame fusion, can adversely effect adherence of deposited thin films by reducing the interfacial oxides necessary for good adhesion.

A radar transmit/receive (T/R) module using Corning 7070 glass with a coefficient of expansion (3.2×10^{-6}/°C) has been reported by Ziegner and Murphy [32]. They first laminated the glass to silicon and then defined the needed circuitry, including vias on the glass. This technology is discussed in the packaging section.

10.15.2 Single-Crystal Substrates

One substrate is uniquely desirable for microwave applications. From Table 10-4, note that sapphire (α-alumina) has an excellent surface, low tan δ, and high thermal conductivity. Sapphire substrates are cut and polished from larger boules, as are most single-crystal materials. Both Verneuil (flame fusion) and Czochralski methods are used to fabricate the larger single crystals. As a single crystal, it has somewhat lower mechanical strength than its polycrystalline counterparts, and its properties are anisotropic. Furthermore, as a single crystal it is inherently expensive and substrate areas $>$9 in^2 are prohibitively expensive. As a result, sapphire's polycrystalline fine-grain analogs are used whenever possible.

10.15.3 Polycrystalline

Polycrystalline substrates have attributes of superior strength, high thermal conductivity, resistance to ion migration, and chemical and thermal shock resistance. For thin film microwave hybrids, the surface is critical. Thinner films are more sensitive to surface irregularities, and consequently most substrates for thin film microwave circuits are lapped and polished to about a 1 μin finish. For most applications, where low or moderate power dissipation is required, alumina is the preferred substrate. Currently, alumina represents about 85% of ceramic substrate sales. Polished 99.6% alumina provides the surface amenable for precision photolithography (fine line and spaces), and the reproducible fabrication of circuit components such as couplers, interdigitated capacitors, coils, and precision resistors. Having few pullouts in high-density material is attributed to its intrinsic smaller average grain size. Even with polishing, pullouts of the larger beryllia grains preclude surfaces with the quality shown by the $99^{+}\%$ alumina. The use of zirconia as a grain-growth inhibitor in beryllia was reported to reduce the average grain size by about 50% and improve electrical and mechanical properties of the body, but at a slight expense in thermal conductivity [33].

10.15.4 High-ϵ Materials

The increasing demand for smaller, lighter weight circuits has spurred renewed interest in high-dielectric-constant ($\epsilon_r > 10$) substrate materials. In the past, high-dielectric losses, low density, poor surfaces, and very negative temperature coefficients of dielectric constant have restricted the commercialization of these materials. Many of the earlier deficiencies were overcome with improvements in powder metallurgy; raw material purity; sintering technology; and cleaning and polishing technology. Table 10-5 [47–51] compares some of the salient properties of the high-dielectric-constant substrates. In many cases the dielectric constant is more temperature dependent (τ_ϵ) than other ceramic substrate materials, but values were reduced to acceptable values.

For the materials in Table 10-5, with high relative permittivities (>10), the temperature coefficient of permittivity (τ_ϵ) is given by

$$\frac{d\epsilon}{\epsilon dT} = A + 0.5 \tan \delta - \alpha\epsilon \tag{45}$$

where A = the temperature coefficient of polarizibility,
α = the linear coefficient of expansion,
ϵ = the permittivity

For some materials, such as TiO_2 and $SrTiO_3$, the first two terms are negligible. However, many of the substrates listed in Table 10-5 have low loss with high permittivity, and with small temperature coefficient *s* of permittivity. As such these materials must have significant *A* values. In alkali halides and high-permittivity glasses, these *a* values may result from expansion effects on ionic countervibration, according to Kell et al. [52].

Barium nanotitanate ($Ba_2Ti_9O_{20}$), with $\epsilon_r \sim 40$, zirconium tin titanate (ZrSn) TiO_4, with $\epsilon_r = 36 - 38$, and barium magnesium tantalate $Ba(MgTa)O_3$, with $\epsilon_r = 25$, have been produced with nearly bulk density by dry pressing. Using roll compaction, shown schematically in Fig. 10-33, Tolino [53] claimed significantly improved surface quality with a reduction in polishing costs. In this method, a mixture of spray-dried powder, binders, and plasticizers are put into a vibrating sieve to break up any agglomerates. The mixture is then put between two rollers, whose rotational speed and separation determine the thickness of the resultant green ceramic. A polyester carrier tape, similar to that used for tape casting, engages the green ceramic as it leaves the rollers. The edges of the tape are trimmed and the ceramic surface is cleaned before being rolled onto a storage roll. Surfaces with 2 to 4 μin finishes for many microwave microcircuit applications arc obtained with this process. Polished zirconium tin titanate surface is characterized by a low incidence of pullouts, but by slightly deeper polishing scratches than found on alumina or beryllia [54]. However, the alkaline earth additives limit the use of process reagents such as mineral acids. Cleaning and pattern etching must be accomplished using precautions not usually necessary with more chemically inert substrates. Care must also be taken when heating many of the titanate-based materials, because they are much more prone to thermal shock than, for example, alumina or beryllia.

TABLE 10-5 Properties of high-k substrates

#	Material	Supplier	Surface roughness, polished (μin)	Dielectric constant	$\tau\epsilon$*	Q (1/Tan δ)	Frequency (approx.) (GHz)	$Q \times f$ ($\times 10^3$)	Thermal expansion (ppm/°C)	Thermal conductivity (W/m·K)	Reference
1	(Zr, Sn) TiO_3	Murata	<1	38	30 ± 30	8000	3	24	6.5	2.3	47
2	(Mg, Ca) TiO_3	Murata	—	21	30 ± 30	9000	3	27	8.5	8.5	47
3	(Ba, P) (Nd, Ti) O_3	Murata	<4	92	30 ± 30	1500	3	4.5	8.5	1.9	47
4	(Zr, Sn) TiO_3	Trans-tech	<1	38	0 ± 30	10000	10	100	6.5	2.5	48
5	(Ba, La) TiOx	Trans-tech	<2	80	0 ± 30	3500	3	10.5	8.5	2.5	48
6	$Ba_2\ Ti_9O_{20}$	Coors	<1	39.5	3 ± 2	5500	4	22	9.1	2.1	49
7	BaO-TiO_2	NTK	<8	13	0 ± 30	18000	13	234	7.6	30	50
8	(Zr, Sn) TiO_3	NTK	<8	36	0 ± 30	8000	8	64	9.5	2.2	50
9	TiO_2	NTK	<8	81	0 ± 30	1500	2.7	4.05	10	3.4	50
10	TiO_2	AlSiMag	<2	87	n/a	2750	0.01	0.03	9	5.0	51

*Temperature coefficient of dielectric constant, ppm/°C.

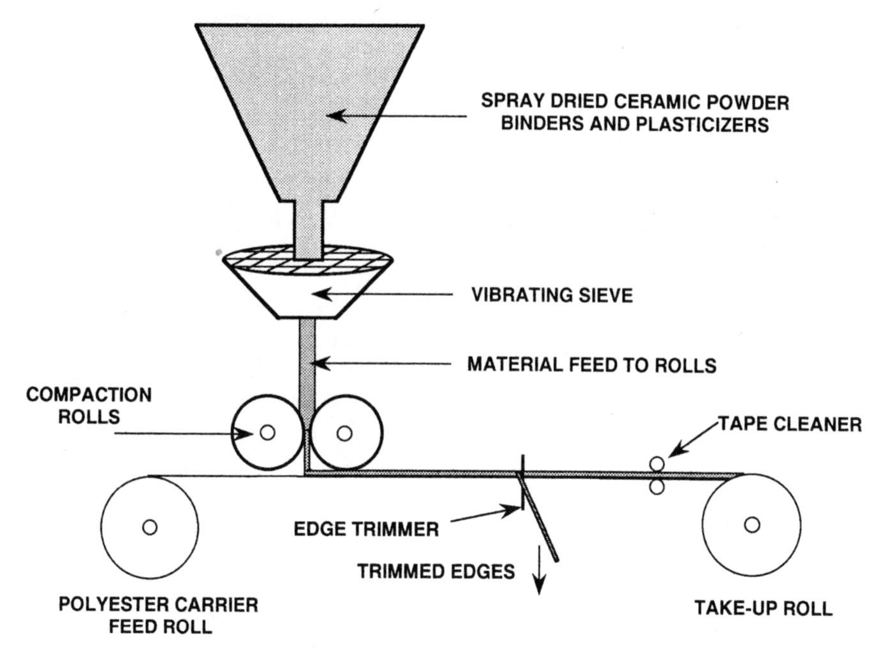

FIGURE 10-33 Schematic of roll compaction manufacturing.

10.15.5 Clad Materials

Glass Transition Temperature (Tg). The glass transition temperature is the temperature at which a relatively brittle material with a small free molecular volume transforms into a material with a high-molecular free volume. With this transformation, individual molecules within the material have increased freedom to move. One important consequence of this is a significant increase in the coefficient of thermal expansion (CTE), shown schematically in Fig. 10-34. T_g essentially defines the upper operating and process range of

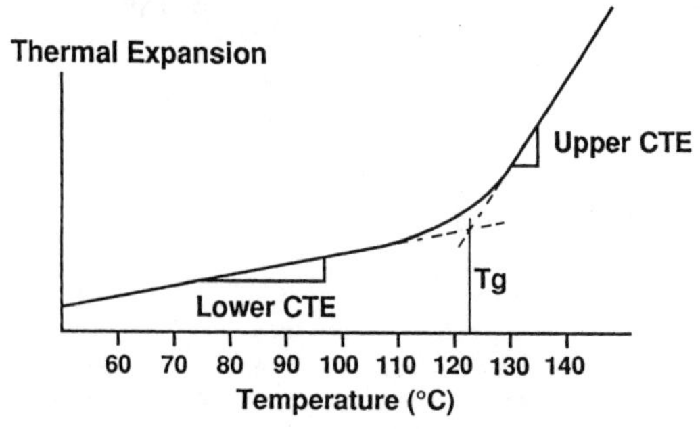

FIGURE 10-34 Schematic of CTE versus temperature showing T_g.

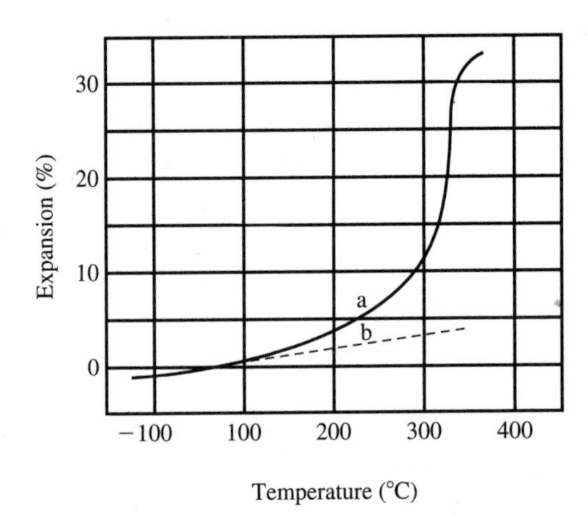

FIGURE 10-35 Comparison of expansion of (*a*) pure PTFE and (*b*) PTFE-composite (*z*-direction), with permission.

polymer-based materials, such as epoxy and cyanate-ester. PTFE, on the other hand, is best described as viscoelastic. Pure PTFE and PTFE composites show nonlinear response with temperature, as shown in Fig. 10-35. Variations in electrical properties are related to density changes in the substrate during heating.

Substrates. For most communications systems operating in the low-megahertz-frequency range, conventional printed circuit boards (PCBs) are adequate. In such systems, the PCB primarily serves as a mechanical support. However, when circuits move into the high-megahertz and gigahertz frequencies, the properties of these boards become as important as other substrates used for high-frequency applications. These soft substrates are laminates, usually consisting of a thermoplastic-resin core with metal foil fused to the top and bottom surfaces. They are particularly attractive for stripline and buried-microstrip geometries, because of the relatively low temperatures required for fabricating multilayer assemblies. Vias and wrap-arounds are more easily incorporated into these substrates compared to hard-substrate materials.

Early materials were nonuniform and characterized by relatively high dielectric losses, poor dimensional stability, and dielectric uniformity. These factors frequently contributed to excessive variations in line impedance and poor circuit performance. Since then many improvements have been made. These improvements include tightening the dimensional thickness of the dielectric and the cladding. Another major advance was in providing uniform, low-loss dielectrics with variations of dielectric constants of ± 0.01 up to millimeter frequencies. Concerns for clad materials include dimensional stability, chemical resistance, anisotropy, and thermal conductivity factors not usually associated with ceramic substrates. The low dielectric loss of PTFE and its chemical resistance has led to its almost universal use as a core material. Nonetheless, in the RF and low-microwave ranges, selected woven, thermoset materials are cost-effective.

Pure PTFE and irradiated polyolefins are preferred as dielectrics for higher frequency applications because they have isotropic dielectric and expansion properties, and low loss over a wide frequency range. However, these materials have two major, although not insurmountable, problems. As pure thermoplastics, they have cold flow; that is, change in dimension or distortion caused by application of pressure, and a higher coefficient of expansion than rigid or semirigid materials. Both of these limitations must be addressed by proper package design, minimizing application of pressure or substrate expansion. At the lower frequency range, glass-filled epoxies, polyimides, and cyanate esters are increasingly used.

To increase rigidity, some manufacturers add fillers to core materials. Some of these additives are woven glass, glass fibers, fiberglass, or ceramics. Figure 10-36 illustrates the effect of adding *E* glass ($\epsilon_r = 6.11$) to epoxy ($\epsilon_r = 3.40$) and PTFE ($\epsilon_r = 2.10$) on the dielectric constant of the composite. Approximately, 60 w/o resin is used to obtain optimal laminate performance for both compositions. Modern laminates are made of PTFE and either woven or nonwoven fiberglass; pressed together under high pressure and temperature. Thin copper plate is then fused to one or both surfaces. Woven fiberglass laminates begin with either fine, medium, or coarse weaves of fiberglass cloth. The cloth, typically 0.002 in thick, is impregnated with PTFE through immersion in a stable PTFE-aqueous dispersion. Silanes are used, either by previous fabric coating or added to the dispersion to provide the necessary moisture resistance. The concentration of PTFE particles in the bath helps determine the dielectric constant of the final laminate. These layers are fused together to the desired thickness.

For RF and wireless applications, the base materials are made by impregnating woven-glass mat with a combination of resin, flame retardant, and hardener or cross-linking agent. The desired thickness is achieved by adjusting the diameter of the glass fibers used for the cloth and the amount of resin. The primary product in this sector is FR-4. When it is made, there are variations in formulations among fabricators to achieve better Cu adhesion, solvent and chemical resistance, and ease of drilling. As a result, wider variations in properties occur with this material than in the PTFE-based substrates.

Nonwoven fiberglass laminates are made in a process similar to making paper. Particles of E glass (about 5 μm in diameter and a 100× aspect ratio) are mixed with PTFE in a slurry. A screen raises a layer of the materials and the liquid is sucked away. The dried and pressed material is layered as with the woven-glass process. Even with this process, the glass fibers still have considerable alignment. The PTFE content is higher, thus lowering the dielectric constant. Alignment of the glass fibers is more important at high microwave frequencies than at the usual RF frequencies.

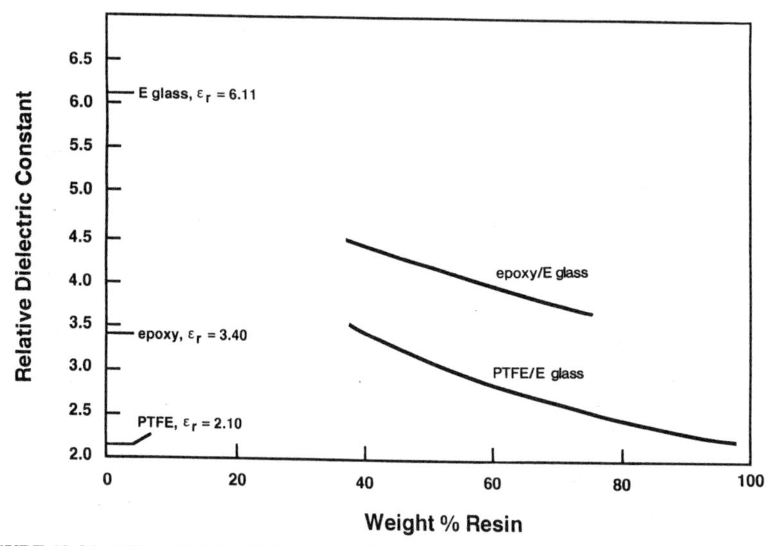

FIGURE 10-36 Effect of adding *E*-glass on the dielectric constant of epoxy and PTFE.

A recently introduced clad substrate is a thermoset ceramic-polymer [55] that combines many of the benefits of both ceramic and PTFE substrates. They are available within a range of dielectric constants, from 3.2 to 9.6. In principle, these laminates with thermoset resins do not have the cold flow and creep problems associated with conventional PTFE composite materials, and have a very low thermal coefficient of dielectric constant. These thermoset materials have a tendency to discolor after prolonged exposure to temperatures of 125°C and above. When temperatures exceed 150°C, the potential exists for delamination of the conductor. This may be corrected by baking out the substrates in oxygen-free ambients.

However, mechanical stability is improved at the expense of electrical performance and other properties such as machinability. The dissipation factor can increase by as much as a factor of 10, depending on the type and amount of additive. This is shown in Table 10-6, which compares the properties of some of the clad substrates. Table 10-6 also shows that moisture absorption can vary by as much as two orders of magnitude. Extreme care must be used when processing these materials to ensure the moisture is baked out, particularly for hermetic packages. This is aided by that fact that the temperature limit for the PTFE/glass core increases from 170°C to about 260°C for the filled materials. However, Cu-etched circuit/lines cannot withstand continuous exposure at these temperatures without bonded-side oxidation. It is recommended that continuous operation of unprotected Cu conductors should be at temperatures $<$50°C.

Although mechanical stability of these composites has improved, significant differences in thermal expansion coefficient between the plastic and cladding remain. This is most evident when Cu is removed from one side of the substrate, causing warping. The degree of mismatch is maximized for pure, unfilled materials, and minimized for woven-fabric materials. Anisotropy in thermal expansion between the *XY* and *Z* axis, as with dielectric properties, is caused by fiber and filler alignment in the core material.

High *Z* expansion causes two major problems. First, the effective dielectric thickness increases with temperature, changing the impedance value. Second, and perhaps most important, the reliability of plated-through-holes (PTH) is directly related to excessive expansion in the *Z* direction. Mismatch between the copper in the barrel and the core material creates electrical opens and conductor chipping and flaking. Recently, a ceramic-filled PTFE was introduced that more closely matches the expansion in the *Z* direction to the expansion of copper, 17 ppm/°C. Table 10-7 summarizes the advantages and disadvantages of some of the dielectrics used for clad substrates.

Another important consideration is the copper foil cladding. It is available in two basic forms: electrodeposited (ED) and rolled (sometimes called wrought). Schematically, the two processes are shown in Figs. 10-37 and 10-38, respectively.

TABLE 10-6 Properties of clad substrates

Laminate type	Dielectric constant	Dissipation factor (1 MHz)	Water absorption (%)	TCE (ppm/°C)	
				X,Y	Z
Polyethylene	2.32 ± 0.005	0.0002	0.01	108	—
PTFE unfilled	2.10	0.0001	0.001	88	1.18
PTFE woven glass	2.4-2.6 ± 04	0.0007	0.01	35	6
PTFE nonwoven glass	2.2-2.4 ± 02	0.0005	0.01	15	20
PTFE glass and ceramic-filled	2.94	0.0012		16	1.5
PTFE ceramic-filled	6-12	0.0020	0.7-1.0	110	1.09

TABLE 10-7 Advantages and disadvantages of clad substrates

Material	Advantages	Disadvantages
Unfilled		
Polyolefin (irradiated)	Good electrical properties Isotropic Easily processed	Poor dimensional stability Limited temperature range Poor solvent resistance
PTFE	Flexible Isotropic Almost universally used	Cladding increases loss Subject to cold flow Most expensive
Polystyrene (cross-linked)	Isotropic over wide temperature and frequency range	Not flexible Poor thermal match Low temperature limit
Polysulfone	Isotropic Low expansion coefficient	Affected by common organic reagents Induced mechanical stresses must be annealed out
Reinforcement		
Woven glass	Improved physical properties and thermal match	Anisotropic Higher dissipation factor, except for GY types
Quartz	Same as glass but with improved dissipation factor	Anisotropic
Ceramic	Extended dielectric constant range Improved electrical properties Better thermal properties Better isotropicity	Slightly poorer physical properties Highest cost High water absorption

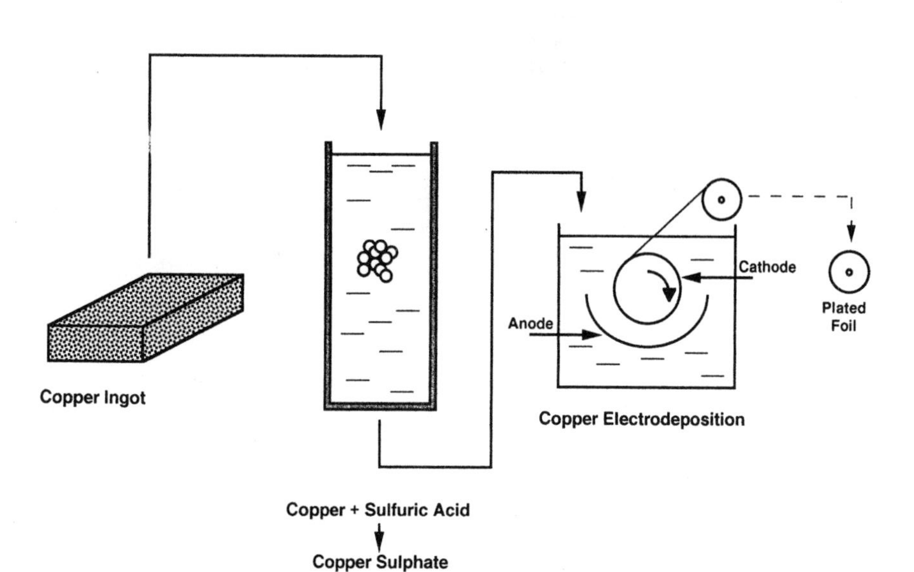

FIGURE 10-37 Schematic for manufacturing electrodeposited (ED) copper.

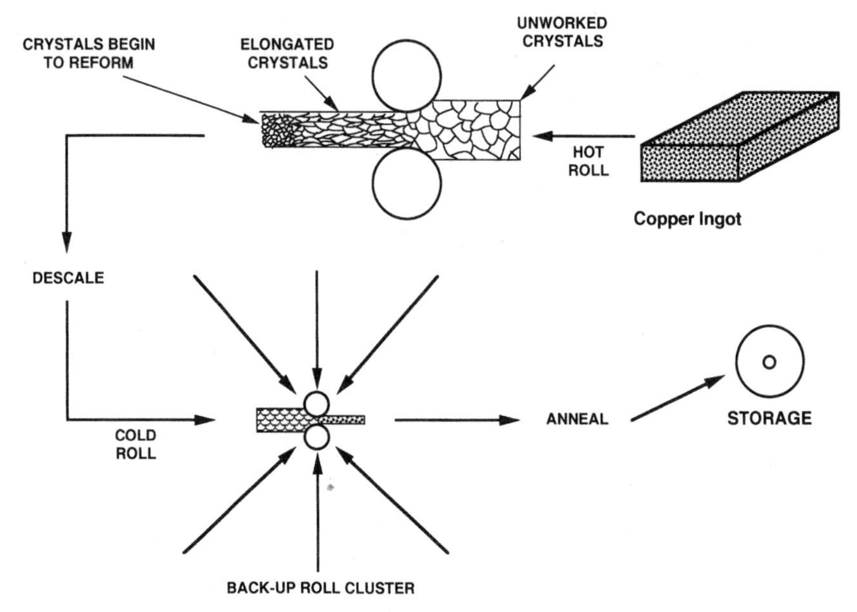

FIGURE 10-38 Schematic for manufacturing rolled or cladded copper.

The ED foil is produced by first dissolving copper metal in sulfuric acid to form copper sulfate. Grain refiners and other additives are combined to make a plating bath. The solution is pumped into a plating tank where the foil is formed in continuous lengths by deposition onto a rotating, nonreactive, polished stainless steel drum partially immersed in the electrolytic copper bath. Typically the plating drum is 50 to 100 in wide. The plated foil is continuously peeled from the drum and coiled. Up to 3500 foot lengths of 1 oz ED copper foil, 8 feet wide, comprise one coil [56]. On one side, the resultant foil replicates the relatively smooth surface of the drum. The other side, where copper growth occurs, typically has a rougher, nodular surface. This effect of having two disparate surfaces is similar to that found in tape-cast ceramics, where the carrier surface is also smoother. The degree of roughness of the growing surface depends to a large extent on the foil manufacturers, who may either roughen or smooth it to optimize subsequent bonding to the core material.

Rolled copper is formed in a process similar to one used for aluminum foil. A raw ingot is hot-rolled into an intermediate gauge (about $\frac{3}{8}$ in). It is then alternately cold-rolled and annealed into a thin foil. Narrow diameter work rollers, typically 12 in diameter, are used to provide maximum force for foil reduction. They are supported either by a second pair of large diameter rolls or a cluster of secondary rollers [57]. These back-up rollers are designed to counteract both vertical and horizontal elements of the rolling forces, enabling minimum-diameter rolls to be used without bending and springing. This process results in a more equiaxed, ductile structure with aligned, elongated copper particles, as shown in Figure 10-39*a*, as opposed to the ED copper dendritic structure shown in Fig. 10-39*b*.

Both types of foil are treated on one side to enhance adhesion. A deposit of copper is used to ensure a nodular, dendritic growth. Oxide-free copper may be bonded directly to the unfilled polymer core materials. An optional thermal barrier treatment with brass, zinc, or nickel, allows the adhesion of the clad laminate to be maintained in spite of the thermal processing conditions. Most foil supplied does not require this additional treatment, ac-

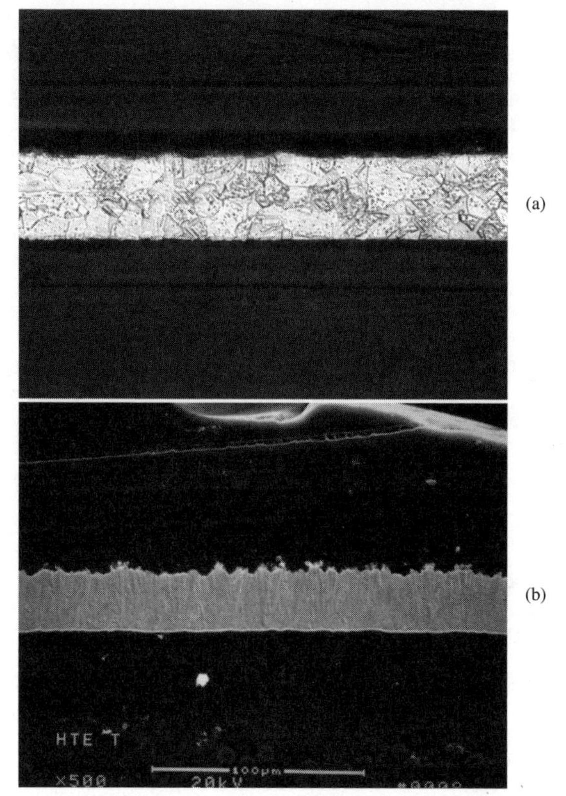

FIGURE 10-39 Cross sections of *(a)* rolled copper, and *(b)* electrodeposited copper. Magnification 500×. *(Courtesy of Somers Thin Strip/Bass Group, Waterbury, CT. The assistance of Steven Struck is greatly appreciated.)*

cording to Savage [58]. To ensure an oxide-free surface, surface passivation treatment with benzotriole or dilute CrO_3 was used. The core material, being thermoplastic, softens sufficiently to wet the roughened copper, providing mechanical anchoring of the copper.

The copper foil is now laminated to the core resin, as depicted in Fig. 10-40. The manufacturing process limits rolled copper to one-half oz thickness. Electrodeposited copper

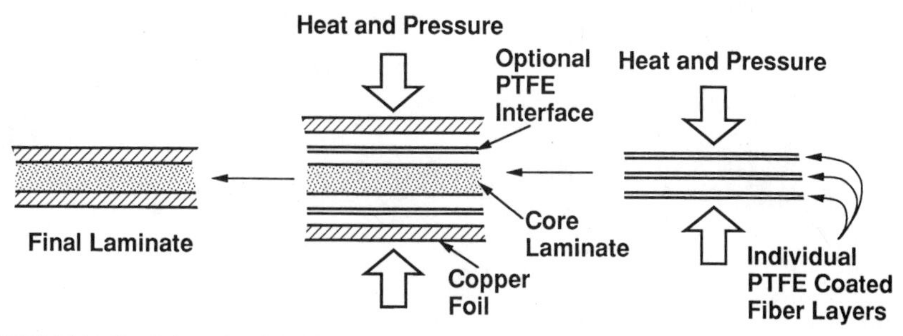

FIGURE 10-40 Schematic of cladding process for soft substrates.

is available down to one-eighth ounce. The initial bond strength of ED copper is higher than rolled, but degrades more quickly at about 260°C, ending with similar values to rolled copper. For reinforced PTFE core materials, sometimes a thin (0.001 in to 0.002 in) film of pure PTFE is used between the copper and core as a bonding layer. The pure-PTFE film beneath the foil, besides its use for bonding, serves other functions; it provides a smooth surface that enhances etching acuity; and also provides a moisture-resistant barrier and protection for the glassy fillers against the hot caustics used for producing subsequent copper oxide bonding treatment.

During lamination, the bonded copper surface is replicated in the melted bonding resin. Photographs of these resin surfaces after etching away the clad copper, as shown in Fig. 10-41, clearly show the effects of copper topography and roughness after lamination on the core-copper interface. The roughness of the ED copper on both the top and substrate increases with increasing deposit thickness, because of the elongated copper growth in the *Z*-axis. Typically, the interface roughness at the substrate for 1 oz copper can reach 95 μin (2.4 μm). On the other hand, the substrate surface of the rolled copper is fairly constant at about 55 μin (about 1.4 μm). During the rolling operation, copper crystals are crushed to small, irregularly shaped spheroids. It is generally preferable to use the rolled copper for microwave applications, where low losses demand smoother, more consistent surfaces. The differences in insertion losses for ED and rolled copper on microstrip line [59] and stripline [60] are shown in Fig. 10-42. As expected, losses are higher for microstrip than for stripline. The flattening of the curves at the higher frequencies for microstrip may result partly from the fact that ground plane is smoother, because in this case it is bonded to a smoother aluminum surface, rather than retaining its original roughness.

The large substrate areas available with soft substrate materials are illustrated by the microwave circuit shown in Fig. 10-43. The panel size of this circuit, operating at about 1 to 2 GHz, is 11 in by 5.6 in. The 0.060 in thick substrate is cladded with 1 oz copper on both sides, and the circuitry is defined by etch-back techniques. This commercial avionics

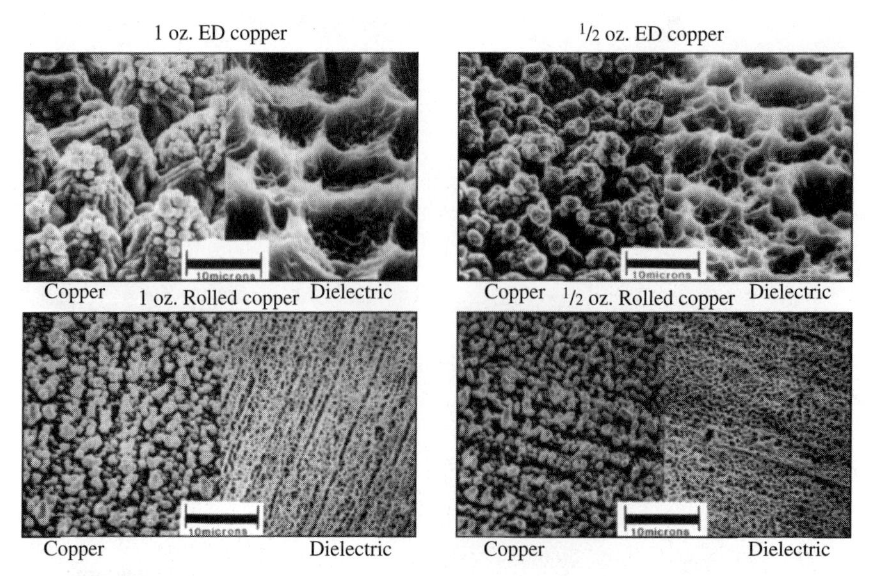

FIGURE 10-41 The upper half of each of these photos shows the bonding side characteristics of 1½ oz electrodeposited copper. The bottom half shows 1 oz rolled copper. The replicated topography of the substrate is shown at the right for each type of copper. *(Courtesy the Rogers Corporation.)*

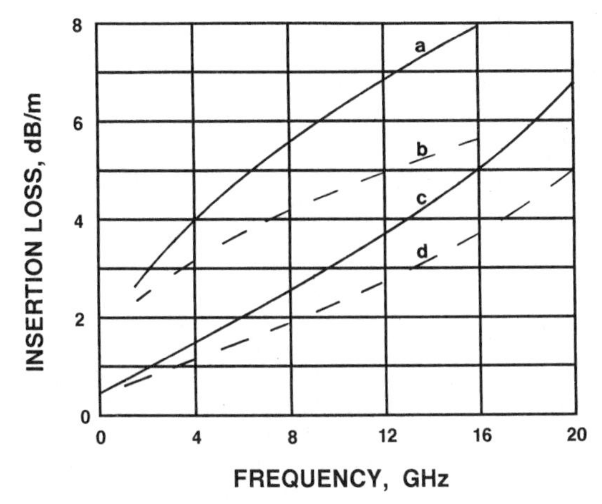

FIGURE 10-42 Insertion loss of electrodeposited (ED) copper and rolled copper. (*a*, *c*) ED copper; (*b*, *d*) rolled copper; (*a*, *b*) microstrip on aluminum backing; (*c*, *d*) stripline, with permission.

circuit is used in cockpits. The flexibility of soft substrates allows for forming complex shapes such as cone antennas.

10.16 CLEANING

A thoroughly cleaned substrate is needed for the preparation of films with proper adhesion and reproducible properties. The choice of cleaning techniques depends on the nature of the substrate, the type of contaminants, and the desired degree of cleanliness. Residues from manufacturing, polishing, fingerprints, oil, and airborne particulate matters are

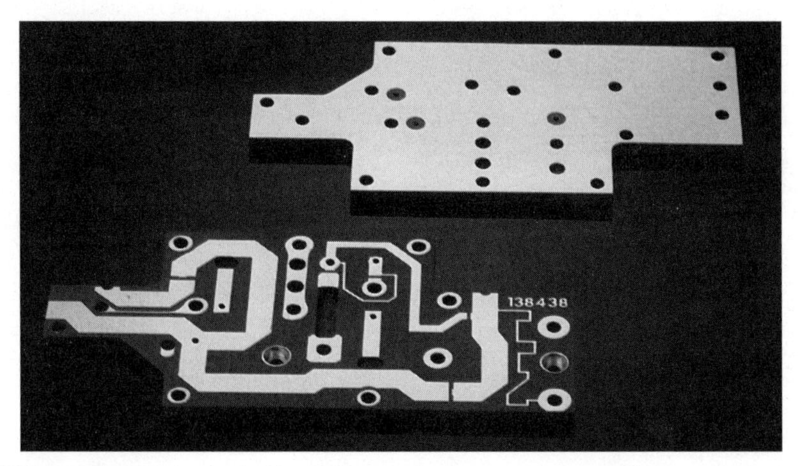

FIGURE 10-43 Avionics circuit of 1 to 2 GHz on a soft substrate. *(Courtesy of the Rogers Corporation.)*

examples of frequently encountered contaminants. Usually, effective cleaning can only be accomplished after identifying the offending material. The reader is directed to an early, but still appropriate, primer by this author on substrate cleaning [61].

Because of their chemical composition, some of the newer substrate materials require special cleaning procedures. Dilute acidic solutions such as phosphoric acid are usually used to clean aluminum nitride, because it is readily attacked by alkali. Chromerge®, a solution of chromium trioxide and sulfuric acid available from Fischer, can also be successful. Titanates are readily attacked by strong acids and should be cleaned in a benign alkaline solution. After etching the copper patterns, soft substrates should be rigorously rinsed with hot water to remove all etching residues.

A technique recently reported for cleaning silicon wafer surfaces may be applicable to hybrid substrates. In this technique, high-purity liquid or gaseous CO_2 is expanded in a special nozzle to form a high-speed jet. The jet contains numerous small-diameter particles of solid CO_2 which strike the surface. On impingement at the surface, this CO_2 "snow" is reported to remove even submicron adherent particles, hydrocarbon stains such as fingerprints, and silicone greases [62,63]. It is the author's experience that, under these special conditions, particles trapped in substrate-surface pores can be removed. One of the major advantages of this technique is its dry nature; that is, the CO_2 has none of the environmental threats and occupational hazards associated with conventional chemical reagents and solvents. The CO_2 spontaneously evaporates, leaving no waste to be disposed of.

10.17 THIN FILMS

It was illustrated how circuit losses are directly affected by the conductors and dielectrics. The materials-deposition techniques have a profound effect on deposit quality of the film. For microwave hybrids, most conductors, and in some cases dielectrics, are deposited by physical vapor deposition. Vapor deposition techniques are described in detail in Chapter 1.

Filament evaporation is not generally used for the fabrication of microwave hybrids because the quality and composition of the deposited films may be questionable. Also, most dielectrics are very difficult to deposit by this method. The reader is referred to Chapter 1 to review thin film deposition techniques.

10.18 MICROWAVE COMPONENTS

10.18.1 Resistors

Resistors are used in high-frequency applications as terminations, attenuators, and for isolation. Depending on the application, resistors should have the following:

- Good stability (particularly important for high-reliability applications, a major MIC market).
- A reproducible temperature coefficient of resistance (TCR), (important when compensation is necessary).
- Good power-handling capability.
- Minimum parasitics.

Table 10-8 lists some of the resistor materials used. Most resistors for higher-frequency applications are thin film; thick film applications are used more often in lower frequency (*S* band) or associated *dc* circuits. Thin film resistive layers are first deposited, defined, and then the contacts (metallizations) are deposited and patterned. Photodefinition of contacts and resistor patterns result in sharp, accurate definition of both geometries.

The series resistance of a strip of resistive material is based on the following equation:

$$R = R_s \frac{l}{w} \tag{46}$$

where: R_s = sheet resistance in Ω/□
l = resistor length
w = resistor width
l/w = number of squares

At high frequencies, the conductor cross section available to carry the current is decreased by the skin effect. Consequently, the high-frequency resistance must be recalculated using incremental inductance.

$$R = 2\pi f\,[L\ (\text{nominal dimensions}) - L\ (\text{dimensions reduced by } \delta/2] \tag{47}$$

For low-thermal resistance, the area of the resistive film should be as large as possible. The width of the resistive film should not be much different than the width of the lines feeding it to minimize discontinuity effects. As such, the length should be as long as possible to keep the thermal resistance as low as possible. The length is specified by rearranging Eq. 10-46 to

$$l = \frac{wR}{R_s} \tag{48}$$

If the load length is increased by decreasing the sheet resistance, the load may show the behavior of a very lossy transmission line, instead of the pure resistance of a lumped resistor. Figure 10-44 shows how the VSWR increases as the length of the load becomes too large, because of low values of sheet resistance. These limitations place some constraints on the processing of resistors. Low-power resistors are designed as short as possible, while still maintaining tight tolerances (±5% or better tolerances in resistor value).

TABLE 10-8 Properties of thin film resistors

Resistor material	Resistance (Ω/□)	TCR, α (ppm/°C)
Nickel-chromium	50-500	<±20 to ±150
Oxidized chromium	100-1000	±25 to ±100
Chromium-silicon monoxide	100-1000	<±20 to ±200
Tantalum nitride	30-200	−60 to −200
Tungsten-ruthenium	25-300	+120 to +300
Titanium	5-2000	100 to −100

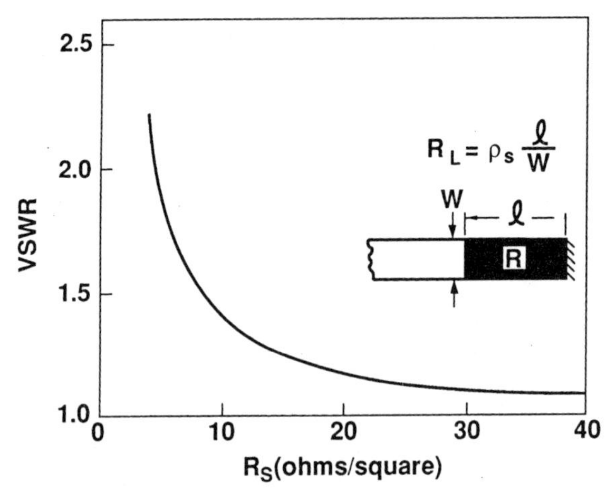

FIGURE 10-44 VSWR as a function of resistor load length.

When designing an MIC resistor, the following rules should be followed:

- To make the resistor lumped, the largest dimension should be at least $\lambda/10$, preferably $\lambda/20$.
- To reduce noise, the substrate should be as smooth as possible because resistors have higher noise values on rougher surfaces.
- Resistors should be as short as possible to minimize inductance value and VSWR. The resistor, as deposited, should be as close to value as possible to minimize trimming. In some cases, thin film resistors can be etch-trimmed without changing their planar dimensions.

Resistor value, geometry, and contact alignment must be tightly controlled. For example, if the tolerance is determined to be ±5%, as is the case for a 50 Ω termination, careful attention must be paid to all aspects of the fabrication process. A 20-mil-wide resistor that is 40 mils long is 2.0 squares. If the ink is 25 Ω/□ the end result is 50 Ω. However, if the width ends up 19 mils, or 5% narrow, the number of squares increases to 2.1, resulting in a resistor that is now 52.5 Ω.

One of the major problems facing the hybrid technologist is providing close access for the resistor to ground. This is accomplished either by via holes through the substrate or wrap-arounds. The former is convenient when only a few vias are needed. Large numbers of vias significantly add to the substrate cost and may weaken the substrate rigidity and mechanical strength. During drilling, using pulsed CO_2 and *Q*-switched Nd:YAG lasers, high-energy pulses (typically 10^6 to 10^7) W/cm^2) are absorbed by the substrate. This leads to intense local heating. Huge thermal gradients are generated, vaporizing some of the material and changing composition, grain size, and flaw distribution. Although the affected total volume of material is small, the edge of the hole greatly affects the residual substrate strength, and holes should be minimized. In addition, molten material remains around the hole lip. This can create discontinuities in narrow lines and damage photomasks, which frequently come in contact with the surface. As a consequence, if polished pieces are used, holes should be realized before polishing. Holes achieved later may require secondary finishing operations, adding

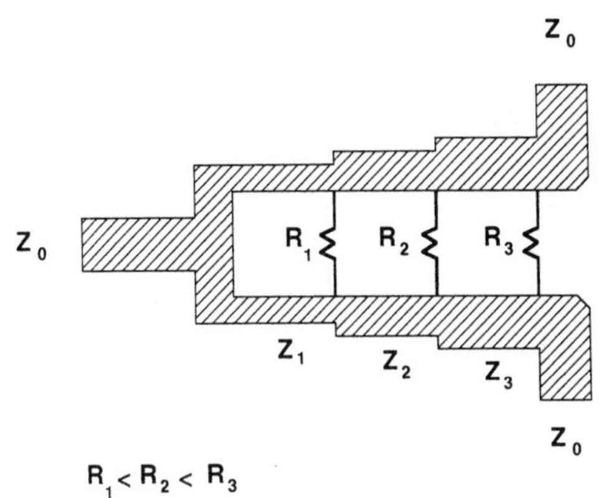

FIGURE 10-45 Schematic of multiple section in-line equal-split power divider.

to the cost. For high volume, holes can be put in the green state. However, location tolerance caused by shrinkage variations may be excessive. Holes may also be realized by ultrasonic material removal; however, this practice is limited to short-run, experimental pieces.

Resistors are frequently used as shunt-to-ground or as isolation resistors in power dividers. A multiple-section in-line, equal-split power divider is shown in Fig. 10-45. This configuration [64] consists of a number of quarter-wave segments with resistive terminations at the end of each segment. When the impedance values are determined, the resistance values are calculated to provide proper isolation. At high frequency, the resistors provide isolation between the shorted arms of the divider. For power dividers containing two sections, the calculations are straightforward; for more than two sections the design equations are more complicated, and both the conductance and the admittance are used instead of resistance and impedance.

Design of the termination conductor is important. The shunt capacitance and inductance leading to the via or wrap-around directly affect resistor-return loss. Two-resistor ground-metallization designs are shown in Fig. 10-46, one straight and the other tapered to account for parasitics. The lower-return loss for the tapered design illustrates the need

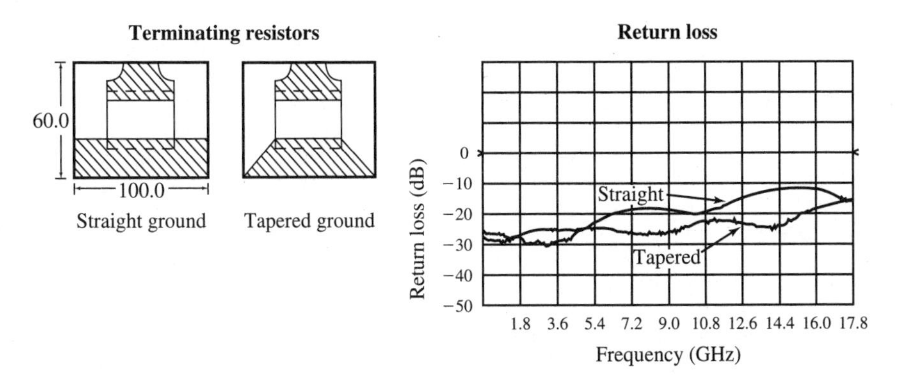

FIGURE 10-46 Return loss versus frequency for two-resistor ground-metallization designs.

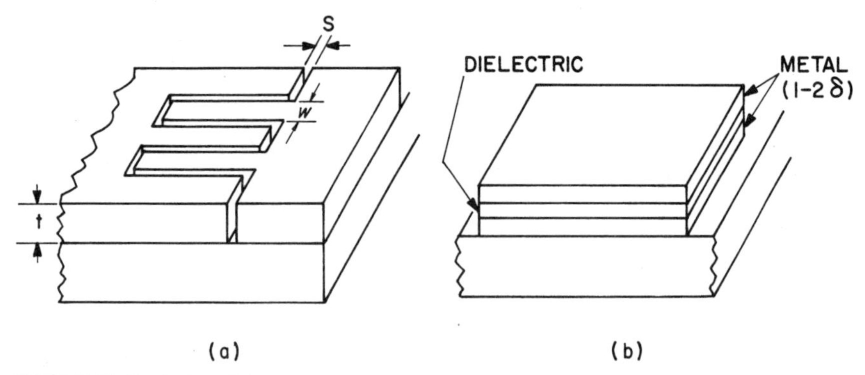

FIGURE 10-47 (*a*) Interdigitated capacitor, and (*b*) metal-insulator-metal structure.

to carefully consider the characteristics of the substrate and the conductor in high-frequency applications.

10.18.2 Capacitors

The basic component for storing electric charge is the capacitor. When a voltage bias is placed across the capacitor, no current flows. Internal polarization of ions and electrons in the dielectric occurs instead. Electrical charge is converted to mechanical energy by the displacement of these electrons and ions from their equilibrium positions within the dielectric material's lattice. There are three types of capacitors used for MICs. Two of these capacitor elements are shown in Fig. 10-47*a* (interdigitated) and Fig. 10-47*b* (parallel plate). The third structure uses the capacitance between conductor and ground, shunt capacitance.

Of the two structures, the interdigitated in Fig. 10-47*a*, is the simpler to construct. This capacitor is fabricated in a single layer using the substrate and air, if unencapsulated, as the dielectric. Because of low-frequency resonance problems resulting from coupling between long lengths of line, only small capacitors (<2 pF) are practical at the higher frequencies, for a gap (s) $=$ 2 mils.

Alley [65] analyzed the tolerance in line width and gap variations on capacitance values. His curve relating the percentage error in capacitance to the width/gap ratio for capacitors fabricated on 99.5% alumina is shown in Fig. 10-48. Under the condition, the sum of line width and space is again a constant; a capacitor tolerance of $\pm 4\%$ is achievable by controlling the line width of a capacitor with nominal 2 mil pitch to ± 0.1 mil, reasonable tolerances for thin film processing. In their work, Taylor and Williams [66] show that for miniaturization, large capacitance values are impractical in the interdigitated form. A 2.5 pF capacitor occupies, for example, an area exceeding 12,000 sq mils, whereas 0.25 pF capacitors use only 1,600 sq mils. Parallel-plate capacitors are more suited for larger-value capacitors because they occupy significantly less area.

The capacitance in this interdigitated structure arises from fringing fields both through the air and the dielectric and by direct fields through the conductor. It is therefore important that the dielectric be uniform and isotropic, the gap (s) between lines be uniform, and the metal has high conductivity. The latter is particularly important because the Q for this type of capacitor is given by

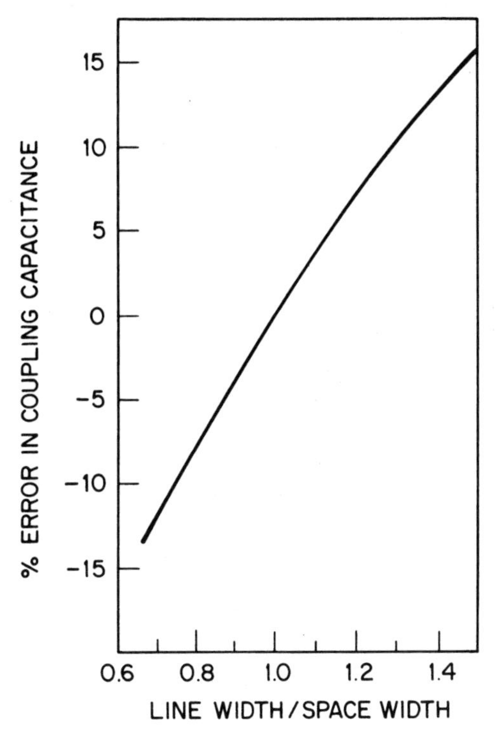

FIGURE 10-48 Interdigitated capacitance error versus *w/s* ratio. $s + w =$ constant.

$$Q = \frac{1}{\omega CR \text{ series}} \tag{49}$$

$$R_{\text{series}} = \frac{4}{3}\frac{1}{XN}R_s \tag{50}$$

where X = cell width
N = number of fingers
C = capacitance in pF
R_s = sheet resistance in $\Omega/\square$

The Q decreases with increasing frequency and the capacitance values can limit this structure to low values.

The other type of capacitor is the three-tiered, parallel-plate structure, or metal-insulator-metal (MIM) as shown in Fig. 10-47*b*. These capacitors, used in tuning, matching, and coupling circuits, require dielectrics with very-low loss at the frequency of interest. The importance of using low-loss dielectric is illustrated in Fig. 10-49, where the capacitor Q is shown to decrease with increasing capacitance values and higher dielectric losses. The increased capacitance values reflect increased values in the electrode aspect ratio, l/w. At low frequencies, the capacitor Q-factor is dominated by the properties of the dielectric, whereas at high frequencies, conductor losses dominate, assuming a frequency-independent loss tangent. The fabrication of small, high-Q capacitors

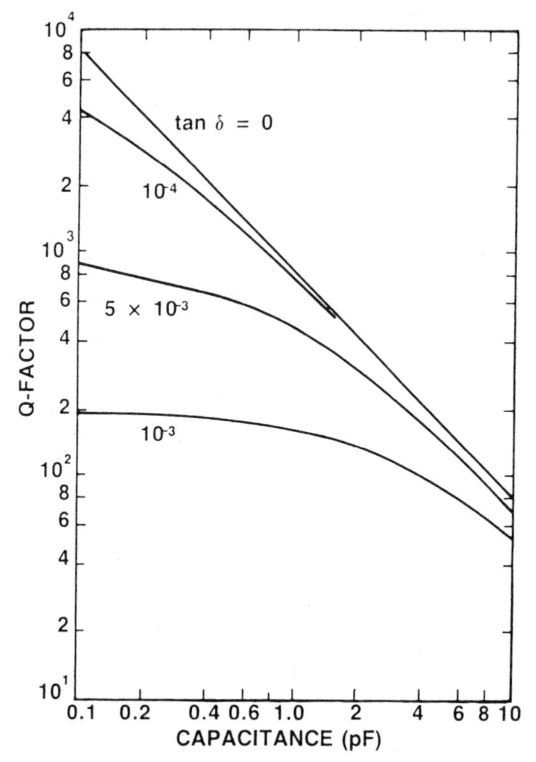

FIGURE 10-49 Capacitor Q-factor versus capacitance for various dielectric loss factors.

requires rigid control over conductor purity, conductor thickness, and the dielectric thickness and its loss.

A number of dielectric materials are available for MIM capacitors. Table 10-9 illustrates the relatively narrow range of the ϵ_r of the thin film dielectrics.

Thin film capacitors are limited in capacitance density and require base conductors with smooth surfaces. For parallel-plate (overlay) capacitors, the value of the capacitance (C) is determined by the overlapping electrode area (A) the thickness of the dielectric (t) and the permittivity (ϵ) or dielectric constant of the dielectric material (ϵ_r). The dielectric constant is a measure of the material's ability to store charge. The capacitance

TABLE 10-9 Properties of capacitor dielectrics

Material	ϵ_r	Dielectric strength (V/mil)	Q at 1 pf
SiO	6-8	4×10^5	30
SiO_2	4-5	10^7	20-500
Al_2O_3	7-10	4×10^6	~800
Ta_2O_5	25	6×10^6	100
Polyimide	3.5	10^4	~100
Benzocyclobutene	2.7	10^4	~800

of any two-plate capacitor may be calculated from Eq. 10-51 if the electrode area, the dielectric constant, and the dielectric thickness are known.

$$C = 225\epsilon_r \frac{A}{t} \times 10^{-6} \tag{51}$$

where C = capacitance in pF
ϵ_r = dielectric constant
A = area of smallest electrode in mils^2
t = dielectric thickness in mils

The total Q of this type of capacitor is of the form

$$\frac{1}{Q} = \frac{1}{Q_C} + \frac{1}{Q_D} \tag{52}$$

where Q_C is the electrode resistance and Q_D is the dielectric loss, and

$$Q_D \cong \frac{1}{\tan\delta} \tag{53}$$

In cases where the conductor losses far exceed those of the dielectric losses, particularly for thin films, the capacitor Q may be expressed as

$$Q = \frac{1}{\omega RC} \tag{54}$$

An MIM capacitor acts as a lossy transmission line. The inductance and the resistance of the base and the counterelectrodes, and the conductance and capacitance of the dielectric all determine the capacitor Q. If the metal thickness of the electrodes exceeds the skin depth, the series resistance is determined by the skin resistance. If the metal thickness is less than the skin depth, the bulk resistance is the determining value. Usually, the bottom electrode is kept thin to maintain a smooth surface, and is typically less than a skin depth, whereas the top electrode is typically realized by electroplating to several microns thickness.

For well-designed overlay capacitors, the length and the width are small compared to the wavelength in the dielectric film. From transmission-line theory, a reasonable equivalent circuit for the capacitor is shown in Fig. 10-50.

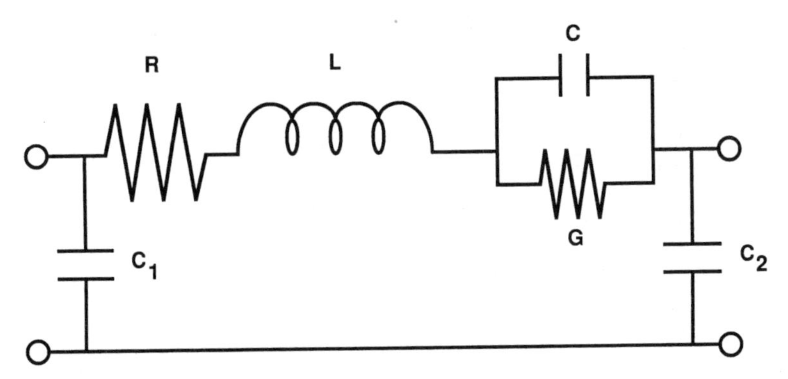

FIGURE 10-50 Equivalent circuit of a thin film capacitor.

When skin losses prevail, the Q_c is given the expression

$$Q_c = \frac{3}{2\omega R_s\,(C/A)l^2} \tag{55}$$

where R_s is surface skin resistance and l is electrode length. The salient point of this expression is the strong dependence of Q_c on conductor length, and to a lesser but important degree, on the unit capacitance, skin resistance, and frequency. An increase in any of these elements will reduce Q_c. However, by significantly increasing the size of the bottom electrode, the conductor Q may be almost doubled.

For a given design frequency and capacitance, the electrode resistance (R) must be kept as low as possible to maximize Q. Within limits, the high-frequency performance of thin film capacitors can be improved by using thicker, highly conductive films for the capacitor electrodes. In the case of tantalum, part of the bottom electrode is thinned as it is anodically converted to its oxide, which serves only to increase the electrode resistance. An underlay of a more-conductive metal like aluminum was suggested [67]. Aluminum has the advantage that it is anodizable and permits oxidation at pinholes in the tantalum layer during the anodization process, fixing an otherwise defective dielectric. The rough surfaces of thick film capacitor electrodes also increase conductor resistance, and these capacitors also have reduced Q.

Generally, it is inadvisable to wire-bond thin film capacitors because thin dielectrics tend to crack during bonding. Sometimes the top electrode is extended to provide a bonding area adjacent to the capacitor. The capacitor dielectric in this structure is also extended over the base electrode, serving as an insulator between the two conductors. Thin film MIM capacitors fabricated this way, Fig. 10-51*a*, have a tendency to short when thin dielectrics are used as insulation. The thin dielectric, typically 0.2 to 1.0 μm, may be stressed over the sharp edge of the bottom conductor and develop hairline cracks. When the top electrode is deposited it shorts through these cracks to the bottom conductor. An alterna-

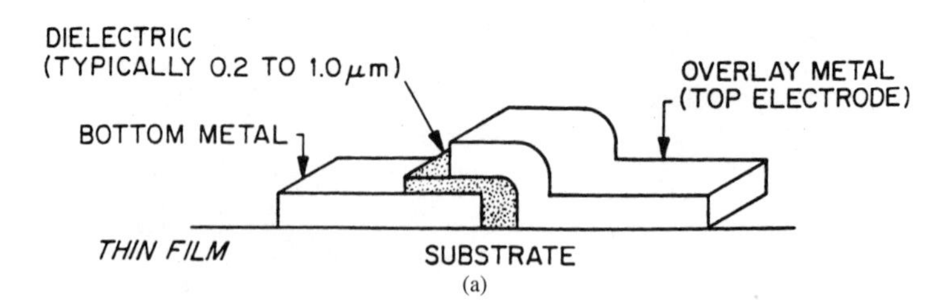

(a)

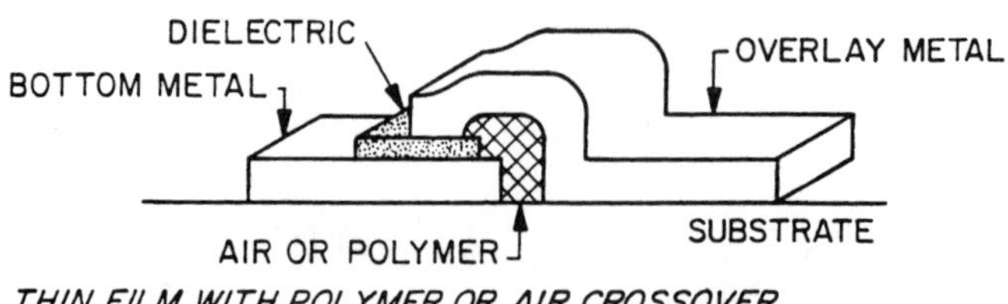

(b)

FIGURE 10-51 (*a*) Thin film parallel-plate capacitor, and (*b*) thin film capacitor with polymer crossover.

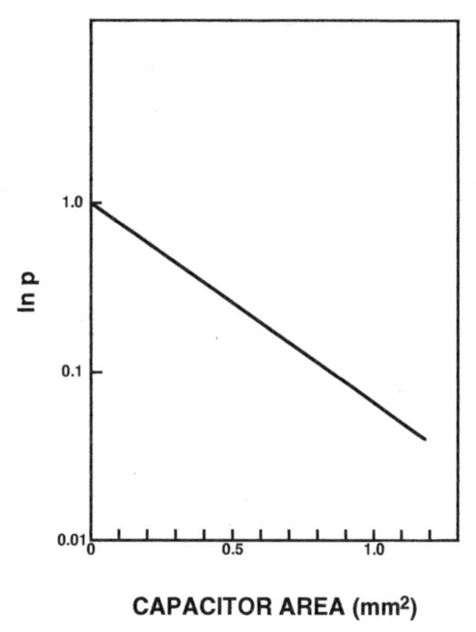

FIGURE 10-52 Fraction of capacitors without defects versus capacitor area.

tive approach is to bring the capacitor dielectric to the end of the bottom conductor, and interpose a thicker, flowable secondary insulator such as polyimide. The top electrode may then be processed and isolated from the base electrode by this redundant insulation, as shown in Fig. 10-51*b*. This technique has the advantage of not having to process and perturb the bottom electrode until after the capacitor dielectric is deposited and defined, maintaining the bottom electrode in an as-deposited state.

At some point, as the bias on a parallel-plate capacitor is increased, the capacitor loses its ability to store charge. Above this maximum-bias point, the capacitor breaks down and the electric current is rapidly transferred between the conductor electrodes. The maximum possible voltage that can be applied across the capacitor is called the dielectric strength of the capacitor, given as volts per unit dielectric thickness. A capacitor with 1.0 μm of silicon nitride breaking down at 400 V would have a breakdown voltage of 4 MV/cm.

Breakdown is usually a consequence of stress, thin spots, cracks, and impurities within the dielectric, rather than an inherent property of the dielectric material. Breakdown reflects more on the processing and the quality of the material. As a consequence, larger-area capacitors are more prone to breakdown than smaller ones. The fraction of capacitors without defects versus the area of the capacitor [68] is shown in Fig. 10-52. The data agree well with

$$\ln P = \rho A \tag{56}$$

For the probability of a capacitor without defects, ρ is the defect density per mm^2 and A is the capacitor area.

Because the capacitor is usually the most sensitive component to processing, circuit yield generally depends on capacitor yield. This is shown in Fig. 10-53, where circuit yield in percent is plotted against the number of capacitors in a circuit. For these tests, 12 pf capacitors that were 14 mil square with 1 μm silicon nitride as the dielectric were used. In effect, circuit yield decreases with increasing capacitor area. High-capacitor yields are

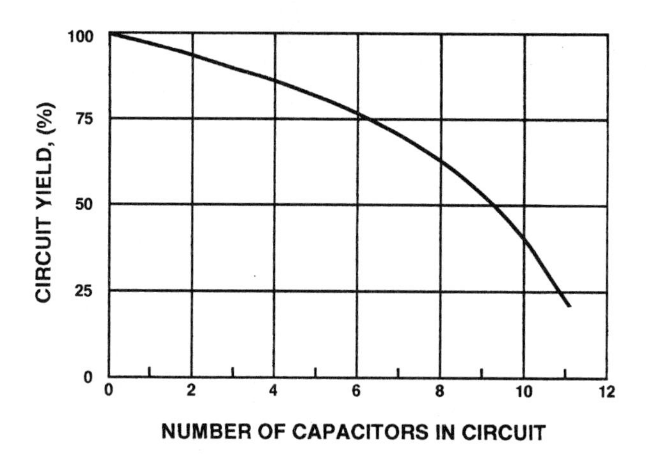

FIGURE 10-53 Circuit yield as a function of number of capacitors in a circuit.

imperative if high-circuit yield are to be realized. Capacitor redundancy is frequently used to ensure circuit survival.

Summarizing, the dielectric for the multilayered structure should

- Be reproducible because trimming is difficult.
- Have a high breakdown voltage, particularly important for thin dielectrics.
- Have low dielectric loss for high Q.
- Have low stress to reduce the sites available for low breakdown.
- Have a dielectric constant independent of frequency, voltage, and humidity.

10.18.3 Coupled-Parallel Microstrip

Couplers. Couplers generally serve as power combiners or splitters. The type discussed in this section are generally four-terminal (port) devices shown schematically in Fig. 10-54*a*. They consist of two or more closely spaced lines, usually of equal width (w) precisely separated by a gap (s). Varying w and s will determine the ratio of energy transferred from one line (leg) to the other. The line length determines the operating frequency.

The signal is introduced into port 1. Part of the signal is then transferred through the substrate and the dielectric (usually air) between the lines to the adjacent leg, where it exits from port 2. The remaining signal then continues out through port 3. Port 4 is usually terminated with a 50 Ω resistor. This value is used because the impedance of the port metallization is normally 50 Ω. For other impedance values the resistor value is adjusted accordingly.

Figure 10-54*b* shows the basic structure of the proximity coupler, comprised of two parallel-coupled microstrip lines. This coupler structure is of special interest for the following reasons:

1. It can be made relatively broadband.
2. Matching and directivity are frequency-independent.

3. The power splitting ratio can be varied over a wide range. The coupling to the two output arms is determined mainly by the distance between the two microstrips in the coupling region. High coupling requires very uniform, narrow gaps, typically 0.2 to 0.5 mils. Even for thin film hybrids, this poses a reproducibility problem. As a result, 3 dB couplers (equal power at both output ports) cannot easily be realized in a pure planar technique. One way to overcome this difficulty is to apply overlay techniques; that is, to increase the coupling by a partial overlapping of the two microstrip lines separated by a thin insulating layer in the coupling region. Couplers of this kind are frequently used in stripline techniques. The coupling ratio can easily be controlled by varying the degree of overlapping.

Interdigitated couplers shown in Fig. 10-54*c* are easier to make, particularly in thin film form because the gaps are at least 0.8 mils. Although wider than the two-line geometries, the gap dimensions nevertheless still dominate. Presser [69] has shown that changing the gap dimension (Fig. 10-55), while keeping the line width constant changes the coupling

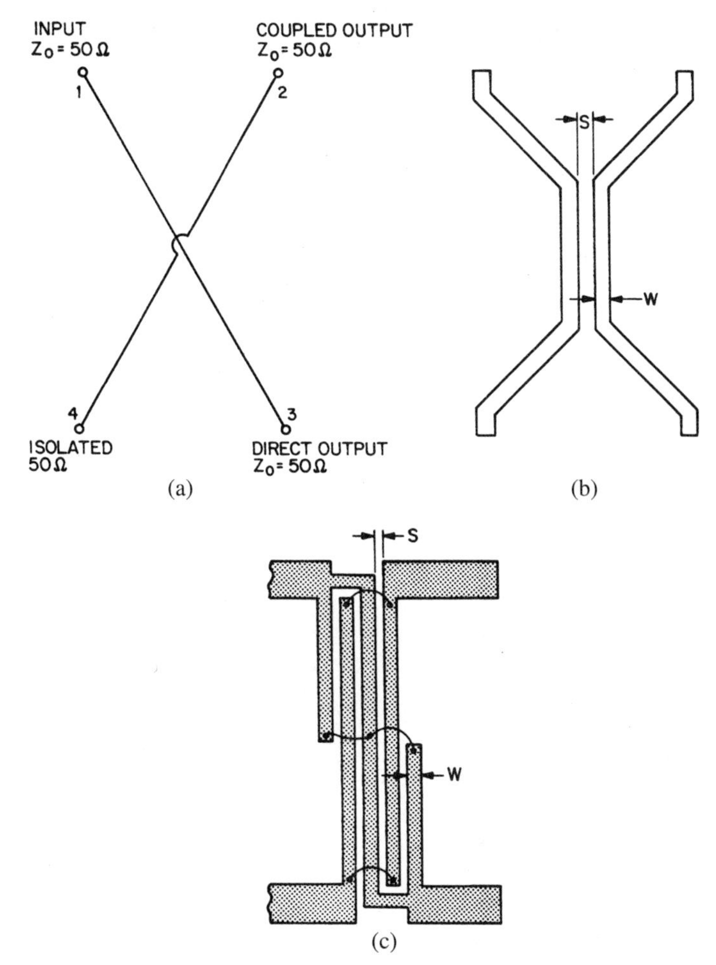

FIGURE 10-54 (*a*) Coupler schematic, (*b*) proximity coupler, and (*c*) interdigitated coupler.

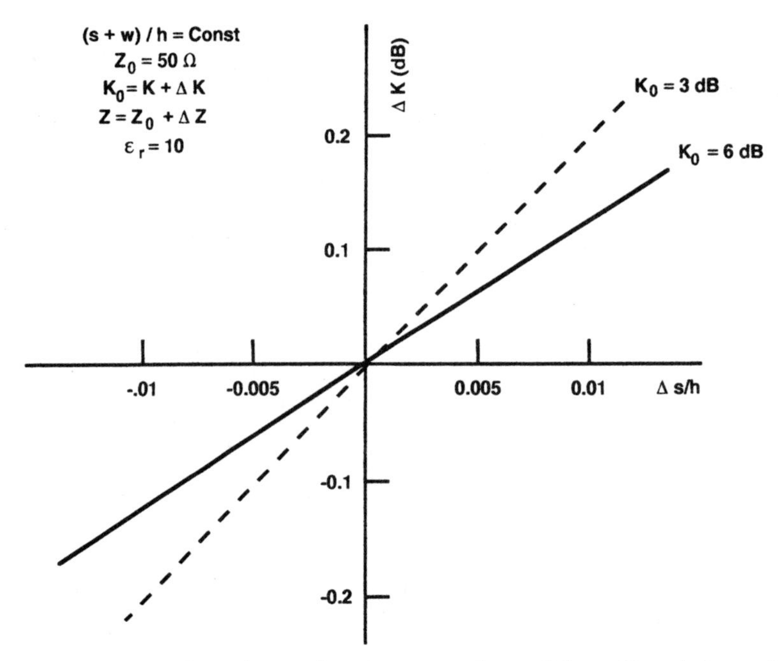

FIGURE 10-55 Effect of changing gap dimensions on coupling coefficient. *(From A. Presser,* IEEE Trans on Microwave Theory and Techniques, *MTT-26, No. 10, October, 1978, p. 801.)*

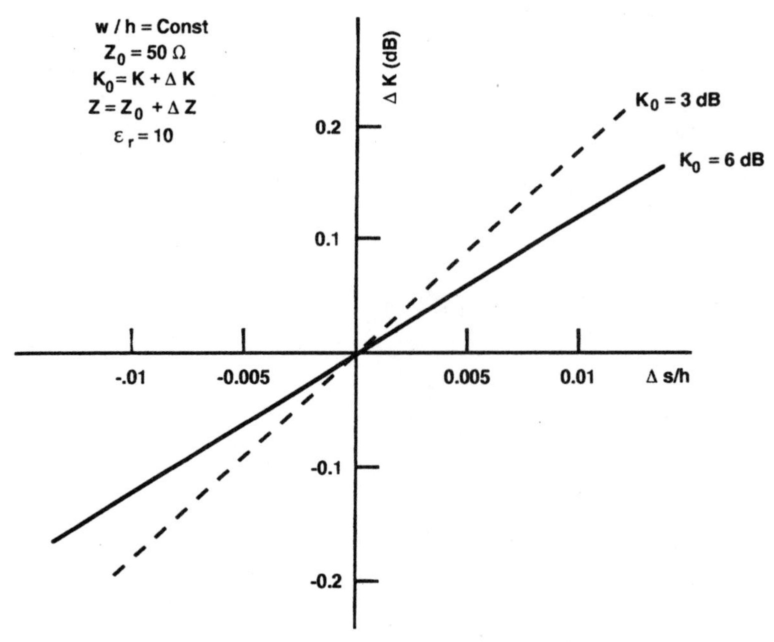

FIGURE 10-56 Coupling deviation versus shape-ratio change (sum ratio constant) (after [144]).

coefficient by as much as three times compared to the same line width change at a constant gap dimension. The line width is nevertheless important. Variations in width change line impedance, causing a mismatch with port impedance.

In practice, the gap and the line width dimensions are interrelated. A wider line results in a narrower gap and vice versa. In Fig. 10-55, Presser shows the effects of changing the gap (Δs) on coupling deviation (s and w/h = constant). Comparing Figs. 10-55 and 10-56, one can immediately see that the gap plays a dominant role in coupler performance. The results of Levy [70] are plotted in Fig. 10-57 where the percentage of coupling change is plotted against gap change for 1 mil lines and spaces. In this figure, both theoretical and actual measurements are plotted. For a 0.25 mil (6.35 μm) increase in gap, a 5% increase in coupling (in excellent agreement with Presser's results) attests to the strict control over artwork, masks, and processes necessary to maintain reproducibility. Others [71] calculated the fabrication tolerances on coupler sensitivity. The complete calculation is beyond the scope of this chapter. However, for coupled microstrip on alumina, a 4% change in coupling factor will generally result from the following:

$$\left.\begin{aligned} \Delta s &= \pm 2.5\ \mu\text{m} \\ \Delta w &= \pm 2.5\ \mu\text{m} \end{aligned}\right\} \quad (\Delta s + \Delta w)/h = \text{constant}$$

$$\Delta h = \pm 25\ \mu\text{m} \qquad h = 0.635\ \text{mm}$$

$$\Delta \epsilon_r = \pm 0.25 \qquad \epsilon_r = 9.7$$

Accordingly, besides maintaining close control over the gap dimensions, comparatively large variations in substrate thickness should be avoided; flat, polished substrates are most desirable. Care must also be taken to monitor the substrate density and the dielectric constant because large variations can occur, particularly between vendors.

Filters. Two types of microstrip filters are generally made in microstrip: (1) end-coupled, and (2) parallel- or edge-coupled, as shown in Fig. 10-58. The problems of dimensional tolerances and sensitivity are unduly severe for end-coupled geometries, so that edge-coupled

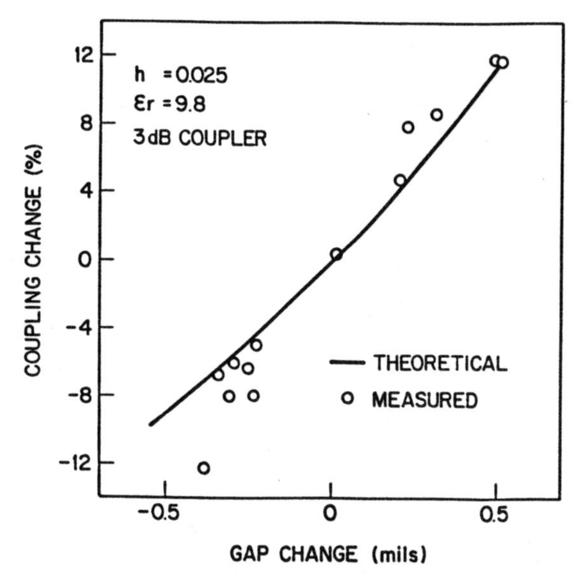

FIGURE 10-57 Coupling change versus gap change (Δs). *(From A. Levy,* Final Report AFWAL-TR-84-0243, *May, 1984, Air Force Materials Laboratory, Wright-Patterson AFB, OH.)*

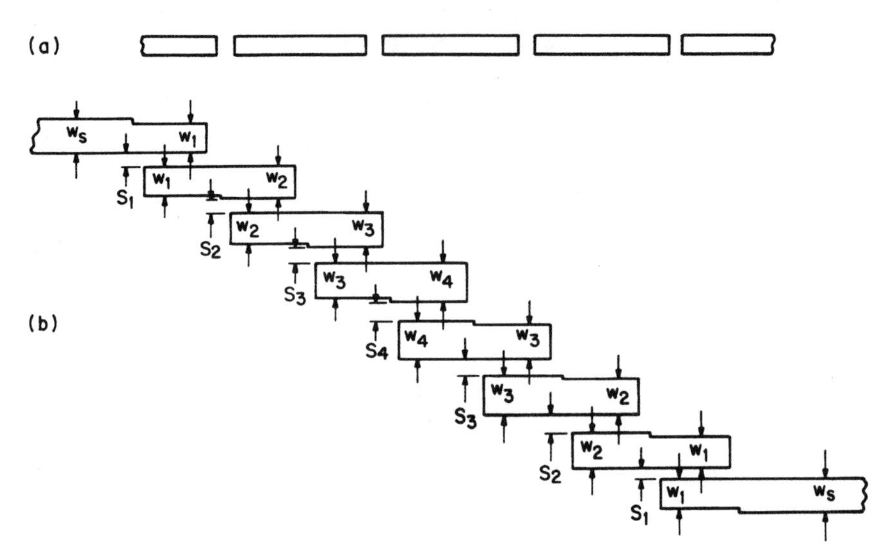

FIGURE 10-58 (*a*) End-coupled filters, and (*b*) edge-coupled filter.

filters, (Fig. 10-58*b*) are generally preferred. The *w* dimension ranges near the 50 Ω line width to slightly less. In the center element, the step is sometimes too narrow to be considered. The step near the ports may be about 5 mils. The gaps are also relatively wide, typically ranging from 5 mils at the outside to 25 mils in the center. These multielement filters are easily fabricated in both thin and thick film forms.

Inductors. There are two basic inductor types used in MICs, wire and printed. Wire inductors, both round and strip, are used for tuning and assembly. Printed inductors will be discussed in this section.

The configuration of a flat, spiral-strip conductor is shown in Fig. 10-59. This type is used when inductances >2 to 3 μH are required. The inductance value increases with increasing number of spirals.

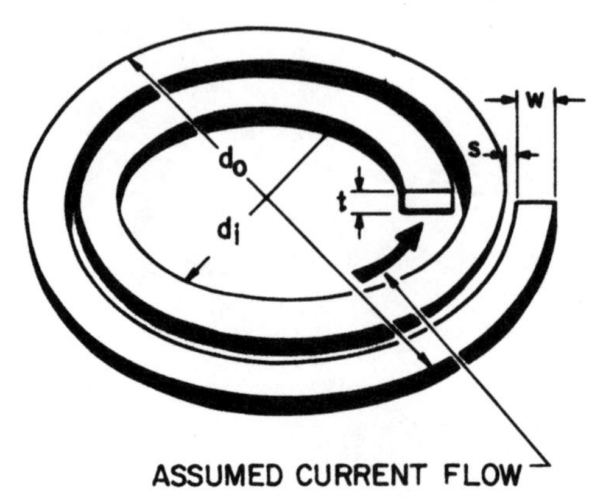

FIGURE 10-59 Printed inductor.

For maximum Q, inductors are usually designed as follows:

1. The spiral should have the widest conductor width possible with the overall diameter (d_o) small. This suggests that the separation between turns should be as small as possible within process limitations, and without introducing unnecessary distributed capacitance. The conductor walls should be clean and straight to provide accurate gap dimensions.
2. There should be some space at the center of the spiral to allow the flux lines to span through, increasing the stored energy per unit length. The accepted value is $d_i/d_o = 1/5$, further decreasing s and w if d_o is fixed.
3. The surface resistance and Q increase as the $\sqrt{2}$. However, it was found experimentally [28] that the conductor Q increases only up to about 6 GHz, and then falls rapidly. This is probably caused by current crowding. Therefore it is very important to minimize inductor *ac* resistance, written as:

$$R_{ac} = k' \pi \, n \, a \, R_s / w \tag{57}$$

where k' = correction factor that accounts for current crowding. It ranges between 1.3 and 2 for w/t between 2 and 100. To minimize current crowding, conductor thicknesses of 5 to 10 skin depths are used.

n = number of turns,

$a = \dfrac{(d_o + d_i)}{4}$ (average radius),

R_s = surface resistance.

4. The current flows within a skin depth on the top and bottom surfaces, and the inner vertical face of the inner turn and the outer vertical face of the outside turn. Thus the smoothness of these surfaces is important, as well, to minimize conductor losses.

The realization of high-Q printed conductors is somewhat limited by conflicting process requirements.

Summarizing planar inductor requirements, it is necessary to use:

1. Wide conductors (within practical limits) for higher Q_s, but width-to-spacing ratio should approach unity.
2. Conductor thickness should be >5 skin depths for low w/t.
3. Conductors should have low-resistance and high-volume conductivity.
4. Gap dimensions are very important. Clean, straight conductor walls are necessary for reproducible results.

10.19 SUPERCONDUCTIVITY

The application of high-temperature superconductivity for microwave circuitry is based on the belief that these new materials will form the foundation of a new technological era. This optimism and enthusiasm not withstanding, the suitable preparation of these materials for practical applications still remains challenging. The more critical areas are reproducibility, quality synthesis, and processing, particularly over large substrate areas. In integration with cooled semiconductor devices, processes, and MCMs, a wide range of problems exist in deposition strategies, insulating materials, patterning, and etching.

The earlier development of planar structures such as microstrip provided significantly improved design flexibility over bulky waveguide. However, the surface resistance of even the best normal conductors, such as Cu and Au, may have unacceptably high-propagation loss. Therefore, among the wide variety of possible applications, those based on the greatly reduced loss properties of superconducting materials will probably have the earliest impact on the form of passive microwave devices. The application of these devices can be grouped into three categories.

- Higher *Q*: performance characteristics vastly superior to conventional components (that is, ultra-high-*Q* values).
- Miniaturization: performance comparable to conventional components, but at greatly reduced volume and weight and with significantly lower power dissipation.
- Unique properties: providing features not available by standard devices (such as switching from resistive to conductive state because of small change in temperature, critical current, or magnetic field).

The general requirements for microwave superconducting thin films are listed in Table 10-10.

Unfortunately, the specific physical and electronic properties of the known cuprate oxides, which all crystallize in layered perovskites, may limit their design and performance. As such, progress in superconductive applications is materials- and process-driven. Nevertheless, significant strides have been made in applying superconductor thin films over a wide range of frequencies. This is illustrated in Table 10-11, which summarizes some of the commercial and military applications of superconductor thin films [72–75]. At present, the applications range over two decades of frequency, from <100 MHz to 11 GHz. It is immediately clear that $LaAlO_3$ (lanthanum aluminate) substrates and YBCO (yttrium-barium-cuprate) films are the materials used almost exclusively.

10.19.1 Properties of High-Tc Films

Superconductivity is characterized by the fact that the dc resistance sharply drops to zero below some transition point. For the dc resistance to become zero, only about 30% of the total electrons need to be superconducting because the current will always take the path of least resistance. Below this transition temperature (T_c), when current is induced into a su-

TABLE 10-10 HTSC thin film requirements

Film formation at the lowest possible processing temperature to minimize substrate structural transitions, and permit sequential two-sided deposition.
Minimal interdiffusion at the substrate interface.
Good thermal expansion match to the substrate.
High transition temperature (at least twice the operating temperature).
Low defect concentration and phase purity.
Chemical stability over time.
Smooth surfaces.
High degree of epitaxy.
High transition temperature (at least twice the operating temperature).
Low RF surface resistance
High current density (at least 10^6 A/cm^2)
Critical field (at the operating temperature it should be much greater than the operating field).

TABLE 10-11 Commercial and military HTSC applications

Vendor	Circuit description	Application	Frequency (GHz)	Material	Substrate	Buffer layer	Tc onset (°K)	Tc Δ (°K)	Jc (10 A/cm)	Reference
Superconductor Technologies, Inc.	Switched filter	Electronic warfare	6 to 11	TBCCO	LaAlO	N/A	102	0.5	0.5	147
Superconductor Technologies, Inc.	Interconnects	MCMs	<100 MHz	YBCO	LaAlO	Proprietary	92	0.3	2.9	147
Westinghouse Electric Corporation	Filter bank	Military	4	YBCO	LaAlO	N/A	89	~2	2	148
Westinghouse Electric Corporation	Delay line	Military	2 to 6	YBCO	LaAlO	N/A	89	~2	2	148
Westinghouse Electric Corporation	Switched filter bank	Military	10	YBCO	LaAlO	N/A	89	~2	2	148
David Sarnoff Research Center/Neocera	Circulator	Military	12	YBCO	YIG	Proprietary	89	0.5	2	149
David Sarnoff Research Center/Neocera		Military	9	YBCO	LaAlO	N/A	89	0.5	2	149
Conductus, Inc.	Delay line	Commercial and military	6	YBCO	LaAlO and		90	0.5 to 1	1 to 2	150
					Sapphire	MgO and CeO*	90	0.5 to 1	1 to 2	
Conductus, Inc.	DIFM	Military	4	YBCO	YBCO/LaAlO	N/A	90	0.5 to 1	1 to 2	150
Conductus, Inc.	Filter	Commercial	1 to 2	YBCO	YBCO/LaAlO	N/A	90	0.5 to 1	2	150
Conductus, Inc.	Interconnects	MCMs	<100 MHz	YBCO	LaAlO	Proprietary				150

*CeO deposited onto sapphire R-plane; MgO deposited onto sapphire M-plane.

perconducting loop, it will circulate almost indefinitely. In regular metals, the penetration of RF fields is governed by the well-known skin depth relationship.

The surface resistance is given by

$$R_s = \frac{1}{\delta\sigma} = \sqrt{\frac{\pi f \mu_o}{\sigma}} \tag{58}$$

It directly follows that losses in TEM transmission lines may be expressed as $\alpha = KR_s$, where K is a geometric factor. For regular conductors, losses increase as $f^{1/2}$.

When conventional metals are cooled, their surface resistance also decreases, but moderately. For example, the surface resistance of a copper conductor may decrease by a factor of three, (depending on material purity) between room temperature and 5°K to 10°K. This relatively small improvement in performance hardly justifies the cooling expense.

Superconductors are fundamentally different. Field penetration is not governed by skin depth, but the London penetration depth (λ) a measure of the distance a magnetic field can penetrate the superconducting material. Gamma is typically 100 to 200 nm, and most importantly, independent of frequency. At low frequencies, the thickness of a superconductor can remain thin and still have very low RF losses. Unfortunately, this behavior does not totally apply to the RF resistance, an important microwave property of a superconducting film.

A good approximation for the ac surface resistance that also holds up well for high-T_c materials is given by Halbritter [76].

$$R_s,\ HTS(T, f) = R_o + f^2 \frac{A(\xi, \lambda)}{T} e^{-\Delta/kT} \quad \text{for } T < 0.5\ T_c \tag{59}$$

$$-f^2\left(\frac{T}{T_c}\right)4\ \frac{1-(T/T_c)^2}{\left[1-(T/T_c)^4\right]}2 \quad \text{for } T \text{ near } T_c \tag{60}$$

where R_o is a temperature-independent residual resistance which incorporates the effects of grain boundary imperfections, surface impurities, surface texture, and other subtle causes. A is a function of the coherence length (ξ) and penetration depth (λ). Δ is the energy gap of the superconductor.

In a superconductor, the oscillating motion of the superconducting electrons generate a reactive voltage. RF losses are induced when this reactive voltage acts on the nonsuperconducting electrons (normal electrons). At higher frequencies, the electrons oscillate faster, increasing the RF losses. This accounts for the f^2 dependence of R_s on a frequency change. Fortunately, losses are reduced as the temperature is lowered below T_c because of the increasing ratio of superconducting to normal electrons with decreasing temperature. For this reason, it is desirable to have the operating temperature as low as possible below the transition temperature. This makes materials with higher transition temperatures, such as thallium-based superconductors, more attractive for microwave applications.

At typical microwave frequencies, R_s is so small that its strong frequency dependence has little effect. However, at some frequency, even the best high-T_c material will become lossy than regular metals whose surface resistance increases only as the square root of the frequency. For practical microwave applications with RF losses well below those of copper, the operating frequency must not be too high and the operating temperature, as discussed above, should be as low as possible below the superconducting transition temperature. Measurements of R_s at 77°K for selected superconducting films [77,78] plotted against frequency clearly illustrate this. The R_s for patterned superconducting film at 77°K is at least one order of magnitude lower at 10 GHz than even Cu at 4°K, but still about one order of magnitude higher than theoretical.

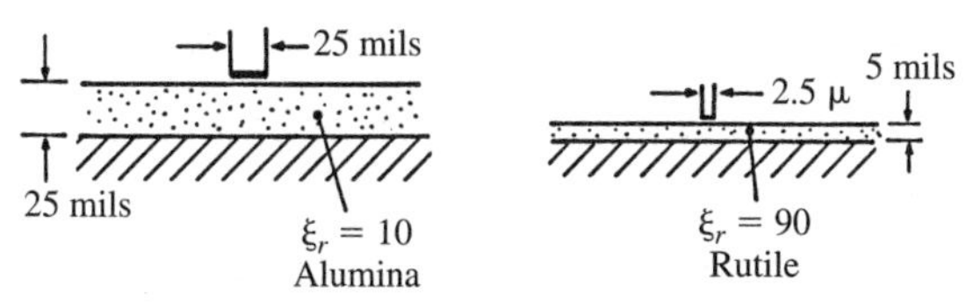

FIGURE 10-60 Comparison of conductor and substrate size for conventional and superconducting circuits on various substrates.

The absence of conductor restraints allows for the use of thinner substrates and increased circuit density with minimum losses. This is shown by the size comparison in Fig. 10-60 between standard and superconductor circuit geometries. High-dielectric-constant substrates reduce both line width and conductor guide length (λg). The result is to shrink the overall circuit volume.

The identical dimensions for conventional and superconducting circuits on 0.010 in thick rutile ($\epsilon_r = 90$) clearly demonstrates that superconducting films by themselves do not generate a size advantage. It is the ability to use very narrow lines that permits the use of thinner, higher-dielectric-constant substrates. However, in the microstrip configuration even extremely narrow, thin superconductor lines on very thin dielectrics can be subject to excessive losses. To reduce losses, other structures such as suspended stripline may be used. The advantage of superconducting circuits for applications such as filters is lack of dispersion and not necessarily size reduction.

10.20 MATERIALS CONSIDERATIONS

One can see with conventional metals and substrates how the choice of materials and processes affects circuit performance, using relatively simple, well-established materials and processes. Three additional factors critically influence the properties of complex superconducting oxide films deposited on various substrate materials: film and substrate crystal structure, film-substrate interactions, and film composition. Substrate interactions can be further subdivided into lattice matching, coefficient of thermal expansion matching, and interdiffusion between film and substrate. These are summarized in Table 10-12. Annealing considerations have not been included because this process is part of film formation.

TABLE 10-12 Superconductor process issues

Substrate choice	To assure proper film structure. To have correct dielectric properties at cryogenic temperatures. To minimize film-substrate interactions.
Film formation	To assure proper film stoichiometry, morphology, and structure. To maximize chemical stability. To assure proper orientation.
Patterning	To provide well-defined, accurate structures. To integrate superconductors without degrading other circuit materials or the superconductor.
Contacts	To provide contacts with resistance commensurate with that of the superconductor.

TABLE 10-13 Desired HTSC substrate properties

Low microwave losses, ($\tan\delta < 10^{-4}$)
Low dielectric constant, ($\epsilon_r \leq 10$) for filters, etc.
Good lattice match to the superconductor of choice
Good thermal expansion match to the superconductor of choice
High surface quality (no twinning or second phases)
No structural changes within the superconductor processing or operational temperature range
Isotropic dielectric properties (or at least predictably anisotropic ones)
High mechanical strength
Chemical inertness

10.20.1 Substrate Materials

Low-loss (high-Q) substrates must be used so that circuit performance based on the low surface resistance of the superconductor is not offset by substrate dielectric losses. At 10 GHz, there is almost a 4-fold increase in α as $\tan\delta$ increases from 10^{-4} to 10^{-6} [79].

The substrate plays an active role in the formation of the superconducting film as well as providing mechanical and electrical support. The important substrate properties identified so far that are most critical in achieving high-quality superconducting films are the crystal structure, lattice and thermal expansion match, and chemical inertness. These substrate properties influence film growth primarily in three ways: lattice matching encourages the proper crystallographic orientation, because the electrical properties are anisotropic; good thermal expansion matching inhibits microcracking in the film; and chemical stability reduces film-substrate interdiffusion and reactions which degrade the superconducting properties of the film. The preparation of high-T_c films involves the deposition of several constituents in a controlled manner and their crystallization in the correct structure, as well as ensuring the proper oxygen content for hole-doping. Thus the starting point for any deposition is the proper substrate and surface to obtain the desired film structure.

A summary of the important parameters for substrates suitable for microwave and millimeter wave high-T_c components is listed in Table 10-13.

Some of the earlier literature covering substrate experimentation was covered by this author [54]. As shown in Table 10-11, lanthanum aluminate is the substrate of choice for most commercial and military superconductive applications. Dilatometry of lanthanum aluminate disclosed neither a first order transition up to ~950°C, nor any abrupt changes in thermal expansion coefficient [80]. A second order rhombohedral-to-cubic transition with no volume change was placed ~500°C, which may account for the twinning observed in the crystal. This twinning, as with the lanthanum gallate, leads to surface steps and an increase in RF surface resistance. The anisotropic dielectric properties are especially bad in this respect because the twins move around, preventing accurate modeling of filters.

10.20.2 Thermal Expansion Coefficient

High T_c superconducting oxides have higher CTE than many of the substrates on which they are deposited. The thermal expansion of YBCO material abruptly changes at about 350°C and 650°C, which reflect changes in oxygen concentration and phase transition, respectively. Mismatch of CTE between the film and the substrate are known to cause cracks in deposited films [81]. This fact is one of the major obstacles to the preparation of superconductor films and integration with dielectrics and semiconductors. Ideally, both expansion and cell dimensions of the film and substrate should be very close. In reality the substrate used is a compromise, as in conventional circuitry.

10.20.3 Buffer (Barrier) Layers

Unfortunately the substrate surface is not always amenable for generating the proper film orientation, and high processing and annealing temperatures promote some interdiffusion between the film and the substrate. Most studies of chemical reactions between high T_c films and substrates were applied to the $RBa_2Cu_3O_{7-x}$ (R = Y, Er, La, etc.) superconductors. Some work is beginning to appear on bismuth and thallium film-substrate reactions because more effort is being applied to these high-temperature systems.

Sapphire and silicon are of interest because of their larger areas and lower dielectric constants compared with the perovskites. Sapphire also has a very-low-loss tangent, making it very attractive for microwave applications. However, Al diffusion from sapphire ($\alpha - Al_2O_3$) severely degrades the film [82], and Cu and Ba both form aluminates at the interface between YBCO and sapphire [83]. MgO and CeO_2 were used on sapphire M and R planes, respectively [84]. Significant substrate interaction was detected with bismuth films on sapphire as well as calcium fluoride and YSZ [85]. Silicon is primarily of interest as a substrate for multichip interconnections. Deposition of superconducting films on silicon is complicated by its very low CTE, large lattice mismatch and the presence of a native oxide. With YBCO, CuO reportedly reacts with the silicon, degrading the superconducting properties of the film [86]. Furthermore, silicon can diffuse directly into YBCO [87], and Cu and Ba have been shown to readily form their respective silicates by reacting with SiO_2 [88]. Rapid thermal annealing may reduce the time-temperature effect on the interface [89].

A number of materials have been used between the substrate and the film to encourage the correct orientation or to suppress chemical interactions. These buffer layers should be passivating, preventing interdiffusion; be unreactive; and grow epitaxially, promoting film epitaxy. Table 10-14 lists some of the single-buffer layer/substrate combinations that have been used to increase onset temperature, current density, and to improve film structure. This table illustrates the wide variety of material combinations available.

In addition to the single layers described above, the use of graded, multiple-buffer layers on silicon (Si) to provide the proper lattice spacing and improve the CTE match

TABLE 10-14 Buffer layer/substrate combinations

Film	Substrate	Buffer
YBCO	YSZ	BaF_2
YBCO	Sapphire	Ag, Au, Pt, MgO, $SrTiO_3$, $BaTiO_3$, CeO_2
YBCO	$SrTiO_3$	Au, Ag
YBCO	MgO	Au, Ag
YBCO	Si	YSZ, Y_2O_3, $SrTiO_3$, SiO_2, MgO
YBCO	99.5% Al_2O_3	Nb
ErBCO	MgO	Pt
BSCCO	β-quartz (SiO_2)	ZrO_2
BSCCO	Si	ZrO_2, ZrSiO
TIBCCO	$LaAlO_3$	MgO, LF_3, ZrO_2, Ta_2O_5
TIBCCO	MgO	BaF_2

deserves special mention. Miura et al. [90] first deposited epitaxial magnesia-spinel ($MgAl_2O_4$) onto Si by CVD technique. The lattice constant of the spinel is only 0.7% smaller than 1.5 times the Si lattice. The lattice constant of the barium titanate used as the upper layer of the compound buffer is only 1.3% smaller than the lattice constant of the spinel. Well-oriented superconducting layers, essentially free of secondary phases and distinct grain boundaries, can be fabricated with excellent transport properties on Si substrates when the proper interface is provided.

10.21 FILM FORMATION

To obtain the proper stoichiometry, a variety of deposition techniques have been used. The general details of these deposition methods were covered earlier. The basic differences involve providing precise stoichiometry, proper phase, and the ability to deposit onto both sides of the substrate. Vacuum-based processes are more amenable to in situ oxidation, which encourages the formation of the superconducting phases (*s*) at lower temperatures.

The special requirements of the superconducting films may require some modifications of thin film deposition techniques. In other cases, new deposition methods are developed. These will be briefly described in the next section.

10.21.1 Off-Axis Sputtering

During the sputtering process using on-axis deposition, energetic ions stripped of their electrons in the plasma, or more often neutrals, bombard the growing film [91]. The film composition is modified by the resultant resputtering of selective components. Second, the film quality is further degraded by differences in angular distribution of sputtering yield. Finally, for the on-axis case, the composition of the film rarely matches that of the target, greatly complicating the process of determining and maintaining the correct target composition. In the off-axis case, the substrate is positioned sufficiently far away from the target axis so that the sputtered material is deposited outside the region of direct on-axis negative-ion flux, but still within the outer edge of the plasma region [92], as discussed in Chapter 1. Atoms that do impinge on the substrate are almost exclusively low-energy, sputtered, neutral atoms that reach the substrate by diffusion. Even with reduced substrate bombardment, slight adjustments to target composition may be necessary to attain proper film stoichiometry. Sputtering rates using off-axis geometries are significantly lower than with on-axis sputtering, typically under 10 nm/min. Long sputtering times are somewhat offset by the fact that most films deposited in this manner may be annealed in situ.

Off-axis sputtering has been reported to improve the surface resistance of YBCO films on buffered sapphire. R_s values as low as 850 $\mu\Omega$ at 10 GHz and 77°K were reported [93]. YBCO sputtered onto heated MgO substrates mounted off-axis at 45° reportedly had significantly lower R_s of only 200 $\mu\Omega$ [94]. The Q of coplanar resonators made from these films were approximately one order of magnitude higher than comparable copper circuits. The authors report also that the microwave performance of their films was equivalent to high-quality films deposited by laser ablation. YBCO ring and stripline resonators fabricated on lanthanum aluminate also showed superior performance compared to copper. R_s values of 300 $\mu\Omega$ were reported. The authors [95] also suggested that improvements in performance could be expected with use of superconducting ground planes instead of the silver they used.

10.21.2 Pulsed-Laser Deposition

The use of high-energy lasers to irradiate and evaporate metals and dielectrics in vacuum is the basis of a technique called pulsed-laser deposition. The interaction of high-powered nanosecond excimer laser ultraviolet pulses with bulk targets results in evaporation and plasma formation of the target species, followed by a plasma expansion and deposition onto the substrate. Although there is some slight spacial compositonal variation if multi-component targets are used, basically the stoichimetry of the target is preserved in the film. This technique is discussed in detail in Chapter 1.

The laser is focused to about 0.5 mm to 1 cm onto a bulk sample of the material to be deposited. The irradiated target is oriented about 30° to the heated receiving substrate. For maximum compositional uniformity, the laser beam can be rastered over the target [96], or the target can be rotated, or both. The substrate can also be rotated. The alkaline earth composition of BSCCO films could be tailored using successive depositions from a multitarget system [97]. Some of the variables that must be controlled include the power to the laser, the angle between the substrate and the target, the background atmospheric pressure, and the composition, the substrate-to-target distance, and the substrate temperature. Film-area coverage depends on spot size and source-material quality, both purity and uniformity. For YBCO and BSCCO, in situ annealing in oxygen generally follows deposition.

Researchers coated both sides of a 5 cm $LaAlO_3$ substrate with YBCO films using direct radiative heating [98]. By using the directed nature of the laser plume, they maintained good stoichiometry and thickness control. The slight degradation in properties of the first side were attributed to the extra time at temperature during processing of the second side.

Laser-ablated YBCO films with low-surface resistance were used by a number of investigators to fabricate passive microwave devices. Meander-line resonators with *Q*s at up to 10 GHz, at least an order of magnitude better than comparable circuits of Cu or Au have been reported [99,100], as have filters with excellent band pass features and low insertion loss [101,102]. Ring resonators of various film thicknesses were fabricated on 0.25 mm thick $LaAlO_3$. The authors [96] determined that at some point film thickness becomes a dominant factor in determining the microstrip losses, just as in normal metals from increased current density. Others have reported on circuits fabricated from TBCCO using ex situ annealing. Straight-line resonators with surface resistance of 100 $\mu\Omega$ have been fabricated [103]. The authors claim that improvements in surface resistance of up to 100 times over conventional metals at almost 10 GHz and 77°K can be achieved using TBCCO, because T/T_c is near 0.5.

10.21.3 Evaporation

One method combines the thermal evaporation of one of the components with a tungsten filament or boat, and the remaining components by electron beam. Ex situ annealed, evaporated films usually have Cu and yttrium metals as sources, and barium fluoride as the barium source. Another method involves the use of electron-beam evaporation only. One investigation reported on the electron-gun evaporation of either YBCO, YbBCO, EBCO or BSCCO from a single sintered source [104] with oxygen directed at the substrate. It was necessary to compensate the sintered charges to allow for differences in vapor pressures. More often, films are deposited in layers, where each component is sequentially deposited as a separate layer forming a basic three-layer stack. Copper is usually deposited first because of its lower affinity for oxygen. Multiple stacks are then formed by repeating the basic stack, which coalesce during annealing. Twelve-layer YbBCO films [105] requiring post-deposition annealing were deposited on MgO with onset temperature of 81°K with ΔT_c of 2.6°K.

Coevaporation is an alternative method where all the constituents are simultaneously deposited, although attainment of the proper stoichiometry over large areas is difficult. Superconducting films produced from both coevaporated and layered films from the same evaporator were compared. Based on process reproducibility, the authors [106] concluded that sequential (layered) evaporation was considerably better than coevaporation. Namordi et al. [107] compared CPW resonators on $LaAlO_3$ and found the surface resistance of the YBCO film at 77°K and 4.75 GHz to be almost 10 times lower than one made of Au with 3 times the thickness.

10.22 METALORGANIC

10.22.1 MOCVD

Metalorganic chemical vapor deposition (MOCVD) is an extremely useful and versatile technique that was successfully used in the growth of many materials such as III/V compounds, optoelectronic materials, and magnetic bubble-memory material. Source materials for MOCVD are metalorganic compounds with high melting points and low vapor pressures. Examples are β-diketonates and acetylacetonates. Both may be fluorinated and combined with a variety of cations, such as the alkaline earths, yttrium, Cu, etc. However, the lack of volatility of the source materials is a problem, and these compounds tend to be unstable and decompose at temperatures very close to those needed to vaporize the material [108]. As a result, run-to-run reproducibility may be difficult to attain if the source material first decomposes. One of the advantages of MOCVD is the almost infinite number of potential organic materials available as MOCVD sources, providing a degree of flexibility unmatched with any other technique [109]. Improvements in source chelates are expected to improve film reproducibility.

In the MOCVD process, the source materials are first volatilized and transported in a hot carrier gas into a reaction chamber. A heterogeneous reaction takes place among the vapor-phase precursors in close proximity to the heated substrate, forming the desired compound and unidentified volatile fragments. When fluorinated metal chelate compounds are used, the decomposition of the fluorinated ligand leaves a residual amount of fluoride ion which must, as with the barium fluoride evaporation, be removed by treating the film with water vapor and oxygen at high temperatures.

YBCO films with an onset temperature of 90°K have been reported [110]. BSCCO films on MgO with onset of 110°K and zero resistance of 77°K [111] illustrate the problem of obtaining single-phase material by MOCVD. In an attempt to reduce the percentage of lower-temperature phases, Pb was added to BSCCO films in a two-step process [112]. Lead toxicity necessitates using a two-step process similar to the one required for MOCVD deposition of Tl-based films [113]. Takemoto et al. [114] were able to fabricate microstrip resonators from ErBCO films deposited on both sides of $LaAlO_3$ substrates with low surface resistance. They reported a two-fold increase in resonator Q compared to superconducting stripline with Ag ground planes. Lu et al. coated both sides of $LaAlO_3$ substrates. They reported that by increasing the substrate temperature by only 5°C during deposition onto the second side, they significantly improved side-to-side uniformity [115].

10.22.2 Spray Pyrolysis

Spray pyrolysis, sometimes referred to as metalorganic deposition (MOD), was successfully applied to high-temperature superconducting films using many of the same ligands as MOCVD. Substrates are flooded with prepared solutions, spin dried at room temperature, and then rapidly heated at 500°C to pyrolyze the metalorganics to their oxides. It was necessary to use rapid thermal annealing to attain 90°K YBCO and YbBCO films with relatively sharp ΔT_c [116], because conventional furnace annealing resulted in substantial interdiffusion with zero resistance only near 37°K. Similarly, sprayed films of YBCO on MgO annealed at 900°C showed marked thickness dependence of onset temperature because of Cu diffusion from the annealed film into the substrate [117]. Sobolewski and Kula [118] attempted to directly pattern BSCCO films on a variety of substrates by spraying aqueous nitrate precursors through a stencil held close to a heated substrate. Only films on MgO showed promise, and microwave detectors subsequently fabricated from YBCO and BSCCO films were considered suitable for operation at 77°K in the submillimeter (terahertz) range [119].

A comparison of the various deposition techniques described above is shown in Table 10-15.

10.23 PATTERNING

10.23.1 Wet Etching

This technique is the most direct patterning method. The simplest in principle, it is complicated by the fact that dilute aqueous-based mineral acids can attack the unprotected portions of the film. The reaction of water with barium (Ba) and barium oxide (BaO) releases oxygen, which promotes the formation of $Ba(OH)_2$, Y_2BaCuO_5, and CuO [120]. CO_2 readily reacts with Ba and BaO to form stable $BaCO_3$. To overcome these problems, barium fluoride (BaF_2), as mentioned earlier, is frequently substituted for Ba or BaO during the evaporation process, or barium alloys which are more environmentally stable are used as sputtering targets. As a result, films are more inert and can be processed and stored for long periods of time without degradation. In one instance, a 60 nm cap of yttrium was used to prevent film reactions with H_2O or CO_2 [121], and the application of a thin layer of Cu (~20 nm) was also reported to minimize environmental degradation [122]. If the

TABLE 10-15 Comparison of superconductor film deposition methods

Technique	Advantages	Disadvantages
Evaporation	Sources easily changed In situ annealing possible Reactive deposition possible	Large area uniformity a problem
Laser ablation	Single source In situ annealing possible High vacuum not necessary Hetereo structures possible	Smaller area coverage
Sputtering	Reproducible Uniform coverage of large areas Two-sided deposition possible Sputter geometry adjustable	On-axis stoichiometry affected by ion and neutrals bombardment Slow deposition rates
Spin pyrolysis	Simplest method Low cost	Contamination possible
MOCVD	Lower temperature process Two-sided deposition possible	Lower current sensitivity Precursers (volatility, etc.)

stability of the film is questionable, it is advisable to pattern films before annealing because many of the reactions with CO_2 and H_2O are reversible.

Aqueous etchants for YBCO include dilute (~1:100) phosphoric acid, (H_3PO_4) [123]. Ethylene-diamine tetra-acetic acid, (EDTA) [124] and dilute nitric (HNO_3) [125] were used for patterning. Various acid solutions were evaluated for high-resolution patterning of YBCO, and it was found that etching in dilute H_3PO_4, HNO_3, or HCl yielded the best results when defining lines and spaces as narrow as 3 μm [126]. CH_3COOH, HCl, and HBr were investigated as selective etchants to remove nonsuperconducting surface layers on BSCCo films [127], with HCl reportedly the most effective. BSCCO films were also selectively cleaned with bromine in absolute ethyl alcohol (EtOH) [128]. Nonaqueous etchants are developed which in principle eliminate problems with water. A phosphate-based etchant, 1,1 diphosphone ethanol in methanol, was used to define 20 μm features in both Y- and Eu-based cuprates [129]. It was necessary to compensate for oxygen secession by posttreatment in an oxygen plasma. More recently, Br/EtOH was also used for patterning [130]. The resolution of wet-etched parts is probably limited to about 6 μm, partly because etching proceeds by granular dissolution. Rosamilia et al. [131] recommended the use of EDTA and $HClO_4$-$NaClO_4$ as etchants for BSSCO.

Microwave circuits are successfully fabricated using wet etchants. YBCO-ring resonators operating at 35 GHz were fabricated using conventional photoresist and dilute phosphoric acid [108]. Resonators were also fabricated using Br/EtOH and dilute H_3PO_4 [132].

10.23.2 Dry Etching

Dry etching was reported to effectively remove YBCO material with good resolution while eliminating the potential of moisture interaction and subsequent degradation of the film. Reactive-ion-beam etching (RIBE) [133], ion-beam milling [134], and laser ablation [135] have all been reported to achieve good resolution without degradation of film properties. One micron features were successfully patterned in YBCO using PMMA resist and deep ultraviolet coupled with ion milling [136]. Heteroepitaxial layers were ion-milled to define junction structures [137].

An excellent example of an application of ion milling is the meander line shown in Fig. 10-61. This line [100], only 1 mil (0.025 mm) wide and 3.3 in (84 mm) long, provides a wide-band delay up to 12 GHz of 1.2 nsec with a maximum loss of only 0.03 dB at 79°K. The extraordinary length-to-width ratio of 3400:1 demands not only continuity, but dimensional accuracy over its entire length.

Filters also demand tight processing tolerances. The precision of definition by ion milling and the improvement in performance of superconducting circuits over normal metals can be illustrated in Fig. 10-62. The filters were fabricated [102] on 20 mil thick $LaAlO_3$ substrate ($\epsilon_r = 26$) using 5 mil wide lines. At the band-pass frequency, the Cu filter typically has losses in the range of 18 dB. In contrast, the superconducting filter typically has losses under 4 dB, almost 20 times better. Figure 10-63 illustrates a commercially available thin film microwave superconductor filter. This circuit [138], a 5-pole filter operating at 2 GHz, has a 0.7% band width. Impedance of the filter elements is 10 Ω.

10.24 MICROWAVE MEASUREMENTS

10.24.1 Frequency Domain Techniques

S-parameters are reflection coefficients directly related to VSWR, impedance and transmission coefficients usually referred to as gain or attenuation. *S*-parameters describe

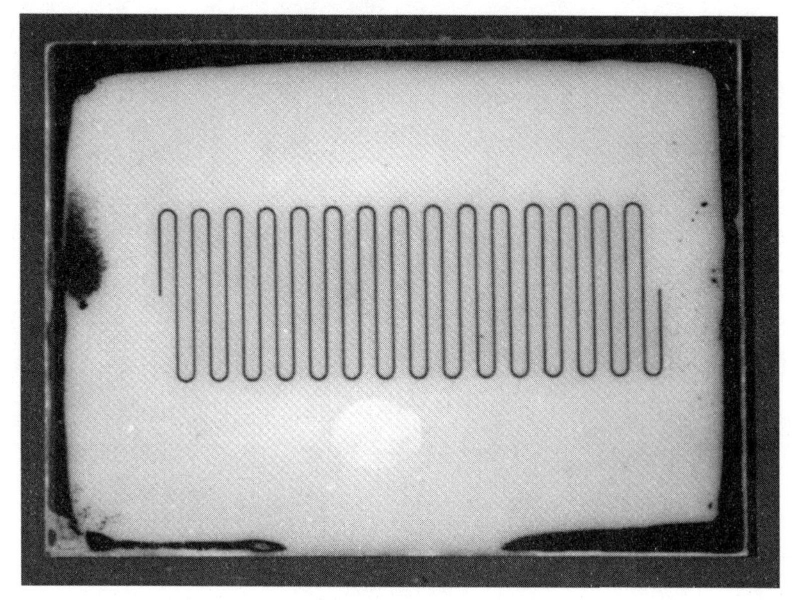

FIGURE 10-61 Ion-milled meander line. *(Courtesy David Sarnoff Research Center, Princeton, NJ. The assistance of V. Pendrick is greatly appreciated.)*

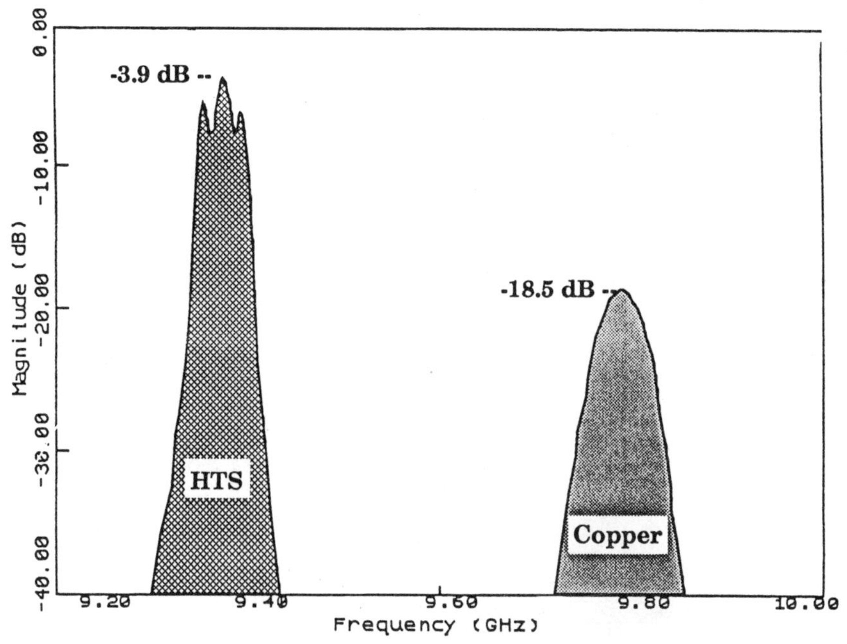

FIGURE 10-62 Measured results for equivalent HTS and copper filters. *(From D. Kalokitis et al.,* IEEE Trans Magn, *vol. 27, no. 2, March, 1991, p. 2540.)*

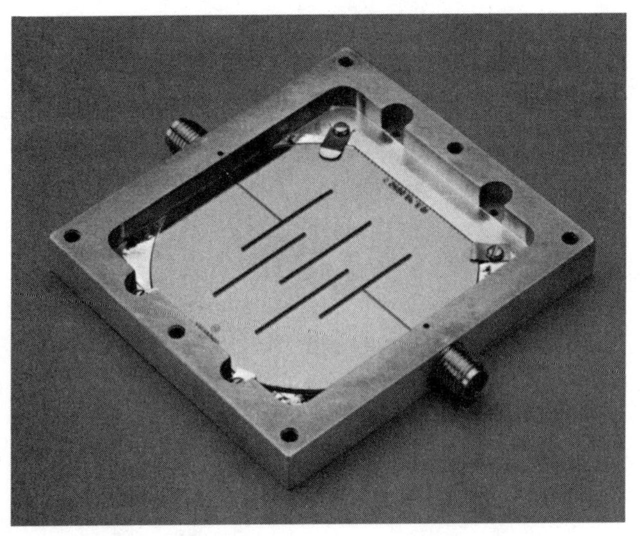

FIGURE 10-63 Five-pole filter, 2 GHz. *(Courtesy of Conductus, Inc., Sunnyvale, CA. The assistance of Dr. G-C. Liang is greatly appreciated.)*

circuit inputs and outputs in terms of power, thus these parameters are a measure of the power transmitted and reflected from a circuit component along a 50-Ω line. *S*-parameters also are measured with all circuits terminated in an actual line impedance of the system. *S*-parameters, being vector quantites, contain both magnitude and phase information. In Figure 10-64, *a* is the input signal into the port and *b* the output signal. Both *a* and *b* are square roots of the power; $(a_1)^2$ is the power incident at port 1, and $(b_2)^2$ is the power exiting port 2. The fraction of a_1 that is reflected at port 1 is defined as S_{11}, whereas the transmitted fraction is S_{21}. The fraction, then, of a_2 that is reflected at port 2 is S_{22}, and in the reverse direction must be S_{12}. The signal b_1 which exits port 1 must by definition be the sum of the fraction of a_1 reflected at port 1 and the part transmitted through port 2. The outputs may be related to the inputs by the relations

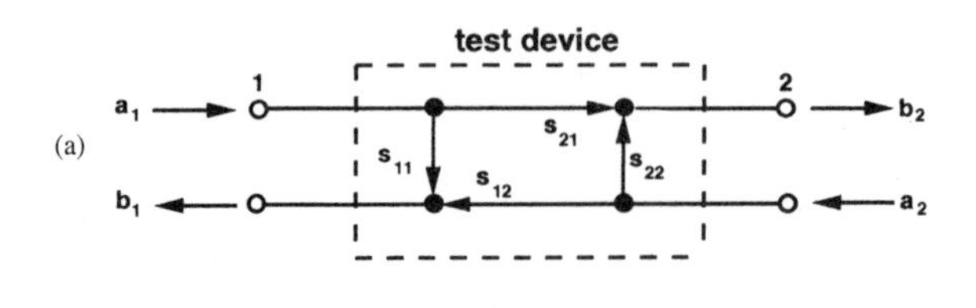

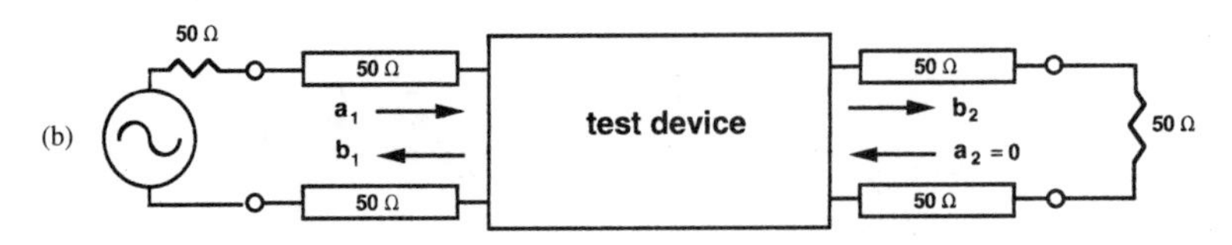

FIGURE 10-64 *S*-parameters.

$$b_1 = S_{11}(a_1) + S_{12}(a_2) \tag{61}$$

$$b_2 = S_{21}(a_1) + S_{22}(a_2) \tag{62}$$

If an RF source is connected to port 1, $a_2 = 0$ by terminating the 50-Ω transmission line from port 2 with its characteristic impedance. A schematic for this set-up is also shown in Fig. 10-64. Setting $a_2 = 0$ in Eqs. 61 and 62, it follows that

$$S_{11} = \frac{b_1}{a_1} \tag{63}$$

$$S_{21} = \frac{b_2}{a_1} \tag{64}$$

By placing the RF source on a_2 and setting $a_1 = 0$, S_{12} and S_{22} can be similarly calculated.

S_{11} and S_{22} are ratios of the reflected power and the incident power, identical to the reflection coefficient (G), commonly used on the Smith chart. As such, for any two-port device, the input and output parameters, as well as its corresponding characteristic impedance, may be extracted from the polar display.

10.24.2 Time-Domain Techniques

Modeling of most high-frequency circuit applications involves design and fabrication before packaging. As such, the parasitic effects of the package are very difficult to predict. Whereas S-parameters yield an overview of the total device-under-test (DUT), it is important to be able to analyze signal integrity through transmission paths. The difficulty in measuring the attributes of high-frequency packages is increasing as the interconnect structure becomes denser and more complex. Isolation is difficult, not only of the effects of discontinuities in the interconnection scheme on package performance, but of the effects of the launchers themselves. A need exists, then, for a technique to characterize all the components making up the package. Time-domain reflectometry (TDR) essentially introduces a fast edge step to a transmission system, and then observes the size and location of the reflected energy from that incident step. In essence, the effects of different discontinuities may be observed as the signal propagates along the interconnect. Additionally, TDR may be used to obtain S-parameters of the interconnect and network. A software library containing circuit parasitics and lossless interconnect components is first developed and then applied to development of an equivalent circuit model. The experimental waveform is then compared to the simulated waveform. Iterative modeling results in an optimization. Figure 10-65 shows the responses from a short, 50-Ω load and open circuit. The real power of TDR is being able to spatially resolve different parts of DUT in terms of impedance and reflections, and the relative simplicity of the necessary equipment. An input step is created in one channel of a sampling oscilloscope and reflections are observed on the same channel. The TDR is calibrated against a standard short and 50-Ω load to reference voltages and time. Figure 10-66 schematically illustrates typical performance characteristics for transmission line discontinuities. Narrowing of the trace width results in an increase in inductance, whereas broadening results in a capacitive discontinuity. Reflections caused by the substrate connection illustrate how this technique can be used to differentiate attachment methods.

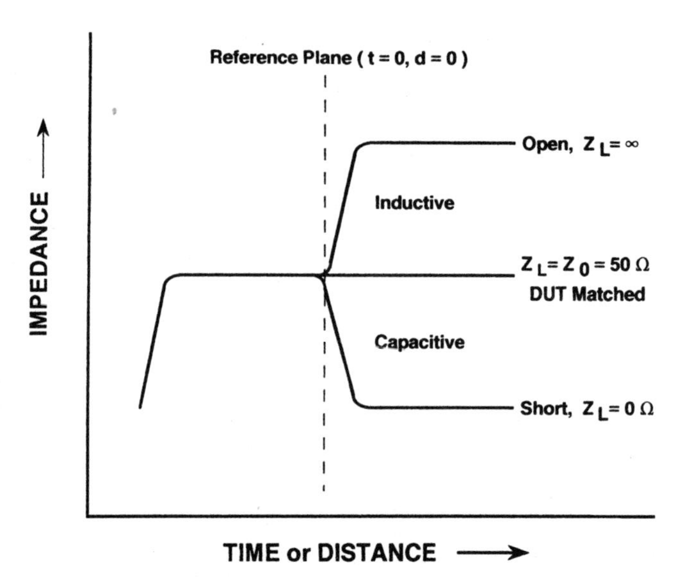

FIGURE 10-65 Time domain reflectometry (TDR).

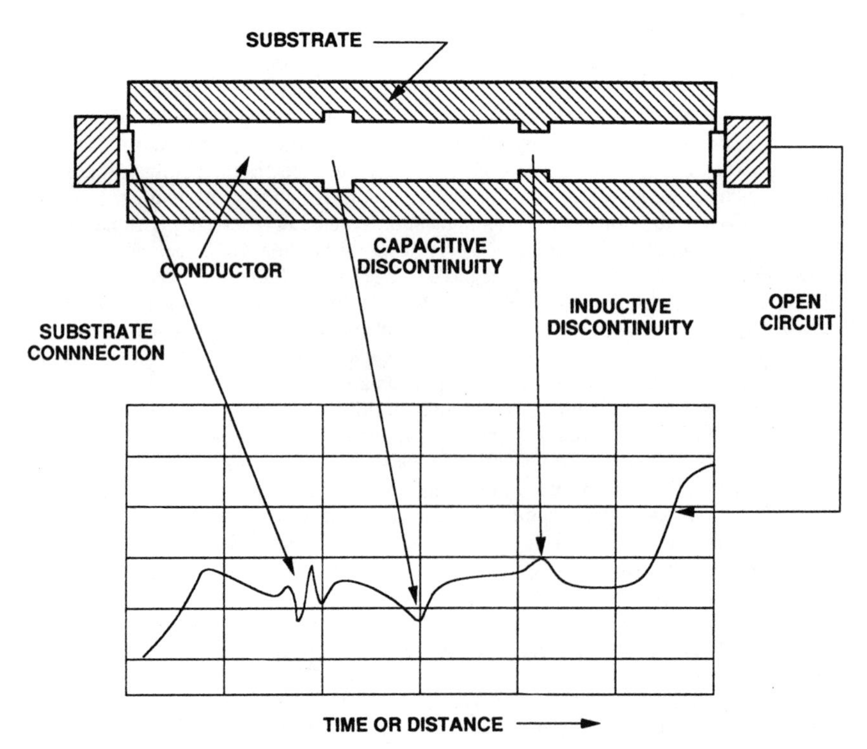

FIGURE 10-66 Typical TDR performance characteristics.

10.25 ELECTRONIC PACKAGING

Probably no subject engenders more controversy than that of electronic packaging, particularly for microwave circuits. The mechanical, thermal, and electrical design of the carrier or package, and the circuit electrical performance are interdependent because of the short wavelengths involved. Before deciding on a package strategy, a number of issues must be addressed.

- Level of integration
- Interconnects
- Enclosure/carrier material selection
- Substrate attachment
- *RF* connector requirements
- Package closure
- Cost
- Electrical requirements (insertion loss, etc.)

An overview of those concerns within the scope of this chapter will be given succeeding sections.

10.26 LEVELS OF INTEGRATION

Why integrate? Current microwave packaging technology is directed mainly toward hermetically packaging specific active devices with limited I/O ports, or the integration of a number of components in a metal housing with many RF connectors and feedthroughs, which may or may not be hemetically sealed. This technology has a high parts count, large numbers of wire interconnects, limited frequency range and chip area, high manual labor, and thus high cost.

The lowest possible cost for a complete functional module, then, depends on the optimization and integration of a large number of interactive components and technologies. Cost reductions in packaging cannot be successfully addressed without taking into account individual chip properties and testing requirements. MMIC yield is the largest cost-driving factor, but significantly, interconnect failures account for about one-third of all circuit failures. Aside from the foregoing yield losses, there are also losses caused by handling, mounting, and interconnecting the relatively large, fragile MMIC chips.

With any technology, there is probably a particular chip size of certain complexity that gives the optimum performance, handling, testing, and interconnect yields. The object is to properly integrate components without introducing parasitic reactances resulting from the connections or package. If the physical size of components such as resistors and inductors is small, compared to the wavelength, they will not be overwhelmed by parasitic reactances. If the components are too large, they will resonate and become ineffective. Table 10-16 compares some of the advantages and disadvantages of integration versus discrete packaging.

Package integration is a multidisciplinary field, requiring innovative inputs from materials, mechanical, electrical, and process engineers. It is important to define advantages, set goals, and to not overintegrate.

TABLE 10-16 Levels of integration

Advantages	Disadvantages
Smaller size	NRE usually higher
Lighter weight	Longer lead time
Greater flexibility	Higher dollar content
Broader band	Specification more difficult
Improved performance	Testing more difficult
Lower cost	Reduced vendor choice

10.29 INTERCONNECTS

An area of vital importance for the electronic packaging of MMIC chips is the interconnection between chips and from chips to the outside world. There are several technologies available, wire bonds, modified TAB, and electroformed interconnects.

10.27.1 Wire

Round. The most common method of chip interconnection is wire bonding. Although wire inductors are used successfully in tuning, as interconnections they require intense

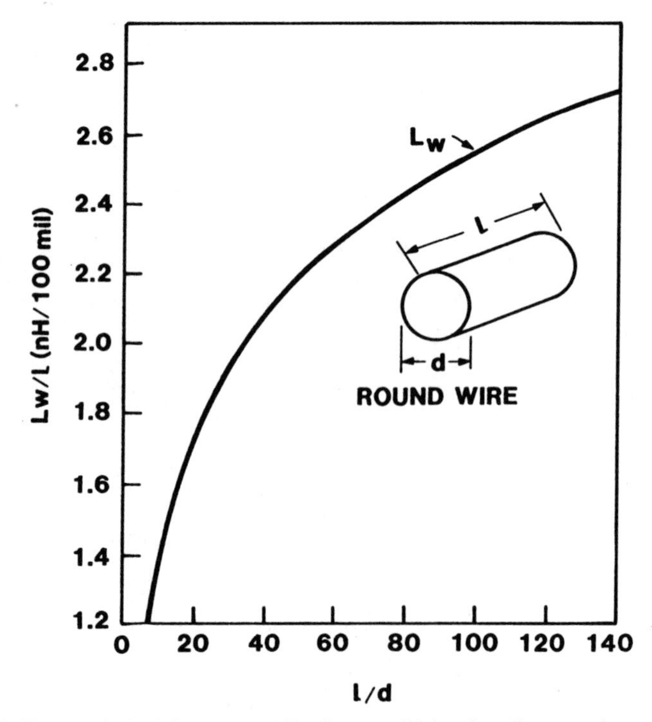

FIGURE 10-67 Round wire inductance per 100 mils versus l/d, length-to-diameter ratio.

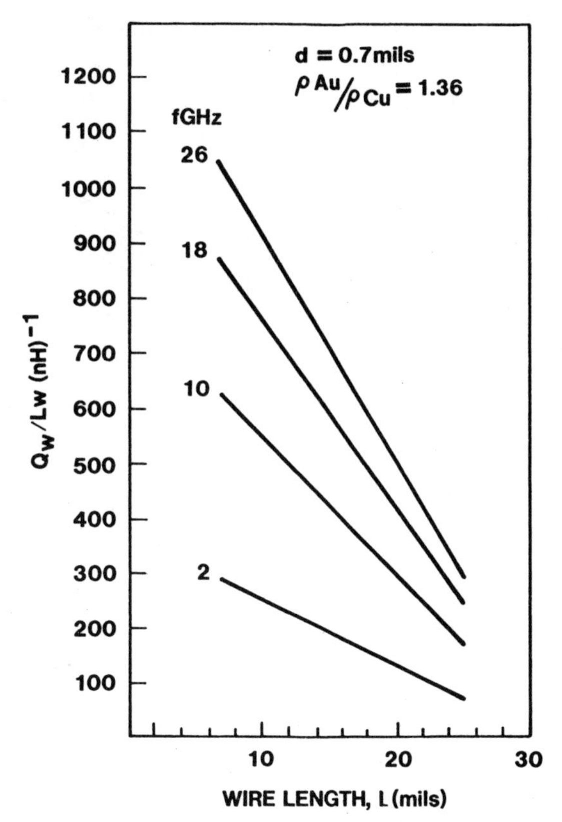

FIGURE 10-68 Q_w/L_w ratio versus wire length for 0.0007 in gold wire.

labor. They have poor reproducibility and, as expected, high parasitic inductance. From Terman [139] the free-space inductance, independent of frequency, of a wire of diameter (d) and length (l) is of the form,

$$L_w(\text{nH}) = 5.08 \times 10^{-3} l[\ln(l/d) + 0.386] \tag{65}$$

where all dimensions are given in mils. The inductance increases commensurately with length. The inductance per 100 mils (L/l) is plotted in Fig. 10-67 as a function of the l/d ratio.

Thus every attempt should be made to keep the wires as short as possible. The diameter of the wire should be as wide as possible because the Q of round wire with resistivity ρ is

$$Q_w = 3.38 \times 10^3 L_w nH \frac{d}{l}\left[\frac{\rho_{\text{Cu}}}{\rho_{\text{metal}}}\right]^{1/2}\left[\frac{\text{GHz}}{2}\right]^{1/2} \tag{66}$$

In practice, however, the wire diameter is limited by the size and spacing of the bonding pads on the active devices. These frequently range about 1 to 2 mils so that it is often necessary to use 0.7 mil wire. Figure 10-68, a plot of Q_w/L_w versus wire length, illustrates two important points. First, for a given wire length, the Q/L ratio increases with increasing frequencies, because of the shallower skin depth reducing the surface

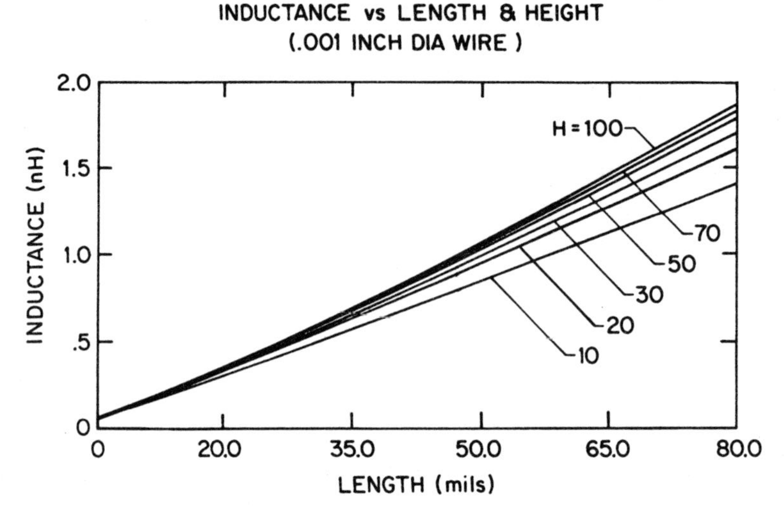

FIGURE 10-69 Inductance versus length and height for 1 mil diameter wire.

resistance. Secondly, this advantage rapidly diminishes as the length of bonding wire is increased.

The height of the wire above the ground plane also has some effect. For short lengths of wire, as shown in Fig. 10-69, there are few differences with height changes, but increasing wire length and increasing height above ground can increase the inductance by as much as 35%. This is another reason for keeping the wire interconnect short and planar.

Finally, wire bonds are usually looped as shown in Fig. 10-70. Thus the additional inductance introduced by the wire bond self-inductance must be considered. This inductance is proportional to the radius of the wire loop,

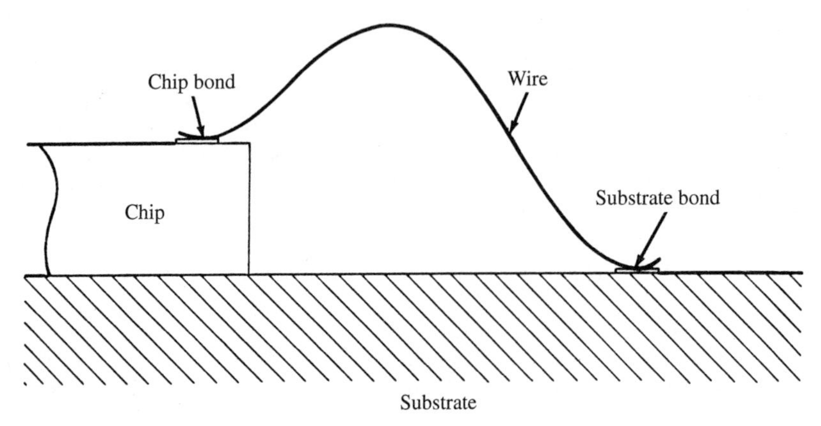

FIGURE 10-70 Schematic of wirebond attachment.

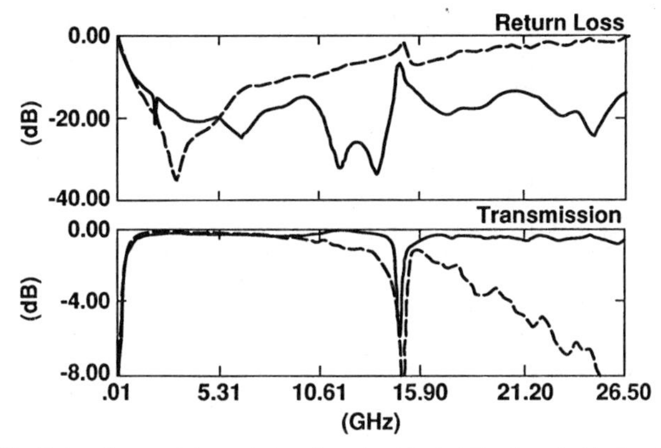

FIGURE 10-71 Return loss and transmission for wire (*dashed lines*) and strip-bonded (*solid lines*) capacitors.

$$L = \frac{\mu_o}{4\pi} \ln r \tag{67}$$

where μ_o is the permeability of free space, $4\pi \times 10^{-7}$H/m, and r is radius of wirebond.

The type of lead and its attachment method directly affect microwave device performance. For example, Kent and Ingalls [130] studied the effect of type of attachment on capacitor performance by varying the location and type of lead. They obtained the best performance using a ribbon lead; the worst using a single wire. They reported only marginal improvement using three wires to try to minimize the inductance effects. Performance curves for single wire and ribbon are shown in Fig. 10-71. The lower return and insertion losses shown by the devices using ribbon leads are directly attributed to the lower inductive parasitics in this type of lead. Whenever possible, wire bonds should be planarized.

Strip Ribbon. Ribbon is frequently used to interconnect substrates, as wrap-arounds, and to connect substrates to the outside world through RF connectors (launchers). Ribbon behaves much like wire, as shown in Fig. 10-72, which plots the ribbon inductance against the l/w ratio.

In summary, the following statements can be concluded with respect to wire and ribbon bonding:

- High conductivity is important for high Q.
- Use as wide a ribbon and wire as possible.
- Keep interconnect as close to ground as possible.
- Keep interconnect as short as possible.
- Use more than one ribbon or wire in parallel to reduce inductance.

Tyler [131] described a controlled-impedance, multilevel structure for GaAs MIMIC packaging, separating the signal and ground planes with a thin layer of polyimide. In this method, metal fingers were etched and plated and windows in the polyimide were opened by laser ablation, as shown in Fig. 10-73. The researcher compared the microwave performance of these structures to interconnects formed with conventional Au wire bonds up

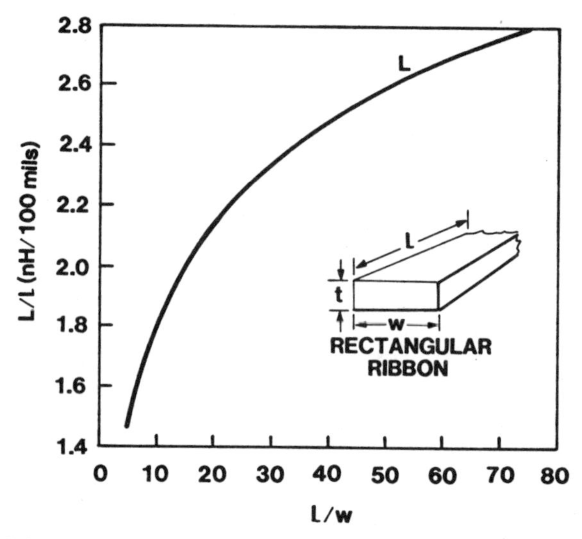

FIGURE 10-72 Inductance (*nH*/100 mil) of rectangular ribbon versus length-to-width ratio *l*/*w*.

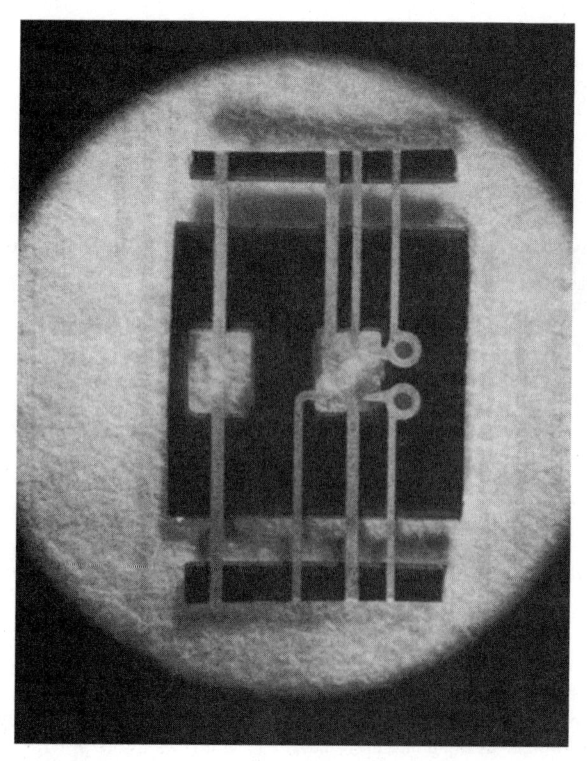

FIGURE 10-73 Controlled impedance TAB interconnect. *(Courtesy of Avantek.)*

TABLE 10-17 Attenuation loss of various interconnect structures [141]

	2 GHz	10 GHz	20 GHz
Single wire bond	−0.22	−2.9	−5.9
Double wire bond	−0.25	−1.9	−3.7
2 × 4 Ribbon bond	−0.23	−1.6	−3.4
Tab interconnect	−0.21	−1.2	−3.1

to 20 GHz. His data are presented in Table 10-17 and clearly show the improvement in attenuation losses of the TAB interconnect relative to wire and ribbon bonds at higher frequencies. *S*-parameter comparisons reinforce these conclusions and that controlled impedance structures would demonstrate even lower losses with a MMIC designed for TAB instead of wire bonds.

Electroformed Interconnect. A novel interconnect technology [132] very effectively overcomes the current drawbacks of wire bonds. Figure 10-74 shows the interconnect bonded to a low-noise FET. The individual fingers have good plated bumps to facilitate bonding to the device bond pads. A polyimide ring, photolithographically defined to precise dimensions, keeps the fingers in place and insulates the fingers from the device, preventing shorting. This also facilitates bonding to internal pads without the looping associated with bond wires. Bond contact areas of 1 ml × 1 mil with spacing of 1 mil or less can be routinely achieved with this technology.

By shaping the width and general geometry, the interconnect medium can become part of a tightly controlled transmission line or matching network. That is something difficult to maintain with wire bonds. Pretesting of devices can also be accomplished with this technology. The relatively large cross section of this interconnect can also act

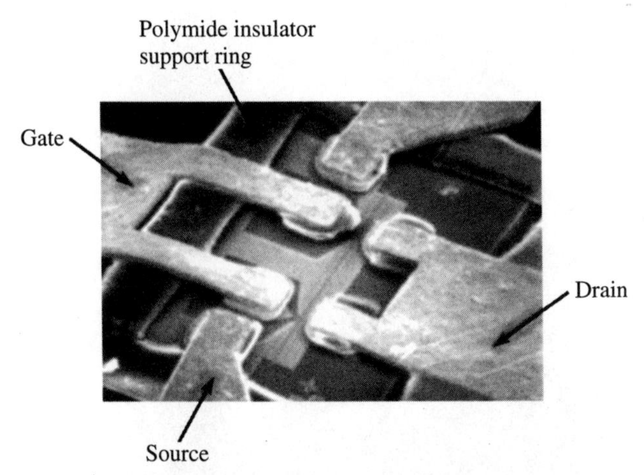

FIGURE 10-74 Bonded electroformed interconnect. *(From S. H. Normann and H. Stripple,* Proc 1985 Intern Symp Hybrid Microelectron, *Anaheim, CA, ISHM, Oct. 19-21, 1992, p. 182.)*

TABLE 10-18 Electroformed interconnect

Easily fabricated	Rapid turn around, low cost
Good reproducibility even for very fine geometries	Suitable for microwave and mmW chips (devices and MMICs)
Larger cross-sectional area than bond wires	Lower parasitics
All-gold construction	Reliable
Integrated insulation	Permits access to internal bond pads
Rugged-beam lead frame and flip-chip mounting	Permits pretesting of device
	Permits preinspection of bonds for chip mounting
Easily customized	Interconnect can be part of matching network

as a thermal conductor in cases where high-power dissipation is not of key importance. Table 10-18 summarizes the key advantages of this interconnect.

Integrated Wiring. Wires are also used to interconnect components such as couplers, as shown in Fig. 10-75*a*. The irreproducible parasitics introduced by the looped wires frequently cause performance shortfalls in these critical components. One alternative is the use of air-insulated crossovers in which the interconnect is electroplated over a sacrificial

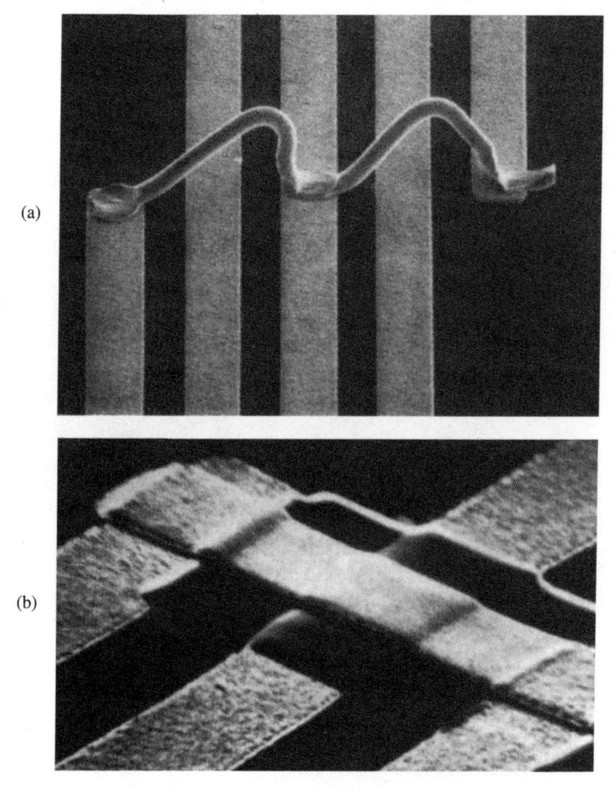

FIGURE 10-75 Lange couplers (*a*) wire-bonded, and (*b*) integrated wiring.

conductive layer. When the conductive support is removed, the resultant air gap separates and insulates the interconnect from the underlying metallization. This structure has the lowest capacitance because the dielectric constant of air is unity. However, these crossovers are subject to shorting out and frequently a drop of polyimide, $\epsilon_r = 3.5$, is put under the bridge to serve as a mechanical support. A more cost-effective approach is to fabricate the interconnect over an accurately defined polyimide insulator. An example of this is shown in Fig. 10-75*b*.

10.28 ENCLOSURE/CARRIER MATERIAL SELECTION

At this point in particular a multidisciplinary approach to packaging is mandated. Table 10-19 lists the problems which must be addressed before any final packaging concept. Low cost is deliberately placed last out of necessity, not choice.

10.28.1 Thermal Expansion Coefficient

The assembly processes and operating environment frequently force the package and its contents to extreme, rapid temperature changes. Examples are illustrated in Table 10-20, which lists some common assembly processes and their typical temperatures.

Naturally, the process is accomplished in descending thermal order, so as not to affect the lower-temperature processed parts. The probem in matching the thermal expansion coefficient (a) of the various materials used in microwave packaging can be appreciated from Fig. 10-76, where the percentage of linear expansion is plotted against the temperature in °C.

Clad substrates are not included in Fig. 10-76 because their thermal expansion coefficient is variable, depending on the cladding thickness and the amount of cladding remaining after pattern etching. As pointed out earlier, expansion values in the X-Y, and Z directions may differ. From Fig. 10-76, it is apparent that many of the metals used for carriers and interconnects expand about 2 to 3 times as much as the ceramic substrates and active devices, an undesirable condition requiring expert package design.

Substrate attachment to the metal enclosure is complicated by the large difference in thermal expansion coefficient between the metal enclosure and the ceramic substrate. This issue, and other issues relative to the substrate attachment, are covered in the next section.

TABLE 10-19 Enclosure/carrier material selection

Compatibility with thermal expansion of substrate
Machinability
Plateability
Weldability
Solderability
Magnetism
Package thermal environment
Package mechanical environment (airborne versus ground-based)
Low cost

TABLE 10-20 Assembly process step temperature

Step	Temperature (°C)
Chip epoxy	150
Sn63 solder	180
96.5 Sn 3.5 Ag solder	220
Eutectic die attach	285
Substrate/carrier attach (Au/Ge)	385

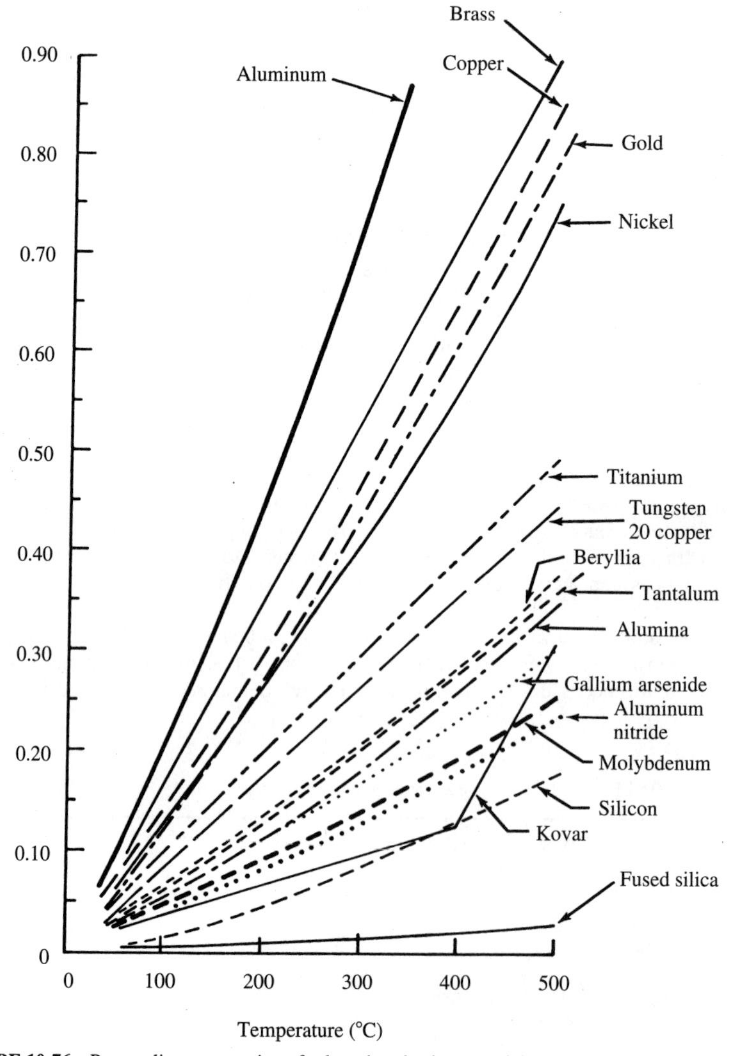

FIGURE 10-76 Percent linear expansion of selected packaging materials versus temperature.

TABLE 10-21 Substrate attachment requirements

Electrical: good RF conductivity.
Mechanical: excellent ground integrity, especially around the substrate perimeter.
Ability to withstand thermal cycling without substrate cracking or bond failure.
Minimum outgasing from flux residues or other organics.
Ability to withstand further package processing.
Bonding materials: solders or epoxies.

10.29 SUBSTRATE ATTACHMENT

Substrate attachment is one of the critical steps that has not received the attention it deserves. At this step, the substrate is populated (expensive), ready to be inserted into the finished enclosure or carrier (expensive). Table 10-21 lists the formidable requirements for good substrate attachment.

To enable the substrate to withstand thermal cycling without cracking, solders are generally used. These are compliant, and have high thermal and electrical conductivities. Table 10-22, lists some representative packaging solders used for substrate attachment. The solder is usually selected based on the metallurgies of the substrate ground, the enclosure or carrier, and the temperature of processes.

10.30 MECHANICAL DESIGN

With increasingly higher frequency requirements, mechanical packaging concerns are affected in ways that are not altogether clear, even to experienced microwave designers. All dimensions become smaller, resulting in tighter tolerance and alignment requirements. Excellent ground continuity is critical to good high-frequency performance, because without proper ground, performance degenerates. The causes of improper grounding are numerous. Any abrupt discontinuity in the ground connection disturbs the field, and power will be redirected in undesirable directions. The importance of good ground continuity cannot be overemphasized. Figures 10-77*a-c* illustrate various situations during substrate mounting where discontinuities exist.

TABLE 10-22 Solders

Type	Composition	Temperature range	Expansion (α) (ppm/°C at 25°C)
Gold-tin eutectic	80Au20Sn	280	16.5
Tin-lead-silver	57Sn40Pb3Ag	180	24
Indalloy #7	50In50Pb	180-200	27
Indalloy #2	60In40Pb	160-180	10
Tin	100% tin	232	23
Indium	100% indium	160	29

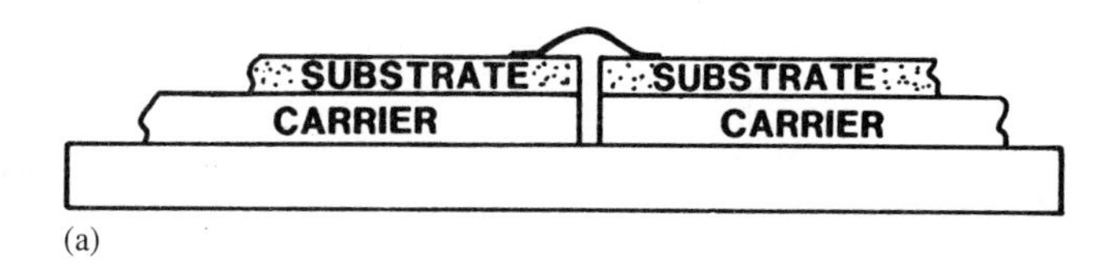

(a)

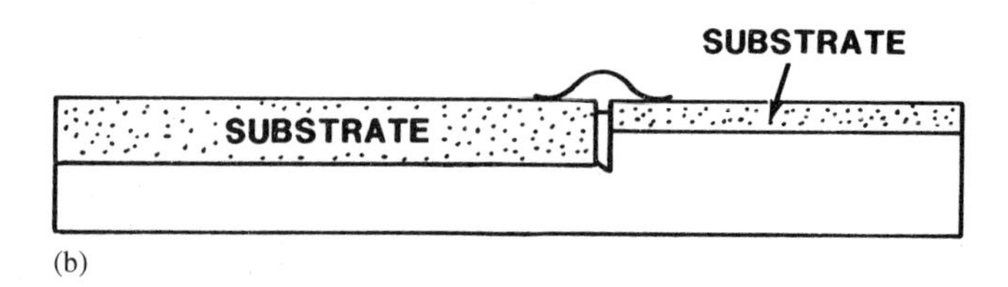

(b)

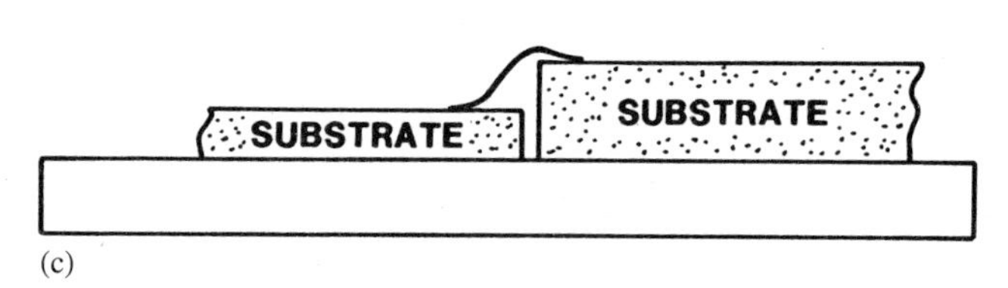

(c)

FIGURE 10-77 Substrate mounting discontinuities. (*a*) Mounting on separate carriers; (*b*) discontinuity in ground plane height, and (*c*) discontinuity in substrate height.

Such discontinuities should be avoided or compensated for. Even so, compensation can introduce other problems, such as excessive inductance from the long ribbon in Fig. 10-77*c*. Another assembly strategy that introduces ground discontinuities is shown in Fig. 10-78. In this structure, two substrates are attached to a common center ground with wire or ribbon. The inductance varies from bond to bond, introducing undesirable parasitics and noticeable variability.

Vias in both hard and soft substrates are generally used to provide either an interlevel connection or close access for the device to ground. When the via is in series with the sig-

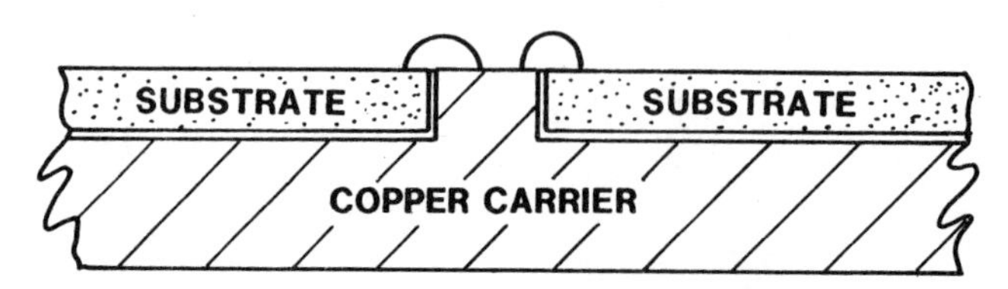

FIGURE 10-78 Circuits on separate carriers attached to a common ground.

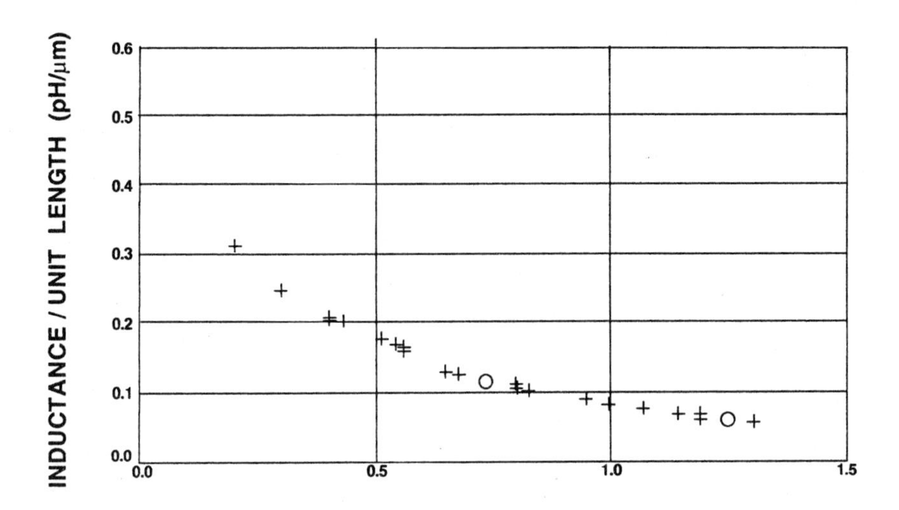

FIGURE 10-79 Via hole inductance per unit length as a function of via diameter/height (*d*/*h*) ratio. Numerical simulation (+), measured data (o).

nal path, it should have the same characteristic impedance as the transmission lines. When used for grounding, vias should have as low an impedance as possible. Two basic problems face the circuit fabricator and designer. First, the added inductance is a function of the diameter-to-height ratio (*d*/*h*) of the via [143], and it decreases with increased *d*/*h* values, as shown in Fig. 10-79. In addition to reducing the inductance, larger *d*/*h* ratios facilitate coating inside the via.

If the hole is totally filled with a dense, low-resistive material, it may be treated as a short piece of round wire, its inductance closely approximated by Eq. 11-65. Gipprich and Grice [144] investigated the ground circuit problem and network equivalent circuits for single and multiple vias. Their table on the effects of via inductance on the insertion loss is plotted in Fig. 10-80. At zero inductance, the via insertion loss is infinite, and the loss for both 20 and 40 dB attenuators are 20 dB and 40 dB, respectively. As the via inductance increases, the losses for the 40 dB attenuator and via increase far more rapidly than for the 20 dB attenuator.

The authors also showed that use of multiple vias (as with multiple wires) reduced circuit losses. They arrived at the following set of via design guidelines:

Small bond pads
- Use large diameter vias.
- Extend the bond pad beyond chip boundaries and add more vias.

Medium bond pads
- Fill the entire pad with vias.
- Smaller vias, closely spaced, are generally more effective than larger vias.

Large bond pads
- Locate via arrays at I/Os.
- A 5 × 5 via array is generally adequate.

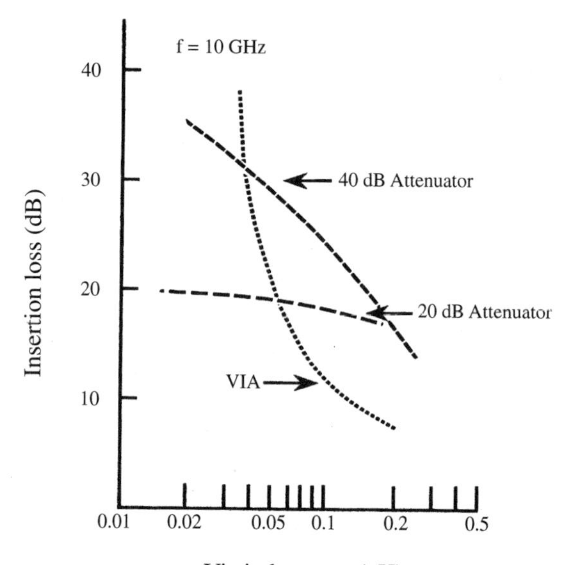

FIGURE 10-80 Effects of via inductance on insertion loss.

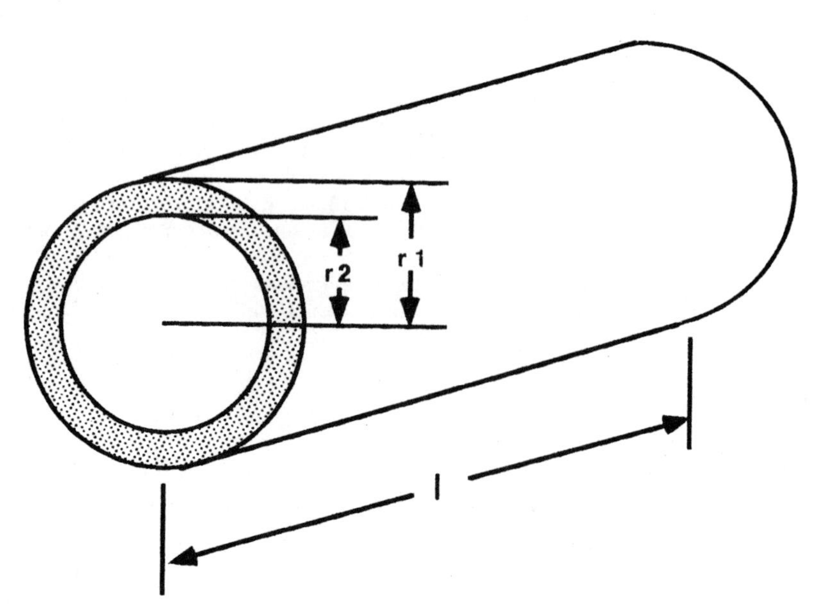

FIGURE 10-81 Schematic of tubular via.

Much of the above work assumes the vias are completely filled. This is not always the case. Tubular inductors described by Grover [145] closely approximate the configuration of a plated-through hole or via. His equations referenced to Fig. 10-81 are

$$L = (0.002)\, l\left(\frac{(2.0)\, l}{\rho_1} + \ln\zeta - 1.0\right) \tag{68}$$

where l, ρ_1 and ρ_2 are in cm.

The value of $\ln\zeta$ is also in Grover [146]. Waddel [147] provides a polynomial for convenience.

$$\ln\zeta = 0.25009128 - 0.0017049618\left(\frac{\rho_2}{\rho_1}\right) - 0.51598981\left(\frac{\rho_2}{\rho_1}\right)^2 + 0.37420782\left(\frac{\rho_2}{\rho_1}\right)^3 - 0.10669571\left(\frac{\rho_2}{\rho_1}\right)^4 \tag{69}$$

The electroless deposition of copper poses severe environmental problems. Treatment of noxious fumes such as formaldehyde and effluent into the waste stream is expensive and poses a serious environmental concern. Many companies considered replacing the electroless process with alternative dry metallization by sputtering. Pargellis [147] presented data that suggests that for holes with aspect ratios ($L/D < 3$), the electroless process may be replaced by sputtering. With holes where $L/D > 3$, only the seeding part of the electroless process may be replaced because the required deposition time becomes excessive for deposition of a functional film. His data is presented in Fig. 10-82. The copper thickness on the substrate surface is about 3 μm, so that even at the inside corner near the top of a hole with an aspect ratio of 1, the thickness only approaches 50%. Sputtering onto both sides of the substrate is usually required to ensure good hole coverage. When evaporation is used, the substrate is angled and rotated to provide better access of the evaporant into the via openings. Ion plating [148] may be used. In this technique, the material to be deposited is conventionally evaporated. A separate glow discharge is maintained near the substrate at a pressure somewhere between 10^{-1} and 10^{-2} torr. The evaporant atoms enter the plasma where they lose much of their directionality and coat the interior of the substrate artifacts such as holes. The conditions must be properly selected so that the rate of deposition is higher than the backsputtering rate.

Partially filled vias may be a problem when spinning on resist and other polymers. It is a fact that with vias present, striations in the polymer appear (localized areas of reduced thickness) according to Besser and Louris [149]. Other methods must be used to provide uniform, liquid-polymer films. These include spray coating [149], dip coating [150], meniscus coating [151], and extrusion [152].

A recently developed proprietary technique uses metal composites for via fill [153]. By varying the composite ratio, the expansion coefficient of the via material can be tailored to match the substrate. A fracturegraph of a 0.010 in via in 0.025 in high density alumina is shown in Fig. 10-83. The taper is typical of laser-drilled holes. The via openings are completely filled with dense metal. As a result, via parasitics are expected to be low and thermal transfer high. Two-level, thin film metallization appears on the substrate surface.

A novel approach to the grounding problem was described by Belohoubek [155]. This approach uses a more sophisticated technology than standard via-based hybrids by laminating ceramic blocks with thin metal foils. These sandwiches are sliced into thin substrates with veins of metal running the substrate length. The technique is the result of attempting to find a solution to several requirements:

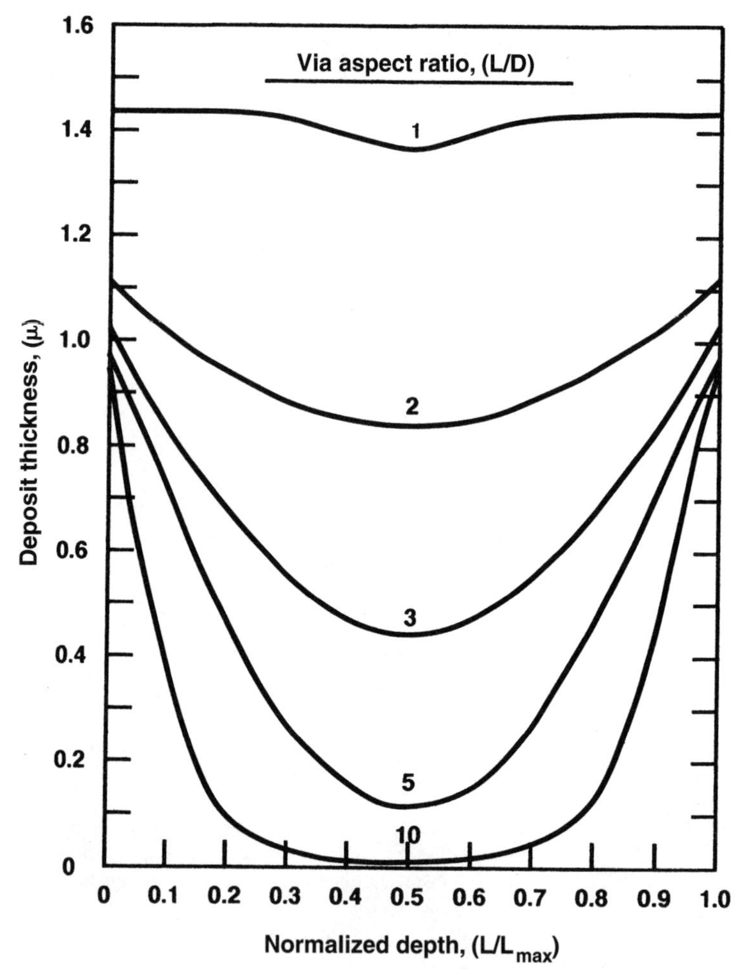

FIGURE 10-82 Deposite thickness of sputtered films for different via aspect ratios.

- A substrate material which can be batch-processed,
- Good RF grounds having low resistance and low inductance,
- A substrate material offering high thermal conductivity,
- A technique which uses thin film deposition of high Q-lumped inductors and capacitors,
- A medium which enables active devices to be mounted onto the circuit without wire bonds, and
- A medium offering small size, low weight, and potentially low cost.

Selectively glazed alumina and beryllia are used so that high Q-lumped components can be defined using thin film techniques. Many of the processes applicable to MMICs are used such as Si_3N_4 overlay capacitors, thin film resistors, and air-bridge interconnections. Figure 10-84 illustrates a cross section of the technique. Ground sep-

FIGURE 10-83 Composite metal-filled via.

tae are provided wherever the circuit topology requires low inductance ground returns. The FETs are flip-mounted onto the substrate to provide good heat conduction and low parasitic interconnections to the RF circuit. X-band and Ku-band power amplifiers have been manufactured. The size of the circuits produced is quite competitive with GaAs MMICs, being about 5 mm × 5 mm. Such a technique is seen as being particularly attractive for power applications at a relatively simple circuit complexity level. Multipe-grounded areas can be achievable using such a construction. The use of the integrated metal ground eliminates many of the problems associated with substrate-to-substrate and circuit-to-ground connections. Significant cost savings are accrued by eliminating wiring, because these circuits can be directly bonded to the enclosure or carrier.

Perko et al. [156] describe a process where a sealing glass $\epsilon_r = 4.1$ (Corning Glass Code 7070) approximately 200 μ thick is bonded to a 75 mm Si carrier. Passive components, interconnections, and via metallization are provided by thin film circuitry on the smooth glass surface. Etched holes in the glass provide the die with a direct,

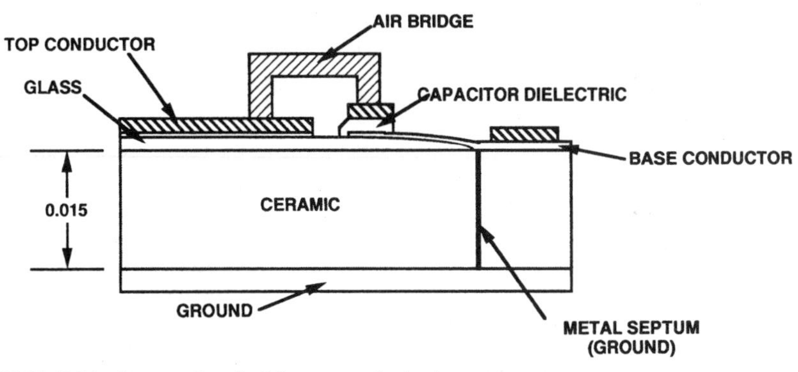

FIGURE 10-84 Cross-section of miniature ceramic circuit.

TABLE 10-23 Surface conversion requirements

Base	Process	Plating required
Aluminum	Soldering	Copper-gold
Brass, copper	Resistance welding	Copper, gold (low-resistance metal)
Kovar	Resistance welding	Gold, nickel
Tungsten-copper	Soldering	Tin

low-inductance connection to the Si carrier. The wafer form factor is a significant advantage in batch manufacturing. Automated cassette-to-cassette handling may be used without modification. Amplifiers up to 18 GHz have been successfully batch-processed and tested.

10.31 PLATABILITY

Plating of the enclosure is needed to convert an otherwise inert surface such as aluminum oxide/aluminum into one which is conductive. Examples are illustrated in Table 10-23.

10.32 CONCLUSION

Hybrid microwave integrated circuit techniques were reviewed in this chapter. The chapter presented the design information necessary to understand the materials and processing requirements of microwave hybrids. Materials technology, including substrates, dielectrics, conductors, and resistive films followed. The various fabrication techniques for thin film circuits were then discussed in light of their applicability for microwave circuitry. Selected packaging requirements for high-speed applications were also reviewed.

REFERENCES

1. H. Hertz, *Electric Waves,* MacMillan, London, 1893.
2. G. R. Cronin and R. W. Haisty, *J Electrochem Soc,* vol. 111, no. 7, 1963, p. 874.
3. C. A. Mead, *Proc IEEE,* vol. 54, 1966, p. 307–8.
4. Photograph courtesy of Varian RF Subsystems, Beverly, MA. The assistance of Dr. David Stevenson is gratefully acknowledged.
5. D. D. Greig and H. F. Englemann, *IRE Proc,* vol. 40, 1952, p. 1644.
6. C. P. Wen, *IEEE Trans on Microwave Theory and Techs,* MTT-17, no. 12, 1969, p. 1087.
7. R. M. Barrett and M. H. Barnes, *IRE National Conference on Airborne Electrics,* Dayton, OH, May 2–5, 1951.
8. D. C. Howe and G. A. Senf, *Proc 1969 Electron Comp Conf,* Washington, DC, April 30–May 2, 1969.
9. A. Presser, *Microwaves,* vol. 53, no. 5, March, 1968.
10. W. Hilberg, *IEEE Trans on Microwave Theory and Techs,* MTT-17, no. 5, 1969, p. 259.

11. Data courtesy of R. E. Stegens, Veritech Microwave Inc., South Plainfield, NJ.
12. P. S. Bachert, *RF Design,* vol. 52, July, 1988.
13. G. Gione and C. Naldi, *Electron Lett,* vol. 20, no. 4, 1984, p. 178.
14. H. Howe Jr., *Stripline Circuit Design,* Artech House, Dedham, MA, 1985.
15. V. Borase, *Microwaves and RF,* vol. 83, Feb., 1983.
16. G. R. Traut, *Microwave J,* no. 11, 1980.
17. B. T. Szentuki, *Electron Lett,* vol. 12, 1976, p. 672.
18. R. Horton et al., *Electron Lett,* vol. 7, no. 17, 1971, p. 490.
19. M. Caulton et al., "Measurement of the Properties of Microwave Transmission Lines for Microwave Integrated Circuits," *RCA Review,* vol. 27, 1966, p. 377.
20. R. A. Pucel et al., *IEEE Trans Microwave Theory Tech,* MTT-16, 1968, p. 342.
21. R. Kulke et al., *Microwaves and RF,* vol. 33, no. 13, 1994, p. 103–144.
22. R. E. Collin, *Foundation for Microwave Engineering,* McGraw-Hill, New York, 1966, p. 314.
23. S. P. Morgan, *J Appl Phys,* vol. 20, 1949, p. 352.
24. R. D. Lending, *Proc National Electronics Conf,* vol. 11, 1955, p. 391.
25. H. Sobol and M. Caulton, "The Technology of Microwave Integrated Circuits," *Advances in Microwaves,* vol. 8, Academic Press, 1974, p. 11–66.
26. F. A. Benson, *Proc Inst Elect Engrs,* (London), Part 3, vol. 100, 1953, p. 85.
27. K. B. Bhasin et al., *J Vac Sci Technol,* vol. A 3, no. 3, 1985, p. 778.
28. E. O. Hammerstadt, *ELAB Report STF44 A74169,* University of Trondheim, Norwegian Institute of Technology, February, 1975.
29. H. van der Peet, *Intern Conf on Thin and Thick Film Technology,* Sept 28–30, 1977, Augsburg, Germany (in English).
30. B. F. Gunshinam et al., "MIC Technology Short Course," *ISHM Monograph Series,* vol. 8, 1185–1, ISHM, Reston, VA, 1985.
31. Kayusa et al., *1993 Proc Intern Symp Microelectron,* Dallas, 1993, p. 193.
32. B.A. Ziegner and T. Murphy, *Appl Microwaves,* Winter, 1991/92, p. 70.
33. A. J. Dawe et al., *1990 Proc Intern Symp Microelectron,* Chicago, Oct. 15–17, 1990, p. 316.
34. J. L. Sepulveda et al., *1992 Proc Intern Symp Microelectron,* San Francisco, Oct. 19–21, 1992, p. 188.
35. E. Foley and G. Rees, *Proc Intern Symp Microelectron,* NY, Oct. 20–22, 1980, p. 99.
36. R. C. Enck et al., *1990 Proc Technical Program,* NEPCON East, Boston, MA, June 1–14, 1990, p. 523.
37. R. Chanchani, *IEEE Trans Comp Hybrids and Manuf Tech,* vol. 10, no. 2, 1987, p. 9.
38. L. J. Bostellar et al., *Hybrid Circuits,* no. 21, January, 1990.
39. E. S. Dettmar and H. K. Charles Jr., *Intern J Hybrid Microelectron,* vol. 10, no. 2, 1987, p. 9.
40. Y. Kurihara et al., *IEEE Trans Comp Hybrids and Manuf Tech,* vol. 14, no. 1, 1991, p. 204.
41. K. J. Lodge et al., *IEEE Trans Comp Hybrids and Manuf Tech,* vol. 13, no. 4, 1960, p. 633.
42. N. Kuramoto et al., *Proc 36th Electron Comp Conf,* Seattle, WA, May 5–7, 1986, p. 424.
43. S. G. Kosnowski et al., *Intern J Hybrid Microelectron,* vol. 10, no. 3, 1987, p. 13.
44. J. H. Harris et al., *J Mater Res,* vol. 5, no. 8, 1990, p. 1763.
45. M. G. Norton, *Hybrid Circuits,* vol. 18, no. 20, January, 1989.
46. M. Feil, *Hybrid Circuits,* vol. 29, no. 29, January, 1989.

47. Product Literature, Murata Corporation, Smyrna, GA. The assistance of Dr. T. Reynolds is greatly appreciated.
48. Product Literature, Trans-tech, Adamstown, MD.
49. Coors Ceramics, Golden, CO. The assistance of R. White is greatly appreciated.
50. Product Literature, NTK Corporation, Mount Prospect, IL. The assistance of Mark Oda is greatly appreciated.
51. AlSiMag Technical Ceramics, Inc., Laurens, SC. The assistance of Ms. B. Crossland is greatly appreciated.
52. R. C. Kell et al., *J Amer Cer Soc,* vol. 56, no. 7, 1973, p. 352.
53. R. Tolino, *Transtech,* Adamstown, DE, private communication.
54. R. Brown in *Materials and Processes for Microwave Hybrids,* ISHM, Reston, VA, 1991.
55. G. R. Traut, *Appl Microwave,* May, 1989, p. 80.
56. R. Savage, Gould Foil Division, Eastlake, OH, personal communication.
57. *Understanding Copper Alloys,* J. H. Mendenhall, (ed.), Krieger Publishing Company, Malabar, FL, 1980.
58. R. Savage, *Electronic Pkg and Prod,* May, 1992, p. 80.
59. J. D. Woermbke, *Microwaves,* Jan, 1982.
60. *Product Bulletin #RT 2.4.3,* Rogers Corporation, Chandler, A. Z. With permission.
61. R. Brown, in L. I. Maissel and R. Glang, (eds.), *Handbook of Thin Film Technology,* McGraw-Hill, NY, 1970, Chapter 6.
62. S. A. Hoenig and K. W. Kinkade, *1992 Proc Intern Sym Hybrid Microelectron,* San Francisco, CA, 1992, p. 29.
63. R. Sherman, *J Vac Sci Techn.,* vol. A, 12, no. 4, 1994, Jul/Aug, p. 1876.
64. S. Cohn, MTT-16, no. 2, 1968, p. 110.
65. G. D. Alley, *IEEE Trans on Microwave Theory and Techniques,* MTT-18, No. 12, December 1970, p. 1028.
66. T. H. Taylor Jr. and M. R. Williams, *27th Electron Compon Conf,* Arlington, VA, May 16–18, 1977, p. 48.
67. D. A. McLean, *IEEE Intern Conv Record,* pt. 7, Electron Devices, Materials, and Microwave Components, 1967, p. 108.
68. S. Asai, et al., Extended Abstracts, vol. 80-2, *Electrochem Soc Fall Meeting,* Hollywood, FL, Abstract no 349, (Oct. 5–10, 1980).
69. A. Presser, *IEEE Trans on Microwave Theory and Techniques,* MTT-26, No. 10, October 1978, p. 801.
70. A. Levy, *Final Report AFWAL-TR-84-4030, Contract #FO 8635-80-C-0243,* May 1984, Air Force Materials Laboratory, Wright Patterson AFB, OH.
71. T. C. Edwards, *Foundations for Microwave Design,* John Wiley & Sons, New York, 1981, p. 155.
72. Superconductor Technologies, Inc., Santa Barbara, CA. The assistance of J. M. Madden is greatly appreciated.
73. Westinghouse Electric Research Center, Pittsburg, PA. The assistance of S. Talese is greatly appreciated.
74. David Sarnoff Research Center, Princeton, NJ. The assistance of E. J. Denlinger is greatly appreciated.
75. Conductus Inc., Sunnyvale, CA. The assistance of G.-C. Liang is greatly appreciated.
76. J. Halbritter, *Z Physik,* vol. 266, 1974, p. 209.
77. A. Fathy et al., *Paper V-3,* IEEE MTT-S Digest, 1990, p. 859.
78. F. A. Miranda et al., *Appl Phys Lett,* vol. 57, 1990, p. 1058.
79. T. E. Van Deventer et al., *1990 IEEE MTS-S Digest,* IEEE, 1990, p. 285.

80. H. M. O'Bryan et al., *J Mater Res,* vol. 5, no. 1, 1990, p. 183.
81. T. Hashimoto et al., *Japan J Appl Phys,* vol. 27, no. 2, February 1988, p. 214.
82. H. C. Li et al., *Appl Phys Lett,* vol. 52(13), March 1988, p. 1098.
83. H. Koinuma et al., *Japan J Appl Phys,* vol. 27, no. 7, July 1988, p. L1216.
84. G.-C. Liang et al., *1994 IEEE MTT-S Digest,* IEEE, 1994, p. 183.
85. D. W. Face et al., *IEEE Trans Mag,* vol. 25, no. 2, Mar. 1989, p. 2341.
86. T. Aida et al., *Japan J Appl Phys,* vol. 26, no. 9, Sept. 1987, p. L1489.
87. G. Koren et al., *Appl Phys Lett,* vol. 53 (23), Dec. 1988, p. 2330.
88. H. Nakajima et al., *Appl Phys Lett,* vol. 53 (15), Oct. 1988, p. 1437.
89. Q. Y. Ma et al., *Appl Phys Lett,* vol. 53 (22), Nov. 1988, p. 2299.
90. S. Miura et al., *Appl Phys Lett,* vol. 53 (20), Nov. 1988, p. 1967.
91. R. L. Sandstrom et al., *Appl Phys Lett,* vol. 53 (5), Aug. 1988, p. 444.
92. M. Muroi, et al., *J Mater Res,* vol. 4, no. 4, July/Aug 1989, p. 781.
93. K. Char et al., *Appl Phys Lett,* vol. 57 (4), July 1990, p. 409.
94. B. B. G. Klopman et al., *IEEE Trans Mag,* vol. 27, no. 2, Mar. 1991, p. 2821.
95. G-C. Liang et al., *IEEE Trans Supercond,* vol. 1, no. 1, Mar. 1991, p. 58.
96. C. M. Chorey et al., *IEEE Trans. Mag,* vol. 27, no. 2, Mar. 1991, p. 2940.
97. H. Tabata et al., *Japan. J. Appl. Phys,* vol. 28, no. 5, May 1989, p. L823.
98. R. Meunchausen, *Superconductor Week,* 5, Apr. 8, 1991.
99. M. Kuhn et al., *IEEE Trans Magn,* vol. 27, no. 2, Mar. 1991, p. 2809.
100. D. Kalokitis et al., *J Electron Matl,* vol. 19, no. 1, 1990, p. 117.
101. H. S. Newman et al., *IEEE Trans Magn,* vol. 27, no. 2, Mar. 1991, p. 2540.
102. D. Kalokitis et al., *Appl Phys Lett,* vol. 58 (5), Feb. 1991, p. 537.
103. G. V. Negrete et al., *Conference Record,* Electro-International, New York City, April 16–18, 1991, p. 564.
104. K. Yoshikawa et al., *5th Intern Workshop on Future Electron Devices, High-Temperature Superconducting Electron Devices,* Miyagi-Zao, June 2–4, p. 51.
105. M. Mukaida et al., *Japan J Appl Phys,* vol. 29, no. 6, June 1990, p. L936.
106. A. Mogro-Camparo et al., *Appl Phys Lett,* vol. 52 (7), Feb. 1988, p. 584.
107. M. R. Namordi et al., *IEEE Trans Microwave Theory and Techn,* Sept. 1991.
108. A. C. Greenwald, *Microelectron Manufact Techn,* May 1991.
109. P. E. Norris and G. W. Orlando, *Supercond Ind,* Spring 1990, p. 14.
110. F. Radpour et al., *J Electrochem Soc,* vol. 137, no. 8, Aug. 1990, p. 2462.
111. H. Yamane et al., *Japan J Appl Phys,* vol. 28, no. 5, May 1989, p. L827.
112. J. M. Zhang et al., *Appl Phys Lett,* vol. 55 (18), Oct. 1989, p. 1906.
113. J. M. Zhang et al., *Appl Phys Lett,* vol. 55 (12), Sept. 1989, p. 1258.
114. J. Takemoto et al., *Appl Phys Lett,* vol. 58 (10), Mar. 1991, p. 1109.
115. Z. Lu et al., submitted to *Appl Phys Lett,* March 6, 1995.
116. Y. L. Chen et al., *J Mater Res,* vol. 4, no. 5, Sept/Oct. 1989, p. 1065.
117. S. J. Golden et al., *J Mater Res,* vol. 5, no. 8, Aug. 1990, p. 1605.
118. R. Sobolewski and W. Kula, in *Book of Abstracts, Conference on the Science and Technology of Thin Film Superconductors,* Nov. 14–18, 1988, Colorado Springs, CO.
119. R. Sobolewski et al., *IEEE Trans Magn,* vol. 25, 1989.
120. P. M. Mankewich et al., *Appl Phys Lett,* vol. 51 (21), Nov. 1987, p. 1753.
121. M. Gurvitch and A. T. Fiory, *Appl Phys Lett,* vol. 51 (13), Sept. 1987, p. 1027.

122. T. S. Kalkur et al., in *Book of Abstracts, Conference on the Science and Technology of Thin Film Superconductors,* Nov. 14–18, 1988, Colorado Springs, CO.
123. Y. Yoshikazu et al., *Japan J Appl Phys,* vol. 26, no. 9, Sept. 1987, p. L1533.
124. F. K. Shokoohi et al., *Appl Phys Lett,* vol. 55 (25), Dec. 1989, p. 2661.
125. A. Mogro-Camparo et al., *Superconducting Sci Technol,* vol. 3, 1990, p. 537.
126. C. X. Qui et al., in *Processing and Applications of High-Tc Superconductors,* William E. Mayo (ed.), The Metallurgical Society, Inc., 1988, p. 83.
127. R. P. Vasquez et al., *Appl Phys Lett,* vol. 54 (11), Mar. 1989, p. 1060.
128. R. Housley, *Superconductor Week,* vol. 4, Apr. 1991.
129. M. Tonouchi et al., *Japan J Appl Phys,* vol. 27, no. 1, Jan. 1988, p. L98.
130. R. P. Vasquez et al., *Appl Phys Lett,* vol. 53 (26), Dec. 1988, p. 2692.
131. J. M. Rosamilia et al., *J Electrochem Soc,* vol. 136, no. 8, Aug 1989, p. 2300.
132. K. B. Bhasin et al., NASA Technical Memorandum 103235, p. 185.
133. S. Matsui et al., *Appl Phys Lett,* vol. 52 (1), Jan. 1988, p. 69.
134. Y. Nishi et al., *J Matl Sci Lett,* vol. 7, 1988, p. 281.
135. A. Inam et al., *Appl Phys Lett,* vol. 51 (10), Oct. 1987, p. 1112.
136. R. Boerstler et al., in *Book of Abstracts, Conference on the Science and Technology of Thin Film Superconductors,* Colorado Springs, CO, Nov. 14–18, 1988, p. 89.
137. C. T. Rogers et al., *Appl Phys Lett,* vol. 55 (19), Jan. 1988, p. 2032.
138. G.-C. Liang et al., *1994 IEEE MTT-S Digest,* IEEE, 1994, p. 183.
139. F. E. Terman, *Radio Engineers Handbook,* McGraw-Hill, New York, 1943.
140. G. Kent and M. Ingalls, *Hybrid Circuit Technol,* Nov. 1988.
141. J. R. Tyler, *Proc 1990 Proc Intern Symp Microelectr,* Oct. 15–17, 1990, Chicago, IL, p. 468.
142. R. Brown, *Connection Technology,* July 1988.
143. M. E. Goldfarb and R. A. Pucel, *IEEE Microwave and Guided Wave Lett,* vol. 1, no. 6, June 1991, p. 135.
144. J. Gipperich and S. Grice, *Proc 1993 Microwave Hybrids Conf,* Oct. 17–20, 1993, Wickenburg, AZ. Sponsored by Rogers Corporation.
145. F. W. Grover, *Inductance Calcualtions,* Van Nostrand, Princeton, NJ, 1946. Reprinted by Dover Publications, NY, 1962.
146. B. C. Waddel, *Transmission Line Design Handbook,* Artech House, Norwood, MA, 1991.
147. A. Pargellis, *Technical Paper IPC-TP-642,* presented at IPC 10th Annual Meeting, Atlanta, GA, Mar. 29–Apr. 3, 1987. IPC, Lincolnwood, IL, 1987.
148. G. M. Mattox, *Electochem Tech,* vol. 2, 1964, p. 295.
149. R. S. Besser and P. J. Louris, *Proc 1993 Intern Symp on Microelectronics,* Nov. 9–11, 1993, Dallas, TX, p. 114.
150. C. J. Brinker et al., *Proc MRS Symp,* vol. 284, 1993, p. 469.
151. H. Bok and H. S. Tong, *SPIE Proceedings,* vol. 1815, Display Technologies, 1992, p. 86.
152. M. Vijan, *U. S. Patent Number 4,696,885,* Sept. 29, 1987, assigned to Energy Conversion Devices, Inc., Troy, MI.
153. E. J. Choinski, *Information Display,* vol. 11, 1991.
154. Microsubstrates Corporation, Tempe, AZ. The assistance of Dr. Ram Panniker is greatly appreciated.
155. E. F. Belohoubek, *RCA Review,* vol. 46, Dec. 1985, p. 464.
156. R. J. Perko et al., *Microwave J,* vol. 11, Nov. 1988, p. 67.

CHAPTER 11
YIELD, TESTING, AND RELIABILITY

David Keezer
Georgia Institute of Technology

This chapter describes the techniques and instrumentation used for assessing the quality and reliability of complete electronic components and subassemblies. An introduction to this topic is provided in section 11.1, including the motivation for testing, yield and fault coverage, screening and reliability, and test economics. In section 11.2, descriptions are provided for design-for-test (DFT), built-in self-test (BIST), boundary scan, automatic test pattern generation (ATPG), and fault dictionary tools. Examples of state-of-the-art test instrumentation are given in section 11.3. The electrical interface between the device-under-test and test instrumentation is described in section 11.4. Some of the commonly used test and evaluation methods for electronics are described in section 11.5. Finally, section 11.6 discusses techniques available for fault isolation and failure analysis.

11.1 INTRODUCTION TO TESTING OF MICROELECTRONIC COMPONENTS AND SUBASSEMBLIES

This section provides motivation for the use of extensive electrical testing. With the rapidly falling cost (per transistor) of electrical systems, one might question whether complete electrical testing is warranted. Could the electronics be treated as dispensable and thereby skip the traditional test process? Unfortunately, as this section shows, such shortcuts are not yet economically feasible.

Some reasons why complete electrical tests remain indespensible are given in section 11.1.1. The closely related issues of yield, fault coverage, and test transparancy are further described in section 11.1.2. When several integrated circuits are assembled into a multichip module (MCM) the yield, test, and quality issues play a dominant role in the production strategy, as illustrated in section 11.1.3. The use of burn-in and accelerated life test to ensure reliability is described in section 11.1.4. Finally, section 11.1.5 discusses the critical factors of cost and economic trade-offs associated with electrical testing.

11.1.1 Introduction—Motivation and Types of Tests

It is understandable, given the relatively large percentage of production costs associated with electrical testing, and given the high yields of electronic components, that one might question the need for extensive testing and detailed diagnostics. After all, how often does one actually find a defective electrical component? Couldn't electrical testing be skipped all together? Unfortunately, as we shall see, the answer today is "no" and we must find economically viable methods for implementing these tests.

In an ideal production environment with near 100% yield, high quality in the delivered product can be achieved by ensuring the near-perfect quality of the manufacturing process. Strictly speaking, if a production yield is actually 100% then no additional testing is necessary. The difficulty lies in qualifying terms such as "almost" perfect and "near" 100%, and in the current rapid pace of microelectronics technology development. First, 100% yield has never been sustained in any commercially viable electronics production line. When yields exceed 90%, technology typically has progressed to the point that either the product becomes obsolete or a new and more aggressive process is adopted (generally with lower initial yields). Therefore, until strict limits are encountered it is unlikely that process technology will evolve to the point where a specific production line can achieve ideal yield and therefore completely bypass testing.

Another point in support for electrical testing is that it actually represents the most critical step of the production process. Not critical in the sense of being overly sensitive, but rather in the sense of being demanding. Electrical testing is usually the final gate through which the product must pass before shipment. It requires that all steps of the production process occur almost perfectly, and that the materials used in the process be almost completely free of defects. When designed and applied correctly, electrical testing identifies and eliminates almost all of the defective or potentially defective components before shipment to a customer. In a sense, design and production can be thought of as simply preparing for testing.

Why is finding defective components so important? First, the degree of importance depends on the application. Clearly a life-critical medical or military application would place a higher premium on ensuring the quality and reliability of the electronic system and components than would a consumer electronics application such as an electronic game. Therefore, there is a wide variation on the importance placed on electrical component quality.

Nevertheless, even ignoring these differences, there is a need to find defective parts at the earliest stage of integration. This is because of the fact that each level of integration results in roughly an order of magnitude (or more) increase in complexity [1–3]. In most cases this increased complexity results in greater difficulty (and cost) in isolating and repairing failed components. Therefore, for example, a defective integrated circuit that escapes detection due to inadequate testing at the component level could cost thousands of dollars in detection and rework at the system level. A complete economic trade-off analysis should therefore consider the entire life-cycle cost for the intended application to avoid simply shifting test cost from one organization to another (and possibly increasing the overall cost).

Given that electrical testing is required, the next question becomes "how much testing is needed?" The short answer, and not coincidentally the economically justifiable answer is, "just enough." As we have seen insufficient testing may result in excessive costs at higher levels of integration. On the other hand, too much testing simply adds to the production cost. How much is just enough is answered in the remainder of this section.

We begin by recognizing that there are many different categories of electrical testing. A successful test strategy will encompass many of these (although rarely will all methods be used by one organization). A sampling of some of the more common categories of electrical testing follows.

Functional Testing. As the name implies, functional testing focuses on verifying that the circuit actually performs its intended functions. If the circuit is a digital binary multiplier, a functional test would input binary words and check the output against the expected product. If the device is an analog-to-digital converter, a functional test would apply some set of analog levels and observe to see if the correct digital outputs are produced. If the device is a radio frequency (RF) amplifier, a functional test would consist of an applied RF input of known power level while measuring the power level (or other RF characteristics) of the output.

If exhaustively performed, under all potential environmental, logical, and electrical conditions, the functional test is sufficient for most applications. However, this extensive combination of conditions is usually impractical even for relatively simple devices. Therefore, a representative subset of functional conditions is usually checked, followed by supplemental tests of the types described below.

Structural Testing. Whereas functional testing aims at exercising the intended device functionality, structural testing is used to verify that the fabricated electrical structure matches the schematic, or computer-aided design (CAD) representation of the device. This is important for two key reasons: (1) as a supplement to partial functional testing and other tests, a structural test can be used as a quantifiable method for checking that the circuit is constructed according to design, and (2) taken together with an accurate and complete circuit model and simulation, structural testing can be used to validate the design where complete functional testing may be impractical. In other words, if an accurate model exists for the circuit, then extensive simulations can be run to verify that it will perform the intended functions. If a structural test is performed that indicates that the physical circuit matches the model, then it can be deduced that the physical device will also perform the intended functions. However, usually there is enough uncertainty in the computer model that structural testing alone is not sufficient.

Built-in Self-Testing (BIST). Sometimes it may be advantageous to incorporate within the circuit a capability to test itself. There are many techniques (mostly used for digital logic) which can be designed into a circuit so that it can perform such a self-test (see section 11.2). Usually such testing is performed off-line during component testing, but may also be useful within an operational system as an on-line self-check. Significant added circuitry is usually required to implement BIST, and the decision to use this technique must be made with careful consideration of the costs and benefits for a particular application. Built-in self-test techniques tend to have the best economic payoff for very large scale integrated (VLSI) circuits where the added BIST circuitry is shared by a large amount of functional circuitry.

Acceptance Testing. These are electrical tests which are performed to ensure compliance with specific customer requirements (specifications). The specific customer requirements may or may not be directly related to the manufacturer's specifications. These tests are usually closely related to the intended application environment. They may be performed on a sample basis (see below) to minimize cost or may be applied to all devices. Such tests are often performed by the component customer and are sometimes referred to as "receiving inspection" testing.

Sample Testing. In this approach to testing, a small random sample of devices is selected from a larger lot and appropriate electrical tests are performed only on the sample. The obvious advantage to this approach is in its cost savings over testing all parts. A variation on this technique is to apply a cursory electrical check to all devices, followed by a more thorough test of a small sample. The probabilities associated with sample testing are well understood and a known confidence level for a given sample size can be predicted (see section 11.2). Nevertheless, when the lot sizes are small or the desired confidence level is nearly 100%, testing of every component may be necessary.

Go/NoGo or Pass/Fail Testing. In this method the device is subjected to a specific set of electrical conditions (stimuli) and is required to produce the expected response. If this response is not observed than the test fails and the part is scrapped (NoGo condition). Otherwise a passing test is called a "Go" condition and the part is accepted. Usualy Go/NoGo testing is performed at the specification limits simply to verify conformance. No parametric measurements are performed (as in characterization testing). Sometimes the term "Go/NoGo" is used to refer to tests performed at nominal conditions, in which case the Go/NoGo test is a prescreen before more extensive tests.

Wafer Probe Testing. These tests are applied to the integrated circuits on a wafer before packaging. Usually wafer probing is performed by the manufacturer to identify those chips which are sufficiently acceptable for packaging. Defective die found during wafer probing are often marked with an ink dot to assure that they are not used in subsequent integration or packaging steps. Sometimes a detailed wafer map is generated instead of or in addition to inking. During wafer probe testing, electrical contact to the integrated circuit (IC) is provided through a probe card supporting fine-pitched needles, or possibly bumps on a flexible membrane. Usually a simple (Go/NoGo) electrical test is applied at this stage, with the understanding that more exhaustive testing will be performed after the devices are packaged. However, in some cases (such as when very expensive packages are used) more extensive testing can be applied at wafer probe testing. In very-high volume, noncritical applications the final electrical test can be applied at the wafer probe stage, assuming that the packaging yield is sufficiently high.

Final Packaged Test. After packaging, the parts are usually tested again at least to verify that no damage has been encountered during packaging. Often the most critical electrical tests are performed at this stage because of the relative ease of providing a clean electrical and thermal test environment. Often the devices will be rated (especially for speed) and sorted into "bins" at this point.

Characterization Test. As distinguished from Go/NoGo testing, characterization tests not only verify conformance to specifications but also measure the margin, or how well the device satisfies the various requirements. For instance, rather than just check that a propagation delay is no more than 3.2 ns, a characterization test would actually measure the delay to be 2.6 ns. In this example, the 0.6 ns margin gives a better understanding of how well the device may perform in adverse electrical and thermal environments. Both the device ac and dc parameters can be measured during characterization testing. Clearly this type of test requires more sophisticated techniques, equipment, time, and associated cost than a Go/NoGo test. For these reasons, characterization testing is often reserved for the initial lot evaluation, sample testing, or selectively applied to critical parameters.

Stress Screening (Burn-in) Test. During these tests the device is subjected to environmental conditions designed to accelerate the onset of failure from latent defects. It is known that if a device is going to fail, it will likely do so within a short time of first operation. The typical time used is 24 to 48 hours at elevated temperature and worst-case bias conditions. This screening for early failure usually reduces the failure rate to acceptable levels over a product life cycle measured in years or decades. After such extended time periods, failure rates again increase. Failure rate as a function of time often exhibits the familiar "bathtub" shape shown in Fig. 11-1.

Reliability (Accelerated Life) Test. This test is used to determine the mean time between failures (MTBF) by extrapolating the burn-in test to find the wear-out time on the bathtub curve. Again, by subjecting the device to elevated temperatures and extreme bias conditions, the onset of "wear-out" can be hastened. Well-known acceleration formulae are available for predicting the acceleration factor for most common electrical failure mechanisms. Generally the higher the temperature, the shorter the time to failure. By measuring time to failure at elevated temperature, the MTBF at nominal temperatures can be deduced.

Diagnostic Test. Although most testing is designed to determine whether a device is functioning as expected, diagnostic testing is used to isolate the specific cause of a failure or degradation. To implement a repair strategy, the defect must first be located. This is usually accomplished through a combination of logical fault isolation tests, possibly followed by physical inspection techniques. Diagnostic tests are often the most difficult to produce because they must generate much more than a Go/NoGo response, or even more than a measured parametric value. In addition, diagnostic tests must also be capable of working in a partially faulty environment. Also, to handle the wide variety of potential defects, the diagnostic test is often very lengthy. In many applications, the cost associated with detailed diagnostic testing and associated repair is too high. In these cases the test and repair strategy is to apply Go/NoGo tests followed by complete module or subsystem replacement.

Quality Test. This term is used to describe tests that require the part to perform beyond the manufacturers specifications. Some devices will be better than others and therefore will provide larger margins and better quality. The degree of economic success that this approach has depends greatly on how the manufacturer has specified and tested the part, and how important such additional margin is to the application. Almost invariably, there will be a trade-off between quality gained and yield lost (and therefore higher cost). Often there is a very narrow spread in the performance characteristics of a given lot of parts, as illustrated in Fig. 11-2*a*. In this situation, using only the very best parts would have a severe impact on cost because of lost yield, not to mention the additional cost of quality testing. In contrast, Fig. 11-2*b* illustrates a distribution of performance where such a quality test may be economically feasible. In this case a substantial fraction of the population will pass the quality test.

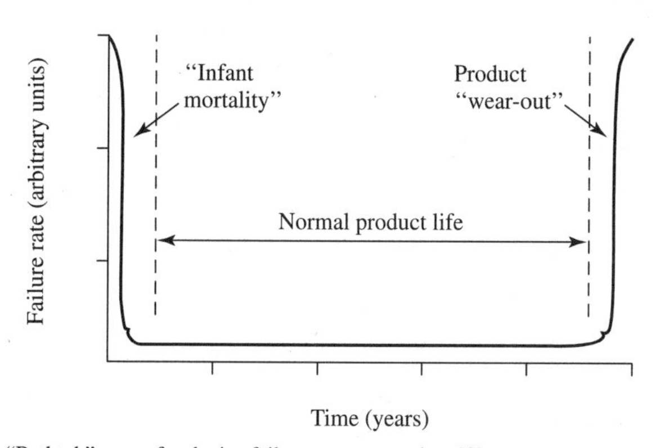

FIGURE 11-1 "Bathtub" curve for device failure rate versus time [3].

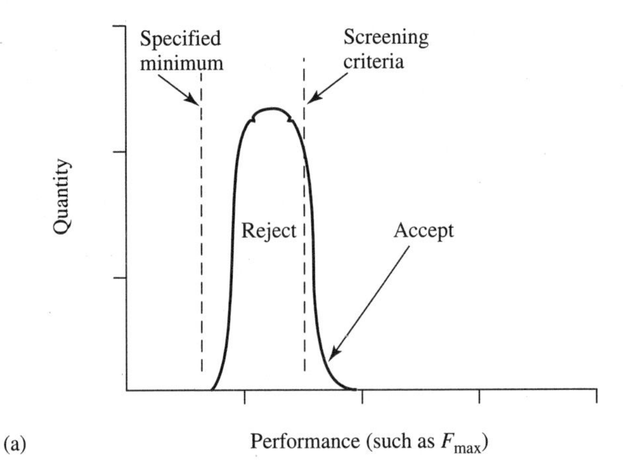

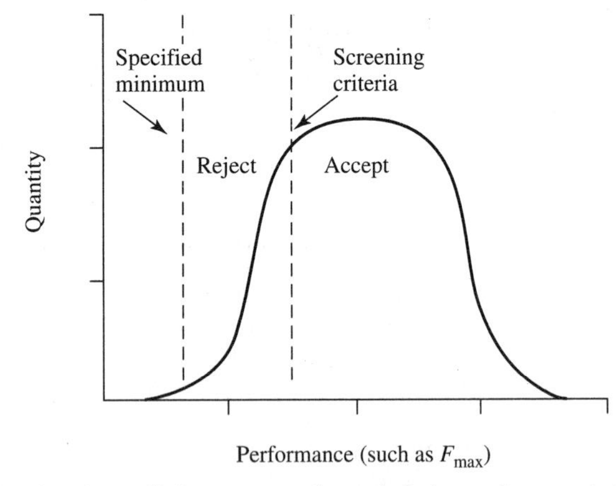

FIGURE 11-2 Illustration of a "quality" test strategy where only the best performance devices are selected for use. *(a)* Ineffective quality test; and *(b)* effective use of quality test.

On-line Test. On-line tests are performed while the system is operational. The possibility to perform testing during system operation clearly depends on the system application, accessability, and intended environment. Its usefulness will also depend on a feasible repair strategy. The application will determine whether the results of such testing are of value to the user of the system. An on-line test may be implemented using external test instrumentation alone, but usually requires some amount of BIST circuitry. There will be costs associated with the design and implementation of this added circuitry. These costs must be carefully weighed against the benefits provided by on-line testing.

Design Verification Testing. Design verification testing is used to check the correctness of the circuit logic design and conformance to design requirements. In a narrow sense, design verification checks just that the fabricated circuit logic agrees with the schematics or CAD database. This is not sufficient to verify the design and only checks the structural integrity of the circuit against an assumed-correct description. In fact, design verification should include elements of almost all types of testing described above. To verify the design means not only to check that the gates are arranged according to the schematic, but also to check that the circuit functions as intended, with sufficient margins to operate in a realistic environment, and with sufficient reliability to produce a quality product. Usually some degree of diagnostic testing is required during design verification to resolve uncertainties and distinguish between repetitive and random defects.

11.1.2 Yield, Fault Coverage, and Quality Level

In section 11.1.1 we saw that a complete testing strategy might be made up of several types of testing. One of the few available methods that provides a quantitative measure of completeness is the structural test technique based on the single-stuck-at (SSA) fault model [1,4–11]. The SSA fault model considers only defects that can be modeled as logical nets fixed at (stuck at) either zero (SA0) or one (SA1) and assumes that no more than one such fault exists within a circuit. This model, of course, is a simplification of the actual situation where many other types of defects can occur and multiple faults can exist. Examples of other types of faults include stuck-at-open faults, bridging faults, gate-delay faults, path-delay faults, and a variety of dc and ac parametric degradation faults.

Figure 11-3 illustrates the concept behind the SSA fault model. In this simple example, the SSA fault model considers the following possibilities:

Net	Fault condition
A	SA0
A	SA1
B	SA0
B	SA1
C	SA0
C	SA1
D	SA0
D	SA1
E	SA0
E	SA1
F	SA0
F	SA1
G	SA0
G	SA1
H	SA0
H	SA1
Z	SA0
Z	SA1

The nets (A, B, C, D, E) are the primary inputs, whereas net (Z) is the circuit output. All tests for this combinational circuit can be expressed as vectors. Each specifies the five input values and the one expected output: (A, B, C, D, E, Z). A test for the fault condition G-SA1 can be written as (A, B, C, D, E, Z) = (0, 0, 0, 0, 0, 0). In other words, inputing

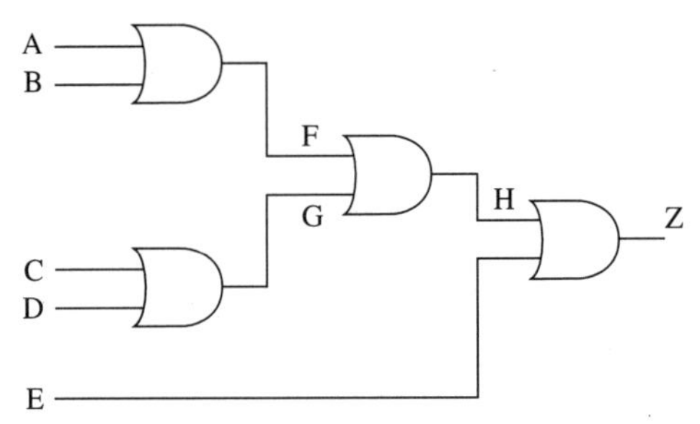

FIGURE 11-3 Example circuit illustrating the SSA fault model.

logic zeros on all inputs should produce a logic zero on the output if the circuit is functioning correctly. However, if net G is stuck at one, then the output will be a logic 1. Notice that the vector (0, 0, 0, 0, 0, 0) also tests for all other nets stuck-at-one for this particular circuit.

To test for the SS0 conditions, different vectors are required. For example, to test for G-SS0 we can apply (0, 0, 0, 1, 0, 1) which also checks for D-SS0, H-SS0, and Z-SS0. A complete set of tests for SSA faults in this circuit would be:

(0,0,0,0,0,0) (1,0,0,0,0,1) (0,1,0,0,0,1) (0,0,1,0,0,1) (0,0,0,1,0,1) (0,0,0,0,1,1)

This set of test vectors has 100% SSA fault coverage for the example circuit. Notice that a fault on one net could be masked by a fault on another net. For instance, G-SA1 could be masked by H-SA0. However, in this case at least H-SA0 would be detected by the 100% SSA test set described above. This example illustrates that many multiple-stuck-at (MSA) faults are often detected by a high-coverage SSA test. It is also easy to see that the number of SSA faults grows linearly with the number of nets in the circuit, whereas the number of MSA faults scales geometrically with gate count. For these reasons, the SSA fault model is widely accepted as a practical method for describing possible faults and for deriving digital tests.

The SSA fault model is used extensively as a measure for structural test fault coverage. Usually fault coverage figures are given as a percentage detected of all possible SSA faults. With modern automated test pattern generation (ATPG) software, 100% coverage of SSA faults for combinational logic is usually possible. However, for very large logic blocks (thousands of gates), 100% coverage may not be practical. In these cases, SSA fault coverage in the upper 90% range is usually achievable and required for adequate testing.

Determining the fault coverage of a set of electrical tests is an important step in predicting the quality level of the resulting product. In addition to the fault coverage (C) of the test, the process yield (Y) must be measured or estimated to determine if an acceptable quality level (AQL) will be achieved. These three measures of the fabrication test process are closely related by the formula [12, 13]

$$QL = Y^{(1 - C)} \tag{1}$$

where QL is the fraction of devices passing the test which are actually good (ideally $QL = 1$). Clearly if the process yield is 100% (that is, $Y = 1$) then by definition all devices are good and no testing is required (that is, $C = 0$). Conversely, if yields are very low ($Y \ll 1$) then 100% fault coverage testing will guarantee that no defective devices escape detection (then $QL = Y^0 = 1$). Usually the result is somewhere in between ($0.05 < Y < 0.99$) and the required fault coverage needed to achieve a desired quality

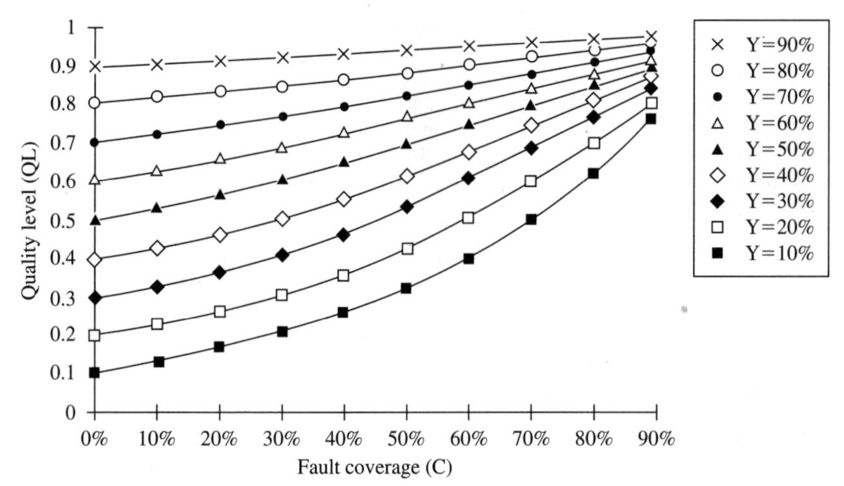

FIGURE 11-4 Quality level (QL) as a function of fault coverage with process yield as a parameter.

level can be calculated by solving the formula above. Some example combinations of the three parameters are plotted in Fig. 11-4. When the quality level is very nearly 100%, the defect level ($DL = 1 - QL$) can be approximated as

$$DL = 1 - QL = (1 - C)(-\ln Y) \quad (\text{for DL} \ll 1) \tag{2}$$

The quantity $(1 - C)$ is called the test transparency and represents the fraction of faults which are not tested. Figure 11-5 illustrates some example combinations of process yield and fault coverage which result in very low defect levels ($0 < DL < 500$ ppm) such as are usually desirable in a manufacturing environment.

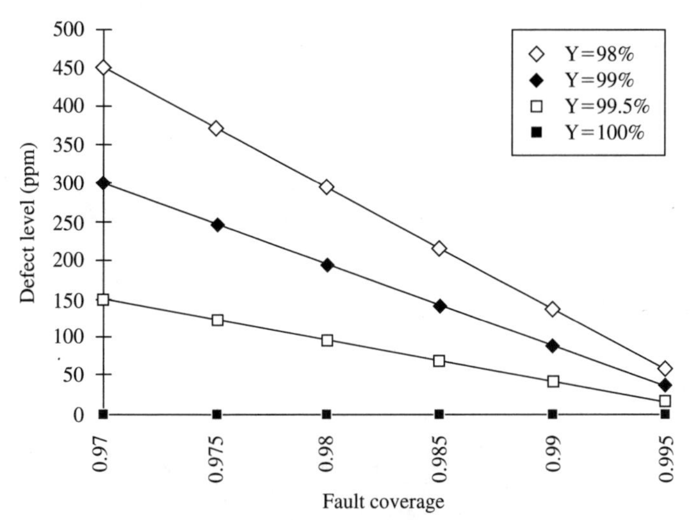

FIGURE 11-5 Defect level ($DL = 1 - QL$) in ppm as a function of fault coverage with process yield as a parameter.

11.1.3 Yield in Multichip Subassemblies

So far we have considered only the probability that a given test will detect the faults in a single integrated circuit. Testing multichip modules (MCMs) has the added difficulty that the uncertainties associated with each of the integrated circuits combine to determine the resulting module yield. Specifically, if there are N die in an MCM, each with yield Y_N, then the resulting module yield is

$$Y_{\text{Module}} = \prod_{i=1}^{N} Y_i \tag{3}$$

For example, if there are three types of chips in the MCM, with yields Y_1, Y_2, and Y_3, then the module yield will be

$$Y_{\text{Module}} = (Y_1)^{N_1}(Y_2)^{N_2}(Y_3)^{N_3} \tag{4}$$

where N_1, N_2, and N_3 are the quantities of the three device types respectively.

There is little difficulty so long as the yields or quality levels of the individual die are relatively high and the number of die in the MCM is not too large. However, as the total number of chips in an MCM increases, the need for extremely high quality levels becomes critical. This fact can be illustrates by considering (for simplicity) MCMs made up of identical chips. In this case, the module yield reduces to:

$$Y_{\text{Module}} = (Y_{IC})^N \tag{5}$$

The probability that an assembled MCM will not have a defect rapidly diminishes as the number of chips (N) increases. This relationship is shown in Fig. 11-6 where module yield (Y_{Module}) is plotted as a function of the number of ICs (N) with the chip yield (Y_{IC}) as a parameter.

This chart clearly illustrates the need for "known-good-die" in assembling even moderate chip-count MCMs. Unless the chips are known-good to a confidence of 99% or better, the chance of producing a defect-free MCM is small. For example, if an 8 × 8 array of memory chips are assembled onto an MCM substrate, even if the die yield is 99% the

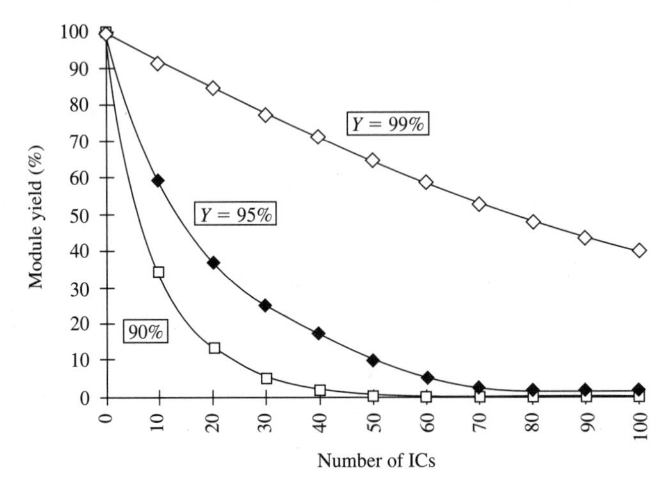

FIGURE 11-6 Module yield as a function of the number of chips in an MCM. (Chip yield is shown as a parameter.)

chances that the entire assembly will be good is only about 50%. If the die yield is only 95%, then the expected module yield drops to below 5%.

Assurance of "known-good die" is achieved by thoroughly testing each chip before MCM assembly. In addition to applying a high fault-coverage electrical test, the chip bias and operating temperature should be adjusted to worst-case conditions. All critical dc parametric values and ac performance characteristics must be checked. Furthermore, an elevated temperature burn-in process should be applied to reduce early part failure (see the next section). When such extensive die screening is accomplished, the confidence that each chip is "known-good" can be well above 99%. However, the assembly process itself can introduce defects even when all the chips are actually "known-good." Therefore the integrated circuit yield as used in the formulae above is never fully 100%. It can be seen from Fig. 11-6 that there is a practical limit of a few tens of die per MCM, above which the expected yields are too small to be economical.

11.1.4 Screening for Reliability

We have seen (section 11.1.1) that during any interval of time there is a finite (although hopefully small) probability that an electronic device will fail. In particular, for relatively new devices (within the first 10,000 hours of operation) a higher rate of failure is expected. This initial dropout is referred to as "infant mortality." The purpose of burn-in testing is to accelerate the failure from infant mortality to the point where the remaining population of devices have a much lower failure rate associated with the normal operating period (see Fig. 11-1). Electrical operation of the device during burn-in can be performed in either a static or dynamic sense. During a static burn-in, electrical input levels to the device are fixed and held in a state that reverses biases of as many junctions as possible. This works reasonably well for discrete components and small-scale integrated circuits. However, for large-scale integrated circuits it is difficult if not impossible to reverse bias of all junctions. In these complex devices the strategy is to toggle the gates in a manner similar to normal operation. This process is called dynamic burn-in and clearly requires more complex instrumentation than that used for static burn-in. Dynamic burn-in is the preferred method for logic devices today.

An obvious approach to reducing early failures in the field is to simply operate all devices for a period of time comparable to the infant-mortality period (typically 1,000 to 10,000 hours) under the same conditions expected during normal operation. Following such operation, a thorough electrical test will identify those devices that have failed. The difficulty with this approach is the associated high cost of operating devices for extended time periods. To reduce cost, the process is accelerated by applying higher electrical bias and subjecting the parts to higher operating temperatures than are expected during normal operation. The effective operating time (t_{eff}) is then given by [14]

$$t_{\text{eff}} = A_T A_v t_{bi} \tag{6}$$

where A_T and A_V are the acceleration factors associated with elevated temperature and voltage, respectively, and t_{bi} is the time duration of the burn-in. The acceleration factors can each be a factor of 10 or more, and can be computed using the following relationships [14–16]:

$$A_T = e^{(Ea/k_B)(1/T_1 - 1/T_2)} \tag{7}$$

$$A_V = e^{(C/t_{\text{ox}})(V_1 - V_2)} \tag{8}$$

where E_a = the activation energy (typically 0.4 eV)
k_B = the Boltzmann constant (8.6×10^{-5} eV/K)
T_1 = the operating temperature
T_2 = the burn-in temperature
C = the voltage acceleration constant (conservatively estimated at 290 Å/V)
t_{ox} = the gate oxide thickness in angstroms
V_1 = the bias voltage during burn-in
V_2 = the bias voltage during normal operation.

As an example, consider a 24-hour burn-in at 125°C with a bias voltage of 7.0 V applied to a 200 Å gate oxide thickness device that normally operates at 25°C and 5.0 V bias. Applying the above formulae, the effective operating time associated with the 24-hour burn-in, t_{eff} is about 21,000 hours (that is, the product of the acceleration factors is about 875). Therefore, it can safely be assumed that the vast majority of device failures will occur during this burn-in procedure. A thorough electrical test after burn-in will identify and eliminate these failing devices before they are placed in the field.

When the parts that have failed in infancy have been eliminated, the failure rate is usually acceptably small for an extended period of time (measured in years or tens of years). After this normal product lifespan, increased failure rates occur during the "wear out" period. Usually the product is taken out of service before this time.

Commonly used measures of failure during the normal operating period include mean time between failures (MTBF), percent failing per 1,000 hours, and device hazard rate (percent of failures per unit time) [14]. The device hazard rate can be expressed in terms of failures in time (FIT). An FIT corresponds to a failure rate of one failure per billion hours (that is, 10^{-9} hr^{-1}) or a probability of failure of 10^{-9} per hour. During the normal operating period, the failure rate is often approximated as a constant, λ. In this case, the exponential distribution is implied where the survivor function is given by [14]:

$$S(t) = e^{-\lambda t} \tag{9}$$

For example, if a device failure rate is 10,000 FITs then $\lambda = 10^{-5}$ per hour. To compute the probability that the device will last one year (8,760 hours) we have

$$S(8{,}760) = e^{-(0.00001)(8{,}760)} = 0.916, \text{ or about } 91.6\% \tag{10}$$

In the same example the chances of surviving 10 years is about 41%, whereas a device with a failure rate of 1,000 FITs would have a 99% chance of surviving one year and a 91% probability of surviving 10 years. Clearly, for long-term use FIT rates <1,000 are required for electronic systems.

To measure the normal life failure rate (life testing), techniques similar to those used in burn-in are used (see above). However, in this case we are not screening out failures but rather are simply measuring the failure rate. The devices that go through a life test must normally be discarded, whether or not they pass the test, because they will have significantly shorter life expectancy after the test. In this sense the life test is destructive. As with burn-in, life testing requires subjecting the parts to operating conditions for extended periods of time. In this case, the time periods are even longer than for burn-in because the normal life failure rates are much lower than infant mortality rates. For this reason, acceleration factors are introduced (as in burn-in) and thousands of parts are required to obtain statistically significant failure rates.

For example, if a device has a failure rate of 1,000 FITs, then testing for 1,000 hours would result in a survival probability of 99.9%. Therefore only one part in 1,000 is expected to fail. By accelerating the test with higher bias voltage and operating temperature, the effective operating time can be greatly lengthened (as during burn-in). Using the same conditions as the burn-in example above, with a combined voltage and temperature acceleration factor of 875, the survival probability during 1,000 hours would be about 41.7%. Therefore, out of 1,000 devices about 583 are expected to fail. Such a test would give a good indication of the actual failure rate.

11.1.5 Test Economics

On the surface, accurately estimating the total cost associated with electrically testing an integrated circuit appears to be simply a matter of multiplying the test time by the hourly labor and equipment rates. However, this simplistic viewpoint overlooks life-cycle costs and may underestimate the value of such features as BIST, which can reduce diagnostic and field testing costs as well as be of use in component and board-level tests. Because there are usually several levels of testing performed on a given component, and the cost associated with each level depends somewhat on the effectiveness of earlier testing, the total life-cycle test cost must be considered to avoid simply transferring test cost from one level to the next. Avoiding such cost transfer is especially important because of the exponential increase in cost associated with testing and repair at higher levels of system integration. Therefore, it is generally advisable to identify failing components as early as possible. Where such a strategy fails is when the incremental cost of improving a component test exceeds the added value in terms of avoiding higher-level testing and repair.

Assuming that there are N steps in the testing process (such as wafer-level test, packaged device preburn-in test, component postburn-in, printed circuit board test, subsystem test, system test, depot-level test, etc.), the total test cost can be expressed as [3]

$$C_{\text{Total}} = \sum_{i=1}^{N} C_i\{[(2 - QL_{i-1})t_i] + [(1 - QL_{i-1})(t_{Di} + t_{Ri})]\} \tag{11}$$

where C_i = cost per hour of the labor and equipment (including overhead, etc.) used in the ith test

QL_{i-1} = quality level of the preceding step

t_i = the average time for testing for the ith

t_{Di} = average time for diagnostics

t_{Ri} = average time for repair for the ith test

This relationship clearly shows how the effectiveness of one test can severely affect the overall cost. Notice that the first term $(2 - QL_{i-1})$ is least when QL_{i-1} is largest. Namely, when $QL_{i-1} = 1$ then the parts are only tested once. In contrast, if the quality level is worst (0) then twice the number of parts must be tested (both the bad ones and the repaired ones). The second term $(1 - QL_{i-1})$ is zero if the preceeding test has 100% quality and the cost of diagnostics and repair would also be zero. Notice that the values for quality levels, QL_i obtained at each step are highly dependent on the amount of effort (and cost) devoted to the test. Therefore, C_{Total} is generally a very complex function of the amount of time spent testing at the various assembly levels.

This formula does not tell the entire story. In addition to the cost of applying the test, there is the material cost and added value, which are lost when a component or subassembly fails. In other words, if a bad component is inserted into an otherwise good assembly, it might not be practical to repair the defective unit. In this case, the product is scrapped and the liability incurred as a result of incomplete testing can be very high. In the final product such liability might even involve the potential for damage, in the familiar sense of product liability. The costs of scrapping units at different stages of production can be added to the equation above by increasing t_{R_i} accordingly or by adding another explicit term, C_{Si}.

The fraction of defective parts which are irreparable, S_i, at each stage of production must also be estimated.

$$C_{\text{Total}} = \sum_{i=1}^{N} C_i\{[(2 - QL_{i-1})t_i] + [(1 - QL_{i-1})(t_{Di} + (1 - S_i)t_{Ri} + S_iC_{Si}/C_i)]\} \qquad (12)$$

The units of C_{Si} in this formula are dollars per part so that S_iC_{Si}/C_i has units of time. Note that this formula does not include the nonrecurring cost of developingthe tests, C_{NREi}. If the total number of production units is known in advance, these costs can be added by amortizing the nonrecurring cost over the total production volume (PV).

$$C_{\text{Total}} = \sum_{i=1}^{N} (C_i\{[2 - QL_{i-1})t_i] + [(1 - QL_{i-1})(t_{Di} + (1 - S_i)t_{Ri} + S_iC_{Si}/C_i)]\} + C_{NREi}/PV) \qquad (13)$$

When an accurate model for the total cost is obtained, then the effects of improving or relaxing a particular test can be ascertained. Only then can an informed decision be made regarding the cost-effectiveness of a test.

11.2 PREPARING FOR TESTING MCMs

The trend toward miniaturization of electronics brings with it a growing use of MCM packaging technology. Although tremendous performance benefits can be realized by eliminating individual chip packages, there are a number of test and fault isolation concerns which are exacerbated by this practice. The most obvious limitation arises from a reduction in controllability and observability. When the chips are assembled into the MCM, electrical access for probing is extremely limited, if not impossible. As compared with older printed circuit boards, the individual chips cannot easily be probed with common test instruments.

Solutions to these problems fall into two categories: (1) DFT methods applied to the chip and MCM electrical/mechanical design [16,18], and (2) new or improved test instrumentation designed for MCM testing [19–37]. The DFT methods are a combination of approaches currently used for both chip-level design and system-level design. For instance, by incorporating boundary scan capability into the I/O pads of the integrated circuits and arranging for the scan paths to be accessible from the MCM primary I/O, each chip can be logically controlled and observed after assembly. Also, many of the techniques used for BIST at the chip level can be incorporated into an MCM design, especially if the chips are custom designed for the MCM application. However, BIST techniques should be applied to partitions of the MCM so that faults that occur can be isolated. This chapter describes the DFT and BIST options available for MCMs and, in particular, the special considerations required by MCMs.

In many cases today it is not possible or economical to include full-boundary-scan capability for all chips in an MCM. Often off-the-shelf components are used which might not have scan capability. Nevertheless it is still desirable to isolate faults within a populated MCM. This is especially true during the development phase when repetitive design errors must be distinguished from random manufacturing defects. Enhanced test instrumentation is required for probing the internal nets within MCMs. For example, in references 19, 22, and 24, the authors have described the use of a modified wafer-probe system together with a low-capacitance, high-bandwidth FET probed for tracing electrical signals within a high-performance MCM. This chapter describes other adaptations of commonly available test equipment for probing the small internal features of MCM circuitry.

Other methods that have been applied to chip-level fault isolation, such as electron-beam and optical-beam probing, can be adapted to MCM fault isolation. Each method has different advantages and limitations.

11.2.1 Scan-Design Methods

Scan design is a general class of structured DFT methods that use existing or added logic storage elements (flip flops) configured during testing as one or more shift registers. The principle advantage gained is a tremendous increase in controllability and observability.

Serial-scan methods have been used since the 1960s when they were first introduced in large mainframe computers. Use in mainframe computers was the principle application until the 1980s, when scan design gained wide acceptance as a chip-level DFT method [17]. Today, with MCMs approaching the complexity of prior systems, the application of scan design techniques to these subsystems is almost universally accepted.

The basic concept of scan design is illustrated in Fig. 11-7. During testing, data are scanned in (usually in a serial manner) to the storage elements to set the state of the system logic. One or more clock pulses are applied to allow the system to process the data and result in another state. This new state is observed by scanning out (serially shifting) the bits of the storage elements in the chain.

The effect is to give control and observation capability for each storage element without adding a great deal of test circuitry. Also, when all or most of the system is scannable the task of generating test patterns is greatly simplified. This is because the logic between the scannable elements is partitioned and can be thought of as purely combinational logic, which is amenable to automated test pattern generation (ATPG) strategies.

The partitioning effect is especially important for fault isolation. Because scan-based testing naturally divides up a large, complex circuit into many smaller logical circuits that are individually testable, we can quickly isolate faults to within one logic block.

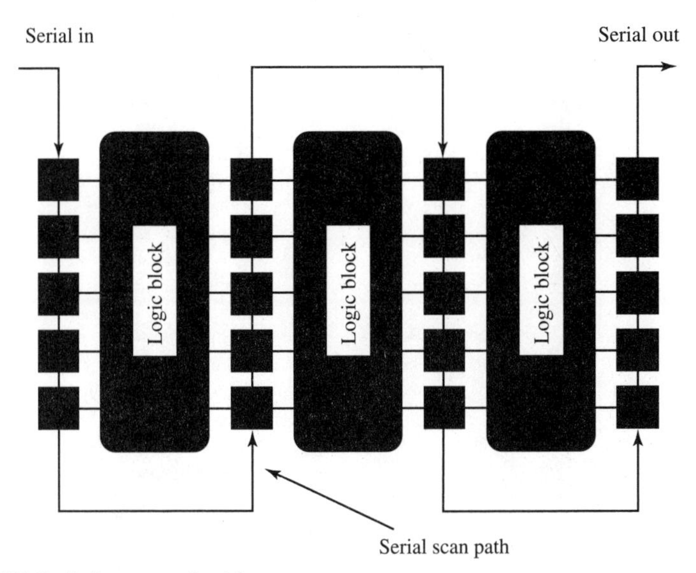

FIGURE 11-7 Basic concept of serial scan.

However, most ATPG software tools take advantage of the scan chain access to many system partitions. This supports parallel testing for improved efficiency. Therefore, the failure of one scan test could result from faults in any of several partitions. Although very efficient at reaching high fault coverage figures, ATPG patterns are not necessarily suitable for fault isolation.

What is also needed is a method for analyzing the test results to isolate the problem. This is accomplished through the use of a fault dictionary, which is generated as part of the ATPG process. The fault dictionary lists all the possible faults that could cause a particular test pattern to fail. Unfortunately, when parallel testing of multiple partitions is applied, there are usually many faults that could cause a pattern to fail. Therefore, further analysis must be conducted. One approach is to look at the overlap in the fault dictionaries for the failing patterns. However, a large degree of parallel testing often results in significant overlap in the fault dictionaires.

A compromise between the efficiency of parallel testing and the resolution needed for fault isolation is required to get around the overlap issue. Fortunately, scan-design methods provide a means of achieving both objectives.

The boundary-scan technique is a subset of the general class of scan-design methods. As depicted in Fig. 11-8, this method incorporates serial-scan register elements into the primary I/O of each integrated circuit. As in the general description of the serial-scan method, boundary scan provides an economic means of accessing each chip input and output (for improved controllability and observability).

When implemented completely throughout an electronic system, the boundary-scan approach partitions the entire system into separate chips and the interconnect network between them. Testing of each chip individually is possible (or multiple chips in parallel). Also, the interconnections between chips can be easily tested simply by setting the outputs of one chip to known states and then scanning out the data received by the other chips. If an interconnect line or chip-to-substrate bond is broken, then data will not be correctly transferred to the receiving chip. The error will show up as a stuck bit in the serial-scan chain and the fault can quickly be isolated to a single line. Similarly,

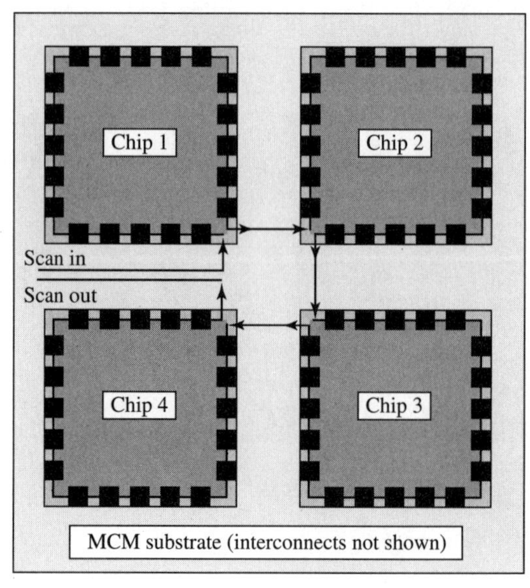

FIGURE 11-8 Boundary scan for MCMs.

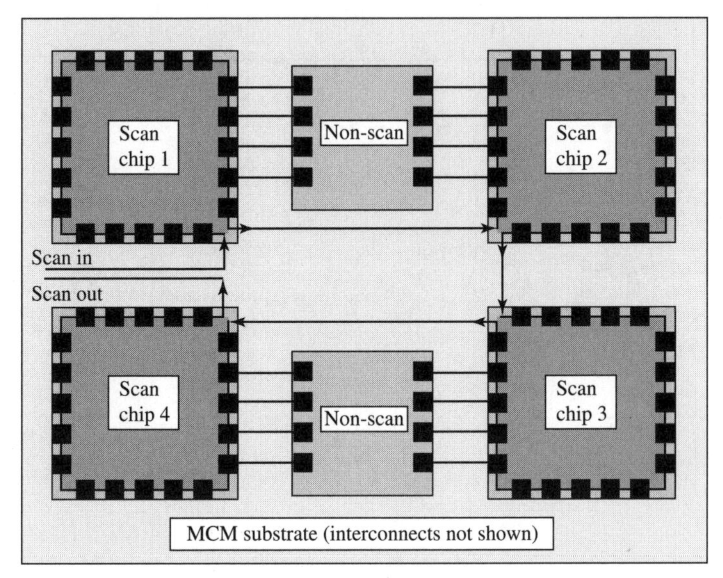

FIGURE 11-9 Boundary scan with nonscannable ICs.

adjacent-line shorts or shorts to power or ground are easily isolated using boundary-scan methods.

For these reasons boundary-scan techniques are widely accepted and a standard (IEEE 1149.1) has been defined. Clearly the approach can be applied to any system of integrated circuits which themselves incorporate boundary-scan registers. Although used at the printed circuit board level, boundary scan is even more essential within MCMs because of the lack of physical access resulting from reduced dimensions and spacing.

One of the limiting factors for applying boundary scan to a PCB or MCM system is that many common chips do not have boundary-scannable registers. Nevertheless, the fault isolation and testability of the circuit can be greatly enhanced if some of the chips are boundary scannable.

Figure 11-9 illustrates that some chips in a partial-scan based circuit can be directly tested using the boundary scan registers in logically adjacent chips. Others, which may be deeply embedded, may be grouped together as a testable partition using available boundary-scan registers on other chips and the module primary I/O.

11.2.2 Built-in Self-Test (BIST)

As described above, scan-based DFT methods can greatly improve the controllability and observability of MCMs. However, because test patterns are usually scanned-in serially, it often requires many millions of clock cycles to apply a sufficient number of tests to the module. In some cases, the number required greatly exceeds the amount that can economically be applied using standard automatic test equipment (ATE). Also, because MCMs are often used to push performance limits, application of at-speed tests can stress the clock rate limits of ATE.

One set of methods used to alleviate this situation uses circuitry built into the device itself to perform some or all of the test functions needed to verify the device operation. These methods are known as built-in self test (BIST).

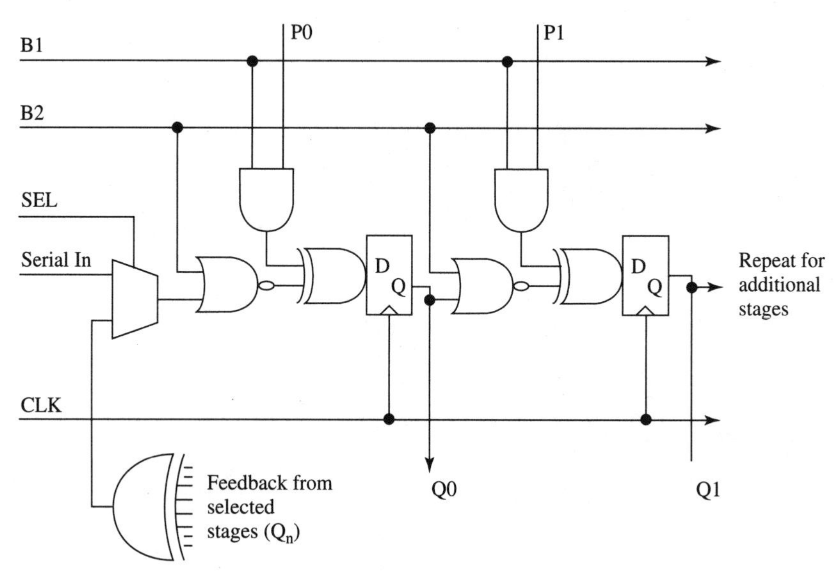

FIGURE 11-10 Built-in logic block observation (BILBO) register.

A common approach to BIST design uses a linear feedback shift register (LSFR) as a pseudo-random number generator (PRNG) to create an effective stimulus pattern within the device. The response (output) of the device is fed into another LFSR configured as a signature generator. After application of many clock cycles, the device is sequenced through enough states to ensure (to a high degree of confidence) that a fault is detected. Evidence of a fault shows up as a deviation from the expected signature, following a given number of clock cycles. The test method can be made autonomous with the addition of a clock counter, a simple sequencer, and a stored "expect" signature.

An efficient method for implementing this type of BIST uses a configurable element known as a built-in logic block observation (BILBO) [17,18] register as shown in Fig. 11-10. This structure can be used in place of functional storage elements (flip-flops and registers) that would otherwise be present (notably scan registers) because it is easily switched between a register mode, a scan mode, a PRNG mode, and a multiple-input signature-generator mode.

By incorporating BILBO-type structures within the ICs of an MCM, self-tests can be used that quickly give an indication of the status of internal circuitry within the chips. This, coupled with the interconnect tests made possible with boundary scan, can be used to isolate faults to a specific chip or interchip signal connection. A top-level schematic for this combined approach is shown in Fig. 11-11.

11.3 AUTOMATED TEST EQUIPMENT (ATE)

Functional and structural testing of digital logic circuits is accomplished using electronic test systems known as ATE. These systems are specialized for handling digital logic components and subsystems, yet are flexible enough to be reprogrammed for a wide variety of different logic functions. For more than 20 years, ATE systems have been computer controlled, permitting automated testing and providing the adaptability needed to cover

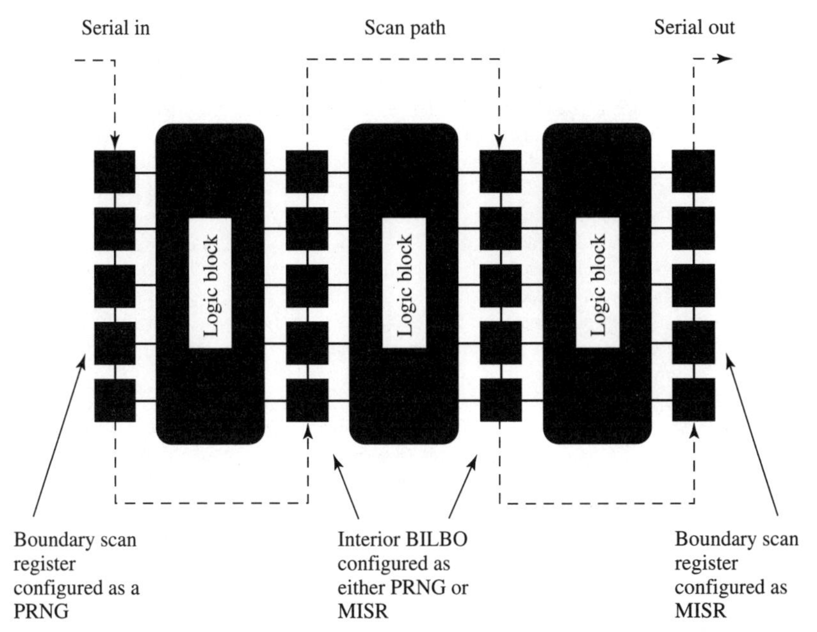

FIGURE 11-11 Self-test using BILBO registers and boundary scan.

the wide variety of logic devices in use. A top-level view of digital ATE architecture and functionality is provided in section 11.3.1 together with a brief overview of the development of ATE. Some detailed characteristics of representative test systems are given in section 11.3.2.

It is likely that an electronic component or subsystem will require testing at several stages during development and production. Many of these will be applied using ATE. This section provides an introduction to the capabilities and operation of this important class of thin film test instrumentation.

11.3.1 ATE Architectures and Capabilities

The fundamental requirements for an ATE system are illustrated in Fig. 11-12. Here a block diagram of an ATE together with the device-under-test (DUT) shows that the main

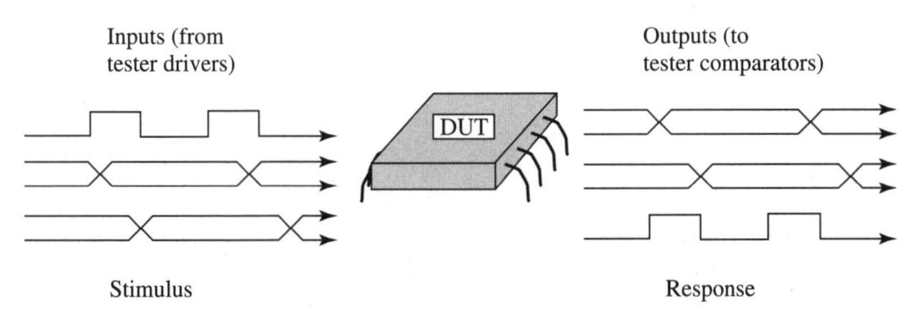

FIGURE 11-12 Simplified ATE functional diagram.

requirements are (1) to produce electrical stimuli (inputs) that cause the DUT to execute its intended functions, and (2) to observe or capture the electrical response (outputs) of the DUT. These electrical responses are compared either in real time with stored, expected response patterns, or are themselves stored for later comparison. An effort is usually made to match the electrical environment during testing with that expected to be encountered within the system application. There are two possibilities: attempt to match the transmission line or parasitic load conditions to those anticipated for the application system; or minimize or optimize the conditions to obtain the intrinsic performance of the DUT. If the second approach is used, then the intrinsic DUT performance is compared with simulation predictions. When the simulation models are validated in this way, they can be used to accurately predict the device performance under the different loading conditions that may be encountered in a system environment. Electrical interface issues that may effect the quality of the test are discussed in more detail in section 11.4. In this section, we concentrate on the ATE design and functionality.

As illustrated in Fig. 11-12, the ATE has two main functions: to produce electrical stimuli; and to capture and compare the DUT output response with expected patterns. At first glance these ATE requirements seem obvious and one would expect their implementation to be straightforward, leaving little room for variety. However, the detailed implementation of these functions results in test systems with a wide range of capabilities. Although digital testing is nothing more than the application and acquisition of simple patterns of ones and zeros, in fact the electrical signals which encode the ones and zeros are complex analog waveforms with voltage and timing characteristics that are not easy to reproduce or capture. How an ATE implements the encoding of data, voltage, and timing can greatly affect the performance and economics of the tester. Some considerations for ATE design (and hence for the selection of an ATE) are described in the remainder of this section.

Shared versus Per-Pin Resources. An ongoing debate centers on the trade-offs between test electronic resources which are either shared among channels or repeated on a per-pin basis, as shown in Fig. 11-13. Because modern ATE can easily accommodate 256 or more channels, the question of whether to repeat electronic hardware for each channel (per-pin)

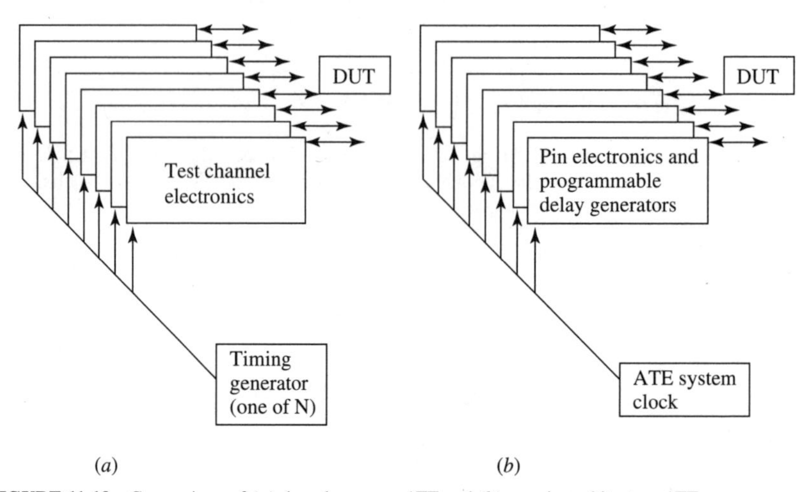

FIGURE 11-13 Comparison of *(a)* shared resource ATE and *(b)* per-pin architecture ATE.

or to share the electronics across multiple channels is an important one. Early shared-resource systems, for example, provided a limited set of timing generators (maybe 16) or reference voltages which were distributed to many or all of the channels. Each channel could then be programmed (using relays, for example) to utilize one of the available resources. After all, the number of different logic levels required by a given device is usually limited, as is the number of different timing delays required. This shared resource approach actually works well under many circumstances. It formed the paradigm for digital test systems in the 1970s and early 1980s.

However, by the mid-1980s this economical approach to ATE design began to exhibit performance limitations. As clock and data rates approached 100 MHz, timing accuracy became a critical requirement. With shared-resource architecture, the ability to distribute several timing signals with subnanosecond accuracy to hundreds of tester channels became an almost insurmountable obstacle for ATE manufacturers. The difficulty stems from the fact that distributed timing signals will usually introduce timing errors. Without an ability to correct for these errors on a per-pin basis, system-wide uncertainties of ± 1 ns were almost inevitable.

By providing a per-pin architecture test system, one or more independently programmable timing generators is incorporated into each channel. Therefore, if a timing error exists on a given channel, it can be corrected by reprogramming the delay for that channel. In addition, because dynamic calibration of logic transitions requires both accurate timing as well as accurate voltage levels, most manufacturers also include programmable pin "driver" and "comparator" levels for each channel. In some cases the process of reproducing tester resources is carried even further with the inclusion of test pattern sequencers on a per-pin basis as well.

Per-pin ATE architecture. When the per-pin architecture is used, most of the electronic resources needed for testing are repeated for each tester channel. Usually each channel can be programmed to act as a driver (for producing test stimulus), a comparator (for acquiring the DUT output response), or as a bidirectional channel (which may switch from driver to comparator depending on the state of the DUT). To understand the basic operation of the tester electronics, it is therefore helpful to discuss the channel function when it is acting as either a driver or as a comparator.

Figure 11.14 shows a simplified functional block diagram for a typical per-pin ATE channel when it is configured as a driver. The system clock is generally provided to all channels and can be used as a timing reference as well as a means for digitally measuring coarse delays. Fine delays (with resolutions of possibly tens of picoseconds) are then generated relative to the reference signal, using dedicated electronics contained within each channel. In the figure, two independent delay generators are shown (although the number available per channel may be more than two, depending on the manufacturer). In this example, the two generators produce a leading-edge (LE) timing marker and a trailing-edge marker. The timing signals from the delay generators are combined with digital pattern data (stored in vector memory) by the formatter. The resulting timed signal is then level-shifted or amplified by a programmable pin driver to produce the desired stimulus waveform. This signal is electrically connected to the DUT, usually with controlled-impedance transmission line structures such as coaxial cables or printed wiring-board striplines. Electrical matching of the transmission line impedance and the driver-output characteristics is crucial for accurate handling of high-speed signals.

Figure 11.15 shows a similar block diagram for the channel electronics when acting as either a receiver or comparator. In this example, an expected response is produced by combining timing signals with the expected data pattern. Here the two timing signals define a window in time where the data are expected to be valid. This information is

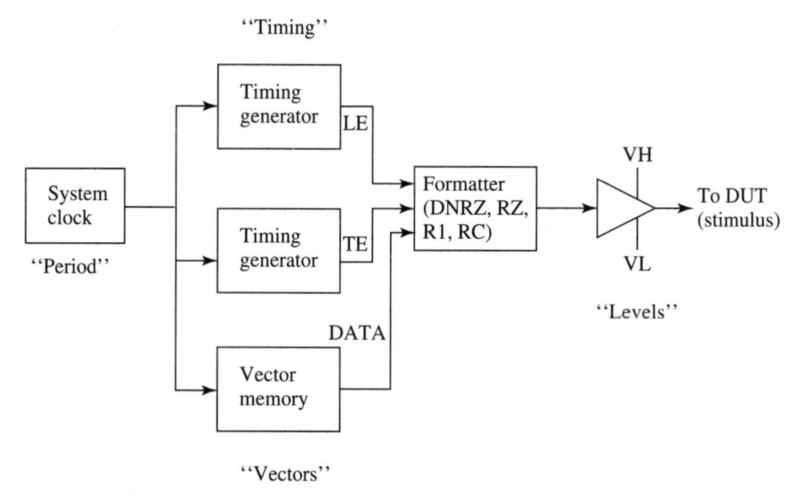

FIGURE 11-14 Simplified block diagram for ATE driver operation.

compared with the digitized response of the DUT output and discrepancies are recorded as errors.

In addition to the functional test capabilities described above, most ATE systems also provide dc parametric test capability. This is accomplished by incorporating relays within each channel that allow a parametric measurement unit (PMU) to be switched onto the signal line (while disconnecting the driver and comparator electronics). Usually the PMU is a central, shared resource, although most modern ATE

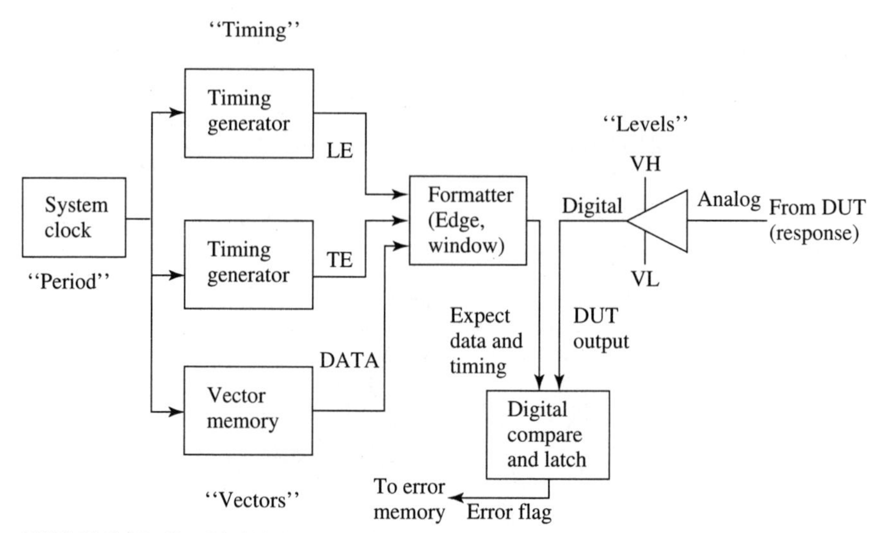

FIGURE 11-15 Simplified block diagram for ATE comparator operation.

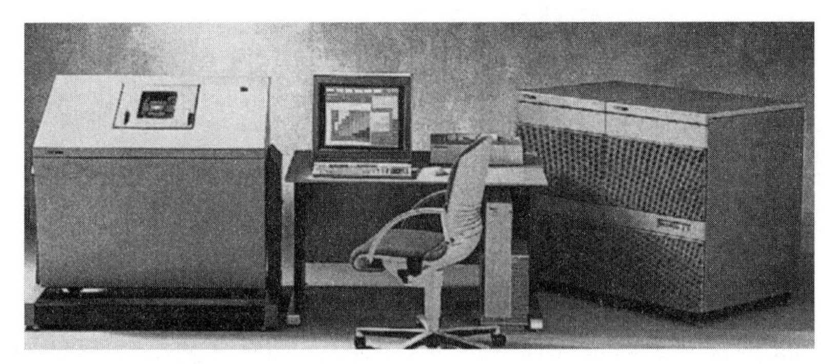

FIGURE 11-16 Photograph of the HP83000-F660 ATE. *(Courtesy of Hewlett-Packard Co.)*

provide multiple PMUs, and some even support a per-pin PMU. The dc measurements tend to require a relatively long time (on the order of milliseconds), so multiple PMUs can reduce the test time by measuring several channels in parallel.

11.3.3 Example ATE Systems

This section provides a brief description of the capabilities of two example test systems, namely the Hewlett-Packard models HP82000-D200 and HP83000-F660. The HP82000 system was introduced in the late 1980s and provided a substantial improvement in performance as a result of its per-pin architecture. With a clock and data rate of 200 MHz (later increased to 400 MHz) this system was about four times faster than previous ATEs (which generally were limited to about 50 MHz). Furthermore, the system timing accuracy of the HP82000-D200 is specified as ±250 ps, which is again a factor of four better than previous ATEs. The HP82000 continues to be widely used and is especially popular in engineering, where accuracy and ease of use are paramount considerations.

The HP83000-F660 introduced in the 1990s is capable of testing at rates as high as 660 MHz. The system uses liquid-cooled, MCM technology and GaAs pin drivers to support these high testing rates. The liquid cooling is especially valuable in maintaining the extremely precise timing accuracy specification of ±50 ps. A photograph of the HP83000-F660 system is shown in Fig. 11-16. The compact design provides the advantage of a small footprint (which is especially important in most production environments). The two systems are compared in Table 11.1.

TABLE 11-1 Comparison of example ATE Systems

	HP82000-D200	HP83000-F660
Maximum clock rate	200 MHz	660 MHz
Timing accuracy	±250 ps	±50 ps
Timing resolution	50 ps	10 ps
Pattern depth(s)	64 k, 256 k, 1 M	1 M, 4 M, 16 M
Maximum channels	512	1024
PMUs	2	One per 16 channels

11.4 ELECTRICAL TEST INTERFACE

11.4.1 Packaged Device Interfaces

The electrical interface between an ATE and a high-performance integrated circuit or subsystem can have a significant effect on the quality of the test, and even the final result. A high-quality interface may have an insignificant effect on the measurements, although a poor interface can result in good devices failing the test.

In the past, when digital clock rates were below 10 MHz or 20 MHz, and logic transitions occurred in tens or hundreds of nanoseconds, the electrical interconnections between the test system and the DUT could be treated as lumped elements. In this case, the main concerns were total (lumped) capacitive load, and in some cases, the total resistive load introduced by the wires of the interface. At low frequencies (typically below 20 MHz) the need for cotrolled-impedance transmission lines was limited to specialized device requirements and analog (RF) or mixed-signal subsystems which might have tight specifications on noise limitations. So, for most applications, the design of the electrical test fixture concentrated mostly on the mechanical requirements of supporting the test socket and providing for the proper number of signal wires. Little regard was generally given to minimizing or matching signal line lengths, controlling impedance, shielding adjacent lines, or minimizing the power supply interconnection impedance.

Today, however, even some of the simplest digital ICs and subsystems support clock rates well in excess of 20 MHz, and may have logic transitions as short as one nanosecond or less. At these rates, a logic transition may occur in a fraction of the time it takes the signal to propagate between the tester and the test device. This point alone generally requires that controlled-impedance transmission lines be used for the interconnections. This is also true because the test sockets must have a mechanism for aligning the device on insertion and for making a mechanically compliant, electrically sound connection to each pin of the device. Usually the contacts are soldered to the printed circuit board and are spring-loaded against the device pins with a clamping mechanism, providing the required pressure to all pins.

An example is shown in Fig. 11-17, which illustrates a 268-pin tape-automated-bond (TAB) device supported by a plastic carrier. In this case, the TAB structure provides a way of fanning-out the electrical contacts from an inner-lead spacing of about 8 mils to a spacing of 20 to 25 mils, which is suitable for contact within the test socket.

Figure 11-18 shows an example of the test socket mounted on an electrical test fixture. The socket sits in the center of an 11″ × 15″ multilayer printed circuit board. In this case a lever-actuated cam is used to supply a controlled amount of pressure to the device leads onto the test socket contact pins. Controlled-impedance microstrip lines are visible in the figure which connect the signals between the test socket and contact points around the board periphery. In this example, the test system contacts through spring-type "pogo" pins which are arranged in ground-signal-ground patterns to the back side of the printed circuit board (PCB).

A close-up view of the back side of the PCB is shown in Fig. 11.19. Here the bottom-side microstrip signal lines are clearly visible. In the center of the board are the test-socket pins. Also visible are surface-mounted decoupling capacitors (located near the socket pins), and surface-mounded chip resistors. The resistors are used in this application to match the device output impedance to the 50-Ω transmission lines.

11.4.2 Wafer-Probe Electrical Interfaces

When the DUT is in a wafer or die form, then electrical contact is more difficult and typically requires an additional level of interconnection and an alignment instrument such as

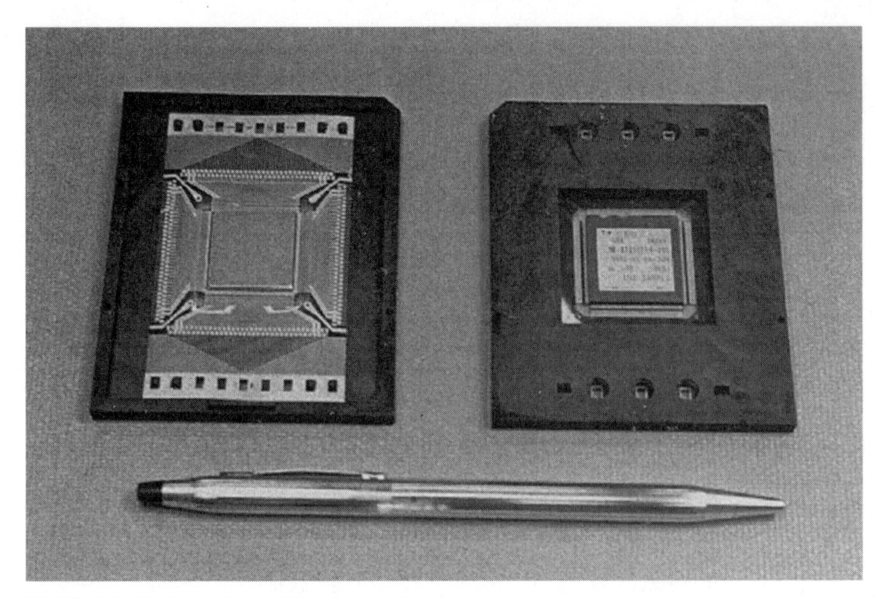

FIGURE 11-17 Example of a TAB-packaged device carrier.

FIGURE 11-18 Electrical interface fixture with test socket for TAB device.

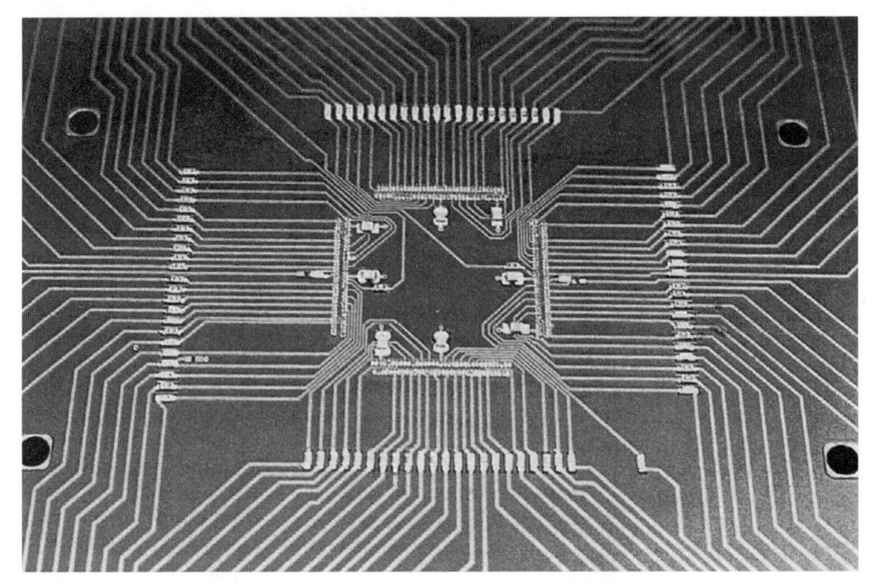

FIGURE 11-19 Backside view of microstrip traces on electrical test fixture.

the wafer prober shown in Fig. 11-20. This tool supports the wafer or die on an automated chuck that is moved in *X*, *Y*, *Z*, and theta (rotation). An optical microscope aids in the alignment of the die-probe pads to needle-type probes on a probe card. Figure 11-21 shows a top view of the circular-probe card and the bundles of coaxial cables used to connect signals to the test system.

11.5 TEST AND EVALUATION METHODS FOR MCMs

11.5.1 Introduction

Most of the difficulty with testing MCMs and their components arises from a lack of electrical access to the ICs at different stages of assembly. At the wafer level, for instance, devices must be nondestructively probed to ensure high speed operation over the required temperature range. These test requirements may also be applied at the bare-die level before commiting the chips to an MCM package. New methods are being developed for electrical burn-in of the bare-chip devices to ensure their reliability (eliminating infant mortality).

This chapter describes many of the techniques currently used for performance-testing MCMs. Also described are several test methods still under development which hold great promise for the future. The need for "known-good die" has driven the development of high-speed bare-die test methods by a number of workers, including this author [38–61]. Existing probe-card technology is adapted for wafer probe at 500 MHz clock rates [62–67]. Recently, one supplier introduced a 360-channel, 2.33 GHz probe card based on wire-over-groundplane geometry [64]. The author has demonstrated a bandwidth above 1 GHz [65] for wafer/die probing using ceramic-blade probes. The so-called "membrane" probe technology [68–70] promises even higher performance. Temporary chip carriers are

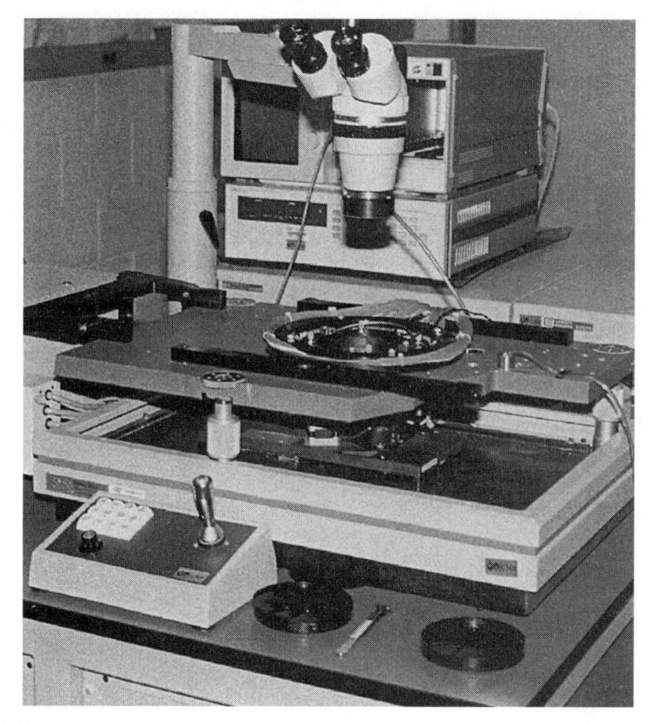

FIGURE 11-20 Automated wafer-probe system (Electroglas 2001×).

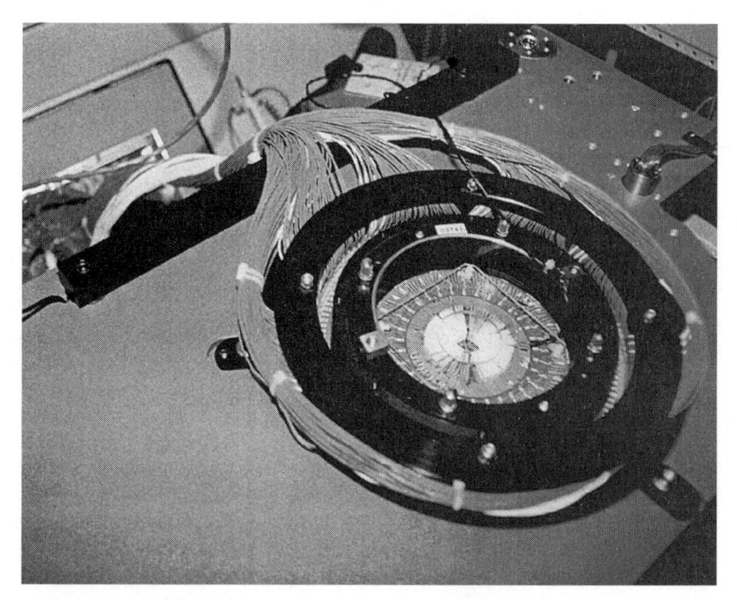

FIGURE 11-21 Electrical interface between wafer prober and ATE systems.

being developed for both performance testing and burn-in of individual chips [39–60,72]. Another technique for ensuring device quality involves rapid temperature cycling of the part while monitoring dc parametrics and ac performance [71]. One system is also under development for internal, noninvasive probing of fully populated MCMs [38,42,45].

When an MCM is assembled, a strategy is needed for isolating defects [73]. Strict use of boundary-scan and BIST techniques greatly simplifies matters [73–78]. However, noninvasive probe techniques such as voltage-contrast electron-beam [79–82] and optical probes [83,84] can be used to trace stuck-at faults, as well as to characterize performance degradations. Test instrument suppliers are beginning to incorporate guided-probe software tools, which have been used in the past for printed circuit board fault isolation [85].

11.5.2 Known Good Die

It is known [52] that the expected yield of an electrical system is based on the product of the probabilities that the individual components are defect-free. Actually this represents the best-case scenario, assuming that the assembly process itself does not introduce faults. Thus the probability of producing a defect-free MCM rapidly approaches zero as the number of ICs increases, unless each chip is known to be good in advance of assembly. At a printed-circuit-board level the requirement is not so stringent because defective chips can often be located and replaced. However, most of the MCM packaging methods presently used do not provide economic methods for rework. Therefore it is widely recognized [38–61,67,71–74] that techniques are required for ensuring "known-good die" for MCMs.

To achieve acceptable MCM yields, the ICs must be good to a level of 99% or better for a typical MCM [52,60,61]. To obtain such levels requires an assessment of both the inherent die yield as well as the actual fault coverage of tests that are applied to the chips before insertion into the MCM. If the die yields are 100%, then no testing is required. Conversely, if die yields are 0% no amount of testing will help the situation. However, because best-case yields are only in the 90% range, a high fault-coverage test must be used to identify the defective chips before MCM assembly, and establish the required assurance of "known-good die."

The same concerns have been shared by developers of high performance printed circuit boards (PCBs). However, ensuring "known-good devices" for PCBs provides the luxury of being able to test the parts separately in standard sockets. Furthermore, a burn-in process can be applied to individually packaged chips to prevent early device failure.

One approach for ensuring "known-good die" for MCMs can be adapted from techniques used to test chips at the wafer level [38–42]. With minor modifications to existing wafer-probe equipment [38,42,45], at-speed tests can be used to measure bare-chip performance over temperature. This approach has the effect of eliminating defective ICs, assuming that a complete (100%) coverage test is used. Except for the need for burn-in, such an approach is adequate and provides the ability to produce prototype MCMs.

Following MCM assembly, the entire module can be burned-in to ensure reliability. However, the undesirable consequence is that MCMs that fail during burn-in must be reworked or discarded.

To accommodate both the need to apply at-speed tests over temperature as well as a burn-in cycle, many groups are developing miniature chip carriers [38–61]. The various approaches provide some method of temporarily connecting electrical signals to the individual chips. The chips can then be treated as packaged devices for high-speed testing and high-temperature burn-in. The difficulty is in developing a low-cost means of making a temporary connection to the chip bond pad that does not damage the pad when removed.

Typical pad dimensions are between 50 and 100 μm; with comparable spaces between pads. These relatively small dimensions present a challenge to test-fixture developers who rarely see packaged devices with lead spacings below 500 μm.

Fortunately MCM technology itself provides the means to make fine-pitch interconnect structures with typical dimensions of 12 to 75 μm. A further difficulty involves the need for a removable contact. The approaches fall into three categories: (1) severable miniature leads, (2) compliant contacts, and (3) removable thin films.

The clearest example of severable miniature leads is provided by TAB. Each chip is mounted onto a dielectric tape film with patterned conductor traces. These traces can be arranged so that the electrical signals fan out from the die-bond pads to more widely spaced test pads located near the edges of the tape. In some configurations the TAB carrier resembles a miniature lead frame, as illustrated in Fig. 11-22.

The TAB film itself is mounted within a larger carrier that can be fitted into a standard socket. Alternatively, the test pads on the tape can be probed using a variety of mechanical probe techniques. After testing, the chip is cut from the tape carrier, leaving the much shorter leads needed for the next assembly step (within the MCM). The TAB approach is effective for both packaging and testing. However, it requires significant tooling and processing costs. It is best used in high-volume applications where the start-up costs can be amortized over a large quantity of devices.

A recent development [56] of a temporary interconnect system for flip-chip devices uses a sacrificial metal film on the solder bumps which is later removed, leaving the bumps undamaged. This method is also in the category of severable, miniature interconnects. However, it is limited to use with bumped die.

The lure of a miniature test socket which would connect directly to the IC has prompted recent work on compliant miniature contacts. One approach [53,67] is adapted from membrane-probe technology [68–70] and is illustrated in Fig. 11-23. Here a film (usually polyimide) is patterned with one or more layers of conductor (usually Cu) to form a fan-out arrangement similar to the TAB tape described above. The contacts are plated to form a raised or bumped area that mates to the chip-bond pad. An array of contacts is formed on the polyimide film that corresponds to the pad arrangement on the chip. A mechanical

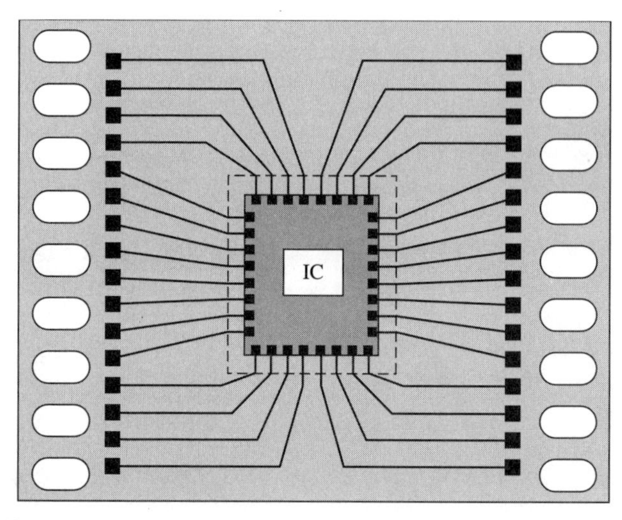

FIGURE 11-22 Bare die test access using TAB.

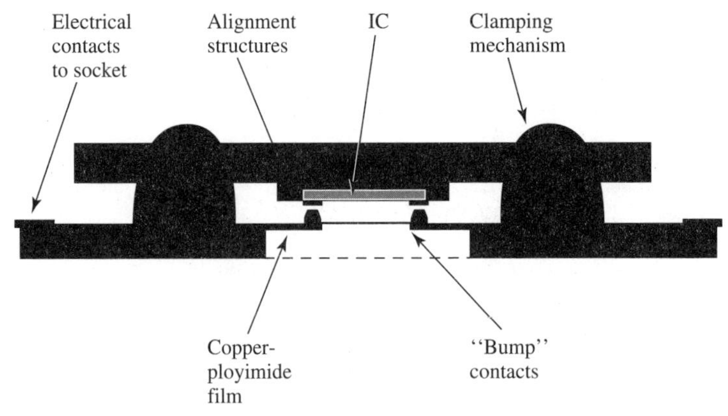

FIGURE 11-23 Membrane-based chip carrier.

fixture or clamping system is used to press the contacts and chip together. Compliancy is provided by the flexibility of the polyimide film.

Another example of a compliant miniature interconnect approach is given in [72] and illustrated in Fig. 11-24. Here an array of Au-plated nickel *z*-axis contacts are arranged throughout a polyimide film with a pitch of as little as 15 μm. The film can be customized to conform to the chip bond pad arrangement and provides a means of temporarily contacting the small pads. The signals are then routed to larger contacts on a customized substrate. The entire assembly is mounted into a test socket as if it were a packaged part.

The third method used for providing a temporary carrier for bare chips uses a removable Kapton film [44], laminated over the surface of a chip or array of chips as shown in Fig. 11-25. Here vias in the film are ablated with a laser to expose the underlying bond pads on the chip. Additional interconnections are formed by sputtering Ti-Cu-Ti and photolithographic patterning. As described in [44] the patterned thin film conductors can connect the underlying chips to standard probe-pad arrangements or form temporary links to

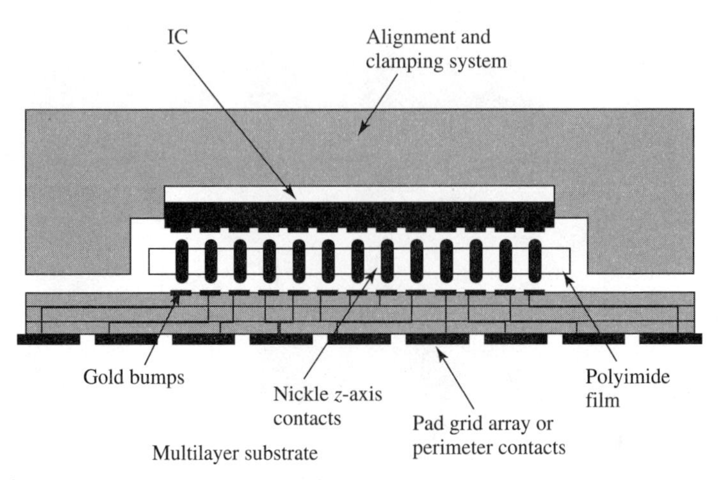

FIGURE 11-24 Chip carrier using compliant, Z-axis contacts.

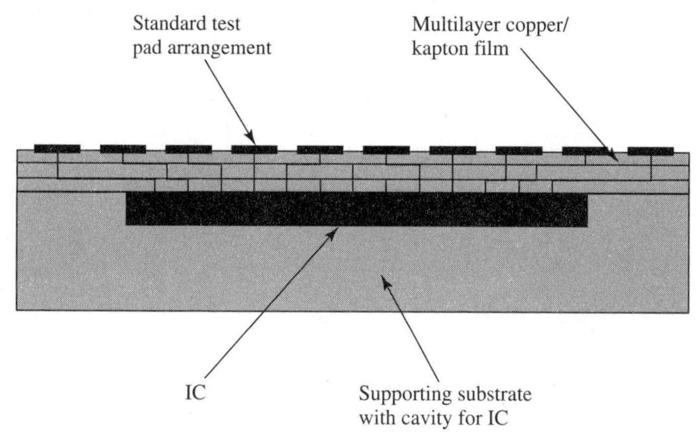

FIGURE 11-25 Kapton film overlay for temporary chip interconnect.

the chip-carrier substrate. Depending on the method chosen, the Kapton film can either be removed after testing or additional layers can be added to interconnect the chips within the MCM. One of the unique advantages to this particular approach is that the overlay films can be removed without damage to the underlying chips.

11.5.3 High-speed Wafer/Die Test

Tremendous advances have been made in the ability to test high-performance devices at the wafer and die levels [62–70]. Until recently, wafer probing was generally applied at low to moderate clock rates (below 20 MHz). In the past this was adequate because many devices did not operate at higher rates and because the parts could be retested (at faster rates if necessary) in packaged form. However, this approach is not adequate for higher-speed devices (50 MHz and higher). Also, as previously discussed, unless a miniature chip carrier is used, performance screening must be performed at the wafer or chip level before insertion into the MCM.

The conventional (needle-type) probe card has several performance-limiting drawbacks. Among these are (1) uncontrolled impedances or impedance discontinuities in the signal lines, and (2) relatively large inductances (3 to 10 nH) in the power and ground connections. These limitations can be minimized by careful design of the probe card. In [62,63] an evolutionary approach is taken whereby performance is improved by shortening the probe needle lengths and providing a controlled-impedance, multilayer PCB for the probe card. In addition,. signal-ground pairs of probes are used to minimize cross talk and to maintain a transmission line structure to the tips of the probe. With this arrangement, testing digital GAs logic devices with 500 MHz clock rates is possible.

Another modification [64] uses a wire-over-groundplane configuration for the probe needles to achieve a bandwidth of over 2.3 GHz on a 360-pin probe card. However, reducing the power and ground supplies' inductance is still a concern for high-current switching devices.

The use of controlled-impedance probe cards and ceramic-blade probes for subnanosecond logic transitions has been described [65]. Furthermore, this arrangement permits the use of surface-mount decoupling capacitors on the power-supply probes to greatly reduce switching noise. This approach is limited in pin count by the thickness of the ceramic blades.

A detailed comparison of several modifications to standard probes and ceramic-blade probes is given in [66]. Here the use of extended ground planes, microstrip ceramic-blade probes, and a secondary ground reference structure above the probe needles, extends the bandwidth to 7.8 GHz with reduced inductance on power connections.

An approach that promises to extend probe performance capabilities even further makes use of a flexible membrane to form the connections to the chip [70–76]. This technique shares many similarities with the MCM packaging methods. Interconnect layers are formed by alternate films of polyimide and patterned metal (usually Cu with barrier metals). Raised areas (bumps) are plated at the contact points and the whole film is pressed in contact with the chip-bond pads. The flexible polyimide film provides the necessary compliance, and signal line impedance can be controlled all the way to the device-bond pad. Ground inductance is minimized through the use of dedicated ground planes and power planes (in multilayer structures).

11.5.4 Bare Die Burn-In

Chip carriers are ensuring known-good die provide not only a way of performing high-speed tests but also provide a means of applying electrical signals to the device during burn-in. Traditionally, packaged devices are subjected to elevated temperatures while under bias (static burn-in) or whiling being functionally activated (dynamic burn-in). The goal is to eliminate those devices that fail within the first few hours of operation. Then the mean time to failure (MTBF) of the remaining population is greatly increased. This method of ensuring the reliability of electronic devices has been widely used for military and medical applications. Variations including system-level burn-in are now widely used, even for commercial applications. Clearly, a comparable method is needed for MCMs and, in particular, for testing the bare die before assembly. When miniature chip carriers are used, the bare chip can be treated as a packaged device and subjected to the same burn-in methods used in the past.

An interesting alternative was recently developed [71] in which integrated circuits are subjected to rapid (<1 min) thermal cycling while electrically monitoring ac performance and dc parametrics. By comparing the measured performance with predicted values, potentially defective devices can be spotted and discarded. This process has far-reaching implications for all levels of device integration. However, when applied at the wafer or bare die level, it has the potential to eliminate the need for a miniature chip carrier. One system combines the membrane-probe technique for high performance with a bare-die alignment mechanism to test complex devices at speeds of 600 to 1000 MHz [67]. Adding a <1 min burn-in capability [71] to this arrangement provides known-good, reliable (burned-in) die for MCMs.

11.5.5 In-Circuit Probe and BIST

Testing assembled MCMs is the subject of the remainder of this chapter. To some extent, the completed MCM can be treated as a very complex component, or alternatively as a very dense subsystem. As such, the methods traditionally used for component or system-level testing can be applied to MCMs with appropriate accommodations made for the typically higher pin count, higher complexity, and higher density.

However, testing MCMs differs from testing of traditional components or systems in the area of fault isolation. The difficulty again arises from the limited accessibility of the internal components and logic elements. Where the test engineer would use an oscilloscope probe in isolating a fault on a PCB, such tools are only now becoming available for probing the interior of an MCM [38,42,45,85].

In this respect, the MCM must be treated more as a component than as a system. Therefore, the techniques used to isolate faults within a component can be applied [73]. These start with DFT techniques built into the device to provide for improved controllability and observability, and potentially BIST [74–78].

Boundary scan [73,76–78] is a standardized technique, (IEEE std. 1149.1) which incorporates a serial shift register into the device I/O pins (see Fig. 11-26). Under test conditions, this shift register takes control of the data presented to the internal logic and effectively isolates an imbedded device from its surroundings. By applying data to this boundary-scan register a test can be run on a particular device without regard to the functions of other devices in the system. Also, if all devices in the system have boundary scan registers, then data can be set up in one device and transferred through the system interconnections to another device boundary-scan register. This allows the interchip connections to be easily checked.

It is clear that incorporating boundary scan on all MCM components would greatly simplify the task of isolating a fault from a defective chip or chip-to-chip interconnection. Although a growing number of devices support boundary scan, the majority of devices do not. Other fault isolation methods are needed for nonscannable MCMs, and they described below.

Although boundary scan provides controllability and observability for the chip inputs and outputs, it does not reduce the number of tests needed to verify the device internal logic. In fact, to achieve a high level of fault coverage (>99%) may require many millions of clock cycles and a prohibitive volume of test data. To get around this problem, one solution uses a linear feedback shift register (LFSR) to generate a pseudorandom test pattern, which is applied to the device internal logic. The output of the logic is fed serially into another LFSR which is used to compress the results into a short signature. At the end of a predetermined number of clock cycles, the residue in the LFSR signature register is compared with an expected value. To a high degree of certainty, a correct signature indicates that the circuit is operational. To provide a BIST capability, both the pattern generator and signature registers are incorporated into the chip design. If the design already contains scan

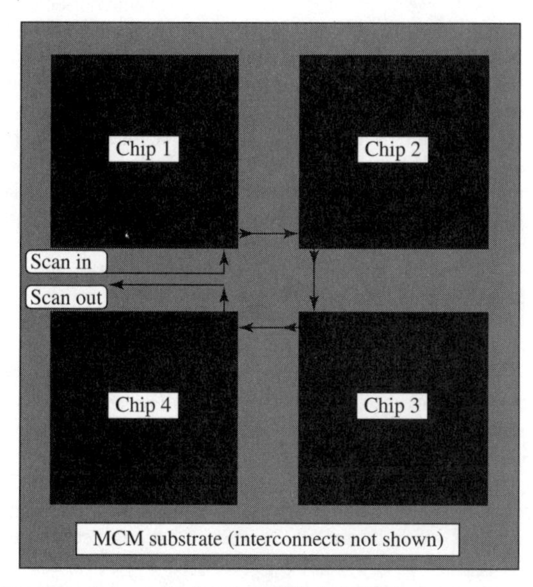

FIGURE 11-26 Using boundary scan for isolating MCM faults [73].

registers, these can be converted to either pattern generators or signature registers by the addition of only a few gates. In fact, one structure known as a built-in logic block observation (BILBO) register can be configured as either a scan register, a simple register, a PRNG, or as a multiple-input scan register (MISR) [74,75].

Even the carefully planned DFT and BIST circuitry may be insufficient during critical testing of the first MCM. When this happens internal probing can be of great value in isolating the fault. The voltage-contrast electron-beam [79–82] and optical-beam probing methods [83,84] provide ways of measuring internal signals without loading the circuit or risking physical damage.

The electron-beam probe has been applied to the measurement of internal logic signals within integrated circuits [79–82]. It is reasonable that this same tool could be applied to fault isolation within MCMs [82]. The basic principle of operation is the same as is used in the familiar scanning electron microscope. Relatively high-energy electrons are focused to a small spot on the device surface (submicron in most cases). The electrons interact with materials on the surface and slightly below the surface. Many types of emissions result from these interactions. However, the most abundant are the secondary electrons with energies between 0 and 50 eV. These carry information about the topology, composition, and electric potential at the surface in the form of variations in the energy and intensity distribution. The topology effects give rise to the familiar scanning electron micrographs.

The variation of the electric potential at the surface also shifts the secondary electron energy distribution and gives rise to the voltage-contrast effect [79–81]. When the electron beam is scanned across the device surface, areas at a positive potential (logic high) appear dark. Lighter areas are at more negative levels (logic low). In this mode, the static-logic state of the device is clearly evident. If the electron beam is pulsed synchronously with the device operation (at a multiple of the clock rate) then a stroboscopic effect results and a picture of the dynamic-logic state can be obtained.

If instead of scanning the surface, the beam is positioned at one small area, then the pulse timing can be adjusted and a sampling oscilloscope or logic analyzer representation can be obtained. In this way the electron-beam probe can be used as a noninvasive (no load) internal probe for ICs or MCMs.

Following the idea of the electron beam as a noninvasive electrical probe, the use of light in a similar fashion shows great promise, especially for extremely high-speed devices [83,84]. In this method, a pulsed beam of light is first linearly polarized, then passed through a quarter-wave plate to produce circular-polarized light. The beam is then focused through an electro-optic crystal in close proximity (a few microns) to the electrical test point. After reflecting from the crystal surface the light is elliptically polarized to a degree which depends on the voltage at the test point. After splitting the beam, the light is passed through crossed polarizers to produce a differential signal pair, which is converted to an electrical signal and amplified. The great advantage that this method has is the potential to use extremely short (femto-second) light pulses to accurately measure signals in the high gigahertz range. A drawback is the need to mechanically position the electro-optic probe very close to the measurement point.

11.6 ELECTRICAL FAULT ISOLATION AND FAILURE ANALYSIS

11.6.1 Electromechanical Probing

As described above, DFT methods are usually the most effective way to ensure that faults can be isolated within a partially functional MCM. However, for a variety of reasons it is not always practical to include boundary scan and BIST into all the chips on an MCM. For

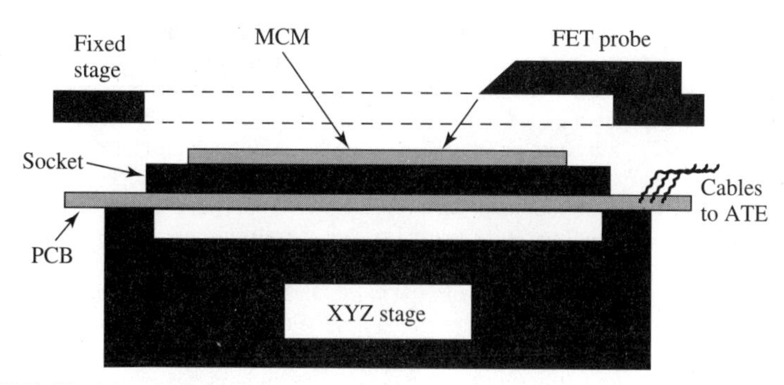

FIGURE 11-27 Adaptation of wafer prober for internal-node probing.

instance, one advantage that MCM packaging has over monolithic wafer-scale technology is the ability to use existing, off-the-shelf components, many of which do not support compatible DFT methods. Therefore additional methods are needed to enhance the fault-isolation capability of nonscan-based MCMs or large imbedded sections of nonscannable logic.

One straightforward approach uses miniature mechanical probes to gain physical access to nodes within an MCM. The mechanical probes are electrically connected to sampling instrumentation such as an oscilloscope, logic analyzer, or ATE. To reduce the added capacitance of the probe, an active device can be used as a buffer. Such FET probes are available with subpicofarad capacitance and several hundred megahertz to gigahertz bandwidth. Figure 11-27 depicts how an existing wafer-probe system can be adapted to position an MCM under such a mechanical probe [86–88]. In this case, the wafer probe system is used as a convenient XYZ positioning system.

Graphical interface software is needed to aid the operator in positioning the FET probe within the MCM. Such an interface is shown in Fig. 11-28. Here the operator can select

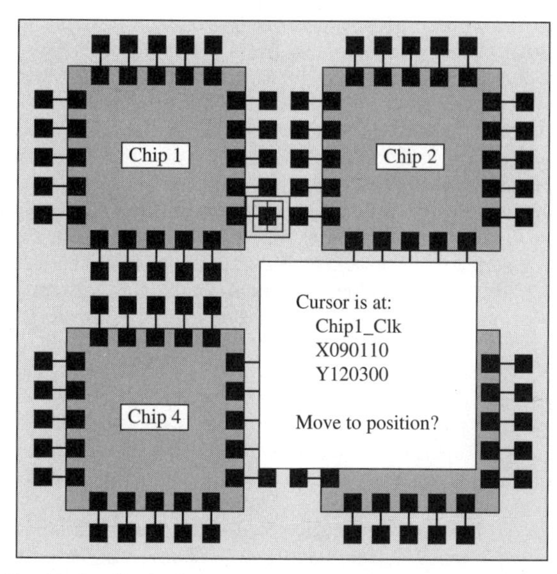

FIGURE 11-28 Graphical navigation.

the physical location for probing by pointing the cursor at the desired location on a drawing of the MCM footprint. The likely probe points are device and substrate wire-bond pads and test pads on the substrate. When the probe point is selected, its coordinates are used to move the MCM. Care must be taken not to collide the probe with three-dimensional structures within the MCM (such as bond wires).

Additional software can greatly improve efficiency in mechanical probing to isolate faults, including direct online links to the CAD database for cross referencing the logic and layout information, and support with guided-probe algorithms. These provide a miniaturized version of the tools used for PCB fault isolation and are available from at least one supplier [89].

11.6.2 Electron-Beam Probing

Introduction. As circuit complexity and density is increased by the use of MCM packaging techniques, tools are required for isolating electrical faults and characterizing signal propagation within the active module. Some work has described the use of BIST and electro-mechanical probes for accomplishing these objectives [90–92]. However, both techniques have limitations, as described in the references. Specifically, BIST requires substantial up-front investment in the chip design and is somewhat, although not entirely, limited to fault isolation and go/no-go testing versus signal delay characterization). On the other hand, mechanical needle probes can be used to extract internal signals within MCMs, but can result in damage and capacitive loading to the probed locations. Mechanical probes are also awkward to position, especially when three-dimensional structures such as wire bonds are present.

To address the need for a noninvasive, high-performance, easily positioned electrical probe for MCMs, we describe the adaptation of a scanning electron microscope (SEM) operated in the voltage-contrast (VC) mode. The VC technique has been successfully used for fault isolation and signal characterization within integrated circuits [92–103]. A similar technique has even been applied to the testing of bare MCM substrates [104]. However, very few applications to active MCMs have been described.

The VC technique uses a beam of relatively low-energy primary electrons (about 1 keV) which are focused to a spot on the surface of the electronic device (IC or MCM). The primary electrons are somewhat insensitive to the few-volt-potential differences found across the surface of most digital logic circuits. However, the secondary electrons emitted from the circuit surface as a result of collisions by the primary electrons have energies between a few volts and about 50 eV. As a result, the secondary electrons are very much affected by the circuit voltage variations. By exploiting this effect, and in some cases enhancing it through the use of extraction and retarding fields, the surface voltage can be deduced. Because the electron beam can be essentially nonloading and is easily positioned by magnetic scan coils, it forms an ideal electrical probe. Furthermore, the use of a beam chopper permits very short (100 to 200 ps) pulses of primary electrons to be synchronized with the clocking of the circuit being tested. The result is comparable to a nonloading sampling oscilloscope with several gigahertz bandwidth.

It should be noted that substantial effort must be directed at providing known-good die (KGD) for use within MCMs. Several techniques are available or under development [105–139]. Even if KGDs are used in an MCM, it is usually required that the complete module be characterized to demonstrate compliance to the specification or to provide feedback to design/simulation. Because multichip packaging is often chosen specifically to gain some performance advantage, the designs often push the limits of the chips in terms of signal propagation delays, setup, and hold times. In these situations, chip-to-chip interconnect delays can be significant to the overall module performance. Furthermore, the use of KGDs may be a necessary but not necessarily sufficient condition for ensuring the performance and functionality of the completed module. Interconnect delays, impedance discontinuities, cross talk, switching noise, and thermal effects are all module-level effects which can limit or destroy the circuit performance, even when all chips are known to be good. Assembly-induced faults can cause module failures even with KGDs in place. Therefore methods are needed to diagnose and isolate such faults. In the early stages of product development, when the design has not been fully validated, it is imperative that assembly-related faults be distinguished from repetitive design errors.

Some of the best techniques used to aid in the characterization and diagnosis of MCMs are the use of boundary scanning and BIST [140–145]. Although the MCM designer often does not have complete control of the individual chip designs, complex ICs with boundary-scan capability are becoming more readily available. Of course full-scan capability within an MCM is a tremendous aid in isolating faults, especially those related to chip-to-chip interconnects. However, even if all chips in the MCM cannot support boundary scan, the use of a partial scan is a step in the right direction. The use of BIST within a particular chip, together with some way of accessing it at the module level (such as through boundary scan), provides a way of checking the chip functionality after MCM assembly.

Even with full boundary scan and BIST implemented throughout an MCM, it may still be necessary at times to electrically probe internal signals within the module. This may be for the purpose of extracting a critical signal to check timing or to measure the dynamic noise characteristics. One approach is to mechanically probe the signals using a fine-tipped needle probe and a high-impedance (usually FET) buffer/amplifier. In some cases it may also be helpful to automate the probe positioning [105,112,140]. With such automation, probe points are selected from the graphic database which represents the MCM layout, or from the electrical schematics (in CAD form). This technique is limited to probing exposed metal lines or pads. It is also a very nonforgiving approach in that damage to the pads or wire bonds is a real possibility.

To reduce, if not eliminate, the mechanical difficulties associated with needle probing, the noncontact method of electron-beam probing can be used [105–119], given the appropriate instrumentation. Electron-beam probing of integrated circuits, using the VC effect, has been fairly common for the past 10 to 15 years. More recently, electron-beam probing of ICs has made the transition from a laboratory research tool to routine engineering application by the automation of signal acquisitions and direct links with CAD systems.

The author has utilized an ISI IC-130 SEM equipped with an applied beam technology (ABT) IL-200 voltage-contrast waveform acquisition system. One of the unique features of this system is that the vacuum chamber is more than 15 inches on a side, and is therefore capable of supporting very large substrates. The large chamber was originally designed to support wafer probing. It was later adapted to support MCMs and PCBs.

Principle of operation. A schematic block diagram of a voltage-contrast SEM is illustrated in Fig. 11-29. The source of electrons is at the top of the SEM column. Typically a tungsten filament is used as a cathode to emit electrons in the "gun" portion of the SEM. However, to achieve brighter (higher-current) beams at low accelerating voltage, LaB_6 or other field emission sources can be used in place of tungsten. However, the field-emission sources require a much better vacuum system than does tungsten.

The electrons are accelerated to an energy of about 1.0 keV by one or more anodes held at the desired potential. In some modes of operation, a beam blanker or chopper is used to pulse the otherwise steady beam of electrons. The pulsed beam can be synchronized to

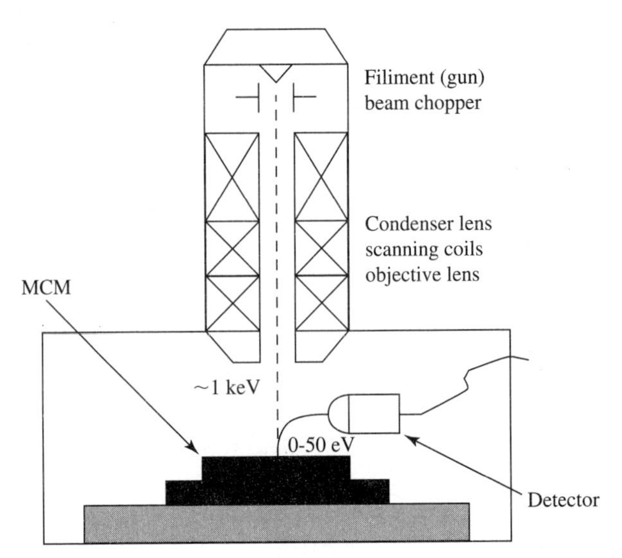

FIGURE 11-29 Schematic block diagram of the voltage-contrast electron-beam system.

the test-circuit clock to produce a stroboscopic effect (described below). The beam blanker is made up of a pair of electrodes through which the beam passes, and a small aperture. A voltage pulse on the electrodes deflects the beam across the aperture so that only a small fraction (pulse) of the beam is passed down the column. The combination of a voltage pulse and a small aperture permits the generation of beam pulses as short as 100 to 200 ps. These are used in a stroboscopic sampling approach to obtain an effective bandwidth of several gigahertz.

The electron-beam column supports one or more magnetic condesing lens. These are used to confine the beam as it travels down the column. Scan coils are used to deflect the beam across the sample (MCM in this case). They can be repetitively indexed to produce a video frame rate, or they can be set at a desired static condition so that the beam is directed to a single spot on the sample surface (spot mode). Typically in SEMs the effective size of the spot, which determines the lateral resolution, is on the order of tens of nanometers. This is how the SEM is able to image such minute structures as are found within today's integrated circuits. For VC testing, extreme lateral resolution is rarely required because it is only necessary to image structures used to carry logic signals. These are about a micron in size within ICs. For MCMs, the dimensions are even larger (tens of microns). Therefore compromises are generally made in VC systems. Image quality (spot size) is often sacrificed so that lower accelerating voltages or higher-beam currents can be used. In MCM testing, where spot size can be an order of magnitude larger than in IC testing, this trade-off can be further extended to speed data acquisition time.

The final control of the beam focus is provided by the objective lens located just above the MCM surface. The quality of the objective lens is one of the primary limits on the overall resolution. To electronically compensate for irregularities in the magnetic imaging system, astigmatism-compensation is usually provided by a set of fine-adjustment coils called stigmators.

After the primary electron beam (at about 1.0 keV) strikes the sample surface, several interactions occur. As a result of these interactions, electrons and radiation of various energies and wavelengths are emitted in all directions. The study of these electron-material interactions and the associated emissions is extensive. To a large degree these effects are fairly well understood. They form the basis for the familiar SEM topological images as well as for other more specialized measurement techniques (including voltage contrast).

The majority of emitted electrons have energies in the range of 0 to 50 eV. These are called secondary electrons. Standard SEM topological images are formed by collecting these electrons with a scintillator and photomultiplier to form a video-type display of signal intensity versus lateral position. The secondary electron signal strength is affected by several variables. These include surface topology, material variations, and variations in surface electric potential. Standard imaging is primarily based on the surface topology effects (with some material sensitivity usually evident). However, for VC testing we concentrate on resolving differences in the secondary electron signal strength, which are a result of electric potential differences across the sample surface. To this end, topological information is actually undesirable and is often suppressed if possible.

The 1.0 keV primary electron beam is only slightly affected by surface voltage variations, which are on the order of a few volts in logic devices. However, the secondary electrons, with energies in the 0 to 50 eV range, are very sensitive to such potential differences. Specifically, the secondary electron energy spectrum is shifted in proportion to the surface potential. By measuring this shift, we can deduce the potential at different locations on the sample surface.

Unfortunately, the secondary electron signal strength itself is not a linear function of surface electric potential. To obtain a linearized signal, an energy analyzer is placed between the sample and the scintillator/PMT. The analyzer is usually a grid or metal mesh held at a controlled potential. By varying this retarding grid potential, the secondary electron signal strength can be adjusted. In qualitative modes, the retarding grid potential is maintained at a fixed value and acts as a filter (only electrons above a certain level can pass through to the scintillator). In quantitative modes, the retarding grid is used to compensate for potential changes on the surface. In this mode the retarding grid potential is adjusted with a closed-loop electronic circuit so as to maintain a constant secondary signal strength. The change in voltage required is then a linear function of the potential change on the surface.

Another potential grid is often used very near to the sample surface (within a few millimeters). This is called the extraction grid. It is used to increase the secondary electron signal strength by attracting the low-energy secondaries and slightly accelerating them as they leave the surface.

Example Modes of Operation. Using different combinations of spot and scan modes, continuous/chopped beam, and qualitative/quantitative modes, a number of useful techniques can be realized. Fig. 11-30 shows a typical static VC image of the surface of an IC. Here the beam is scanned across the sample surface while holding the IC at a particular logic state. A continuous beam is used and the effects are observed qualitatively. The retarding-grid and extraction-grid potentials are fixed so that only electrons from surfaces at about 0 V are transmitted to the scintillator. These form the light regions in the photomicrograph. The areas of the sample at a high logic level (5 V) have secondary electron energies shifted down by about 5 V. These have lower energy and are therefore not able to pass across the retarding-grid potential barrier. As a result, these areas appear dark in the image. Logic zeros and ones are clearly discernable in this image.

In Fig. 11-31 the sample clock is synchronized with the video-scan rate to create a stroboscopic effect. In this image, the beam is still continuous. However, because the device is synchronized with the video scan, each point on the screen corresponds to a particular phase of the clock. In this example a data bus with a number of lines is arranged to run from the video screen top to bottom. Because the video scans from top to bottom, the

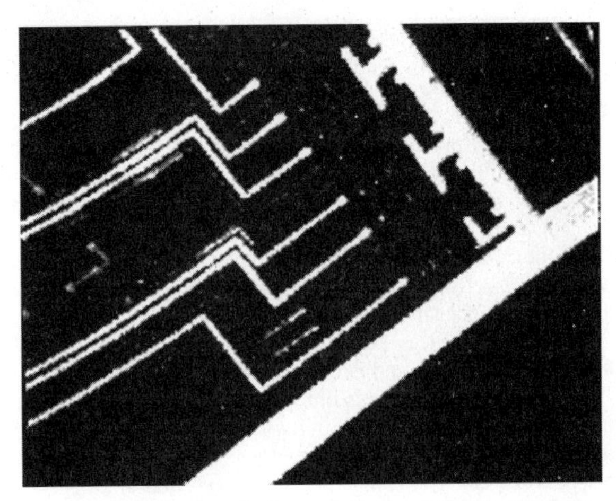

FIGURE 11-30 Continuous, scan, and qualitative mode image of IC static-logic states.

vertical position also corresponds to advancing phase (or time) relative to the clock. Therefore the dynamic states of the signals on the lines are visualized in a logic-stage diagram. In this example, the bus data are counting in binary so that each signal is either two times or half the rate of its neighbor.

In Fig. 11-32 the IC signal is clocked at a rate far exceeding that of the video scan. However, close inspection of the image reveals a stripe pattern on the signal line. This is a result of a stroboscopic-beat rate between the video-scan rate and the IC-clock rate. In this example, the IC surface has the common dielectric passivation coating that normally suppresses static voltage contrast signals. This is why most of the image is a neutral grey

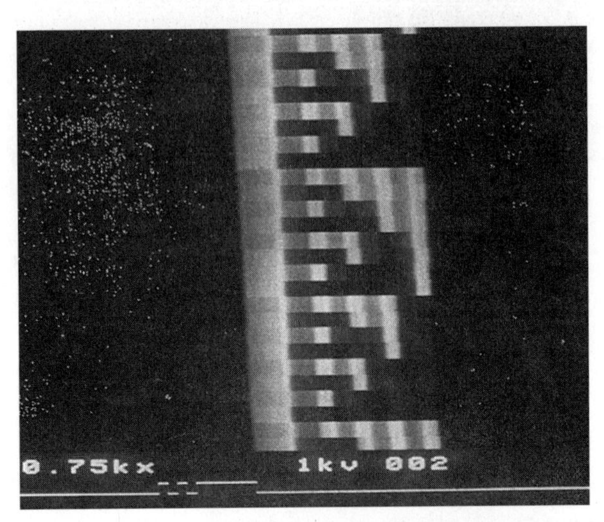

FIGURE 11-31 Continuous, scan, and qualitative mode image of IC dynamic states synchronized with the video scan.

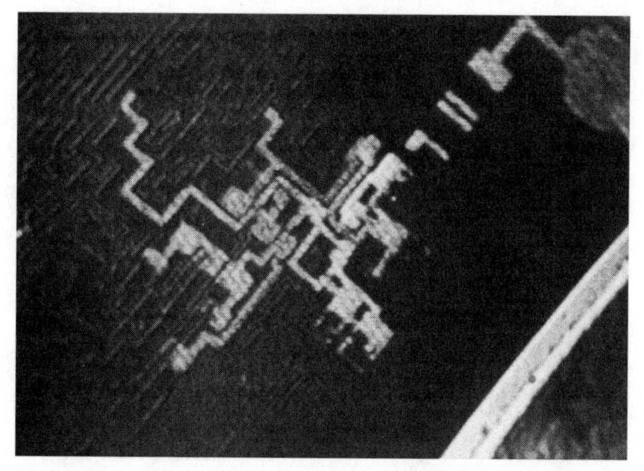

FIGURE 11-32 Dynamic imaging of a high-frequency signal capacitively coupled through the passivation layer.

shade (not logic low or high). However, dynamic signals can capacitively couple to the dielectric surface and become visible with VC. In this example, only one signal is clocking and it is easily traced throughout the circuit.

If the beam is positioned at a single location within the circuit (spot mode) then a voltage versus time plot can be generated, as shown in Fig. 11-33. In addition to using the spot mode, the beam is chopped with the beam blanker. The sample pulse is synchronized with the IC-clock rate so that the voltage can be quantitatively measured at a particular phase (relative to the IC clock). The sampling-pulse phase is systematically advanced so that the

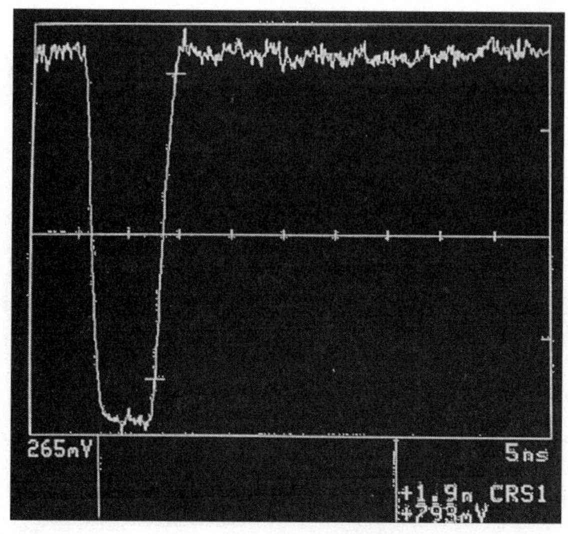

FIGURE 11-33 Quantitative, chopped, and spot-mode measurement of a 1.9 ns risetime signal with 200 ps resolution.

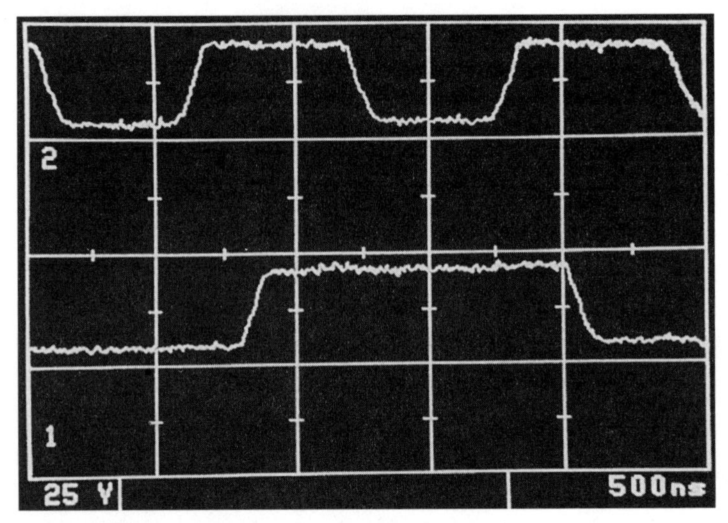

FIGURE 11-34 Example of multiple waveform and qualitative comparison.

IC voltage can be measured as a function of time. In this example, the chopper produced a 200 ps beam pulse and was delayed in 50 ps steps to produce a waveform illustrated on a 5 ns per division scale. The example illustrates measurement of the IC signal risetime of 1.9 ns (10% to 90%).

A similar technique is used to produce the waveform plots shown in Fig. 11-34. Here two different signals were measured (from two different locations). This example clearly illustrates how internal signals can be compared. Parameters such as propagation delays, set up times, and hold times can be accurately measured. Furthermore, as shown in Fig. 11-34, rise and fall times, pulse widths, duty cycles, overshoot, undershoot, and coupling noise are measurable on a subnanosecond scale. Under ideal conditions, the voltage resolution can be as fine as about 1 mV. However, in most situations a few tens of millivolts resolution is all that is required.

Implications for MCM probing. The examples shown illustrate the tremendous accuracy and flexibility that VC electron-beam testing provides. The techniques demonstrated on micron-scale dimensions of ICs are easily applied to structures within MCMs such as bond pads, exposed signal traces, or even buried signal traces. In fact, application to MCMs has some advantages. These are associated primarily with the larger dimensions involved. As a result, a larger spot size can be used (sacrificing extreme high resolution) to produce a higher current beam and decrease the time required for signal acquisition. Furthermore, because the beam is spread over a wider surface area, undesirable effects of charging, sample vibration, and topology are minimized.

The main additional requirement imposed by MCMs is the need for a large chamber with a wide-range stage. To meet this challenge, the author's system is adapted from an SEM designed for probing 6-in wafers. The inner-chamber dimensions exceed 15 inches. Better than 2-μm stage positioning repeatability is achieved using optical encoders and computer-driven motors. A photograph of the inner chamber and stage is shown in Fig. 11-35.

An electron micrograph of a probed IC is shown in Fig. 11-36. This example illustrates the scale difference between IC internal probing and MCM electron-beam testing. The chip-bond pads are 100 μm on a side and are clearly visible on this scale. However, the

FIGURE 11-35 Photograph of the large-chamber SEM interior and stage.

internal (micron-scale) structures are not discernable. In MCM testing we are interested in extracting signals from structures the size of the bond pads illustrated.

The use of an electron beam to provide a safe, high-performance method of electrically probing logic signals within MCMs has been described. The basic VC technique can be varied to determine static voltage levels, dynamic logic states, detailed voltage-time

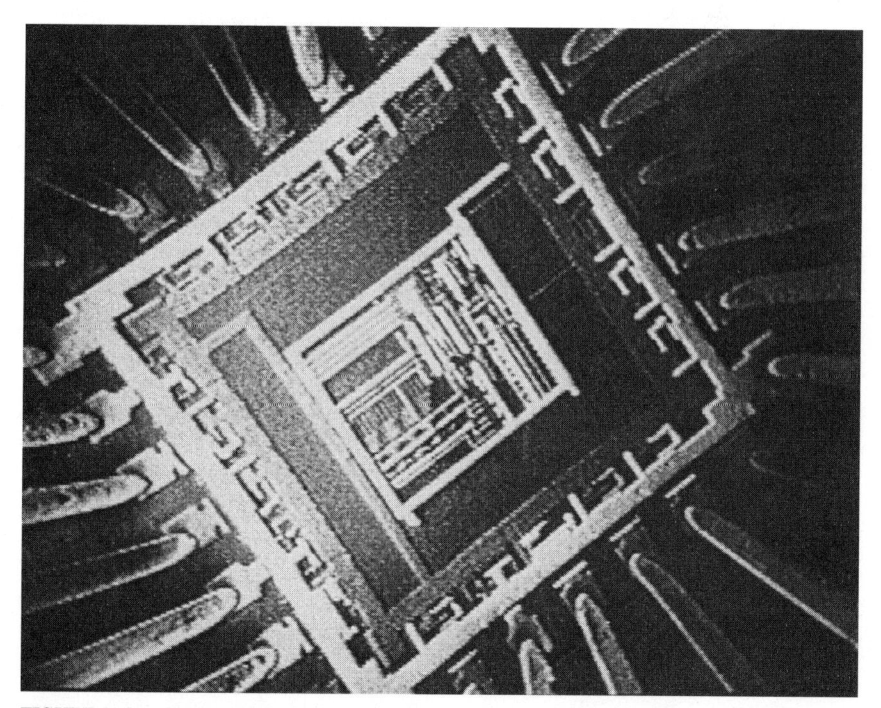

FIGURE 11-36 Photomicrograph of a probed IC.

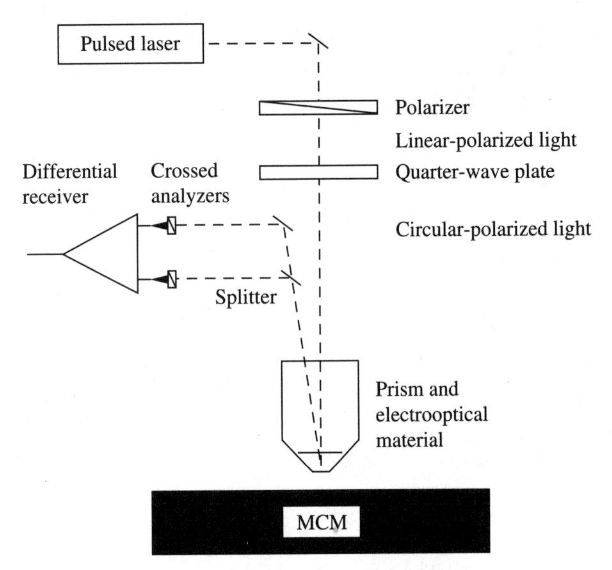

FIGURE 11-37 Electro-optical probe showing principle of operation.

waveforms, and comparison of good and faulty behavior (dynamic fault imaging). These techniques can prove invaluable in characterizing the performance of an MCM or in diagnosing faulty behavior.

11.6.3 Optical-Beam Probing

Another noninvasive probing technique that has been applied to integrated circuits and may have application to MCMs is optical-beam probing [146,147]. This technique is based on the electro-optical effect and the probe is a beam of light passed through an electro-optical material (usually a crystal) placed in close proximity to the probe point. As the voltage on the device surface changes, the polarization of the light is rotated by a corresponding degree. The light is reflected from the probe tip, back through the electro-optical material, through an analyzer (orthogonal polarizers) to a detector. This arrangement is illustrated in Fig. 11-37.

The electro-optic effect has promise because of the extremely short (femtosecond) light pulses which are possible. The general sampling and stroboscopic techniques are otherwise similar to the VC method described above. However, this technique is not nearly as well developed as the electron-beam method. It also requires mechanical positioning of the probe in close proximity to the device surface.

REFERENCES

1. T. W. Williams and K. P. Parker, "Design for Testability—A Survey," *Proc of the IEEE,* vol. 71, no. 1, January 1983, pp. 98–112.
2. B. Davis, *The Economics of Automatic Testing,* McGraw-Hill, U.K., 1982, p. 12.

3. J. M. Cortner, *Digital Test Engineering,* John Wiley & Sons, Inc., New York, 1987.
4. M. Abramovici, M. A. Beuer, and A. D. Friedman, *Digital Systems Testing and Testable Design,* AT&T Bell Laboratories and W. H. Freeman and Co., New York, 1990.
5. C. Timoc et al., "Logical Models of Physical Failures," *Proc of the Intl Test Conf (ITC),* Oct. 1983, pp. 546–553.
6. A. Goundan, "Fault Equivalence in Logic Networks," *Ph.D. Dissertation,* Univ. of Southern Calif., Mar. 1978.
7. E. J. McCluskey and F. W. Clegg, "Fault Equivalence in Combinational Logic Networks," *IEEE Trans on Computers,* vol. C-20, no. 11, Nov. 1971, pp. 1286–1293.
8. M. A. Breuer, (ed.), *Diagnosis and Reliable Design of Digital Systems,* Computer Science Press, Rockville, MD, 1976.
9. H. Y. Chang, E. G. Manning, and G. Metze, *Fault Diagnosis of Digital Systems,* John Wiley & Sons, New York, 1970.
10. A. D. Friedman and P. R. Menon, *Fault Detection in Digital Circuits,* Prentice-Hall, Englewood Cliffs, NJ, 1971.
11. A. K. Susskind, "Diagnostic for Logic Networks," *IEEE Spectrum,* vol. 10, Oct. 1973, pp. 40–47.
12. T. W. Williams and N. C. Brown, "Defect Level as a Function of Fault Coverage," *IEEE Trans on Computers,* vol. C-30(12), Dec. 1981, pp. 987.
13. E. S. Park, M. R. Mercer, and T. W. Williams, "Statistical Delay Fault Coverage and Defect Level for Delay Faults," *Proc of the International Test Conf (ITC),* 1988, pp. 492–499.
14. D. J. Klinger, Y. Nakada, and M. A. Mendendez, *AT&T Reliability Manual,* Van Nostrand Reinhold, New York, 1990.
15. S. Arrhenius, *Z Physik Chemie,* vol. 4, 1889, pp. 226.
16. D. Crook, "Method of Determining Reliability Screens for Time-Dependent Dielectric Breakdown," *Proc of the Intl Reliability Physics Symposium, IEEE,* 1979, p. 1.
17. T. W. Williams and K. P. Parker, "Design for Testability—A Survey," *Proc of the IEEE,* vol. 71, no. 1, January 1983, pp. 98–112.
18. B. Koenemann, J. Mucha, and G. Zwiehoff, "Built-in Logic Block Observation Techniques," *Digest of the 1979 Test Conf,* October 1979, pp. 37–41.
19. D. C. Keezer, "Bare Die Testing and MCM Probing Techniques" *Proc of the IEEE Multichip Module Conf,* March 1992, pp. 20–23.
20. G. A. Forman, J. A. Nieznanski, and J. Rose, "Die for MCMs: IC Preparation for Testing, Analysis, and Assembly," *Proceedings of the IEEE Multichip Module Conf,* March 1992, pp. 32–35.
21. Meeting Minutes from the MCM Test Workshop and the STAR Packaging Workshop Test Session, October 1991 and March 1992.
22. D. C. Keezer, "Assessment of High Performance ICs Prior to MCM Insertion," *Government Microelectronics Applications Conf (GOMAC),* Nov. 1992, pp. 341–344.
23. L. E. Roszel and W. Daum, "MCM Prototyping Using Overlay Interconnect Process," *Proc of the IEEE Multichip Module Conf,* March 1992, pp. 36–39.
24. D. C. Keezer, "MCM Testing Using Available Technology," *Proc of the 1992 Intl Test Conf (ITC'92),* Sept. 20–24, 1992, p. 253.
25. M. Taylor and W. W. M. Dai, "Tiny MCM," *Proc of the IEEE Multichip Module Conf,* March 1991, pp. 143–147.
26. R. W. Bassett, P. S. Gillis, and J. J. Shushereba, "Testing and Diagnosis of High-Density CMOS Multichip Modules," *Proc of the IEEE Multichip Module Conf,* March 1991, pp. 108–113.
27. J. K. Hagge and R. J. Wagner, in Messner, Turlik, Balde, Garron (eds.), *Thin Film Multichip Modules,* Intl. Society for Hybrid Microelectronics (ISHM) Publ., Chapter 13.2, 1992, pp. 501–544.
28. S. P. Athan, D. C. Keezer, and J. McKinley, "High-Frequency Wafer Probing and Power Supply Resonance Effects," *Proceedings of the Intl Test Conf (ITC'91),* Oct. 1991, pp. 1069–1077.

29. M. Beiley et al., "Array Probe Card," *Proc of the IEEE Multichip Module Conf,* March 1992, pp. 28–31.
30. T. Tada et al., "A Fine-Pitch Probe Technology for VLSI Wafer Testing," *Proc of the Intl Test Conf (ITC'90),* Sept. 1990, pp. 900–906.
31. B. Leslie and F. Matta, "Membrane Probe Card Technology (The Future of High-Performance Wafer Test)" *Proc of the Intl Test Conf (ITC'88),* Sept. 1988, pp. 601–607.
32. *"A New HP 82000 System for Testing MCMs to 200 MHz,"* Hewlett Packard, *Pathways,* Spring 1993, pp. 3–8.
33. E. Menzel and E. Kubalek, "Fundamentals of Electron Beam Testing of Integrated Circuits," *SCANNING,* vol. 5, 1983, pp. 103–122.
34. E. Menzel and E. Kubalek "Secondary Electron Detection Systems for Quantitative Voltage Measurement," *SCANNING,* vol. 5, 1983, pp. 151–171.
35. A Gopinath and C. C. Sangen "A Technique for Linearization of Voltage Contrast in the Scanning Electron Microscope," *J Phys E4,* 1971, pp. 334–336.
36. F. J. Henley and H. J. Choi, "Test-Head Design using Electro-Optic Receivers and GaAs Pin Electronics for a Gigahertz Production Test System," *Proc of the Intl Test Conf (ITC'88),* 1988, pp. 700–709.
37. M. Shinagawa and T. Nagatsuma, "An Automated Optical On-Wafer Probing System for Ultra-High-Speed ICs," *Proc of the Intl Test Conf (ITC'92),* 1992, pp. 834–839.
38. D. C. Keezer, "Bare Die Testing and MCM Probing Techniques," *Proc of the IEEE Multichip Module Conf,* March 1992, pp. 20–23.
39. R. H. Parker, "Bare Die Test," *Proceedings of the IEEE Multichip Module Conf,* March 1992, pp. 24–27.
40. G. A. Forman, J. A. Nieznanski, and J. Rose, "Die for MCMs: IC Preparation for Testing, Analysis, and Assembly," *Proceedings of the IEEE Multichip Module Conf,* March 1992, pp. 32–35.
41. Meeting Minutes from the MCM Test Workshop and the STAR Packaging Workshop Test Session, October 1991 and March 1992.
42. D. C. Keezer, "Assessment of High-Performance ICs Prior to MCM Insertion," *Government Microelectronics Applications Conf (GOMAC),* Nov. 1992, pp. 341–344.
43. R. A. Fillion, R. J. Wojnarowski, and W. Daum, "Bare Chip Test Techniques for Multichip Modules," *ECTC Conf Proceedings,* 1990, p. 554.
44. L. E. Roszel, W. Daum, "MCM Prototyping Using Overlay Interconnect Process," *Proc of the IEEE Multichip Module Conf,* March 1992, pp. 36–39.
45. D. C. Keezer, "MCM Testing Using Available Technology," *Proc of the 1992 Intl Test Conf (ITC'92),* Sept. 1992, p. 253.
46. R. A. Barrett, "Testing Tasks and Protocols for University MCMs," *Proceedings of the IEEE Multichip Module Conf,* March 1991, pp. 164–171.
47. M. Taylor and W. W. M. Dai, "Tiny MCM," *Proc of the IEEE Multichip Module Conf,* March 1991, pp. 143–147.
48. R. W. Bassett, P. S. Gillis and J. J. Shushereba, "Testing and Diagnosis of High-Density CMOS Multichip Modules," *Proc of the IEEE Multichip Module Conf,* March 1991, pp. 108–113.
49. R. Lang, "Cost-Effectiveness of nCHIP's MCM Technology," *Proc of the IEEE Multichip Module Conf,* March 1991, pp. 16–23.
50. K. P. Shambrook, "Overview of Multichip Module Technologies," *Proc of the IEEE Multichip Module Conf,* March 1991, pp. 1–9.
51. J. K. Hagge and R. J. Wagner, et al., (eds.), *Thin Film Multichip Modules, Messner,* Intl. Society of Hybrid Microelectronics (ISHM), 1992, Chap. 13.2, pp. 501–544.
52. J. K. Hagge and R. J. Wagner, "High Yield Assembly of Multichip Modules Through Known-Good IC's and Effective Test Strategies," *Proc of the IEEE,* December 1992, pp. 1965–1994.

53. J. Godfrey, "Interfacing/Fixturing for the HP82000/HP83000," *Proc of the Intl HP82000 Users Group Meeting,* May 1993, pp. 268–276.

54. T. Costlow, "Motorola Shipping Bare Die," *Electronic Engineering Times,* June 21, 1993, p. 54.

55. G. Derman, "MCMs Make Right Connections," *Electronic Engineering Times,* June 28, 1993, p. 54.

56. S. Crum, "MCNC Develops Flip Chip Test Method," *Electronic Packaging and Production,* May 1993, pp. 11–12.

57. J. Tuck, "On Board the MCM-L Train" *Circuits, Assembly,* March 1993, pp. 20–21.

58. Z. Sekulic, "Testing Multichip Modules," *IMS-Test Solutions,* November 1992, pp. 4–5.

59. R. K. Scannell and J. K. Hagge, "Development of a Multichip Module DSP," *IEEE Computer,* April 1993, pp. 13–21.

60. W. Daum, W. E. Burdick, and R. A. Fillion, "Overlay High-Density Interconnect: A Chips-First Multichip Module Technology," *IEEE Computer,* April 1993, pp. 23–29.

61. C. M. Habiger and R. M. Lea, "Hybrid-WSI: A Massively Parallel Computing Technology?" *IEEE Computer,* April 1993, pp. 50–61.

62. J. Huppenthal, "500 MHz Functional Test System Development," *Proc of the Rocky Mountain Test Conf,* Colorado Springs, Oct. 1989, pp. 35–44.

63. J. Peoples, "Probe Card Design for High Speed Digital Testing," *Proc of the Rocky Mountain Test Conf,* Denver, May 1991, pp. 16–20.

64. R. Nelson and M. Bonham, "High-Density/High-Speed Wafer Probing: A 360 Pin-2.33 GHz Probe Card for Production," *Proc of the Intl HP82000 Users Group Meeting,* San Francisco, June 1992, pp. 188–213.

65. S. P. Athan, D. C. Keezer and J. McKinley, "High-Frequency Wafer Probing and Power Supply Resonance Effects," *Proc of the Intl Test Conference (ITC'91),* Oct. 1991, pp. 1069–1077.

66. P. Warwick and P. Giard, "Enhancing Standard Probe Technology for UHF Probing," *Proc of the Intl HP82000 Users Group Meeting,* May 1993, pp. 32–51.

67. A. Barber, K. Lee, and H. Obermaier, "A Bare Chip Probe for High I/O, High-Speed Testing," *Proc of the Intl HP82000 Users Group Meeting,* May 1993, pp. 19–31.

68. M. Beiley et al., "Array Probe Card," *Proc of the IEEE Multichip Module Conf,* March 1992, pp. 28–31.

69. T. Tada et al., "A Fine-Pitch Probe Technology for VLSI Wafer Testing," *Proc of the Intl Test Conf (ITC'90),* Sept. 1990, pp. 900–906.

70. B. Leslie and F. Matta, "Membrane Probe Card Technology (The Future of High Performance Wafer Test)" *Proc of the Intl Test Conf (ITC'88),* Sept. 1988, pp. 601–607.

71. T. E. Figal, "Below A Minute Burn-In," *U.S. Patent 5,030,905,* July 9, 1991.

72. T. Costlow, "Two Team in MCM Test," *Electronic Engineering Times,* July 19, 1993, p. 64.

73. D. C. Keezer, "Fault Isolation Methods for Multichip Modules," to appear in *Proc of the Intl Symposium on Testing and Failure Analysis (ISTFA'93),* Nov. 1993.

74. T. W. Williams and K. P. Parker, "Design for Testability—A Survey," *Proc of the IEEE,* vol. 71, no. 1, January 1983, pp. 98–112.

75. B. Koenemann, J. Mucha, and G. Zwiehoff, "Built-in Logic Block Observation Techniques," *Digest of the 1979 Test Conf,* October 1979, pp. 37–41.

76. R. Vemuri et al., "An Integrated Multicomponent Synthesis Environment for MCMs," *IEEE Computer,* April 1993, pp. 62–74.

77. J. Novellino, "ITC'91 Focuses on Testability Issues," *Electronics Design,* October 24, 1991, pp. 38–45.

78. K. E. Posse, "A Design-for-testability Architecture for Multichip Modules," *Proc of the Intl Test Conf (ITC'91),* October 1991, pp. 113–120.

79. E. Menzel and E. Kubalek, "Fundamentals of Electron Beam Testing of Integrated Circuits," *SCANNING,* vol. 5, 1983, pp. 103–122.
80. E. Menzel and E. Kubalek, "Secondary Electron Detection Systems for Quantitative Voltage Measurements," *SCANNING,* vol. 5, 1983, pp. 151–171.
81. A. Gopinath and C. C. Sangen, "A Technique for Linearization of Voltage Contrast in the Scanning Electron Microscope," *J Phys E4,* 1971, pp. 334–336.
82. R. Woolnough, "E-Beam Checks LCDs, MCMs," *Electronic Engineering Times,* August 9, 1993, p. 31.
83. F. J. Henley and H. J. Choi, "Test Head Design using Electro-Optic Receivers and GaAs Pin Electronics for a Gigahertz Production Test System," *Proc of the Intl Test Conf (ITC'88),* 1988, pp. 700–709.
84. M. Shinagawa and T. Nagatsuma, "An Automated Optical On-Wafer Probing System for Ultra-High-Speed ICs," *Proc of the Intl Test Conf (ITC'92),* 1992, pp. 834–839.
85. "A New HP 82000 System for Testing MCMs to 200 MHz," Hewlett Packard, *Pathways,* Spring 1993, pp. 3–8.
86. D. C. Keezer, "Bare Die Testing and MCM Probing Techniques," *Proc of the IEEE Multichip Module Conf,* March 1992, pp. 20–23.
87. D. C. Keezer, "Assessment of High Performance ICs Prior to MCM Insertion," *Government Microelectronics Applications Conf (GOMAC),* Nov. 1992, pp. 341–344.
88. D. C. Keezer, "MCM Testing Using Available Technology," *Proc of the Intl Test Conf (ITC'92),* Sept. 20–24, 1992, p. 253.
89. "A New HP 82000 System for Testing MCMs to 200 MHz," Hewlett Packard, *Pathways,* Spring 1993, pp. 3–8.
90. D. C. Keezer, "Bare Die Testing and MCM Probing Techniques," *Proc of the IEEE Multichip Module Conf,* March 1992, pp. 20–23.
91. D. C. Keezer, "MCM Testing Using Available Technology," *Proc of the Intl Test Conf (ITC'92),* Sept. 20–24, 1992, p. 253.
92. D. C. Keezer, "Fault Isolation Methods for Multichip Modules," *Proc of the Intl Symposium for Testing & Failure Analysis (ISTFA'93),* Los Angeles, Calif., November 15–18, 1993, pp. 135–141.
93. E. Menzel and E. Kubalek, "Fundamentals of Electron Beam Testing of Integrated Circuits," *SCANNING,* vol. 5, 1983, pp. 103–122.
94. E. Menzel and E. Kubalek, "Electron Beam Test Techniques for Integrated Circuits," *Scanning Electron Microscopy 1981/I,* SEM Inc, AMF O'Hare, Chicago, pp. 305–318.
95. E. Menzel and E. Kubalek, "Electron Beam Chopping Systems in the SEM," *Scanning Electron Microscopy 1979/I,* SEM Inc, AMF O'Hare, Chicago, pp. 305–318.
96. M. Ostrow et al., "IC-Internal Electron Beam Logic State Analysis," *Scanning Electron Microscopy 1982/II,* SEM Inc, AMF O'Hare, Chicago, pp. 563–572.
97. E. Menzel and M. Brunner, "Secondary Electron Analyzers for Voltage Measurements," *Scanning Electron Microscopy 1983,* SEM Inc, AMF O'Hare, Chicago, pp. 65–75.
98. E. Menzel and E. Kubalek, "Secondary Electron Detection Systems for Quantitative Voltage Measurements," *SCANNING,* vol. 5, 1983, pp. 151–171.
99. J. P. Flemming and E. W. Ward, "A Technique for Accurate Measurement and Display of Applied Potential Distributions Using the SEM," *Scanning Electron Microscopy 1970,* IITRI, Chicago, pp. 465–470.
100. A. Gopinath and C. C. Sangen, "A Technique for Linearization of Voltage Contrast in the Scanning Electron Microscope," *J Phys E 4,* 1971, pp. 334–336.
101. J. L. Balk et al., "Quantitative Voltage Contrast at High-Frequencies in the SEM," *Scanning Electron Microscopy 1976/I,* IITRI, Chicago, pp. 615–624.
102. H. P. Feuerbaum, "VLSI Testing Using the Electron Probe," *Scanning Electron Microscopy 1979/I,* SEM Inc, AMF O'Hare, pp. 285–296.

103. L. Kotorman, "Non-Charging Electron Beam Pulse Prober on FET Wafers," *Scanning Electron Microscopy 1980/IV,* SEM Inc, AMF O'Hare, pp. 77–84.
104. S. D. Golladay, "A Voltage Contrast Detector for Electrical Testing of Multichip Substrates," *2nd European Conf on Electron and Optical Beam Testing of Integrated Circuits,* Poster P2, 1989.
105. D. C. Keezer, "Bare Die Testing and MCM Probing Techniques," *Proc of the IEEE Multichip Module Conf,* March 1992, pp. 20–23.
106. R. H. Parker, "Bare Die Test," *Proc of the IEEE Multichip Module Conf,* March 1992, pp. 24–27.
107. G. A. Forman, J. A. Nieznanski, and J. Rose, "Die for MCMs: IC Preparation for Testing, Analysis and Assembly," *Proc of the IEEE Multichip Module Conf,* March 1992, pp. 32–35.
108. Meeting Minutes from the MCM Test Workshop and the STAR Packaging Workshop Test Session, October 1991 and March 1992.
109. D. C. Keezer, "Assessment of High Performance ICs Prior to MCM Insertion," *Government Microelectronics Applications Conf (GOMAC),* Nov. 1992, pp. 341–344.
110. R. A. Fillion, R. J. Wojnarowski, and W. Daum, "Bare Chip Test Techniques for Multichip Modules," *ECTC Conf Proceedings,* 1990, p. 554.
111. L. E. Roszel and W. Daum "MCM Prototyping Using Overlay Interconnect Process," *Proc of the IEEE Multichip Module Conf,* March 1992, pp. 36–39.
112. D. C. Keezer, "MCM Testing Using Available Technology," *Proc of the 1992 Intl Test Conf (ITC'92),* Sept. 20–24, 1992, p. 253.
113. R. A. Barrett, "Testing Tasks and Protocols for University MCMs," *Proc of the IEEE Multichip Module Conf,* March 1991, pp. 164–171.
114. M. Taylor and W. W. M. Dai, "Tiny MCM," *Proc of the IEEE Multichip Module Conf,* March 1991, pp. 143–147.
115. R. W. Bassett, P. S. Gillis, and J. J. Shushereba, "Testing and Diagnosis of High-Density CMOS Multichip Modules," *Proc of the IEEE Multichip Module Conf,* March 1991, pp. 108–113.
116. R. Lang, "Cost Effectiveness of nCHIP's MCM Technology," *Proc of the IEEE Multichip Module Conf,* March 1991, pp. 16–23.
117. K.P. Shambrook, "Overview of Multichip Module Technologies," *Proc of the IEEE Multichip Module Conf,* March 1991, pp. 1–9.
118. J. K. Hagge and R. J. Wagner, *Thin Film Multichip Modules,* Ed. Messner et al., (eds.), Intl. Society for Hybrid Microelectronics (ISHM) Publ., Chapter 13.2, 1992, pp. 501–544.
119. J. K. Hagge and R. J. Wagner, "High-Yield Assembly of Multichip Modules Through Known-Good ICs and Effective Test Strategies," *Proc of the IEEE,* December 1992, pp. 1965–1994.
120. J. Godfrey, "Interfacing/Fixturing for the HP82000/HP83000," *Proc of the Intl HP82000 Users Group Meeting,* May 1993, pp. 268–276.
121. T. Costlow, "Motorola Shipping Bare Die," *Electronic Engineering Times,* 1993, p. 72.
122. G. Derman, "MCMs Make Right Connections," *Electronic Engineering Times,* 1993, p. 54.
123. S. Crum, "MCNC Develops Flip Chip Test Method," *Electronic Packaging and Production,* May 1993, pp. 11–12.
124. J. Tuck, "On Board the MCM-L Train," *Circuits, Assembly,* March 1993, pp. 20–21.
125. Z. Sekulic, "Testing Multichip Modules," *IMS-Test Solutions,* November 1992, pp. 4–5.
126. R. K. Scannell and J. K. Hagge, "Development of a Multichip Module DSP," *IEEE Computer,* April 1993, pp. 13–21.
127. W. Daum, W. E. Burdick, and R. A. Fillion, "Overlay High-Density Interconnect: A Chips-First Multichip Module Technology," *IEEE Computer,* April 1993, pp. 23–29.
128. C. M. Habiger and R. M. Lea, "Hybrid-WSI: A Massively Parallel Computing Technology?" *IEEE Computer,* April 1993, pp. 50–61.

129. J. Huppenthal, "500 MHz Functional Test System Development," *Proc of the Rocky Mountain Test Conf,* Colorado Springs, Oct. 1989, pp. 4-35 – 4-44.

130. J. Peoples, "Probe Card Design for High-Speed Digital Testing," *Proc of the Rocky Mountain Test Conf,* Denver, May 1991, pp. 16–20.

131. R. Nelson, M. Bonham, "High-Density/High-Speed Wafer Probing: A 360 Pin-2.33 GHz Probe Card for Production," *Proc Intl HP82000 Users Group Mtg,* June 1992, pp. 188–213.

132. S. P. Athan, D. C. Keezer, J. McKinley, "High-Frequency Wafer Probing and Power Supply Resonance Effects," *Proc of the Intl Test Conf (ITC'91),* October 1991, pp. 1069–1077.

133. P. Warwick and P. Giard, "Enhancing Standard Probe Technology for UHF Probing," *Proc of the Intl HP82000 Users Group Meeting,* May 1993, pp. 32–51.

134. A. Barber, K. Lee, and H. Obermaier, "A Bare Chip Probe for High I/O, High Speed Testing," *Proc of the Intl HP82000 Users Group Meeting,* May 1993, pp. 19–31.

135. M. Beiley et al., "Array Probe Card," *Proc of the IEEE Multichip Module Conf,* March 1992, pp. 28–31.

136. T. Tada et al., "A Fine Pitch Probe Technology for VLSI Wafer Testing," *Proc of the Intl Test Conf (ITC'90),* September 1990, pp. 900–906.

137. B. Leslie and F. Matta, "Membrane Probe Card Technology (The Future of High Performance Wafer Test)," *Proc of the Intl Test Conf (ITC'88),* September 1988, pp. 601–607.

138. T. E. Figal, "Below A Minute Burn-in," *U.S. Patent 5,030,905,* July 9, 1991.

139. T. Costlow, "Two Team in MCM Test," *Electronic Engineering Times,* July 19, 1993, p. 64.

140. D. C. Keezer, "Fault Isolation Methods for Multichip Modules," *Proc of the Intl Symposium for Testing & Failure Analysis (ISTFA'93),* Nov. 1993, pp. 135–141.

141. T. W. Williams and K. P. Parker, "Design for Testability—A Survey," *Proc of the IEEE,* vol. 71, no. 1, January 1983, pp. 98–112.

142. B. Koenemann, J. Mucha, and G. Zwiehoff, "Built-in Logic Block Observation Techniques," *Digest of the 1979 Test Conf,* October 1979, pp. 37–41.

143. R. Vemuri et al., "An Integrated Multicomponent Synthesis Environment for MCMs," *IEEE Computer,* April 1993, pp. 62–74.

144. J. Novellino, "ITC'91 Focuses on Testability Issues," *Electronics Design,* October 24, 1991, pp. 38–45.

145. K. E. Posse, "A Design-for-Testability Architecture for Multichip Modules," *Proc of the Intl Test Conf (ITC'91),* October 1991, pp. 113–120.

146. F. J. Henley and H. J. Choi, "Test Head Design using Electro-Optic Receivers and GaAs Pin Electronics for a Gigahertz Production Test System," *Proc of the Intl Test Conf (ITC'88),* 1988, pp. 700–709.

147. M. Shinagawa and T. Nagatsuma, "An Automated Optical On-Wafer Probing System for Ultra-High-Speed ICs," *Proc of the Intl Test Conf (ITC'92),* 1992, pp. 834–839.

INDEX

D

E

ABOUT THE EDITORS

Aicha Elshabini is Professor in the Bradley Department of Electrical and Computer Engineering, and Director of the Microelectronics Laboratory at Virginia Tech. She is a Fellow member of both the IMAPS society and the IEEE/CPMT society, and she is editor of the *International Journal of Microcircuits & Electronic Packaging.*

Fred Barlow is a member of the Research Faculty in the Bradley Department of Electrical and Computer Engineering at Virginia Tech, where he is involved in research in electronic packaging. He is currently a member of the IMAPS Power Packaging Subcommittee, as well as the Chairman of the IMAPS Internet Subcommittee.